Theory of Mechanisms and Machines

Theory of Mechanisms and Machines

C.S. SHARMA
Formerly Professor
Department of Mechanical Engineering
Jai Narain Vyas University
Jodhpur

KAMLESH PUROHIT
Professor
Department of Mechanical Engineering
Jai Narain Vyas University
Jodhpur

PHI Learning Private Limited
Delhi-110092
2015

₹ 425.00

THEORY OF MECHANISMS AND MACHINES
C.S. Sharma and Kamlesh Purohit

© 2006 by PHI Learning Private Limited, Delhi. All rights reserved. No part of this book may be reproduced in any form, by mimeograph or any other means, without permission in writing from the publisher.

ISBN-978-81-203-2901-0

Sixth Printing ... ... **August, 2015**

Published by Asoke K. Ghosh, PHI Learning Private Limited, Rimjhim House, 111, Patparganj Industrial Estate, Delhi-110092 and Printed by Rajkamal Electric Press, Plot No. 2, Phase IV, HSIDC, Kundli-131028, Sonepat, Haryana.

Contents

Preface .... *xiii*

Chapter 1   Introduction to Mechanisms and Machines   1–15

    1.1   Introduction 1
    1.2   Kinematic Link 2
    1.3   Kinematic Pairs 2
            1.3.1   Nature of Contact 3
            1.3.2   Nature of Relative Motion 3
            1.3.3   Nature of Mechanical Constraints 4
    1.4   Kinematic Chain 5
    1.5   Mechanism 6
    1.6   Four Bar Chain 7
    1.7   Inversions of Kinematic Chain 9
            1.7.1   Four Bar Chain 9
            1.7.2   Slider-crank Chain 10
            1.7.3   Double Slider-crank Chain 13
    Exercises 15

Chapter 2   Kinematic Analysis of Mechanism   16–71

    2.1   Introduction 16
    2.2   Velocity Analysis 17
    2.3   Relative Velocity Method 17
            2.3.1   Slider-crank Mechanism 19
            2.3.2   Four Bar Mechanism 19
            2.3.3   Crank and Slotted Lever Mechanism 20
    2.4   Instantaneous Centre Method 32
            2.4.1   Arnold Kennedy Theorem 34
            2.4.2   Locating Instantaneous Centres 35
    2.5   Acceleration Analysis 40
            2.5.1   Procedure to Draw Acceleration Polygon of a Mechanism 42
    2.6   Klein's Construction 49

- 2.7 Coriolis Acceleration 51
- 2.8 Analytical Method 58
- Exercises 63
- Multiple Choice Questions 69

Chapter 3 Synthesis of Mechanisms 72–103

- 3.1 Introduction 72
- 3.2 Number Synthesis 72
  - 3.2.1 Degree of Freedom of a Mechanism 72
  - 3.2.2 Grashof's Criterion of Movability 77
  - 3.2.3 Limit Positions, Dead Centres and Transmission Angle 80
- 3.3 Dimensional Synthesis 82
- 3.4 Synthesis by Relative Pole Method 82
  - 3.4.1 Synthesis of Four Bar Mechanism 84
  - 3.4.2 Synthesis of Slider-Crank Mechanism 86
- 3.5 Synthesis of Mechanism by Inversion Method 87
  - 3.5.1 Synthesis of Four Bar Mechanism 88
  - 3.5.2 Synthesis of Slider-Crank Mechanism 89
- 3.6 Slider-Crank Mechanism for Given Quick Return Ratio 90
- 3.7 Synthesis of Mechanism by Analytical Method 91
  - 3.7.1 Freudenstein's Method of Synthesis 92
  - 3.7.2 Approximate Synthesis Using Least Square Technique 93
  - 3.7.3 Bloch's Method 94
- Exercises 99
- Multiple Choice Questions 101

Chapter 4 Lower Pair Mechanisms 104–135

- 4.1 Introduction 104
- 4.2 Straight Line Motion Mechanism 104
  - 4.2.1 Exact Straight Line Motion Mechanism 104
  - 4.2.2 Approximate Straight Line Motion Mechanism 109
- 4.3 Pantograph 112
- 4.4 Engine Indicator Mechanisms 114
- 4.5 Automobile Steering Mechanism 120
  - 4.5.1 Davis Steering Mechanism 122
  - 4.5.2 Ackermann Steering Mechanism 124
- 4.6 Hooke's Joint 126
- Exercises 133

Chapter 5 Friction 136–183

- 5.1 Introduction 136
- 5.2 Laws of Dry Friction 137

5.3	Angle of Friction	137
5.4	Inclined Plane Friction	138
	5.4.1 Motion Up the Plane	139
	5.4.2 Motion Down the Plane	141
5.5	Inclined Plane with Guide Friction	145
5.6	Wedge Friction	147
5.7	Screw Friction	153
5.8	Pivot and Collar Friction	158
	5.8.1 Collar Bearings	160
	5.8.2 Pivot Bearing	163
5.9	Friction Clutches	165
	5.9.1 Plate Clutches	166
	5.9.2 Cone Clutches	168
5.10	Friction in Turning Pairs	175
5.11	Friction Axis	176
Exercises		178
Multiple Choice Questions		182

Chapter 6 Belts, Ropes and Chains 184–234

6.1	Introduction	184
6.2	Belt Drive	184
6.3	Mechanics of Belt Drive	188
6.4	Velocity Ratio	190
6.5	Length of Belt	192
	6.5.1 Open-belt Drives	192
	6.5.2 Cross-belt Drives	194
6.6	Ratio of Belt Tensions	195
	6.6.1 Ratio of Tensions in V-Belt	197
6.7	Effect of Centrifugal Force	199
6.8	Power Transmission Capacity	200
6.9	Stresses in Belt	201
6.10	Elastic Creep	203
6.11	Belt Materials	204
6.12	Stepped Pulley Drive	214
6.13	Rope Drive	218
	6.13.1 Wire Rope Construction	218
	6.13.2 Forces on a Wire Rope	219
	6.13.3 Breaking Strength	221
6.14	Chain Drive	224
6.15	Types of Chain	225
6.16	Chain Length	228
6.17	Chordal Action	229
Exercises		230
Multiple Choice Questions		233

Chapter 7 Brakes and Dynamometers 235–291

- 7.1 Introduction 235
- 7.2 Block Brakes 235
  - 7.2.1 Block Brakes with Small Angle of Contact 236
  - 7.2.2 Block Brake with Large Angle of Contact 239
- 7.3 Band Brakes 250
  - 7.3.1 Simple Band Brake 251
  - 7.3.2 Differential Band Brake 252
- 7.4 Band and Block Brake 262
- 7.5 Internal Expanding Shoe Brake 265
- 7.6 Effect of Braking on Vehicle 270
- 7.7 Dynamometers 274
  - 7.7.1 Absorption Dynamometers 274
  - 7.7.2 Transmission Dynamometers 278
- Exercises 285
- Multiple Choice Questions 291

Chapter 8 Governors 292–330

- 8.1 Introduction 292
- 8.2 Watt Governor 292
- 8.3 Porter Governor 295
- 8.4 Proell Governor 300
- 8.5 Hartnell Governor 305
- 8.6 Wilson-Hartnell Governor 312
- 8.7 Spring-Controlled Gravity Governor 316
- 8.8 Pickering Governor 317
- 8.9 Characteristics of a Governor 318
- 8.10 Effort of a Governor 320
- 8.11 Power of a Governor 321
- 8.12 Controlling Force 322
- Exercises 326
- Multiple Choice Questions 329

Chapter 9 Cams 331–387

- 9.1 Introduction 331
- 9.2 Classification of Cams and Followers 331
- 9.3 Cam Terminology 334
- 9.4 Analysis of Follower Motions 335
  - 9.4.1 Constant Velocity Motion 336
  - 9.4.2 Simple Harmonic Motion 338
  - 9.4.3 Constant Acceleration and Deceleration 339
  - 9.4.4 Cycloidal Motion 343
  - 9.4.5 Polynomial Motion 344

9.5	Graphical Synthesis of Cam Profile	346
9.6	Analytical Synthesis of Cam Profile	355
9.7	Pressure Angle	359
9.8	Cams with Specified Profile	362
	9.8.1 Tangent Cam with Roller Follower	362
	9.8.2 Circular Cam with Roller Follower	372
	9.8.3 Circular Cam with Flat Faced Follower	374
	Exercises	384
	Multiple Choice Questions	386

Chapter 10 Gears   388–440

10.1	Introduction	388
10.2	Gear Terminology	389
10.3	Law of Gearing	391
10.4	Velocity of Sliding	393
10.5	Forms of Gear Teeth	394
	10.5.1 Involute Profile	394
	10.5.2 Cycloidal Profile	396
10.6	Arc of Contact	397
10.7	Interference	405
10.8	Minimum Number of Teeth to Avoid Interference	406
10.9	Minimum Number of Teeth on Pinion with Rack	408
10.10	Force Analysis of a Spur Gear	415
10.11	Helical Gear	417
	10.11.1 Formative Number of Teeth	419
	10.11.2 Forces on Helical Gear	419
10.12	Spiral Gear	420
	10.12.1 Efficiency of Spiral Gear	422
10.13	Worm Gear Set	427
	10.13.1 Efficiency of Worm Gear Set	430
10.14	Bevel Gears	432
	Exercises	436
	Multiple Choice Questions	439

Chapter 11 Gear Train   441–470

11.1	Introduction	441
11.2	Simple Gear Train	441
11.3	Compound Gear Train	443
11.4	Epicyclic Gear Train	445
	11.4.1 Tabulation Method	445
	11.4.2 Relative Velocity Method	446
11.5	Torque Transmitted by Epicyclic Gear Train	447
	Exercises	464
	Multiple Choice Questions	469

Chapter 12 Force Analysis — 471–541

- 12.1 Introduction 471
- 12.2 Static Force Analysis 471
  - 12.2.1 Static Equilibrium 471
  - 12.2.2 Free Body Diagram 474
  - 12.2.3 Effect of Friction 475
  - 12.2.4 Principle of Virtual Work 476
- 12.3 Dynamic Force Analysis 486
- 12.4 Inertia Force Analysis 488
- 12.5 Dynamics of Slider–Crank Mechanism 495
  - 12.5.1 Displacement, Velocity and Acceleration of Piston 495
  - 12.5.2 Velocity and Acceleration of Connecting Rod 497
  - 12.5.3 Crank Effort 498
- 12.6 Dynamically Equivalent Link 500
- 12.7 Inertia Force in Reciprocating Engine (Graphical Method) 512
- 12.8 Turning Moment Diagram 515
- 12.9 Flywheel 517
- Exercises 534
- Multiple Choice Questions 541

Chapter 13 Balancing — 542–591

- 13.1 Introduction 542
- 13.2 Static Balancing of Rotating Masses 542
- 13.3 Dynamic Balancing 547
- 13.4 Two Plane Balancing 548
- 13.5 Balancing Machines 558
  - 13.5.1 Pivoted-Cradle Balancing Machine 559
  - 13.5.2 Nodal Point Balancing Machine 560
  - 13.5.3 Field Balancing 561
- 13.6 Balancing of Reciprocating Mass 564
- 13.7 Balancing of Locomotive 567
- 13.8 Balancing of In-line Engine 571
- 13.9 Balancing of V-Engines 577
- 13.10 Direct and Reverse Crank Method 581
- Exercises 586
- Multiple Choice Questions 590

Chapter 14 Mechanical Vibrations — 592–666

- 14.1 Introduction 592
- 14.2 Elements of Vibratory System 592
- 14.3 Undamped Free Vibration 593
- 14.4 Torsional Vibration 597

14.5	Free Damped Vibration		605
	14.5.1	Logarithmic Decrement	608
14.6	Forced Damped Vibration		616
	14.6.1	Forced Vibration Due to Rotating Unbalance	620
	14.6.2	Excitation of the Support	621
	14.6.3	Vibration Isolation and Transmissibility	623
14.7	Transverse Vibration		633
	14.7.1	Vibration due to Single Point Load	633
	14.7.2	Several Point Loads	634
14.8	Whirling of Shaft		637
14.9	Torsional Vibration of Rotor System		640
	14.9.1	Two Rotor System	640
	14.9.2	Three Rotor System	642
	14.9.3	Geared System	651
14.10	Electrical Analogy		656
Exercises			660
Multiple Choice Questions			665

Chapter 15 Gyroscope 667–697

15.1	Introduction	667
15.2	Gyroscopic Couple	668
15.3	Gyroscopic Stabilization of Ship	675
15.4	Gyroscopic Effect in Grinding Mill	682
15.5	Stability of Automotive Vehicle	684
	15.5.1 Stability of Four Wheels Vehicle	684
	15.5.2 Stability of Two Wheeler Vehicle	690
Exercises		694
Multiple Choice Questions		696

Bibliography 699–700

Answers to Multiple Choice Questions 701–702

Index 703–706

Preface

Theory of mechanisms and machines is one of the principal subjects being taught to undergraduate students in mechanical, production and mining engineering disciplines. It is an applied science, which deals with developing relationship between geometry and motion of the parts of a machine or mechanism and the forces which produce these motions. Therefore, the subject is naturally divided into three parts—the kinematics (the analysis of motion of the parts), synthesis of mechanism and machine components and the study of kinetics, the time varying forces in machine and resulting dynamic phenomena, which are required to be considered prior to any design.

This book is intended to cover syllabi of most of the Indian universities, technical institutions and professional bodies. It contains fifteen chapters. The first chapter gives introduction to mechanisms and machines including their kinematic inversions. The second chapter deals with the kinematic analysis of mechanism along with graphical and analytical methods for velocity and acceleration analysis. The instantaneous centre method and Coriolis acceleration are also discussed. The third chapter deals with synthesis of mechanism. Freudenstein's method along with approximate synthesis using the least square technique is also discussed. Various lower pair mechanisms including automobile steering mechanisms are discussed in fourth chapter. The influence of friction in various elements such as screws, wedge, pivot and collar bearings and clutches has been discussed in fifth chapter. The mechanical power transmission elements and measuring devices namely belts, ropes and chain drive, brakes, dynamometers, cam and followers, gears and gear trains are discussed in Chapters 6 to 11. The force analysis, including inertia force and dynamics of reciprocating parts, is discussed in twelfth chapter. The balancing of rotating and reciprocating masses is covered in thirteenth chapter. The fourteenth chapter deals with vibration of mechanical system. The final chapter introduces the concept of gyroscopic stabilization used for navigational purpose.

It is but natural that some errors might have crept into a work of such volume. We shall appreciate if such errors and shortcomings are brought to our notice. Suggestions for improvement to the book would also be welcome.

We wish to acknowledge indebtedness to the authors and publishers whose books, articles and research papers have been consulted during the preparation of the manuscript.

Finally, we thank the editorial and production staff of Prentice-Hall of India, New Delhi, for their continuous cooperation and help in publishing this book.

C.S. Sharma
Kamlesh Purohit

CHAPTER 1

Introduction to Mechanisms and Machines

1.1 INTRODUCTION

Mechanism is an assemblage of several rigid bodies connected in such a way that they produce constrained motion between them. For example, the assembly of piston, cylinder, connecting rod and crank forms a mechanism commonly referred to as **slider-crank mechanism,** which converts reciprocating motion of piston (or slider) into a rotary motion of crank or vice-versa (Figure 1.1). Some other examples of mechanisms are mechanical clocks, toys, drafter, valve operating mechanism of IC engine and so forth.

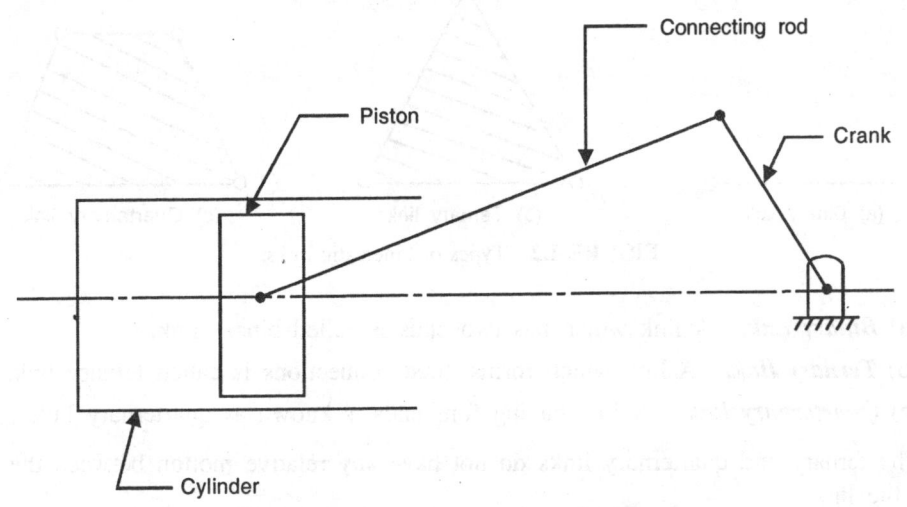

FIGURE 1.1 Slider-crank mechanism.

On the other hand, a machine is defined as combination of mechanisms which apart from producing constrained motion also transmits forces and does useful work. In other words, a machine is a contrivance which transforms energy available in one form into another to do certain type of desired work. The lathe, drill, shaper and planer are common examples of machine in which electrical energy is transformed into the motion of cutting tool which performs machining operation.

The theory of mechanisms and machines is an applied science which comprises the study of relationship between the geometry and motion of the parts of a mechanism or machine and the study of forces which act on those parts. Thus the subject of theory of mechanisms and machines can be broadly classified into the following two categories:

(a) *Kinematics.* It is that branch which deals with the study of geometric relationship and the relative motion between the parts.

(b) *Dynamics.* It is that branch which deals with the study of forces acting on various parts of a machine. The forces can be either static or dynamic due to the inertia of parts of machine.

1.2 KINEMATIC LINK

A kinematic link is a resistant body or assembly of resistant bodies which constitutes the part of a mechanism. It is the smallest element of the machine which transmits motion to other links. A **resistant body** is one which does not change its physical form by the application of forces. In a mechanism, three types of links—rigid link, flexible link and fluid link—are the most widely used. A rigid link, depending upon its ends with which connection between the links is formed, can be classified into the following three categories (Figure 1.2):

(a) Binary link   (b) Ternary link   (c) Quarternary link

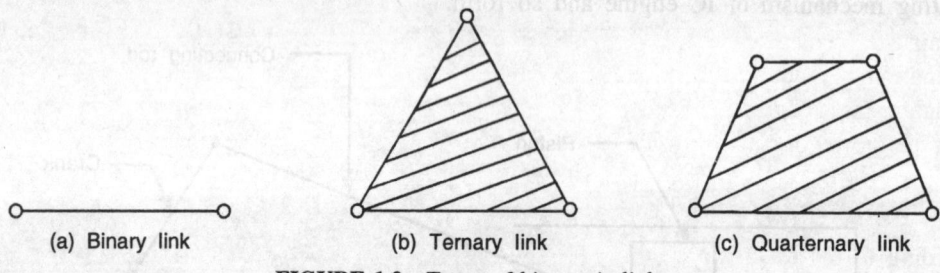

FIGURE 1.2 Types of kinematic links.

(a) *Binary link.* A link which has two ends is called binary link.

(b) *Ternary link.* A link which forms three connections is called ternary link.

(c) *Quarternary link.* A link having four ends is known as quarternary link.

The ternary and quarternary links do not have any relative motion between the joints within the link.

1.3 KINEMATIC PAIRS

If two kinematic links are joined in such a way that the relative motion between them is completely or successfully constrained, such a joint is termed as **kinematic pair.** The term successfully constrained means that connection between the links is not in itself such as to give complete constrained but the constraint is achieved by some other means. For example, a pair is formed between the shaft of turbine and a thrust bearing in which, besides rotation, a vertical axial movement of the shaft is also permissible. However, the axial movement is prevented due to the weight of the turbine which is far in excess of any vertical upward force. Thus it forms a successfully-constrained pair.

A kinematic pair can be classified according to the following criteria:

(i) Nature of contact between the links
(ii) Nature of relative motion between the links
(iii) Nature of mechanical constraint

A brief discussion about these is given further.

1.3.1 Nature of Contact

According to the nature of contact formed between two links, a kinematic pair is classified as lower pair and higher pair.

Lower pair. When two links have surface or area contact between them while in motion, such a pair is known as **lower pair.** The relative motion in a lower pair is either purely turning or sliding. The contact surfaces of lower pair are complimentary to each other, e.g. shaft rotating in a journal bearing, automobile steering, universal joint, all the pairs of a slider-crank mechanism.

Higher pair. When two links have point or line contact between them while in motion, the pair so formed is termed as **higher pair.** The relative motion in higher pair is a combination of sliding and turning. The contact surfaces of higher pair are dissimilar. The kinematic pairs formed between ball, roller bearing, gears, cam and follower, etc. belong to higher pair category.

1.3.2 Nature of Relative Motion

According to the nature of relative motion between two links, a kinematic pair is classified into the following types:

Sliding pair. A kinematic pair is said to be **sliding pair** when two links are so connected that one is constrained to have a sliding motion relative to other. Figure 1.3(a) shows a prismatic sliding pair in which a prismatic bar slides in a rectangular hole. The connection between piston and cylinder of a slider-crank mechanism is another example of a sliding pair.

Turning pair. When two links are connected in such a way that only a constrained motion of rotation is possible between them, the kinematic pair so formed is called **turning pair** or **revolute pair.** A circular shaft with two collars rotating in a bearing as shown in Figure 1.3(b), the connection between piston, connecting rod and crank are common examples of turning pair.

Rolling pair. A pair of two links having a rolling motion relative to each other is called **rolling pair,** e.g. contact between ball or roller with the cage [Figure 1.3(c)] and a pulley in a belt drive.

Spherical pair. When one link is in the form of a sphere which turns inside a fixed link, such a pair is called a **spherical pair.** A ball and socket joint is an example of spherical pair [Figure 1.3(d)].

Screw pair. A pair of screw and threads which has both turning and sliding motion between them is called a **screw pair**. It is sometimes also referred to as **helical pair**. See Figure 1.3(e).

(a) Sliding pair
(b) Turning pair
(c) Rolling pair
(d) Spherical pair
(e) Screw pair

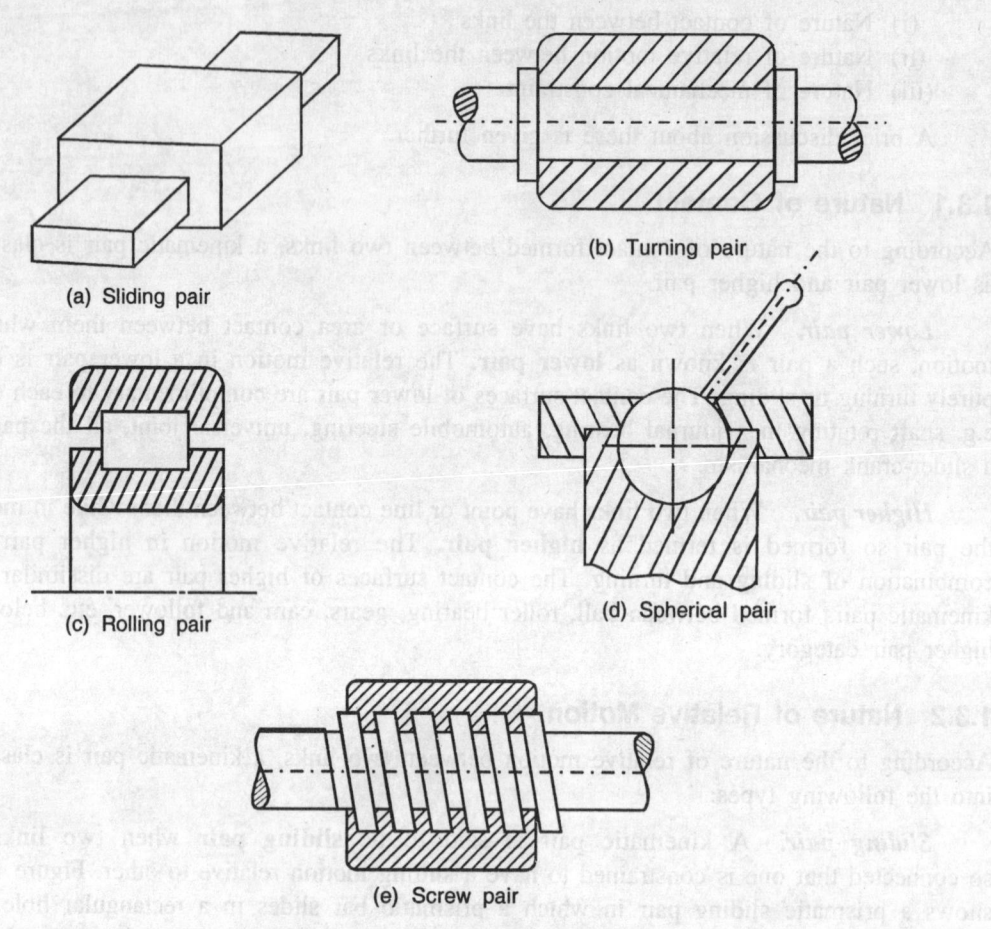

FIGURE 1.3 Types of kinematic pair.

1.3.3 Nature of Mechanical Constraints

Depending upon the nature of mechanical constraints, the kinematic pair is classified into the following two categories:

Closed pair. When two links, which form a kinematic pair are held together mechanically such a pair is called a **closed pair**. The screw pair and spherical pair are examples of closed pair.

Open pair. When two links forming a pair are held together by means of some external force namely gravity force, spring force, etc., they constitute **open pair** or **unclosed pair**. The cam-follower of IC engine valve operating mechanism is an example of open pair.

1.4 KINEMATIC CHAIN

A kinematic chain is a closed network of links connected by kinematic pairs such that the relative motion between them is completely constrained motion.

For a kinematic chain to give constrained motion, certain conditions are required to be satisfied which are:

(i) A kinematic chain must have at least four links with four lower pairs.
(ii) The motion of each link relative to other should be definite.

Mathematically, these conditions can be fulfilled by satisfying Grubler's equation*, which is:

$$j = \frac{3n}{2} - 2 \quad (1.1)$$

where
  j is the number of binary joints
    [= $(2n_2 + 3n_3 + 4n_4)/2$]
    with n_2, n_3 and n_4 as number of binary, ternary and quarternary links respectively
and n is the number of links in a kinematic chain (= $n_2 + n_3 + n_4$).

EXAMPLE 1.1 Examine the chains shown in Figure 1.4 below and indicate whether these are kinematic chains or not.

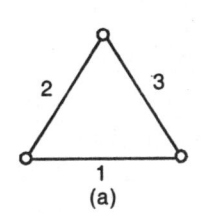

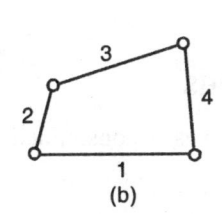

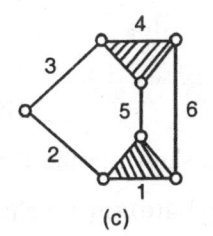

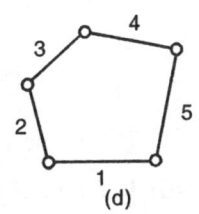

FIGURE 1.4

Solution

(a) Refer to Figure 1.4(a), in which:
  the number of links, $n = 3$
  and the number of binary joint, $j = 3$
  Now check whether Eq. (1.1) is satisfied or not.

$$j = \frac{3n}{2} - 2$$

or

$$3 = \frac{3 \times 3}{2} - 2$$

---

*Grubler's equation is discussed in detail in Chapter 3 (Synthesis of Mechanism).

or $\qquad 3 \neq 2.5$

Here Left Hand Side (LHS) parameter is not equal to Right Hand Side (RHS), LHS > RHS, so chain is locked and it is called structure.

(b) Refer to Figure 1.4(b), in which:

$$n = 4 \quad \text{and} \quad j = 4$$

Applying Eq. (1.1),

$$4 = \frac{3 \times 4}{2} - 2$$

or $\qquad 4 = 4$

or $\qquad$ LHS = RHS

Hence the chain shown in the Figure (b) is a constrained kinematic chain.

(c) Refer to Figure 1.4(c), in which:

$$n = 6$$

and $\qquad j = \frac{(2n_2 + 3n_3)}{2} = \frac{(2 \times 4 + 3 \times 2)}{2} = 7$

Applying Eq. (1.1),

$$7 = \frac{3 \times 6}{2} - 2$$

or $\qquad 7 = 7$

or $\qquad$ LHS = RHS

Hence the chain shown in the Figure (c) is a constrained kinematic chain.

(d) Refer to Figure 1.4(d), in which:

$$n = 5 \text{ and } j = 5$$

Applying Eq. (1.1),

$$5 = \frac{3 \times 5}{2} - 5$$

or $\qquad 5 \neq 2.5$

Since LHS > RHS, so chain is unconstrained. Though such type of chain is non-kinematic chain, some authors prefer to call such chain unconstrained kinematic chain.

1.5 MECHANISM

A mechanism, as defined earlier, is an assemblage of links in which each link produces constrained motion relative to other. However, in the context of theory of mechanism, it is defined as that kinematic chain in which one link is fixed. In a mechanism, when we say that one link is fixed, we mean that it is chosen as frame of reference for all other links and the motion of all other points on the links are measured with respect to this fixed link.

Mechanisms are of two types—simple and compound. A simple mechanism has four links whereas a compound mechanism has more than four links. It may be a combination of two or more simple mechanisms.

When a mechanism is used to transmit forces or to do some useful work, it is called a **machine.** Figure 1.5 illustrates how kinematic links make up a machine.

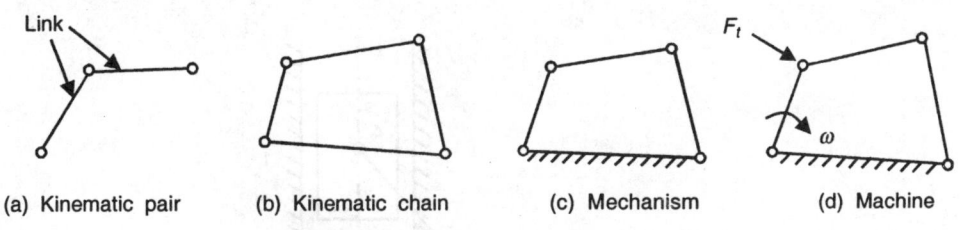

(a) Kinematic pair (b) Kinematic chain (c) Mechanism (d) Machine

FIGURE 1.5 Journey from links to machine.

1.6 FOUR BAR CHAIN

A four bar (link) chain consists of four kinematic links of different lengths which are connected by turning type lower pair [Figure 1.6(a)]. In a four bar chain, the fixed link is called **frame.** The link, which is not connected to frame, is called **coupler link.** Among other two links hinged to frame, one which receives input is termed as **crank** and the output link is known as **follower** or **rocker link.**

(a) Four bar chain

(b) Four bar chain with curved slider (c) Slider-crank mechanism

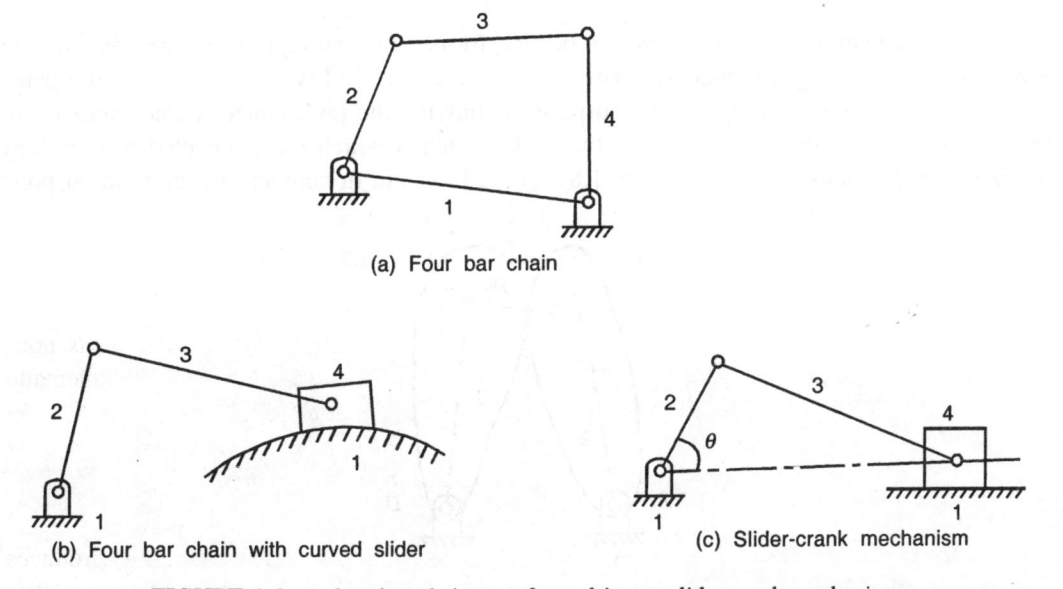

FIGURE 1.6 A four bar chain transformed into a slider-crank mechanism.

In a four bar chain, if one of the turning pairs is replaced by a sliding pair, the resulting mechanism is known as **slider-crank mechanism.** To demonstrate, let us consider a four bar chain as shown in the Figure 1.6(a). If the turning pair between the links 1 and 4 is replaced by a curved slider as shown in the Figure 1.6(b), it still remains turning pair as the relative motion between the links 1 and 4 is represented by angular displacement. However, if the radius of curvature of the slider becomes infinitely large, the turning pair is transformed into a sliding pair and the resulting mechanism [Figure 1.6(c)] is a slider-crank mechanism.

In a four bar chain, if two turning pairs are replaced by sliding pairs, the resulting mechanism is termed as **double slider-crank mechanism** (Figure 1.7).

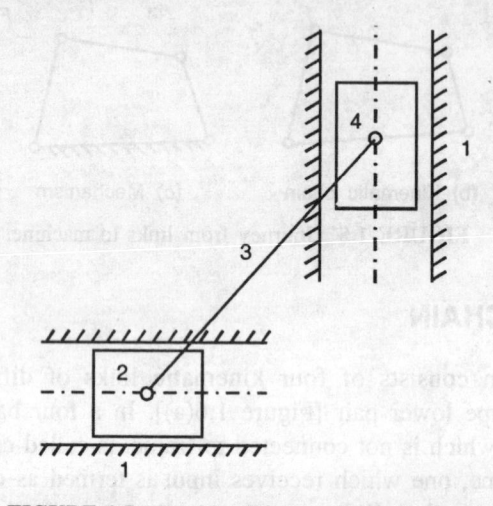

FIGURE 1.7 Double slider-crank mechanism.

A mechanism with higher pairs can be transformed into an equivalent mechanism with lower pair by replacing a higher pair with an additional link and two turning or sliding pairs. This equivalence is valid only for the purpose of studying the instantaneous characteristics of the mechanism. For example, consider two links 2 and 3 which are connected to link 1 by turning pairs at points A and D (Figure 1.8). These links are in contact to each other at point

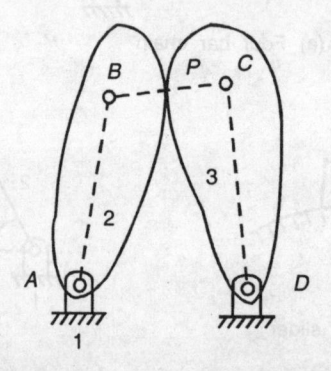

FIGURE 1.8 Higher pair mechanism.

P forming a higher pair. This higher pair can be replaced by a link BC with turning pair at B and C such that the centre of curvature of contact point P, at this instant, lies at B and C respectively. The resulting mechanism $A\ B\ C\ D$ is a mechanism with lower pair. Another example is shown in Figure 1.9.

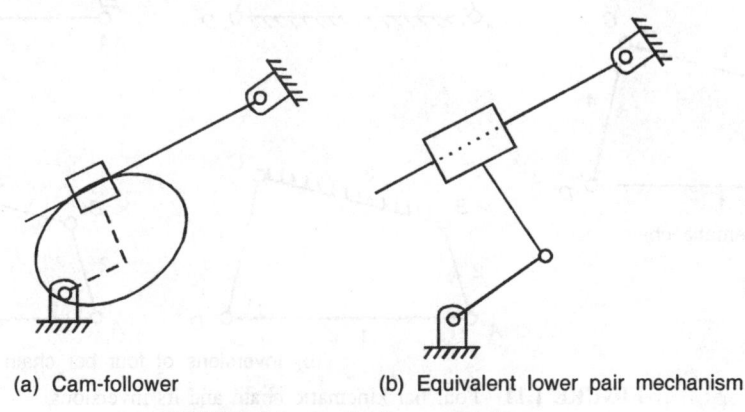

(a) Cam-follower (b) Equivalent lower pair mechanism

FIGURE 1.9 Equivalent mechanism.

A spring connection in a mechanism simulates the action of two binary links joined by turning pair so it can be replaced by two binary links (See Figure 1.10).

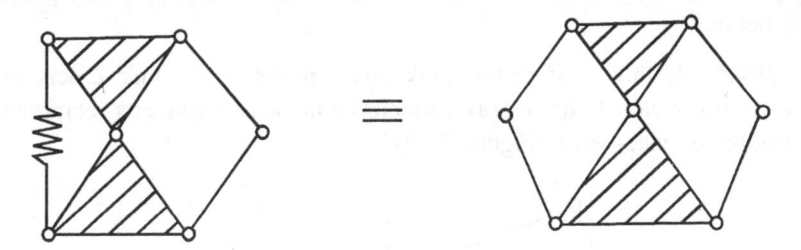

FIGURE 1.10 Simulation of spring connection.

1.7 INVERSIONS OF KINEMATIC CHAIN

A mechanism is defined as a kinematic chain in which one of the links is fixed. Thus by fixing the links of a kinematic chain one at a time, we get as many number of variants as the number of links. These variants of kinematic chain are called **inversions of kinematic chain** or **inversions of mechanism.**

A few inversions of some of the popular kinematic chains are discussed further.

1.7.1 Four Bar Chain

The simplest kinematic chain which results into four different types of mechanism is a four bar kinematic chain. The four different mechanisms obtained by fixing one link at a time are shown in Figure 1.11. Though these four inversions look identical, their mobility can be varied by suitably altering the proportions of lengths of various links.

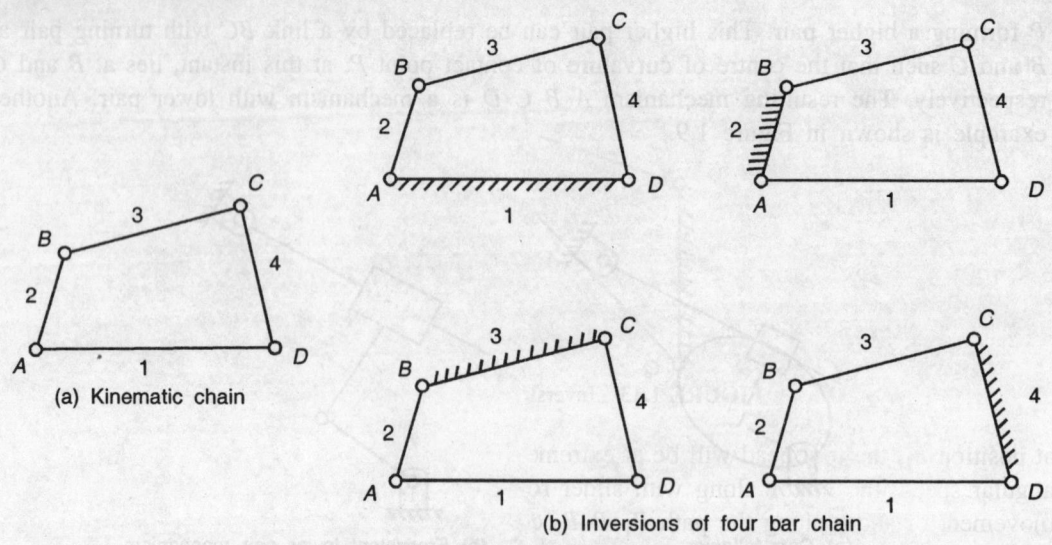

FIGURE 1.11 Four bar kinematic chain and its inversions.

(a) Kinematic chain

(b) Inversions of four bar chain

1.7.2 Slider-crank Chain

A slider-crank chain is a special form of four bar chain in which one turning pair is replaced by a sliding pair. Various inversions of slider-crank chain obtained by fixing different links are discussed below:

Link 1 fixed. If link 1 of slider-crank chain is fixed and link 2 acts as crank, the resulting mechanism is called **slider-crank mechanism**, which converts reciprocating motion into rotary motion or vice-versa (Figure 1.12).

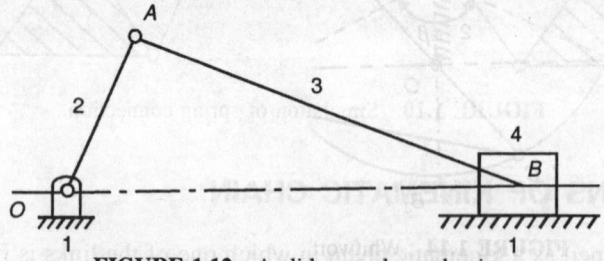

FIGURE 1.12 A slider-crank mechanism.

Link 2 fixed. When link 2 of slider-crank chain is fixed and link 3 acts as crank, link 1 rotates about centre O along with slider 4 which reciprocates on it (Figure 1.13).

This inversion gives two most popular forms of mechanism—whitworth quick return mechanism and rotary engine.

Whitworth quick return mechanism. A whitworth quick return mechanism, used in slotting machine, is shown in Figure 1.14. The link 2 is fixed and link 3 acts as input crank. The link 1 which rotates about centre O, is extended upto point C. An additional link 5 is pivoted at point C. The other end of link 5 is pivoted to tool head. The axis of tool movement passes through centre O and is perpendicular to link 2. In this mechanism, when slider 4 is

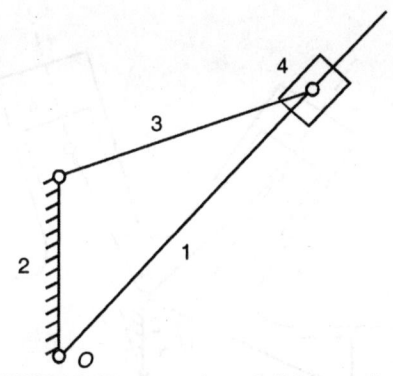

FIGURE 1.13 Inversion of slider-crank chain.

at position B_1, the tool head will be at extreme left position. When crank 3 rotates at constant angular speed, the link 1 along with slider rotates at the circumference of outer circle. The movement of slider along the path B_1–B–B_2 completes the forward movement of tool head. Further rotation of crank 3 causes slider to move along the path B_2–B_3–B_1 during which the tool head takes return stroke. It is evident from Figure 1.14 that angle for forward stroke is

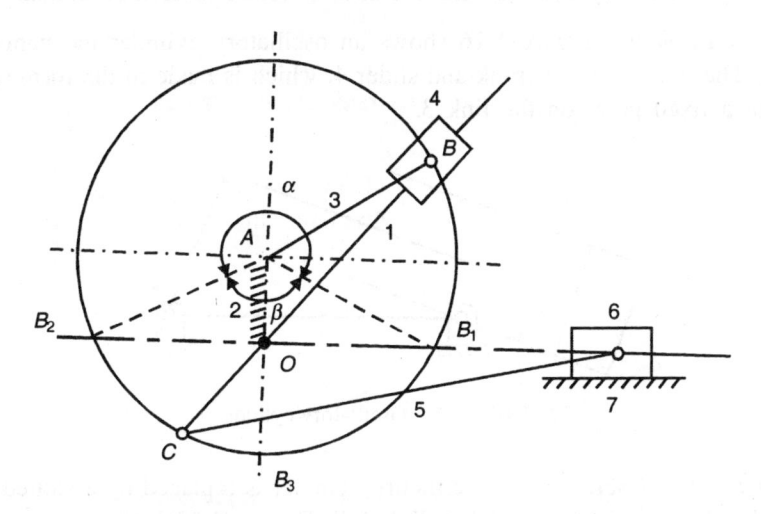

FIGURE 1.14 Whitworth quick return mechanism.

an obtuse angle α which is greater than angle for return stroke β, thus tool traverses rapidly during the return stroke. The ratio of forward to return stroke timing is given as:

$$r = \frac{\text{Time of cutting}}{\text{Time of return}} = \frac{\alpha}{\beta} \tag{1.2}$$

Rotary engine. Figure 1.15 shows a multi-cylinder rotary engine mechanism. In this inversion, link 2 is fixed and link 3 acts as crank. The slider 4 is represented by piston and link 1, which rotates about point O, is represented by a cylinder.

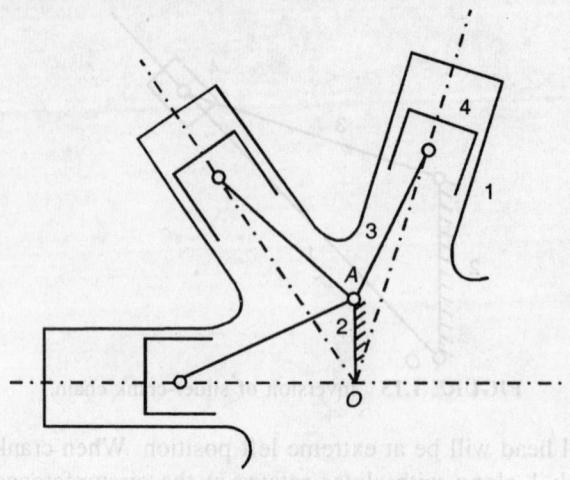

FIGURE 1.15 Rotary engine.

Link 3 fixed. By fixing the link 3 of the slider crank chain, it gives two different inversions namely oscillatory cylinder and crank and slotted lever mechanism.

Oscillatory cylinder. Figure 1.16 shows an oscillatory cylinder mechanism in which link 3 is fixed. The link 2 acts as crank and slider 4, which is made in the form of a cylinder, oscillates about a fixed point on the link 3.

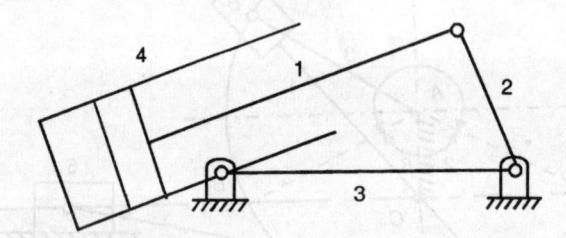

FIGURE 1.16 Oscillatory cylinder.

Crank and slotted lever. If the oscillatory cylinder is replaced by a slotted lever which guides the slider, [i.e. piston connected to link 1 in Figure (1.17)], the resulting mechanism is called **crank and slotted lever mechanism.** In this mechanism also, link 3 is fixed and link 2 acts as crank. The link 4 acts as an oscillatory link whose one end is pivoted to link 3 and the other end is pivoted to another link 5 which is connected to tool head through turning pair. The two extreme positions of oscillating slotted lever along with the slider is shown in Figure 1.17(b). This mechanism is also called a quick return mechanism, as the time for forward stroke is proportional to angle α which is greater than the time for angle β for return stroke.

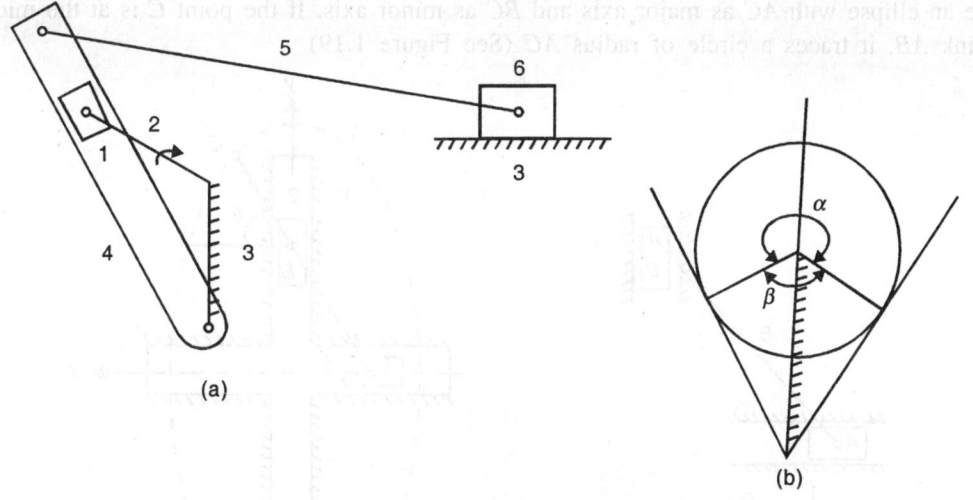

FIGURE 1.17 Inversion of slider-crank chain with link 3 fixed.

Link 4 fixed. When link 4 (slider) of the slider-crank chain is fixed and link 2 acts as rocker whose one end *A* oscillates about *B* and link 1 reciprocates along the axis of fixed link 4, it results into a hand pump mechanism as shown in Figure 1.18.

(a) Kinematic Inversion   (b) Hand pump mechanism

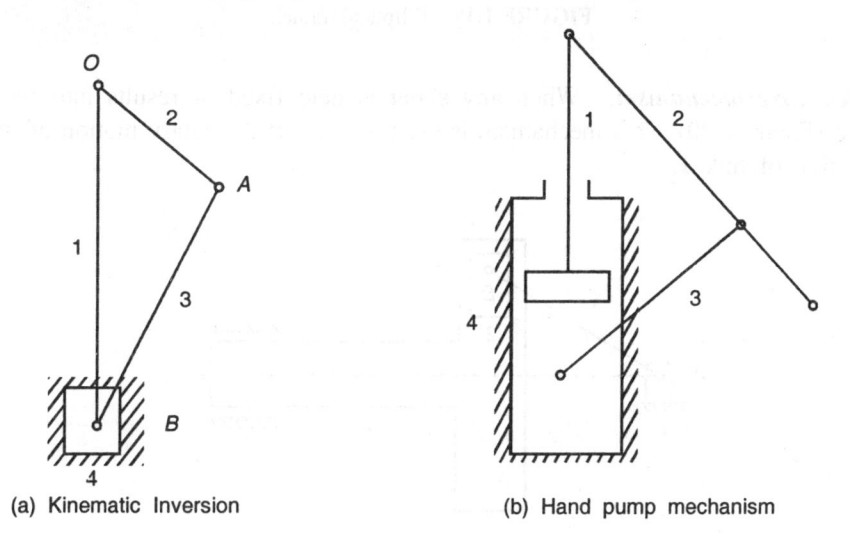

FIGURE 1.18 Inversion of slider-crank chain with link 4 fixed.

1.7.3 Double Slider-crank Chain

A double slider-crank chain, as shown in Figure 1.6 has three popular inversions which are discussed below:

Elliptical tramel. If the link 1 acts as a guideway for sliders 2 and 4, it results into an elliptical tramel in which any point *C* on the link 3, except at the mid-point of *AB*, will trace an ellipse with *AC* as major axis and *BC* as minor axis. If the point *C* is at the middle of link *AB*, it traces a circle of radius *AC* (See Figure 1.19).

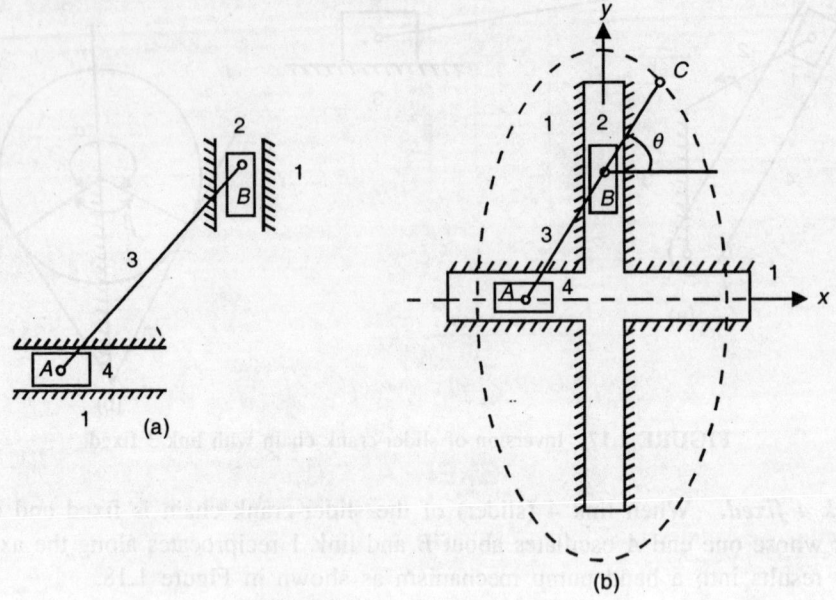

FIGURE 1.19 Elliptical tramel.

Scotch yoke mechanism. When any slider is held fixed, it results into scotch yoke mechanism (Figure 1.20). This mechanism is used to convert the rotary motion of link 3 into sliding motion of link 1.

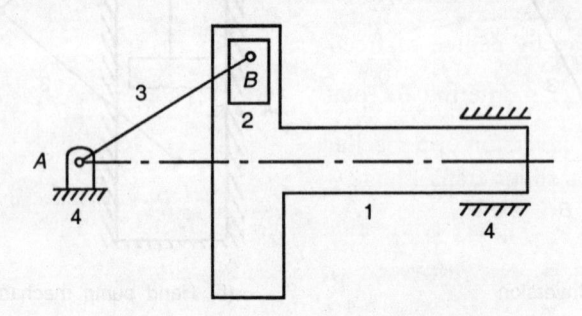

FIGURE 1.20 Scotch yoke mechanism.

Oldham's coupling. The third inversion of double slider crank chain is Oldham's coupling which is used to connect two parallel shafts whose axes are not in perfect alignment. In this inversion, the coupler link 3 is held fixed and the links 2 and 4 have slotted grooves to form sliding pairs with link 1 (Figure 1.21).

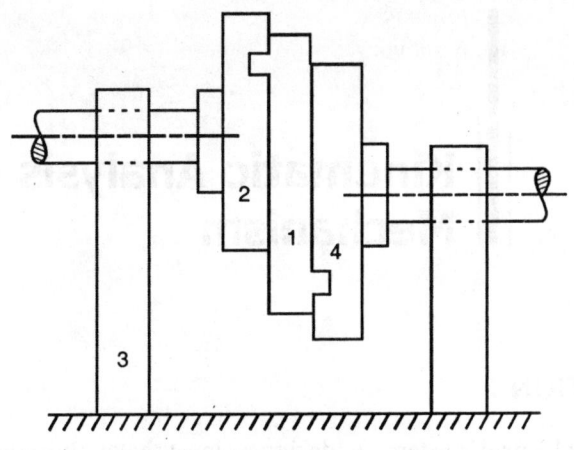

FIGURE 1.21 Oldham's coupling.

EXERCISES

1. What do you understand by terms machine and mechanism? State how these differ from each other.
2. Differentiate between analysis and synthesis.
3. What is a kinematic pair? How can it be classified?
4. What is the difference between lower and higher pairs? Give examples of each type.
5. What is a kinematic chain? What conditions are required to be satisfied by a kinematic chain to give constrained motion?
6. What do you mean by degree of freedom of a mechanism?
7. Explain the Grubler's criterion for plane mechanism to obtain the degree of freedom.
8. What is meant by inversion of a mechanism? Describe with the help of suitable sketches the inversion of a slider-crank chain.
9. Describe Oldham's coupling with suitable sketch.
10. Describe elliptical tramel. How does it enable us to describe a true ellipse?

CHAPTER 2

Kinematic Analysis of Mechanism

2.1 INTRODUCTION

In the design of mechanical system, a designer must have thorough understanding of kinematics of mechanism. **Kinematics** is the study of displacement, rotation, speed, velocity and acceleration of each link at various positions during the operating cycle. Using these information, a designer can compute forces and thereby dimensions of all the links.

The kinematic analysis of the mechanism can be performed either by graphical or by analytical method. The graphical method, though less accurate, is preferred as it gives the motion characteristics of all the links. In this chapter, the graphical method is restricted only for determining velocity and acceleration of mechanism with lower pairs. However, a mechanism with higher pair can be first converted into an equivalent lower pair mechanism and then can be analysed. The analytical method of analysis using vector algebra is discussed later.

In kinematic analysis, the motion of a point on the link relative to the fixed frame of reference is called **absolute motion**. However, the motion of a point relative to some other moving link or moving frame of reference is said to be **relative motion**. In graphical method, the relative motion of a link can be represented as a vector. The position of a point on the link can be represented as a vector from the origin of a specified reference coordinate system to the point. Generally, the symbol R_{PO} is used to denote the vector position of point P relative to point O (Figure 2.1). Graphically, the length of vector $\overrightarrow{op}$ is drawn to a convenient scale to represent the magnitude and the direction sense is given by the tail end to the head end p.

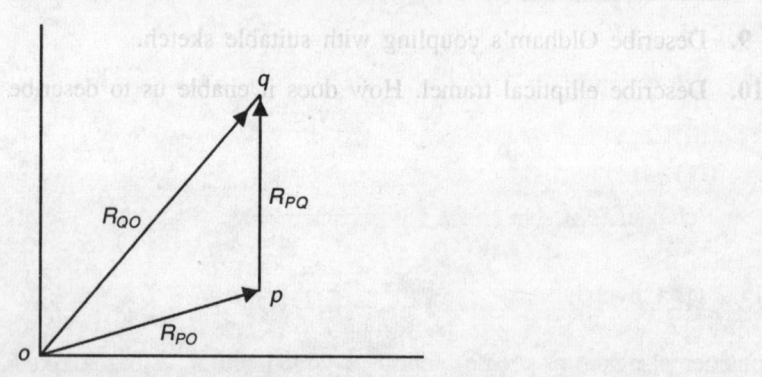

FIGURE 2.1 Vector representation of two points P and Q.

Kinematic Analysis of Mechanism

The position of point P relative to another point Q on the link is represented by a position difference between the points P and Q as shown in Figure 2.1. In vector notations, it is called subtraction of vector and is represented as:

$$R_{PQ} = R_{QO} - R_{PO}$$

or

$$\vec{qp} = \vec{oq} - \vec{op} \tag{2.1}$$

The law of vector addition states that the position of point Q relative to the origin O is equal to the vectorial sum of position difference between the points P and Q and the position of point P relative to the origin (Refer Figure 2.1).

$$R_{QO} = R_{PQ} + R_{PO}$$

or

$$\vec{oq} = \vec{qp} + \vec{op} \tag{2.2}$$

2.2 VELOCITY ANALYSIS

The change of position of a link with reference to some fixed frame of coordinates is called **displacement**. The rate of change of displacement of a link with reference to time, i.e. the time derivative of displacement, is commonly referred as **velocity of link**. Depending upon the type of motion, the velocity is classified into two types, namely linear velocity and angular velocity.

Linear velocity:
$$v = \frac{ds}{dt} \tag{2.3}$$

Angular velocity:
$$\omega = \frac{d\theta}{dt} \tag{2.4}$$

where
s = linear displacement of the link
θ = angular displacement of the link and
t = time.

In kinematic analysis, the velocities of various links can be determined by the following methods:

(i) Relative velocity method or velocity polygon method
(ii) Instantaneous centre method
(iii) Auxiliary point method
(iv) Analytical method

In this chapter, some of these methods are discussed.

2.3 RELATIVE VELOCITY METHOD

Consider a rigid link PQ of length r which rotates about a fixed point P with a uniform angular velocity ω rad/s in clockwise direction [See Figure 2.2(a)]. When the link PQ takes a small turn through an angle $\delta\theta$ in a time δt, the point Q will travel along the arc QQ' as shown in Figure 2.2(b). The velocity of point Q relative to the fixed point P can be expressed as:

$$v_{qp} = \frac{\delta\theta}{\delta t} \times r$$

or
$$v_{qp} = \frac{d\theta}{dt} \times r \quad \text{when } \delta t \to 0$$

or
$$v_{qp} = \omega r \tag{2.5}$$

This shows that as the time δt approaches zero, the arc QQ' will be perpendicular to link PQ. The magnitude of velocity of point Q is ωr and its direction is perpendicular to the axis of the link in the direction of rotation. Therefore, velocity vector of link PQ can be represented as shown in Figure 2.2(c).

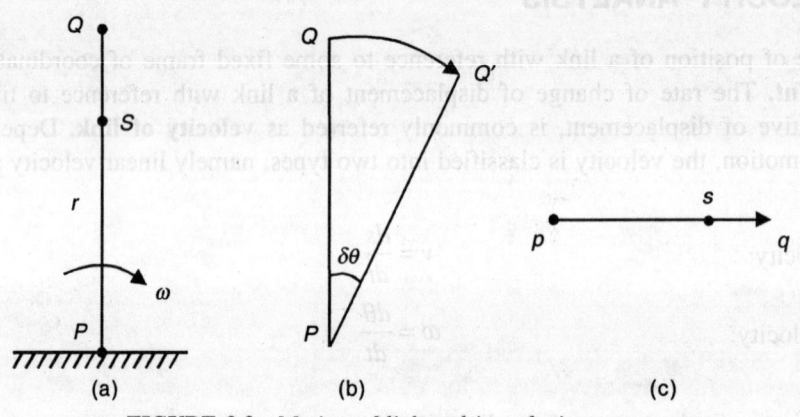

FIGURE 2.2 Motion of link and its velocity vector.

Consider an intermediate point S on the link PQ. The velocity of point S relative to the point P is:

$$v_{sp} = \omega \cdot SP \tag{2.6}$$

and velocity v_{sp} can be represented by vector $\overrightarrow{ps}$ on vector diagram such that:

$$\frac{v_{sp}}{v_{qp}} = \frac{\overrightarrow{ps}}{\overrightarrow{pq}} = \frac{\omega \times SP}{\omega \times QP} = \frac{SP}{QP} \tag{2.7}$$

Hence the point s divides the vector $\overrightarrow{pq}$ in the same ratio as the point S divides the link PQ. Thus, this law of proportionality is useful in drawing the velocity polygon and finding the relative velocities of points on the link.

The application of relative velocity method is illustrated through the following three most popular mechanisms:

2.3.1 Slider-crank Mechanism

Consider a slider-crank mechanism used in reciprocating engine in which the crank OC rotates with uniform angular velocity ω rad/s in clockwise direction [Figure 2.3(a)]. The slider P moves on a fixed guide G. The crank and slider are joined by a coupler link CP, commonly referred to as connecting rod. The velocities of various links of a slider-crank mechanism can be determined by drawing velocity polygon as discussed below:

1. Compute velocity of point C relative to the fixed point O. $v_{co} = \omega \times OC$
2. Choose a convenient point o and draw a vector $\overrightarrow{oc}$ perpendicular to link OC to some suitable scale as shown in Figure 2.3(b).

(a) Slider-crank mechanism (b) Velocity polygon

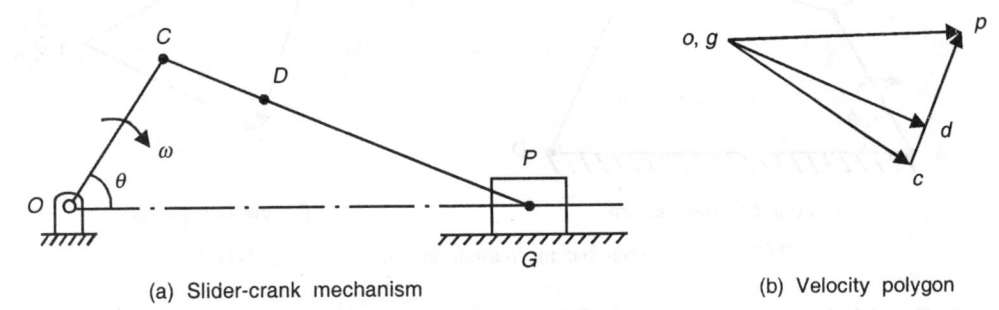

FIGURE 2.3 Velocity analysis of slider-crank mechanism.

3. Considering the coupler CP as a rigid link, the velocity of slider P relative to the point C v_{pc} is perpendicular to the link CP. Thus from point c, draw a vector $\overrightarrow{cp}$ of unknown magnitude, perpendicular to the link CP.
4. The velocity of slider P relative to guide G v_{pg} is along the line of stroke. Since there is no relative motion between the guide G and the fixed point O, the point g on the velocity polygon can be assumed to be at the point o. Thus from point g, draw a line parallel to OP to intersect vector $\overrightarrow{cp}$ at point p. The vector $\overrightarrow{gp}$ represents the velocity of the slider P.

The velocity of any point D located on the connecting rod CP can be determined with the help of the following relation:

$$\frac{\overrightarrow{cd}}{\overrightarrow{cp}} = \frac{CD}{CP} \quad \text{or} \quad \overrightarrow{cd} = \overrightarrow{cp} \times \frac{CD}{CP} \tag{2.8}$$

Therefore, the point d on the velocity vector $\overrightarrow{cp}$ can be located as shown in Figure 2.3(b). The vector $\overrightarrow{od}$ represents the absolute velocity of point D.

2.3.2 Four Bar Mechanism

Consider a four bar mechanism in which the driver link AB rotates at uniform angular velocity ω rad/s in counter clockwise direction [Figure 2.4(a)]. The link AD is a fixed link and BC

is coupler link. The velocity of the point B relative to point A is given as $v_{ba} = \omega \times AB$ in counterclockwise sense and perpendicular to link AB. The velocity polygon of a four bar mechanism can be constructed as follows:

1. Compute the velocity of point B relative to point A, $v_{ba} = \omega \times AB$
2. Choose a convenient point a and draw a vector $\overrightarrow{ab}$ perpendicular to link AB in counterclockwise sense to some suitable scale as shown in Figure 2.4(b).

(a) Four bar mechanism   (b) Velocity polygon

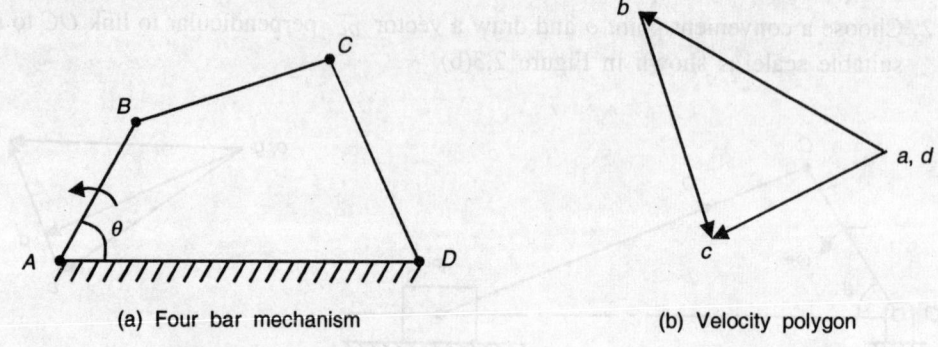

FIGURE 2.4 Four bar mechanism and its velocity polygon.

3. Considering the coupler link BC as a rigid link, the velocity of point C relative to the point B, v_{cb}, is perpendicular to the link BC. Thus from point b, draw a vector $\overrightarrow{bc}$ of unknown magnitude, perpendicular to link BC.
4. From point d, which is a fixed point lying at point a, draw a vector $\overrightarrow{dc}$ perpendicular to link CD. The vectors $\overrightarrow{dc}$ and $\overrightarrow{bc}$ intersect at point c. The vector $\overrightarrow{dc}$ represents the velocity of point C relative to point D, v_{cd}.

The velocity of any point on coupler link BC and follower link CD can be found by law of proportionality as discussed in previous section.

2.3.3 Crank and Slotted Lever Mechanism

A crank and slotted lever mechanism, which is one type of quick return motion mechanism, is widely used in shaper machines. It executes the return stroke in the shortest possible time (Figure 2.5). In this mechanism, the crank OB rotates with uniform angular velocity ω rad/s about the fixed point O. At the end B of the crank, a slider is pivoted such that C be a point on the link AD immediately below the slider B. As the crank OB rotates about the point O, the slider reciprocates along the lever AD and there is a relative motion between the points B and C. The lever AD oscillates about the fulcrum point A. The other end of the lever is connected to ram R through a link DR. A cutting tool is attached to the ram which performs cutting and idle stroke.

The velocities of various links of a crank slotted lever mechanism can be determined by drawing a velocity polygon as discussed below:

1. Compute tangential velocity of crank OB, $v_{bo} = \omega \times OB$

Kinematic Analysis of Mechanism

2. Choose a convenient point o and draw a vector $\overrightarrow{ob}$ perpendicular to crank OB in counterclockwise direction to some suitable scale as shown in Figure 2.5(b).

(a) Crank and slotted lever mechanism

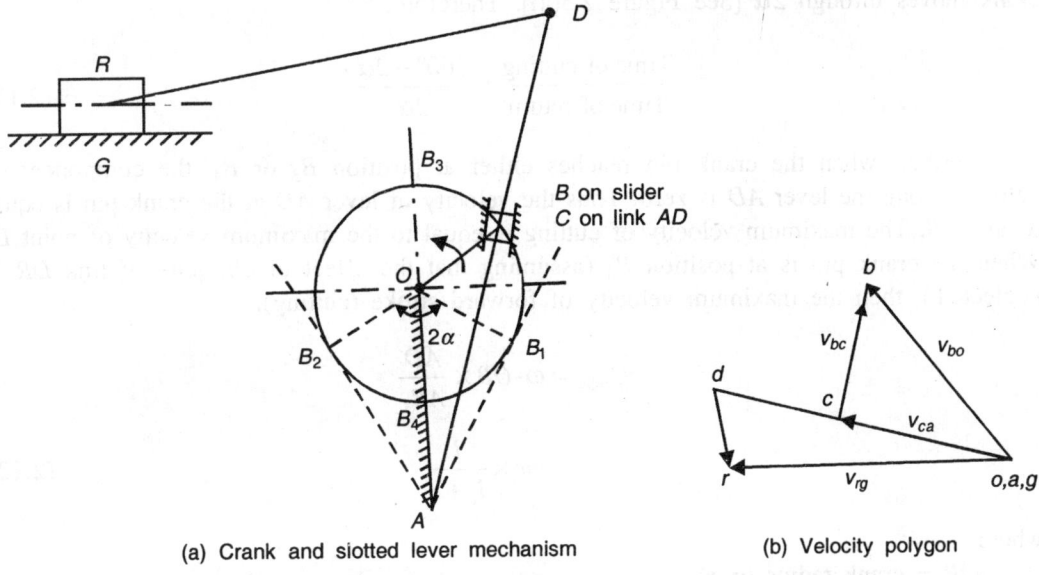

(b) Velocity polygon

FIGURE 2.5 Velocity analysis of crank and slotted lever mechanism.

The vector $\overrightarrow{ob}$ has two components, one parallel to the lever AD and another perpendicular to it, such that the following vector equation is satisfied:

$$v_{bo} = v_{ca} + v_{bc}$$

or
$$\overrightarrow{ob} = \overrightarrow{ac} + \overrightarrow{cb} \qquad (2.9)$$

where velocity of point C relative to point A, v_{ca} is perpendicular to lever AD and v_{bc} is parallel to the lever AD.

3. Therefore, draw a line ac perpendicular to the lever AD. Also draw another line bc from point b which is parallel to the lever AD. These lines intersect at point c.

4. The velocity of point C on the lever AD is represented by vector $\overrightarrow{ac}$. Extend this line up to point d in proportion such that it satisfies the following relation:

$$\frac{\overrightarrow{ac}}{\overrightarrow{ad}} = \frac{AC}{AD} \qquad (2.10)$$

5. Draw a vector $\overrightarrow{dr}$ of unknown magnitude perpendicular to the link DR.

6. The velocity of the ram R relative to the fixed guide G, v_{rg} is along the line of stroke. Thus from point g, draw a line parallel to the line of stroke to intersect vector $\overrightarrow{dr}$ at point r. The vector $\overrightarrow{gr}$ represents the velocity of the ram relative to the fixed point G.

In a crank and slotted lever mechanism, when the crank takes the position OB_1 and OB_2, the component of velocity of crank pin perpendicular to the lever is zero. Thus during forward stroke, the crank moves through $(360° - 2\alpha)$ and during return stroke, it executes when the crank moves through 2α [See Figure 2.5(a)]. Therefore,

$$\frac{\text{Time of cutting}}{\text{Time of return}} = \frac{360° - 2\alpha}{2\alpha} \qquad (2.11)$$

Further, when the crank pin reaches either at position B_3 or B_4, the component of velocity along the lever AD is zero. Thus the velocity of lever AD at the crank pin is equal to $\omega \cdot OB$. The maximum velocity of cutting is equal to the maximum velocity of point D. When the crank pin is at position B_3 (assuming that the effect of obliquity of link DR is neglected.), then the maximum velocity of forward stroke (cutting):

$$v_{c,\max} = \omega \cdot OB \times \frac{AD}{AB_3}$$

$$= \omega r \times \frac{l}{l_1 + r} \qquad (2.12)$$

where
OB = crank radius $(= r)$
AD = length of the lever $(= l)$ and
l_1 = length of the fixed link OA.

Similarly, the maximum velocity of return stroke:

$$v_{r,\max} = \omega \cdot OB \times \frac{AD}{AB_4}$$

$$= \omega r \times \frac{l}{l_1 - r} \qquad (2.13)$$

Therefore, the ratio of velocities is:

$$\frac{v_{r,\max}}{v_{c,\max}} = \frac{\text{Maximum velocity of return}}{\text{Maximum velocity of cutting}} = \frac{l_1 + r}{l_1 - r} \qquad (2.14)$$

EXAMPLE 2.1 In a slider-crank mechanism, the crank and the connecting rod are 150 mm and 400 mm long respectively. The crank rotates at 360 rpm in clockwise direction. When crank has turned through 60°, determine:

(i) velocity of the slider
(ii) angular velocity of the connecting rod
(iii) velocity of point C on the connecting rod which is 300 mm away from point A
(iv) position of any point on connecting rod which has the minimum velocity.

Solution

Angular velocity of crank, $\omega_{ao} = \dfrac{2\pi N}{60}$

or
$$\omega_{ao} = \dfrac{2\pi \times 360}{60} = 37.7 \text{ rad/s}$$

Linear velocity of crank, $v_{ao} = \omega_{ao} \times OA$

$$= 37.7 \times 0.15 = 5.655 \text{ m/s}$$

The procedure to draw velocity polygon is given below:

1. Draw configuration diagram of slider-crank mechanism as shown in Figure 2.6(a).

(a) Configuration diagram

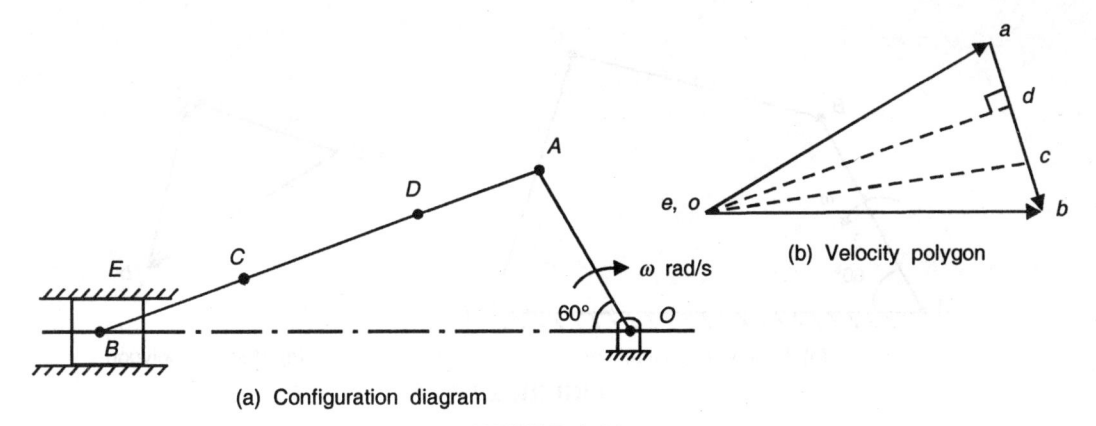

(b) Velocity polygon

FIGURE 2.6

2. Draw velocity vector $\overrightarrow{oa}$ perpendicular to link OA equal to 5.655 m/s to convenient scale (1 m/s = 10 mm).
3. From point a, draw vector $\overrightarrow{ab}$ perpendicular to link AB. From point o, draw vector $\overrightarrow{ob}$ parallel to the path of slider B, both meeting at b. The velocity polygon is oab. [Figure 2.6(b)].
4. Mark point c on vector $\overrightarrow{ab}$, such that

$$\dfrac{AC}{AB} = \dfrac{\overrightarrow{ac}}{\overrightarrow{ab}}$$

   (i) Velocity of slider: $v_b = \overrightarrow{ob} = $ **5.9 m/s**   Ans.
   (ii) Angular velocity, of connecting rod:

$$\omega_{ab} = \dfrac{\overrightarrow{ab}}{AB} = \dfrac{2.9}{0.4} = 7.25 \text{ rad/s} \qquad \text{Ans.}$$

   (iii) Velocity of point C: $v_c = \overrightarrow{oc} = $ **5.6 m/s**   Ans.
   (iv) For minimum velocity, draw a line od perpendicular to vector $\overrightarrow{ab}$ at point d. The position of point d decides the location of point D on connecting rod.

$$\frac{AD}{AB} = \frac{\overrightarrow{ad}}{\overrightarrow{ab}}$$

or

$$AD = \frac{\overrightarrow{ad}}{\overrightarrow{ab}} \times AB = \frac{11}{29} \times 400 = \mathbf{151.7 \text{ mm}} \qquad \text{Ans.}$$

EXAMPLE 2.2 The dimensions and configuration of the four bar mechanism shown in Figure 2.7(a) are as follows:

$AB = 300$ mm, $BC = 360$ mm, $CD = 360$ mm, $AD = 600$ mm and $\angle BAD = 60°$

The crank AB has an angular velocity of 10 rad/s in clockwise direction. Determine the angular velocities of links BC and CD and the velocity of the joint C.

(a) Four bar mechanism   (a) Velocity polygon

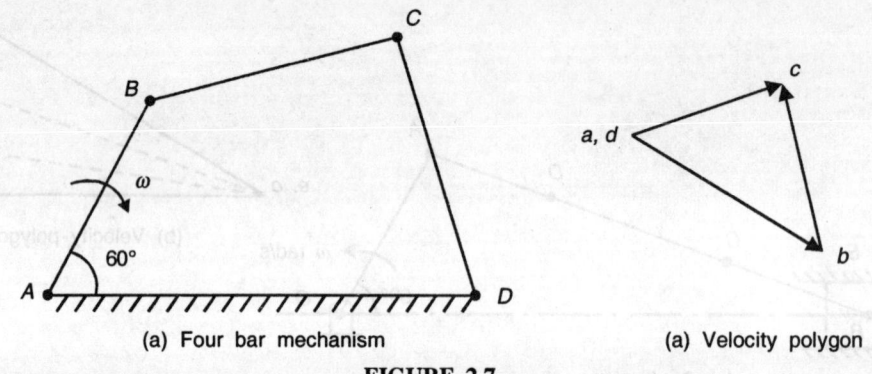

FIGURE 2.7

Solution Linear velocity of crank point B relative to A is:

$$v_{ba} = \omega \times AB = 10 \times 0.3 = 3 \text{ m/s}$$

Draw the configuration diagram and obtain the velocity polygon abc, to a convenient scale as shown in Figure 2.7(b).

Measure vectors $\overrightarrow{bc}$ and $\overrightarrow{cd}$.

(i) $\omega_{bc} = \dfrac{\overrightarrow{bc}}{BC} = \dfrac{2.25}{0.36} = \mathbf{6.25 \text{ rad/s}}$ \hfill Ans.

(ii) $\omega_{cd} = \dfrac{\overrightarrow{cd}}{CD} = \dfrac{2.1}{0.36} = \mathbf{5.83 \text{ rad/s}}$ \hfill Ans.

(iii) $v_c = \overrightarrow{cd} = \mathbf{2.1 \text{ m/s}}$ \hfill Ans.

EXAMPLE 2.3 In a mechanism shown in Figure 2.8, the crank AB rotates about point A at uniform speed of 240 rpm in clockwise direction. The link CD oscillates about the fixed point D, which is connected to link AB by a coupler link BC. The slider F moves in horizontal guides, being driven by the link EF.

Kinematic Analysis of Mechanism

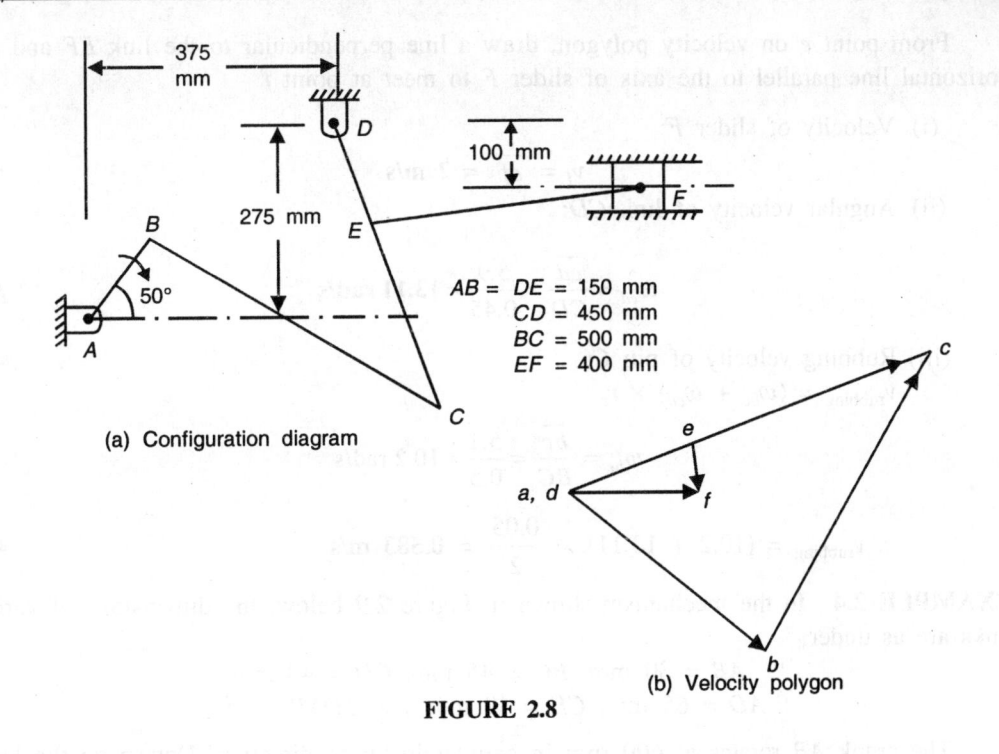

FIGURE 2.8
(a) Configuration diagram
(b) Velocity polygon

AB = DE = 150 mm
CD = 450 mm
BC = 500 mm
EF = 400 mm

Determine:

(i) velocity of slider F
(ii) angular velocity of link CD
(iii) rubbing velocity of pin C which is 50 mm in diameter.

Solution

Angular velocity of link AB: $\omega_{ba} = \dfrac{2\pi N}{60}$

or

$$\omega_{ba} = \dfrac{2\pi \times 240}{60} = 25.13 \text{ rad/s}$$

Velocity of point B on link AB:

$$v_{ba} = \omega_{ba} \times AB = 25.13 \times 0.15 = 3.77 \text{ m/s}$$

Draw configuration diagram (to the scale) as shown in Figure 2.8(a).
Draw velocity polygon $abcd$ to a convenient scale as shown in Figure 2.8(b).
In velocity polygon, locate point e on vector $\vec{cd}$ such that

$$\dfrac{ED}{CD} = \dfrac{\vec{ed}}{\vec{cd}} \quad \text{or} \quad \vec{ed} = \dfrac{150}{450} \times 60.0 = 20 \text{ mm}$$

From point *e* on velocity polygon, draw a line perpendicular to the link *EF* and a horizontal line parallel to the axis of slider *F* to meet at point *f*.

(i) Velocity of slider *F*:
$$v_f = \overline{af} = 2 \text{ m/s} \qquad \text{Ans.}$$

(ii) Angular velocity of link *CD*:
$$\omega_{cd} = \frac{\overline{cd}}{CD} = \frac{5.9}{0.45} = 13.11 \text{ rad/s} \qquad \text{Ans.}$$

(iii) Rubbing velocity of pin *C*:
$$v_{\text{rubbing}} = (\omega_{bc} + \omega_{cd}) \times r_c$$
$$\omega_{bc} = \frac{\overline{bc}}{BC} = \frac{5.1}{0.5} = 10.2 \text{ rad/s}$$
$$\therefore v_{\text{rubbing}} = (10.2 + 13.11) \times \frac{0.05}{2} = 0.583 \text{ m/s} \qquad \text{Ans.}$$

EXAMPLE 2.4 In the mechanism shown in Figure 2.9 below, the dimensions of various links are as under:
$$AB = 30 \text{ mm}, BC = 45 \text{ mm}, CD = 40 \text{ mm}$$
$$AD = 65 \text{ mm}, CE = 40 \text{ mm and } \angle DAB = 75°$$

The crank *AB* rotates at 600 rpm in counterclockwise direction. Determine the linear velocity of slider *E* and angular velocity of link *CE*.

(a) Configuration diagram

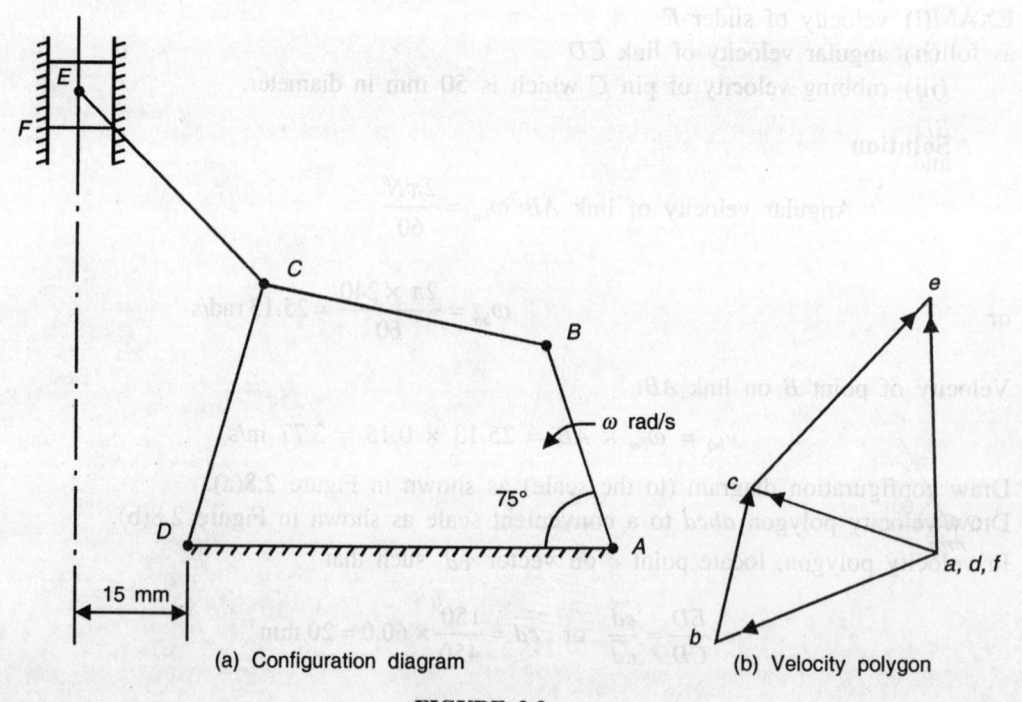

(b) Velocity polygon

FIGURE 2.9

Solution Angular velocity of crank AB:

$$\omega_{ba} = \frac{2\pi N}{60}$$

Linear velocity of point B on link AB:

$$v_{ba} = \omega_{ba} \times AB$$

$$= \frac{2\pi \times 600}{60} \times \frac{30}{1000} = 1.885 \text{ m/s}$$

Draw the configuration diagram to a convenient scale as shown in Figure 2.9(a). To draw the velocity polygon as shown in Figure 2.9(b), draw vector $\overline{ab}$ perpendicular to AB and equal to 1.885 m/s to convenient scale.

From point b, draw vector $\overline{bc}$ perpendicular to link BC and from point d, draw another vector $\overline{cd}$ perpendicular to link CD, both meeting at point c. Then from point c, draw a vector $\overline{ce}$ perpendicular to link CE and from point f, draw a line parallel to the axis of slider E, both meeting at point e.

(i) Linear velocity of slider: $v_e = \overline{fe} = \mathbf{1.85}$ **m/s** Ans.
(ii) Angular velocity of link CE:

$$\omega_{ce} = \frac{\overline{ce}}{CE} = \frac{1.95}{0.04} = \mathbf{48.75 \text{ rad/s}} \qquad \text{Ans.}$$

EXAMPLE 2.5 Figure 2.10 below shows a mechanism in which lengths of various links are as follows:

$OA = 100$ mm, $AB = 350$ mm, $BC = 170$ mm
$BD = CD = 140$ mm, $OD = 400$ mm, $CE = 350$ mm
and $\angle DOA = 60°$

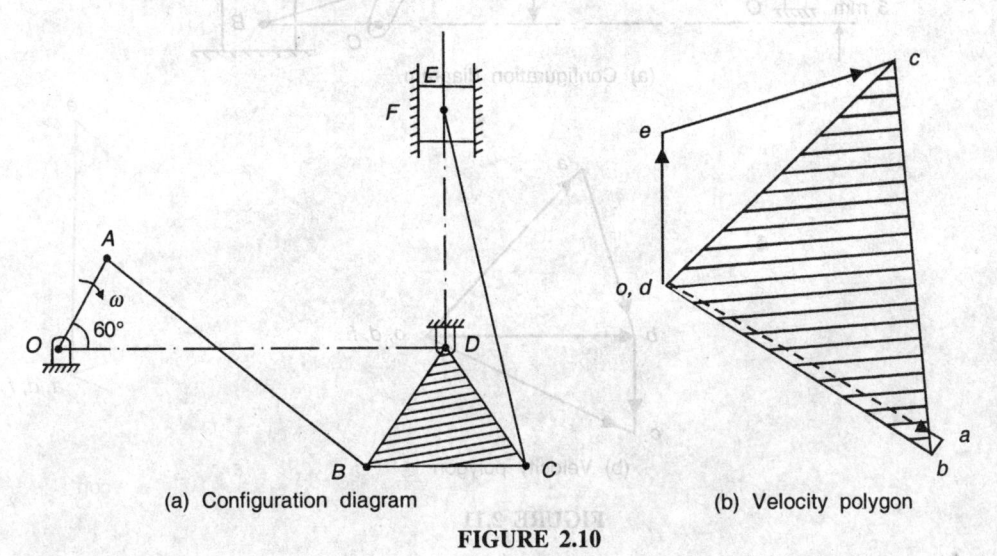

(a) Configuration diagram (b) Velocity polygon

FIGURE 2.10

If the slider E moves up with a velocity of 5 m/s when the crank OA rotates at 60°, determine the speed and direction of rotation of the crank OA.

Solution Draw configuration diagram as shown in Figure 2.10(a) to a convenient scale. To draw the velocity polygon as shown in Figure 2.10(b), draw vector $\overline{de}$ in vertical upward direction equal to 5 m/s to a convenient scale as slider E moves vertical upward direction. Draw vector $\overline{ec}$ perpendicular to EC and vector $\overline{dc}$ perpendicular to DC to meet at point c.

Draw velocity image of ternary link BCD. Also draw vectors $\overline{ab}$ and $\overline{oa}$ perpendicular to AB and OA respectively to meet at point a. Measure the length of vector $\overline{oa}$.

Angular velocity of link OA: $\omega_{oa} = \dfrac{\overline{oa}}{OA} = \dfrac{10.6}{0.1} = 106$ rad/s

Speed: $N = \dfrac{60\omega}{2\pi} = \dfrac{60 \times 106}{2\pi} = $ **1012.2 rpm in clockwise direction.** **Ans.**

EXAMPLE 2.6 In the mechanism shown in Figure 2.11 below, the dimensions of various links are as follows:

$OA = 25$ mm, $AB = 80$ mm, $BC = 20$ mm, $CD = 33$ mm, $DE = 20$ mm

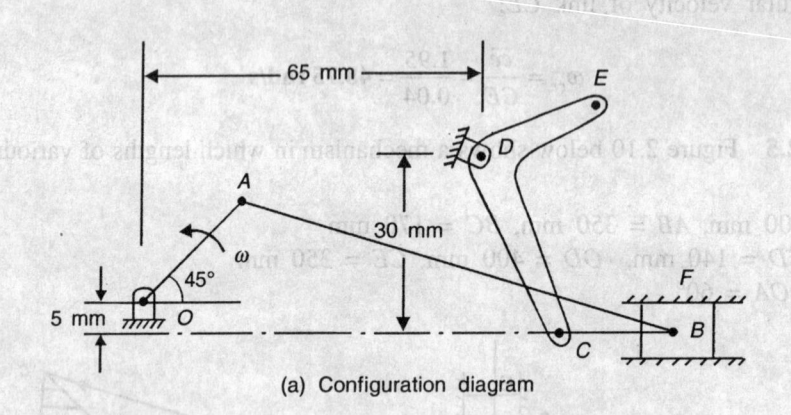

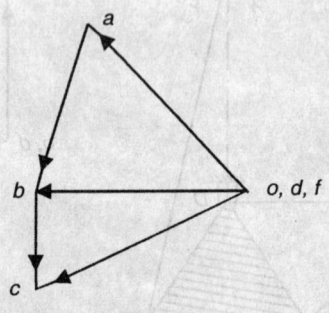

(a) Configuration diagram

(b) Velocity polygon

FIGURE 2.11

If the crank OA is inclined at 45° and revolves at uniform speed of 120 rpm in counter clockwise direction, determine the angular velocities of links AB, BC and CD. Also find the linear velocity of slider B.

Solution Linear velocity of point A on link OA:

$$v_{ao} = \frac{2\pi N}{60} \times OA = \frac{2\pi \times 120}{60} \times \frac{25}{1000} = 0.314 \text{ m/s}$$

Construct configuration diagram to the convenient scale as shown in Figure 2.11(a). Draw vector $\overrightarrow{oa}$ perpendicular to link OA equal to 0.314 m/s at convenient scale. Draw vector $\overrightarrow{ab}$ perpendicular to link AB and $\overrightarrow{ob}$ parallel to the axis of the slider to meet at point b.

Further, draw vector $\overrightarrow{bc}$ perpendicular to link BC and $\overrightarrow{cd}$ perpendicular to link CD meeting at point c [Figure 2.11(b)].

(i) Angular velocity of link AB:

$$\omega_{ab} = \frac{\overrightarrow{ab}}{AB} = \frac{0.23}{0.08} = 2.875 \text{ rad/s} \quad \text{Ans.}$$

(ii) Angular velocity of link BC:

$$\omega_{bc} = \frac{\overrightarrow{bc}}{BC} = \frac{0.13}{0.02} = 6.5 \text{ rad/s} \quad \text{Ans.}$$

(iii) Angular velocity of link CD:

$$\omega_{cd} = \frac{\overrightarrow{cd}}{CD} = \frac{0.31}{0.033} = 9.4 \text{ rad/s} \quad \text{Ans.}$$

(iv) Linear velocity of slider B:

$$v_b = \overrightarrow{ob} = 0.29 \text{ m/s} \quad \text{Ans.}$$

EXAMPLE 2.7 The following data refer to a crank and slotted lever type quick return mechanism (Figure 2.12).

Distance between the fixed centres O and A = 200 mm
Length of driver link OB = 75 mm
Length of link AD = 400 mm
Length of link DE = 200 mm

If the crank OB rotates at 300 rpm in counterclockwise direction and makes an angle AOB = 120°, determine the velocity of the slider E and angular velocities of the links AD and DE.

(a) Configuration diagram

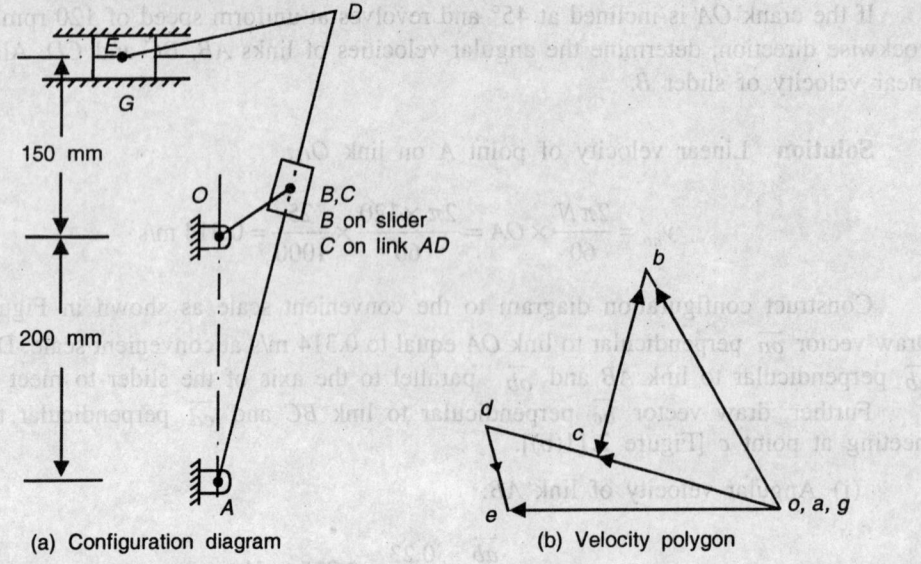

(b) Velocity polygon

FIGURE 2.12

Solution Draw configuration diagram to a convenient scale as shown in Figure 2.12(a).
Angular velocity of crank:

$$\omega_{bo} = \frac{2\pi N}{60} = \frac{2\pi \times 300}{60} = 31.41 \text{ rad/s}$$

Linear velocity of point B relative to O:

$$v_{bo} = \omega_{bo} \times BO = 31.41 \times 0.075 = 2.356 \text{ m/s}$$

Draw vector $\overrightarrow{ob}$ perpendicular to link OB equal to 2.356 m/s to convenient scale. Draw a vector $\overrightarrow{ac}$ perpendicular to link AD and another vector $\overrightarrow{bc}$ parallel to link AD to meet at point c.

Extend the vector $\overrightarrow{ac}$ up to $\overrightarrow{ad}$ such that

$$\frac{\overrightarrow{ac}}{\overrightarrow{ad}} = \frac{AC}{AD}$$

From point d, draw a vector $\overrightarrow{de}$ perpendicular to link DE and another vector $\overrightarrow{ge}$ parallel to the axis of slider to meet at point e. See Figure 2.12(b).

(i) Velocity of slider E:

$$v_e = \overrightarrow{ge} = 2.4 \text{ m/s} \qquad \text{Ans.}$$

(ii) Angular velocity of link AD:

$$\omega_{ad} = \frac{\overrightarrow{ad}}{AD} = \frac{2.65}{0.4} = 6.625 \text{ rad/s} \qquad \text{Ans.}$$

(iii) Angular velocity of link DE:

$$\omega_{de} = \frac{\overrightarrow{de}}{DE} = \frac{0.75}{0.2} = 3.75 \text{ rad/s} \qquad \text{Ans.}$$

EXAMPLE 2.8 In a quick return mechanism used in shaping machine, the ratio of maximum velocities is 5/3. If the length of the stroke be 200 mm, find:

  (i) the length of the slotted lever
  (ii) quick return ratio
  (iii) maximum cutting velocity per second, if the crank rotates at 25 rpm.

Solution Ratio of velocities:

$$\frac{\text{Maximum velocity during return stroke}}{\text{Maximum velocity during forward stroke}} = \frac{l_1 + r}{l_1 - r} = \frac{5}{3}$$

or
$$3(l_1 + r) = 5(l_1 - r)$$

or
$$l_1 = 4r$$

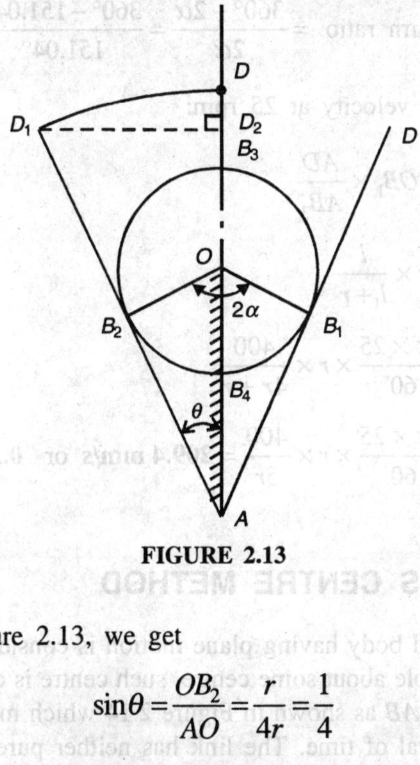

FIGURE 2.13

(i) Referring to Figure 2.13, we get

$$\sin\theta = \frac{OB_2}{AO} = \frac{r}{4r} = \frac{1}{4}$$

or
$$\theta = 14.48°$$

Also in triangle D_2AD_1,

$$\sin\theta = \frac{D_1D_2}{AD_1} = \frac{\frac{1}{2} \times \text{Stroke length}}{l}$$

or
$$l = \frac{\text{Stroke length}}{2\sin\theta} = \frac{200}{2\times\sin 14.48°}$$
$$= 400 \text{ mm}$$

(ii) Quick return ratio $= \dfrac{\text{Time for cutting}}{\text{Time for return}} = \dfrac{360° - 2\alpha}{2\alpha}$

In triangle OB_2A,
$$\alpha = 180° - 90° - 14.48° = 75.52°$$
or
$$2\alpha = 151.04°$$
Therefore,
$$\text{Quick return ratio} = \frac{360° - 2\alpha}{2\alpha} = \frac{360° - 151.04°}{151.04°} = 1.383 \qquad \text{Ans.}$$

(iii) Maximum cutting velocity at 25 rpm:
$$v_{max} = \omega \cdot OB_1 \times \frac{AD}{AB_3}$$
$$= \omega r \times \frac{l}{l_1 + r}$$
$$= \frac{2\pi \times 25}{60} \times r \times \frac{400}{4r + r}$$
$$= \frac{2\pi \times 25}{60} \times r \times \frac{400}{5r} = 209.4 \text{ mm/s} \quad \text{or} \quad 0.2094 \text{ m/s} \qquad \text{Ans.}$$

2.4 INSTANTANEOUS CENTRE METHOD

If the displacement of a rigid body having plane motion is considered as equivalent to a pure rotation of the body as a whole about some centre, such centre is called **instantaneous centre**. To illustrate, consider a link AB as shown in Figure 2.14 which moves from its initial position AB to A_1B_1 in a short interval of time. The link has neither pure translatory nor pure rotary motion but a combination of these two motions.

The combined motion of translation and rotation of the link from its initial position AB to the position A_1B_1 may be assumed to be a motion of pure rotation about a certain centre I which is the point of intersection of the perpendicular bisector of AA_1 and BB_1. Mathematically, it can be shown that the angular rotation of point A about the centre of

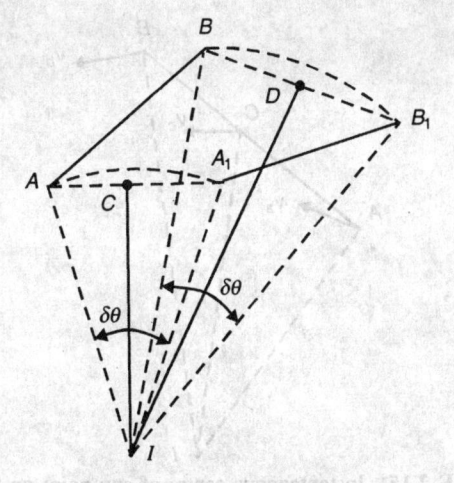

FIGURE 2.14 Motion of a link about instantaneous centre.

rotation I is equal to the angular rotation of point B about the centre I. The displacements of points A and B can be given by:

$$\delta x_a = IA \sin \delta\theta/2 \qquad (2.15)$$

and

$$\delta x_b = IB \sin \delta\theta/2 \qquad (2.16)$$

For a small displacement, the angle $\delta\theta$ becomes infinitesimal and the rigid body AB is said to be rotating instantaneously about the point I as an instantaneous centre. For each configuration of a mechanism, there is one instantaneous centre. The position of the instantaneous centre changes with the alteration of configuration of mechanism.

Dividing both sides of Eqs. (2.15) and (2.16) by δt and taking the limits, the magnitude of instantaneous velocities of points A and B can be represented as:

$$v_a = \omega \cdot IA \quad \text{and} \quad v_b = \omega \cdot IB \qquad (2.17)$$

where ω is the angular velocity of IA and IB ($= d\theta/dt$).

The direction of velocities of points A and B are obviously perpendicular to the radii IA and IB.

The usefulness of the instantaneous centre of link lies in the fact that velocity of any point in the body is proportional to its distance from the instantaneous centre and the direction of velocity is in a line perpendicular to the line joining that point to the instantaneous centre. Therefore, referring to Figure 2.15, the following relation holds good:

$$\frac{v_c}{IC} = \frac{v_a}{IA} = \frac{v_b}{IB} = \text{angular velocity of link } AB \qquad (2.18)$$

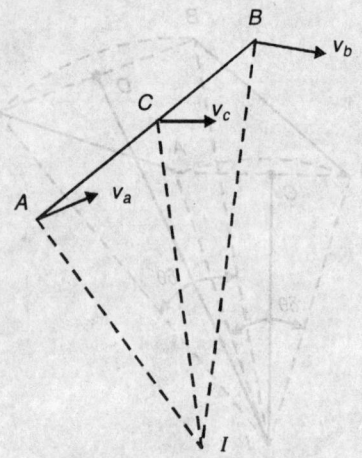

FIGURE 2.15 Instantaneous centre of any point on the link.

Thus an instantaneous centre makes it possible to determine the velocity of any point on a link if velocities of at least two points on the link or location of its instantaneous centre are known.

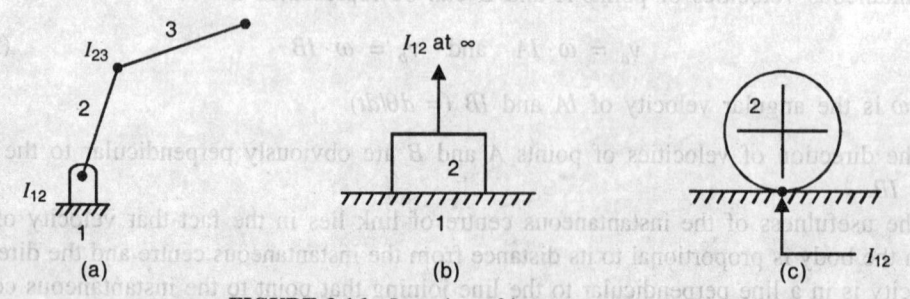

The location of instantaneous centres for various types of motion in a mechanism can be decided on the basis of the following rules:

1. When two links form a turning pair, the instantaneous centre is assumed to be located at the centre of the pair [Figure 2.16(a)].
2. In case of sliding pair, the instantaneous centre lies at infinity in the direction perpendicular to the path of motion of the slider [Figure 2.16(b)].
3. When two links make a pure rolling contact, the point of contact at a given instant is taken as instantaneous centre [Figure 2.16(c)].

FIGURE 2.16 Location of instantaneous centre.

2.4.1 Arnold Kennedy Theorem

The Arnold Kennedy theorem states that if three links have relative motion with respect to each other, their relative instantaneous centre lies on a straight line.

To prove this theorem, let us consider three planar links 1, 2 and 3 having relative motion with respect to each other in a plane. Let the links 1 and 2 rotate about the centres

I_{13} and I_{23} respectively relative to the fixed link 3. Thus the instantaneous centre of the links 1 and 3 is I_{13} and that of the links 2 and 3 is I_{23} (Figure 2.17).

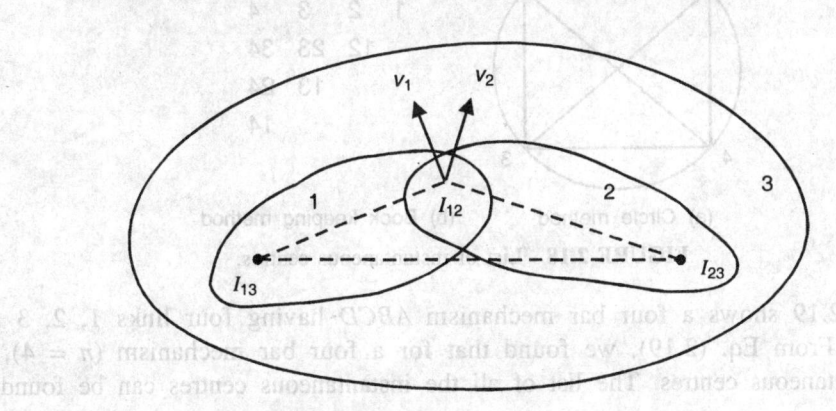

FIGURE 2.17 Arnold Kennedy theorem.

Now let us consider that the links 1 and 2 both are moving relative to the fixed link 3 such that their mutual instantaneous centre I_{12} lies outside the line joining I_{13} and I_{23} on either link 1 or link 2. If the point I_{12} is considered on the link 1, its velocity v_1 will be perpendicular to the line joining I_{13} and I_{12}. Similarly, if the point I_{12} is considered on the link 2, its velocity v_2 will be perpendicular to the line joining I_{23} and I_{12}. By definition, we know that the velocity of point I_{12} should be same whether it is on link 1 or on the link 2. This is possible only if point I_{12} lies on the line joining I_{13} and I_{23}. This proves the Arnold Kennedy theorem. The actual location of point I_{12} on this line depends upon the magnitude of angular velocities of links 1 and 2 relative to link 3.

2.4.2 Locating Instantaneous Centres

The following procedure may be used for locating instantaneous centres on the mechanism:

1. Determine the number of instantaneous centres. For two links having relative motion between them, there is one instantaneous centre. Therefore, total number of instantaneous centres is:

$$N = \frac{n(n-1)}{2} \qquad (2.19)$$

where n is the number of links

2. Make a list of all the instantaneous centres with the help of circle diagram or by book keeping method, as shown in Figure 2.18.
3. Locate primary centres at the kinematic pairs of the link.
4. Locate remaining centres by the use of kennedy theorem.

The above procedure is illustrated further with the help of an example of four bar mechanism.

(a) Circle method   (b) Book keeping method

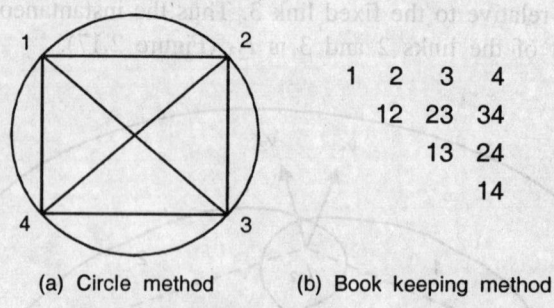

FIGURE 2.18 List of instantaneous centres.

Figure 2.19 shows a four bar mechanism *ABCD* having four links 1, 2, 3 and 4 respectively. From Eq. (2.19), we found that for a four bar mechanism ($n = 4$), there are six instantaneous centres. The list of all the instantaneous centres can be found from Figure 2.18.

Now the centre of rotation of link 2 relative to the link 1, which is fixed at point *A*, is I_{12}. Since the link 1 is fixed, the location of this instantaneous centre is not going to change with rotation of link 2. Therefore, this instantaneous centre is called **fixed instantaneous centre.** Similarly, another fixed instantaneous centre I_{14} between links 1 and 4 is located at point *D*.

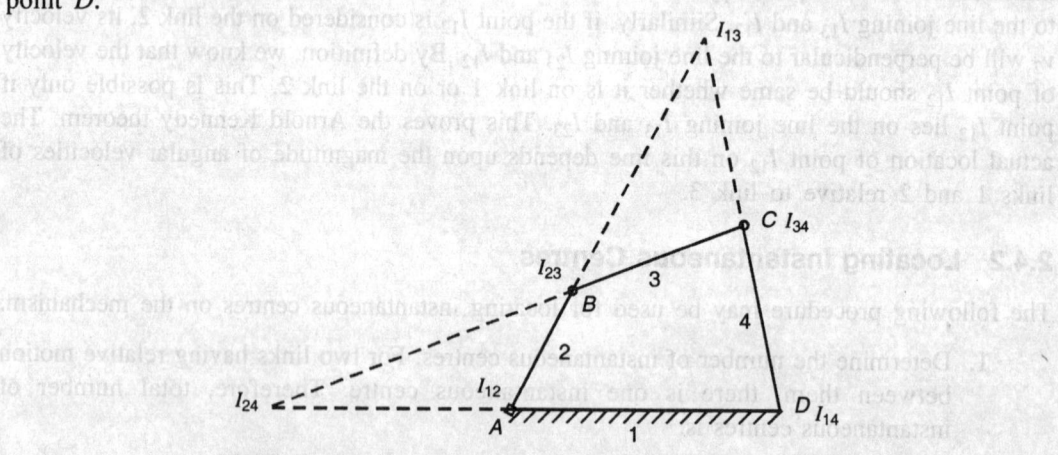

FIGURE 2.19 Instantaneous centre of a four bar mechanism.

The instantaneous centre for the links 2 and 3, I_{23}, lies at the point *B*. Although the position of instantaneous centre I_{23} changes with rotation of either link 2 or link 3, it will always be located at the centre of turning pair. Such type of instantaneous centres are called **permanent instantaneous centres.** Similarly, another permanent instantaneous centre I_{34} is located at the point *C*. These fixed and permanent instantaneous centres on the mechanism can be located by inspection only. However, the third type of instantaneous centres which are neither fixed nor permanent can be located by Arnold Kennedy theorem.

Now the instantaneous centre I_{13} is the point about which the link 3 rotates with reference to link 1 for a given position. Since the links 1 and 3 do not have any connection, the instantaneous centre I_{13} is neither fixed nor permanent. To locate this point, consider three links 1, 2 and 3 for which instantaneous centres I_{12} and I_{23} are already known. Therefore, according to Kennedy theorem, instantaneous centre I_{13} must lie on the line joining the points I_{12} and I_{23}. Similarly, for another set of links 1, 4 and 3, the positions of instantaneous centres I_{14} and I_{34} are already known. Therefore, the location of instantaneous centre I_{13} can be marked on the line joining the centres I_{14} and I_{34}. The intersection of these two lines locate the position of instantaneous centre I_{13}.

The instantaneous centre I_{24} can be located in similar way to instantaneous centre I_{13} (Figure 2.19).

EXAMPLE 2.9 Find all the instantaneous centres of the slider-crank mechanism shown in Figure 2.20 below and find the velocity of the slider when the crank OA rotates with an angular velocity of 10 rad/s. Also determine the angular velocity of the connecting rod. The lengths of the crank and the connecting rod are 240 mm and 800 mm and the crank makes an angle of 45° from the inner dead centre.

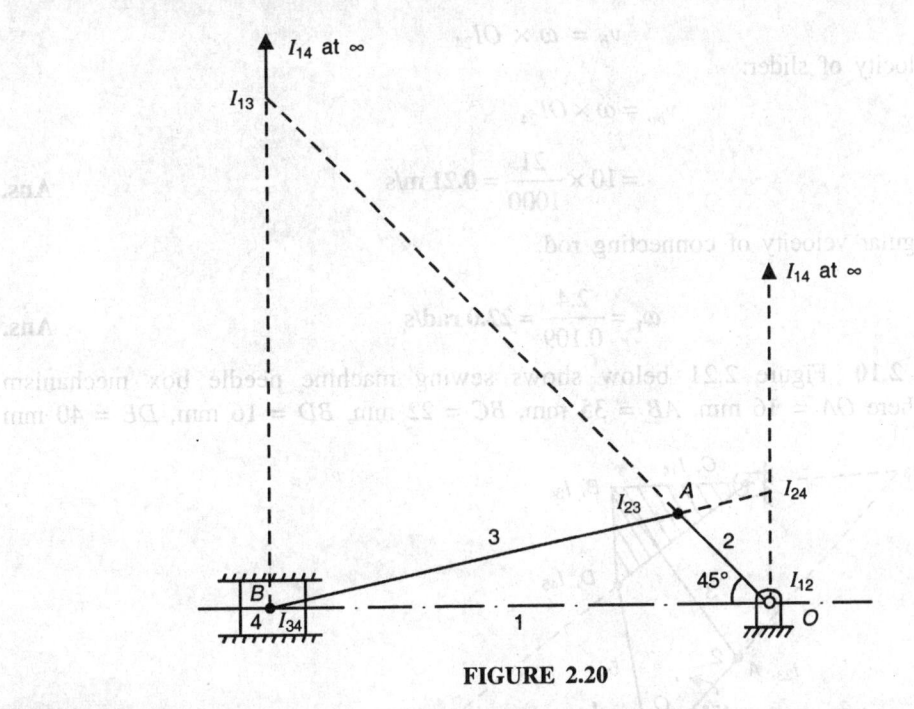

FIGURE 2.20

Solution Velocity of point A on link OA:

$$v_a = \omega \times OA = 10 \times 0.24 = 2.4 \text{ m/s}$$

Draw configuration diagram of the slider-crank mechanism.

Number of instantaneous centres: $N = \dfrac{n(n-1)}{2} = \dfrac{4(4-1)}{2} = 6$

1   2   3   4
  12 23 34
    13 24
      14

Locate the instantaneous centres with Arnold Kennedy theorem as shown in Figure 2.20.

Let ω_1 be the angular velocity of the connecting rod AB, then

$$\omega_1 = \frac{v_a}{AI_{13}} = \frac{v_b}{BI_{13}}$$

where

$$v_a = \omega \cdot OA \quad \text{and}$$

$$v_b = \omega \times OI_{24}$$

(i) Velocity of slider:

$$v_{bo} = \omega \times OI_{24}$$

$$= 10 \times \frac{21}{1000} = 0.21 \text{ m/s} \qquad \textbf{Ans.}$$

(ii) Angular velocity of connecting rod:

$$\omega_1 = \frac{2.4}{0.109} = \textbf{22.0 rad/s} \qquad \textbf{Ans.}$$

EXAMPLE 2.10 Figure 2.21 below shows sewing machine needle box mechanism $OABCDE$, where OA = 16 mm, AB = 35 mm, BC = 22 mm, BD = 16 mm, DE = 40 mm

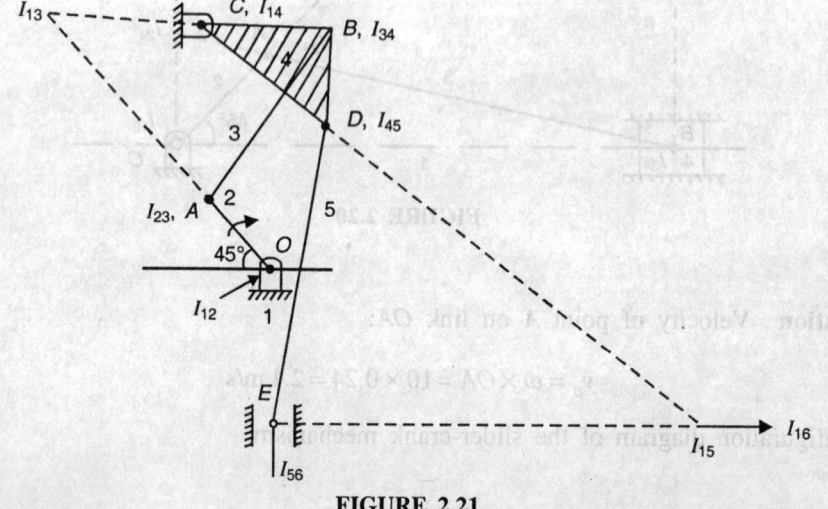

FIGURE 2.21

and angle at $B = 90°$. The crank OA makes $45°$ angle and rotates at 400 rpm in clockwise direction. The horizontal and vertical distances between OC are 13 mm and 40 mm respectively. The point E lies below the point O. Find the velocity of the needle.

Solution Velocity of point A:

$$v_a = \frac{2\pi N}{60} \times OA$$

$$= \frac{2\pi \times 400}{60} \times 0.016$$

$$= 0.67 \text{ m/s}$$

Now
$$\frac{v_a}{I_{13}A} = \frac{v_b}{I_{13}B}$$

$$v_b = v_a \times \frac{I_{13}B}{I_{13}A} = 0.67 \times \frac{50}{42} = 0.797 \text{ m/s}$$

Further
$$\frac{v_d}{CD} = \frac{v_b}{CB}$$

$$v_d = 0.797 \times \frac{28}{22} = 1.015 \text{ m/s}$$

Again
$$\frac{v_e}{I_{15}E} = \frac{v_d}{I_{15}D}$$

Velocity of needle: $v_e = v_d \times \frac{I_{15}E}{I_{15}D} = 1.015 \times \frac{74}{81} = \mathbf{0.927}$ **m/s** Ans.

EXAMPLE 2.11 Locate the instantaneous centres of the mechanisms shown in Figure 2.22.

Solution

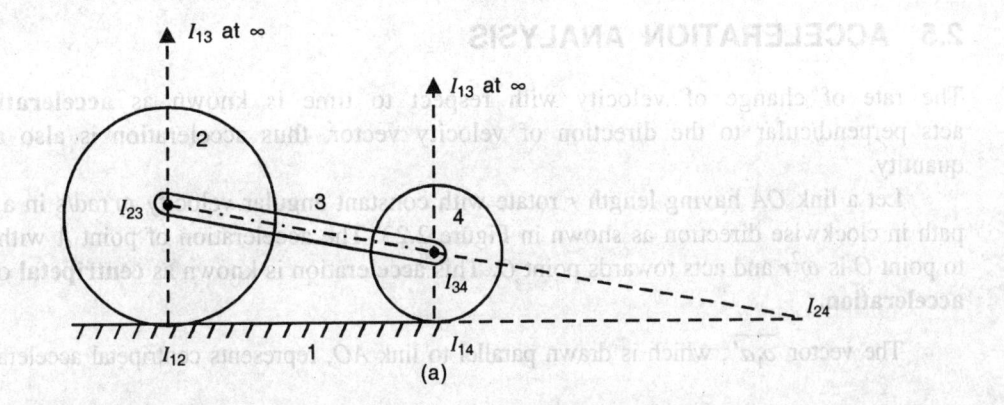

FIGURE 2.22

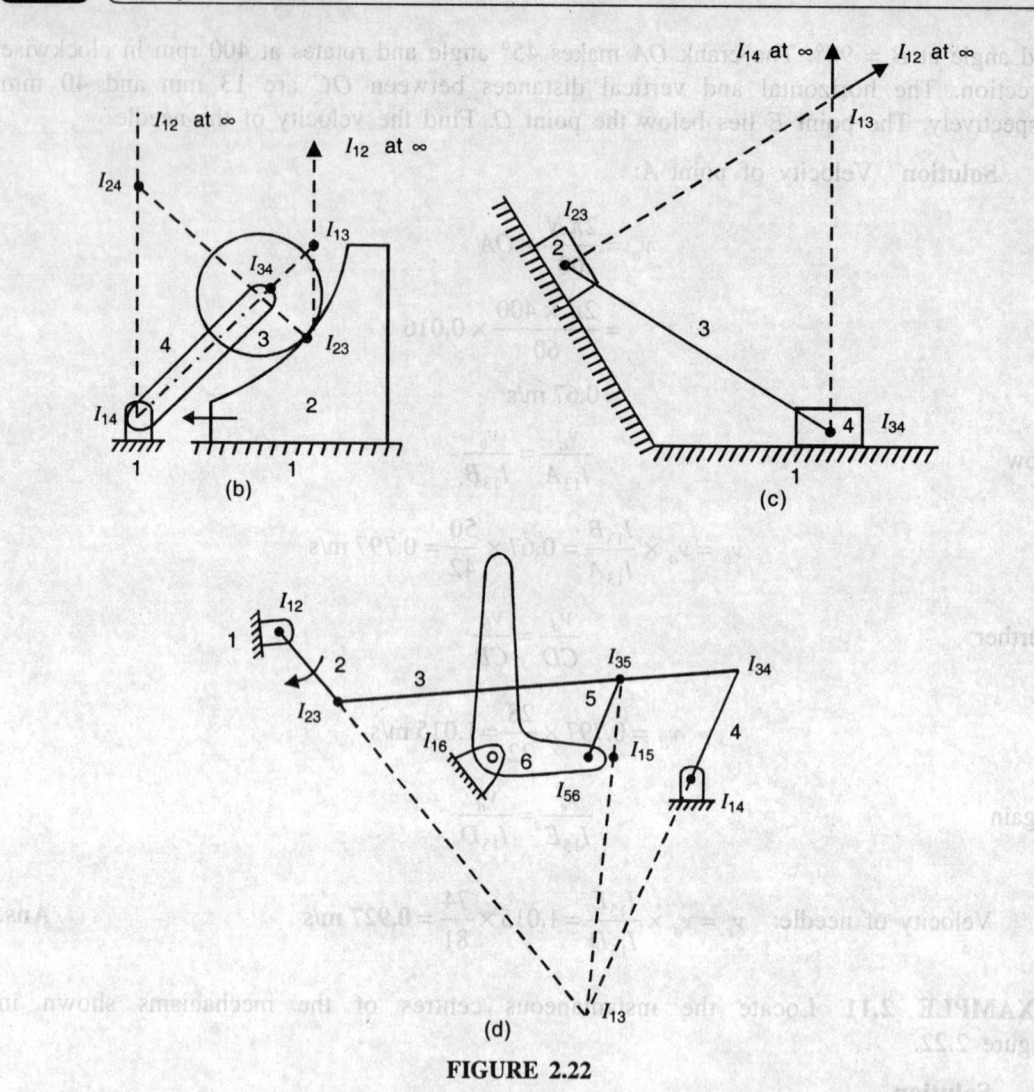

2.5 ACCELERATION ANALYSIS

The rate of change of velocity with respect to time is known as **acceleration** and acts perpendicular to the direction of velocity vector, thus acceleration is also a vector quantity.

Let a link OA having length r rotate with constant angular velocity ω rad/s in a circular path in clockwise direction as shown in Figure 2.23. The acceleration of point A with respect to point O is $\omega^2 r$ and acts towards point O. This acceleration is known as **centripetal** or **radial acceleration.**

The vector $\overrightarrow{o_1 a'}$, which is drawn parallel to link AO, represents centripetal acceleration, f^c.

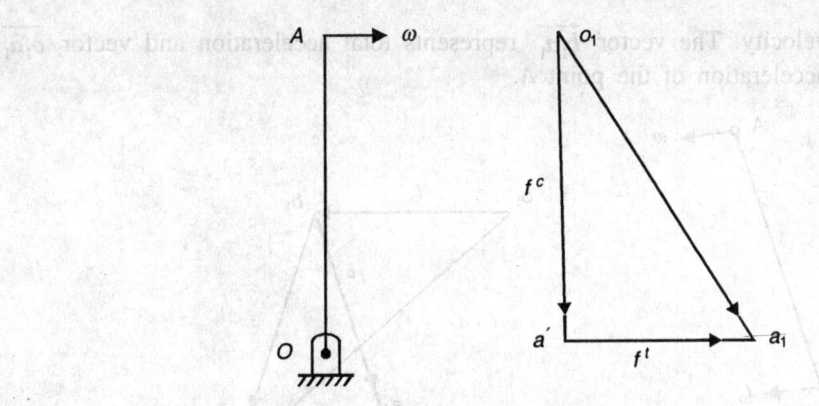

FIGURE 2.23 Total acceleration of a point on link.

Centripetal acceleration, $\quad f^c = \omega^2 r \quad$ (2.20)

$$= \frac{v_{ao}^2}{r}$$

where v_{ao} is velocity of point A with respect to point O.

Now suppose that the link OA is rotating with non-uniform angular velocity having angular acceleration α rad/s^2. In such a case, the total acceleration of point A is vector sum of (a) centripetal acceleration, f^c and (b) tangential acceleration, f^t.

Tangential acceleration, $\quad f^t = \alpha \cdot r \quad$ (2.21)

In vector diagram, the tangential acceleration, f^t is represented by a vector $\overrightarrow{a'a_1}$ drawn perpendicular to link OA, i.e. parallel to the direction of velocity (Figure 2.23). Thus, total acceleration:

$$\overrightarrow{f_{ao}} = \overrightarrow{f^c} + \overrightarrow{f^t}$$

or $\quad\quad \overrightarrow{f_{ao}} = \frac{v_{ao}^2}{r} + \alpha \cdot r \quad$ (2.22)

In case the acceleration, f_o, of end O of the link OA is given in the direction and magnitude as shown in Figure 2.24, then acceleration of point A with respect to point O is determined by vector sum of all the accelerations.

$$\overrightarrow{f} = \overrightarrow{f_o} + \overrightarrow{f^c} + \overrightarrow{f^t} \quad (2.23)$$

To construct an acceleration diagram, take a point o_1 and draw a vector $\overrightarrow{o_1 b_1}$ equal to acceleration f_o in the direction and magnitude. Draw the centripetal acceleration f^c, vector $\overrightarrow{b_1 a'}$, parallel to the link and tangential acceleration f^t, vector $\overrightarrow{a'a_1}$, perpendicular to the link in the direction of velocity. The vector $\overrightarrow{b_1 a_1}$ represents total acceleration and vector $\overrightarrow{o_1 a_1}$ represents absolute acceleration of the point A.

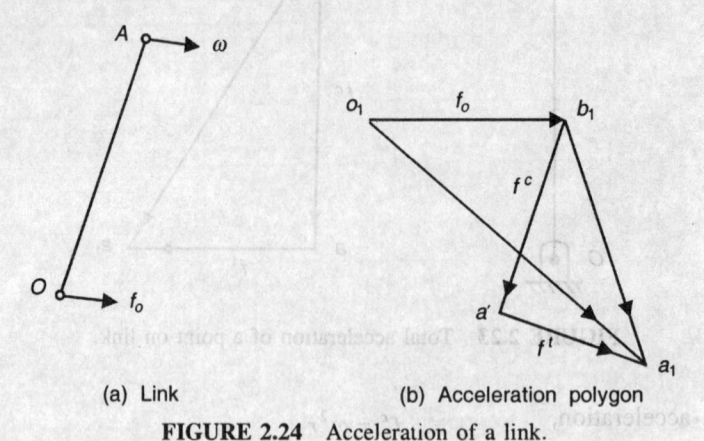

(a) Link (b) Acceleration polygon

FIGURE 2.24 Acceleration of a link.

2.5.1 Procedure to Draw Acceleration Polygon of a Mechanism

The following procedure may be followed to construct an acceleration polygon of a given mechanism:

1. Draw the configuration diagram of the given mechanism to a suitable scale. To illustrate, let us consider a slider-crank mechanism as shown in Figure 2.25(a).
2. Draw the velocity polygon by relative velocity method as discussed in Section 2.3 [Figure 2.25(b)].

(a) Configuration diagram (b) Velocity polygon

(c) Acceleration polygon

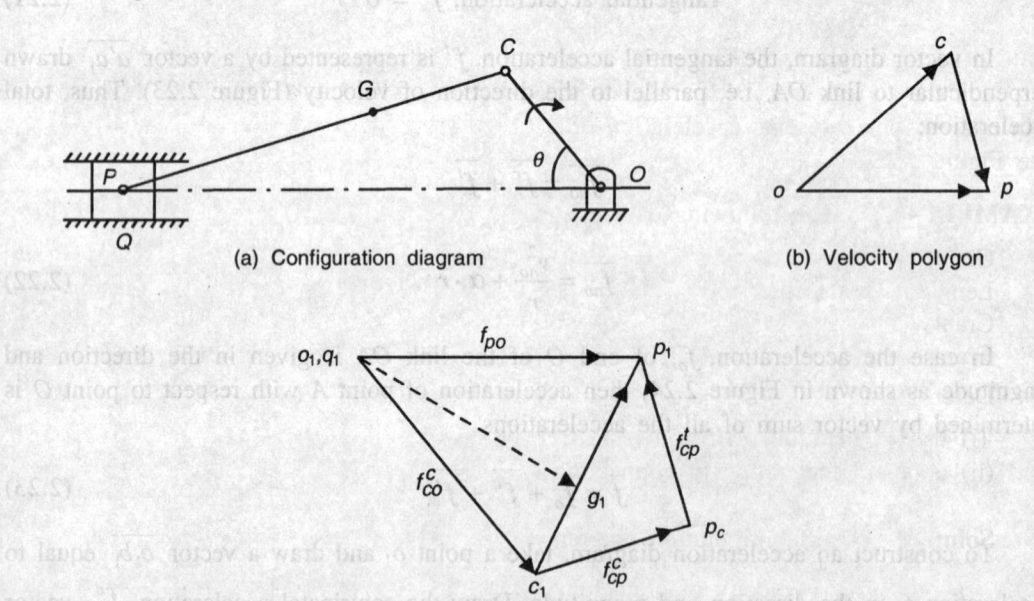

FIGURE 2.25 Velocity and acceleration analysis of a slider-crank mechanism.

3. Prepare a table of centripetal acceleration in the following format as given below:

Link	Length m	Velocity v m/s	Centripetal acceleration $f^c = v^2/\text{length}$ m/s²	Remarks
OC	r_1	v_{co} ($\overrightarrow{oc}$)	$f^c_{co} = v^2_{co}/r_1$	$f^t_{co} = 0$
CP	r_2	v_{cp} ($\overrightarrow{cp}$)	$f^c_{cp} = v^2_{cp}/r_2$	—
PO	—	v_{po} ($\overrightarrow{op}$)	—	$f^c_{po} = 0$

4. Construct acceleration polygon as described below:
   (a) Draw vector $\overrightarrow{o_1 c_1}$, centripetal acceleration parallel to link CO and equal to f^c_{co} at suitable scale.
   (b) Draw vector $\overrightarrow{c_1 p_c}$ representing centripetal acceleration of link CP ($= v^2_{cp}/r_2$).
   (c) Draw $\overline{p_c p_1}$ perpendicular to vector $\overrightarrow{c_1 p_c}$ representing tangential acceleration of link CP f^t_{cp}.
   (d) From point o_1, draw vector $\overrightarrow{o_1 p_1}$ representing the direction of motion of piston P, intersecting vector $\overline{p_c p_1}$ at point p_1.
   (e) The vectors $\overline{p_c p_1}$ and $\overline{c_1 p_1}$ represent tangential and total acceleration of link CP respectively and $\overline{o_1 p_1}$ represent acceleration of piston P.
   (f) To obtain acceleration of point G on link CP, divide $\overline{c_1 p_1}$ at g_1 such that:

$$\frac{\overline{c_1 g_1}}{\overline{c_1 p_1}} = \frac{CG}{CP}$$

Join $o_1 g_1$ which represents acceleration of point G on link CP.
See Figure 2.25(c).

EXAMPLE 2.12 The following data relate to a slider-crank mechanism:

Crank radius	150 mm
Length of connecting rod	600 mm
Crank angle	45°
Crank speed	300 rpm in clockwise direction

Determine:
(i) the acceleration of piston
(ii) angular acceleration of connecting rod

Solution Draw configuration diagram to a convenient scale [Figure 2.26(a)].

$$\text{Velocity of crank pin} = \omega \cdot OC$$
$$= \frac{2\pi \times 300}{60} \times 0.15 = 4.71 \text{ m/s}$$

(a) Configuration diagram

(b) Velocity polygon

(c) Acceleration polygon.

FIGURE 2.26

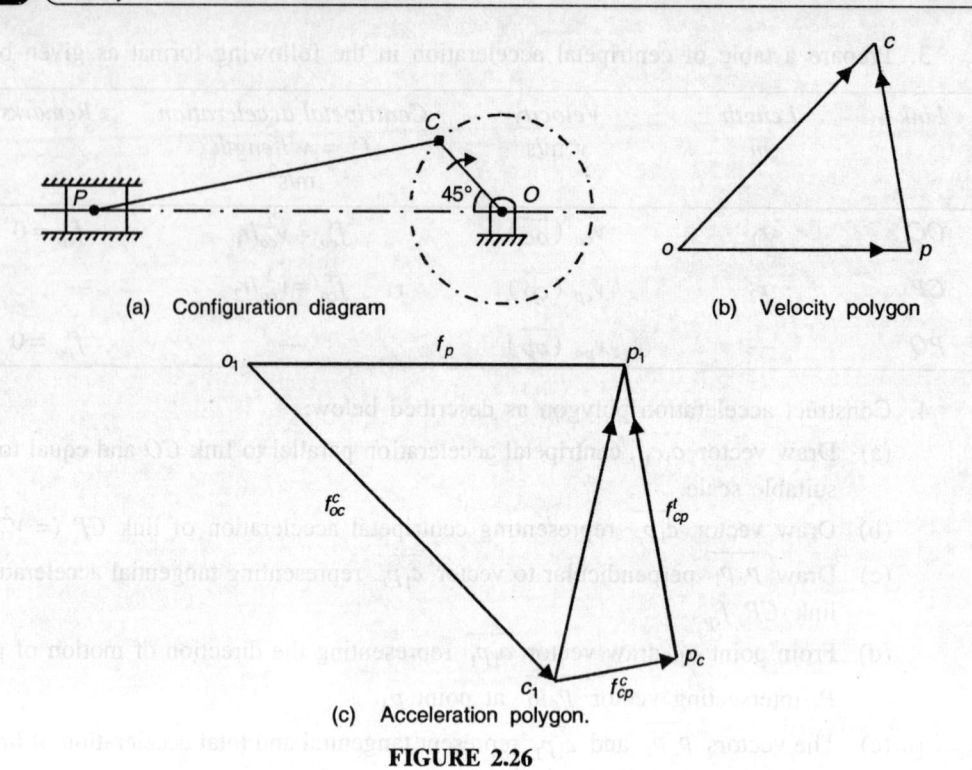

Draw vector $\overrightarrow{oc}$ equal to $v_c = 4.71$ m/s to convenient scale. Draw vector $\overrightarrow{cp}$ perpendicular to link CP and vector $\overrightarrow{op}$ parallel to the axis of slider to meet at point p. The polygon ocp represents velocity polygon [Figure 2.26(b)].

Prepare table of centripetal acceleration.

Link	Length (m)	Velocity (v) (from velocity polygon)	Centripetal acceleration $f^c = v^2/\text{length}$ m/s^2	Remarks
OC	0.15	$\overrightarrow{oc} = 4.71$ m/s	$4.71^2/0.15 = 147.9$	Tangential acceleration is zero
CP	0.60	$\overrightarrow{cp} = 3.4$ m/s	$3.4^2/0.6 = 19.26$	
OP	—	$\overrightarrow{op} = 3.9$ m/s	0	Centripetal acceleration is zero

(i) Draw vector $\overrightarrow{o_1 c_1}$ parallel to link OC and equal to 147.9 m/s^2 to convenient scale.
(ii) Draw vector $\overrightarrow{c_1 p_c}$ representing centripetal acceleration of link CP equal to 19.26 m/s^2.
(iii) Draw $p_c p_1$ perpendicular to vector $\overrightarrow{c_1 p_c}$ to represent tangential acceleration of link CP.
(iv) From point o_1, draw a vector $\overrightarrow{o_1 p_1}$ parallel to axis of slider and join $\overrightarrow{c_1 p_1}$ representing total acceleration of link CP.
See Figure 2.26(c).

Acceleration of slider: $f_p = \overline{o_1 p_1} = 128 \text{ m/s}^2$ **Ans.**

Angular acceleration of link CP:

$$\alpha_{cp} = \frac{f_{cp}^t}{CP} = \frac{\overline{P_c P_1}}{CP}$$

or $\alpha_{cp} = \dfrac{98}{0.6} = 163.34 \text{ rad/s}^2$ **Ans.**

EXAMPLE 2.13 The link AB of a four bar mechanism as shown in Figure 2.27, revolves uniformly at 120 rpm in clockwise direction. Find the angular acceleration of links BC and CD. The dimensions of various links are as given below:

$AB = 40$ mm, $BC = 100$ mm, $CD = 80$ mm, $AD = 60$ mm and $\angle DAB = 90°$

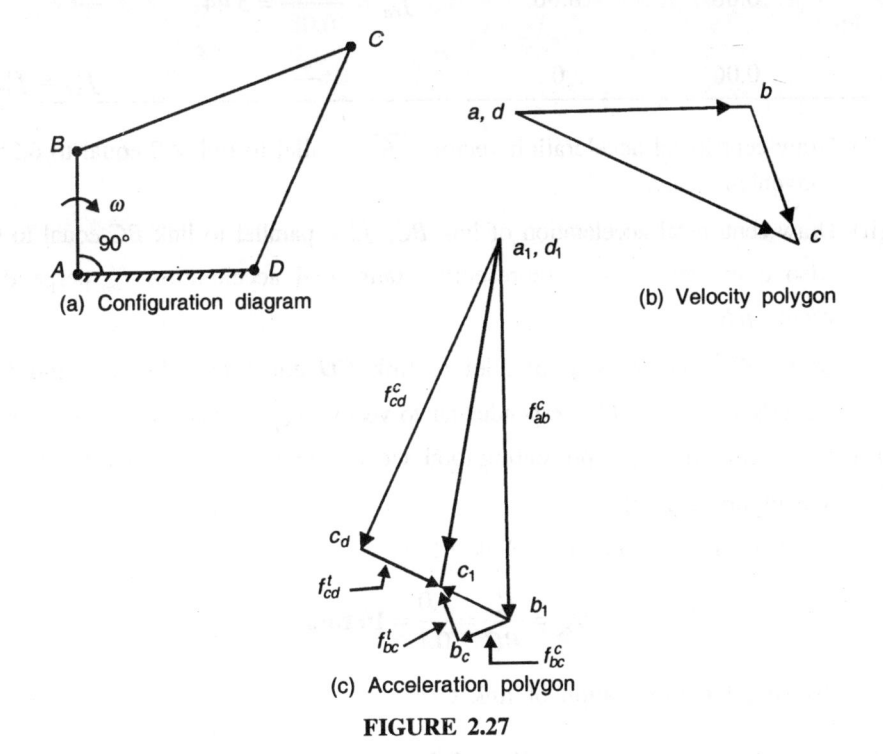

(a) Configuration diagram

(b) Velocity polygon

(c) Acceleration polygon

FIGURE 2.27

Solution   Draw the configuration diagram to a convenient scale [Figure 2.27(a)].

Angular velocity of link AB:

$$\omega = \frac{2\pi N}{60} = \frac{2\pi \times 120}{60} = 12.56 \text{ rad/s}$$

Linear velocity of point B relative to A:

$$v_{ba} = \omega \times AB = 12.56 \times 0.04 = 0.5 \text{ m/s}$$

Draw velocity polygon $a_3b_3c_3$ with vector $\overrightarrow{ab}$ equal to 0.5 m/s drawn perpendicular to link AB. Draw vectors $\overrightarrow{bc}$ and $\overrightarrow{cd}$ perpendicular to links BC and CD respectively. See Figure 2.27(b). Prepare centripetal acceleration by measuring velocities from velocity polygon.

Link	Length m	Velocity m/s	Centripetal acceleration $f^c = v^2/r$ m/s^2	Remarks
AB	0.04	0.5	$f^c_{ab} = \dfrac{0.5^2}{0.04} = 6.25$	$f^t_{ab} = 0$
BC	0.1	0.3	$f^c_{bc} = \dfrac{0.3^2}{0.1} = 0.9$	—
CD	0.08	0.66	$f^c_{cd} = \dfrac{0.66^2}{0.08} = 5.44$	—
AD	0.06	0	—	$f^c_{ad} = f^t_{ab} = 0$

(i) Draw centripetal acceleration vector $\overrightarrow{a_1b_1}$ parallel to link AB equal to 6.25 m/s^2 in convenient scale.

(ii) Draw centripetal acceleration of link BC, f^c_{bc}, parallel to link BC equal to 0.9 m/s^2. Also draw vector $\overrightarrow{b_cc_1}$ representing tangential acceleration f^t_{bc} perpendicular to vector $\overrightarrow{b_1b_c}$.

(iii) Draw f^c_{cd} vector $\overrightarrow{a_1c_d}$ parallel to link CD equal to 5.44 m/s^2 and tangential acceleration vector f^t_{cd} perpendicular to vector $\overrightarrow{a_1c_d}$ to meet vector $\overrightarrow{b_cc_1}$ at point c_1.

(iv) Join $\overrightarrow{b_1c_1}$ and $\overrightarrow{c_1d_1}$ representing total accelerations of link BC and CD respectively. See Figure 2.27(c).

(a) Angular acceleration of link BC:

$$\alpha_{bc} = \frac{f^t_{bc}}{BC} = \frac{1.0}{0.1} = 10 \text{ rad/s}^2 \qquad \textbf{Ans.}$$

(b) Angular acceleration of link CD:

$$\alpha_{cd} = \frac{f^t_{cd}}{CD} = \frac{1.4}{0.08} = 17.5 \text{ rad/s}^2 \qquad \textbf{Ans.}$$

EXAMPLE 2.14 For the mechanism given in Example 2.4, determine the linear acceleration of slider E and angular acceleration of link BC.

Solution The configuration diagram and velocity polygon are drawn in Example 2.4. However, for the sake of continuity, these are produced here again in Figure 2.28.

Based on velocities of various links measured from velocity polygon Figure 2.28(b), a table for centripetal acceleration is prepared as follows:

Link	Length m	Velocity m/s	Centripetal acceleration $f = v^2/r$ m/s^2	Remarks
AB	0.03	1.885	$f_{ab}^c = \dfrac{1.885^2}{0.03} = 118.4$	$f_{ab}^t = 0$
BC	0.045	1.15	$f_{bc}^c = \dfrac{1.15^2}{0.045} = 29.39$	—
CD	0.04	1.5	$f_{cd}^c = \dfrac{1.5^2}{0.04} = 56.25$	—
CE	0.04	1.95	$f_{ce}^c = \dfrac{1.95^2}{0.04} = 95.06$	—
E	—	1.85	—	—

(a) Configuration diagram

(b) Velocity polygon

(c) Acceleration polygon

FIGURE 2.28

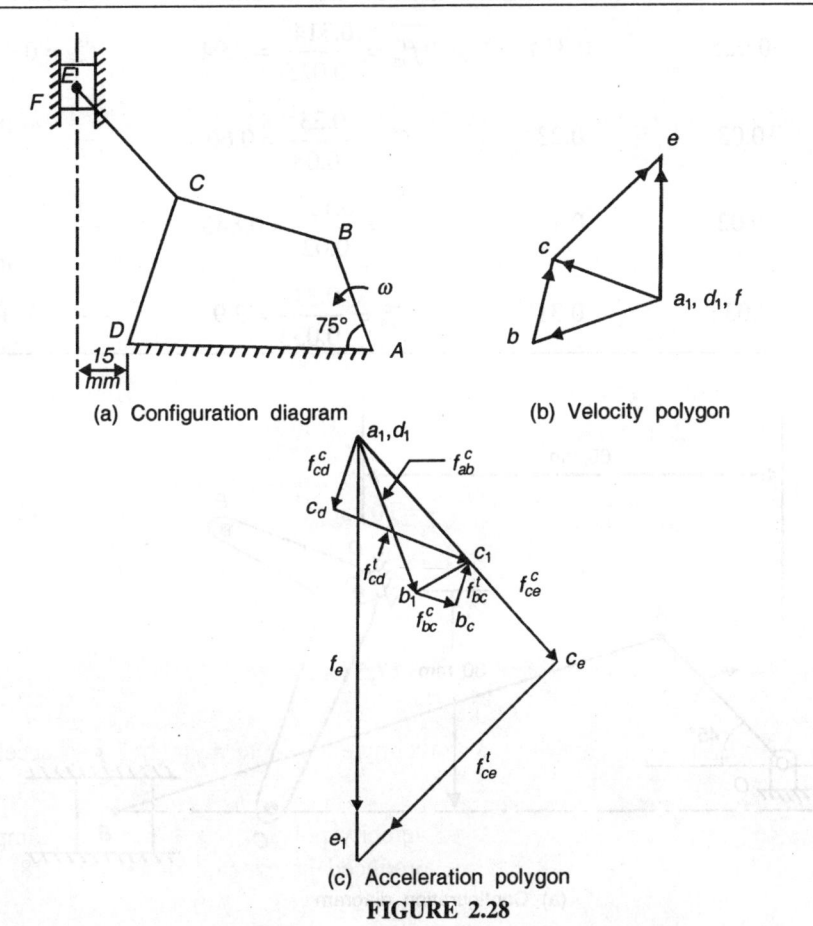

Draw acceleration polygon in convenient scale as shown in Figure 2.28(c).

(a) Linear acceleration of slider: $f_e = \overrightarrow{a_1 e_1} = 296$ m/s² **Ans.**

(b) Angular acceleration of link BC:

$$\alpha_{bc} = \frac{f_{bc}^t}{BC} = \frac{28}{0.045} = 622.22 \text{ rad/s} \quad \textbf{Ans.}$$

EXAMPLE 2.15 For the mechanism given in Example 2.6, determine the linear acceleration of slider and angular acceleration of link AB.

Solution The configuration diagram and velocity polygon are drawn in Example 2.6. However, for the sake of continuity, these are produced here again in Figure 2.29.

Based on velocities of various links measured from velocity polygon, a table for centripetal acceleration is prepared as follows:

Link	Length m	Velocity m/s	Centripetal acceleration $f = v^2/r$ m/s²	Remarks
OA	0.025	0.314	$f_{oa}^c = \dfrac{0.314^2}{0.025} = 3.94$	$f_{oa}^t = 0$
AB	0.08	0.23	$f_{ab}^c = \dfrac{0.23^2}{0.08} = 0.66$	—
BC	0.02	0.13	$f_{bc}^c = \dfrac{0.13^2}{0.02} = 0.845$	—
CD	0.033	0.31	$f_{cd}^c = \dfrac{0.31^2}{0.033} = 2.9$	—

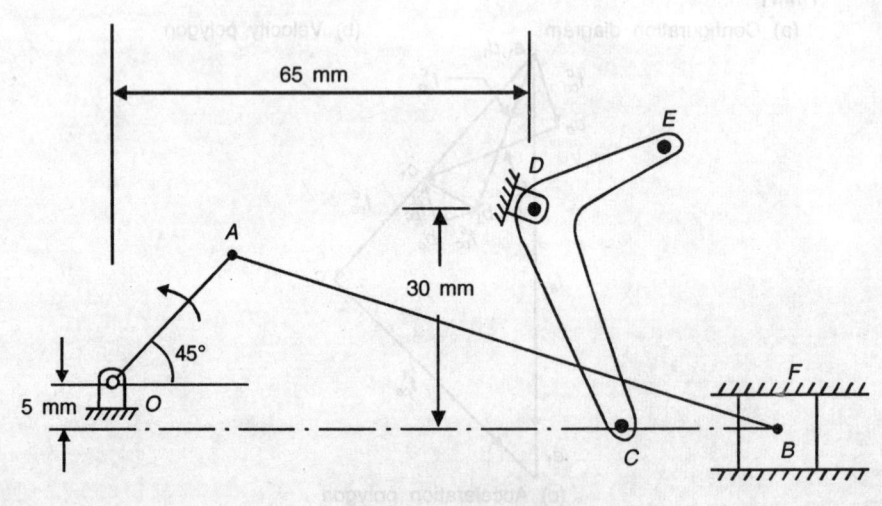

(a) Configuration diagram

Kinematic Analysis of Mechanism 49

(b) Velocity polygon

(c) Acceleration polygon

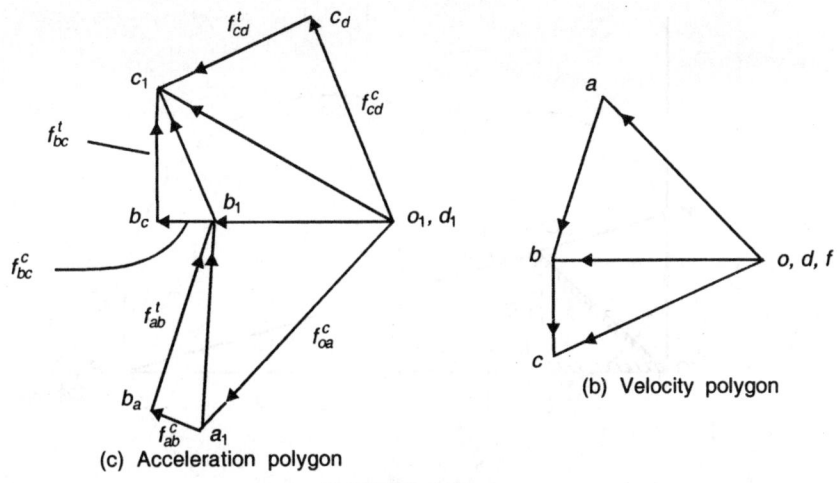

FIGURE 2.29

Draw acceleration diagram to a convenient scale as shown in Figure 2.29(c).

(i) Linear acceleration of slider:

$$f_b = \overrightarrow{o_1 b_1} = 2.5 \text{ m/s}^2 \quad \text{Ans.}$$

(ii) Angular acceleration of link AB:

$$\alpha_{ab} = \frac{f_{ab}^t}{AB} = \frac{2.7}{0.08} = 33.75 \text{ rad/s}^2 \quad \text{Ans.}$$

2.6 KLEIN'S CONSTRUCTION

For a slider-crank mechanism, a simpler method to construct velocity and acceleration polygon on the configuration diagram itself has been suggested by Klein. Accordingly, the line representing crank in the configuration diagram also represents velocity and acceleration of the moving end.

Consider a configuration diagram of a slider-crank mechanism OCP as shown in Figure 2.30. Let r be the length of crank OC.

Velocity polygon. Let r represents velocity v_{co}. Extend the line PC to intersect line drawn perpendicular to OP through O at point M. The triangle OCM represents velocity polygon which is similar to the velocity polygon shown in Figure 2.3(b) rotated through 90° in the direction opposite to that of the crank.

In Figure 2.30:

(i) OC represents magnitude of velocity of point C relative to point O, v_{co}
(ii) CM represents magnitude of velocity of point C relative to point P, v_{cp} and
(iii) OM represents magnitude of slider velocity, v_{po}

Acceleration polygon. For acceleration polygon, let r represents acceleration of point C with respect to point O, f_{co} (= $\omega^2 \times r$). This assumption gives the scale for acceleration polygon. For construction of acceleration polygon, the following procedure may be used:

1. Draw a circle of radius CM with point C as centre of the circle.

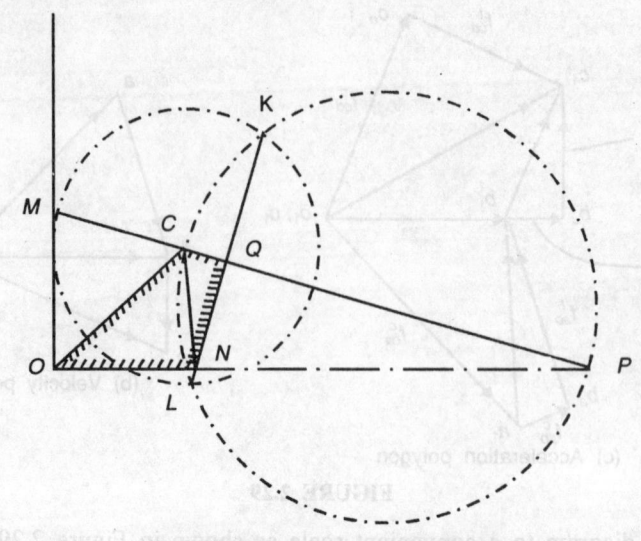

FIGURE 2.30 Klein's construction.

2. Draw another circle with link *CP* as diameter and mid-point of the line *CP* as the centre of the circle.
3. Join the points of intersection of these two circles by a straight line *KL*, which is chord common to both circles. Let this common chord cuts the links *CP* and *OP* at points *Q* and *N* respectively.

The quadrilateral *OCQN* is the required acceleration polygon which is similar to the acceleration polygon $o_1 c_1\, p_c\, p_1$ shown in Figure 2.25(c), rotated through 180°. Referring to Figure 2.30:

ON represents linear acceleration of slider *P* (= $o_1 p_1$)
OC represents centripetal acceleration of crank *OC* (= $o_1 c_1$)
CQ represents centripetal acceleration of link *CP* (= $c_1 p_c$)
QN represents tangential acceleration of link *CP* (= $p_c p_1$) and
CN represents total acceleration of link *CP*

The scale of acceleration diagram is equal to linear scale of configuration diagram multiplied by square of angular velocity of crank.

EXAMPLE 2.16 Find the velocity and acceleration of piston of a slider-crank mechanism, whose data are given in Example 2.12.

Solution Angular velocity of crank:

$$\omega = \frac{2\pi N}{60} = \frac{2\pi \times 300}{60} = 31.415 \text{ rad/s}$$

$$v_{co} = \omega \times OC = 31.415 \times 0.15 = 4.71 \text{ m/s}$$

Velocity polygon. Draw configuration diagram to a convenient scale. Extend line *PC* up to point *M*. Triangle *OCM* represents velocity polygon (Figure 2.31).

15 mm long crank OC drawn on configuration diagram (scale 1:10) represents v_{co} (= 4.71 mm). Therefore,

$$1 \text{ mm length on velocity polygon } OCM = \frac{4.71}{15} = 0.314 \text{ m/s}$$

Velocity of slider: $v_{po} = OM = 13 \times 0.314 = \textbf{4.08 m/s}$ **Ans.**

FIGURE 2.31

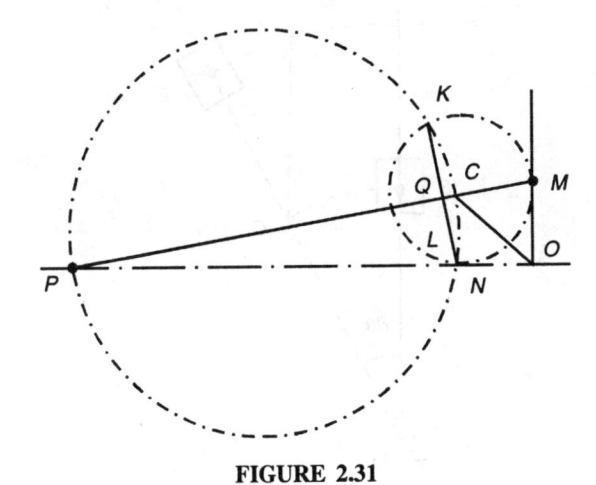

Acceleration polygon. The following procedure may be used to construct acceleration polygon:

(i) Draw a circle of radius CM at point C.
(ii) Draw another circle with CP as diameter.
(iii) Join intersection points K and L to meet on OP at point N.

Triangle $OCQN$ represents acceleration polygon.

$$\text{Acceleration scale} = \text{Scale of configuration diagram} \times \omega^2$$

or $\quad 1 \text{ mm length on acceleration polygon } = \dfrac{10 \times 31.415^2}{1000} = 9.869 \text{ m/s}^2$

Acceleration of slider: $f_{po} = ON = 13 \times 9.869 = \textbf{128.3 m/s}^2$ **Ans.**

2.7 CORIOLIS ACCELERATION

The total acceleration of a point with respect to another point on a rigid link is vector sum of the centripetal and tangential acceleration provided that the distance between these two points is fixed and relative acceleration of the two points on the link has been considered. However, if the distance between these two points varies, as found in crank and slotted lever mechanism, the total acceleration will contain one additional acceleration, called **Coriolis acceleration.**

Consider link OA which oscillates about the fixed centre O with constant angular velocity ω rad/s (Figure 2.32). The link OA also consists of a slider B which is free to slide with velocity v m/s on the link OA. When link OA is rotated by angle $d\theta$ to a new position OA'

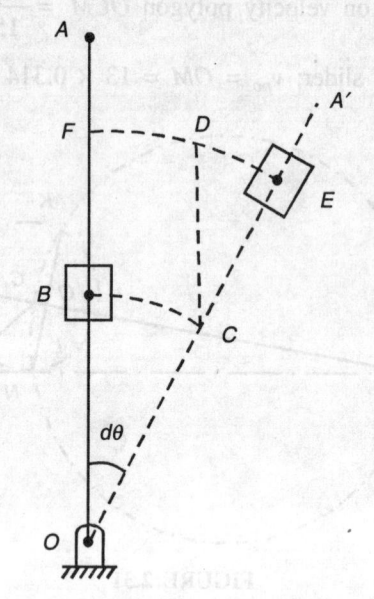

FIGURE 2.32 Coriolis acceleration.

in time dt, the slider moves outwardly from position B to E during the same interval of time. The slider is assumed to have moved from position B to E as follows:

(i) From point B to C due to rotation of link
(ii) From point C to D due to outward velocity of the slider v m/s
(iii) From point D to E due to acceleration perpendicular to the link. This additional acceleration is called **Coriolis acceleration.**

The angular displacement:

$$\text{arc } DE = \text{arc } EF - \text{arc } FD$$
$$= \text{arc } EF - \text{arc } BC$$
$$= OF \cdot d\theta - OB \cdot d\theta$$
$$= BF \cdot d\theta$$
$$= CD \cdot d\theta$$

where

CD = linear displacement ($= v \cdot dt$)
$d\theta$ = angular displacement ($\omega \cdot dt$), and

Therefore, $\quad\quad\quad\quad \text{arc } DE = (v \cdot dt) \times (\omega \cdot dt)$
$$= v \cdot \omega dt^2 \quad\quad\quad\quad (2.24)$$

We know that tangential component of velocity is perpendicular to the link and is given by $v = \omega r$. In the present case, the angular velocity ω rad/s is constant and the slider is moving on the link with constant velocity, therefore velocity is increasing proportional to distance r. This means that there is a constant acceleration which acts perpendicular to the link.

Let f^{cr} be the Coriolis acceleration. The displacement arc DE, which is approximately equal to chord DE for small angle of rotation $d\theta$, can be found as:

$$DE = \frac{1}{2} f^{cr} \times (dt)^2 \tag{2.25}$$

Equating Eqs. (2.24) and (2.25), we get the expression for Coriolis acceleration as

$$f^{cr} = 2\omega v \tag{2.26}$$

The direction of Coriolis acceleration is such that it rotates the slider velocity vector by 90° in the direction of angular velocity of the link, as shown in Figure 2.33.

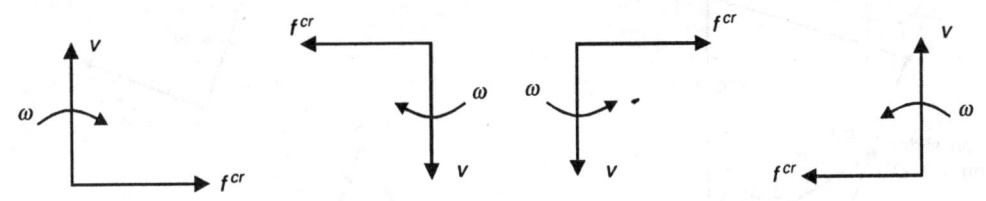

FIGURE 2.33 Direction of Coriolis acceleration.

EXAMPLE 2.17 In a crank and slotted lever type quick return motion mechanism, the crank AB rotates at 90 rpm. Determine:

(a) velocity of ram E
(b) magnitude of Coriolis acceleration
(c) acceleration of ram E.

The length of links are:
$OA = 300$ mm, $AB = 200$ mm, $OD = 600$ mm and $DE = 500$ mm. The crank AB makes an angle of 60° and rotate in clockwise direction.

Solution Draw the configuration diagram to a convenient scale [Figure 2.34(a)].

Angular velocity of crank, $\omega = \dfrac{2\pi N}{60} = \dfrac{2\pi \times 90}{60} = 9.42$ rad/s

Linear velocity, $v_{ba} = \omega \times AB = 9.42 \times 0.2 = 1.88$ m/s

To construct velocity polygon [Figure 2.34(b)], draw vector $\overrightarrow{ab}$ perpendicular to crank AB equal to 1.88 m/s to a convenient scale. Draw another vector $\overrightarrow{oc}$ perpendicular to link OC and vector $\overrightarrow{bc}$ to meet at right angle to vector $\overrightarrow{oc}$. Extend vector $\overrightarrow{oc}$ to $\overrightarrow{od}$ such that

$$\frac{OC}{OD} = \frac{\overrightarrow{oc}}{\overrightarrow{od}}$$

From point d, draw vector $\overrightarrow{de}$ perpendicular to link DE and from point o, draw a line parallel to axis of ram E to meet at point e.

Vector $\overline{bc}$ represents velocity of slider, $v_b = \overline{bc} = 1.1$ m/s

Angular velocity of link OC, $\omega_1 = \dfrac{\overline{oc}}{OC} = \dfrac{1.5}{0.39} = 3.842$ rad/s

Coriolis acceleration, $f_{bc}^{cr} = 2v_b\omega_1 = 2 \times 1.1 \times 3.842 = \mathbf{8.45}$ **m/s** **Ans.**

Velocity of ram, $v_e = \overline{oe} = \mathbf{2.4}$ **m/s** **Ans.**

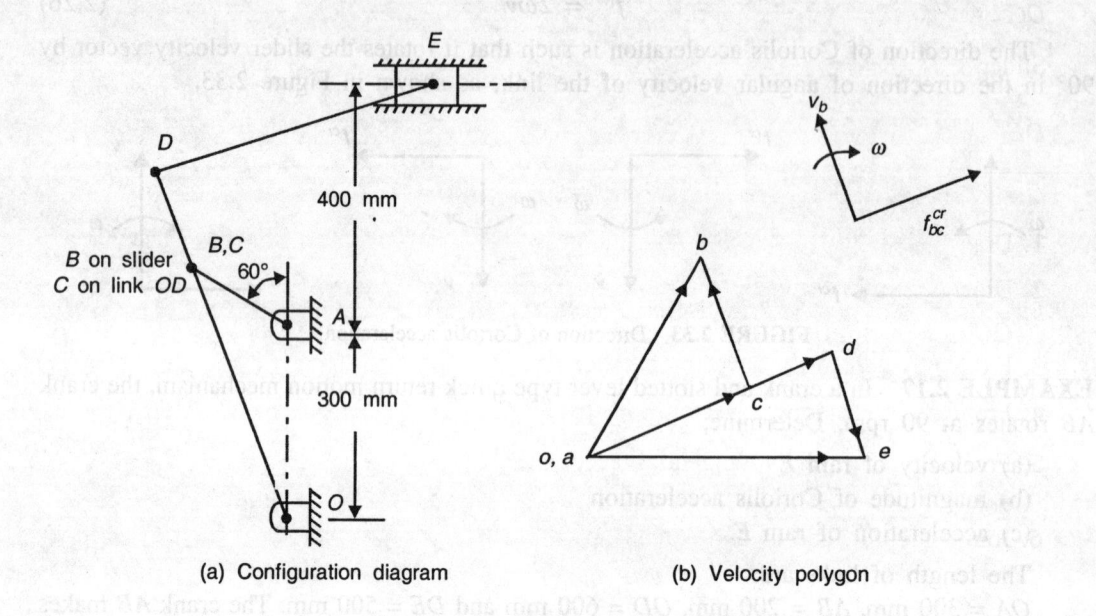

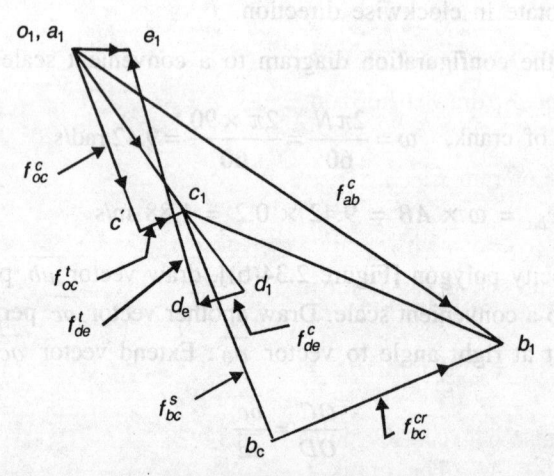

(a) Configuration diagram

(b) Velocity polygon

(c) Acceleration polygon

FIGURE 2.34

Kinematic Analysis of Mechanism **55**

To construct acceleration polygon [Figure 2.34(c)], prepare table for centripetal acceleration as given under:

Link	Length m	Velocity m/s	Centripetal acceleration $f = v^2/r$ m/s^2	Remarks
AB	0.2	1.88	$f^c_{ab} = \dfrac{1.88^2}{0.2} = 17.67$	$f^t_{ab} = 0$
OC	0.39	1.55	$f^c_{oc} = \dfrac{1.55^2}{0.39} = 6.16$	—
Slider B	—	1.1	—	$f^{cr} = 8.45 \perp OC$
DE	0.5	0.9	$f^c_{de} = \dfrac{0.9^2}{0.5} = 1.62$	—

1. Draw vector $\overrightarrow{a_1 b_1}$ equal to $f^c_{ab} = 17.67$ m/s^2 to a convenient scale.
2. Draw vector $\overrightarrow{b_c b_1}$ equal to coriolis acceleration $f^{cr}_{bc} = 8.45$ m/s^2 perpendicular to link OC as shown in the direction.
3. From point b_c, draw a vector parallel to link OC.
4. From point o_1, draw a vector $\overrightarrow{o_1 c'}$ parallel to link OC equals to $f^c_{oc} = 6.16$ m/s^2.
5. From point c', draw tangential acceleration vector $\overrightarrow{c' c_1}$ to meet vector $\overrightarrow{b_c c_1}$ at point c_1.
6. Join points o_1, and c_1 and extend up to d_1 such that:

$$\frac{OC}{OD} = \frac{\overrightarrow{o_1 c_1}}{\overrightarrow{o_1 d_1}}$$

7. At point d_1, draw vector $\overrightarrow{d_1 d_e}$ equal to f^c_{de} (= 1.62 m/s^2) and vector $\overrightarrow{d_e e_1}$ equal to f^t_{de}. From point o_1, draw a horizontal line parallel to axis of ram E to meet vector $\overrightarrow{d_e e_1}$ at point e_1.

(c) Acceleration of ram E:

$$f_e = \overrightarrow{o_1 e_1} = 0.45 \text{ m/s}^2 \qquad \text{Ans.}$$

EXAMPLE 2.18 In the swivelling point mechanism shown in Figure 2.35, the dimensions of various links are:

OA = 120 mm, AB = 900 mm, BC = 300 mm, DE = 600 mm, EF = 500 mm and OC = 750 mm

The vertical distance between the point C and axis of slider F is 650 mm. The point D is the mid-point of link AB. The crank OA rotates at 120 rpm in clockwise direction. Determine the acceleration of sliding link DE in the trunnion.

Theory of Mechanisms and Machines

(a) Configuration diagram

(b) Velocity polygon

(c) Acceleration polygon

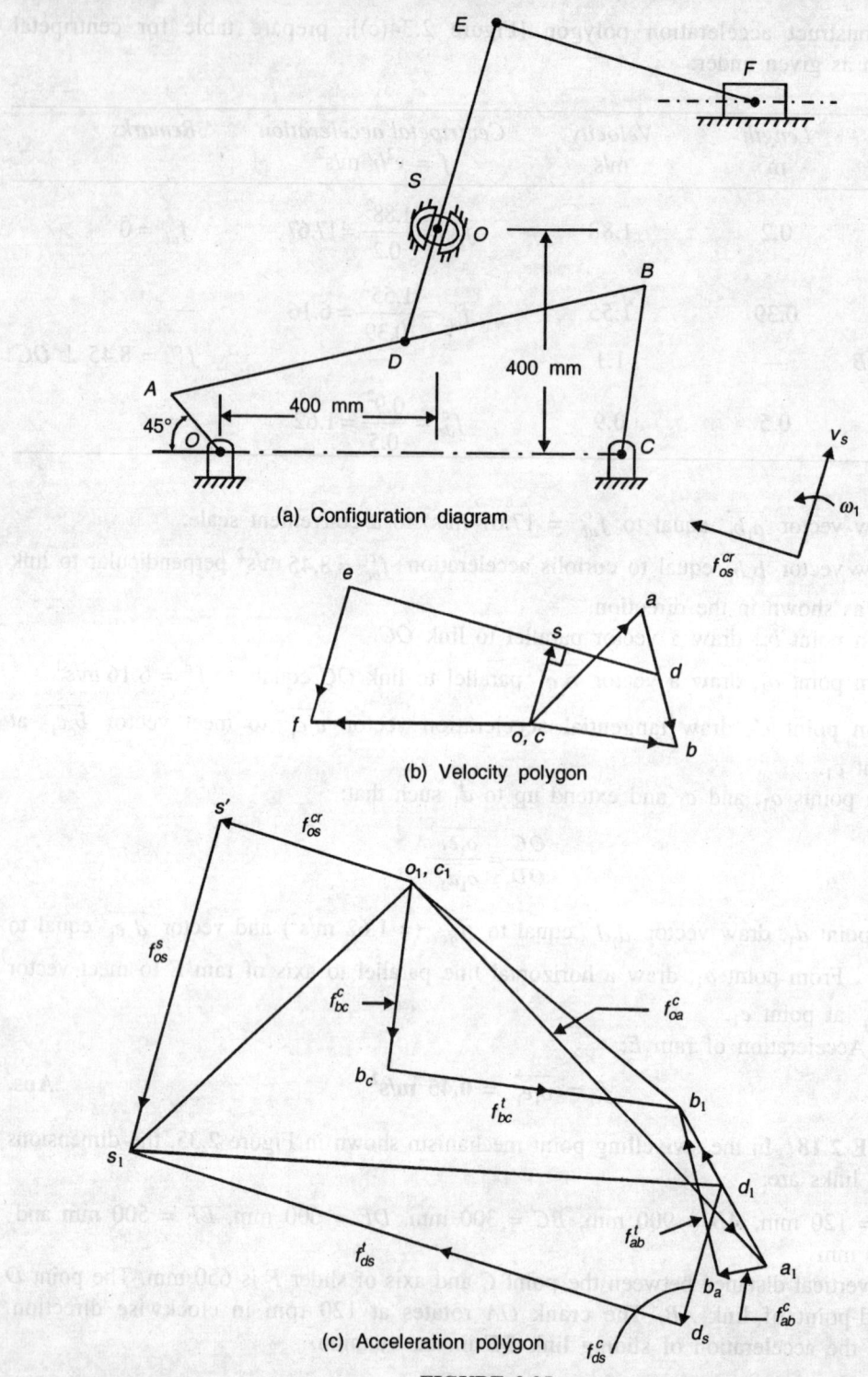

FIGURE 2.35

Solution Draw configuration diagram to a convenient scale as shown in Figure 2.35(a).

Angular velocity of link OA: $\omega = \dfrac{2\pi N}{60} = \dfrac{2\pi \times 120}{60} = 12.56$ rad/s

Velocity of point A relative to O: $v_{ao} = \omega \times OA$
$$= 12.56 \times 0.12$$
$$= 1.5 \text{ m/s}$$

Velocity polygon [Figure 2.35(b)]

(i) Draw vector $\overrightarrow{oa}$ perpendicular to link OA equal to 1.5 m/s to a convenient scale.

(ii) From point a, draw vector $\overrightarrow{ab}$ perpendicular to link AB and vector $\overrightarrow{cd}$ perpendicular to link BC to meet at point b.

(iii) Locate point d on vector $\overrightarrow{ab}$ such that:
$$\dfrac{AD}{AB} = \dfrac{\overrightarrow{ad}}{\overrightarrow{ab}}$$

(iv) Draw a line perpendicular to link DE from point d and drop a perpendicular from point o to meet at point s. The vector $\overrightarrow{os}$ represents velocity of slider in trunnion.

(v) Extend vector $\overrightarrow{ds}$ to $\overrightarrow{de}$ such that:
$$\dfrac{DS}{DE} = \dfrac{\overrightarrow{ds}}{\overrightarrow{de}}$$

(vi) From point e, draw a line perpendicular to link EF and a horizontal line from point o, parallel to slider F to meet at point f.

Velocity of slider S in trunnion: $v_s = \overrightarrow{os} = 0.75$ m/s

Angular velocity of link DE: $\omega_1 = \dfrac{\overrightarrow{de}}{DE} = \dfrac{3}{0.6} = 5$ rad/s

Coriolis acceleration: $f_{os}^{cr} = 2 v_s \omega_1$
$$= 2 \times 0.75 \times 5 = 7.5 \text{ m/s}^2$$

The direction of Coriolis acceleration is shown in Figure 2.35.

Acceleration polygon [Figure 2.35(c)]

(i) Draw vector $\overrightarrow{o_1 a_1}$ parallel to link OA equal to f_{oa}^c ($= 18.75$ m/s^2) to a convenient scale.

(ii) Draw vector $\overrightarrow{a_1 b_a}$ parallel to link AB equal to f_{ba}^c ($= 1.73$ m/s^2) and at point b_a, draw a perpendicular line $\overrightarrow{b_a b_1}$ representing tangential acceleration of link AB, f_{ab}^t.

(iii) From point c_1, draw vector $\overrightarrow{c_1 b_c}$ equal to f_{bc}^c (= 6.53 m/s^2) and at point b_c, draw a perpendicular, representing f_{bc}^t, to meet f_{ab}^t at point b_1.

(iv) Join vectors $\overrightarrow{a_1 b_1}$ and $\overrightarrow{b_1 c_1}$. Locate point d_1 such that:

$$\frac{AD}{AB} = \frac{\overrightarrow{a_1 d_1}}{\overrightarrow{a_1 b_1}}$$

(v) From point d_1, draw vector $\overrightarrow{d_1 d_s}$ equal to f_{ds}^c (= 5.2 m/s) and draw a perpendicular at point d_s to represent f_{ds}^t.

(vi) At point o_1, draw vector $\overrightarrow{o_1 s'}$ equal to f_{os}^{cr} (= 7.5 m/s^2) and at point s', draw a perpendicular to meet at point s_1.

(vii) Join $o_1 s_1$ and $d_1 s_1$ representing total acceleration of trunnion and link DS respectively.

Prepare acceleration table as given below:

Link	Length m	Velocity m/s	Centripetal acceleration $f = v^2/r$ m/s^2	Remarks
OA	0.12	1.5	$f_{oa}^c = \frac{1.5^2}{0.12} = 18.75$	$f_{oa}^t = 0$
AB	0.9	1.25	$f_{ab}^c = \frac{1.25^2}{0.9} = 1.73$	—
BC	0.3	1.4	$f_{bc}^c = \frac{1.4^2}{0.3} = 6.53$	—
DE	0.6	3	$f_{de}^c = \frac{3^2}{0.6} = 15$	—
Slider S	—	0.75	—	$f_{oc}^{cr} = 7.5$ m/s^2 $\perp$ DS

The acceleration of sliding link DE in trunnion:

$$\text{Vector } \overrightarrow{s' s_1} = 3.05 \text{ m/s}^2 \qquad \text{Ans.}$$

2.8 ANALYTICAL METHOD

Consider a four bar mechanism ABCD. Let the link AB rotate with uniform angular velocity ω_2 rad/s, as shown in Figure 2.36. Let us compute the velocity and acceleration of point B. Since the links are rigid, the lengths of the links remain unchanged.

Kinematic Analysis of Mechanism 59

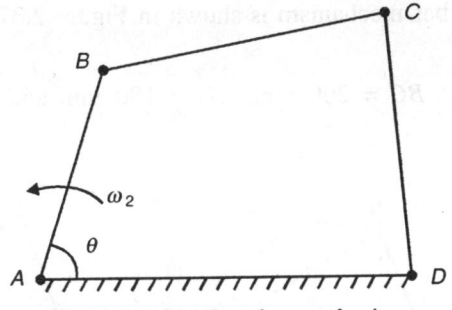

FIGURE 2.36 Four bar mechanism.

Velocity analysis. Let the vectors $\overrightarrow{AB}$, $\overrightarrow{BC}$ and $\overrightarrow{CD}$ are predetermined if the angle of input link made with link *AD* is known.

From the concept of relative velocity, the absolute velocity of point *B* in vector notation is:

$$\vec{v}_b = \vec{v}_{ba} + \vec{v}_{bc} + \vec{v}_{cd} \tag{2.27}$$

The above equation is true as velocity of points *A* and *D* is zero.

Velocity of *B* relative to $A = \vec{\omega}_2 \times \overrightarrow{AB}$, which is completely known.

Velocity of *B* relative to $C = \vec{\omega}_3 \times \overrightarrow{CB}$ and

Velocity of *C* relative to $D = \vec{\omega}_4 \times \overrightarrow{DC}$

Therefore,
$$\vec{\omega}_2 \times \overrightarrow{AB} = (\vec{\omega}_3 \times \overrightarrow{CB}) + (\vec{\omega}_4 \times \overrightarrow{DC}) \tag{2.28}$$

In Eq. (2.28), the magnitude of angular velocities of links *BC* and *CD*, ω_3 and ω_4 respectively, are unknown. Since this equation can be decoupled into two equations, the angular velocities ω_3 and ω_4 can be computed.

Acceleration analysis

Acceleration of *B* relative to *A* = Acceleration of *B* relative to *C*
    + Acceleration of *C* relative to *D*   (2.29)

Acceleration of *B* relative to $A = \vec{\omega}_2 \times (\vec{\omega}_2 \times \overrightarrow{AB})$

Acceleration of *B* relative to $C = \vec{\omega}_3 \times (\vec{\omega}_3 \times \overrightarrow{CB}) + (\vec{\alpha}_3 \times \overrightarrow{CB})$

Similarly,

Acceleration of *C* relative to $D = \vec{\omega}_4 \times (\vec{\omega}_4 \times \overrightarrow{DC}) + (\vec{\alpha}_4 \times \overrightarrow{DC})$

Hence Eq. (2.29) can be written as:

$$\vec{\omega}_2 \times (\vec{\omega}_2 \times \overrightarrow{AB}) = \vec{\omega}_3 \times (\vec{\omega}_3 \times \overrightarrow{CB}) + (\vec{\alpha}_3 \times \overrightarrow{CB}) + \vec{\omega}_4 \times (\vec{\omega}_4 \times \overrightarrow{DC}) + (\vec{\alpha}_4 \times \overrightarrow{DC}) \tag{2.30}$$

The above procedure of analytical analysis is demonstrated in Example 2.19.

EXAMPLE 2.19 A four bar mechanism is shown in Figure 2.37 below. The dimensions of various links are:

$$AB = 150 \text{ mm}, \quad BC = 200 \text{ mm}, \quad CD = 180 \text{ mm and } AD = 220 \text{ mm}$$

FIGURE 2.37

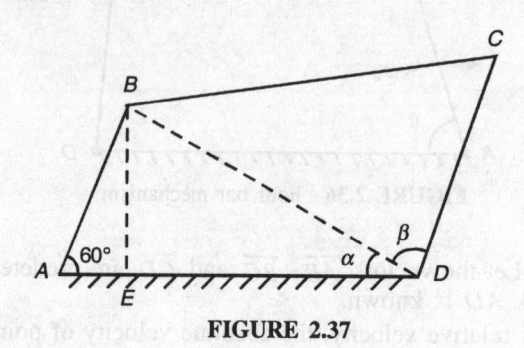

The link AD is fixed. The crank link AB rotates with uniform angular velocity of 10 rad/s in counterclockwise direction and makes an angle of 60° with link AD at an instant. Compute the angular velocity and acceleration of point C.

Solution Let the coordinates of point A be (0, 0). Referring to Figure 2.37, the coordinates of point B:

$$(x_b, y_b) = (150 \times \cos 60°, 150 \times \sin 60°) = (75, 129.9)$$

From the geometry of the figure,

$$BD^2 = BE^2 + ED^2$$
$$= (150 \times \sin 60°)^2 + (220 - 150 \times \cos 60°)^2$$
$$= 37900$$

or
$$BD = 194.68$$

$$AB^2 = BD^2 + AD^2 - 2BD \cdot AD \cos \alpha$$

or
$$\cos \alpha = \frac{BD^2 + AD^2 - AB^2}{2BD \cdot AD} = \frac{37900 + 220^2 - 150^2}{2 \times 194.68 \times 220} = 0.7448$$

or
$$\alpha = \cos^{-1}(0.7448) = 41.86°$$

Similarly,

$$\cos \beta = \frac{CD^2 + BD^2 - BC^2}{2CD \cdot BD} = \frac{180^2 + 37900 - 200^2}{2 \times 180 \times 194.68} = 0.4323$$

or

$$\beta = \cos^{-1}(0.4323) = 64.38°$$

Coordinates of point C:

$$x_c = AD + CD \cos[180° - (\alpha + \beta)]$$
$$= 200 + 180 \times \cos[180° - (41.86 + 64.38)]$$
$$= 270.3 \text{ mm}$$

$$y_c = CD \sin[180° - (\alpha + \beta)]$$
$$= 180 \times \sin[180° - (41.86 + 64.38)]$$
$$= 172.8 \text{ mm}$$

$$(x_c, y_c) = (270.3, 172.8)$$

Coordinates of point D are $(220, 0)$. Therefore,

Vectors: $\overrightarrow{AB} = (x_b - x_a)i + (y_b - y_a)j$ (i)
$$= (75 - 0)i + (129.9 - 0)j$$
$$= 75i + 129.9j$$

$\overrightarrow{CB} = (x_b - x_c)i + (y_b - y_c)j$ (ii)
$$= (75 - 270.3)i + (129.9 - 172.8)j$$
$$= -195.3i - 42.9j$$

$\overrightarrow{DC} = (x_c - x_d)i + (y_c - y_d)j$ (iii)
$$= (270.3 - 220)i + (172.8 - 0)j$$
$$= 50.3i + 172.8j$$

Velocity analysis

$$\overrightarrow{\omega_2} \times \overrightarrow{AB} = \overrightarrow{\omega_3} \times \overrightarrow{CB} + \overrightarrow{\omega_4} \times \overrightarrow{DC}$$

Let $\overrightarrow{\omega_2} = 10k, \; \overrightarrow{\omega_3} = \omega_3 k \text{ and } \overrightarrow{\omega_4} = \omega_4 k$

Substituting,

$$10k(75i + 129.9j) = \omega_3 k \times (-195.3i - 42.9j) + \omega_4 k \times (50.3i + 172.8j)$$
$$750j - 129i = -195.3\omega_3 j + 42.9\omega_3 i + 50.3\omega_4 j - 172.8\omega_4 i$$

Equating i and j components, we have

$$42.9\omega_3 - 172.8\omega_4 = -1299 \quad \text{(iv)}$$
$$-195.3\omega_3 + 50.3\omega_4 = 750 \quad \text{(v)}$$

Solving Eqs. (iv) and (v), we get

$$\omega_3 = -2.03 \text{ rad/s}$$

$$\omega_4 = 7.0 \text{ rad/s}$$

Velocity of point $C = \overrightarrow{\omega_4} \times \overrightarrow{DC}$
$$= 7k \times (50.3i + 172.8j)$$
$$= 352.1j - 1209.6i$$

or

$$v_c = \sqrt{1209.6^2 + 352.1^2} = 1259.8 \text{ mm/s}$$

or

$$v_c = 1.26 \text{ m/s} \quad \text{Ans.}$$

Acceleration analysis

Acceleration of B relative to A = Acceleration of B relative to C
+ Acceleration of C relative to D

$$\overrightarrow{\omega_2} \times (\overrightarrow{\omega_2} \times \overrightarrow{AB}) = \overrightarrow{\omega_3} \times (\overrightarrow{\omega_3} \times \overrightarrow{CB}) + (\overrightarrow{\alpha_3} \times \overrightarrow{CB}) + \overrightarrow{\omega_4} \times (\overrightarrow{\omega_4} \times \overrightarrow{DC}) + (\overrightarrow{\alpha_4} \times \overrightarrow{DC})$$

Let $\overrightarrow{\alpha_3} = \alpha_3 k$ and $\overrightarrow{\alpha_4} = \alpha_4 k$

$$\overrightarrow{\omega_2} \times (\overrightarrow{\omega_2} \times \overrightarrow{AB}) = 10k \times (750j - 1299i) = -7500i - 12990j$$

$$\overrightarrow{\omega_3} \times (\overrightarrow{\omega_3} \times \overrightarrow{CB}) = -2.03k \times (-2.03k \times (-195.3i - 42.9j))$$
$$= 4.12(195.3i + 42.9j)$$
$$= 804.6i + 176.75j$$

$$\overrightarrow{\omega_4} \times (\overrightarrow{\omega_4} \times \overrightarrow{DC}) = 7k \times (7k \times (50.3i + 172.8j))$$
$$= 49(-50.3i - 172.8j)$$
$$= -2464.7i - 8467.2j$$

Therefore,

$$-7500i - 12990j = 804.6i + 176.75j + \alpha_3 k \times (-195.3i - 42.9j)$$
$$- 2464.7i - 8467.2j + \alpha_4 k \times (50.3i + 172.8j)$$
$$= -1660.1i - 8290.45j - 195.3\alpha_3 j + 42.9\alpha_3 i$$
$$+ 50.3\alpha_4 j - 172.8\alpha_4 i$$

Separating components of i and j, we have

$$-7500 = -1660.1 + 42.9\alpha_3 - 172.8\alpha_4$$

$$-12990 = -8290.45 - 195.3\alpha_3 + 50.3\alpha_4$$

or

$$42.9\alpha_3 - 172.8\alpha_4 = -5839.9 \quad \text{(vi)}$$

$$-195.3\alpha_3 + 50.3\alpha_4 = 4699.55 \qquad \text{(vii)}$$

Solving Equations (vi) and (vii), we get

$$\alpha_3 = 35.1 \text{ rad/s}^2, \ \alpha_4 = 42.28 \text{ rad/s}^2$$

Acceleration of point C:

$$\vec{\alpha_c} = \vec{\omega_4} \times (\vec{\omega_4} \times \vec{DC}) + (\vec{\alpha_4} \times \vec{DC})$$

$$= -2464.7i - 8467.2j + 42.28k \times (50.3i + 172.8j)$$

$$= -2464.7i - 8467.2j + 2126.68j - 7306i$$

$$= -9770.7i - 6340.52j$$

or

$$\alpha_c = \sqrt{9.7707^2 + 6.34052^2} = \mathbf{11.64 \ m/s^2} \qquad \text{Ans.}$$

EXERCISES

1. Explain the concept of relative velocity method and write the procedure to draw velocity polygon of a slider-crank mechanism.

2. Explain the procedure to draw velocity polygon of a crank and slotted lever mechanism.

3. For a crank and slotted lever mechanism, show that the ratio of velocity of return stroke to cutting stroke is $(l_1 + r)/(l_1 - r)$, where l_1 is length of the fixed link and r is radius of the crank.

4. What do you mean by instantaneous centre? Explain the properties of instantaneous centre.

5. State and prove the Arnold Kennedy theorem.

6. What do you mean by fixed, permanent and moving instantaneous centres? Explain the procedure to locate the instantaneous centre of a mechanism with suitable example.

7. Explain, with suitable example, the procedure to draw acceleration polygon of a mechanism.

8. What do you mean by coriolis acceleration? Show that coriolis acceleration is $2\omega v$. Also suggest the procedure to find the direction of coriolis acceleration.

9. In a four bar mechanism $ABCD$, the link AD is fixed link of 120 mm long, crank AB is 30 mm long and rotate uniformly at 100 rpm clockwise, while the link CD is 60 mm long and oscillates about fixed point D. The link BC is equal to length AD. Find the angular velocity of link DC when angle BAD is 60°.

[**Ans:** 3.77 rad/s]

10. In a slider-crank mechanism shown in Figure E2.1, the line of stroke of slider P is offset by a distance 50 mm from the centre O. The links OC and CP are 200 mm and 750 mm long respectively. The crank OC is rotating clockwise at a uniform speed of 200 rpm. Determine:

    (i) Acceleration of slider and
    (ii) Angular acceleration of link CP when angle AOC is 135°.

    [Ans: 5.1 m/s^2, 83 rad/s^2]

FIGURE E2.1

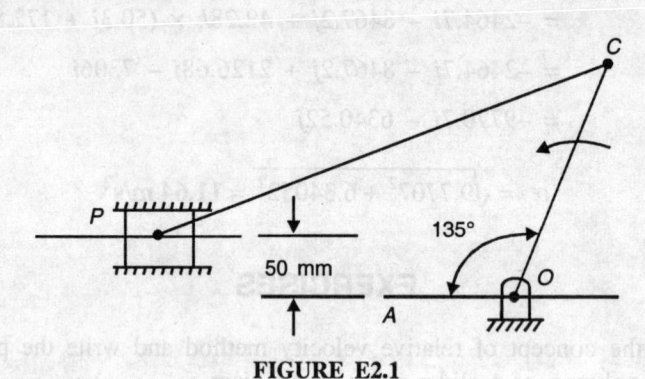

11. In the mechanism shown in Figure E2.2 the crank AB rotates uniformly in clockwise direction at 240 rpm. The dimensions of various links are:

    $AB = 60$ mm, $BC = 160$ mm, $CD = 100$ mm, $AD = 200$ mm, $EF = 200$ mm and $CE = 40$ mm.

    Determine, for the position shown, the linear acceleration of slider F. Also determine the angular velocity and acceleration of link EF.

    [Ans: 20 m/s^2, 3.95 rad/s, 199.5 rad/s^2]

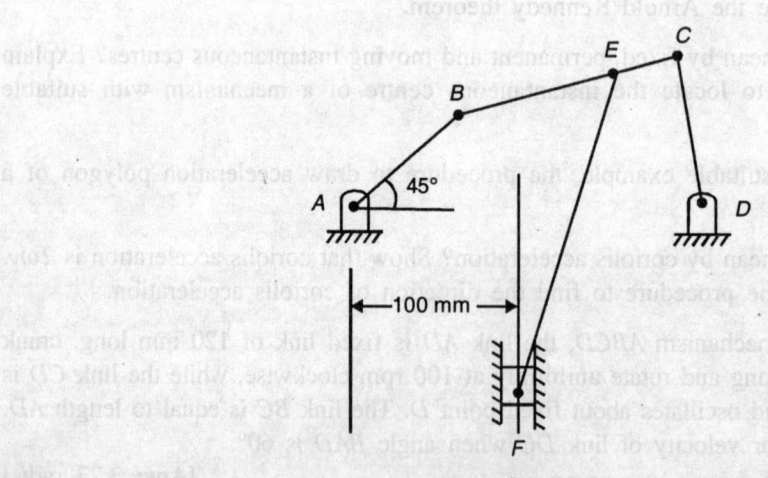

FIGURE E2.2

Kinematic Analysis of Mechanism 65

12. An Andrew variable stroke engine mechanism is shown in Figure E2.3. The crank OA rotates at 100 rpm. Find
    (a) the linear acceleration of slider D
    (b) the angular acceleration of links AC, BC and CD
    [Ans: 10.65 m/s^2, 142.7 rad/s^2 (cw); 47.8 rad/s^2 (ccw), 46.7 rad/s^2 (cw)]

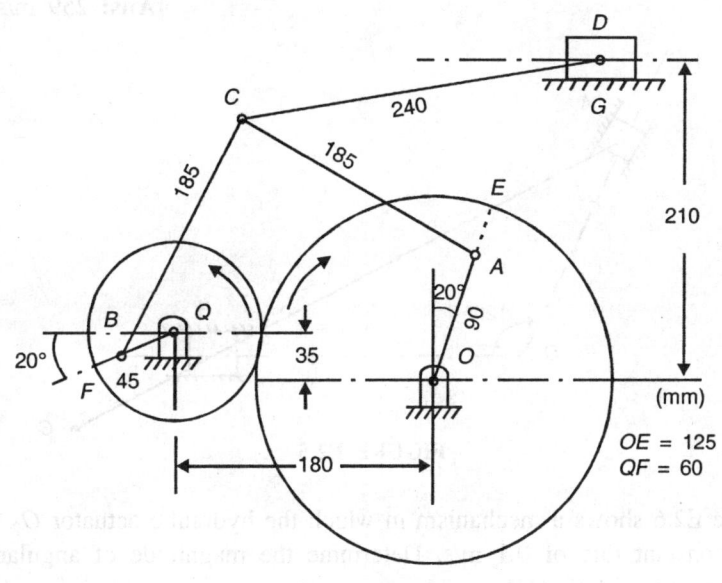

FIGURE E2.3

13. Figure E2.4 shows a Peaucellier inversor comprising six links. Links OU and OV are 1 m long and the four remaining links are 0.5 m long. The point P is guided so as to move along a straight line through OQ. At the instant when the angle PUQ is 60°, the point P is moving towards O with a velocity of 0.5 m/s and an acceleration of 1 m/s^2.

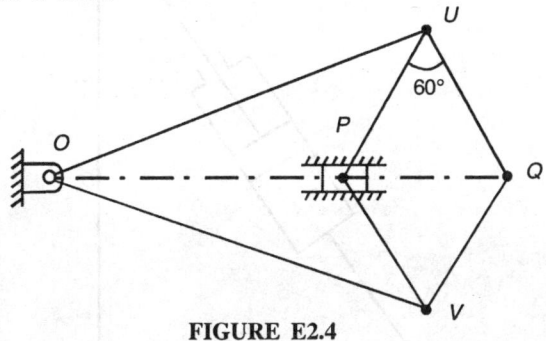

FIGURE E2.4

Draw velocity and acceleration polygon for the configuration given and find the velocity and acceleration of point Q.

[Ans: 0.88 m/s, 3.2 m/s^2]

66 Theory of Mechanisms and Machines

14. In the link *ABC* as shown in Figure E2.5, the lengths of links *AB* and *BC* are 600 mm and 225 mm respectively. The points *A* and *B* are attached by pin joints to the slider blocks. If, for the position where *BD* = 375 mm, *A* is sliding towards *D* with velocity of 6 m/s and a retardation of 150 m/s^2, find the acceleration of point *C* and the angular acceleration of link.

[Ans: 259 m/s^2, 294 rad/s^2]

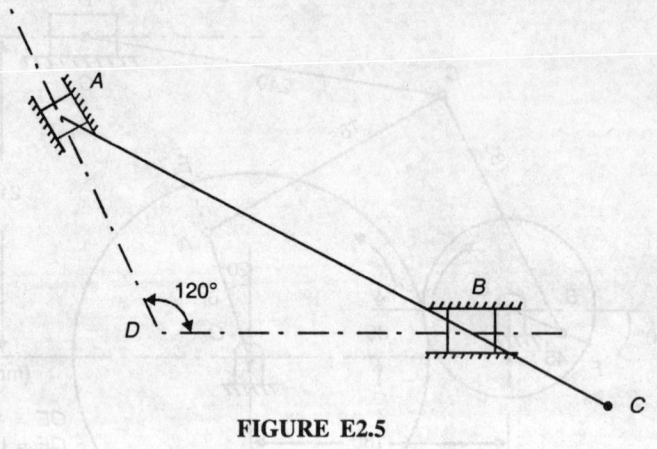

FIGURE E2.5

15. Figure E2.6 shows a mechanism in which the hydraulic actuator O_2A is expanding at a constant rate of 0.1 m/s. Determine the magnitude of angular velocity and acceleration of link O_4A.

[Ans: 0.56 rad/s, 0.136 rad/s^2]

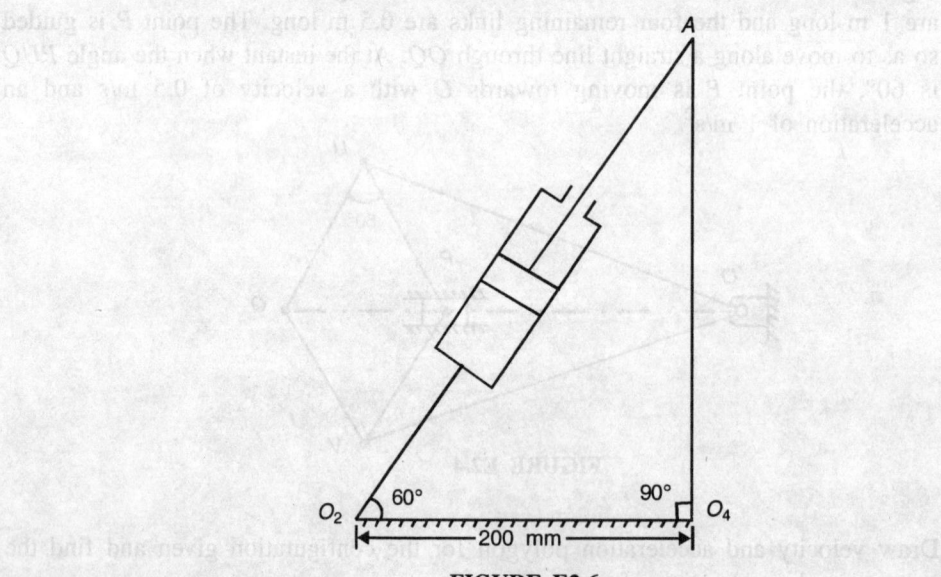

FIGURE E2.6

Kinematic Analysis of Mechanism 67

16. In the mechanism shown in Figure E2.7 given below, the slider D is constrained to move in horizontal direction. The dimensions of various links are:

    $OC = 150$ mm, $CB = 300$ mm, $AB = 260$ mm, $BD = 450$ mm

    If the crank OC makes an angle of 45° from horizontal axis and rotates in counter-clockwise direction at a speed of 180 rpm increasing at the rate of 50 rad/s^2, determine the velocity and acceleration of slider D and the angular velocity and angular acceleration of link BD.

    [Ans: 1.45 m/s, 12.6 m/s^2, 4.89 rad/s, 52.2 rad/s^2]

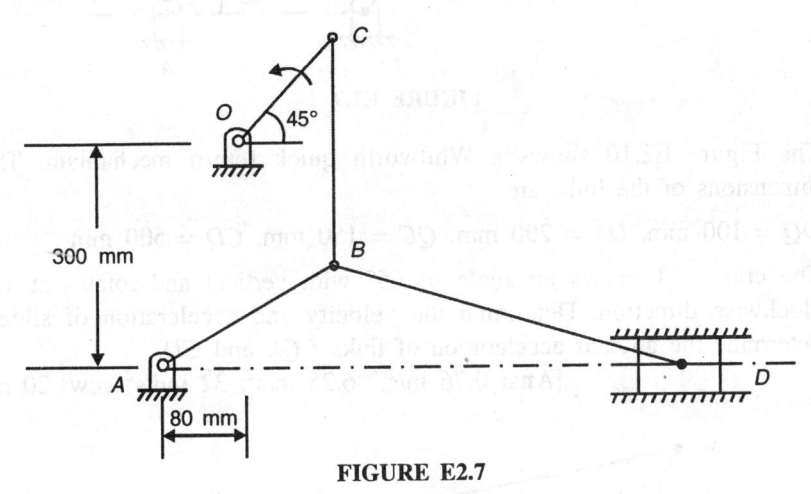

    FIGURE E2.7

17. In the mechanism shown in Figure E2.8, the crank AC makes an angle of 30° with horizontal axis and rotates at 300 rpm in counterclockwise direction. Find the angular velocity of link BC and velocity of slider.

    [Ans: 8.5 m/s, 2.6 m/s]

    FIGURE E2.8

18. A single cylinder rotary engine is shown in Figure E2.9. OA is the fixed link, 200 mm long. OB is the connecting rod and is 520 mm long. The line of stroke is along AD and at the instant is inclined at 30° to the vertical. The body of the engine consisting of cylinders rotates at a uniform speed of 400 rpm about fixed centre A. Determine the acceleration of slider B and angular acceleration of connecting rod.

    [Ans: 390 m/s^2, 288.5 rad/s^2]

Theory of Mechanisms and Machines

FIGURE E2.9

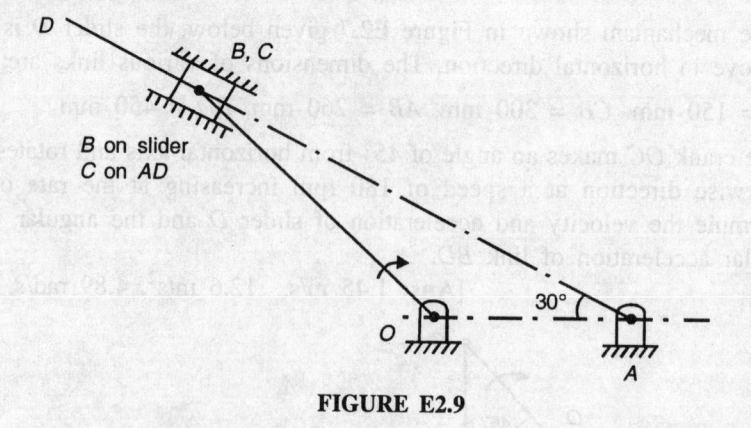

19. The Figure E2.10 shows a Whitworth quick return mechanism. The various dimensions of the links are:

OQ = 100 mm, OA = 200 mm, QC = 150 mm, CD = 500 mm

The crank OA makes an angle of 65° with vertical and rotates at 120 rpm in clockwise direction. Determine the velocity and acceleration of slider D. Also determine the angular acceleration of links CQB and CD.

[Ans: 0.76 m/s, 6.25 m/s^2, 32 rad/s^2 ccw, 20 rad/s^2 ccw]

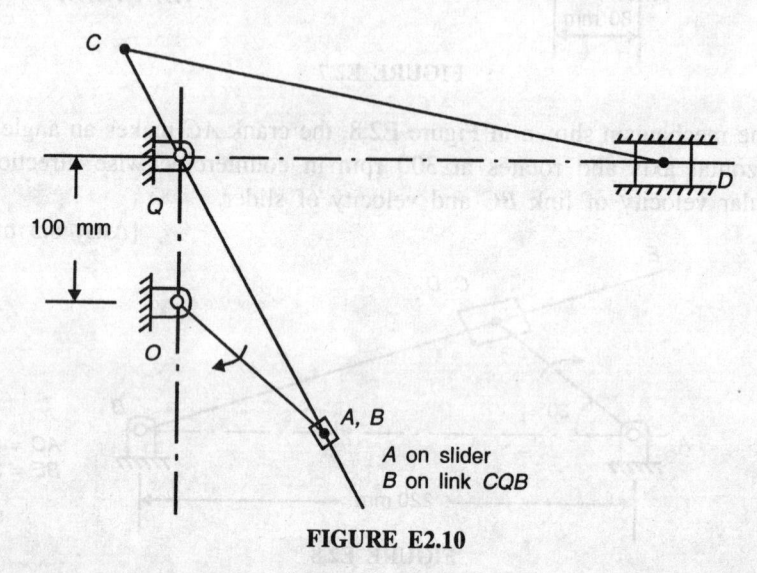

FIGURE E2.10

20. Figure E2.11 shows a quick return mechanism with the dimensions. Find:
    (i) velocity and acceleration of slider E
    (ii) the tangential accelerations of links AC and DE
    [Ans: 3.1 m/s, 49 m/s^2, 50 rad/s^2, 12 rad/s^2]

FIGURE E2.11

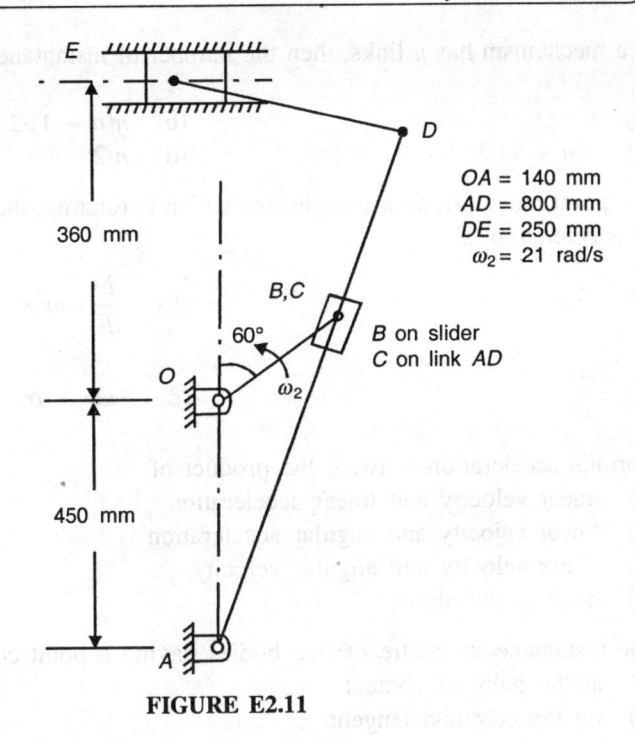

OA = 140 mm
AD = 800 mm
DE = 250 mm
ω_2 = 21 rad/s

B on slider
C on link AD

MULTIPLE CHOICE QUESTIONS

1. The number of instantaneous centres for a 8 link kinematic chain is
   - (a) 16
   - (b) 8
   - (c) 28
   - (d) 24

2. In a four bar mechanism, the mechanical advantage is maximum when velocity ratio is
   - (a) maximum
   - (b) minimum
   - (c) 1
   - (d) 1/2

3. Coriolis acceleration is considered if
   - (a) the point considered moves on a path that rotates
   - (b) the point considered moves along a path that is stationary
   - (c) the point considered moves along a circular path
   - (d) the point considered moves in any curvilinear path

4. The direction of coriolis acceleration
   - (a) is always directed to the centre of curvature of the path
   - (b) is sometimes directed to the centre of curvature
   - (c) depends upon the direction of the velocity of the point on the path
   - (d) none of the above

5. If a mechanism has n links, then the number of instantaneous centres would be equal to
   (a) n
   (b) $n(n-1)/2$
   (c) $n(n+1)/2$
   (d) $n/2$

6. If a point moves along a straight line which is rotating, then the tangential component of acceleration is
   (a) v^2/r
   (b) $\dfrac{dv}{dt} - \omega^2 r$
   (c) $\dfrac{dv}{dt}$
   (d) $2\omega v + \alpha r$

7. Coriolis acceleration is twice the product of
   (a) linear velocity and linear acceleration
   (b) linear velocity and angular acceleration
   (c) linear velocity and angular velocity
   (d) none of the above

8. The instantaneous centre of two bodies having a point contact lies
   (a) at the point of contact
   (b) on the common tangent
   (c) on the common normal
   (d) none of the above

9. If three bodies move relatively to each other, then, according to Kennedy's theorem, their instantaneous centre will lie on
   (a) parabolic curve
   (b) elliptical path
   (c) circle
   (d) straight line

10. The magnitude of linear velocity of a point B on a link AB, relative to A is
    (a) $\omega \cdot AB$
    (b) $\omega \cdot AB^2$
    (c) $\omega^2 \cdot AB$
    (d) $(\omega \cdot AB)^2$

11. When a point is moving in a curved path, the tangential acceleration is equal to
    (a) rate of change in speed
    (b) rate of change of square of speed
    (c) square of speed/radius of curvature
    (d) speed/radius of curvature

12. When a point is moving in a curved path, the normal (centripetal) acceleration is equal to
    (a) rate of change in velocity
    (b) rate of change of square of velocity
    (c) square of velocity/radius of curvature
    (d) velocity/radius

13. In which of the following mechanism, the coriolis acceleration exists?
    (a) Whitworth quick return mechanism
    (b) tangent cam mechanism
    (c) both (a) and (b)
    (d) one of the above

14. Klein's construction is used when
    (a) crank has uniform acceleration.
    (b) crank has uniform angular velocity.
    (c) crank has non-uniform angular velocity.
    (d) angular velocity of crank is zero.

CHAPTER 3

Synthesis of Mechanisms

3.1 INTRODUCTION

The study of theory of mechanisms and machines has two aspects namely kinematic synthesis and analysis. The term **kinematic synthesis** means a process of contriving a scheme through which a machine or mechanism is developed which satisfies the given kinematic specifications. It is also referred to as **kinematic design.** The **kinematic analysis** is an evaluation technique which allows the designer to critically examine an already existing or a proposed mechanism to judge its suitability for the given task. In this chapter, various techniques of kinematic synthesis are discussed.

The kinematic synthesis of a mechanism is accomplished in the following three stages:

 (i) Type synthesis
 (ii) Number synthesis
(iii) Dimensional synthesis

The type synthesis is the first step of kinematic synthesis in which the designer usually choose type of mechanism suitable for a given kinematic specification, for example cams, gears, belts and pulleys or a linkage. This selection is usually based upon availability of space, technically and manufacturing feasibility, economic viability, reliability and safety. The detailed discussion on type synthesis is beyond the scope of this chapter. However, number synthesis and dimensional synthesis are discussed in the following sections.

3.2 NUMBER SYNTHESIS

The second step in the kinematic synthesis is number synthesis. It deals with determining the number of links, number of kinematic pairs and movability of the mechanism in terms of degrees of freedom.

3.2.1 Degree of Freedom of a Mechanism

The term **degree of freedom of a mechanism** (dof) refers to number of independent input parameters which are required to specify the relative position of all the links.

A two dimensional planar link as shown in Figure 3.1 can be specified by three independent parameters namely coordinates of a point (x, y) on the link and its orientation angle θ. Therefore, one unconstrained planar link has three degrees of freedom.

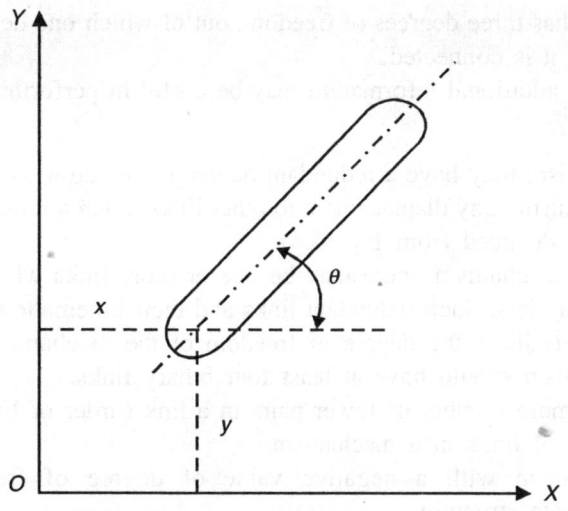

FIGURE 3.1 Degree of freedom of a link.

A mechanism consists of n links out of which one link is fixed; so there are $(n-1)$ number of movable links. The total number of degrees of freedom in the absence of connections is $3(n-1)$. When two links are connected by a turning pair or a sliding pair (lower pair), its two degrees of freedom per joint are suppressed. Thus, the number of degrees of freedom of a mechanism is given as:

$$\text{dof} = 3(n-1) - 2j \tag{3.1}$$

where
n = number of links, and
j = number of lower pairs in the mechanism

If a mechanism includes higher order links namely ternary and quarternary links besides binary links, the number of lower pairs can be determined by the following relation:

$$j = (2n_2 + 3n_3 + 4n_4)/2 \tag{3.2}$$

where n_2, n_3 and n_4 denote number of binary, ternary and quarternary links respectively.

A mechanism with one degree of freedom is said to be a **constrained mechanism** as only one coordinate or one input is sufficient to define the movement of all other links. Similarly, a mechanism with two degrees of freedom needs two inputs to produce a unique output. However, an assemblage of links having zero degree of freedom is called **structure** as it does not permit any motion between the links. For a constrained mechanism (with dof equal to unity), Eq. (3.1) reduces to the following, which is called Grubler's equation.

$$3(n-1) - 2j = 1$$

or
$$2j - 3n + 4 = 0 \tag{3.3}$$

The degree of freedom of a mechanism having h number of higher pairs can be found by the following equation:

$$\text{dof} = 3(n-1) - 2j - h \tag{3.4}$$

A higher pair has three degrees of freedom, out of which one degree of freedom per pair is suppressed when it is connected.

The following additional information may be useful in performing the number synthesis of a mechanism:

1. A mechanism may have a redundant degree of freedom, i.e. a link can be displaced without causing any displacement to other links. Such a redundant degree of freedom should be deduced from Eq. (3.4).
2. In certain mechanism, there may be one or more links which do not introduce any extra constraints. Such redundant links and their kinematic pairs should be removed while determining the degree of freedom of the mechanism.
3. A mechanism should have at least four binary links.
4. The maximum number of lower pairs in a link (order of link) can be $n/2$, where n is number of links in a mechanism.
5. A mechanism with a negative value of degree of freedom forms statically indeterminate structure.

EXAMPLE 3.1 Find the degrees of freedom in each of the mechanisms shown in Figure 3.2.

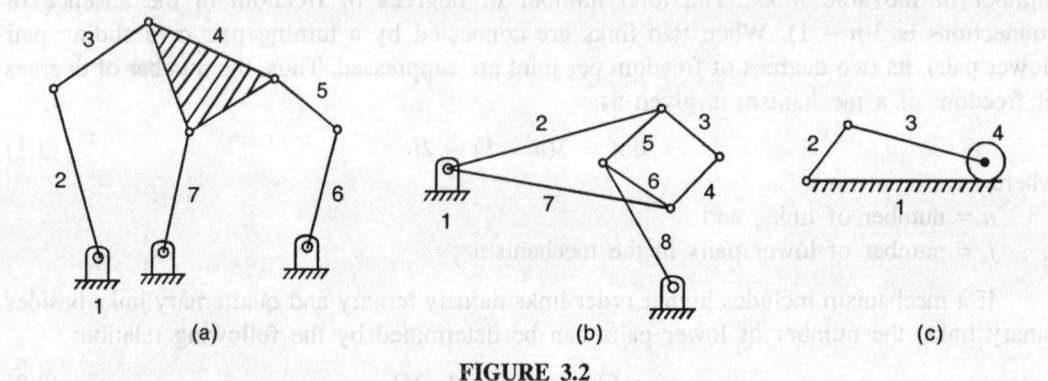

FIGURE 3.2

Solution

(i) Referring to Figure 3.2(a),
Number of links, $n = 7$
Number of lower pairs, $j = 8$
∴ dof = $3(n - 1) - 2j$
     = $3(7 - 1) - 2 \times 8$
     = 2 **Ans.**

(ii) Referring to Figure 3.2(b),
Number of links, $n = 8$
Number of lower pairs, $j = 10$
∴ dof = $3(n - 1) - 2j$
     = $3(8 - 1) - 2 \times 10$
     = 1 **Ans.**

The mechanism is completely constrained.

(iii) Referring to Figure 3.2(c), in which the roller having slipping motion forms a higher pair,
Number of links, $n = 4$
Number of lower pairs, $j = 3$
Number of higher pairs, $h = 1$
$\therefore$ dof $= 3(n - 1) - 2j - h$
$= 3(4 - 1) - 2 \times 3 - 1$
$= 2$ Ans.

EXAMPLE 3.2 Determine the degrees of freedom of the mechanisms shown in Figure 3.3.

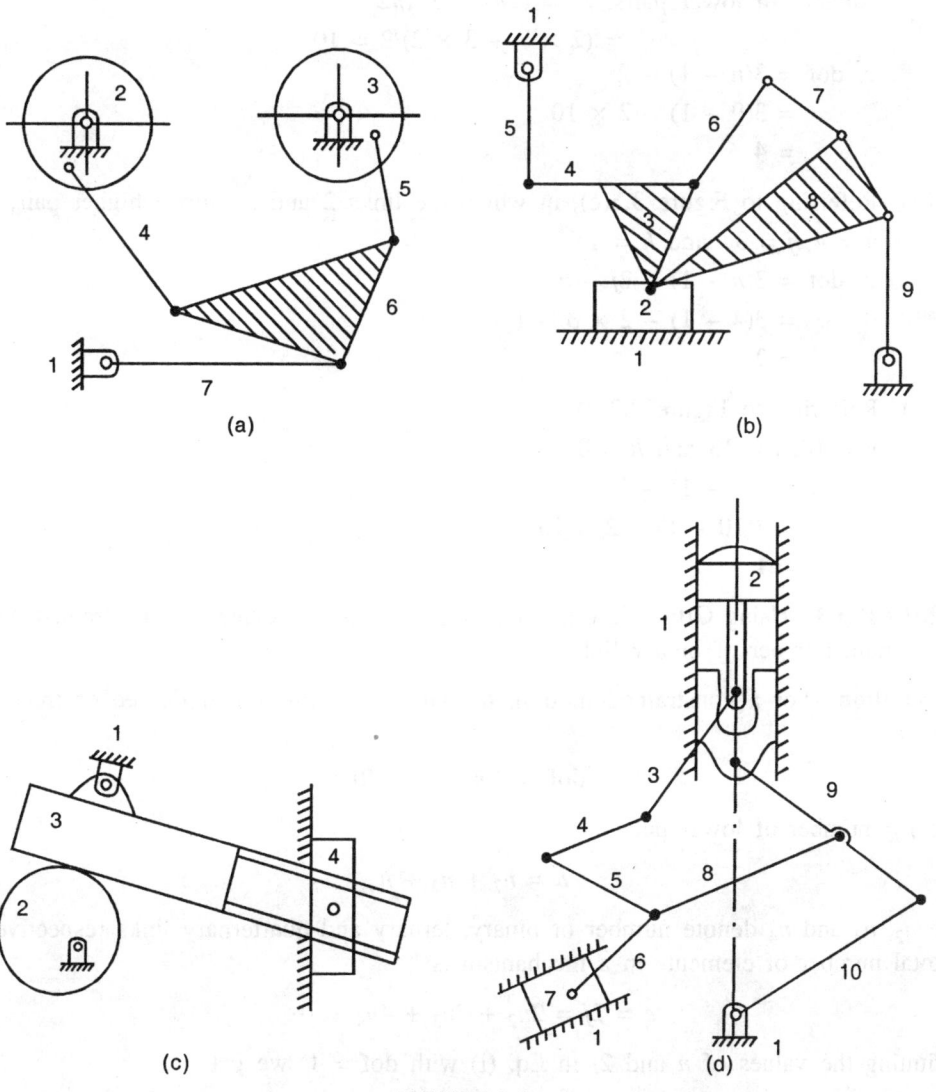

FIGURE 3.3

Solution

(i) Referring to Figure 3.3(a),
Number of links, $n = 7$
Number of lower pairs, $j = 8$
$\therefore$ dof $= 3(n - 1) - 2j$
$= 3(7 - 1) - 2 \times 8$
$= 2$ **Ans.**

(ii) Referring to Figure 3.3(b),
Number of links, $n = n_2 + n_3 = 7 + 2 = 9$
Number of lower pairs, $j = (2n_2 + 3n_3)/2$
$= (2 \times 7 + 3 \times 2)/2 = 10$
$\therefore$ dof $= 3(n - 1) - 2j$
$= 3(9 - 1) - 2 \times 10$
$= 4$ **Ans.**

(iii) Referring to Figure 3.3(c), in which the links 2 and 3 form a higher pair,
$n = 4$, $j = 3$ and $h = 1$
$\therefore$ dof $= 3(n - 1) - 2j - h$
$= 3(4 - 1) - 2 \times 3 - 1$
$= 2$ **Ans.**

(iv) Referring to Figure 3.3(d),
$n = 10$, $j = 13$ and $h = 0$
$\therefore$ dof $= 3(n - 1) - 2j$
$= 3(10 - 1) - 2 \times 13$
$= 1$ **Ans.**

EXAMPLE 3.3 Using Grubler's criterion, show that for achieving the constrained motion, the minimum number of binary links is four.

Solution For a constrained motion, the Grubler's criterion of degree of freedom is written as:

$$\text{dof} = 3(n - 1) - 2j \qquad (i)$$

where j = number of lower pairs

Also
$$n = n_2 + n_3 + n_4 \qquad (ii)$$

where n_2, n_3 and n_4 denote number of binary, ternary and quarternary links respectively. The total number of elements in a mechanism is:

$$e = 2j = 2n_2 + 3n_3 + 4n_4 + \cdots \qquad (iii)$$

Substituting the values of n and $2j$ in Eq. (i) with dof = 1, we get

$$1 = 3[(n_2 + n_3 + n_4 + \cdots) - 1] - (2n_2 + 3n_3 + 4n_4 + \cdots)$$

or
$$n_2 = 4 + n_4 + \cdots$$

Since n_4 is number of quarternary links, the smallest number of binary links n_2 is 4.

EXAMPLE 3.4 A mechanism is formed by six links. Determine the order of links and their numbers. Also suggest the possible layout.

Solution The maximum order of links,
$$i_{max} = \frac{n}{2} = \frac{6}{2} = 3,$$
i.e. ternary links may be used. Therefore,
$$n = n_2 + n_3$$
or
$$6 = n_2 + n_3 \tag{i}$$
Using Grubler's equation,
$$dof = 3(n-1) - 2j$$
or
$$1 = 3(6-1) - 2j$$
∴ The number of lower pairs, $j = 7$
Total number of elements,
$$2j = 2n_2 + 3n_3$$
or
$$14 = 2n_2 + 3n_3 \tag{ii}$$
Solving Equations (i) and (ii), we get
$$n_2 = 4 \quad \text{and} \quad n_3 = 2 \qquad \textbf{Ans.}$$
The possible layout are as shown in Figure 3.4 below:

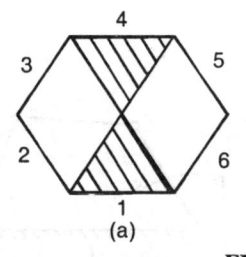

FIGURE 3.4

3.2.2 Grashof's Criterion of Movability

According to Grashof's criterion of movability, a four bar mechanism can be classified into the following three types:

(i) *Double crank mechanism.* A mechanism in which both the driving and driven links make a complete rotation is called a **double crank mechanism**. These driving and driven links are usually referred to as crank and follower links respectively.

(ii) *Double rocker mechanism.* If both crank and follower links make only oscillation and none of them makes a complete rotation, such a mechanism is termed as **double rocker mechanism**.

(iii) *Crank-rocker mechanism.* A mechanism in which the driving link makes complete rotation and the driven link makes only oscillation, such a mechanism is commonly referred to as **crank-rocker mechanism**.

Grashof suggested that the above classification of mechanism is based upon relative dimensions of various links and the choice of a link which acts as a frame. A four bar planar chain satisfying the following relation is known as **Grashof's chain**.

$$l + s \leq p + q \qquad (3.5)$$

where l and s are lengths of the longest and the shortest links and p and q are lengths of other two links (Figure 3.5).

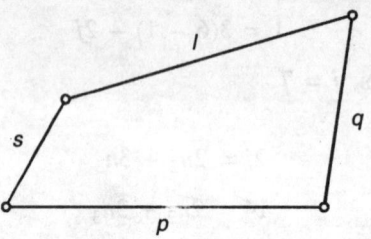

FIGURE 3.5 Grashof's chain.

The inversions of Grashof's chain result into double crank, double rocker and crank-rocker type mechanisms.

(a) If the shortest link is treated as frame, the resulting mechanism is double crank mechanism [Figure 3.6(a)].

(a) Double crank mechanism

(b) Double rocker mechanism

(c) Crank-rocker mechanism

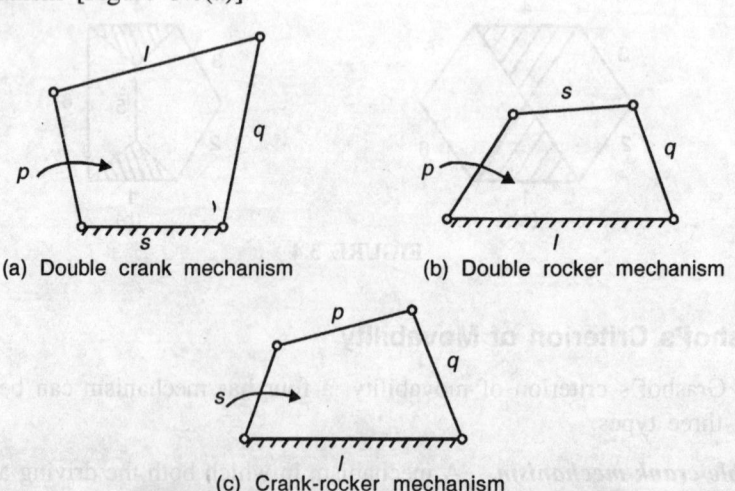

FIGURE 3.6 Inversions of Grashof's chain.

(b) If the shortest link is coupler link, it results into a double rocker mechanism [Figure 3.6(b)].
(c) When the shortest link is crank and any other link is frame, the resulting mechanism is called **crank-rocker mechanism** [Figure 3.6(c)].

In a four bar chain, if the sum of the longest and the shortest links is greater than the sum of the lengths of two other links, $l + s > p + q$, all the four inversions of such a chain results in a double rocker mechanism. Further, a four bar chain in the situation $l + s = p + q$, with l equal to p and s equals to q, results in the following three special mechanisms:

(i) Parallelogram mechanism, in which two equal length links are not adjacent [Figure 3.7(a)]. All four inversions of this mechanism yield double crank mechanisms.
(ii) Deltoid mechanism, in which two equal length links are placed adjacent and the longer link is fixed [Figure 3.7(b)]. Its inversions are crank-rocker mechanisms.
(iii) Galloway mechanism, which is a Deltoid mechanism with a shorter link fixed [Figure 3.7(c)].

(a) Parallelogram mechanism   (b) Deltoid mechanism

(c) Galloway mechanism

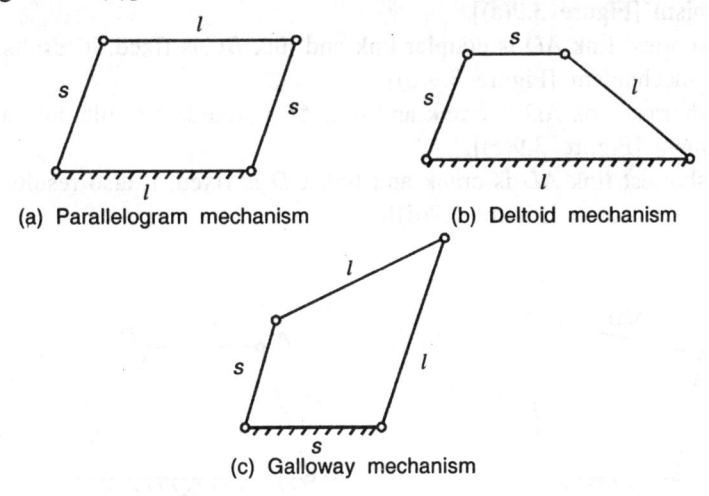

FIGURE 3.7 Other inversions of four bar chain.

EXAMPLE 3.5 A four bar kinematic chain is shown in Figure 3.8. Draw its inversions and identify the nature of each mechanism.

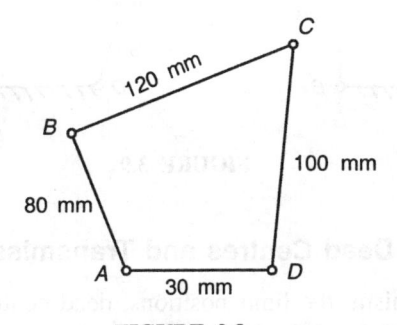

FIGURE 3.8

Solution A kinematic chain is called Grashof's chain if it satisfies the following relation:

$$l + s \leq p + q,$$

where

l = length of the longest link (120 mm)

s = length of the shortest link (30 mm)

Given: $\quad p = 80$ mm

$\quad q = 100$ mm

Therefore, $\quad 120 + 30 \leq 80 + 100$

So the kinematic chain satisfies the Grashof's law.
There are four possible inversions:

(i) If link AD, the shortest link, is fixed, the resulting mechanism is double crank mechanism [Figure 3.9(a)].

(ii) If the shortest link AD is coupler link and link BC is fixed, it results into a double rocker mechanism [Figure 3.9(b)].

(iii) If the shortest link AD is crank and link AB is fixed, it results into a crank–rocker mechanism [Figure 3.9(c)].

(iv) If the shortest link AD is crank and link CD is fixed, it also results into a crank–rocker mechanism [Figure 3.9(d)].

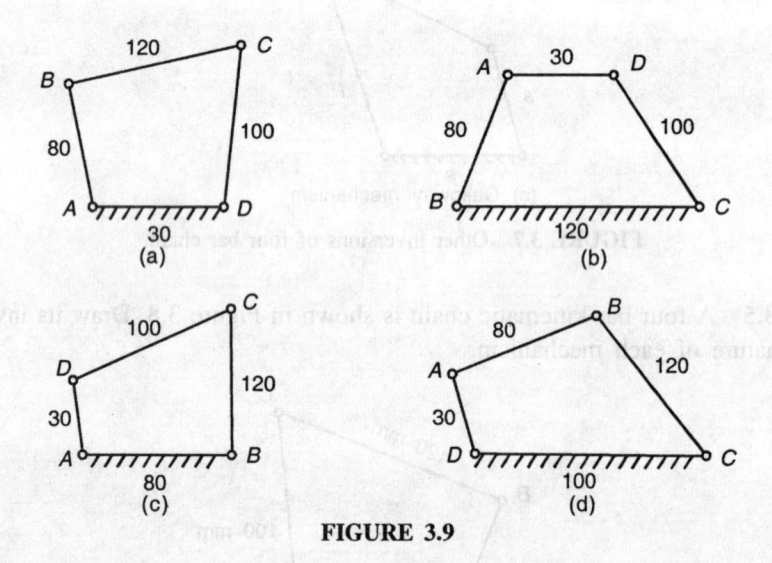

FIGURE 3.9

3.2.3 Limit Positions, Dead Centres and Transmission Angle

For the synthesis of mechanism, the limit positions, dead centres and transmission angle are important terms which are defined further.

Limit positions. It is a position of mechanism in which interior angle between the coupler and crank links is either 180° or 360°. Figures 3.10(b) and (c) show two extreme limit positions of a four bar mechanism.

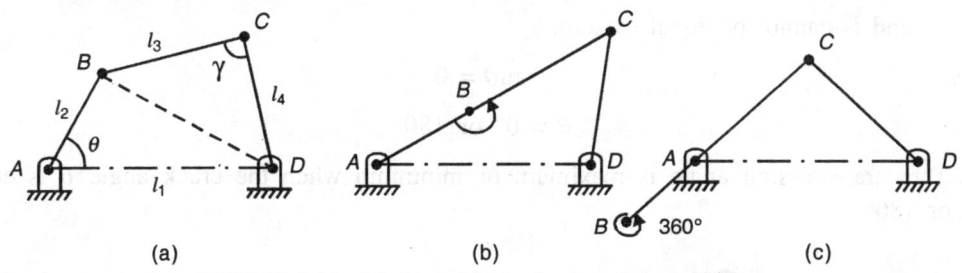

FIGURE 3.10 Limit positions of a four bar mechanism.

Dead centres. It is that position of mechanism in which the interior angle between the coupler and the follower links is either 180° or 360° as shown in Figure 3.11.

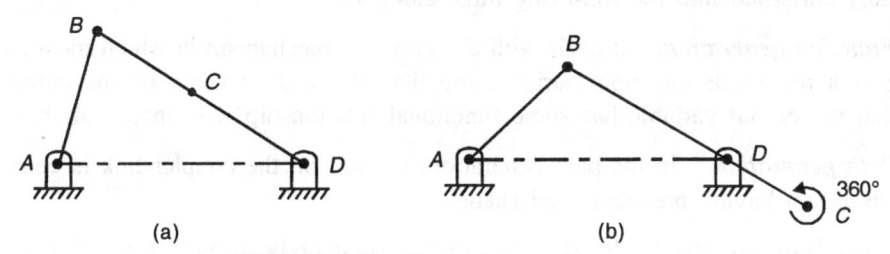

FIGURE 3.11 Dead centre positions.

Transmission angle. An interior angle between the coupler and the follower links at any position other than the dead centre is called **transmission angle** γ [Figure 3.10(a)]. The transmission angle is important from the point of view of transmission efficiency. For a high speed mechanism, the value of transmission angle less than 20° is not acceptable.

Applying cosine law to triangle *ABD* [Refer Figure 3.10(a)],

$$AB^2 + AD^2 - 2AB \times AD \cos\theta = BD^2$$

or

$$l_2^2 + l_1^2 - 2l_2 l_1 \cos\theta = BD^2 \tag{3.6}$$

Similarly, for triangle *BCD*,

$$l_3^2 + l_4^2 - 2l_3 l_4 \cos\gamma = BD^2 \tag{3.7}$$

Equating Eqs. (3.6) and (3.7), we get

$$l_2^2 + l_1^2 - 2l_2 l_1 \cos\theta = l_3^2 + l_4^2 - 2l_3 l_4 \cos\gamma$$

or

$$\cos\gamma = -\frac{l_1^2 + l_2^2 - l_3^2 - l_4^2 - 2l_2 l_1 \cos\theta}{2l_3 l_4} \tag{3.8}$$

For minimum or maximum value of transmission angle, differentiate Eq. (3.8) with respect to θ and put it equal to zero to get the following condition:

$$l_1 l_2 \sin\theta = 0 \qquad (3.9)$$

Since l_1 and l_2 cannot be equal to zero,

$$\sin\theta = 0$$

or

$$\theta = 0° \text{ or } 180°$$

Thus transmission angle is maximum or minimum when the crank angle θ is either zero or 180°.

3.3 DIMENSIONAL SYNTHESIS

The dimensional synthesis aims at determining dimensions of links and starting position of the links in a mechanism to satisfy the kinematic motion constraints. The dimensional synthesis is broadly classified into the following three categories:

Function generation. It deals with designing of mechanism in which movement of the driving link represents the input variable and that of the driven link by the output variable such that the output variable has some functional relationship with input variable.

Path generation. In the path generation, a point on the coupler link is constrained to generate a path having predetermined shape.

Rigid body guidance. It refers to guide a rigid body through a number of specified positions.

A large number of methods, both graphical and analytical, are available for the synthesis of mechanism. The graphical method being a direct approach is widely used. These methods produce reasonable accurate result with much less effort. However, in analytical method, a rigorous mathematical treatment is essential. Analytical methods are based on generalised equations, so these can be easily implemented to digital computer for higher accuracy.

In the following sections, we intend to cover both graphical and analytical methods of synthesis of mechanism.

3.4 SYNTHESIS BY RELATIVE POLE METHOD

The pole is a centre of rotation of moving link relative to a fixed link of the mechanism. To locate the pole point, let us consider a four bar mechanism *ABCD* in which link *AD* is fixed and crank *AB* is at initial rotation angle θ. If the crank *AB* is rotated by angle α in anti-clockwise direction to take new position *AB'*, the follower link *CD* rotates by angle β to reorient the coupler link *BC* to the new position *B'C'* as shown in Figure 3.12. The normals drawn at the mid-points of *BB'* and *CC'* intersect at point *P*. This point is centre of rotation of coupler relative to the fixed link *AD* and is called **pole**.

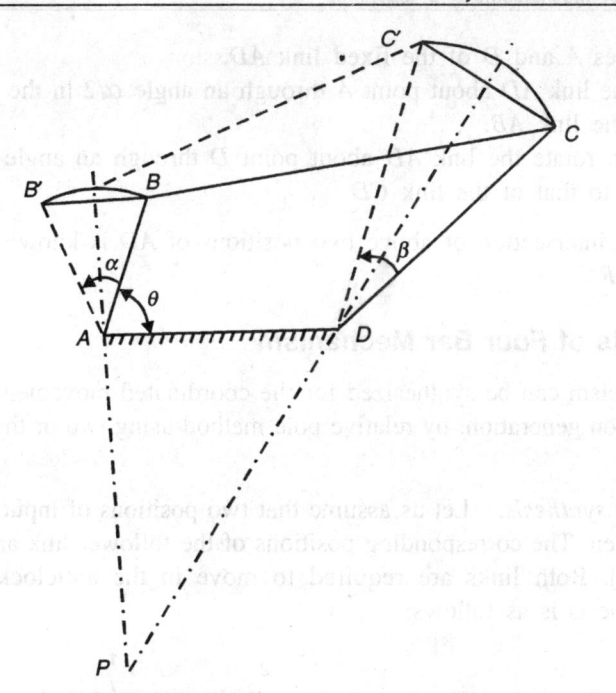

FIGURE 3.12 Pole of a four bar mechanism.

A relative pole is a centre of rotation of a link relative to other moving links such as crank and follower. In a four bar mechanism, if the crank and the follower links rotate by angles α and β respectively from their initial positions θ and ϕ, then the relative pole of coupler link relative to crank and follower can be found by the following procedure (Figure 3.13):

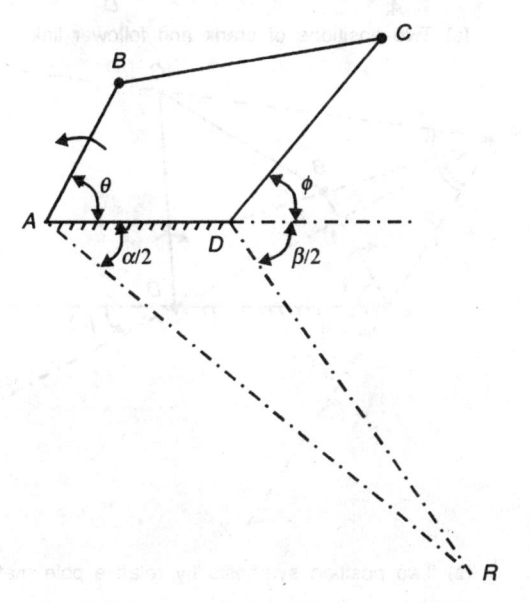

FIGURE 3.13 Relative pole of coupler link.

1. Join nodes *A* and *D* of the fixed link *AD*.
2. Rotate the link *AD* about point *A* through an angle $\alpha/2$ in the direction opposite to that of the link *AB*.
3. Similarly, rotate the link *AD* about point *D* through an angle $\beta/2$ in the direction opposite to that of the link *CD*.

The point of intersection of above two positions of *AD* is known as relative pole and is represented by *R*.

3.4.1 Synthesis of Four Bar Mechanism

A four bar mechanism can be synthesized for the coordinated movement of input and output links, called function generation, by relative pole method using two or three positions of input and output links.

Two position synthesis. Let us assume that two positions of input link crank separated by angle α are given. The corresponding positions of the follower link are separated by angle β [Figure 3.14(a)]. Both links are required to move in the anticlockwise direction. The procedure of synthesis is as follows:

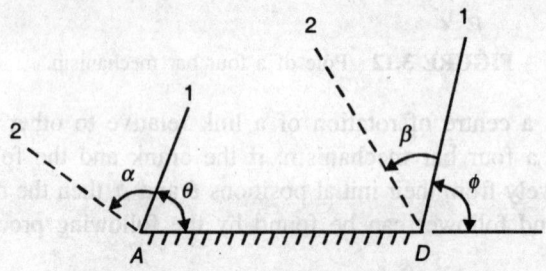

(a) Two positions of crank and follower link

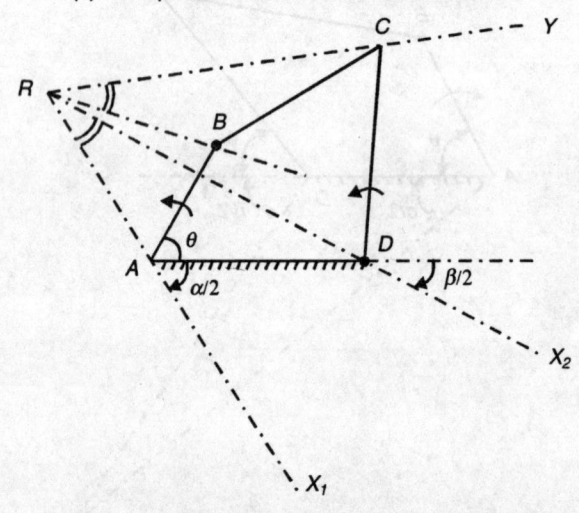

(b) Two position synthesis by relative pole method

FIGURE 3.14

1. Choose the frame *AD* arbitrarily at certain distance apart [See Figure 3.14(b)].
2. Draw a line AX_1 passing through point *A* and inclined at an angle $\alpha/2$ from link *AD* in the direction opposite to the direction of rotation of the link *AB*.
3. Similarly, draw another line DX_2 passing through point *D* inclined at angle $\beta/2$ from link *AD* in the direction opposite to the direction of rotation of link *CD*.
4. The intersection of lines AX_1 and DX_2 at point *R* is known as **relative pole**.
5. Choose suitable length for crank *AB* and mark the first position of the crank.
6. Join the points *B* and *R* by a straight line.
7. Draw a line *RY* from point *R* such that angle *ARD* is equal to angle *BRY*, i.e. the angle subtended by the frame is equal to the angle subtended by the coupler link.
8. Select the suitable location of point *C* on line *RY* and join *BC* and *CD* to form a mechanism *ABCD*.

Three position synthesis. In three position synthesis, two pairs of the crank and follower rotations are required to be coordinated. Let the two pairs of coordinates indicating three positions of crank and follower are (α_1, β_1) and (α_2, β_2) respectively [See Figure 3.15(a)]. The stepwise procedure is as given below (refer Figure 3.15).

1. Choose the frame *AD* arbitrarily at certain distance apart.

(a) Three positions of crank and follower

(b) Three position synthesis by relative pole method

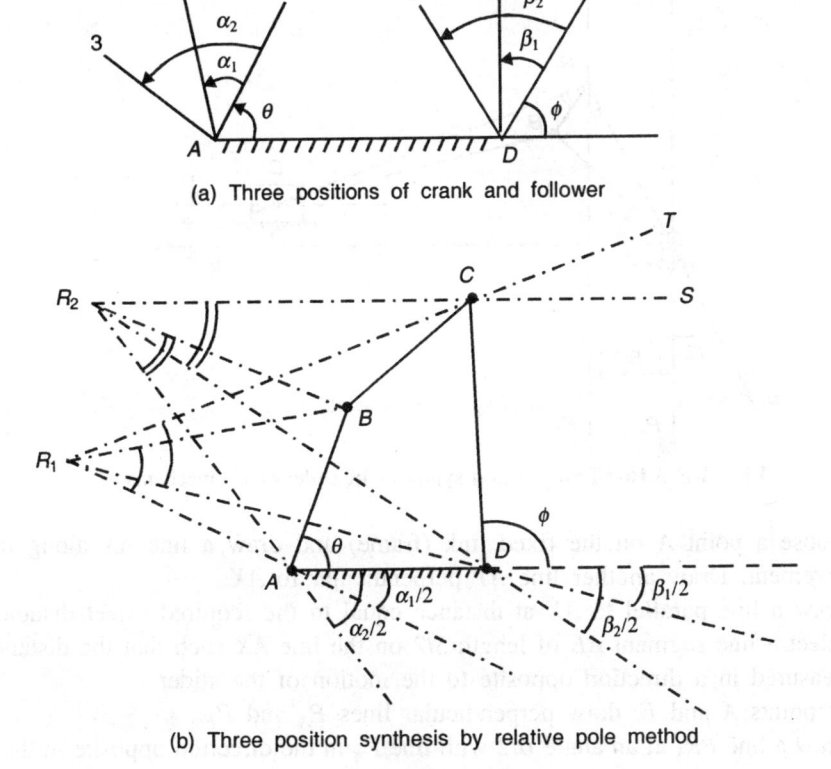

FIGURE 3.15

2. Locate the relative poles R_1 and R_2 as described previously for two pairs of coordinates (α_1, β_1) and (α_2, β_2).
3. Choose suitable length of crank AB and mark the first position of the crank.
4. As the crank and follower subtend equal angles at the relative pole, so draw a line R_1T such that angle DR_1T is equal to angle AR_1B.
5. Join R_2 and B by a straight line.
6. Draw a line R_2S such that angle AR_2B is equal to angle DR_2S.
7. Locate the point C at the intersection of lines R_1T and R_2S.
8. Join links BC and CD to complete the mechanism $ABCD$.

3.4.2 Synthesis of Slider-Crank Mechanism

The synthesis of slider-crank mechanism is presented below to coordinate the movement of input crank and slider by the relative pole method using two and three positions of slider and crank.

Two position synthesis. Consider a slider-crank mechanism in which when crank takes turn by angle α in anticlockwise direction from the first position, the slider changes its position by a distance S. Let the slider axis is offset by distance e from the axis of fixed link. The detailed procedure of synthesis is as follows (Refer Figure 3.16):

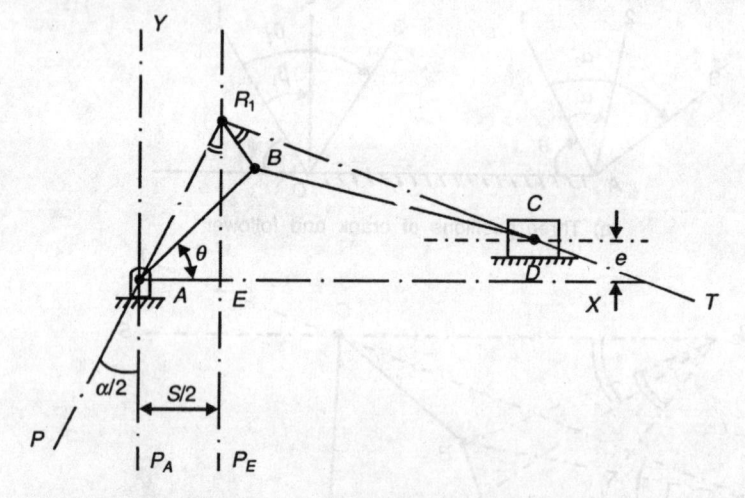

FIGURE 3.16 Two position synthesis of slider-crank mechanism.

1. Choose a point A on the fixed link (frame) and draw a line AX along the slider movement. Draw another line AY perpendicular to AX.
2. Draw a line parallel to AX at distance equal to the required offset distance e.
3. Select a line segment AE of length $S/2$ on the line AX such that the distance AE is measured in a direction opposite to the motion of the slider.
4. At points A and E, draw perpendicular lines P_A and P_E.
5. Draw a line PR_1 at an angle $\alpha/2$ with line P_A in the direction opposite to the rotation of crank.

6. The intersection of line PR_1 with line P_E is relative pole R_1.
7. Choose suitable length of crank AB and mark its first position.
8. Join B with R_1 by a straight line.
9. Construct angle BR_1T equals to angle AR_1E. The line R_1T intersects the offset line at point C. Join BC to complete the mechanism $ABCD$.

Three position synthesis. In the three position synthesis, the two pairs of movement of crank and slider are coordinated. Let (α_1, S_{12}) and (α_2, S_{13}) be coordinates of the crank and the slider. The crank is assumed to rotate in anticlockwise direction. The stepwise procedure of synthesis is as under (See Figure 3.17):

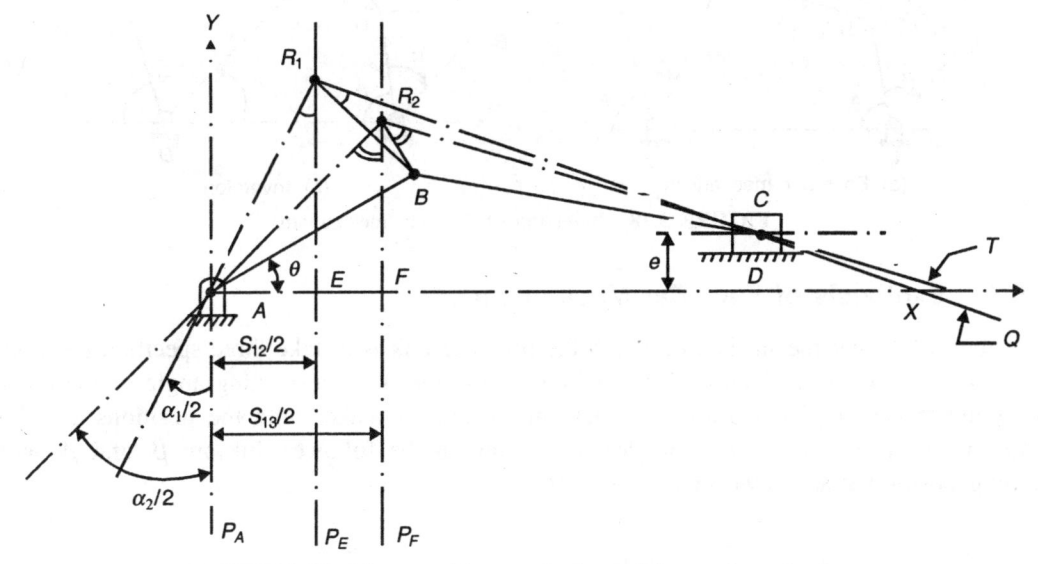

FIGURE 3.17 Three position synthesis of slider-crank mechanism.

1. Choose point A on the fixed link and draw a line AX along the movement of slider.
2. Locate the relative poles R_1 and R_2 for two pairs of coordinates (α_1, S_{12}) and (α_2, S_{13}) respectively (as described in two position synthesis method).
3. Choose the length of crank AB and mark its first position.
4. Join B with R_1 and R_2 by straight lines.
5. Draw a line R_1Q making an angle BR_1Q equals to angle AR_1E.
6. Similarly, draw another line R_2T which makes an angle BR_2T equals to angle AR_2F.
7. Mark the position of slider C at the intersection of these two lines.
8. Join BC to form a slider-crank mechanism.

3.5 SYNTHESIS OF MECHANISM BY INVERSION METHOD

A four bar mechanism $ABCD$ is shown in Figure 3.18(a) in which the rotation of the crank AB through angle α causes rotation of the follower link CD by the angle β. To get the inversion of mechanism, let us hold the link CD stationary and permit the other links including frame AD to take new positions by moving backward through the angle β such that the same relative motion is obtained [Figure 3.18(b)]. The resulting mechanism $A'B'_1 CD$ is called **inversion of mechanism**. This technique of inverting on the follower link is widely used in the design of four bar and slider-crank mechanisms.

(a) Four bar mechanism (b) Inversion

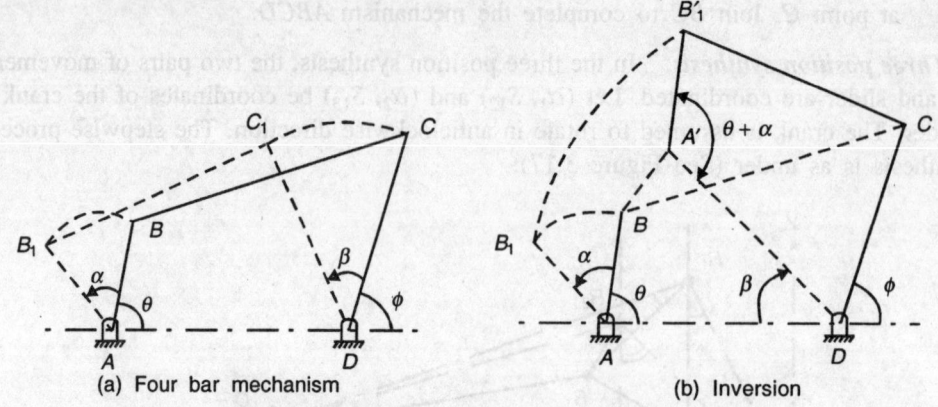

FIGURE 3.18 Inversion of four bar mechanism.

3.5.1 Synthesis of Four Bar Mechanism

Consider a four bar mechanism in which the follower link is to take three specified positions corresponding to three given positions of the crank link. Let the starting angle of the crank be θ and α_1 and α_2 be two angles which the position 1 makes with the positions 2 and 3 respectively. The corresponding angles of rotation of the follower link are β_1 and β_2 with starting position ϕ as shown in Figure 3.19.

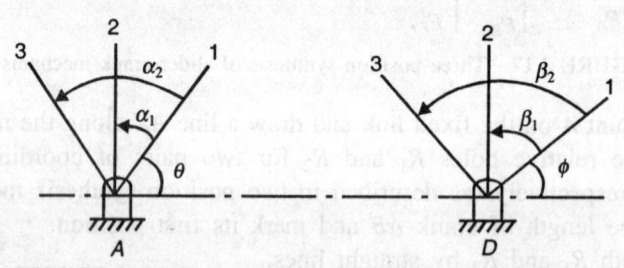

FIGURE 3.19 Three positions of crank and follower links.

The stepwise procedure to synthesize a mechanism is given below (Refer Figure 3.20):

1. Choose the frame AD arbitrarily at certain distance apart.
2. Select the appropriate length of the crank link AB and mark three positions AB, AB_1, AB_2 as shown in Figure 3.20.
3. Join points B_1 and D by a straight line. Draw an arc $B_1B'_1$ subtending angle β_1 with point D as centre and radius DB_1.
4. Similarly, join points B_2 and D. Draw an arc $B_2B'_2$ which subtends an angle β_2 at fixed point D.
5. Join BB'_1 and $B'_1B'_2$ by straight lines.

6. Draw two lines P and Q which are normal bisectors to BB'_1 and $B'_1B'_2$ respectively.
7. Intersection of lines P and Q at the point C forms a mechanism ABCD.

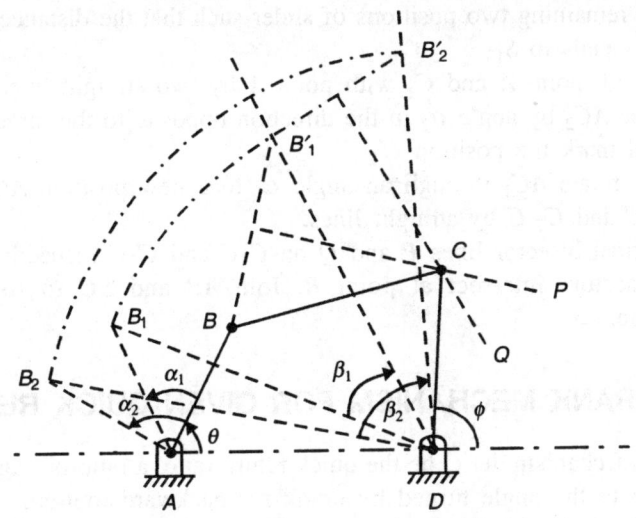

FIGURE 3.20 Synthesis of four bar mechanism by inversion method.

3.5.2 Synthesis of Slider-Crank Mechanism

For a slider-crank mechanism, let the two pairs of coordinates which denote the three specified positions of the crank and the slider are (α_1, S_{12}) and (α_2, S_{13}) respectively. The stepwise procedure of synthesis is as follows (Refer Figure 3.21).

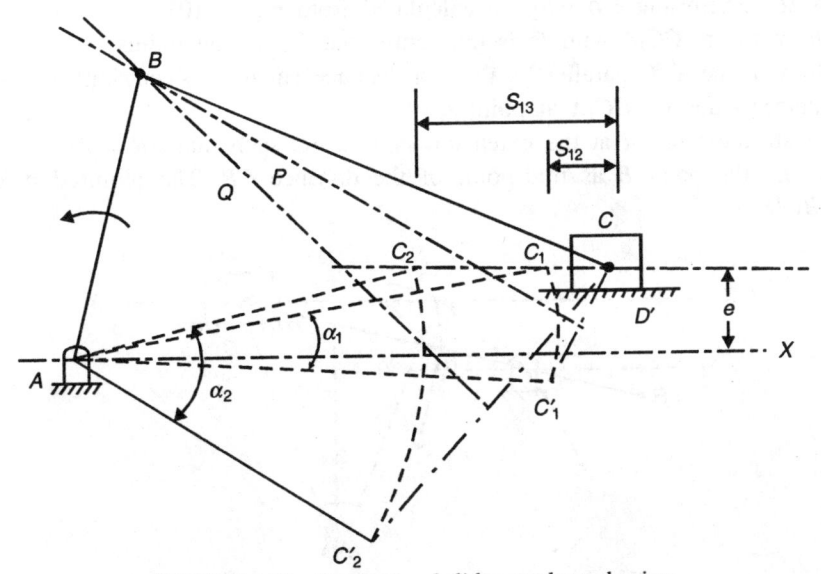

FIGURE 3.21 Synthesis of slider-crank mechanism.

1. Locate the point A on AX-axis and draw the slider axis parallel to AX at a given offset distance e.
2. Locate the first position of the slider C arbitrarily but conveniently.
3. Mark the remaining two positions of slider such that the distance CC_1 equals to S_{12} and CC_2 equals to S_{13}.
4. Join C_2 with point A and C_1 with point A by two straight lines.
5. Rotate line AC_2 by angle α_2 in the direction opposite to the direction of rotation of crank and mark the position C'_2.
6. Similarly, rotate AC_1 through an angle α_1 to a new position AC'_1.
7. Join $C'_1 C$ and $C'_2 C$ by straight lines.
8. Draw normal bisector lines P and Q on $C'_1 C$ and $C'_2 C$ respectively.
9. These bisectors intersect at point B. Join AB and BC to form a slider-crank mechanism.

3.6 SLIDER-CRANK MECHANISM FOR GIVEN QUICK RETURN RATIO

For a slider-crank mechanism, let Q be the quick return ratio, a ratio of angle turned by crank for forward motion to the angle turned by crank for backward motion.

$$Q = \frac{180° + \alpha}{180° - \alpha} \qquad (3.10)$$

Let S be the linear displacement of the slider. The procedure of synthesis is as under [Refer Figure 3.22]:

1. Draw a line XY and mark two positions C and C_1 of the slider at distance S apart.
2. Locate the point P on the perpendicular bisector of CC_1 such that angle CPC_1 is equal to 2α, where angle α may be calculated from Eq. (3.10).
3. Draw an arc $CC_1 A$ with P as the centre and PC as the radius.
4. Draw a line $X_1 Y_1$ parallel to XY at a distance equal to eccentricity (e), such that it intersects the arc $CC_1 A$ at point A.
5. Locate the point R at the extension of line AC such that $AR = AC_1$.
6. Obtain the point B as mid-point of the distance CR. The required mechanism is ABCD.

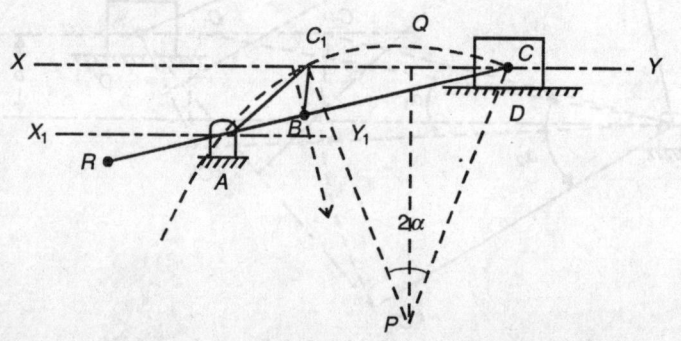

FIGURE 3.22 Slider-crank mechanism for a given quick return ratio (Q).

3.7 SYNTHESIS OF MECHANISM BY ANALYTICAL METHOD

A mechanism in which the output and input variables are proportionally related to some function $y = f(x)$ is called **function generator**. In synthesis of these mechanisms, the number of points, where the output generated by mechanism matches with the output computed from mathematical function, are called **precision points**. The selection of number of such points depends upon the design accuracy which generally vary between three to six. However, as the number of precision points increases, the computational difficulty increases. The best choice for the spacing of precision point is the one which gives minimum value of error between two adjacent precision points. As a first approximation, the Chebyshev spacing of precision points is the most widely used. Accordingly, the precision point is:

$$x_j = \bar{x} - r \cos\left\{\frac{(2j-1)\pi}{2n}\right\} \qquad (3.11)$$

where

$\bar{x}$ = mean value of input variable $\left(= \dfrac{x_i + x_f}{2}\right)$

r = range value $\left(= \dfrac{x_f - x_i}{2}\right)$

j = jth number of spacing point
x_i = initial value of input variable
x_f = final value of input variable and
n = number of spacing points

Graphically, the Chebyshev spacing can be obtained by drawing a circle of radius r with the centre at distance $\bar{x}$ from the origin O. Now construct a regular inscribed polygon of sides $2n$ such that two sides are perpendicular to the X-axis, as shown in Figure 3.23. The projections of vertices of the polygon on X-axis gives location of n spacing points.

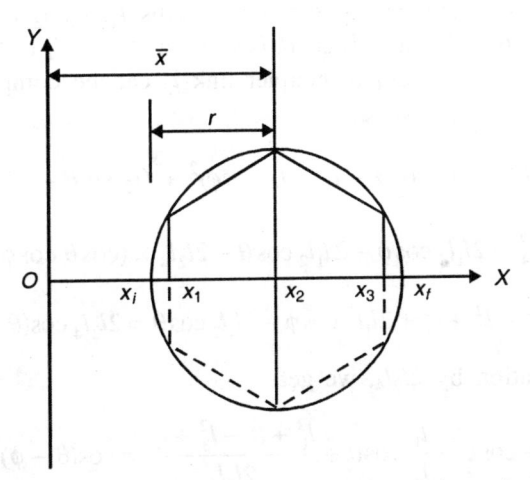

FIGURE 3.23 Chebyshev spacing of precision points.

In a function generator mechanism, the input variable x represents rotation of the driving link θ and the output variable y represents rotation of the driven link ϕ (See Figure 3.24). The relation between x and θ and that between y and ϕ is usually assumed to be linear as given below:

$$\frac{\theta_f - \theta_i}{x_f - x_i} = k_x = \frac{\theta - \theta_i}{x - x_i} \tag{3.12}$$

and

$$\frac{\phi_f - \phi_i}{y_f - y_i} = k_y = \frac{\phi - \phi_i}{y - y_i} \tag{3.13}$$

where k_x and k_y are called scale factors and the subscripts i and f represent initial and final values respectively.

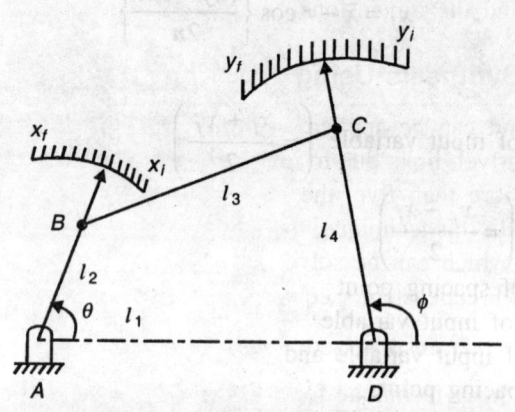

FIGURE 3.24 Function generator mechanism.

3.7.1 Freudenstein's Method of Synthesis

Consider a four bar mechanism having the link lengths l_1, l_2, l_3 and l_4 respectively. The coordinates of the points B and C are $(l_2 \cos\theta, l_2 \sin\theta)$ and $(l_1 + l_4 \cos\phi, l_4 \sin\phi)$ respectively. From geometric relations, the length of coupler link l_3 can be computed as follows (Refer to Figure 3.24):

$$BC^2 = l_3^2 = (l_2 \cos\theta - l_1 - l_4 \cos\phi)^2 + (l_2 \sin\theta - l_4 \sin\phi)^2$$

or

$$l_3^2 = l_1^2 + l_2^2 + l_4^2 + 2l_1 l_4 \cos\phi - 2l_1 l_2 \cos\theta - 2l_2 l_4 \times (\cos\theta \cos\phi + \sin\theta \sin\phi)$$

or

$$l_1^2 + l_2^2 - l_3^2 + l_4^2 + 2l_1 l_4 \cos\phi - 2l_1 l_2 \cos\theta = 2l_2 l_4 \cos(\theta - \phi)$$

Dividing the above equation by $2l_2 l_4$, we get

$$\frac{l_1}{l_2} \cos\phi - \frac{l_1}{l_4} \cos\theta + \frac{l_1^2 + l_2^2 - l_3^2 + l_4^2}{2l_2 l_4} = \cos(\theta - \phi)$$

or
$$k_1 \cos\phi - k_2 \cos\theta + k_3 = \cos(\theta - \phi) \tag{3.14}$$

where
$$k_1 = \frac{l_1}{l_2}, \quad k_2 = \frac{l_1}{l_4} \quad \text{and} \quad k_3 = \frac{l_1^2 + l_2^2 - l_3^2 + l_4^2}{2l_2 l_4}$$

Equation (3.14) is known as **Freudenstein's equation of displacement**.

Let θ_1, θ_2 and θ_3 be the three positions of driving link and ϕ_1, ϕ_2 and ϕ_3 three positions of driven link corresponding to Chebyshev spacing of three precision points.

Substituting these three positions of driving and driven links into Eq. (3.14), we get three linear simultaneous equations which can be solved either by Gauss elimination method or any other method to determine the constants k_1, k_2 and k_3. By selecting appropriate length of either link, one can find the dimensions of other links.

3.7.2 Approximate Synthesis Using Least Square Technique

The Freudenstein's method can be applied to synthesize a mechanism up to five precision positions of driving and driven links due to computational limitation. However, if the number of precision points are more than five, the approximate solution can be obtained by Least Square Technique (LST). In this technique, the sum of square of errors is minimized to obtain a set of linear equations which can be solved for unknown variables.

Let us consider the Freudenstein's equation for displacement.

$$k_1 \cos\phi - k_2 \cos\theta + k_3 = \cos(\theta - \phi)$$

To apply the least square technique for n number of precision points, the sum of square of the error is given as:

$$E = \sum_{i=1}^{n} \left[k_1 \cos\phi_i - k_2 \cos\theta_i + k_3 - \cos(\theta_i - \phi_i) \right]^2 \tag{3.15}$$

For minimum error, the partial derivative of error E with respect to k_1, k_2 and k_3 must be equal to zero.

$$\frac{\partial E}{\partial k_1} = \frac{\partial E}{\partial k_2} = \frac{\partial E}{\partial k_3} = 0$$

or

$$\frac{\partial E}{\partial k_1} = \sum_{i=1}^{n} 2\left[k_1 \cos\phi_i - k_2 \cos\theta_i + k_3 - \cos(\theta_i - \phi_i) \right] \cos\phi_i = 0 \tag{3.16}$$

$$\frac{\partial E}{\partial k_2} = \sum_{i=1}^{n} 2\left[k_1 \cos\phi_i - k_2 \cos\theta_i + k_3 - \cos(\theta_i - \phi_i) \right] \cos\theta_i = 0 \tag{3.17}$$

$$\frac{\partial E}{\partial k_3} = \sum_{i=1}^{n} 2\left[k_1 \cos\phi_i - k_2 \cos\theta_i + k_3 - \cos(\theta_i - \phi_i) \right] = 0 \tag{3.18}$$

The above three equations can be written in simplified form as:

$$k_1 \Sigma \cos^2 \phi_i - k_2 \Sigma \cos \theta_i \cos \phi_i + k_3 \Sigma \cos \phi_i = \Sigma \cos(\theta_i - \phi_i) \cdot \cos \phi_i \qquad (3.19)$$

$$k_1 \Sigma \cos \phi_i \cos \theta_i - k_2 \Sigma \cos^2 \theta_i + k_3 \Sigma \cos \theta_i = \Sigma \cos(\theta_i - \phi_i) \cos \theta_i \qquad (3.20)$$

$$k_1 \Sigma \cos \phi_i - k_2 \Sigma \cos \theta_i + k_3 = \Sigma \cos(\theta_i - \phi_i) \qquad (3.21)$$

where Σ represents summation of n precision points.

The Eqs. (3.19), (3.20) and (3.21) are simultaneous linear non-homogeneous equations which can be solved for three unknowns k_1, k_2 and k_3.

3.7.3 Bloch's Method

A Russian kinematician Bloch developed a method using complex algebra for synthesis of mechanism. According to him, if the links of a four bar mechanism are replaced by position vector, the vector equation is written as:

$$r_1 + r_2 + r_3 + r_4 = 0 \qquad (3.22)$$

where $\quad r_1 = l_1 e^{j\theta_1}, \quad r_3 = l_3 e^{j\theta_3}, \quad r_2 = l_2 e^{j\theta_2}, \quad r_4 = l_4 e^{j\theta_4}$

l_1, l_2, l_3 and l_4 are lengths of the links and θ_1, θ_2, θ_3 and θ_4 are angles made by them (See Figure 3.25).

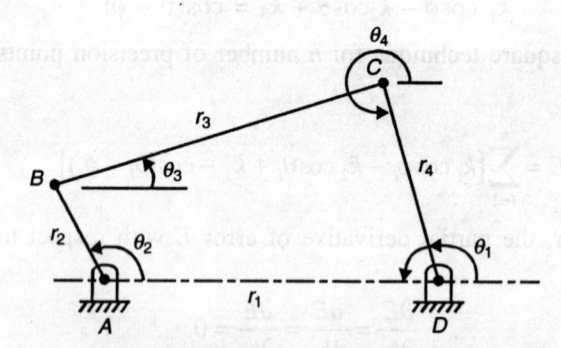

FIGURE 3.25 Four bar mechanism.

In complex polar notations, the vector equation can be written as:

$$l_1 e^{j\theta_1} + l_2 e^{j\theta_2} + l_3 e^{j\theta_3} + l_4 e^{j\theta_4} = 0 \qquad (3.23)$$

Let ω be the angular velocity $\left(= \dfrac{d\theta}{dt} \right)$

and α be the angular acceleration $\left(= \dfrac{d\omega}{dt} = \dfrac{d^2\theta}{dt^2} \right)$

Considering the vector r_1 as frame and taking first and second derivatives of Eq. (3.23), we get

$$l_2\omega_2 e^{j\theta_2} + l_3\omega_3 e^{j\theta_3} + l_4\omega_4 e^{j\theta_4} = 0 \qquad (3.24)$$

$$l_2(\alpha_2 + j\omega_2^2)e^{j\theta_2} + l_3(\alpha_3 + j\omega_3^2)e^{j\theta_3} + l_4(\alpha_4 + j\omega_4^2)e^{j\theta_4} = 0 \qquad (3.25)$$

Now transforming Eqs. (3.23), (3.24) and (3.25) back into vector notations, we get a set of the following homogeneous vector equations having complex coefficients:

$$r_1 + r_2 + r_3 + r_4 = 0 \qquad (3.26)$$

$$\omega_2 r_2 + \omega_3 r_3 + \omega_4 r_4 = 0 \qquad (3.27)$$

$$(\alpha_2 + j\omega_2^2)r_2 + (\alpha_3 + j\omega_3^2)r_3 + (\alpha_4 + j\omega_4^2)r_4 = 0 \qquad (3.28)$$

In the above three equations, there are four unknowns. So one unknown may be chosen arbitrarily, say a frame of length l_1 with angle θ_1.

Solving the above equations for vector r_2, we get

$$r_2 = \frac{\begin{vmatrix} -1 & 1 & 1 \\ 0 & \omega_3 & \omega_4 \\ 0 & \alpha_3 + j\omega_3^2 & \alpha_4 + j\omega_4^2 \end{vmatrix}}{|D|} \qquad (3.29)$$

where

$$|D| = \begin{vmatrix} 1 & 1 & 1 \\ \omega_2 & \omega_3 & \omega_4 \\ \alpha_2 + j\omega_2^2 & \alpha_3 + j\omega_3^2 & \alpha_4 + j\omega_4^2 \end{vmatrix}$$

Similar expressions can be obtained for vectors r_3 and r_4. Since the determinant $|D|$ is common in all the expressions, the effect of this is to change the magnitude of vector. Thus, we may assume the value of determinant to be unity. The solution then obtained gives dimensions of vectors as follows:

$$r_2 = \omega_4(\alpha_3 + j\omega_3^2) - \omega_3(\alpha_4 + j\omega_4^2)$$

$$r_3 = \omega_2(\alpha_4 + j\omega_4^2) - \omega_4(\alpha_2 + j\omega_2^2)$$

$$r_4 = \omega_3(\alpha_2 + j\omega_2^2) - \omega_2(\alpha_3 + j\omega_3^2)$$

and

$$r_1 = -r_2 - r_3 - r_4$$

EXAMPLE 3.6 Synthesize a four bar mechanism to generate a function $y = \log_{10} x$ in the interval $1 \le x \le 10$. The input crank length is 50 mm. The input crank is to rotate from 45° to 105° while the output link moves from 135° to 225°. Use three accuracy points with Chebyshev's spacing.

Solution Using Chebyshev's three accuracy point,

$$x_j = \bar{x} - r\cos\left[\frac{(2j-1)\pi}{2n}\right], \quad j = 1, 2, \ldots, n$$

where

$$\bar{x} = \frac{x_i + x_f}{2} = \frac{1 + 10}{2} = 5.5$$

$$r = \frac{x_f - x_i}{2} = \frac{10 - 1}{2} = 4.5$$

$$n = 3$$

$$x_1 = 5.5 - 4.5 \times \cos\left[\frac{(2 \times 1 - 1)\pi}{2 \times 3}\right] = 1.603$$

$$x_2 = 5.5 - 4.5 \times \cos\left[\frac{(2 \times 2 - 1)\pi}{2 \times 3}\right] = 5.5$$

$$x_3 = 5.5 - 4.5 \times \cos\left[\frac{(3 \times 2 - 1)\pi}{2 \times 3}\right] = 9.397$$

Corresponding function values are:

$$y_1 = \log_{10}(x_1) = \log_{10}(1.603) = 0.2049$$
$$y_2 = \log_{10}(x_2) = \log_{10}(5.5) = 0.7403$$
$$y_3 = \log_{10}(x_3) = \log_{10}(9.397) = 0.9730$$

Let θ_1, θ_2 and θ_3 be the three positions of the driving link and ϕ_1, ϕ_2 and ϕ_3 be the three positions of the driven link.

Assuming that the relations between x and θ and that between y and ϕ are linear, we get

$$\frac{\theta_f - \theta_i}{x_f - x_i} = k_x = \frac{\theta_j - \theta_i}{x_j - x_i}$$

or

$$\theta_j = \theta_i + k_x(x_j - x_i)$$

where

$$k_x = \frac{105 - 45}{10 - 1} = 6.667$$

$$\theta_1 = 45° + 6.667\,(1.603 - 1) = 49.02°$$
$$\theta_2 = 45° + 6.667\,(5.5 - 1) = 75°$$
$$\theta_3 = 45° + 6.667\,(9.397 - 1) = 100.98°$$

Similarly, the position of the output link:
$$\phi_j = \phi_i + k_y(y_j - y_i)$$

where
$$k_y = \frac{\phi_f - \phi_i}{y_f - y_i} = \frac{225° - 135°}{1 - 0} = 90$$

$$\phi_1 = 135° + 90(0.2049 - 0) = 153.44°$$
$$\phi_2 = 135° + 90(0.7403 - 0) = 201.62°$$
$$\phi_3 = 135° + 90(0.973 - 0) = 222.57°$$

Substituting the values of (θ_1, ϕ_1), (θ_2, ϕ_2) and (θ_3, ϕ_3) in the following Freudenstein's equation to form three simultaneous equations:
$$k_1 \cos \phi - k_2 \cos \theta + k_3 = \cos(\theta - \phi)$$

For pair (θ_1, ϕ_1),
$$k_1 \cos 153.44° - k_2 \cos 49.02° = \cos(49.02 - 153.44)$$

or
$$-0.8944 k_1 - 0.6557 k_2 + k_3 = -0.2490 \qquad (i)$$

For pair (θ_2, ϕ_2),
$$k_1 \cos 201.62° - k_2 \cos 75° + k_3 = \cos(75 - 201.62)$$

or
$$-0.9296 k_1 - 0.2588 k_2 + k_3 = -0.5965 \qquad (ii)$$

For pair (θ_3, ϕ_3),
$$k_1 \cos 222.57° - k_2 \cos 100.98° + k_3 = \cos(100.98 - 222.57)$$

or
$$-0.7364 k_1 + 0.1904 k_2 + k_3 = -0.5238 \qquad (iii)$$

Solving Eqs. (i), (ii) and (iii) for k_1, k_2 and k_3, we get
$$k_1 = 2, k_2 = -0.783 \text{ and } k_3 = 1.078.$$

So
$$k_1 = \frac{l_1}{l_2} = 2 \text{ and } k_2 = \frac{l_1}{l_4} = -0.703$$

Assume that the smallest link is:
$$l_2 = 50 \text{ mm (given)}$$

Therefore,
$$l_1 = 2l_2 = \mathbf{100 \text{ mm}} \qquad \text{Ans.}$$

$$l_4 = \frac{l_1}{k_2} = -\frac{100}{0.703} = \mathbf{142.2 \text{ mm (negative)}} \qquad \text{Ans.}$$

$$k_3 = \frac{l_1^2 + l_2^2 - l_3^2 + l_4^2}{2 l_2 l_4} = 1.078$$

or
$$\frac{100^2 + 50^2 - l_3^2 + 142.2^2}{-2 \times 50 \times 142.2} = 1.078$$

or $\qquad l_3 = 219.2$ mm $\qquad$ **Ans.**

EXAMPLE 3.7 Design a four bar mechanism which can coordinate the input and output angles as given below:

Input crank angle: $\qquad$ 30° $\qquad$ 50° $\qquad$ 80°

Output crank angle: $\qquad$ 0° $\qquad$ 30° $\qquad$ 60°

Solution Writing Freudenstein's equation for each of the three positions:

$$k_1 \cos 0° - k_2 \cos 30° + k_3 = \cos(30 - 0)$$

$$k_1 \cos 30° - k_2 \cos 50° + k_3 = \cos(50 - 30)$$

$$k_1 \cos 60° - k_2 \cos 80° + k_3 = \cos(80 - 60)$$

After simplifying,

$$k_1 - 0.866 \, k_2 + k_3 = 0.866 \qquad \text{(i)}$$

$$0.866 \, k_1 - 0.6428 \, k_2 + k_3 = 0.9397 \qquad \text{(ii)}$$

$$0.5 \, k_1 - 0.1736 \, k_2 + k_3 = 0.9397 \qquad \text{(iii)}$$

Solving Eqs. (i), (ii) and (iii) for the values of k_1, k_2 and k_3, we get

$$k_1 = 1.831, \quad k_2 = 1.4294 \quad \text{and} \quad k_3 = 0.2718$$

Let the length of the smallest link, $l_2 = 1$ unit

$$l_1 = k_1 l_2 = 1.831$$

$$l_4 = \frac{l_1}{k_2} = \frac{1.831}{1.4294} = 1.281$$

$$k_3 = \frac{l_1^2 + l_2^2 - l_3^2 + l_4^2}{2 l_2 l_4} = 0.2718$$

or

$$\frac{1.831^2 + 1^2 - l_3^2 + 1.281^2}{2 \times 1 \times 1.281} = 0.2718$$

or $\qquad l_3 = 2.155$

The lengths of the links are:

$\qquad l_1 = 1.831, \quad l_2 = 1, \quad l_3 = 2.155 \quad \text{and} \quad l_4 = 1.281$ $\qquad$ **Ans.**

EXERCISES

1. What do you mean by kinematic synthesis? Explain three stages in which kinematic synthesis is accomplished.
2. What do you mean by degree of freedom of a mechanism? Explain Grubler's equation for determining dof of a mechanism.
3. Prove that for a constrained motion, the minimum number of binary links in a mechanism is four.
4. What is meant by movability of a mechanism? Explain the Grashof's criterion of deciding movability of a mechanism.
5. Define the following terms related to synthesis of mechanism:
   (a) Limit position
   (b) Dead centre
   (c) Transmission angle
   (d) Function generator
   (e) Path generator
6. Explain the graphical procedure to synthesize four bar mechanism by (a) two position synthesis and (b) three position synthesis methods.
7. Explain the three position synthesis of a slider-crank mechanism.
8. Explain the Freudenstein's method of synthesis of a four bar mechanism.
9. Explain the procedure to design a four bar mechanism using the least square technique to coordinate motion of input and output links for n number of positions governed by a function.
10. Determine the movability of the mechanisms shown in Figure E3.1 by Grashof's criterion (numbers indicate the respective link length in mm).

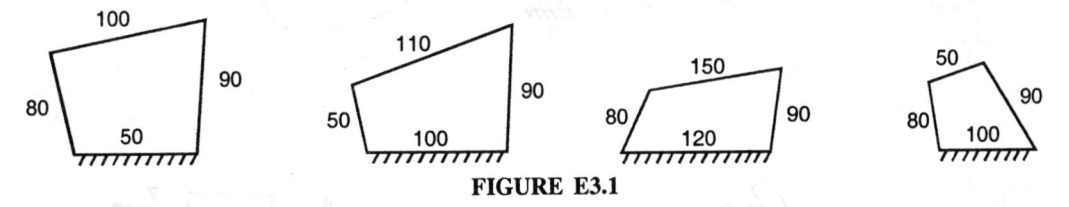

FIGURE E3.1

[Ans: (a) C–C (b) C–R, (c) C–R, (d) R–C]

11. The dimensions of a four bar mechanism are 100 mm, 400 mm, 700 mm and 800 mm. Draw the inversions of the mechanism and decide their movability.
12. The length of links of a four bar mechanism are in Geometric Progression Series. Discuss the movability of the mechanism.
13. In a four bar mechanism, the lengths of driver crank, coupler, follower and fixed link are 100 mm, 200 mm, 300 mm and L_0 respectively. Find the range of values for L_0 so as to make it (a) crank-rocker mechanism and (b) crank-crank mechanism.
    [Ans: 200 mm $\leq L_0 \leq$ 400 mm]
14. The mechanism shown in Figure E3.2 is driven by link AB. Find out the maximum and minimum transmission angles.
    [Ans: θ = 46.5°, 124.2°]

100 Theory of Mechanisms and Machines

FIGURE E3.2

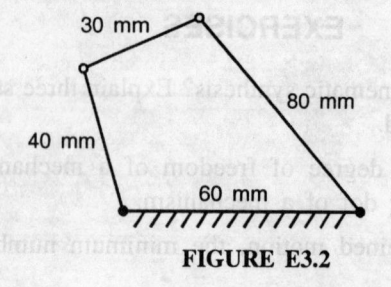

15. Examine the mechanisms shown in Figure E3.3 and determine the degree of freedom of each mechanism.

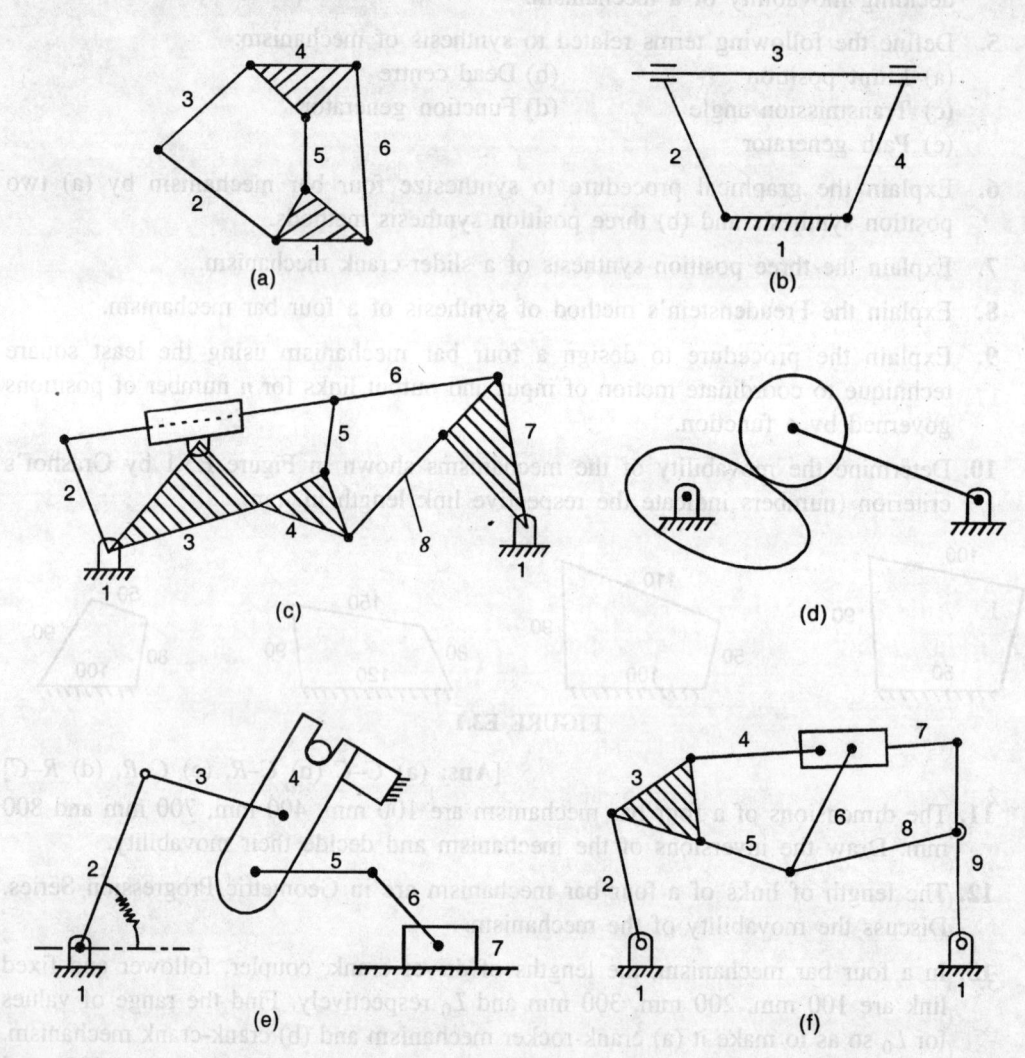

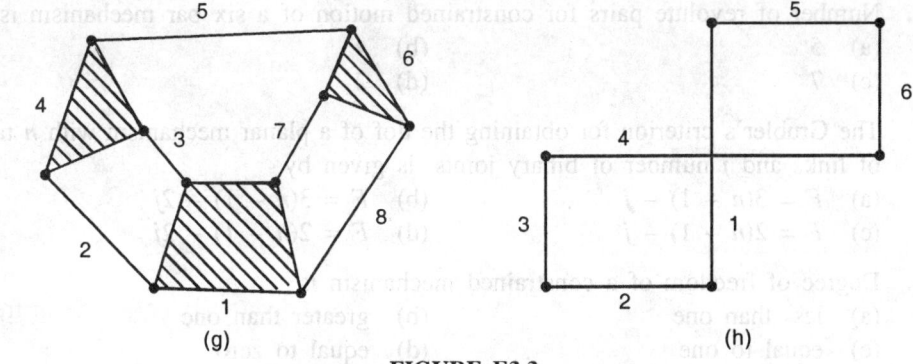

FIGURE E3.3

[Ans: (a) 1, (b) 0, (c) 3, (d) 1, (e) 3, (f) 2, (g) 1, (h) 1]

16. Synthesize a function generator to solve the following equation:

$$y = \frac{1}{x} \quad 1 \leq x \leq 2$$

The input link is to rotate from 30° to 120° while the output link rotates from 240° to 330°. Use three accuracy points.

[Ans: $l_1 = 0.968$, $l_2 = 1$, $l_3 = l_4 = 2.48$]

17. Design a four bar mechanism to generate function $y = x^{1.5}$ where x varies from 1 to 4. The length of one link is 20 mm. Use three accuracy points.

[Ans: 20, 119.5, 211.7, 240 mm]

MULTIPLE CHOICE QUESTIONS

1. A negative dof for a mechanism means
   (a) constrained motion mechanism
   (b) unconstrained motion mechanism
   (c) any motion
   (d) statically indeterminate structure

2. Zero dof for a mechanism means
   (a) statically indeterminate structure
   (b) statically determinate structure
   (c) constrained motion mechanism
   (d) unconstrained motion mechanism

3. For a four bar mechanism to act as a double crank mechanism
   (a) the largest link should be follower
   (b) the shortest link should be crank
   (c) the shortest link should be frame
   (d) none of the above

4. Number of revolute pairs for constrained motion of a six bar mechanism is
   (a) 5
   (b) 4
   (c) 7
   (d) 2

5. The Grubler's criterion for obtaining the dof of a planar mechanism with n number of links and j number of binary joints, is given by
   (a) $F = 3(n - 1) - j$
   (b) $F = 3(n - 1) - 2j$
   (c) $F = 2(n - 1) - j$
   (d) $F = 2(n - 1) - 2j$

6. Degree of freedom of a constrained mechanism is
   (a) less than one
   (b) greater than one
   (c) equal to one
   (d) equal to zero

7. The dof of a rigid link placed on a table is
   (a) 0
   (b) 1
   (c) 2
   (d) 3

8. In a four bar chain, if l is length of the longest link, s is length of the shortest link and p and q are lengths of the remaining two links. Grashof's law states that
   (a) $l + s < p + q$
   (b) $l + s > p + q$
   (c) $l - s < p - q$
   (d) $l - s > p + q$

9. The maximum number of lower pairs in a link can be (n is number of links in a mechanism)
   (a) n
   (b) $2n$
   (c) $\dfrac{n}{2}$
   (d) $\dfrac{n}{4}$

10. When the shortest link is coupler link, it results into
    (a) crank-crank mechanism
    (b) crank-rocker mechanism
    (c) double rocker mechanism
    (d) none of the above

11. When the shortest link is crank and any other link is frame, it results into
    (a) crank-crank mechanism
    (b) crank-rocker mechanism
    (c) double rocker mechanism
    (d) none of the above

12. A mechanism in which two links of equal length are placed adjacent and the longer link is fixed, is called
    (a) parallelogram mechanism
    (b) deltoid mechanism
    (c) Galloway mechanism
    (d) crank-rocker mechanism

13. A mechanism in which two links of equal length are not adjacent is called
    (a) parallelogram mechanism
    (b) deltoid mechanism
    (c) Galloway mechanism
    (d) crank-rocker mechanism

14. Dead centre is that position of mechanism in which the interior angle between coupler and follower links is
    (a) 0°
    (b) 90°
    (c) 180°
    (d) 270°

15. A mechanism in which two links of equal length are placed adjacent and the shorter link is fixed is called
    (a) parallelogram mechanism
    (b) deltoid mechanism
    (c) Galloway mechanism
    (d) crank-rocker mechanism

CHAPTER 4

Lower Pair Mechanisms

4.1 INTRODUCTION

When two links have surface or area contact while in motion, the pair so formed is known as a **lower pair.** In such kinematic pair, the relative motion is purely turning (pivoted) or sliding. A mechanism formed with turning or sliding pairs is called **lower pair mechanism.** In mechanical system, a large number of mechanisms with lower pairs are used to produce different types of motion such as tracing a point, straight line motion, automobile steering, rotation with angular misalignment and so forth.

In this chapter, a brief discussion on some of the most commonly used mechanisms with lower pairs is given.

4.2 STRAIGHT LINE MOTION MECHANISM

Mechanisms which are designed to produce a straight line motion are known as **straight line motion mechanisms.** A mechanism with sliding pair is the most common type of straight line motion mechanism. But sliding pair is bulky and is subjected to comparatively rapid wear so a mechanism which is constrained by the use of turning pair is often called a straight line motion mechanism. These mechanisms are divided into three categories:

(i) Exact straight line motion mechanism
(ii) Approximate straight line motion mechanism
(iii) Straight line copying mechanism

4.2.1 Exact Straight Line Motion Mechanism

Exact straight line motion mechanism follows a mathematical relation which holds true for all positions of input link such that output link follows straight line.

Let a link OA turn about centre O and the position of the point A be such that the product $OA \cdot OB$ is constant. Then the path of the point A will be straight line perpendicular to diameter OC of the circle along with circumference of which B moves (Figure 4.1).

To prove mathematical relation, draw AD perpendicular to OC or OC produced. Join CB to form triangle OBC. Then the triangles OBC and ODA are similar as angle BOC is common and $\angle OBC = \angle ODA$ (each right angle).

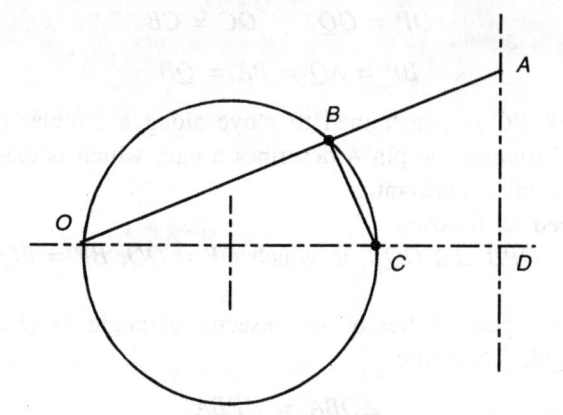

FIGURE 4.1 Straight line motion.

Therefore
$$\frac{OB}{OC} = \frac{OD}{OA}$$

or
$$OD = OA \cdot \frac{OB}{OC} \qquad (4.1)$$

But *OC* is constant as it is the diameter of the circle. If the product $OA \cdot OB$ is constant, then the distance *OD* will be constant and the point *A* moves along the straight line *AD* which is perpendicular to *OD*.

Several mechanisms have been devised by connecting points *O*, *B* and *A* in such a way as to satisfy the above condition. A few of them are described further.

Peaucellier mechanism. Consider a mechanism as shown in Figure 4.2 in which various links are having the following proportions:

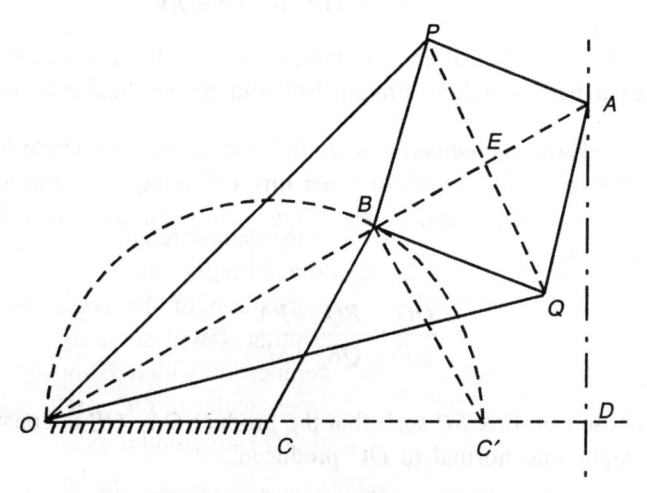

FIGURE 4.2 Peaucellier mechanism.

$$OP = OQ, \quad OC = CB$$
and
$$BP = AQ = PA = QB$$

The pin B on link BC is constrained to move along a circular path $C'BO$ with C as centre. As the link BC rotates, the pin A describes a path which is exactly a straight line as the product $OB \cdot OA$ remains constant.

This can be proved as follows:

Consider triangles OPB and OQB, in which $OP = OQ$, $BP = BQ$ and OB is common. Hence $\angle POB = \angle QOB$

This means that the point B lies on the bisector of angle POQ and $\angle OBP = \angle OBQ$. Further, $\triangle APB \equiv \triangle AQB$. Therefore,

$$\angle QBA = \angle PBA$$
and
$$\angle OBP + \angle PBA = \angle QBO + \angle QBA$$

But OB is a straight line. Thus

$$\angle OBP + \angle PBA = \angle QBO + \angle QBA = 180°$$

Hence OBA is a straight line.
Join PQ to bisect BA at point E.
From $\triangle OPE$,

$$OP^2 = OE^2 + PE^2 \tag{4.2}$$

and from $\triangle AEP$,

$$AP^2 = EA^2 + PE^2 \tag{4.3}$$

Subtracting Eq. (4.3) from Eq. (4.2), we get

$$OP^2 - AP^2 = OE^2 - EA^2$$
$$= (OE + EA)(OE - EA)$$
$$= OA \cdot OB \quad \text{as} \quad EA = BE \tag{4.4}$$

In Eq. (4.4), lengths OP and AP are constant, therefore the product $OA \cdot OB$ is constant and the point A traces a path which is straight line and perpendicular to OD.

Hart mechanism. Hart mechanism is a six link mechanism in which link PQ equals to RS and link QR equals to PS. The length of fixed link OC is equal to the length of link BC (Figure 4.3). Further, points O, B and A are three points on the links PQ, QR and PS respectively such that:

$$\frac{OQ}{QP} = \frac{BQ}{QR} = \frac{PA}{SP} \tag{4.5}$$

The point A is chosen on link SP such that the product $OA \cdot OB$ is constant and it traces a path which is a straight line normal to OC produced.

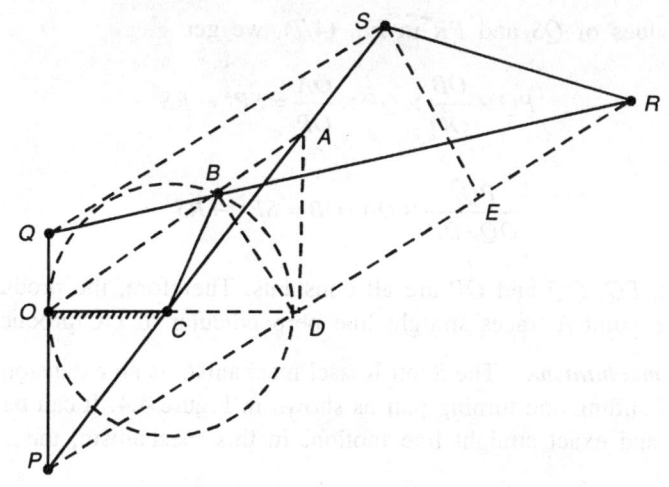

FIGURE 4.3 Hart mechanism.

The lines joining QS and PR are parallel for all positions of the mechanism. Hence, if three points O, B and A lie on a line parallel to QS or PR for one position, then they will lie on straight line parallel to them for all positions of the mechanism.

From $\triangle SPR$,

$$SP^2 = PR^2 + RS^2 - 2PR \times RS \times \cos\angle SRP \tag{4.6}$$

But

$$\cos\angle SRP = \frac{PR - QS}{2RS}$$

or

$$SP^2 = PR^2 + RS^2 - 2PR \times RS \times \frac{PR - QS}{2RS}$$

$$= PR^2 + RS^2 - PR(PR - QS)$$

or

$$PR \times QS = SP^2 - RS^2 \tag{4.7}$$

$$= \text{constant, as } SP \text{ and } RS \text{ both are constant.}$$

From similar triangles QPS and OPA,

$$\frac{QS}{QP} = \frac{OA}{OP} \quad \text{or} \quad QS = QP \times \frac{OA}{OP} \tag{4.8}$$

Similarly, from triangles PQR and OQB,

$$\frac{PR}{PQ} = \frac{OB}{OQ} \quad \text{or} \quad PR = PQ \times \frac{OB}{OQ} \tag{4.9}$$

Substituting the values of QS and PR in Eq. (4.7), we get

$$PQ \times \frac{OB}{OQ} \times QP \times \frac{OA}{OP} = SP^2 - RS^2$$

or

$$\frac{PQ^2}{OQ \cdot OP} \times OA \cdot OB = SP^2 - RS^2 \tag{4.10}$$

In Eq. (4.10), PQ, OQ and OP are all constants. Therefore, the product $OA \cdot OB$ must be constant. Hence point A traces straight line perpendicular to OC produced.

Scott Russel mechanism. The Scott Russel mechanism is an extension of a slider-crank chain consisting of sliding and turning pair as shown in Figure 4.4. It can be used to generate both approximate and exact straight line motion. In this mechanism, the crank AB and the

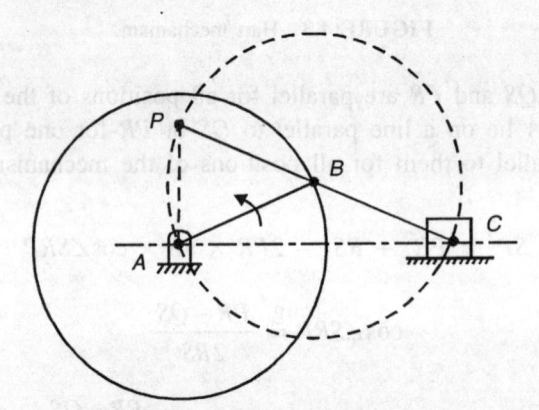

FIGURE 4.4 Scott Russel mechanism.

connecting rod BC are connected through a pin joint. The point C is constrained to move in a straight line by a sliding pair. The extension of connecting rod BC up to point P permits the point P to move along a straight line perpendicular to AC for the following two conditions:

(a) When lengths of the links AB, BC and BP are equal, the point P describes a straight line AP perpendicular to AC provided the slider C moves in a straight line along AC. As the lengths of the links AB, BC and BP are equal, the angle PAC is right angle and PA is perpendicular to axis of slider AC. This is true for all positions of the slider C and is possible only if P moves in a straight line perpendicular to AC at point A.

(b) When length of the link BC is not equal to that of BP, then the point P describes an approximate straight line perpendicular to AC provided the slider C moves along a straight line and the links are proportional in such a way that length of the link BC is mean proportional to the lengths of the links AB and BP, i.e.

$$\frac{AB}{BC} = \frac{BC}{BP}$$

or
$$AB = \frac{BC^2}{BP}$$

which is equation of elliptical trammel. This mechanism is called **modified Scott Russel mechanism** (Figure 4.5).

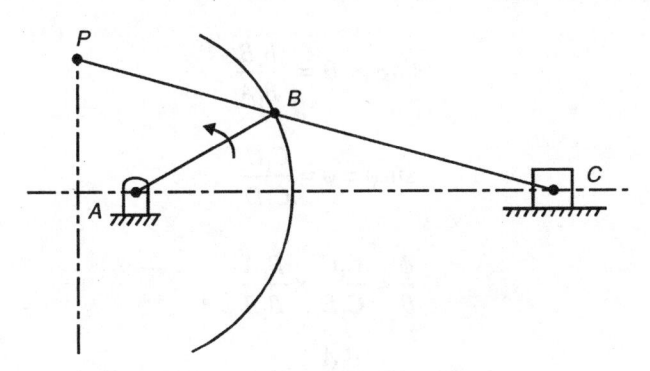

FIGURE 4.5 Modified Scott Russel mechanism.

4.2.2 Approximate Straight Line Motion Mechanism

Some mechanisms describe an approximate straight line motion. Usually a four bar chain or its modified variant gives approximate straight line motion. A few of them are described further.

Watt mechanism. Watt mechanism is one of the simplest straight line motion mechanisms. It consists of a four bar chain *ABCD* with fixed link *AD* (Figure 4.6). In the mean position, the crank *AB* and the follower link *CD* are parallel and the coupler link *BC* is perpendicular to the links *AB* and *CD*. The point *P* on the coupler link traces a curve called **'Lemniscoid'**; part of the path traced by it is approximately a straight line.

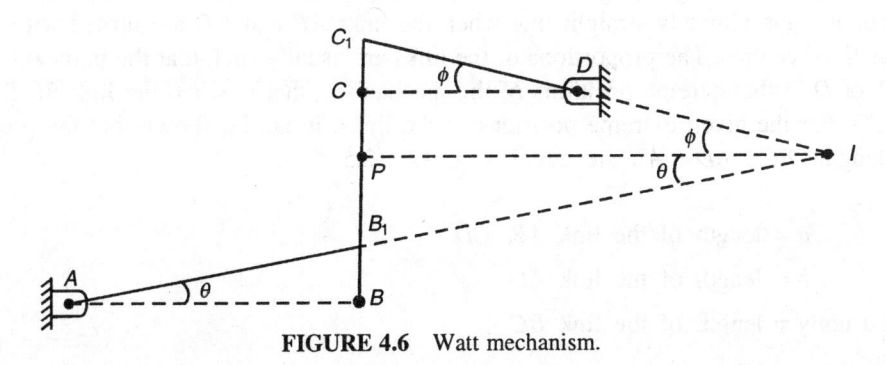

FIGURE 4.6 Watt mechanism.

The position of the point *P* can be found by making use of instantaneous centre of link *BC*. For the position of the mechanism shown, the instantaneous centre of the link *BC* is given by point *I*.

Let the angle B_1AB be equal to θ and the angle C_1DC be equal to ϕ.
Since the angles θ and ϕ are small angles, the following relation holds true:

$$\frac{C_1P}{B_1P} = \frac{\phi}{\theta} \tag{4.11}$$

From the geometry of the Figure 4.6, we know that

$$\sin\theta \simeq \theta = \frac{B_1B}{B_1A}$$

and

$$\sin\phi \simeq \phi = \frac{C_1C}{C_1D}$$

Therefore,

$$\frac{\phi}{\theta} = \frac{C_1C}{C_1D} \times \frac{B_1A}{B_1B} \tag{4.12}$$

$$= \frac{B_1A}{C_1D}$$

Equating Eqs. (4.11) and (4.12), we get

$$\frac{C_1P}{B_1P} = \frac{B_1A}{C_1D}$$

or

$$C_1P : B_1P :: B_1A : C_1D \tag{4.13}$$

Thus the point P divides the coupler link BC into two parts, which are inversely proportional to the lengths of the links AB and CD.

Techebicheff mechanism. This is a four bar mechanism in which the crossed links AB and CD are equal in length (Figure 4.7). The tracing point P which is mid-point of the link BC traces an approximately straight line when the links AB and CD are turned with points A and D as fixed centres. The proportions of the links are usually such that the point P is directly above A or D in the extreme positions of the mechanism, that is when the link BC lies along AB or CD. For the given extreme positions of the links, it can be shown that the proportions of the lengths $BC : AD : AB$ are equal to $1 : 2 : 2.5$.

Let

$a =$ length of the link AB, CD

$b =$ length of the link AD

and unity = length of the link BC

From triangle C_2AD,

$$AC_2^2 = C_2D^2 - AD^2$$

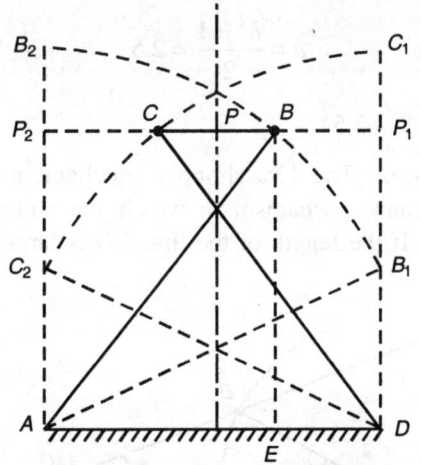

FIGURE 4.7 Techebicheff mechanism.

$$(AB_2 - B_2C_2)^2 = a^2 - b^2$$

or
$$(a-1)^2 = a^2 - b^2$$

or
$$a = \frac{b^2 + 1}{2} \tag{4.14}$$

Let *BE* be perpendicular to the link *AD*.
Then
$$AB^2 = BE^2 + AE^2$$

or
$$AB^2 = AP_2^2 + (AD - ED)^2$$

$$AB^2 = (AB_2 - B_2P_2)^2 + [AD - (PP_1 - PB)]^2$$

$$a^2 = \left(a - \frac{1}{2}\right)^2 + \left[b - \left(\frac{b}{2} - \frac{1}{2}\right)\right]^2$$

or
$$a = \frac{b^2}{4} + \frac{b}{2} + \frac{1}{2} \tag{4.15}$$

Equating Eqs. (4.14) and (4.15), we get

$$\frac{b^2 + 1}{2} = \frac{b^2}{4} + \frac{b}{2} + \frac{1}{2}$$

or
$$b = 2$$

and
$$a = \frac{b^2 + 1}{2} = 2.5$$

Hence $BC : AD : AB = 1 : 2 : 2.5$

Grasshopper mechanism. The Grasshopper mechanism shown in Figure 4.8 is a modified version of slider-crank mechanism in which the sliding pair C is replaced by a turning pair using a link CD. If the length of the link CD is large enough, the point C moves

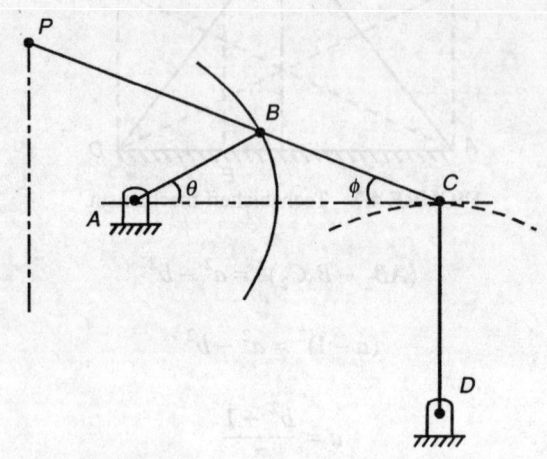

FIGURE 4.8 Grasshopper mechanism.

in an approximately straight line for small angular displacement of the link AB. In such mechanism, the point P on an extension of the link BC describes an approximately straight line path for small angular displacement of the link AB on each side of the axis AC. The extended length of the link BC can be found from the principle of ellipse trammel. Accordingly,

$$PB = \frac{BC^2}{AB} \tag{4.16}$$

Robert's mechanism. It is a simple four bar chain in which the links AB and CD are of equal length and the tracing point P is rigidly attached to the link BC on a line which bisects the link BC at right angle.

The best position for the point P may be found by making use of instantaneous centre of the link BC as shown in Figure 4.9. The point P describes approximately horizontal line for small angular displacement of the driving link AB.

4.3 PANTOGRAPH

A pantograph is a four bar mechanism with lower pairs used to produce the path exactly similar to one traced out by a point on the mechanism. The produced path may be on an

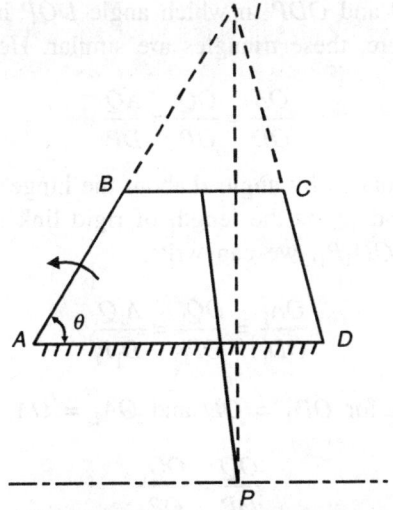

FIGURE 4.9 Robert's mechanism.

enlarged scale or reduced scale as required. It is used as a geometrical instrument to reproduce the geometrical figures and plane area of irregular shapes such as maps on enlarged or reduced scale.

Figure 4.10 shows a pantograph mechanism in which length of the link *AB* is equal to that of *CD* and length of the link *AD* is equal to that of *BC*. The links are pin jointed at points

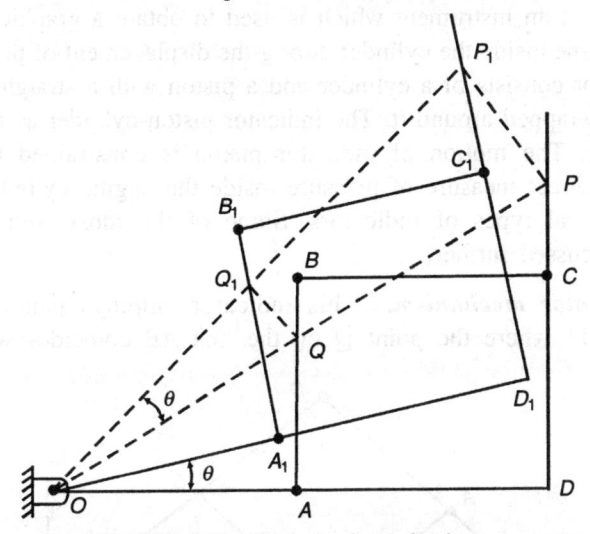

FIGURE 4.10 Pantograph mechanism.

A, *B*, *C* and *D*. The link *AB* is parallel to the link *CD* and the link *AD* is parallel to *BC*. The link *AD* is extended to the fixed point *O*.

Let *Q* be the point on the link *AB* and the point *P* on the extension of the link *CD*. Both the points lie on a straight line which passes through the point *O*. It can be shown that under these circumstances, the point *P* will reproduce the motion of the point *Q* to an enlarged scale or the point *Q* will reproduce the motion of the point *P* to a reduced scale.

Consider triangles OAQ and ODP in which angle DOP is common and angle AQO is equal to angle DPO. Therefore, these triangles are similar. Hence

$$\frac{OA}{OD} = \frac{OQ}{OP} = \frac{AQ}{DP} \qquad (4.17)$$

Let the mechanism be rotated by angle θ about the hinge point O, the new positions of the points P and Q are P_1 and Q_1 as the length of rigid link remains same. Hence for two similar triangles OA_1Q_1 and OD_1P_1, we can write

$$\frac{OA_1}{OD_1} = \frac{OQ_1}{OP_1} = \frac{A_1Q_1}{D_1P_1} \qquad (4.18)$$

From Eqs. (4.17) and (4.18), for $OD_1 = OD$ and $OA_1 = OA$,

$$\frac{OQ}{OP} = \frac{OQ_1}{OP_1} \qquad (4.19)$$

Therefore, the triangles OPP_1 and OQQ_1 are similar and PP_1 and QQ_1 are parallel. Hence the point P describes a path similar to that described by the point Q.

4.4 ENGINE INDICATOR MECHANISMS

An engine indicator is an instrument which is used to obtain a graphical record of variation of pressure and volume inside the cylinder during the displacement of piston in a reciprocating engine. The indicator consists of a cylinder and a piston with a straight line mechanism and a drum with paper wrapped around it. The indicator piston-cylinder assembly is connected to the engine cylinder. The motion of indicator piston is constrained by the spring, so its displacement gives direct measure of pressure inside the engine cylinder.

There are several types of indicators. Some of the most commonly used indicator mechanisms are discussed further.

Simplex indicator mechanism. This indicator employs pantograph mechanism as shown in Figure 4.11 where the point Q on the link AB coincides with the point B. The

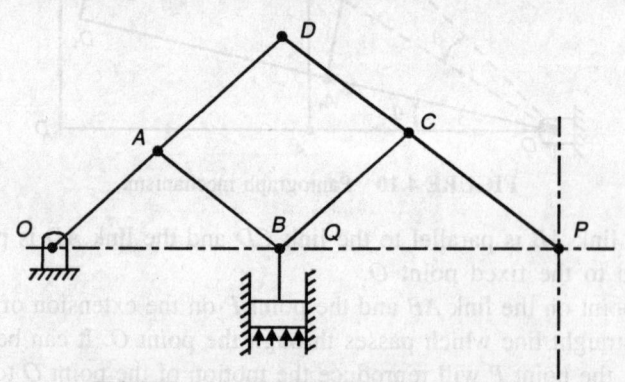

FIGURE 4.11 Simplex indicator mechanism.

point P on the link DC produced is such that OQP is a straight line and the links AB, BC, CD and DA form a parallelogram. The indicator piston is connected at the point Q through a piston rod.

The displacement of the indicator piston at the point Q is recorded at the point P on enlarged scale as it describes a path similar to that described by the point Q.

Crosby indicator mechanism. A crosby indicator mechanism, as shown in the Figure 4.12, is a modified version of pantograph in which the points O and D are fulcrum points fixed to the body of the indicator. The links OA, AB, BC and CD are joined through lower pair joints. The link AB is extended to the point P such that the points O, Q and P lie on a straight line.

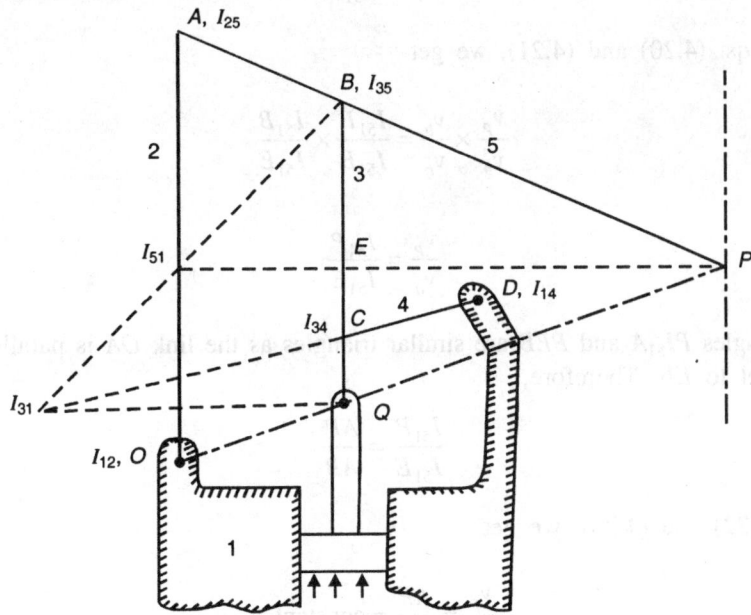

FIGURE 4.12 Crosby indicator mechanism.

This mechanism is required to satisfy the following conditions:
(i) link OA is parallel to BQ.
(ii) the point P travels along a straight line.
(iii) the velocity ratio between P and Q is constant, i.e. $\dfrac{v_p}{v_q}$ = constant.

Locate the various instantaneous centres as shown in Figure 4.12. The centre I_{31} is obtained by drawing a horizontal line from the fulcrum of indicator piston Q which meets the link DC produced at I_{31}. Similarly, instantaneous centre between the links 3 and 5 can be located according to Kennedy theorem.

$$\text{Velocity ratio, } \frac{v_b}{v_q} = \frac{I_{31}B}{I_{31}Q}$$

But

$$\frac{I_{31}B}{I_{31}Q} = \frac{I_{51}B}{I_{51}E}$$

Therefore,

$$\frac{v_b}{v_q} = \frac{I_{31}B}{I_{31}Q} = \frac{I_{51}B}{I_{51}E} \tag{4.20}$$

Similarly,

$$\frac{v_p}{v_b} = \frac{I_{51}P}{I_{51}B} \tag{4.21}$$

Multiplying Eqs. (4.20) and (4.21), we get

$$\frac{v_p}{v_b} \times \frac{v_b}{v_q} = \frac{I_{51}P}{I_{51}B} \times \frac{I_{51}B}{I_{51}E}$$

or

$$\frac{v_p}{v_q} = \frac{I_{51}P}{I_{51}E} \tag{4.22}$$

The triangles $PI_{51}A$ and PEB are similar triangles as the link OA is parallel to QB and $I_{51}A$ is parallel to EB. Therefore,

$$\frac{I_{51}P}{I_{51}E} = \frac{AP}{AB} \tag{4.23}$$

From Eqs. (4.22) and (4.23), we get

$$\frac{v_p}{v_q} = \frac{AP}{AB} = \text{constant} \tag{4.24}$$

Hence the tracing point P moves in an approximate straight line and traces the movement of the indicator piston.

Thomson indicator mechanism. Figure 4.13 shows a Thomson indicator mechanism in which the links OA, AC, CD and OD constitute a Grasshopper-type straight line mechanism. The link OD is a fixed link. The tracing point P is on the link AC produced. The link AC gets the motion from the displacement of piston rod of the indicator through the link BQ which is approximately parallel to the link OA, like crosby indicator mechanism.

It holds the ratio of velocities of the points P and Q as constant. To prove this, locate all instantaneous centres as shown in Figure 4.13. Let the link QB intersect $I_{51}P$ at point E.

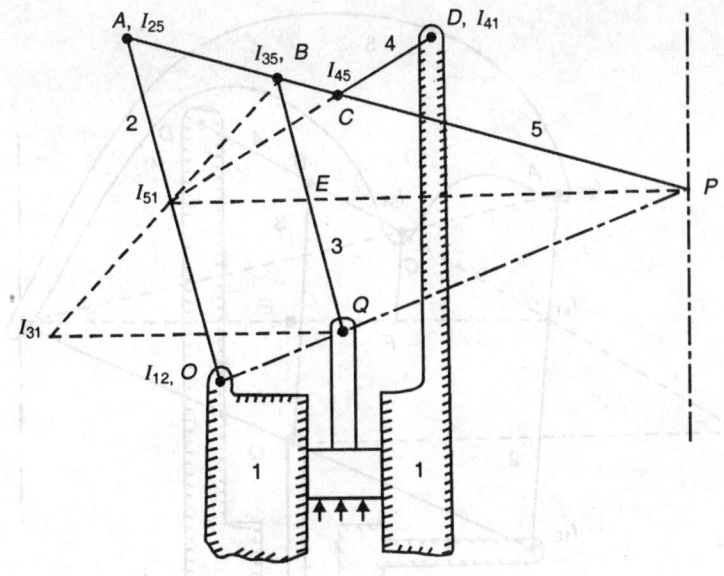

FIGURE 4.13 Thomson indicator mechanism.

Then
$$\frac{v_b}{v_q} = \frac{I_{31}B}{I_{31}Q} = \frac{I_{51}B}{I_{51}E} \qquad (4.25)$$

Similarly,
$$\frac{v_p}{v_b} = \frac{I_{51}P}{I_{51}B} \qquad (4.26)$$

From Eqs. (4.25) and (4.26), we get
$$\frac{v_p}{v_q} = \frac{I_{51}P}{I_{51}E} \qquad (4.27)$$

If the links QB and OA are parallel, the triangles PEB and $PI_{51}A$ are similar triangles.

Therefore,
$$\frac{v_p}{v_q} = \frac{I_{51}P}{I_{51}E} = \frac{AP}{AB} = \text{constant} \qquad (4.28)$$

Hence the tracing point P moves in an approximate straight line and traces the movement of the indicator piston.

Dobbie McInne indicator mechanism. The Dobbie McInne indicator mechanism is similar to Thomson indicator mechanism with a difference that the link BQ is connected to the link CD instead of the link AC (Figure 4.14). Thus the motion of the indicator piston is transmitted from the point Q to the pin C on the link CD. However, the links OA and BQ are approximately parallel.

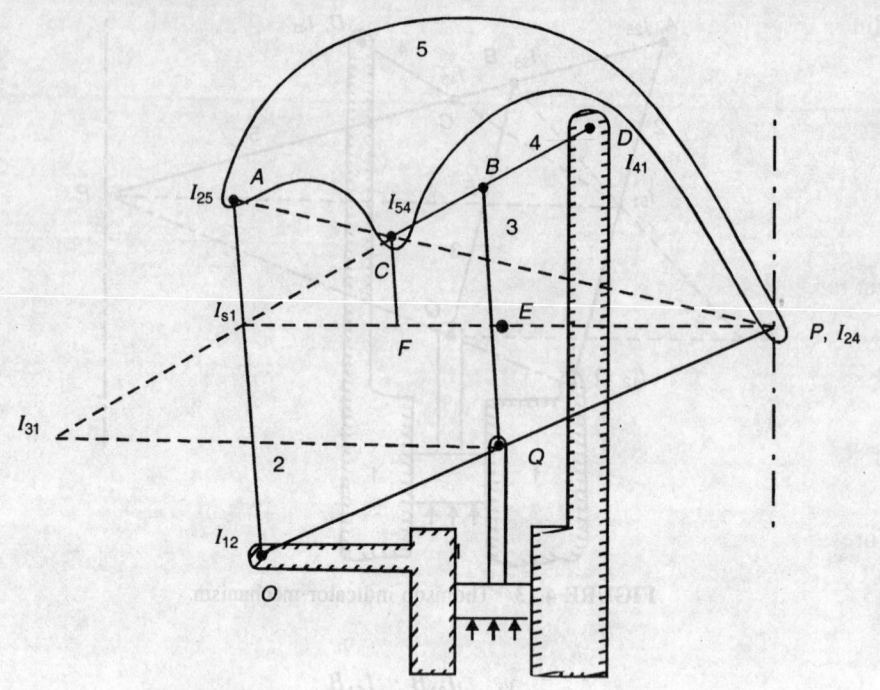

FIGURE 4.14 Dobbie McInne indicator mechanism.

In order to verify that velocity ratio between the points P and Q is constant, locate all the instantaneous centres as shown in Figure 4.14.

The velocity ratio,
$$\frac{v_b}{v_q} = \frac{I_{31}B}{I_{31}Q} = \frac{I_{51}B}{I_{51}E} \tag{4.29}$$

Also
$$\frac{v_c}{v_b} = \frac{DC}{DB} \tag{4.30}$$

and
$$\frac{v_p}{v_c} = \frac{I_{51}P}{I_{51}C} \tag{4.31}$$

Therefore
$$\frac{v_p}{v_q} = \frac{v_p}{v_c} \times \frac{v_c}{v_b} \times \frac{v_b}{v_q} \tag{4.32}$$

$$= \frac{I_{51}P}{I_{51}C} \times \frac{DC}{DB} \times \frac{I_{51}B}{I_{51}E}$$

Let a line CF be drawn parallel to BQ. Then

$$\frac{I_{51}F}{I_{51}C} = \frac{I_{51}E}{I_{51}B}$$

or

$$I_{51}B = I_{51}C \times \frac{I_{51}E}{I_{51}F}$$

Substituting $I_{51}B$ in Eq. (4.32), we get

$$\frac{v_p}{v_q} = \frac{I_{51}P}{I_{51}F} \times \frac{DC}{DB} \qquad (4.33)$$

If the links QB and OA are parallel, the triangles PCF and PAI_{51} are similar.

Therefore,

$$\frac{I_{51}P}{I_{51}F} = \frac{AP}{AC}$$

so that

$$\frac{v_p}{v_q} = \frac{AP}{AC} \times \frac{DC}{DB} = \text{constant} \qquad (4.34)$$

This expression gives approximately the ratio of the displacement of the point P to the displacement of the point Q.

EXAMPLE 4.1 Design a pantograph mechanism to be used in simplex indicator mechanism which should represent four times the gas pressure inside the cylinder of the engine. The distance between the fixed point and the tracing point of the indicator diagram is 200 mm.

Solution Referring to Figure 4.15,

$$\frac{OP}{OQ} = 4, \quad OP = 200 \text{ mm}$$

Therefore, $OQ = 50$ mm

Also

$$\frac{OQ}{OP} = \frac{AQ}{DP} = \frac{OA}{OD} = \frac{50}{200} = \frac{1}{4}$$

In triangle ODP,

$$OD + DP > OP$$

As the triangle ODP is an isosceles triangle, it should satisfy the following condition:

$$OD > \frac{OP}{2} > 100 \text{ mm}$$

FIGURE 4.15

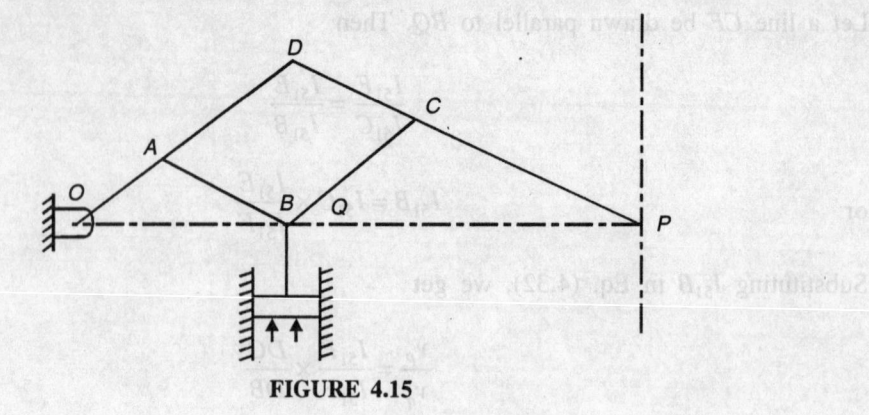

Let OD be equal to 120 mm.

We have
$$\frac{OA}{OD} = \frac{1}{4}$$

or
$$OA = \frac{120}{4} = 30 \text{ mm}$$

$$AD = OD - OA = 120 - 30 = \textbf{90 mm} \qquad \textbf{Ans.}$$

$AD = BC = 90$ mm ($\because$ $ABCD$ is a parallelogram.)

Again
$$\frac{AQ}{DP} = \frac{1}{4}$$

or
$$AQ = \frac{DP}{4} = \frac{120}{4} = 30 \text{ mm} \quad (\because OD = DP \text{ as } \Delta ODP \text{ is an isosceles triangle.})$$

Therefore $DC = AQ = \textbf{30 mm}$ \qquad \textbf{Ans.}

4.5 AUTOMOBILE STEERING MECHANISM

The relative motion between the wheels of an automotive vehicle and the road surface should be pure rolling so that there is line contact between the road surface and the vehicle tyres which prevent uneven wear of tyres. In order to satisfy this condition when the vehicle is moving along a curved path, the steering mechanism must be so designed that the paths of the points of contact of each wheel with the road surface are concentric circular arcs. This is usually effected by turning the axes of rotation of the two front wheels relative to vehicle chassis and to satisfy the above condition, the axis of the wheel on the inside of the curve must be turned through a larger angle than the axis of the wheel on the outside of the curve. The axes of these two wheels, when produced, intersect on the common axis of rear wheels called instantaneous centre of rotation (Figure 4.16).

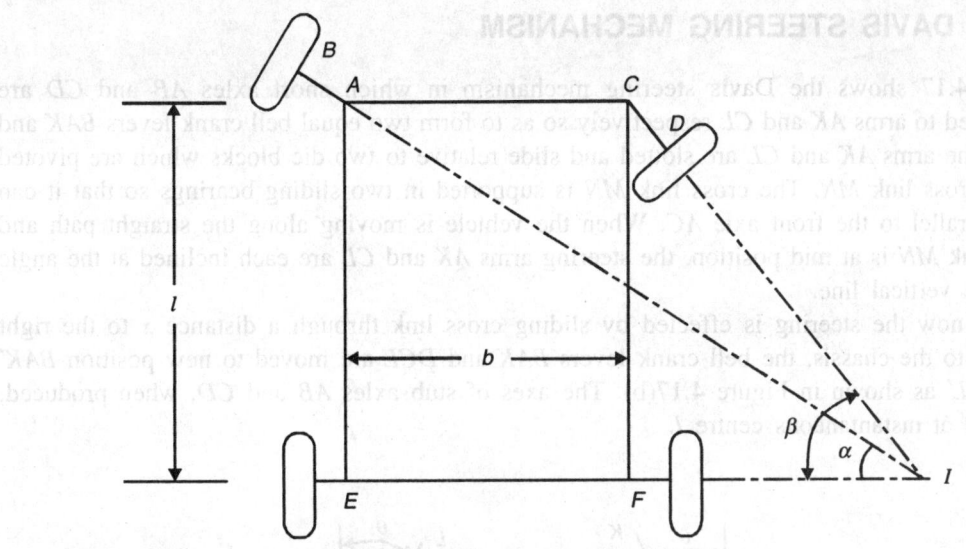

FIGURE 4.16 Automobile steering principle.

Figure 4.16 shows a plan view of an automotive vehicle in which AB and CD are two short axles called **stub axles.** When vehicle takes right turn, the axes AB and CD intersect the common axis of rear wheel axle EF at point I so the path of contact of each wheel with the road is a circular arc with centre I.
Let

b = distance between pivots of front axle
l = wheel base and
α, β = angles turned by the stub axles.

From the geometry of Figure 4.16,

$$AC = EF = EI - FI$$
$$= AE \cot \alpha - CF \cot \beta$$
$$= AE (\cot \alpha - \cot \beta) \qquad (\because AE = CF)$$

or
$$\cot \alpha - \cot \beta = \frac{AC}{AE} = \frac{b}{l} \qquad (4.35)$$

This is called the fundamental equation for correct steering and if it is satisfied, there will not be any lateral slip of wheel when automotive vehicle is taking turn.

A mechanism which satisfies the above fundamental equation of steering at all radii of curvature of the path followed by the vehicle is called **steering mechanism.** Generally two types of steering mechanism namely *Davis steering mechanism* and *Ackermann steering mechanism* are most commonly used. The main distinction between these two types is that the Davis mechanism has sliding pairs compared to turning pairs used in Ackermann steering mechanism. A brief description about these steering mechanisms is given further.

4.5.1 DAVIS STEERING MECHANISM

Figure 4.17 shows the Davis steering mechanism in which short axles *AB* and *CD* are connected to arms *AK* and *CL* respectively so as to form two equal bell crank levers *BAK* and *DCL*. The arms *AK* and *CL* are slotted and slide relative to two die blocks which are pivoted to the cross link *MN*. The cross link *MN* is supported in two sliding bearings so that it can slide parallel to the front axle *AC*. When the vehicle is moving along the straight path and cross link *MN* is at mid position, the steering arms *AK* and *CL* are each inclined at the angle θ to the vertical line.

If now the steering is effected by sliding cross link through a distance x to the right relative to the chassis, the bell crank levers *BAK* and *DCL* are moved to new position *BAK'* and *DCL'* as shown in Figure 4.17(b). The axes of stub axles *AB* and *CD*, when produced, intersect at instantaneous centre *I*.

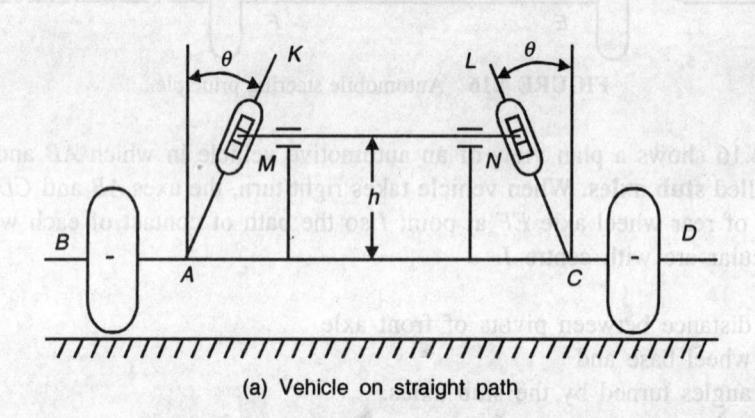

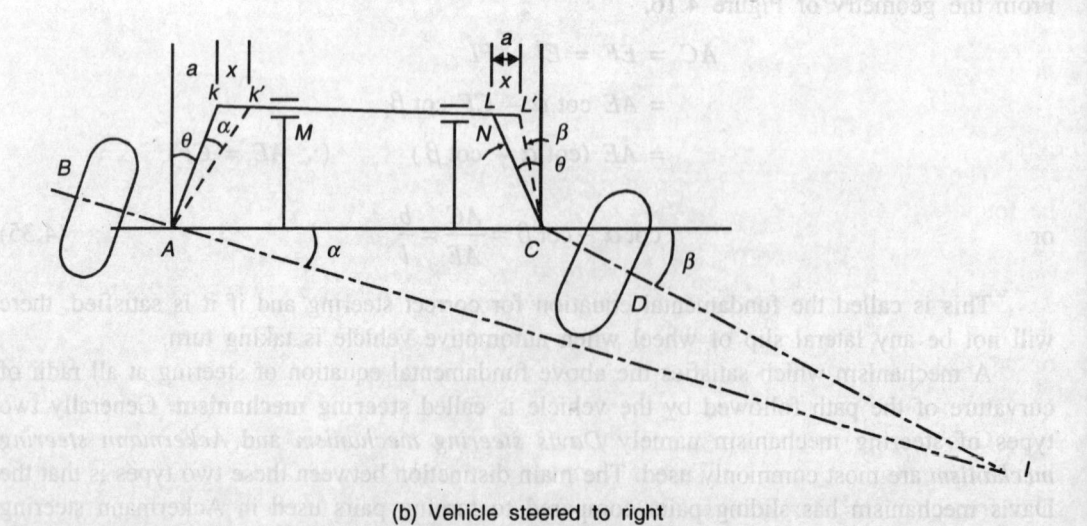

(a) Vehicle on straight path

(b) Vehicle steered to right

FIGURE 4.17 Davis steering mechanism.

Let

 x = horizontal displacement of cross link MN when steering is effected (= KK' or LL')
 h = distance between the cross link MN and front axle AC
 $2a$ = difference between AC and KL
 α, β = angles turned by the stub axles and
 θ = angle of inclination of links AK and CL to the vertical line.

Referring to Figure 4.17(b),

$$\tan(\theta + \alpha) = \frac{\tan\theta + \tan\alpha}{1 - \tan\theta \tan\alpha} = \frac{a + x}{h}$$

Substituting the value of $\tan\theta$ (= a/h) and simplifying, we get

$$\tan\alpha = \frac{xh}{h^2 + a^2 + ax} \qquad (4.36)$$

Also

$$\tan(\theta - \beta) = \frac{\tan\theta - \tan\beta}{1 + \tan\theta \tan\beta} = \frac{a - x}{h}$$

Substituting the value of $\tan\theta$ and simplifying, we get

$$\tan\beta = \frac{xh}{h^2 + a^2 - ax} \qquad (4.37)$$

From Eqs. (4.36) and (4.37), the condition of correct steering is:

$$\cot\alpha - \cot\beta = \frac{h^2 + a^2 + ax}{xh} - \frac{h^2 + a^2 - ax}{xh}$$

$$= \frac{2a}{h} = 2\tan\theta \qquad (4.38)$$

Thus for correct steering, the required inclination angle θ of the links AK and CL can be found from Eq. (4.35) as

$$\cot\alpha - \cot\beta = \frac{2a}{h} = 2\tan\theta = \frac{b}{l}$$

or

$$\tan\theta = \frac{b}{2l} \qquad (4.39)$$

Generally, the value of b/l is kept between 0.4 and 0.5. Hence the inclination angle θ lies in the range of 11°–14°. The main limitation of Davis steering mechanism is that due to high friction at the sliding pairs, the contact surfaces wear out rapidly and impairs the accuracy of the mechanism.

EXAMPLE 4.2 In a Davis steering mechanism, the distance between the pivots of the front axle is 1.2 m and wheel base is 3 m. When the automotive vehicle is moving along a straight path, find the inclination of the track arms to longitudinal axis of the vehicle.

If the vehicle takes turn and the angle turned through by the inner wheel is 30°, find the angle turned through by the outer wheels.

Solution Given: $b = 1.2$ m and wheel base, $l = 3$ m

(i) Inclination angle of the track arms is given by

$$\tan\theta = \frac{b}{2l}$$

or
$$\tan\theta = \frac{1.2}{2 \times 3} = 0.2$$

or
$$\theta = 11.31°$$ Ans.

(ii) Equation of correct steering is:

$$\cot\alpha - \cot\beta = \frac{b}{l}$$

or
$$\cot\alpha - \cot 30° = \frac{1.2}{3}$$

or
$$\cot\alpha - 1.732 = 0.4$$

or
$$\text{angle } \alpha = 25.13°$$ Ans.

4.5.2 Ackermann Steering Mechanism

The Ackermann steering mechanism is based on a four bar chain in which two longer links AC and KL are unequal in length while the shorter links AK and CL are equal in length. All links are connected through turning pairs. During the straight path motion of the vehicle, the longer links are parallel and each shorter link is inclined at an angle θ to the longitudinal axis of the chassis. When the vehicle is to steer to right curve path, the shorter link CL is turned to new position CL' so as to increase the angle θ while the other shorter link AK is to turn to new position AK' so as to reduce the angle θ (Figure 4.18).

From the Figure 4.18(b), it is clear that the angle turned by left stub axle AB is smaller than the angle turned by right stub axle CD ($\alpha < \beta$). Further, the value of angle α, for the given value of angle β, depends upon the ratio AK/AC and the angle θ. For the given value of the ratio AK/AC and the angle θ, corresponding values of angles β and α can be obtained either by graphical method or by analytical procedure.

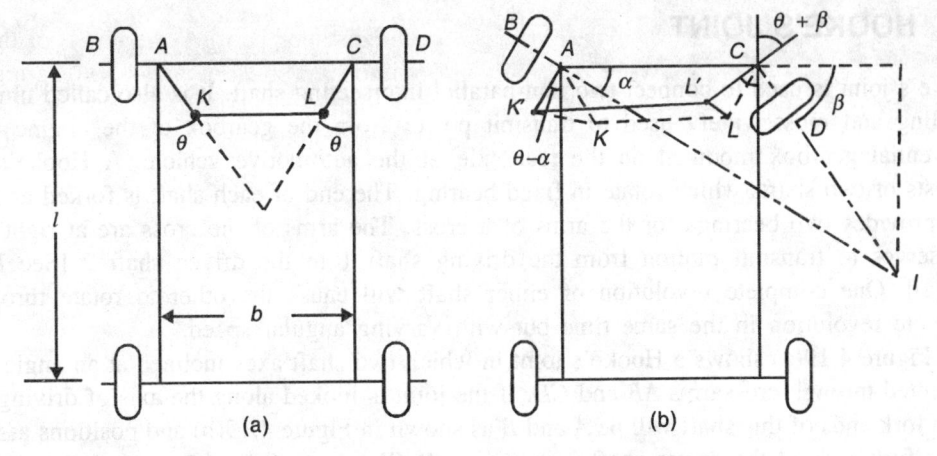

FIGURE 4.18 Ackermann steering mechanism.

For correct steering, besides satisfying the fundamental equation of correct steering Eq. (4.35), the instantaneous centre I must lie on the common axis of the rear wheels but in actual practice, it is found to lie on a line parallel to the rear axis at an approximate distance of 0.3 times wheel base above it. In Ackermann steering mechanism, if the distance $AC(=b)$ is 0.45 times the wheel base of the vehicle, it gives correct steering. In fact, there are three conditions where Ackermann steering mechanism gives correct steering of an automobile:

(i) when the vehicle moves straight
(ii) when the vehicle turns at a correct angle to the right
(iii) when the vehicle turns at a correct angle to the left

In other positions, pure rolling is not possible due to slipping of the wheel. Hence correct steering is not possible.

In order to determine the inclination angle of track arms θ, let us assume that projections of KK' and LL' on the link AC are approximately equal. Therefore,

projection of arc LL' on AC = projection of arc KK' on AC

or $\quad CL\,[\sin(\theta + \beta) - \sin\theta] = AK\,[\sin\theta - \sin(\theta - \alpha)]$

Since lengths of the links AK and CL are equal,

$\sin\theta\cos\beta + \sin\beta\cos\theta - \sin\theta = \sin\theta - \sin\theta\cos\alpha + \cos\theta\sin\alpha$

or $$\tan\theta = \frac{\sin\alpha - \sin\beta}{\cos\alpha + \cos\beta - 2} \qquad (4.40)$$

The values of angles α and β to be taken in this equation are those found for correct steering. Further, when value of angle θ is known, the length of track arm can easily be determined.

4.6 HOOKE'S JOINT

Hooke's joint is used to connect two non-parallel intersecting shaft. It is also called universal coupling and most widely used to transmit power from the gearbox of the engine to the differential gearbox mounted on the rear axle of the automotive vehicle. A Hooke's joint consists of two shafts which rotate in fixed bearing. The end of each shaft is forked and each fork provides two bearings for the arms of a cross. The arms of the cross are at right angle and serves to transmit motion from the driving shaft 1 to the driven shaft 2 [See Figure 4.19(a)]. One complete revolution of either shaft will cause the other to rotate through a complete revolution in the same time but with varying angular speed.

Figure 4.19(a) shows a Hooke's joint in which two shaft axes inclined at an angle α are connected through crossarms AB and CD. If the joint is looked along the axis of driving shaft 1, the fork ends of this shaft will be A and B as shown in Figure 4.19(b) and positions assumed by the fork ends of the driven shaft 2 are C and D. The axis of the driven shaft 2 is inclined at an angle α from the axis of the driving shaft.

(a) Hooke's joint

(b) Axes of shafts

FIGURE 4.19

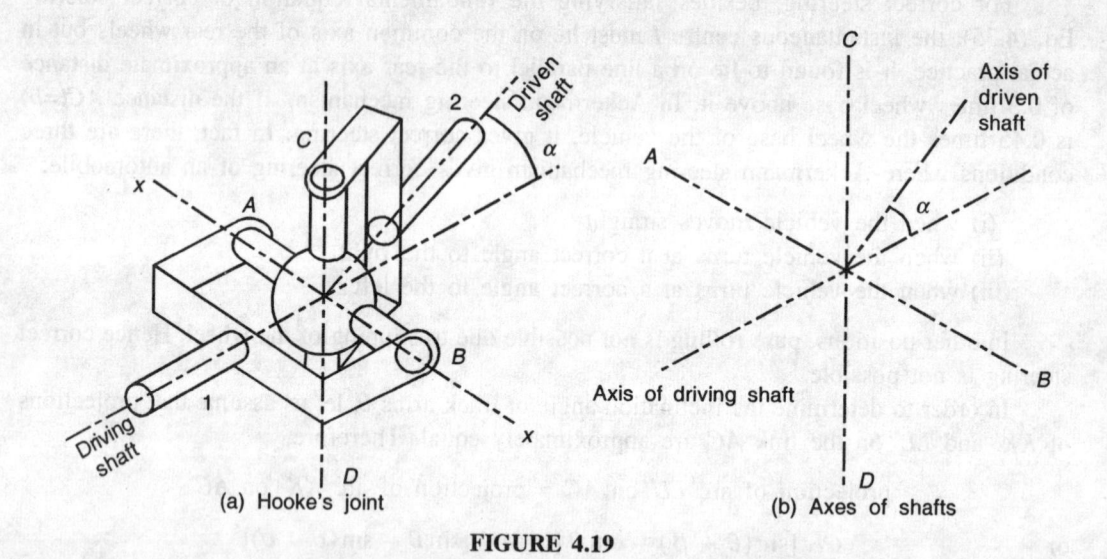

Figure 4.20 shows the path of rotation of driving shaft which is circle; however, the driven shaft moves along the path of an ellipse if looked along the axis of the driving shaft. However, true path of the driven shaft is also circular when viewed along its own axis.

Let the driving shaft rotate through an angle θ so that fork ends assume the positions A_1 and B_1. The driven shaft also rotates by same angle but the true angle of rotation of driven shaft is larger when it is looked along its own axis. If point C_1 is extended horizontally upto circle at C_2, the angle C_2OE is actual angle turned by the driven shaft.

Referring to geometry of Figure 4.20,

$$\tan \phi = \frac{C_2 E}{OE} \quad \text{and} \quad \tan \theta = \frac{C_1 E}{OE}$$

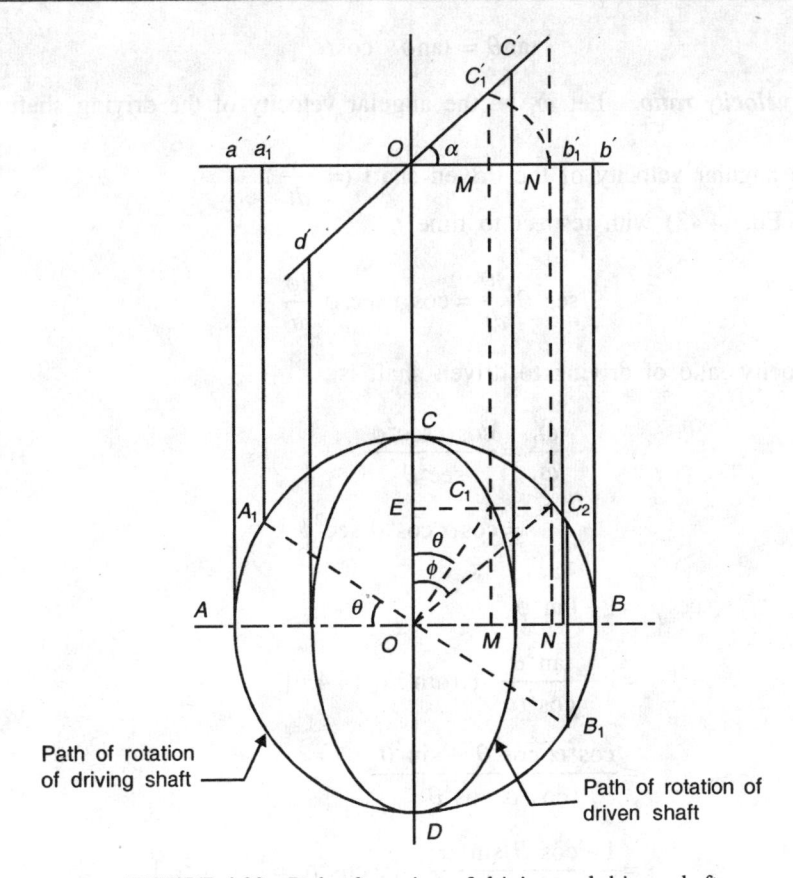

FIGURE 4.20 Path of rotation of driving and driven shafts

or
$$\frac{\tan\phi}{\tan\theta} = \frac{C_2E/OE}{C_1E/OE} = \frac{C_2E}{C_1E}$$

or
$$\frac{\tan\phi}{\tan\theta} = \frac{ON}{OM} \quad (4.41)$$

Let α be the angle of inclination of the driven shaft with respect to the driving shaft. Referring to plan view in Figure 4.20,

$$\cos\alpha = \frac{OM}{OC_1'} = \frac{OM}{ON} \quad (4.42)$$

Equating Eqs. (4.41) and (4.42), we get

$$\frac{\tan\phi}{\tan\theta} = \frac{ON}{OM} = \frac{1}{\cos\alpha}$$

or
$$\tan\theta = \tan\phi \cdot \cos\alpha \tag{4.43}$$

Angular velocity ratio. Let ω_1 be the angular velocity of the driving shaft ($= \dfrac{d\theta}{dt}$) and ω_2 be the angular velocity of the driven shaft ($= \dfrac{d\phi}{dt}$). Differentiating Eq. (4.43) with respect to time t,

$$\sec^2\theta \frac{d\theta}{dt} = \cos\alpha \sec^2\phi \frac{d\phi}{dt}$$

or angular velocity ratio of driving to driven shaft is:

$$\frac{\omega_1}{\omega_2} = \frac{\cos\alpha \sec^2\phi}{\sec^2\theta}$$

$$= \cos\alpha \cos^2\theta \sec^2\phi \tag{4.44}$$

But
$$\sec^2\phi = 1 + \tan^2\phi$$

$$= 1 + \frac{\tan^2\theta}{\cos^2\alpha} \quad \text{[from Eq. (4.43)]}$$

$$= \frac{\cos^2\alpha \cos^2\theta + \sin^2\theta}{\cos^2\alpha \cos^2\theta}$$

$$= \frac{1 - \cos^2\theta \sin^2\alpha}{\cos^2\alpha \cos^2\theta} \quad [\because \cos^2\alpha = 1 - \sin^2\alpha]$$

Substituting the value of $\sec^2\phi$ in Eq. (4.44), we get

$$\frac{\omega_1}{\omega_2} = \frac{\cos\alpha \cos^2\theta \times (1 - \cos^2\theta \sin^2\alpha)}{\cos^2\alpha \cdot \cos^2\theta}$$

$$= \frac{1 - \cos^2\theta \sin^2\alpha}{\cos\alpha} \tag{4.45}$$

For the given values of angle α, this ratio is maximum when $\cos\theta = 0$, i.e. when $\theta = 90°$ or $270°$, and it is minimum when $\cos\theta = \pm 1$, i.e. when $\theta = 0°$ or $180°$.

Maximum angular velocity of driven shaft:

$$\omega_{2\max} = \frac{\omega_1}{\cos\alpha} \tag{4.46}$$

Minimum angular velocity of driven shaft:

$$\omega_{2\min} = \omega_1 \cos\alpha \tag{4.47}$$

The value of angle θ for which the speed of driving and driven shafts are equal is given by

$$\frac{\omega_1}{\omega_2} = \frac{1 - \cos^2\theta \sin^2\alpha}{\cos\alpha} = 1$$

or
$$1 - \cos^2\theta \sin^2\alpha = \cos\alpha$$

or
$$\cos^2\theta = \frac{1 - \cos\alpha}{\sin^2\alpha} = \frac{1}{1 + \cos\alpha} \tag{4.48}$$

Further
$$\sin^2\theta = 1 - \cos^2\theta$$

$$= 1 - \frac{1}{1 + \cos\alpha} = \frac{\cos\alpha}{1 + \cos\alpha} \tag{4.49}$$

From Eqs. (4.48) and (4.49), we get

$$\tan^2\theta = \cos\alpha$$

The condition of equal velocities of driving and driven shaft is:

$$\tan\theta = \sqrt{\cos\alpha} \tag{4.50}$$

Angular acceleration of driven shaft. The angular acceleration of the driven shaft is:

$$\alpha_2 = \frac{d\omega_2}{dt} = \frac{d\omega_2}{d\theta} \cdot \frac{d\theta}{dt}$$

$$= \frac{-\omega_1^2 \cos\alpha \sin^2\alpha \sin 2\theta}{(1 - \cos^2\theta \sin^2\alpha)^2} \tag{4.51}$$

The value of angle θ for which angular acceleration of the driven shaft is maximum may be found by differentiating Eq. (4.51) with respect to angle θ and equating it to zero.

$$\frac{d\alpha_2}{d\theta} = \frac{d}{d\theta}\left[\frac{-\omega_1^2 \cos\alpha \sin^2\alpha \sin 2\theta}{(1 - \cos^2\theta \sin^2\alpha)^2}\right] = 0$$

or
$$\cos 2\theta = \frac{2\sin^2\alpha}{2 - \sin^2\alpha} \tag{4.52}$$

This gives maximum angular acceleration when angle θ is approximately 45° or 135°, i.e. when the arms of cross are inclined at 45° to the plane containing the axes of the two shafts.

For the shafts rotating at high speeds, the angle between their axes and mass moment of inertia of joint should be as low as possible, otherwise very high alternating stresses may be set up in the parts of joint owing to the alternate angular acceleration and retardation. In some drives, to keep the angle α small, a double universal joint is used.

EXAMPLE 4.3 A Hooke's joint is used to connect two shafts. If the driving shaft rotates at 1000 rpm and the permissible variation in speed of driven shaft is not to exceed ±4 per cent of the mean speed, determine the angle between the axes of the shafts. Also calculate the maximum and minimum speed of the driven shaft.

Solution We have

$$\omega_1 = \frac{2\pi N_1}{60} = \frac{2\pi \times 1000}{60} = 104.72 \text{ rad/s}$$

$$\frac{\omega_2}{\omega_1} = \frac{1}{\cos \alpha}$$

Maximum angular speed of driven shaft: $\omega_{2\max} = \dfrac{\omega_1}{\cos \alpha}$

Minimum angular speed of driven shaft: $\omega_{2\min} = \omega_1 \cos \alpha$

Therefore, $\quad$ variation of speed $= \omega_1 \left(\dfrac{1}{\cos \alpha} - \cos \alpha \right)$

Let ω_1 be the mean speed.

Permissible variation of speed = ±4% of mean speed

Therefore,

$$\omega_1 \left(\frac{1}{\cos \alpha} - \cos \alpha \right) = 0.08 \omega_1$$

or $\quad \cos^2 \alpha + 0.08 \cos \alpha - 1 = 0$

or $\quad \cos \alpha = +0.96$

or $\quad \alpha = 16.1°$ $\hfill$ Ans.

Maximum speed of driven shaft:

$$N_{2\max} = \frac{N_1}{\cos \alpha} = \frac{1000}{0.96}$$

$$= 1041.67 \text{ rpm} \hfill \text{Ans.}$$

Minimum speed of driven shaft: $N_{2min} = N_1 \cos\alpha$ **Ans.**

$$= 1000 \times \cos 16.1°$$

$$= 960 \text{ rpm}$$

EXAMPLE 4.4 A universal joint is used to connect shaft whose axes are inclined at 20° and the speed of driving shaft is 500 rpm. Find the extreme angular velocities of the driven shaft and its maximum angular acceleration.

Solution We have

$$\omega_1 = \frac{2\pi N}{60} = \frac{2\pi \times 500}{60} = 52.36 \text{ rad/s}$$

Maximum angular speed:

$$\omega_{2max} = \frac{\omega_1}{\cos\alpha}$$

$$= \frac{52.36}{\cos 20°} = \mathbf{55.72 \text{ rad/s}} \quad \text{**Ans.**}$$

Minimum angular speed: $\omega_{2min} = \omega_1 \cos\alpha = 52.36 \cos 20°$

$$= \mathbf{49.2 \text{ rad/s}} \quad \text{**Ans.**}$$

The acceleration of the driven shaft is maximum when

$$\cos 2\theta \simeq \frac{2\sin^2\alpha}{2-\sin^2\alpha}$$

$$= \frac{2\sin^2 20°}{2-\sin^2 20°}$$

$$= 0.1242$$

or $2\theta = 82.86°$ or $277.14°$

or $\theta = 41.43°$ or $137.57°$

Substituting the value of ω_1, α and θ in the following expression,

$$\alpha_2 = \left[\frac{-\omega_1^2 \cos\alpha \sin^2\alpha \sin 2\theta}{(1-\cos^2\theta \sin^2\alpha)^2}\right]$$

$$= -52.36^2 \times \frac{\cos 20° \times \sin^2 20° \times \sin 82.86°}{(1-\cos^2 41.43° \sin^2 20°)^2}$$

$$= \mathbf{-342.6 \text{ rad/s}^2} \quad \text{**Ans.**}$$

The negative sign indicates that the acceleration is in the opposite direction to the velocity, i.e. the driven shaft has the maximum retardation when $\theta = 41.43°$ or $221.43°$ and it is in the same direction as the velocity when $\theta = 137.57°$ or $317.57°$.

EXAMPLE 4.5 The axes of two shafts, connected through Hooke's joint, are inclined at 15°. At what positions of their driving shaft, the velocities of the two shafts are equal? State whether the accelerations are positive or negative at these positions.

If the driving shaft rotates at 750 rpm, find the acceleration of the driven shaft at any one of the above positions.

Solution For $\dfrac{\omega_1}{\omega_2} = 1$, the required condition is:

$$\tan \theta = \sqrt{\cos \alpha}$$

or $$\tan \theta = \sqrt{\cos 15°} = \pm 0.9828$$

or $$\theta = 44.5°,\ 135.5°,\ 224.5°\ \text{and}\ 315.5° \qquad \textbf{Ans.}$$

$$\text{Acceleration, } \alpha_2 = \frac{-\omega_1^2 \cos \alpha \sin^2 \alpha \sin 2\theta}{(1 - \sin^2 \alpha \cos^2 \theta)^2}$$

If $\sin 2\theta$ in the above equation is negative, acceleration is positive and vice-versa.

Hence acceleration is positive for $\theta = 135.5°$ and $315.5°$

and acceleration is negative for $\theta = 44.5°$ and $224.5°$ **Ans.**

$$\text{Acceleration at } \theta = 135.5°,\ \alpha_2 = -\left(\frac{2\pi \times 750}{60}\right)^2 \times \left\{\frac{\cos 15° \times \sin^2 15° \times \sin 271°}{(1 - \sin^2 15° \cos^2 135.5°)^2}\right\}$$

$$= \frac{-6168.5 \times -0.0647}{0.933}$$

$$= 427.76 \text{ rad/s}^2 \qquad \textbf{Ans.}$$

EXERCISES

1. What do you mean by straight line mechanism? Name the different mechanisms which are used for exact straight line motion.
2. Draw a neat sketch of Peaucellier straight line mechanism. Explain with a proof how the tracing point describes a straight line path.
3. Describe Hart's mechanism with a neat sketch and prove that the tracing point describes a straight line path.

4. With a neat sketch, explain the working principle of following mechanisms:
   (i) Scott Russel mechanism
   (ii) Watt mechanism
   (iii) Crosby indicator mechanism
   (iv) Dobbie McInne indicator mechanism

5. What do you mean by pantograph and discuss its uses? Describe with a neat sketch the principle and working of pantograph.

6. What is condition for correct steering? Sketch and describe the Davis and Ackermann steering mechanisms and discuss their relative advantages.

7. Sketch the Dobbie McInne indicator mechanism and find the ratio of the displacement of the pencil to the displacement of piston of the indicator mechanism.

8. What is a Hooke's joint? With a sketch, describe the working of Hooke's joint. Also show for a Hooke's joint that
$$\tan \theta = \tan \phi \cos \alpha$$
where the angles θ, ϕ and α have their usual meanings.

9. In a modified Scott Russel mechanism shown in Figure 4.5, the lengths of the links AB and BC are 30 mm and 40 mm respectively. Find the length of extended link BP so that the point P is having approximate straight line motion.

[Ans: 53.33 mm]

10. Design a pantograph for the simplex indicator mechanism in which the tracing point is located at a distance of 160 mm from the fixed point O. The indicator diagram should be magnified by four times the displacement of the piston.

[Ans: OA = 20 mm, AB = 40 mm, OB = 80 mm, and OP = 160 mm]

11. In a Watt mechanism, the links AB and CD are perpendicular to the link BC in the mean position. The lengths of the moving links are:

AB = 120 mm, CD = 200 mm and BC = 175 mm.

Locate the position of point P on link BC to trace approximately a straight line motion. Also plot the path of point P and mark and measure the straight line segment of the path of point P.

[Ans: 109.3 mm]

12. In a Grasshopper mechanism as shown in Figure 4.8, if $BC^2 = PB \cdot AB$, determine the vertical force at point P necessary to resist a torque T applied to the link AB.

$$\left[\text{Ans}: F_P = T \times \frac{PB}{BC \cdot CP \cos \phi} \right]$$

13. The connecting rod BC of a slider-crank mechanism is 0.6 m long. The piston C is constrained to move in horizontal line AC as shown in Figure E4.1. If a tracing point

P is taken on BC produced such that CP is equal to 0.96 m and its locus is a straight line, find the mean radius of the crank AB for its travel between 0° to 20° with the line of stroke.

[Ans: 0.944 m]

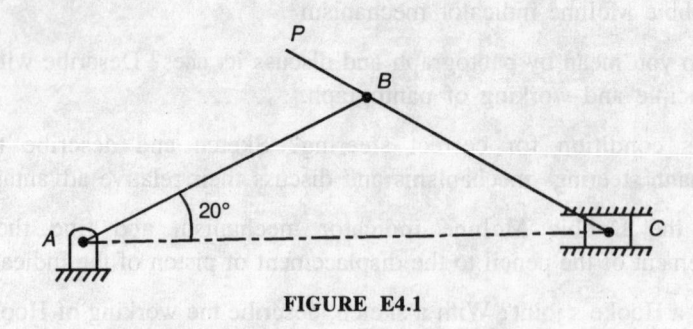

FIGURE E4.1

14. In a Davis steering mechanism, find the inclination of the track arm to the longitudinal axis of the car if the length of car between axles is 2.3 m and the steering pivots are 1.3 m apart. The car is moving in a straight path.

[Ans: 15.8°]

15. The distance between the pivots of the front stub axles of an automobile is 1.3 m, the length of track rod is 1.2 m, the wheel track is 1.45 m and the wheel base is 2.8 m. What should be the length of track arm if the steering mechanism is to give a correct steering while rounding a corner of 6 m radius?

[Ans: 0.145 m]

16. An automotive vehicle using Ackermann steering mechanism has a wheel base of 2.7 m and a track of 1.4 m. The track rod is 1.15 m and each track arm is 0.14 m long. The distance between the pivot of front stub axles is 1.23 m. If the car is turning to the right, find the radius of curvature of the path followed by the front inner wheel for the correct steering.

[Ans: 5.6 m]

17. The two shafts are connected by a Hooke's joint and the driving shaft rotates at 500 rpm. If the speed of driven shaft lies between 475 and 525 rpm, determine the maximum permissible angle between the shafts.

[Ans: 17.72°]

18. A Hooke's joint is used to connect two shafts whose axes are inclined at 20°. The driving shaft rotates uniformly at 6000 rpm. What are the extreme angular velocities of the driven shaft? Find the maximum value of acceleration or retardation and state the angle where both will occur.

[Ans: ω_{2max} = 667 rad/s, ω_{2min} = 590 rad/s
θ_{max} = 138.56° and 318.56°, θ_{min} = 41.43° and 221.43°, α_2 = 50,800 rad/s^2]

19. Two shafts are connected by a Hooke's joint. The driving shaft is rotating at a uniform speed of 1000 rpm. Find the greatest permissible angle between the axes of the two shafts so that the total fluctuation of speed does not exceed 90 rpm.

[**Ans:** 17.06°]

20. A Hooke's joint connects two shafts which are having 165° as the included angle. The driving shaft rotates uniformly at 1000 rpm. Find the maximum angular acceleration of the driven shaft and the maximum torque required if the driven shaft carries a flywheel of mass 10 kg and radius of gyration 100 mm.

[**Ans:** 761.18 rad/s^2, 76.12 Nm]

CHAPTER 5

Friction

5.1 INTRODUCTION

On every machined surface, some marks of imperfection are bound to be there which take the form of hills and valleys that vary both in height and spacing. The resulting texture of the surface is called **surface roughness**. The characteristics of surface roughness depend upon the method of machining. In mechanical components, when two such surfaces are in contact, there is always a certain amount of interlocking owing to the hills on one fitting into the valley of the other. If one surface slides over the other, a force, commonly referred to as **friction force** is introduced which retards the motion. Friction force acts in the direction opposite to that of the sliding surface and is tangential (parallel) to the sliding surface.

In mechanical assemblies such as shaft bearing, lathe spindle, slideways and so forth, the friction force is harmful as it consumes certain amount of energy which is converted into heat and therefore, is waste. Thus the friction between these surfaces need to be reduced by the use of lubrication. However, in other applications like belt drive, clutches, etc., the power transmission capacity is a function of friction force between the contacting surfaces. Thus in these applications, designer usually attempts to increase friction force between contacting surfaces so as to increase the power transmission capacity.

Depending upon the condition of surfaces, friction is classified into the following three categories:

Dry friction. When the relative motion between two contacting surfaces is completely without lubrication, the friction between the surfaces is called **dry friction**. If the relative motion is sliding in nature, it is referred to as **sliding friction;** otherwise the friction due to rolling of one surface over another is called **rolling friction.**

Greasy friction. When two contacting surfaces have a skin thin layer of lubrication between them, the friction is known as **greasy friction**. In theory of lubrication parlance, it is referred to as **boundary lubrication.**

Film friction. When two contacting surfaces are completely separated by a thin layer of lubricant and the friction occurs only due to resistance of the motion of lubricant layers, the friction between the surfaces is termed as **film friction.**

5.2 LAWS OF DRY FRICTION

Experimental evidences have shown that when two mechanical members with smooth and dry surfaces are in contact, the minimum amount of force required to cause one member to slide over the other obeys the following laws:

(i) The frictional force is directly proportional to the normal reaction between the two surfaces.
(ii) Friction force opposes the motion between the surfaces.
(iii) It depends upon the material of contacting surfaces.
(iv) It is independent of area of the contacting surfaces for a given normal force.
(v) The friction force is largely independent of the velocity of sliding, though at high velocity it slightly decreases.

5.3 ANGLE OF FRICTION

Consider a body, which is at rest on a smooth and dry horizontal plane surface as shown in Figure 5.1. The weight of body W acts vertically downward and the plane surface exerts a reaction force R_n on the body which is normal to the surface. Thus under the equilibrium condition of the body, weight of the body is equal to normal reaction.

$$W = R_n \tag{5.1}$$

Now if a horizontal force F is applied to the body, no relative motion will take place till the applied force F is greater than or at least equal to the friction force F'. The body will be in equilibrium under the action of three forces—applied force F, weight W and reaction force R between the plane surface and the body which is equal and opposite to the resultant of the forces F and W. It is inclined at an angle ϕ to the normal reaction R_n. As the applied force F increases, the angle ϕ also increases till the body begins to slide on the plane surface. The angle ϕ at this point is called **angle of friction**.

(a) Body at rest   (b) Body at sliding

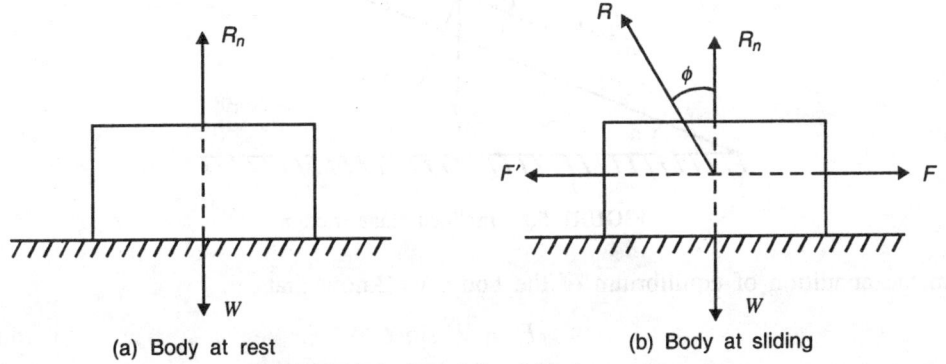

FIGURE 5.1 Principle of friction angle.

According to the laws of dry friction, friction force is directly proportional to the normal reaction, i.e.

$$F' \propto R_n$$

or
$$F' = \mu R_n \tag{5.2}$$

where μ is a constant of proportionality, also termed as **coefficient of friction**.

or
$$\mu = \frac{F'}{R_n} \tag{5.3}$$

Also referring to Figure 5.1,
$$\tan\phi = \frac{F'}{W} = \frac{F'}{R_n} = \mu \tag{5.4}$$

Hence when a body slides over a plane surface, the true reaction between them is always inclined at the friction angle ϕ to the normal reaction.

5.4 INCLINED PLANE FRICTION

Consider a body at rest on an inclined dry plane which makes an angle α with the horizontal plane. The body under such circumstance is subjected to the following forces as shown in Figure 5.2:

(i) Weight of the body W, acting in downward direction
(ii) Normal reaction R_n
(iii) Friction force F' ($= \mu R_n$), resisting the motion of the body

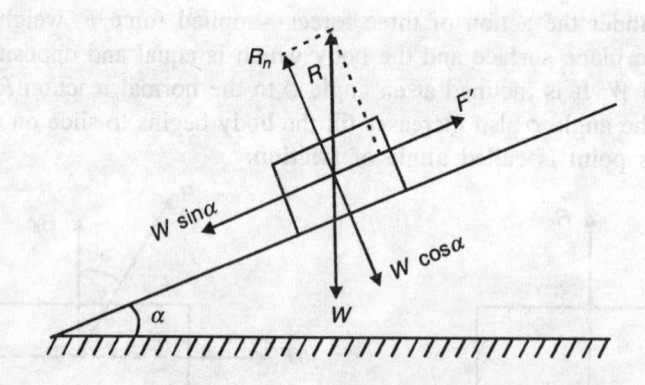

FIGURE 5.2 Inclined plane friction.

From the condition of equilibrium of the body, we know that
$$F' = W \sin\alpha$$
and
$$R_n = W \cos\alpha$$

If the angle of inclined plane α is increased gradually to a point where the body begins to slide down the plane of its own, this angle is known as **angle of repose** or **limiting angle of inclination**.

Referring to Figure 5.2,
$$W \sin \alpha = F' = \mu R_n$$
or
$$W \sin \alpha = \mu W \cos \alpha$$
or
$$\tan \alpha = \mu \tag{5.5}$$

However, the coefficient of friction μ is equal to $\tan \phi$. Therefore, the limiting angle of inclination is equal to friction angle, i.e. $\alpha = \phi$.

In the problem of inclined plane friction, there are two cases namely motion up the plane and motion down the plane. These are discussed in the following sub-sections.

5.4.1 Motion Up the Plane

Consider a body of weight W moving up the inclined plane with uniform velocity under the action of external force F. Let the plane be inclined at an angle α to the horizontal plane and normal reaction R_n acts normal to the inclined plane as shown in Figure 5.3. When the body moves up the plane, the friction force F' $(= \mu R_n)$, which opposes the motion, acts down the plane. Under the condition of equilibrium, the following three forces act on the body [Figure 5.3(b)]:

(i) Weight of the body W, acting vertically downward
(ii) External force F, acting up the plane at an angle β from vertical plane
(iii) Reaction force R, which is equal to the resultant of F' and R_n.

(a) Motion up the plane (b) Force diagram

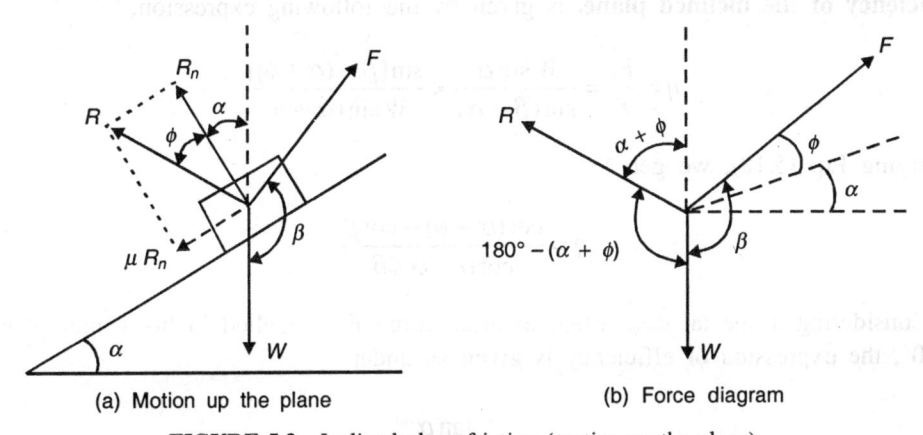

FIGURE 5.3 Inclined plane friction (motion up the plane).

For equilibrium of forces, as shown in Figure 5.3(b), applying Lami's theorem, we get

$$\frac{W}{\sin[(\pi - \beta) + (\alpha + \phi)]} = \frac{F}{\sin[\pi - (\alpha + \phi)]} = \frac{R}{\sin \beta}$$

or
$$\frac{W}{\sin[\beta - (\alpha + \phi)]} = \frac{F}{\sin(\alpha + \phi)} = \frac{R}{\sin \beta}$$

or
$$F = \frac{W \sin(\alpha + \phi)}{\sin[\beta - (\alpha + \phi)]} \qquad (5.6)$$

The condition for the minimum force required to move up the body is given as

$$\sin[\beta - (\alpha + \phi)] = 1$$

or
$$\beta - (\alpha + \phi) = 90°$$

or
$$\phi = \beta - (90° + \alpha) \qquad (5.7)$$

Thus for the minimum force condition, the external force should be inclined to the inclined plane by an angle equal to the friction angle ϕ. [See Figure 5.3(b)].

Therefore,
$$F_{min} = W \sin(\alpha + \phi) \qquad (5.8)$$

Now consider a case when a body moves up the plane without any friction between them. Let F_0 be the force required to move up the plane without any friction. In such a circumstance, the friction angle ϕ becomes zero and reaction force R coincides with normal reaction R_n. Thus Eq. (5.6) reduces to

$$F_0 = \frac{W \sin \alpha}{\sin(\beta - \alpha)} \qquad (5.9)$$

The ratio of the force required without friction to the force required with friction, known as **efficiency of the inclined plane,** is given by the following expression:

$$\eta = \frac{F_0}{F} = \frac{W \sin \alpha}{\sin(\beta - \alpha)} \times \frac{\sin[\beta - (\alpha + \phi)]}{W \sin(\alpha + \phi)} \qquad (5.10)$$

Simplifying Eq. (5.10), we get

$$\eta = \frac{\cot(\alpha + \phi) - \cot \beta}{\cot \alpha - \cos \beta} \qquad (5.11)$$

Considering a special case when external force F is applied in horizontal plane, i.e. $\beta = 90°$, the expression of efficiency is given as under:

$$\eta = \frac{\tan \alpha}{\tan(\alpha + \phi)} \qquad (5.12)$$

For the maximum efficiency, Eq. (5.10) can be simplified using trigonometrical identity

$$2 \sin A \sin B = \cos(A + B) - \cos(A - B)$$

Therefore,
$$\eta = \frac{\cos(\beta - \phi - 2\alpha) - \cos(\beta - \phi)}{\cos(\beta - \phi - 2\alpha) - \cos(\beta + \phi)} \qquad (5.13)$$

For given values of angles β and ϕ, which are constant, there is one value of angle α which gives the maximum efficiency. Therefore, the condition of maximum efficiency is given as

$$\cos(\beta - \phi - 2\alpha) = 1$$

or

$$\beta - \phi - 2\alpha = 0°$$

or

$$\alpha = \frac{\beta - \phi}{2} \quad (5.14)$$

Substituting the value of angle α in Eq. (5.13), we get

$$\eta_{max} = \frac{1 - \cos(\beta - \phi)}{1 - \cos(\beta + \phi)} \quad (5.15)$$

If the external force F is horizontal, i.e. when $\beta = 90°$, the expression for maximum efficiency is given as

$$\eta = \frac{1 - \sin\phi}{1 + \sin\phi} \quad (5.16)$$

5.4.2 Motion Down the Plane

When a body moves down the plane, the friction force $F'(= \mu R_n)$ acts in the upward direction, i.e. the direction opposite to the direction of motion. In such a case, the reaction force R is inclined at an angle ϕ to the normal reaction R_n towards right side as shown in Figure 5.4. The body remains in the equilibrium under the action of three forces—the weight W, the external force F and the reaction force R.

(a) Motion down the plane

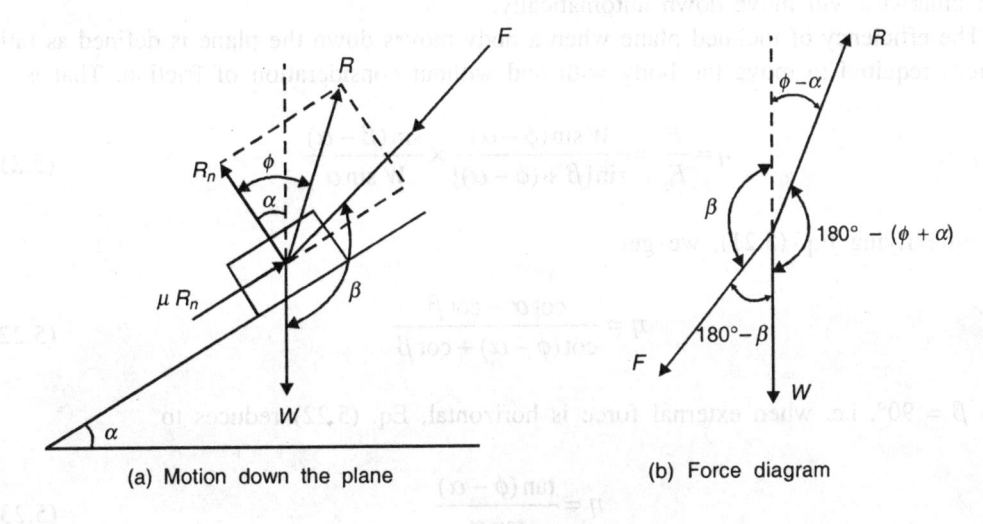

(b) Force diagram

FIGURE 5.4 Inclined plane friction (motion down the plane).

Applying Lami's theorem, we have [refer Figure 5.4(b)]

$$\frac{W}{\sin[\beta + (\phi - \alpha)]} = \frac{F}{\sin[\pi - (\phi - \alpha)]} = \frac{R}{\sin(\pi - \beta)} \quad (5.17)$$

or the external force, $F = \dfrac{W \sin(\phi - \alpha)}{\sin[\beta + (\phi - \alpha)]}$ (5.18)

From Eq. (5.18), the following interpretation can be drawn:

(i) When the friction angle is greater than the angle of inclination ($\phi > \alpha$), the external force is required to move the body down the plane.
(ii) When the friction angle is equal to the angle of inclination ($\phi = \alpha$), the body is on the point of moving down under its own weight without any external force.
(iii) When the friction angle is less than the angle of inclination ($\phi < \alpha$), the external force will be negative. This means that the external force is required to be applied in the opposite direction so as to resist the motion.
(iv) The minimum force required to move the body down the plane is:

$$F_{min} = W \sin(\phi - \alpha) \quad (5.19)$$

When a body moves down the inclined plane without any friction, i.e. $\phi = 0$, Eq. (5.18) reduces to

$$F_0 = \frac{-W \sin \alpha}{\sin(\beta - \alpha)} \quad (5.20)$$

where F_0 is the force required to move the body without any friction. The negative sign indicates that a force in the opposite direction is required to stop the motion down the plane, which otherwise will move down automatically.

The efficiency of inclined plane when a body moves down the plane is defined as ratio of forces required to move the body with and without consideration of friction. That is

$$\eta = \frac{F}{F_0} = \frac{W \sin(\phi - \alpha)}{\sin[\beta + (\phi - \alpha)]} \times \frac{\sin(\beta - \alpha)}{W \sin \alpha} \quad (5.21)$$

After simplifying Eq. (5.21), we get

$$\eta = \frac{\cot \alpha - \cot \beta}{\cot(\phi - \alpha) + \cot \beta} \quad (5.22)$$

When $\beta = 90°$, i.e. when external force is horizontal, Eq. (5.22) reduces to

$$\eta = \frac{\tan(\phi - \alpha)}{\tan \alpha} \quad (5.23)$$

EXAMPLE 5.1 A force of 240 N inclined at 30° to the horizontal plane is required to pull a body weighing W N. It is found that if the direction of force is reversed to push the body, the force required is 300 N. Determine the following:
  (i) weight of the body
  (ii) coefficient of friction
  (iii) minimum amount of force required to pull the body

Solution Refer to Figure 5.5(a), when the body is being pulled.

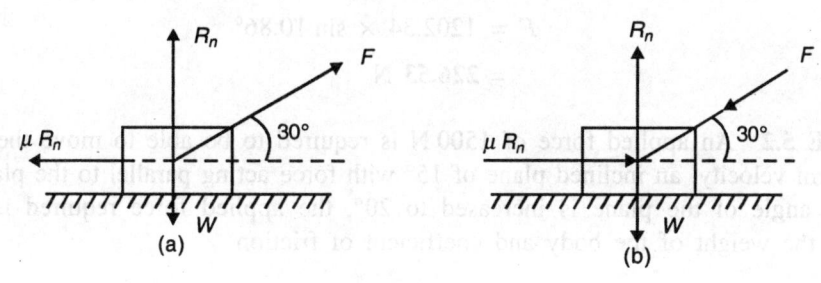

FIGURE 5.5

For equilibrium of the body [Figure 5.5(a)], we must have

$$F \cos 30° = \mu R_n$$

or
$$240 \cos 30° = \mu R_n \quad \text{(i)}$$

Also
$$R_n + F \sin 30° = W$$

or
$$R_n = W - 240 \sin 30° \quad \text{(ii)}$$

Substituting the value of R_n in Eq. (i), we get

$$240 \cos 30° = \mu(W - 240 \sin 30°)$$

or
$$207.84 = \mu(W - 120) \quad \text{(iii)}$$

Similarly, referring to Figure 5.5(b), when body is being pushed, for equilibrium

$$300 \cos 30° = \mu R_n \quad \text{(iv)}$$

and
$$W + 300 \sin 30° = R_n \quad \text{(v)}$$

Substituting the value of R_n from Eq. (v) in Eq. (iv), we get

$$300 \cos 30° = \mu(W + 300 \sin 30°)$$

or
$$259.8 = \mu(W + 150) \quad \text{(vi)}$$

Solving Eqs. (iii) and (vi), we get

(i) Weight of the body, $W = 1202.34$ N **Ans.**
(ii) Coefficient of friction, $\mu = 0.192$ **Ans.**
(iii) The minimum force required to pull the body:

$$F = W \sin \phi$$

where

angle ϕ = friction angle $[= \tan^{-1} \mu]$
$= \tan^{-1} (0.192)$ or $10.86°$

Thus
$$F = 1202.34 \times \sin 10.86°$$
$$= 226.53 \text{ N}$$ **Ans.**

EXAMPLE 5.2 An applied force of 1500 N is required to be able to move the body up, with uniform velocity, an inclined plane of 15° with force acting parallel to the plane. When inclination angle of the plane is increased to 20°, the applied force required is 1800 N. Determine the weight of the body and coefficient of friction.

Solution Referring to Figure 5.3, the applied force is:

$$F = \frac{W \sin(\alpha + \phi)}{\sin[\beta - (\alpha + \phi)]}$$

where $\beta = 90° + \alpha$, as it is given that the applied force is parallel to the plane.

or
$$F = \frac{W(\sin\alpha\cos\phi + \cos\alpha\sin\phi)}{\sin[(90° + \alpha) - (\alpha + \phi)]}$$
$$= W(\sin\alpha + \mu\cos\alpha)$$

where
$$\mu = \tan\phi$$

Substituting the given values,

$$1500 = W(\sin 15° + \mu\cos 15°)$$
$$1800 = W(\sin 20° + \mu\cos 20°)$$

or
$$\frac{1500}{1800} = \frac{\sin 15° + \mu\cos 15°}{\sin 20° + \mu\cos 20°}$$

or
$$0.833 = \frac{0.2588 + \mu \times 0.966}{0.342 + \mu \times 0.94}$$

or
$$\mu = 0.142$$ **Ans.**

Weight, $W = \dfrac{F}{\sin\alpha + \mu\cos\alpha} = \dfrac{1500}{\sin 15° + 0.142 \times \cos 15°}$

$$= 3788.0 \text{ N}$$ **Ans.**

5.5 INCLINED PLANE WITH GUIDE FRICTION

Consider an inclined plane with guide friction arrangement as shown in Figure 5.6 in which two elements A and B are free to slide in the guides and are in contact along inclined plane XX. The angle between the inclined plane XX and axis of slider A is $90° - \alpha$ and that between XX and the axis of slider B is $90° - \beta$.

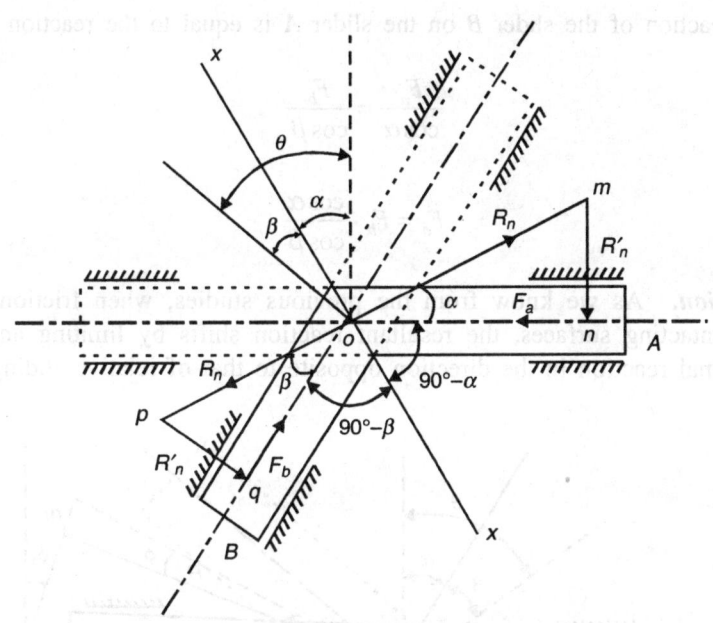

FIGURE 5.6 Inclined plane with guides without friction.

Let

F_a = applied force to the slider A to overcome the resistance offered by the slider B
F_b = applied force to the slider B to overcome the resistance offered by the slider A
ϕ = limiting angle of friction between the contact surfaces of the sliders A and B
and
ϕ_1 = limiting angle of friction between the contacting surfaces of each of the sliders A and B with their respective guides

The relation between the forces F_a and F_b for the cases of without and with friction is discussed further.

Without friction. The slider B as shown in Figure 5.6 is in equilibrium under the action of the following three forces:

(a) Applied force, F_b
(b) Normal reaction R_n of the slider A on the slider B, acting perpendicular to the plane XX
(c) The sum of normal reactions of the two guides on the slider B, R'_n.

The above three forces form a triangle of forces, Δopq, for which

$$F_b = R_n \cos \beta \tag{5.24}$$

Similarly, the slider A is in equilibrium under the action of three forces which form a triangle of forces Δomn for which

$$F_a = R_n \cos \alpha \tag{5.25}$$

However, the reaction of the slider B on the slider A is equal to the reaction of A on B.

Therefore,
$$\frac{F_a}{\cos \alpha} = \frac{F_b}{\cos \beta} \tag{5.26}$$

or
$$F_a = F_b \cdot \frac{\cos \alpha}{\cos \beta} \tag{5.27}$$

With friction. As we know from the previous studies, when friction is considered between the contacting surfaces, the resultant reaction shifts by limiting angle of friction ϕ, with the normal reaction in the direction opposite to that of relative sliding, as shown in Figure 5.7.

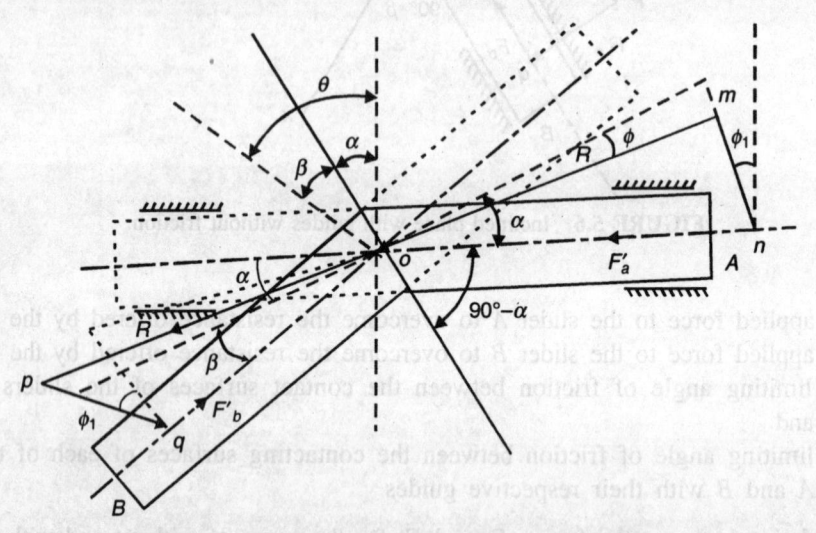

FIGURE 5.7 Inclined plane guides with friction.

Using sine rule for Δopq, we have

$$\frac{R}{\sin(90° + \phi_1)} = \frac{F_b'}{\sin[90° - (\beta + \phi + \phi_1)]}$$

or
$$R = F_b' \times \frac{\cos \phi_1}{\cos(\beta + \phi + \phi_1)} \tag{5.28}$$

Similarly, for Δomn, we have

$$\frac{R}{\sin(90° - \phi_1)} = \frac{F_a'}{\sin[90° - (\alpha - \phi - \phi_1)]}$$

or
$$R = F_a' \times \frac{\cos\phi_1}{\cos(\alpha - \phi - \phi_1)} \qquad (5.29)$$

Equating Eqs. (5.28) and (5.29), we get

$$\frac{F_a'}{F_b'} = \frac{\cos(\alpha - \phi - \phi_1)}{\cos(\beta + \phi + \phi_1)} \qquad (5.30)$$

The ratio of force F_a required without friction to that required with friction (when F_b = constant) is known as **efficiency** and is expressed as

$$\eta = \frac{F_a}{F_a'} = \frac{\cos\alpha}{\cos\beta} \times \frac{\cos(\beta + \phi + \phi_1)}{\cos(\alpha - \phi - \phi_1)} \qquad (5.31)$$

Let θ be the angle between the two guides ($= \alpha + \beta$).

$$\eta = \frac{\cos\alpha}{\cos(\theta - \alpha)} \times \frac{\cos(\theta - \alpha + \phi + \phi_1)}{\cos(\alpha - \phi - \phi_1)}$$

Using trigonometrical identity $2\cos A \cos B = \cos(A + B) + \cos(A - B)$, we get

$$\eta = \frac{\cos(\theta + \phi + \phi_1) + \cos(2\alpha - \theta - \phi - \phi_1)}{\cos(\theta - \phi - \phi_1) + \cos(2\alpha - \theta - \phi - \phi_1)} \qquad (5.32)$$

For given values of angles θ, ϕ and ϕ_1, the efficiency will be maximum when $\cos(2\alpha - \theta - \phi - \phi_1)$ is maximum, i.e. when $2\alpha - \theta - \phi - \phi_1 = 0$.

or
$$\alpha = \frac{\theta + \phi + \phi_1}{2}$$

Substituting the value of angle α in Eq. (5.32), the maximum efficiency is given by

$$\eta_{max} = \frac{1 + \cos(\theta + \phi + \phi_1)}{1 + \cos(\theta - \phi - \phi_1)} \qquad (5.33)$$

5.6 WEDGE FRICTION

A wedge is usually of a triangular or trapezoidal in cross section. It consists of sliding pair. It is, generally, used for slight adjustment in the position of a body or to lift heavy load like a screw jack. A typical wedge system used for lifting loads is shown in Figure 5.8. This system consists of three elements—frame A, wedge B and slider C. When a force is applied to the wedge, the slider is raised in the guided frame. Under the equilibrium of wedge, when friction is neglected, it is subjected to three forces:

(i) applied force F
(ii) normal reaction between the frame A and the wedge, R_{n1}
(iii) normal reaction between the wedge and the slider, R_{n2}.

A component of normal reaction R_{n2}, ($R_{n2} \cos\alpha$) holds the external weight W. Therefore,

$$W = R_{n2} \cos\alpha$$

and

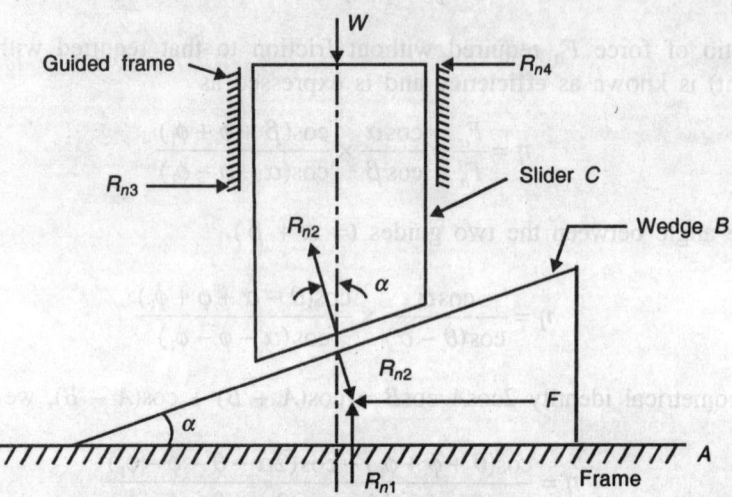

$$R_{n1} = F \cot\alpha \tag{5.34}$$

FIGURE 5.8 A typical wedge system.

The slider is acted upon by the three forces:
(i) weight W
(ii) normal reaction R_{n2}
(iii) reaction between the guide frame and the slider.

The horizontal component of normal reaction R_{n2} (= $R_{n2} \sin \alpha$) forms a clockwise couple which tries to tilt the slider. The slider remains balanced only if the normal reactions R_{n3} and R_{n4} act as shown in Figure 5.8. When friction between the frame, wedge, slider and guide frame is taken into account, the resultant reaction will be inclined to the normal reaction at the angle of friction in the direction opposite to the relative motion, as shown in Figure 5.9.
Let

μ = coefficient of friction between the guide frame and the slider (= $\tan \phi$)

and μ_1 = coefficient of friction between the frame, wedge and slider (= $\tan \phi_1$)

Consider the equilibrium of the wedge under the action of three forces:

(i) external force F
(ii) resultant reaction between the frame and the wedge R_1
(iii) resultant reaction between the wedge and the slider R_2

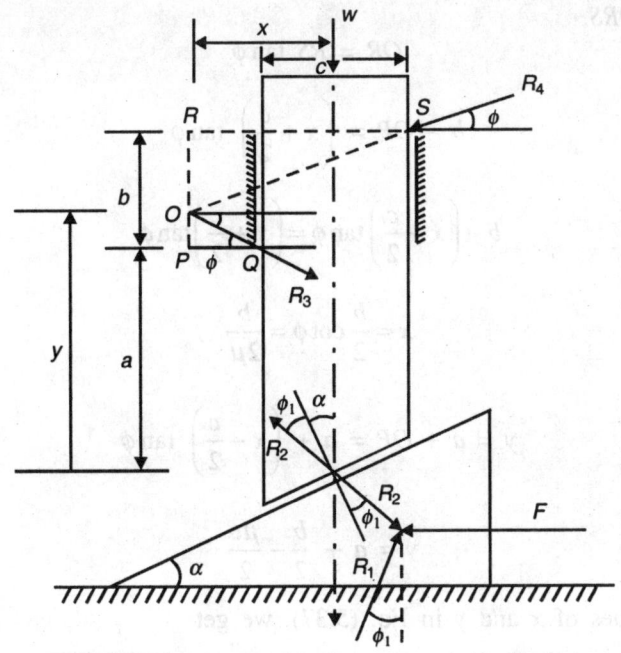

FIGURE 5.9 A typical wedge system (with friction).

According to Lami's theorem,

$$\frac{F}{\sin[180° - (\alpha + 2\phi_1)]} = \frac{R_1}{\sin[90° + (\alpha + \phi_1)]} = \frac{R_2}{\sin(90° + \phi_1)}$$

Therefore,
$$R_1 = F \times \frac{\cos(\alpha + \phi_1)}{\sin(\alpha + 2\phi_1)} \qquad (5.35)$$

and
$$R_2 = F \times \frac{\cos\phi_1}{\sin(\alpha + 2\phi_1)} \qquad (5.36)$$

The slider is in equilibrium under the four forces—W, R_2, R_3 and R_4. Resolving R_2 into two components, parallel and perpendicular to the axis of guide, and taking moments about O, the point of intersection of reactions R_3 and R_4, we get

$$[R_2\cos(\alpha + \phi_1) - W]x = R_2 \sin(\alpha + \phi_1)y \qquad (5.37)$$

where x and y are distances as shown in Figure 5.9.

From $\triangle OPQ$,

$$OP = PQ \tan\phi = \left(x - \frac{c}{2}\right)\tan\phi$$

Similarly, from $\triangle ORS$,

$$OR = RS \tan \phi$$

or
$$b - OP = \left(x + \frac{c}{2}\right) \tan \phi$$

or
$$b - \left(x - \frac{c}{2}\right) \tan \phi = \left(x + \frac{c}{2}\right) \tan \phi$$

or
$$x = \frac{b}{2} \cot \phi = \frac{b}{2\mu} \qquad (5.38)$$

and distance:
$$y = a + OP = a + \left(x - \frac{c}{2}\right) \tan \phi$$

or
$$y = a + \frac{b}{2} - \frac{\mu c}{2} \qquad (5.39)$$

Substituting the values of x and y in Eq. (5.37), we get

$$[R_2 \cos(\alpha + \phi_1) - W] \frac{b}{2\mu} = R_2 \sin(\alpha + \phi_1) \times \left(a + \frac{b}{2} - \frac{\mu c}{2}\right)$$

Therefore, the load–carrying capacity of the wedge is:

$$W = R_2 \left[\cos(\alpha + \phi_1) - \mu \left(1 + \frac{2a - \mu c}{b} \sin(\alpha + \phi_1) \right) \right] \qquad (5.40)$$

EXAMPLE 5.3 A block weighing 3 kN is lifted up by a driving wedge against a vertical wall as shown in Figure 5.10. If the wedge angle is 15° and coefficient of friction between

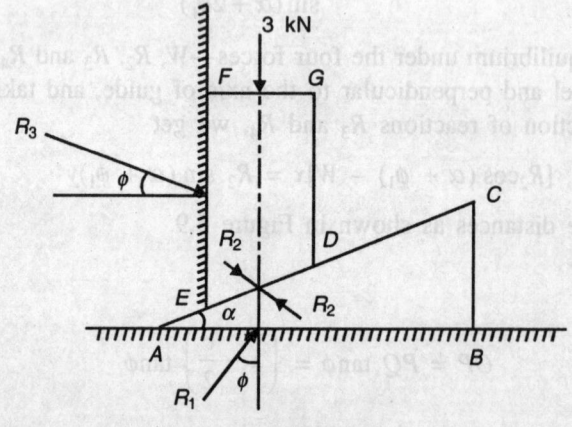

FIGURE 5.10

all the surfaces in contact is 0.35, determine the minimum horizontal force to be applied to raise the block.

Solution Consider the equilibrium of the block *defg*, which is subjected to the following forces as shown in figure 5.11:

(i) weight, $W = 3000$ N
(ii) reaction, R_3
(iii) reaction, R_2

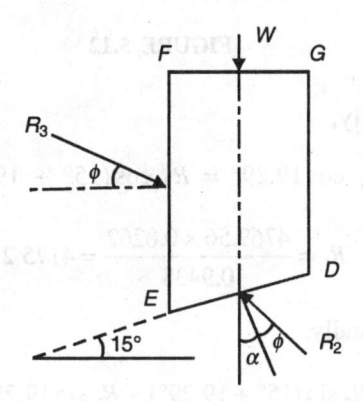

FIGURE 5.11

$$\phi = \tan^{-1} \mu = \tan^{-1} 0.35$$
$$= 19.29°$$

Resolving forces horizontally,

$$R_3 \times \cos 19.29° = R_2 \times \sin(15° + 19.29°)$$

or

$$R_2 = 1.675 \, R_3 \qquad \text{(i)}$$

Now, resolving forces vertically,

$$3000 + R_3 \sin 19.29° = R_2 \times \cos(15° + 19.29°)$$

or

$$3000 + 0.33 \, R_3 = 0.826 \, R_2 \qquad \text{(ii)}$$

Solving Eqs. (i) and (ii), we get

Reaction, $R_3 = 2847.5$ N and $R_2 = 4769.56$ N

Now consider the equilibrium of the wedge, which is subjected to the following forces as shown in Figure 5.12:

(i) reaction R_2
(ii) reaction R_1
(iii) horizontal force F required to drive wedge

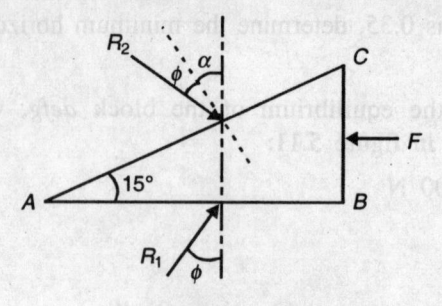

FIGURE 5.12

Resolving the forces vertically,

$$R_1 \cos 19.29° = R_2 \cos(15° + 19.29°)$$

or

$$R_1 = \frac{4769.56 \times 0.8262}{0.9438} = 4175.2 \text{ N}$$

Resolving the forces horizontally,

$$F = R_2 \sin(15° + 19.29°) + R_1 \sin 19.29°$$
$$= 4769.5 \times \sin 34.29° + 4175.2° \times \sin 19.29°$$
$$= 4066.3 \text{ N} \qquad \text{Ans.}$$

EXAMPLE 5.4 Two blocks A and B connected by a horizontal rod by frictionless hinges are supported on two rough surfaces as shown in Figure 5.13.

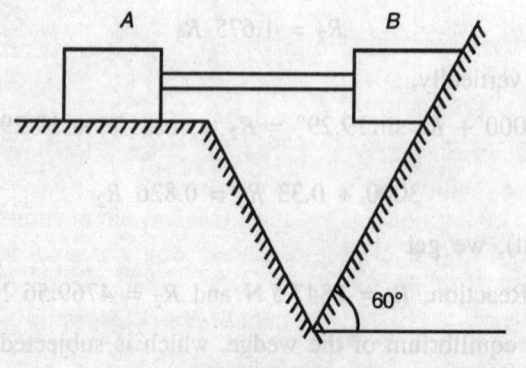

FIGURE 5.13

The coefficients of friction are 0.25, between the block A and the horizontal surface, and 0.3, between the block B and the inclined surface. If the block B weighs 200 N, determine the smallest weight of block A that will hold the system in equilibrium.

Solution Consider the equilibrium of the block B under the action of the following forces, as shown in Figure 5.14:

  (i) weight of the block B, W_b
 (ii) normal reaction, R_n
(iii) force of friction of block A acting horizontally on the block $B = \mu_a \times W_a = 0.25 W_a$
(iv) force of friction between the block B and the inclined plane $= \mu_b \times R_n = 0.3 R_n$.

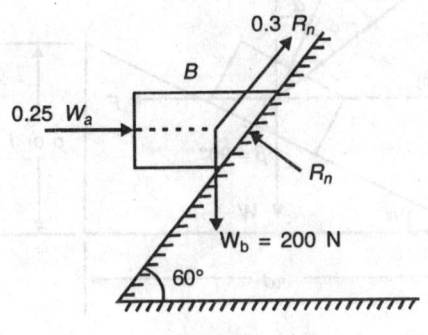

FIGURE 5.14

Resolving the forces along the plane,

$$0.3 R_n + 0.25 W_a \times \cos 60° = 200 \times \sin 60°$$

or
$$0.3 R_n + 0.125 W_a = 173.2 \qquad (i)$$

and resolving the forces perpendicular to the inclined plane,

$$R_n = 0.25 W_a \times \sin 30° + 200 \times \cos 60°$$

or
$$R_n = 0.216 W_a + 100 \qquad (ii)$$

Solving Eqs. (i) and (ii) for weight W_a, we get

Weight of the block A, $W_a = 754.4$ N **Ans.**

5.7 SCREW FRICTION

The screw, bolts, studs, nuts are widely used in various machines and structures for fastening. These fastenings have screw threads which are made by cutting a continuous helical grooves on a cylindrical surface. A screw thread is obtained when hypotenuse of a right angled triangle is wrapped around the circumference of a cylinder.

The screw threads are mainly of two types—square threads and V-threads. V-threads are stronger and offer more frictional resistance to the motion. It is because of this reason that V-threads are more commonly used for fastening purpose.

Kinematically, the development of a screw thread, when unwound from the body of screw, is an inclined plane, the angle of inclination is equal to helix angle. Thus the movement between nut and screw against the applied axial force is analogous to the movement of a weight on an inclined plane as shown in Figure 5.15. The load on the screw, when transferred to the nut, acts as a distributed load on the surface of the thread in contact. For simplicity,

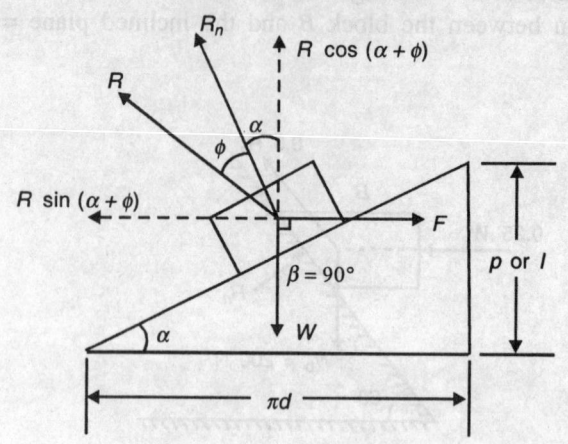

FIGURE 5.15 Screw friction (square thread).

it is assumed that the distributed load is concentrated at a point on the mean circumference of the thread for square threaded screw used in screw jack, the faces of the square threads are normal to the axis of the spindle.

Let

α = helix or lead angle of the thread
p = pitch of the thread
d = mean diameter of the screw
μ = coefficient of friction and
ϕ = angle of friction (= $\tan^{-1} \mu$)

Referring to Figure 5.15 and Eq. (5.6), the expression for horizontal applied force F (with $\beta = 90°$) required to raise the weight up on the inclined plane is given as

$$F = W \tan(\alpha + \phi) \qquad (5.41)$$

and the frictional torque:

$$T = \frac{Wd}{2} \tan(\alpha + \phi) \qquad (5.42)$$

Similarly, the force required to lower the weight is:

$$F = W \tan(\phi - \alpha)$$

The transmission efficiency of the screw nut can be expressed as

$$\eta = \frac{\tan \alpha}{\tan(\alpha + \phi)} \qquad (5.43)$$

In case of V-threads, the faces of the V-groove are inclined to the axis of the screw. The normal reaction R_n, therefore, acts perpendicular to the inclined face as shown in Figure 5.16. The axial component of normal reaction bears the external applied force W.

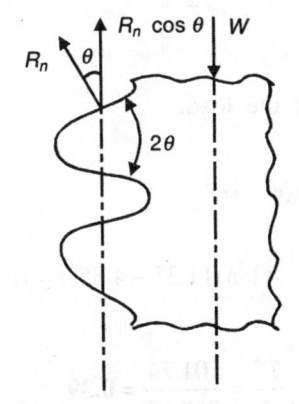

FIGURE 5.16 Friction in V-threads.

Therefore,
$$W = R_n \cos\theta$$
or
$$R_n = \frac{W}{\cos\theta}$$
$$\text{Friction force} = \mu R_n$$
$$= \frac{\mu W}{\cos\theta}$$
$$= \mu' W,$$

where μ' = equivalent coefficient of friction (= $\mu/\cos\theta$).

Thus the conditions for the V-thread, so far as friction is concerned, are identical with those for square thread in which the coefficient of friction is μ'.

EXAMPLE 5.5 A screw jack with square threads having 60 mm pitch diameter is used to raise a load of 30 kN. The pitch of the thread is 16 mm and coefficient of friction between screw and nut is 0.2. Determine the ratio of torque required to lower the load to the torque required to raise the load and also determine the efficiency of the screw jack when the load is raised.

Solution Helix angle: $\alpha = \tan^{-1}\left(\dfrac{p}{\pi d}\right) = \tan^{-1}\left(\dfrac{16.0}{\pi \times 60}\right) = 4.85°$

Friction angle: $\phi = \tan^{-1}\mu = \tan^{-1}0.2 = 11.3°$

Torque required to raise the load:

$$T = \frac{Wd}{2}\tan(\alpha + \phi) = \frac{30 \times 60}{2} \times \tan(4.85° + 11.3°)$$

$$= 260.62 \text{ Nm}$$

Similarly, torque required to lower the load:

$$T' = \frac{Wd}{2}\tan(\phi - \alpha)$$

$$= \frac{30 \times 60}{2} \times \tan(11.3° - 4.85°) = 101.74 \text{ Nm}$$

The required ratio: $\dfrac{T'}{T} = \dfrac{101.74}{260.62} = \mathbf{0.39}$ **Ans.**

$$\eta = \frac{\tan\alpha}{\tan(\alpha + \phi)} = \frac{\tan 4.85°}{\tan(4.85° + 11.3°)} \times 100 = \mathbf{29.3\%} \quad \textbf{Ans.}$$

EXAMPLE 5.6 A turnbuckle consists of a box nut connecting two rods, one screwed right handed and other left handed, each having a pitch of 4 mm and a mean diameter of 23 mm. The thread is of V form with included angle of 55° and coefficient of friction may be taken as 0.18. Assuming that rods do not turn, calculate

(i) torque required on the nut to produce a pull of 50 kN
(ii) efficiency of the screw

Solution The effective coefficient of friction of the V-thread:

$$\mu' = \frac{\mu}{\cos\theta} = \frac{0.18}{\cos 27.5°} = 0.203$$

Friction angle: $\phi = \tan^{-1}\mu' = \tan^{-1}0.203 = 11.47°$

Helix angle: $\alpha = \tan^{-1}\left(\dfrac{p}{\pi d}\right) = \tan^{-1}\left(\dfrac{4}{\pi \times 23}\right) = 3.17°$

Torque:
$$T = \frac{Wd}{2}\tan(\alpha + \phi)$$

$$= \frac{50 \times 23}{2} \times \tan(3.17° + 11.47°)$$

$$= \mathbf{150.2 \text{ Nm}} \quad \textbf{Ans.}$$

Transmission efficiency: $\eta = \dfrac{\tan\alpha}{\tan(\alpha+\phi)}$

$= \dfrac{\tan 3.17°}{\tan(3.17°+11.47°)} \times 100$

$= 21.2\%$ Ans.

EXAMPLE 5.7 A lifting jack with differential screw threads is shown diagrammatically in Figure 5.17. The portion B screws into a fixed base C and carries a right handed square thread of pitch 10 mm and mean diameter 60 mm. The square thread of 6 mm pitch on a mean diameter of 30 mm, screwing into B. If the coefficient of friction for each thread is 0.2, find the overall efficiency and the torque to be applied to B to raise a load of 5 kN.

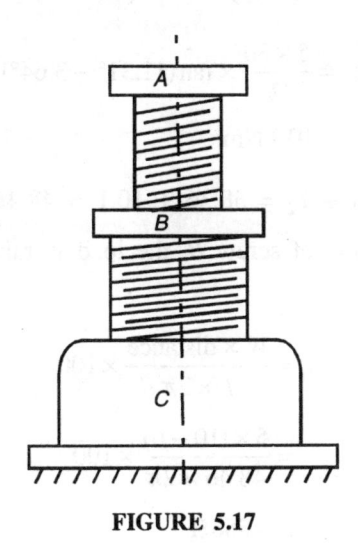

FIGURE 5.17

Solution Between the screw B and body C, the relative motion is against the external load, therefore

Torque: $T_1 = \dfrac{Wd_1}{2}\tan(\alpha_1+\phi)$

where $\tan\alpha_1 = \dfrac{p_1}{\pi d_1}$

or $\alpha_1 = \tan^{-1}\left(\dfrac{10}{\pi \times 60}\right) = 3.036°$

Also $\phi = \tan^{-1} 0.2 = 11.31°$

$$T_1 = \frac{5 \times 60}{2} \times \tan(3.036 + 11.31°)$$

$$= 38.36 \text{ Nm}$$

Similarly, between the screws B and A, the resistive torque is

$$T_2 = \frac{W d_2}{2} \tan(\phi - \alpha_2)$$

where

$$\alpha_2 = \tan^{-1}\left(\frac{p_2}{\pi d_2}\right) = \tan^{-1}\left(\frac{6}{\pi \times 30}\right) = 3.64°$$

$$T_2 = \frac{5 \times 30}{2} \times \tan(11.31° - 3.64°)$$

$$= 10.1 \text{ Nm}$$

Total torque: $T = T_1 + T_2 = 38.36 + 10.1 = $ **48.46 Nm** Ans.

For one complete revolution of screw B, the load is raised by a distance equal to the difference in pitch.

Overall efficiency:

$$\eta = \frac{W \times \text{distance}}{T \times 2\pi} \times 100$$

$$= \frac{5 \times (10 - 6)}{48.46 \times 2\pi} \times 100$$

$$= \mathbf{6.568\%} \qquad \text{Ans.}$$

5.8 PIVOT AND COLLAR FRICTION

In some mechanical machines such as propellar shaft of ship, shaft of steam turbine and so forth, a rotating shaft is subjected to axial thrust. A bearing surface has to be provided to take this thrust and to maintain alignment of shaft in its correct axial position. These surfaces may be either flat or conical in shape. The bearing surface provided at the end of shaft is known as **pivot**. It is also sometimes known as **foot-step bearing** (Figure 5.18). When a collar is provided alongwith the length of shaft to bear the axial thrust on the mating surface, as shown in Figure 5.19, it is called **collar bearing**. In order to increase the thrust–carrying capacity, a number of collars may be provided on a single shaft.

FIGURE 5.18 Pivot bearings.

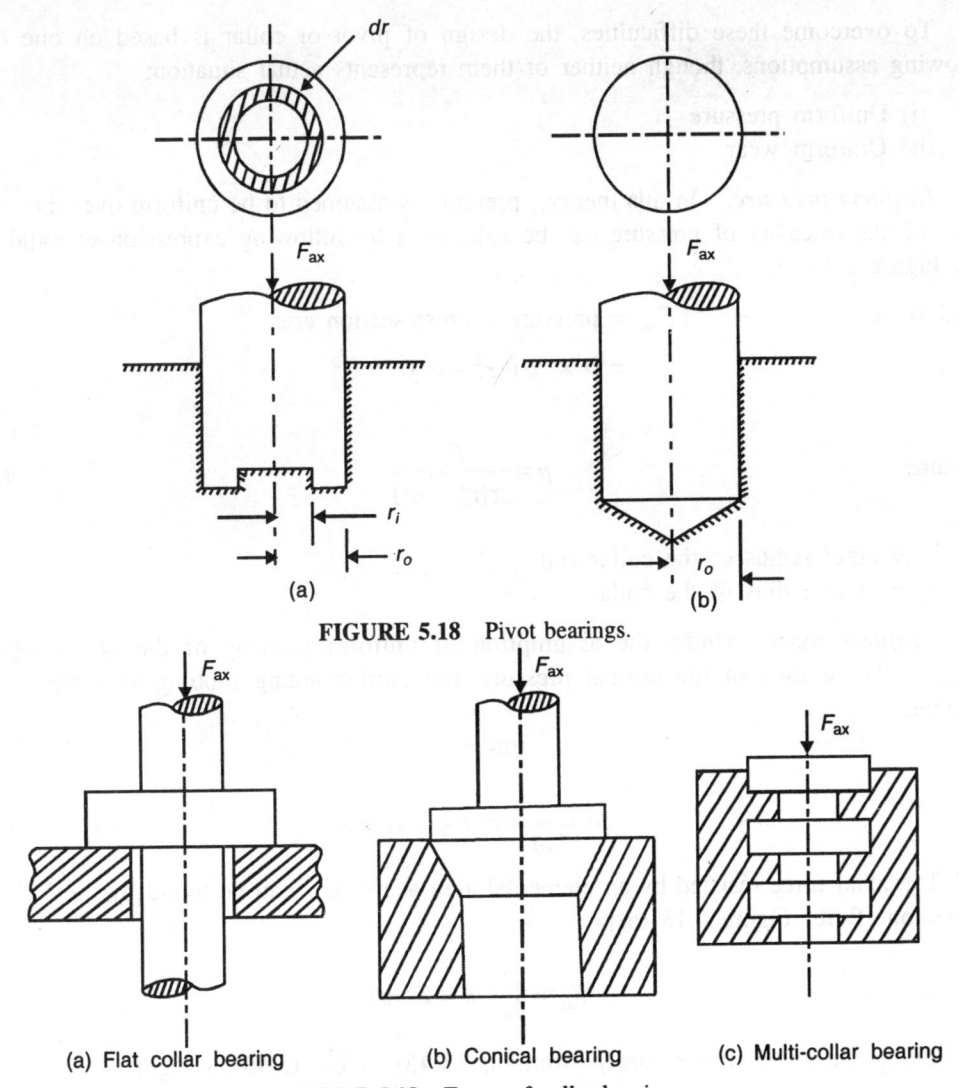

(a) Flat collar bearing   (b) Conical bearing   (c) Multi-collar bearing

FIGURE 5.19 Types of collar bearing.

The relative motion between the contact surfaces of a thrust bearing is resisted by the friction between the contacting surfaces which results into loss of power. The friction torque on a pivot or collar bearing is usually calculated either on the basis of uniform pressure on the bearing surface or uniform rate of wearing of the bearing surfaces. The present knowledge of friction is insufficient to state which of these two theories are more accurate as the actual physical changes which take place are complex in nature. In a new bearing, the intensity of pressure can be assumed uniform at various radii because of exact tolerance to which bearing is manufactured. However, after the shaft has run in the bearing for sufficient time, the intensity of pressure is no more uniformly distributed as the points at different radii move at different rubbing velocities. The rate of wear will be greater at those points where the velocity of rubbing is greater.

To overcome these difficulties, the design of pivot or collar is based on one of the following assumptions, though neither of them represents actual situation:

(i) Uniform pressure
(ii) Uniform wear

Uniform pressure. In this theory, pressure is assumed to be uniform over the surface area and the intensity of pressure can be calculated by following expression of axial force [see Figure 5.18(a)]:

Axial force:
$$F_{ax} = \text{pressure} \times \text{cross-section area}$$
$$= p \times \pi((r_o^2 - r_i^2))$$

or

Pressure:
$$p = \frac{F_{ax}}{\pi(r_o^2 - r_i^2)} \tag{5.44}$$

where
r_o = outer radius of the collar and
r_i = inner radius of the collar

Uniform wear. Under the assumption of uniform wearing of the two contacting surfaces, the product of the normal pressure and corresponding rubbing velocity must be constant.

$$pv = c_1$$

or
$$pr = \frac{c_1}{\omega} = c \quad (\text{as } v = \omega r) \tag{5.45}$$

The axial force resisted by an elemental area at radius r can be found by the following expression [Refer Figure 5.18 (a)]:

Axial force:
$$F_{ax} = \int_{r_i}^{r_o} p \times 2\pi r \, dr \tag{5.46}$$

Substituting the value of pressure p from Eq. (5.45) in Eq. (5.46), we get

$$F_{ax} = \int_{r_i}^{r_o} \frac{c}{r} \times 2\pi r \, dr$$
$$= 2\pi p r(r_o - r_i) \tag{5.47}$$

or the intensity of pressure for uniform wear rate at a radius r is:

$$p = \frac{F_{ax}}{2\pi r(r_o - r_i)} \tag{5.48}$$

5.8.1 Collar Bearings

In collar bearing a collar is provided alongwith the length of shaft to bear the axial thrust on a mating surface. If the surface of the collar is flat plane normal to the axis of shaft, it is known as **flat collar bearing** and if the surface is conical such as frustum of cone, it is called **conical collar bearing**. The expressions for frictional torque of these two types of bearings are discussed further.

Flat collar bearing. Consider a flat single collar bearing as shown in Figure 5.20. The bearing is subjected to axial thrust force of magnitude F_{ax}.

Let

p = normal pressure over an area
r_o = outer radius of the collar
r_i = inner radius of the collar
N = speed of the shaft, and
μ = coefficient of friction between the two surfaces

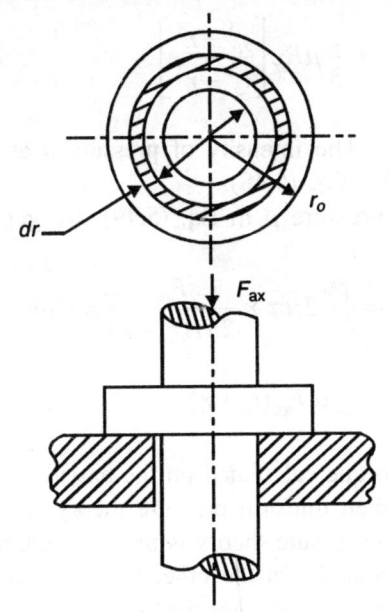

FIGURE 5.20 Flat collar bearing.

Consider an element of width dr of the collar at radius r. The area of the surface of the elemental ring is $2\pi r dr$.

The frictional force: $\quad F' = \mu \times$ pressure $\times$ elemental area

$$= \mu p \times 2\pi r dr$$

and frictional torque about the shaft axis:

$$dT = dF' \times r$$
$$= 2\mu p \pi r^2 dr$$

or total frictional torque:
$$T = \int_{r_i}^{r_o} 2\mu p \pi r^2 dr \qquad (5.49)$$

The total frictional torque can be calculated for the following two cases:

(a) Uniform pressure theory. Considering the assumption that pressure is uniform over the contacting surface. The intensity of pressure is given as

$$p = \frac{F_{ax}}{\pi(r_o^2 - r_i^2)} \quad (5.50)$$

Substituting the value of pressure p in Eq. (5.49),

Frictional torque:
$$T = \int_{r_i}^{r_o} 2\mu\pi \times \frac{F_{ax}}{\pi(r_o^2 - r_i^2)} \times r^2 dr \quad (5.51)$$

$$= \frac{2}{3}\mu F_{ax}\left[\frac{r_o^3 - r_i^3}{r_o^2 - r_i^2}\right]$$

(b) Uniform wear theory. The intensity of pressure p at a radius r of the collar for uniform wear theory is given by Eq. (5.48).

Substituting the value of pressure p in Eq. (5.49), we get

Frictional torque:
$$T = \int_{r_i}^{r_o} 2\mu\pi \times \frac{F_{ax}}{2\pi r(r_o - r_i)} \times r^2 dr \quad (5.52)$$

$$= \frac{1}{2}\mu F_{ax}(r_o + r_i)$$

Comparison of frictional torque calculated on the basis of the above two theories reveal that frictional torque is more when uniform pressure theory is used. Thus while calculating power loss in a bearing, uniform pressure theory is mostly used, though the actual power loss will be little less than that calculated. On the other hand, in friction clutches, uniform wear theory predicts lower torque transmission capacity.

Conical collar bearing. A conical collar bearing, which is in the shape of frustum of cone, is shown in Figure 5.21. The bearing is subjected to an axial thrust force of magnitude F_{ax}. Consider an element of width dr of the collar at a radius r. The area of the surface of the collar element ring is $2\pi r dr/\sin\alpha$.

The frictional force:
$$dF' = \mu p \times 2\pi r \frac{dr}{\sin\alpha}$$

and the frictional torque:
$$dT = dF' \times r = \mu p \times 2\pi r^2 \frac{dr}{\sin\alpha}$$

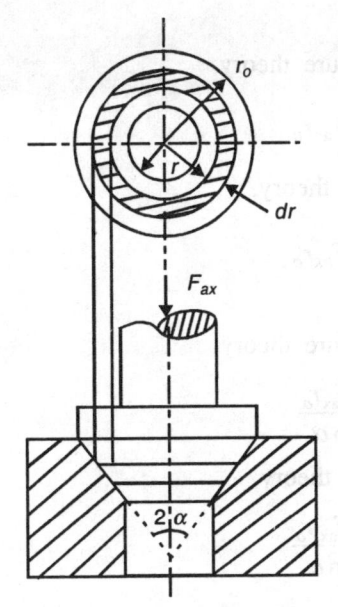

FIGURE 5.21 Conical collar bearing.

Total frictional torque:
$$T = \int_{r_i}^{r_o} \mu p \times \frac{2\pi r^2}{\sin\alpha} dr \tag{5.53}$$

where α = semi-cone angle of the conical bearing.

The expressions of frictional torque can be obtained for the following two cases:

(a) Uniform pressure theory

$$\text{Frictional torque: } T = \frac{2\mu F_{ax}}{3\sin\alpha}\left[\frac{r_o^3 - r_i^3}{r_o^2 - r_i^2}\right] \tag{5.54}$$

(b) Uniform wear theory

$$\text{Frictional torque: } T = \frac{\mu F_{ax}}{2\sin\alpha}[r_o + r_i] \tag{5.55}$$

Comparing the frictional torque expressions of flat collar bearing and conical collar bearing, we find that the frictional torque of conical collar bearing is $1/\sin\alpha$ times that of a flat collar bearing.

5.8.2 Pivot Bearing

A flat or conical bearing surface provided at the end of the shaft without any collar is known as **pivot bearing** (See Figure 5.18). The expression of frictional torque can be derived on the lines similar to those of the collar bearing except that inner radius r_i is equal to zero.

1. *Flat pivot*
   (a) For uniform pressure theory,

   Torque: $T = \dfrac{2}{3}\mu F_{ax} r_o$

   (b) For uniform wear theory,

   Torque: $T = \dfrac{1}{2}\mu F_{ax} r_o$

2. *Conical pivot*
   (a) For uniform pressure theory,

   Torque $T = \dfrac{2}{3}\dfrac{\mu F_{ax} r_o}{\sin \alpha}$

   (b) For uniform wear theory,

   Torque $T = \dfrac{1}{2}\dfrac{\mu F_{ax} r_o}{\sin \alpha}$

EXAMPLE 5.8 The thrust of a propeller shaft in a marine engine is taken up by a number of collars integral with the shaft which is 300 mm in diameter. The thrust on the shaft is 200 kN and the speed is 75 rpm. Taking coefficient of friction equal to 0.05 and assuming intensity of pressure as uniform and equal to 0.3 MPa, find the external diameter of the collar and the number of collars required, if the power lost in friction is not to exceed 16 kW.

Solution

Torque:
$$T = \dfrac{60 \times P}{2\pi N} = \dfrac{60 \times 16 \times 10^3}{2\pi \times 75} = 2037.18 \text{ Nm}$$

Also
$$T = \dfrac{2}{3}\mu F_{ax}\left[\dfrac{r_o^3 - r_i^3}{r_o^2 - r_i^2}\right]$$

or
$$2037.18 = \dfrac{2}{3} \times 0.05 \times 200 \times 10^3 \left[\dfrac{(r_o - r_i)(r_o^2 + r_o r_i + r_i^2)}{(r_o + r_i)(r_o - r_i)}\right]$$

or $\quad 0.3055(r_o + r_i) = (r_o^2 + r_o r_i + r_i^2)$

or $\quad 0.3055(r_o + 0.15) = (r_o^2 + 0.15 r_o + 0.0225)$

or $\quad r_o^2 - 0.1555 r_o - 0.0233 = 0$

or $\quad r_o = 0.249$ m or 249 mm

External diameter: $\quad d_o = 2r_o = 498 \text{ mm}$ Ans.

Number of collars:
$$n = \frac{F_{ax}}{p\pi(r_o^2 - r_i^2)}$$

$$= \frac{200 \times 10^3}{0.3 \times 10^6 \pi (0.249^2 - 0.15^2)}$$

$= 5.37$, say **6 numbers** Ans.

EXAMPLE 5.9 In a flat collar thrust bearing the inner and outer radii are 120 mm and 72 mm respectively. The total axial thrust is 60 kN and the intensity of pressure is 0.25 MPa. If the coefficient of friction is 0.05 and shaft rotates at 600 rpm, determine the power lost in overcoming the friction. Also determine the number of collars required to withstand the axial thrust.

Solution Assuming uniform pressure theory,

Frictional torque:
$$T = \frac{2}{3}\mu F_{ax}\left[\frac{r_o^3 - r_i^3}{r_o^2 - r_i^2}\right]$$

$$= \frac{2}{3} \times 0.05 \times 60 \times 10^3 \left[\frac{0.12^3 - 0.072^3}{0.12^2 - 0.072^2}\right]$$

$= 294$ Nm

Power lost:
$$P = \frac{2\pi NT}{60 \times 1000} = \frac{2\pi \times 600 \times 294}{60 \times 1000} = 18.47 \text{ kW} \quad \text{Ans.}$$

Number of collars:
$$n = \frac{F_{ax}}{p \times \pi(r_o^2 - r_i^2)}$$

$$= \frac{60 \times 10^3}{0.25 \times 10^6 \times \pi (0.12^2 - 0.072^2)}$$

$= 8.29$, say **9 numbers** Ans.

5.9 FRICTION CLUTCHES

A friction clutch transmits power under the influence of friction contact between two or more members. It generally has two or more rotating concentric surfaces with at least one surface lined with a friction material. When these surfaces are faced against each other firmly, a tangential friction force is produced between them which transmits torque from the input shaft to the output shaft. When a friction clutch is engaged, its members tend to rotate as a single unit. However, under certain conditions, namely when these members are not in equilibrium, they may have relative motion. This phenomenon is called **slip.** In a power transmission system, the slippage of clutch is undesirable for obvious reasons.

The friction clutch offers the following advantages:

(i) It can be easily engaged or disengaged at high speed.
(ii) It enables the driver to pick up and accelerate gradually without any major shock.
(iii) The friction element gets slipped when the torque transmitted through it exceeds a safe value, thus acting as a safety device as well.

The selection of a suitable material for friction lining primarily depends upon the torque capacity and operating conditions. For a clutch, the friction material must meet the following requirements:

(i) It must have a high coefficient of friction. It should be between 0.2 and 0.35 for a dry clutch and between 0.08 and 0.15 for a lubricated clutch.
(ii) It should not get affected by moisture or oil.
(iii) It should have the requisite strength and wear resistance.
(iv) It must have low thermal expansion and high heat-soaking capacity to prevent thermal distortion and heat spotting.

The most commonly used friction materials are wood, woven and moulded asbestos, corks, cast iron, leather, felt, phosphor bronze and powder metals.

5.9.1 Plate Clutches

In plate clutches, the friction surfaces are always kept normal to the axis of the connecting shaft and are brought in contact by an axial force.

Figure 5.22 shows the schematic diagrams of a single plate and multiple plate clutches consisting of flywheel, clutch plate with lining, pressure plate, and thrust springs. In a plate clutch, on the shaft of a prime mover, a flywheel or disc is mounted. The clutch plate is generally free to slide on the splined shaft. In an ordinary situation, a clutch always remains engaged. The axial force, which is introduced through springs, is applied to the pressure plate which keeps the pressure plate, the clutch plate and the flywheel surfaces in contact to transmit power.

When the clutch pedal is pressed down, the pressure on the pressure plate is released and it moves back slightly and relieves pressure on the clutch plate. The clutch plate then slips back from the flywheel and the driven shaft is disconnected.

When pressure on the clutch pedal is released, it builds up pressure on the pressure plate which ultimately forces the pressure plate and the clutch plate to bring them in contact with the flywheel surface and the power transmission is restored.

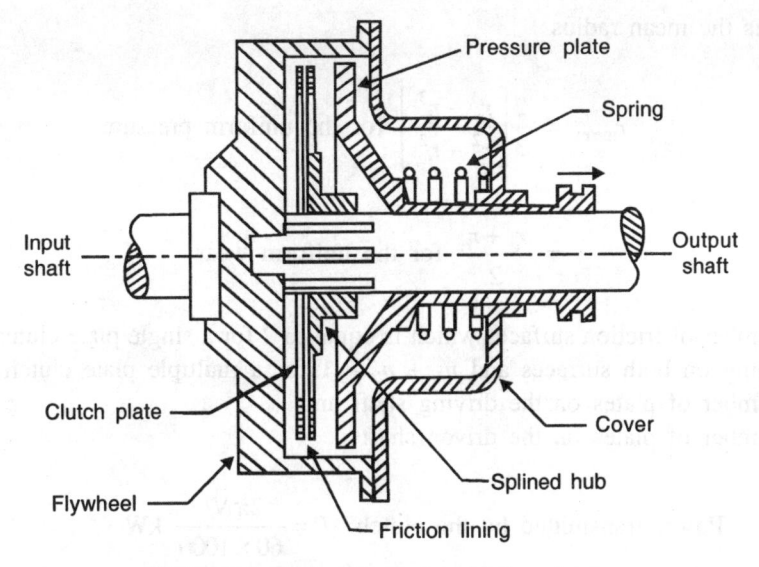

FIGURE 5.22(a) Single plate friction clutch.

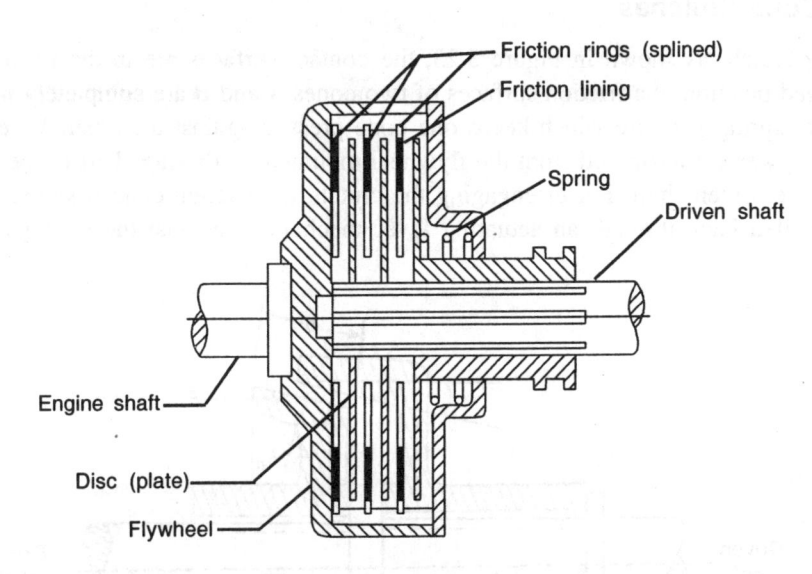

FIGURE 5.22(b) Multiple plate friction clutch.

The power or torque transmission capacity of a friction clutch depends upon the friction force, radius at which it acts and the number of friction surfaces. Kinematically, a friction clutch is equivalent to a multi-collar friction problem; hence the friction torque can be computed on lines similar to collar friction.

Therefore, friction torque:

$$T = \mu n F_{ax} r_{mean} \tag{5.56}$$

where r_{mean} is the mean radius

$$r_{mean} = \frac{2}{3}\left[\frac{r_o^3 - r_i^3}{r_o^2 - r_i^2}\right] \text{ for the uniform pressure}$$

$$= \frac{r_o + r_i}{2} \text{ for the uniform wear}$$

and n = number of friction surfaces which is equal to 2 for a single plate clutch with friction lining on both surfaces and $n_1 + n_2 - 1$ for a multiple plate clutch.

with n_1 = number of plates on the driving shaft and
n_2 = number of plates on the driven shaft

Power transmitted by the clutch: $P = \dfrac{2\pi NT}{60 \times 1000}$ kW (5.57)

5.9.2 Cone Clutches

In a cone clutch, as shown in Figure 5.23, the contact surfaces are in the form of cones. In the engaged position, the friction surfaces of two cones A and B are completely in contact due to external spring pressure which keeps one cone pressed against the other. When a clutch is engaged, power is transmitted from the flywheel mounted on driving shaft to the friction cone keyed to the driven shaft. For disengaging the clutch, the friction cone B keyed to the driven shaft is pulled back through an actuating lever mechanism against the spring force.

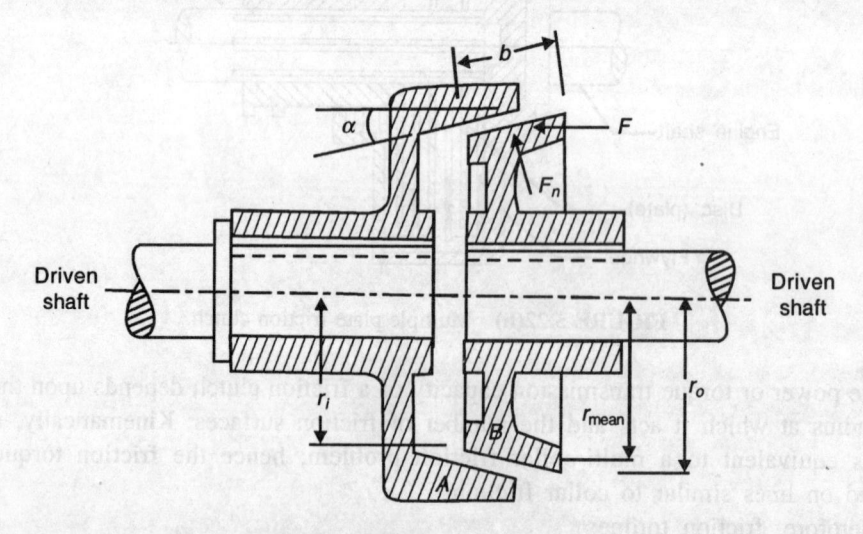

FIGURE 5.23 Cone clutch.

Friction 169

In cone clutches, the normal force on the contact surfaces is larger than that on the plate clutches. Therefore, for the same axial thrust, the cone clutches can transmit more power. However, a cone clutch exposed to dust and dirt may create some difficulty in disengagement. Further, score marks on the friction lining surface may impede sliding.

The power transmission capacity can be computed by treating a cone clutch as a conical collar bearing. In uniform wear theory, the torque transmitted by clutch is:

$$T = \mu' F_{ax} r_{mean} \tag{5.58}$$

where
μ' = equivalent coefficient of friction and ($= \mu/\sin\alpha$) and
r_{mean} = mean radius [$= (r_o + r_i)/2$]

The equation of axial force $F_{ax} = F_n \sin\alpha$ is valid under steady operation after the clutch is engaged (See Figure 5.24). However, during engagement, there is an additional component of force which is mainly due to friction force as shown in Figure 5.24.

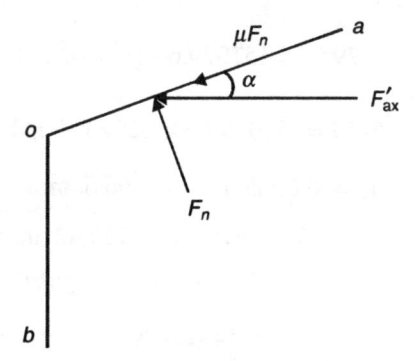

FIGURE 5.24 Forces on cone clutch.

This component resists the action of engagement. Therefore, the axial force required for engaging the clutch increases. Thus

$$F'_{ax} = F_n \sin\alpha + \mu F_n \cos\alpha$$

However, various experimental results have indicated that $\mu F_n \cos\alpha$ is only 25 per cent effective. Therefore, the actual axial force required to keep the clutch engaged is:

$$F'_{ax} = F_n \left(\sin\alpha + \frac{\mu \cos\alpha}{4}\right) \tag{5.59}$$

EXAMPLE 5.10 A single plate clutch, effective on both sides, is required to transmit 25 kW at 3000 rpm. Determine the outer and inner radii of frictional surface, if the coefficient of friction is 0.25, the ratio of radii is 1.25 and the maximum pressure is not to exceed 0.1 MPa. Also determine the axial thrust to be provided by the springs. Assume uniform wear condition.

Solution In case of uniform wear, the pressure is maximum at the inner radius, therefore

axial force: $F_{ax} = 2\pi p\, r_i\, (r_o - r_i)$

Angular speed: $$\omega = \frac{2\pi N}{60} = \frac{2\pi \times 3000}{60} = 100\pi \text{ rad/s}$$

Torque: $$T = \frac{P \times 10^3}{\omega} = \frac{25 \times 10^3}{100\pi} = 79.57 \text{ Nm}$$

Frictional torque transmitted by the clutch:

$$T = \frac{\mu F_{ax}}{2} \times n \times (r_o + r_i)$$

where n = number of friction surfaces (= 2)

or $$79.57 = \frac{0.25}{2} \times 2\pi \times 0.1 \times 10^6 \times r_i(r_o - r_i) \times 2 \times (r_o + r_i)$$

or $$79.57 = 157079.6 \, r_i \, (r_o^2 - r_i^2)$$

or $$79.57 = 157079.6 \, r_i[(1.25 \, r_i)^2 - r_i^2]$$

or r_i = 0.0965 m or **96.5 mm** Ans.

r_o = 1.25 × 96.5 or **120.62 mm** Ans.

Axial thrust: $F_{ax} = 2\pi \times 0.1 \times 10^6 \times 0.0965 \, (0.12062 - 0.0965)$

= **1462.4 N** Ans.

EXAMPLE 5.11 A plate clutch has three discs on the driving shaft and two discs on the driven shaft. The outside diameter of the contact surfaces is 240 mm and inside diameter is 120 mm. Assuming uniform pressure and coefficient of friction 0.3, determine the total axial force on the springs to transmit 25 kW at 1500 rpm. If there are 6 springs each of stiffness 10 kN/m and each of contact surface has worn away by 0.5 mm, what is the maximum power that can be transmitted at the same speed with uniform wear?

Solution

Torque: $$T = \frac{P \times 10^3}{\omega}$$

Angular speed: $$\omega = \frac{2\pi N}{60} = \frac{2\pi \times 1500}{60} = 157.0 \text{ rad/s}$$

$$\therefore T = \frac{25 \times 10^3}{157} = 159.23 \text{ Nm}$$

Number of pairs of friction surfaces:
$$n = n_1 + n_2 - 1$$
$$= 3 + 2 - 1$$
$$= 4$$

For uniform pressure,

Torque:
$$T = \frac{2}{3} \mu F_{ax} \times n \times \left[\frac{r_o^3 - r_i^3}{r_o^2 - r_i^2} \right]$$

$$159.23 = \frac{2}{3} \times 0.3 \times F_{ax} \times 4 \times \left[\frac{0.12^3 - 0.06^3}{0.12^2 - 0.06^2} \right]$$

or

Axial thrust: F_{ax} = **1421.7 N**   Ans.

Total wear = 0.5 × number of friction surfaces
$$= 0.5 \times 4 \times 2$$
$$= 4 \text{ mm}$$

Reduction in spring force = $\dfrac{4}{1000} \times 10 \times 10^3 \times 6 = 240$ N

New axial thrust: F'_{ax} = 1421.7 − 240 = 1181.7 N

For uniform wear, torque $T' = \dfrac{\mu n F'_{ax}}{2} (r_o + r_i)$

$$= \frac{0.3 \times 4 \times 1181.7}{2} \times (0.12 + 0.06)$$

$$= 127.62 \text{ Nm}$$

∴ Power: $P' = T' \times \omega = \dfrac{127.62 \times 157}{1000} =$ **20 kW**   Ans.

EXAMPLE 5.12 A multiple plate clutch transmits 55 kW of power at 1800 rpm. The coefficient of friction is 0.1. The inner radius is 80 mm and is 0.65 times the outer radius. If the intensity of pressure is not to exceed 160 kN/m², determine the number of pair of friction surfaces needed to transmit the required torque.

Solution

Angular speed:
$$\omega = \frac{2\pi N}{60} = \frac{2\pi \times 1800}{60} = 60\pi \text{ rad/s}$$

Torque:
$$T = \frac{P}{\omega} = \frac{55 \times 10^3}{60\pi} = 291.78 \text{ Nm}$$

$$r_i = 0.65 \, r_o \quad \text{or} \quad r_o = \frac{r_i}{0.65} = \frac{80}{0.65} = 123 \text{ mm}$$

Axial force:
$$F_{ax} = 2\pi p r_i (r_o - r_i) \quad \text{(for maximum pressure, } r = r_i\text{)}$$
$$= 2\pi \times 160 \times 10^3 \times 0.08 \times (0.123 - 0.08)$$
$$= 3458.26 \text{ N}$$

Assuming uniform wear, the torque capacity is:

$$T = \frac{1}{2} \mu n F_{ax} (r_o + r_i)$$

$$291.78 = \frac{1}{2} \times 0.1 \times n \times 3458.26 \times (0.123 + 0.08)$$

or Number of friction surfaces: $n = 8.3$, **say 9** Ans.

EXAMPLE 5.13 A cone clutch with semi-cone angle 12.5° transmits 15 kW at 500 rpm. The width of the friction surface is 40 per cent of mean diameter. If the normal pressure between the surfaces in contact is not to exceed 120 kN/m², determine

(i) the outer and inner radii of the cone
(ii) the axial force required to engage the clutch.

Solution

Face width:
$$b = \frac{r_o - r_i}{\sin \alpha} = 0.4 \, d_m = 0.4 (r_o + r_i)$$

or
$$r_o - r_i = 0.4 (r_o + r_i) \sin 12.5°$$

or
$$r_o = 1.189 \, r_i \qquad \text{(i)}$$

For uniform wear theory, the intensity of pressure is maximum at inner radius. Therefore, the torque capacity:

$$T = \frac{\mu F_{ax}}{2 \sin \alpha} (r_o + r_i)$$

where
$$F_{ax} = \text{axial force}$$
$$= 2\pi p r_i (r_o - r_i)$$

or
$$T = \frac{\mu}{2\sin\alpha} \times 2\pi p r_i (r_o - r_i) \times (r_o + r_i)$$

or
$$\frac{60P}{2\pi N} = \frac{\mu\pi p}{\sin\alpha} \times r_i \times (r_o^2 - r_i^2)$$

or
$$\frac{60 \times 15 \times 10^3}{2\pi \times 500} = \frac{0.25 \times \pi \times 120 \times 10^3}{\sin 12.5°} \times r_i (r_o^2 - r_i^2)$$

or
$$5.4825 \times 10^{-4} = r_i (r_o^2 - r_i^2) \qquad \text{(ii)}$$

Solving Eqs. (i) and (ii), we get

$r_i = 0.1098$ m or **109.8 mm** Ans.

$r_o = 1.189 \times 109.8$ or **130.5 mm** Ans.

Axial force: $F_{ax} = 2\pi p r_i (r_o - r_i)$

$= 2\pi \times 120 \times 0.1098 \times (0.1305 - 0.1098)$

$= \mathbf{1.7137\ kN}$ Ans.

EXAMPLE 5.14 A cone clutch is to transmit 20 kW at 900 rpm. The mean diameter is limited to 250 mm and the slope of the face is 15°. Determine the necessary axial force to engage the clutch and face width if the coefficient of friction is 0.25 (Assume a contact pressure of 80 kN/m²).

Solution

Torque:
$$T = \frac{P \times 60}{2\pi N} = \frac{20 \times 10^3 \times 60}{2\pi \times 900} = 212.2\ \text{Nm}$$

Also
$$T = \mu F_n\, r_{mean}$$
or

Normal force:
$$F_n = \frac{T}{\mu r_{mean}} = \frac{212.2}{0.25 \times 0.125} = 6790.4\ \text{N}$$

Axial force during engagement:

$$F_{ax} = F_n (\sin\alpha + \mu \cos\alpha)$$

$= 6790.4\,(\sin 15° + 0.25 \times \cos 15°)$

$= \mathbf{3397.2\ N}$ Ans.

Face width: $$b = \frac{F_n}{2\pi r_{mean} p} = \frac{6790.4}{2\pi \times 0.125 \times 80 \times 10^3}$$
$$= 0.108 \text{ m} \quad \text{or} \quad \mathbf{108 \text{ mm}} \qquad \text{Ans.}$$

EXAMPLE 5.15 In a cone clutch, the contact surfaces have an effective mean diameter of 80 mm. The semi–cone angle is 15° and coefficient of friction is 0.25. Find the torque required to produce slipping of the clutch if the axial force applied is 200 N.

This clutch is employed to connect an electric motor running uniformly at 1000 rpm with a flywheel which is initially stationary. The flywheel has a mass of 14 kg and its radius of gyration is 150 mm. Calculate the time required for the flywheel to attain full speed and also the energy lost in slippage of the clutch.

Solution For uniform wear theory,

Torque:
$$T = \frac{\mu F_{ax}}{\sin \alpha} \times \frac{r_o + r_i}{2}$$

$$= \frac{0.25 \times 200}{\sin 15°} \times 0.04$$

$$= 7.72 \text{ Nm}$$

For flywheel, $T = I\alpha = mk^2 \times \alpha$

or

Angular acceleration, $\alpha = \dfrac{7.72}{14 \times 0.15^2} = 24.5 \text{ rad/s}^2$

or

Time taken, $t = \dfrac{\dfrac{2\pi}{60} \times 1000}{24.5} = 4.27 \text{ s}$

Let

θ_m = angle turned by the motor before slippage

and θ_f = angle turned by the flywheel before slippage

Then
$$\theta_m = \left(\frac{2\pi}{60} \times 1000\right) \times 4.27 = 447.15 \text{ rad}$$

Since flywheel accelerates from rest to 1000 rpm,

$$\theta_f = \frac{0 + \theta_m}{2} = \frac{447.15}{2} = 223.575 \text{ rad}$$

Energy lost = $T(\theta_m - \theta_f)$

$$= \frac{7.72 \times (447.15 - 223.575)}{1000}$$

$$= 1.726 \text{ kJ}$$ Ans.

5.10 FRICTION IN TURNING PAIRS

Mechanism is an assemblage of kinematic links which are joined by kinematic pair. The turning pair is most commonly used kinematic pair in which a circular pin supported in a bearing allows turning or revolving motion between two links. When the pin is at rest in bearing and contact surfaces are frictionless, the reaction of the bearing on the pin will act at point A in upward direction, as shown in Figure 5.25(a). However, when the pin rotates, say in counter clockwise direction, the point of contact is shifted towards left to point B [Figure 5.25(b)]. At this point, the pin is subjected to two forces besides external force F:

(i) normal reaction R_n
(ii) frictional force μR_n, which acts in the direction opposite to that of rotation and is tangential at point B.

The resultant reaction R is inclined at an angle ϕ with normal reaction R_n.

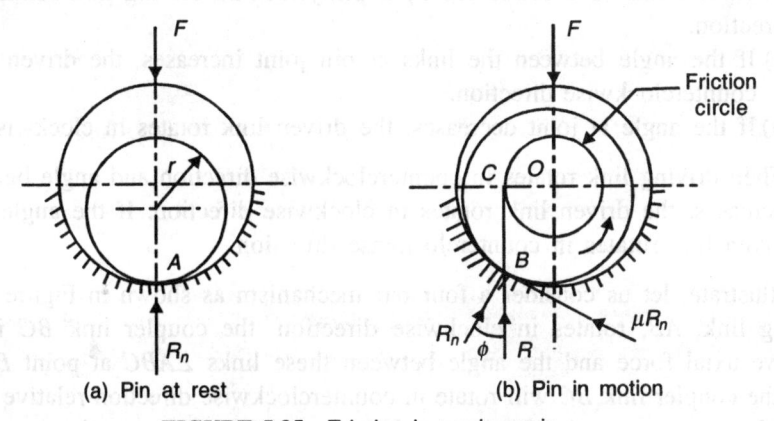

(a) Pin at rest   (b) Pin in motion

FIGURE 5.25 Friction in turning pair.

Therefore, the pin is subjected to two forces—external force F and reaction force R. For equilibrium, the forces F and R must be equal in magnitude. However, since these two forces are parallel to each other at distance OC, they constitute a couple.

Frictional torque: $T = F \times OC$

$$= F \times r \sin \phi$$

For small angle, $\sin \phi \simeq \tan \phi = \mu$

Therefore, $$T = \mu F r \tag{5.60}$$

Thus the effect of friction on a turning pair is to displace the reaction force through a distance equal to $r \sin \phi$ $(= \mu r)$ such that it is tangential to circle drawn with radius OC. This circle is commonly known as **friction circle**.

5.11 FRICTION AXIS

When a link of a mechanism is joined to another link by a frictionless turning pair, the line of action of the force passes through the line joining the centres of pins. However, if the friction is taken into account, the force does not pass through the centre of the pin but it is tangential to the friction circle of the pin. Hence the line of action of the force is common tangent to the friction circles of two pins. Since there are four possible common tangents to a given pair of friction circles, it is necessary to determine the following information:

(i) The direction of external force acting on the link
(ii) The direction of rotation of one link relative to other connecting link

Based upon the above information, a common tangent, which gives a friction couple in the direction opposite to that of a couple producing the motion, is selected. This line of action of the force is called **friction axis** of the link.

The direction of motion of the link relative to another link can be determined by following criteria:

1. When two links are connected by a pin joint and driving link rotates in clockwise direction.
    (a) If the angle between the links at pin joint increases, the driven link rotates in counterclockwise direction.
    (b) If the angle at joint decreases, the driven link rotates in clockwise direction.
2. When driving link rotates in counterclockwise direction and angle between the links increases, the driven link rotates in clockwise direction. If the angle decreases, the driven link rotates in counterclockwise direction.

To illustrate, let us consider a four bar mechanism as shown in Figure 5.26(a). When the driving link, AB, rotates in clockwise direction, the coupler link BC is subjected to compressive axial force and the angle between these links $\angle ABC$ at point B is increasing; therefore the coupler link BC will rotate in counterclockwise direction relative to driving link AB. Therefore tangent to the friction circle at point B will be on the upper side to give clockwise friction couple. Similarly, when link CD is rotating in clockwise direction, the angle $\angle DCB$ decreases and the link BC relative to CD will rotate in clockwise direction. Therefore, tangent to the friction circle will be on the lower side to give counterclockwise friction couple. The tangent to these two circles is known as **friction axis**.

Referring to the another position of a four bar mechanism as shown in Figure 5.26(b), the coupler link BC is subjected to tension force and angle $\angle ABC$ at point B is increasing, so link BC will rotate in counterclockwise direction and the friction couple will be on the

Friction

Friction axis

Friction circle

(a)

Friction axis

(b)

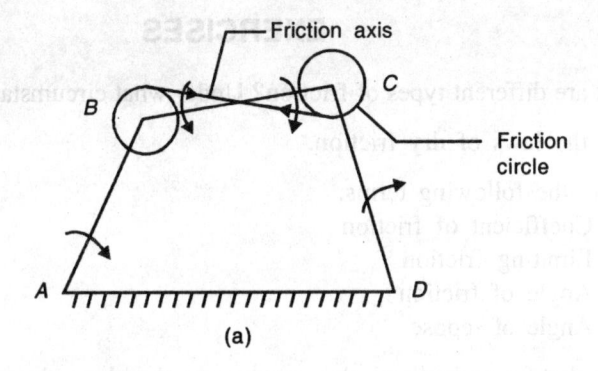

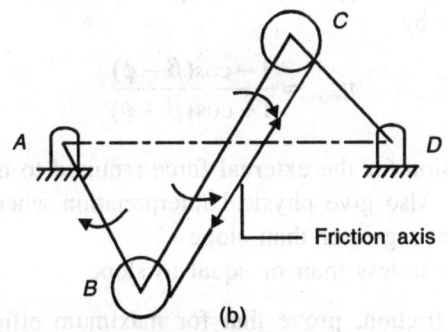

FIGURE 5.26 Friction axis of four bar mechanism.

lower side of the link. Similarly, with reference to link *CD*, the coupler *BC* rotates in clockwise direction such that the friction couple at pin *C* opposes the motion. The resulting friction axis of the link *BC* is shown in Figure 5.26(b).

The friction axes of a slider-crank mechanism for various positions of crank are shown in Figure 5.27. In slider-crank mechanism, there is another method to know the position of friction axis. Accordingly, out of four possible tangents one that gives the least intercept (OC) from origin *O* is the required friction axis of the coupler link.

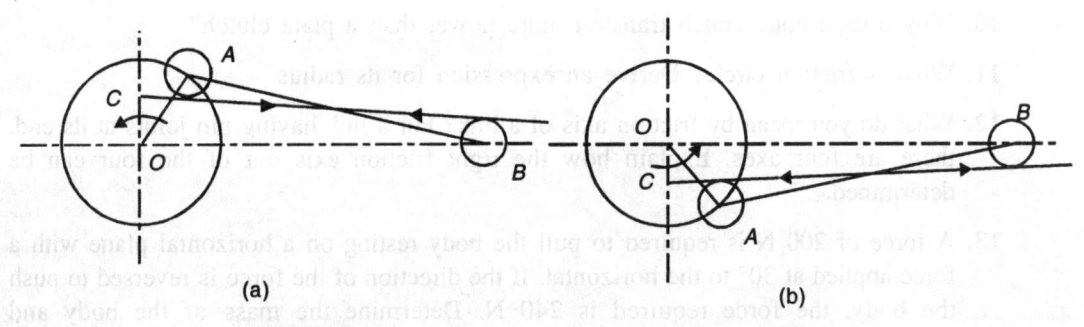

(a)   (b)

FIGURE 5.27 Friction axes of slider-crank mechanism.

EXERCISES

1. What are different types of friction? Under what circumstances each type will occur?

2. State the laws of dry friction.

3. Define the following terms:
   (i) Coefficient of friction
   (ii) Limiting friction
   (iii) Angle of friction
   (iv) Angle of repose

4. Prove that for an inclined plane motion up the plane, the condition for the maximum efficiency is given by

$$\eta_{max} = \frac{1-\cos(\beta - \phi)}{1-\cos(\beta + \phi)}$$

5. Derive the expression for the external force required to move down a weight W on an inclined plane. Also give physical interpretation when
   (i) friction angle is greater than slope
   (ii) friction angle is less than or equal to slope

6. Neglecting collar friction, prove that for maximum efficiency of square threaded screw jack, the helix angle α is given by

$$\alpha = \frac{\pi}{4} - \frac{\phi}{2}$$

7. Derive an expression for frictional torque of a truncated conical pivot bearing assuming uniform wear.

8. What do you understand by uniform pressure theory and uniform wear theory? Which theory is the most suitable for (i) bearing design and (ii) clutch design?

9. Discuss the suitability of friction clutch for high torque transmission.

10. Why does a cone clutch transmit more power than a plate clutch?

11. What is friction circle? Derive an expression for its radius.

12. What do you mean by friction axis of a link? For a link having pin joints at its end, there are four axes. Explain how the right friction axis out of the four can be determined.

13. A force of 200 N is required to pull the body resting on a horizontal plane with a force applied at 30° to the horizontal. If the direction of the force is reversed to push the body, the force required is 240 N. Determine the mass of the body and coefficient of friction between the body and the surface. Also determine minimum force required to pull the body if the inclination can be varied.

[**Ans:** 122.3 kg,   0.1575,   186.65 N]

14. A block weighing 10 kN is to be raised against a surface which is inclined at 60° with the horizontal force by means of a 15° inclined wedge as shown in Figure E5.1. Find the magnitude of horizontal force which will just start the block to move, if the coefficient of friction between all surfaces of contact is 0.2.

[Ans: 6 kN]

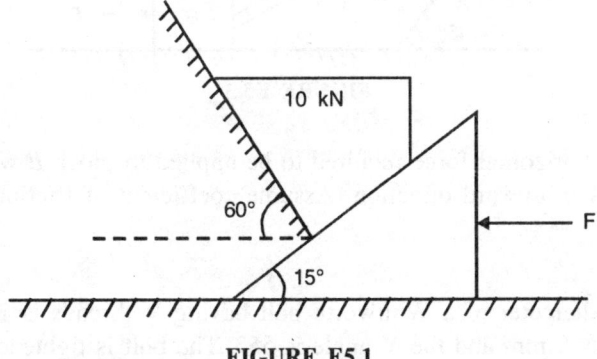

FIGURE E5.1

15. A body weighing 3 kN is required to move up a rough plane by a force being applied parallel to the plane. The inclination of the plane is such that when same body is kept on a perfectly smooth plane inclined at the same angle, a force of 600 N applied at an inclination of 30° to the plane keeps the body in equilibrium. Determine the force required to move up the body on rough plane (Assume $\mu = 0.3$).

[Ans: 1406 N]

16. Two bodies C and D weighing 1 kN and W kN respectively are resting on two inclined rough planes OA and OB. These bodies are connected by a horizontal link CD as shown in Figure E5.2. Determine the maximum and minimum values of weight W for which the equilibrium of body exist (Take $\mu = 0.364$).

[Ans: 12.16 kN, 391 N]

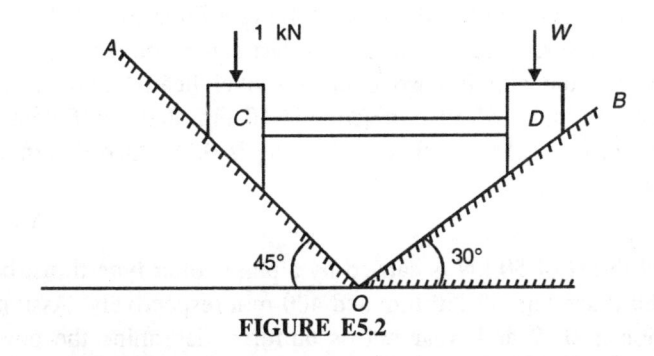

FIGURE E5.2

17. A block A weighing 100 N rests on a rough inclined plane whose inclination to the horizontal is 45°. This block is connected to other block B weighing 300 N which rests on a rough horizontal plane by a weightless rigid bar inclined at 30° to the horizontal as shown in Figure E5.3.

FIGURE E5.3

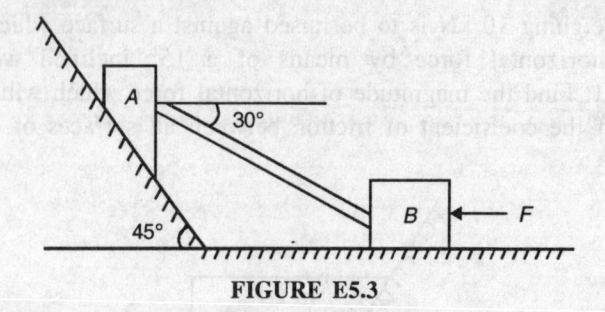

Find the horizontal force required to be applied to block *B* which causes to move the block *A* in upward direction (Assume coefficient of friction at both surfaces as 0.268.).

[Ans: 180.4 N]

18. The mean diameter of a Whitworth bolt having V threads is 25 mm. The pitch of the thread is 5 mm and the V angle is 55°. The bolt is tightened by screwing a nut whose mean radius of the bearing surface is 25 mm. If the coefficient of friction for nut and bolt is 0.1 and that for nut and bearing surface is 0.16, find the torque required and the force to be applied at the end of 500 mm long spanner when the load on the bolt is 10 kN.

[Ans: 62.205 Nm, 124.41 N]

19. A vertical screw with single start square threads 50 mm mean diameter and 10 mm pitch is raised against a load of 5500 N by means of a hand wheel, the boss of which is threaded to act as a nut. The axial load is taken up by a thrust collar which supports the wheel boss and has a mean diameter of 65 mm. If the coefficient of friction is 0.15 for the screw and 0.18 for the collar and tangential force applied by each hand to the wheel is 140 N, find suitable diameter of the hand wheel.

[Ans: 441 mm]

20. Find the load that can be lifted by applying a force of 220 N at the end of 500 mm long lever of screw jack using single start square threads. The load does not rotate with the spindle and is carried on a swivel head having a bearing of 100 mm diameter. The pitch of 50 mm screw is 10 mm and coefficient of friction between nut and screw is 0.18 and that for bearing is 0.15. Also determine the efficiency of the jack.

[Ans: 7.8 kN, 11.28%]

21. An axial thrust of 50 kN is carried by a plain collar type thrust bearing having inner and outer diameters of 250 mm and 400 mm respectively. Assuming that coefficient of friction is 0.02 and wear rate is uniform, determine the power absorbed in the friction at a speed of 120 rpm.

[Ans: 2.04 kW]

22. A thrust bearing has contact surfaces 200 mm external diameter and 150 mm internal diameter. The coefficient of friction is 0.08, the total axial load is 3000 N and the intensity of pressure is 0.35 N/mm². Calculate the number of collars required and the power lost in the friction at 420 rpm.

[Ans: 7, 9.45 kW]

23. The thrust of a propeller shaft in a marine engine is taken up by a number of collars, integral with the shaft of 300 mm diameter. The axial thrust is 20 kN and the speed is 75 rpm. Taking $\mu = 0.05$ and uniform intensity of pressure as 0.3 N/mm², find the external diameter of the collar and number of collars required if the power lost in friction is not to exceed 16.5 kW.

[Ans: 210 mm, 10 collars]

24. A conical pivot supports a load of 2kN. The cone angle is 120° and intensity of pressure is not to exceed 0.3 N/mm². The external diameter is three times the internal diameter of the bearing surface. If $\mu = 0.06$ and the shaft rotates at 120 rpm, what will be the power lost? Also find the breadth of the cone.

[Ans: 2.53 kW, 161.7 mm]

25. A multiple plate clutch is to transmit 12 kW at 1500 rpm. The inner and outer radii for the plates are to be 50 mm and 100 mm respectively. The maximum axial spring force is restricted to 1 kN. Calculate the necessary number of pairs of surfaces if $\mu = 0.35$, assuming constant wear. What will be the necessary axial force?

[Ans: 3,970 N]

26. A plate clutch consists of a flat driven plate gripped between a driving plate and a pressure plate so that there are two active driving surfaces each having an inner diameter 200 mm and outer diameter 350 mm. Also $\mu = 0.4$ and the working pressure is limited to 170 kN/m². Assuming that the pressure is uniform, calculate the power which can be transmitted at 1000 rpm. If the clutch becomes worn so that the intensity of pressure is inversely proportional to the radius, the total axial force remains unchanged, calculate the revised power transmission capacity and the greatest intensity of pressure.

[Ans: 130 kW, 127 kW, 234 kN/m²]

27. A multiple plate friction clutch is designed to transmit 75 kW from an engine running at 2000 rpm. Assuming that the pressure distribution is uniform and limited to 150 kN/m² and that the inner and outer diameters of lining are 100 mm and 150 mm respectively, determine the necessary end thrust and number of plate. Take $\mu = 0.25$. If this clutch is used to transmit the power from the engine to a rotor which has mass of 1150 kg and radius of gyration of 200 mm, determine the time required for this rotor to reach 1500 rpm from standstill.

[Ans: 1.414 kN, 8 plates, $p = 144$ kN/m², 20.2 s]

28. A cone clutch is required to transmit 30 kW at 1200 rpm. The mean diameter of the surface is 250 mm and the cone angle is 25°. Assuming μ as 0.3 and normal pressure as 140 kN/m², determine the axial width of the conical surface and the required axial force.

[Ans: 54.1 mm, 1.32 kN]

29. A cone clutch with a cone angle 24° is required to transmit 10 kW at 750 rpm. The width of the face is one–fourth of the mean diameter and normal pressure between surfaces is 250 kN/m². Take $\mu = 0.2$ with uniform wear. Determine the main dimensions of the clutch.

[Ans: r_i = 115.7 mm, r_o = 121.83 mm, b = 29.69 mm]

30. The crank of a steam engine is 300 mm and the ratio of connecting rod to crank is 4.5. The journals at the cross head, crank pin and crank shafts are of diameters 80 mm, 120 mm, and 160 mm respectively. The coefficient of friction between the cross head and guides is 0.08 and for journal is 0.06. Draw the friction axis of connecting rod and crank if crank makes an angle of 45°.

MULTIPLE CHOICE QUESTIONS

1. The friction force is independent of
   (a) material
   (b) area of contact
   (c) normal reaction
   (d) surface finish

2. In an inclined plane friction with motion down the plane, the body moves down without any force for
   (a) $\phi > \alpha$
   (b) $\phi = \alpha$
   (c) $\phi < \alpha$
   (d) any value of ϕ

3. For the same pitch, the efficiency of a square threaded screw jack is
   (a) more than that of V thread
   (b) same as that of V thread
   (c) less than that of V thread
   (d) dependent of load on the jack

4. In case of pivot bearing, the wear is
   (a) maximum at the centre
   (b) zero at the centre
   (c) uniform throughout the contact area
   (d) minimum at maximum radius

5. The frictional torque produced due to friction for same shaft diameter, in conical bearing is
   (a) less than flat bearing
   (b) equal to that for flat bearing
   (c) more than that for flat bearing
   (d) be less or more depending upon cone angle

6. A friction clutch should be designed on the assumption of
   (a) uniform wear
   (b) uniform pressure
   (c) constant friction
   (d) any one

7. The ratio of frictional torque for conical collar bearing to that for flat collar bearing is proportional to
   (a) $\sin\alpha$
   (b) $\cos\alpha$
   (c) $\tan\alpha$
   (d) $1/\sin\alpha$

Friction

8. The equivalent coefficient of friction for a cone clutch is
   (a) $\mu \sin\alpha$
   (b) $\mu \tan\alpha$
   (c) $\mu/\sin\alpha$
   (d) $\mu/\tan\alpha$

9. The ratio of frictional torque produced for uniform wear to that for uniform pressure is
   (a) 1
   (b) $\dfrac{2}{3}$
   (c) $\dfrac{4}{3}$
   (d) $\dfrac{3}{4}$

10. The number of collars are provided to carry a fixed axial force in a flat collar bearing
    (a) to increase torque
    (b) to decrease torque
    (c) to decrease intensity of pressure
    (d) to increase intensity of pressure

11. Angle of repose as referred to friction is equal to
    (a) $\alpha = \tan^{-1}\mu$
    (b) $\alpha = \sin^{-1}\mu$
    (c) $\alpha = \cot^{-1}\mu$
    (d) $\alpha = \operatorname{cosec}^{-1}\mu$

12. The efficiency of screw jack depends on
    (a) pitch
    (b) coefficient of friction
    (c) load
    (d) all of the above

CHAPTER 6

Belts, Ropes and Chains

6.1 INTRODUCTION

The transmission of power from the prime mover to drive a machine is made either through flexible drive or a positive drive element. The flexible drive includes the following types:

(i) belt drive
(ii) rope drive
(iii) chain drive

These drives greatly simplify the construction of machine and allow a considerable flexibility in location of the driving and driven machine elements. The location tolerances are not as critical as found in the positive drives such as gears, power screws, clutches and so forth.

Flexible drive elements are simple in construction, run quieter and suitable for long distances. These drives require low maintenance and cost less compared to gear drive. Flexible drive also plays an important role in absorbing shock load and damping out the effect of vibration due to its long length and elastic properties. However, the important limitations of these drives are the following:

(i) low velocity ratio
(ii) Velocity ratio is not constant.
(iii) They require regular readjustment of centre distance to keep certain amount of initial tension.

The uses of belts and chains are grouped into three categories:

(i) power transmission
(ii) conveyor service
(iii) elevator service

The ropes are widely used for haulage purpose.

6.2 BELT DRIVE

A belt drive, used for the power transmission, usually consists of three elements—driving pulley, driven pulley and belt. Belt is wrapped around the pulleys with a certain amount of initial tension as shown in Figure 6.1. In a belt drive, the tangential force is transmitted from one pulley to another through a belt, thereby it transmits mechanical power between two shafts. When an unstretched belt is wrapped around the two pulleys located at a certain distance apart, the outer and inner faces of the belt are subjected to tension and compression stresses respectively. In between these two faces, there is a neutral section which remains neutral and is free from either type of stress. While analysing a belt drive the effective radius of the pulley is taken at this neutral section which is sum of the radius of the pulley and half of the belt thickness.

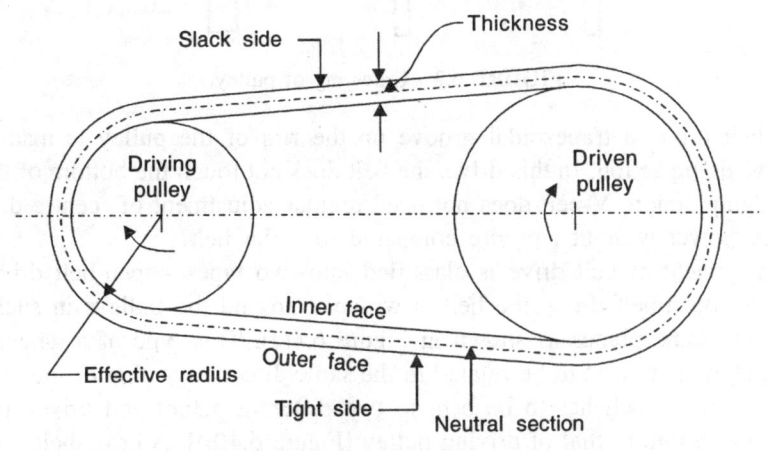

FIGURE 6.1 An open-belt drive.

In a belt drive, the belt section is either of rectangular section or a trapezoidal section (Figure 6.2). A belt of rectangular section is known as **flat belt** whereas that of a trapezoidal section is called **V-belt.**

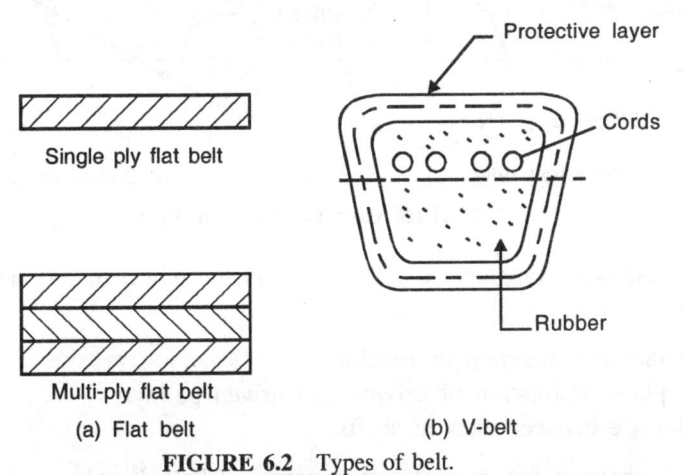

FIGURE 6.2 Types of belt.

In a flat belt drive, the rim of the pulley is slightly crowned to keep the belt running centrally on the pulley rim (Figure 6.3). The amount of crown height is approximately 1 per cent of the face width of pulley.

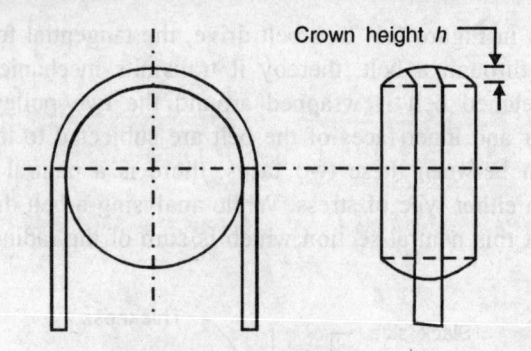

FIGURE 6.3 Crowning of pulley.

In a V-belt drive, a trapezoidal groove on the rim of the pulley is made to take the advantage of wedging action. In this drive, the belt does not touch the bottom of the V-groove. Owing to wedging action, V-belt does not need regular adjustment of centre distance and it transmits more power without any slip compared to a flat belt.

The arrangement of belt drive is classified into two types—open-belt drive and cross-belt drive. In an open-belt drive, the belt is wrapped around the pulleys in such a way that belt bends in the same planes as shown in Figure 6.4(a). This type of arrangement is used when driven pulley is desired to be rotated in the same direction as that of the driving pulley. In cross-belt drive, the belt has to be bent in two different planes and driven pulley rotates in the direction opposite to that of driving pulley [Figure 6.4(b)]. A cross-belt drive transmits more power compared to open-belt drive due to larger angle of contact. However, the life of the cross belt is shorter as it wears out more rapidly.

(a) Open-belt drive   (b) Cross-belt drive

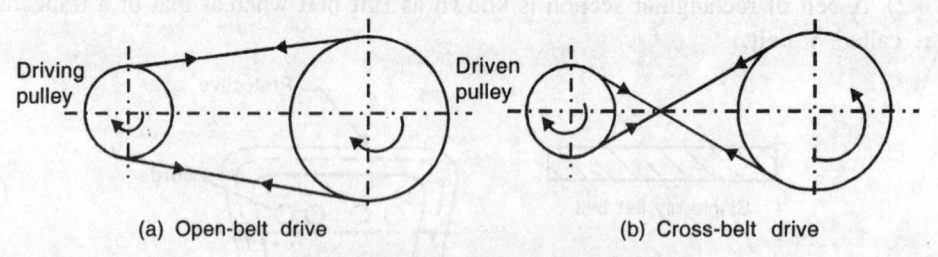

FIGURE 6.4 Types of belt drive.

Besides these two arrangements, a belt can transmit power in various other arrangements which primarily depend upon the following:

  (i) the required direction of rotation
 (ii) the plane of rotation of driving and driven pulley
(iii) the angle between axes of shafts

Figure 6.5 shows a few typical arrangements of belt drive.

One of the important limitations of the belt drive is loosening of the belt. Belts after some period of service may be stretched permanently which may lead to loosening of initial tension. To restore the initial tension, the centre distance between two shafts is increased.

(a) Quarter turn drive **(b) Right angled drive**

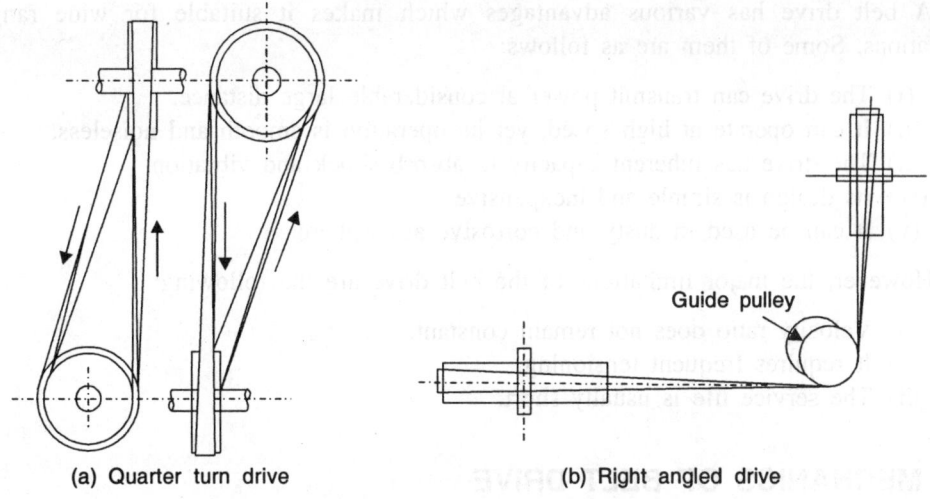

FIGURE 6.5 Typical arrangements of belt drive.

However, in those circumstances where the centre distance can not be increased due to physical limitation of the drive layout, the use of idler pulley is suggested. Such belts are subjected to reversed bending stresses when they pass through idler pulley. Some typical arrangements are shown in Figure 6.6.

(a) Gravity idler pulley

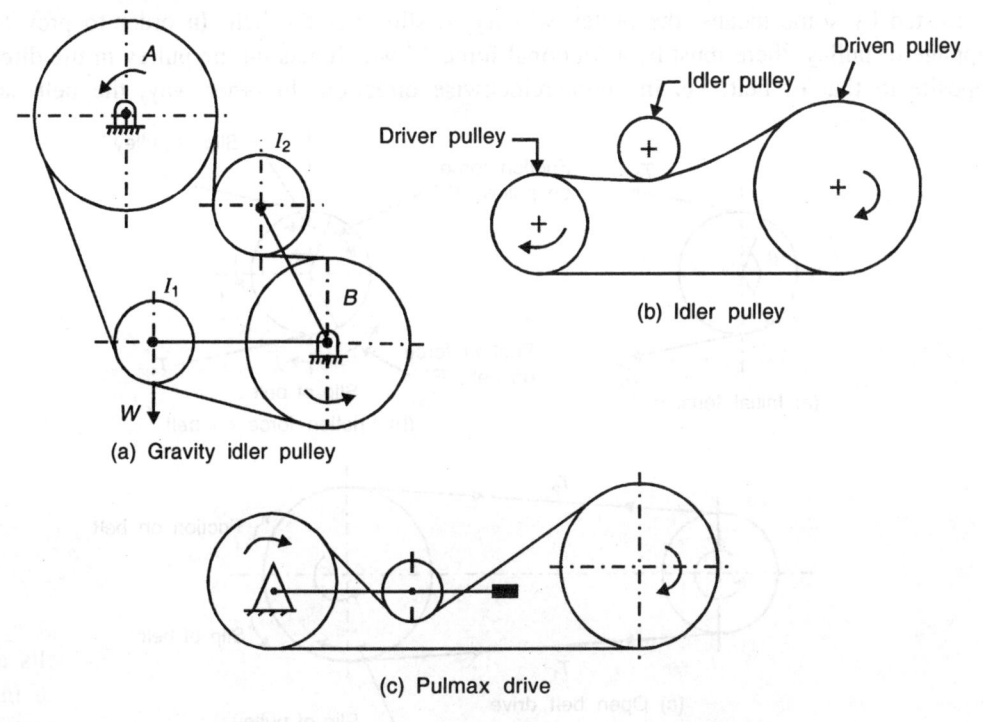

(b) Idler pulley

(c) Pulmax drive

FIGURE 6.6 Some typical arrangements of belt drive.

A belt drive has various advantages which makes it suitable for wide range of applications. Some of them are as follows:

(i) The drive can transmit power at considerable large distance.
(ii) It can operate at high speed, yet its operation is smooth and noiseless.
(iii) The drive has inherent capacity to absorb shock and vibration.
(iv) Its design is simple and inexpensive.
(v) It can be used in dusty and corrosive atmosphere.

However, the major limitations of the belt drive are the following:

(i) Velocity ratio does not remain constant.
(ii) It requires frequent tensioning.
(iii) The service life is usually short.

6.3 MECHANICS OF BELT DRIVE

In a belt drive, one of the primary requirements is that the belt should not slip over the pulley. To understand the mechanics of belt drive, let us consider a belt wrapped around the pulley subtending an angle θ at its centre. It is assumed that when drive does not operate or when it is at idle condition, the belt is subjected to initial tension T_0 only [Figure 6.7(a)].

When belt starts transmission of power, a tangential force F_t is applied in the direction of rotation, say in clockwise direction as shown in Figure 6.7(b). If the motion of the belt is resisted by some means, the pulley will try to slip over the belt. In order to prevent the slippage of pulley, there must be a frictional force F' which acts on the pulley in the direction opposite to that of belt, i.e. in counterclockwise direction. In other way, the belt can be

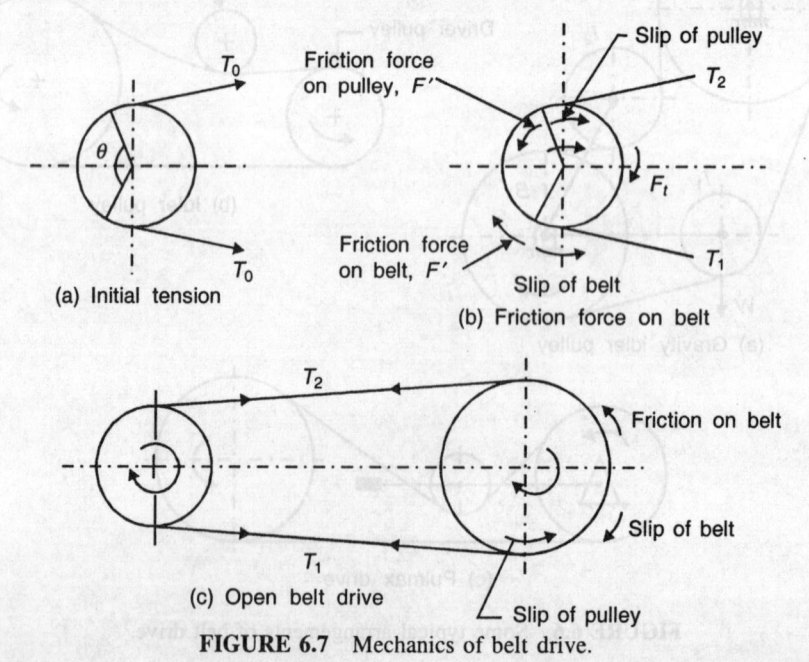

FIGURE 6.7 Mechanics of belt drive.

prevented from slipping if the friction force acts on the belt in the direction opposite to that of pulley.

In a belt drive, if the tangential force on the pulley is increased, a stage will come where it equalizes the frictional force. Any further increase of tangential force may cause a slip between the belt and the pulley.

The frictional force F' in the belt results in an increased tension on one side called **tight side** and decreased tension on other side called **slack side**. The magnitudes of tight side tension T_1 and slack side tension T_2 are such that the difference between these two is equal to tangential force.

Tension in the tight side: $T_1 = T_0 + \delta T$

Tension in the slack side: $T_2 = T_0 - \delta T$

Initial tension: $T_0 = \dfrac{T_1 + T_2}{2}$

Tangential force: $F_t = T_1 - T_2$

Further, a belt drive must satisfy a condition called **law of belting**. According to it, the centre line of belt, when it approaches a pulley, must lie in a plane perpendicular to the axis of the pulley or in the plane of the pulley, otherwise the belt will run off the pulley.

Figure 6.8 shows a quarter turn drive in which two non-parallel shafts are connected by means of a belt. In such a drive, when the driven pulley rotates in anticlockwise direction, the centre line of the belt approaching the pulley lies in the plane perpendicular to the axis of pulley. Thus the belt transmits power. However, if the direction of the drive is reversed, the belt will run off as the centre line of the belt would not approach in the plane of the pulley.

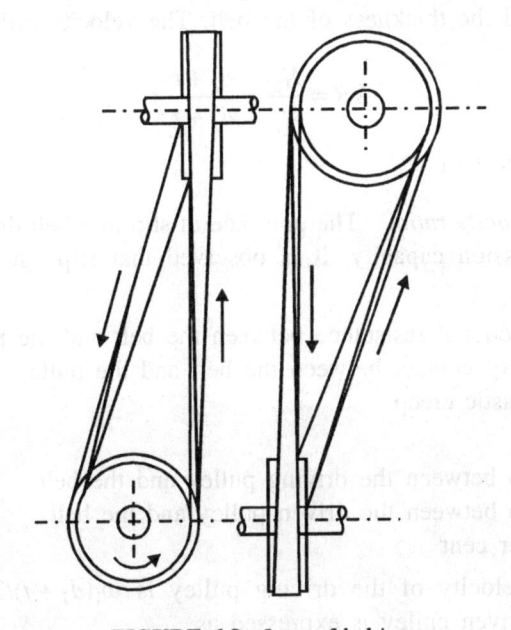

FIGURE 6.8 Law of belting.

6.4 VELOCITY RATIO

In a belt drive, the **velocity ratio** is defined as a ratio of angular velocity of the driven pulley to that of the driving pulley.

Let
N_1 = speed of the driving pulley
N_2 = speed of the driven pulley
ω_1 = angular velocity of the driving pulley
ω_2 = angular velocity of the driven pulley
d_1 = pitch diameter of the driving pulley
and d_2 = pitch diameter of the driven pulley

If the belt material is elastic and slip between the belt and the pulley is neglected, the kinematic equilibrium of the drive is obtained by equating the peripheral velocities of driving and driven pulleys as follows:

$$\pi d_1 N_1 = \pi d_2 N_2$$

or

$$\frac{N_2}{N_1} = \frac{d_1}{d_2}$$

Velocity ratio:
$$VR = \frac{\omega_2}{\omega_1} = \frac{N_2}{N_1} = \frac{d_1}{d_2} \tag{6.1}$$

where ω_1 and ω_2 are angular velocities of driving and driven pulleys. However, for the case when thickness of the belt is sufficiently large, the effective diameter of the pulley is sum of the pitch diameter and the thickness of the belt. The velocity ratio is then defined as

$$VR = \frac{N_2}{N_1} = \frac{d_1 + t}{d_2 + t} \tag{6.2}$$

where t = thickness of the belt

Effect of slip on velocity ratio. The presence of slip in a belt drive reduces its velocity ratio and power transmission capacity. It is observed that slip usually occurs due to the following reasons:

(i) Insufficient frictional resistance between the belt and the pulley
(ii) Smaller angle of contact between the belt and the pulley
(iii) Presence of elastic creep

Let
S_1 = percentage slip between the driving pulley and the belt.
S_2 = percentage slip between the driven pulley and the belt
and S = total slip in per cent

If the peripheral velocity of the driving pulley is $\omega_1(d_1 + t)/2$, then the peripheral velocity of belt on the driven pulley is expressed as

$$v = \frac{\omega_1(d_1 + t)}{2} \times \frac{100 - S_1}{100}$$

The belt moves with this velocity up to driven pulley and there it again slips by S_2 percentage. Thus the peripheral velocity of the driven pulley is:

$$\frac{\omega_1(d_1 + t)}{2} \times \left(\frac{100 - S_1}{100}\right) \times \left(\frac{100 - S_2}{100}\right)$$

If the total slip in the drive is S, the following condition must hold true:

$$\frac{\omega_1(d_1 + t)}{2} \times \left(\frac{100 - S}{100}\right) = \omega_1 \left(\frac{d_1 + t}{2}\right) \times \left(\frac{100 - S_1}{100}\right) \times \left(\frac{100 - S_2}{100}\right)$$

or

$$\frac{100 - S}{100} = \frac{100 - S_1}{100} \times \frac{100 - S_2}{100}$$

or

$$S = S_1 + S_2 - 0.01 S_1 S_2 \qquad (6.3)$$

The effect of the slip on velocity ratio is accounted by the following relation:

$$VR = \frac{N_2}{N_1} = \left(\frac{d_1 + t}{d_2 + t}\right)\left(\frac{100 - S}{100}\right) \qquad (6.4)$$

EXAMPLE 6.1 A motor shaft operates at 1440 rpm and drives a machine at 720 rpm by means of a flat belt. The diameter of driving pulley is 300 mm. If there is a slip of 2 per cent on driving pulley and 3 per cent on the driven pulley, determine the diameter of driven pulley. Belt thickness is 6 mm.

Solution

Total slip of the drive: $\qquad S = S_1 + S_2 - 0.01 S_1 S_2$

where S_1 = slip on driving pulley and

S_2 = slip on driven pulley

$$S = 2 + 3 - 0.01 \times 2 \times 3 = 4.94 \text{ per cent}$$

Velocity ratio:

$$VR = \frac{N_2}{N_1} = \left(\frac{d_1 + t}{d_2 + t}\right)\left(\frac{100 - S}{100}\right)$$

or

$$\frac{720}{1440} = \left(\frac{300 + 6}{d_2 + 6}\right) \times \frac{100 - 4.94}{100}$$

or

Diameter of driven pulley: $d_2 = 575.76$ mm   **Ans.**

6.5 LENGTH OF BELT

The length of belt required to be wrapped around the pulley primarily depends upon the following factors:

    (i) diameters of driving and driven pulleys
    (ii) distance between the centres of power transmitting shafts
    (iii) type of drive arrangement

In the following sub-section, the formulae for computing length of open-belt and cross-belt drives are derived. However, the same logic may be extended for other types of drive arrangement.

6.5.1 Open-belt Drives

Consider an open-belt drive with O_1 and O_2 as the centres of the driving and driven pulleys respectively. Let AB and CD be the common tangents to the pulley circles representing the centre line of the belt (Figure 6.9). Belt makes contact with driving pulley at point C and leaves the contact at point A. If the common tangent AF subtends an angle $\angle AO_1F$ (or angle ϕ) with the centre line at pulley centre O_1, the angle of contact on the driving (usually a smaller pulley) and driven pulleys are defined as

Angle of contact on the driving pulley: $\theta_1 = \pi - 2\phi$
Angle of contact on the driven pulley: $\theta_2 = \pi + 2\phi$
with

$$\angle\phi = \sin^{-1}\left(\frac{d_2 - d_1}{2c}\right) \qquad (6.5)$$

Now construct a line O_1K parallel to the line AB such that $\angle O_2O_1K$ equals to angle ϕ and O_2K represents $(d_2 - d_1)/2$ (See Figure 6.9).

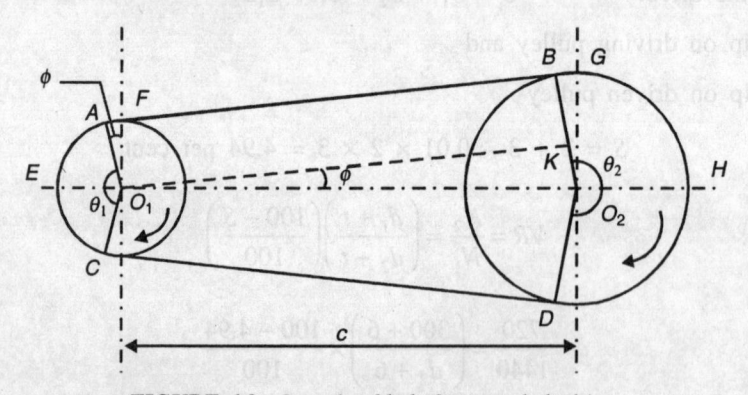

FIGURE 6.9 Length of belt for open-belt drive.

The length of belt for the open-belt drive is sum of the following:

(i) The arc length between the points C and A
(ii) Length between the points A and B
(iii) The arc length between the points B and D
(iv) Length between the points D and C

Therefore,

length:
$$L_o = 2[\text{Arc } EA + AB + \text{Arc } BH]$$

$$= 2\left[\left(\frac{\pi}{2} - \phi\right)\frac{d_1}{2} + O_1K + \left(\frac{\pi}{2} + \phi\right)\frac{d_2}{2}\right]$$

$$= 2\left[\left(\frac{\pi}{2} - \phi\right)\frac{d_1}{2} + c\cos\phi + \left(\frac{\pi}{2} + \phi\right)\frac{d_2}{2}\right]$$

or
$$L_o = \pi\frac{(d_1 + d_2)}{2} + \phi(d_2 - d_1) + 2c\cos\phi \qquad (6.6)$$

If the centre distance c is large compared to difference in the radii of the two pulleys, the angle ϕ may be assumed as small angle and for that $\sin\phi = \phi$.

We have
$$\phi = \frac{d_2 - d_1}{2c}$$

$$\cos\phi = (1 - \sin^2\phi)^{0.5}$$

and
$$= \left(1 - \frac{1}{2}\sin^2\phi + \cdots\right) \text{ (Using binomial theorem)}$$

or
$$\cos\phi = 1 - \frac{1}{2}\phi^2 = 1 - \frac{1}{2}\left(\frac{d_2 - d_1}{2c}\right)^2$$

Substituting the values of angle ϕ and $\cos\phi$ in Eq. (6.6), we get the approximate length of the belt for open-belt drive:

$$L_o = \frac{\pi(d_1 + d_2)}{2} + \frac{d_2 - d_1}{2c} \times (d_2 - d_1) + 2c\left[1 - \frac{1}{2}\left(\frac{d_2 - d_1}{2c}\right)^2\right]$$

or
$$L_o = \frac{\pi(d_1 + d_2)}{2} + \frac{(d_2 - d_1)^2}{4c} + 2c \qquad (6.7)$$

6.5.2 Cross-belt Drives

Let O_1 and O_2 be the centres of two pulleys connected by a cross-belt and tangents AB and CD represent the centre line of the belt (See Figure 6.10).

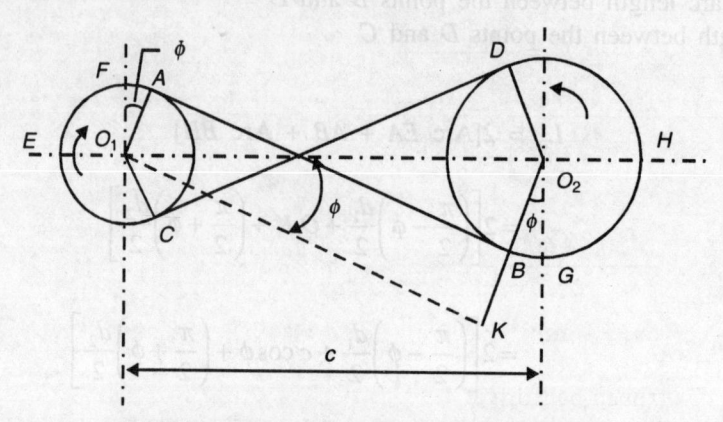

FIGURE 6.10 Length of belt for cross-belt drive.

Draw a line O_1K parallel to the line AB such that it intersects the extended portion of line O_2B at the point K. The angle O_2O_1K equals angle ϕ and O_2K represents sum of radii of driving and driven pulleys.

The length of belt for cross-belt:

$$L_c = 2(\text{Arc } EA + AB + \text{arc } BH)$$

$$= 2\left[\left(\frac{\pi}{2}+\phi\right)\frac{d_1}{2} + c\cos\phi + \left(\frac{\pi}{2}+\phi\right)\frac{d_2}{2}\right]$$

$$= \frac{\pi(d_1+d_2)}{2} + 2c\cos\phi + \phi(d_1+d_2)$$

With $\phi = (d_2+d_1)/2c$ and $\cos\phi = \left[1 - \frac{1}{2}\left(\frac{d_2+d_1}{2c}\right)^2\right]$, the approximate length of the cross-belt drive is given as

$$L_c = \frac{\pi(d_1+d_2)}{2} + \frac{(d_1+d_2)^2}{4c} + 2c \tag{6.8}$$

The length of the belt calculated by above formula need to be reduced by 2 to 5 per cent to compensate for elastic extension due to initial tension.

EXAMPLE 6.2 A compressor is driven by an electric motor using cross-belt drive. Pulleys mounted on the motor and compressor are of diameters 300 mm and 600 mm respectively. The centre distance is 2m. Later on, in order to change the direction of rotation of the compressor, the drive arrangement is changed to open-belt drive. State whether the same belt can be used or not. If not, what is the remedy?

Solution The exact length of cross-belt drive is:

$$L_c = \frac{\pi(d_1 + d_2)}{2} + 2c \cos \phi + \phi(d_1 + d_2)$$

where
$$\phi = \sin^{-1}\left(\frac{d_1 + d_2}{2c}\right) = \sin^{-1}\left(\frac{300 + 600}{2 \times 2000}\right) = 13°$$

$$\therefore \quad L_c = \frac{\pi(300 + 600)}{2} + 2 \times 2000 \times \cos 13° + \frac{13 \times \pi}{180} \times (300 + 600)$$

$$= 5515.4 \text{ mm} \qquad \text{Ans.}$$

The exact length of open-belt drive:

$$L_o = \frac{\pi(d_1 + d_2)}{2} + 2c \times \cos\phi + \phi(d_2 - d_1)$$

$$\phi = \sin^{-1}\left(\frac{d_2 - d_1}{2c}\right) = \sin^{-1}\left(\frac{600 - 300}{2 \times 2000}\right) = 4.3°$$

where
$$L_o = \frac{\pi(300 + 600)}{2} + 2 \times 2000 \times \cos 4.3° + \frac{\pi}{180} \times 4.3° \times (600 - 300)$$

$$= 5424.9 \text{ mm} \qquad \text{Ans.}$$

It is obvious that for a fixed centre distance, the belt length should be reduced by $(L_c - L_o)$, i.e. 90.5 mm or the centre distance between the motor and the compressor may be increased.

6.6 RATIO OF BELT TENSIONS

The power transmission capacity of a belt drive depends upon the difference in tensions of tight and slack sides of the belt. These tensions are also known as **friction tensions.**
Let

T_1 = tension on tight side of the belt
T_2 = tension on slack side of the belt
θ = angle of contact of belt with the pulley and
μ = coefficient of friction between the belt and the pulley

Consider a driving pulley rotating in clockwise direction in which the belt on left hand side is stretched more than the belt on right hand side. Thus the belt on left hand side is called **tight side** and that on right hand side is termed as **slack side.** The ratio of tensions may be

196 *Theory of Mechanisms and Machines*

found by considering an elemental strip AB of belt subtending an angle $\delta\theta$ at the centre of pulley (Figure 6.11). Various forces acting on the elemental strip AB are the following:

   (i) tension $T + \delta T$ in the belt at point A on tight side
  (ii) tension T in the belt at point B on slack side
 (iii) normal reaction R acting radially upward between elemental length of the belt and the pulley
  (iv) the friction force μR acting tangentially to the pulley rim, which resists slipping of the elemental strip of belt on the pulley.

Under the action of above forces, the elemental strip will remain in the equilibrium, if the force polygon is closed [Figure 6.11(b)]. Mathematically, the algebraic sum of forces acting in horizontal and vertical directions must be zero.

(a) Forces acting on the belt   (b) Force polygon

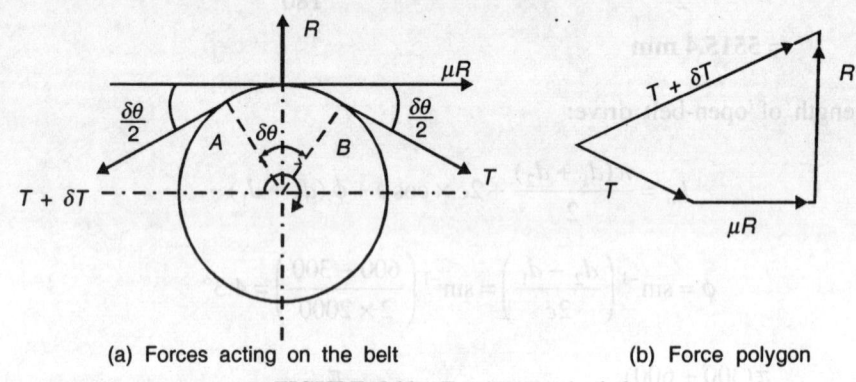

FIGURE 6.11 Tensions in the belt.

Resolving the forces in horizontal direction,

$$\mu R + T \cos\frac{\delta\theta}{2} - (T + \delta T) \cos\frac{\delta\theta}{2} = 0$$

For a small angle $\delta\theta$, $\cos\dfrac{\delta\theta}{2} = 1$ and $\sin\dfrac{\delta\theta}{2} = \dfrac{\delta\theta}{2}$.

Therefore, $\quad\quad\quad\quad\quad\quad\quad\quad \mu R + T - (T + \delta T) = 0$

or $\quad\quad\quad\quad\quad\quad\quad\quad\quad\quad\quad \delta T = \mu R \quad\quad\quad\quad\quad\quad\quad\quad\quad\quad\quad (6.9)$

Resolving the forces in vertical direction,

$$R - T \sin\frac{\delta\theta}{2} - (T + \delta T) \sin\frac{\delta\theta}{2} = 0$$

or $\quad\quad\quad\quad\quad\quad\quad R - T \times \dfrac{\delta\theta}{2} - (T + \delta T) \times \dfrac{\delta\theta}{2} = 0$

Neglecting the product of small quantities, we get

$$R = T\,\delta\theta \tag{6.10}$$

Substituting the value of R in Eq. (6.9), we get

$$\frac{\delta T}{T} = \mu\delta\theta \tag{6.11}$$

Integrating Eq. (6.11) between appropriate limits, we get

$$\int_{T_2}^{T_1} \frac{dT}{T} = \int_0^\theta \mu\,d\theta$$

or

$$\log_e\left(\frac{T_1}{T_2}\right) = \mu\theta$$

or

$$\text{Ratio of tensions, } \frac{T_1}{T_2} = e^{\mu\theta} \tag{6.12}$$

6.6.1 Ratio of Tensions in V-Belt

When a trapezoidal section belt is designed to run in a V-shaped grooved pulley, it is known as **V-belt drive**. In a V-belt drive, owing to the wedging action of belt with V-groove, the normal reactions act perpendicular to the face of the groove [See Figure 6.12(a)].

Let
  2α = included angle of V-groove and
  R = normal reaction between each side of the groove and the corresponding side of the belt
  Radial reaction = $2R \sin \alpha$
  Frictional force = $2\mu R$

Now consider an element strip of V-belt subtending an angle $\delta\theta$ at the centre of the pulley. The strip is subjected to the following forces [Figure 6.12(b)]:

  (i) Tension $T + \delta T$ at the point A
  (ii) Tension T at the point B
  (iii) Normal reaction $2R \sin \alpha$ acting radially outward
  (iv) Friction force $2\mu R$ acting tangentially to the pulley

For equilibrium of the elemental strip, the algebraic sum of forces acting in horizontal and vertical directions must be zero.

Resolving the forces in horizontal directions with the assumption that $\delta\theta$ is small angle for which $\cos\dfrac{\delta\theta}{2} = 1$, we get

198 Theory of Mechanisms and Machines

(a) Normal reaction on V-belt

(b) Forces acting on the V-belt

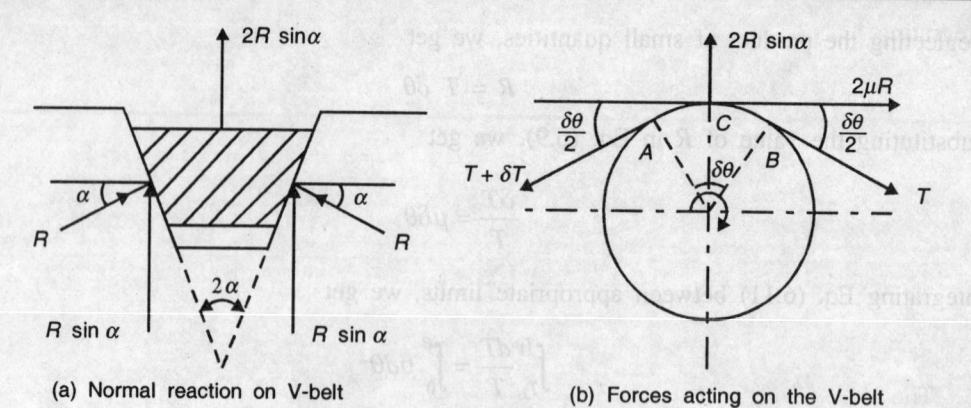

FIGURE 6.12 Tensions in V-belt.

$$2\mu R + T \cos \frac{\delta\theta}{2} - (T + \delta T) \cos \frac{\delta\theta}{2} = 0$$

or
$$\delta T = 2\mu R \tag{6.13}$$

Similarly, resolving the forces in vertical directions

$$2R \sin \alpha - T \sin \frac{\delta\theta}{2} - (T + \delta T) \sin \frac{\delta\theta}{2} = 0$$

For small angle $\delta\theta$, $\sin(\delta\theta/2) = \delta\theta/2$

Therefore,
$$2R \sin \alpha - T \times \frac{\delta\theta}{2} - (T + \delta T) \times \frac{\delta\theta}{2} = 0$$

Neglecting the product of small terms in the above equation, we get

$$2R \sin \alpha - 2T \times \frac{\delta\theta}{2} = 0$$

or
$$R = \frac{T \delta\theta}{2 \sin \alpha} \tag{6.14}$$

Substituting the value of reaction R in Eq. (6.13), we get

$$\delta T = 2\mu \times \frac{T \delta\theta}{2 \sin \alpha}$$

or
$$\frac{\delta T}{T} = \frac{\mu \delta\theta}{\sin \alpha} \tag{6.15}$$

Integrating Eq. (6.15) between appropriate limits, we get

$$\int_{T_2}^{T_1} \frac{dT}{T} = \int_0^{\theta} \frac{\mu \delta\theta}{\sin\alpha}$$

or

$$\log_e\left(\frac{T_1}{T_2}\right) = \frac{\mu\theta}{\sin\alpha}$$

or

$$\frac{T_1}{T_2} = e^{\frac{\mu\theta}{\sin\alpha}} \tag{6.16}$$

The above expression of ratio of tensions is just similar to that of flat belt drive except that coefficient of friction μ here is increased to $(\mu/\sin\alpha)$ times. Therefore, the ratio of tensions of the V-belt is greater than that of the flat belt. Hence it transmits more power compared to flat belt.

6.7 EFFECT OF CENTRIFUGAL FORCE

When a belt operates at high speed, there is a considerable inertial force acting on the belt. This inertial force is produced due to centrifugal action of the belt weight which tries to lift the belt from the pulley. On account of this inertial force, two tensions of equal magnitude are induced on the tight and slack sides of the belt.

Consider a small element of belt which is in equilibrium condition under the action of the following forces (refer Figure 6.13):

(i) Centrifugal force F_c
(ii) Centrifugal tension T_c acting on tight and slack sides of the belt

Let
m = mass of the belt per unit length
r = radius of the pulley
v = peripheral velocity of the belt
and $\delta\theta$ = angle of contact of the element over the pulley

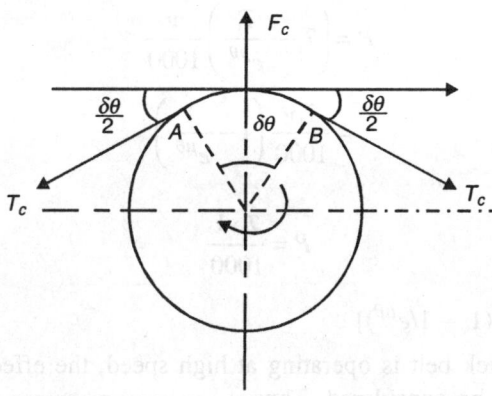

FIGURE 6.13 Centrifugal force on the belt.

The centrifugal force on the element:

$$F_c = \text{mass of the element} \times \text{acceleration}$$

$$= mr\delta\theta \times \frac{v^2}{r}$$

or

$$F_c = mv^2\delta\theta \tag{6.17}$$

For the equilibrium of element AB, let us resolve the force in vertical direction.

$$\therefore F_c = 2T_c \sin\frac{\delta\theta}{2}$$

or

$$F_c = T_c \times \delta\theta \quad \text{(For small angle, } \sin\frac{\delta\theta}{2} = \frac{\delta\theta}{2}\text{)} \tag{6.18}$$

Equating Eqs. (6.17) and (6.18), we get

$$\text{Centrifugal tension, } T_c = mv^2 \tag{6.19}$$

Centrifugal tension on the belt is a function of the mass of belt and peripheral velocity and it is independent of tensions on tight and slack sides.

6.8 POWER TRANSMISSION CAPACITY

The power transmission capacity of the belt is given by the following relation:

$$\dot{P} = \frac{(T_1 - T_2)v}{1000} \text{ kW} \tag{6.20}$$

From the ratio of tensions [Eq. (6.12)], we know that $T_2 = T_1/e^{\mu\theta}$. Substituting in Eq. (6.20), we get

$$P = \left(T_1 - \frac{T_1}{e^{\mu\theta}}\right)\frac{v}{1000}$$

$$= \frac{T_1 v}{1000}\left(1 - \frac{1}{e^{\mu\theta}}\right)$$

or

$$P = \frac{T_1 vk}{1000} \tag{6.21}$$

where k is a constant [$= (1 - 1/e^{\mu\theta})$]

However, when a thick belt is operating at high speed, the effect of centrifugal tension on the belt is required to be considered. Thus total tension on tight side: $T = T_1 + T_c$

total tension on slack side: $T_s = T_2 + T_c$

or

Frictional tension on tight side: $T_1 = T - T_c$ (6.22)

Substituting the value of T_1 in Eq. (6.21), we get

Power: $$P = \frac{(T - T_c)vk}{1000}$$

or $$P = \frac{vk(T - mv^2)}{1000}$$ (6.23)

For the condition of maximum power transmission capacity, let us differentiate Eq. (6.23) with respect to velocity v; and equating it to zero, we get

$$\frac{\partial P}{\partial v} = kT - 3kmv^2 = 0$$

or $$T = 3mv^2 = 3T_c$$ (6.24)

Therefore, for maximum power to be transmitted, the maximum allowable belt tension should be equal to three times the centrifugal tension. The threshold speed of the belt is given as

$$v_{max} = \sqrt{\frac{T}{3m}}$$ (6.25)

The ratio of tensions on tight and slack sides, when the effect of centrifugal tension is considered, is given by

$$\frac{T_1 - T_c}{T_2 - T_c} = e^{\mu\theta}$$ (6.26)

6.9 STRESSES IN BELT

When a belt transmits power from the driving pulley to driven pulley, a fibre on it, which travels along the complete path, is subjected to the following forces at different positions:

(i) Tension on the tight side T_1
(ii) Tension on the slack side T_2
(iii) Centrifugal tension T_c
(iv) Bending of the belt over the pulley

These forces produce three types of stresses:
  (a) Static stress due to tension on the tight side:

$$\sigma_1 = \frac{T_1}{bt} \tag{6.27}$$

where
  b = width of the belt
  and t = thickness of the belt

  (b) The centrifugal tension due to weight of the belt produces centrifugal stresses, which are tensile in nature.

$$\sigma_c = \frac{T_c}{bt} = \frac{\rho v^2}{g \times 10^6} \tag{6.28}$$

where
  ρ = weight density of the belt material (N/m^3)

  (c) Bending stress due to curvature effect of the belt around the pulley is:

$$\frac{\sigma_b}{y} = \frac{E}{R}$$

where
  y = distance of fibre from neutral axis (= $t/2$)
  E = modulus of elasticity
  and R = radius of curvature, i.e. radius of pulley
  Therefore,

$$\sigma_b = \frac{E \times t/2}{R} = \frac{Et}{d} \tag{6.29}$$

The distribution of these stresses at various points in a belt is shown in Figure 6.14 which indicates that the maximum stress, which is the sum of these three stresses, occurs at the beginning of the arc of the driving pulley. For safe design of belt, the maximum stress should be less than or equal to allowable strength of belt material.

$$\sigma_{max} = \sigma_1 + \sigma_c + \sigma_b \leq \sigma_d \tag{6.30}$$

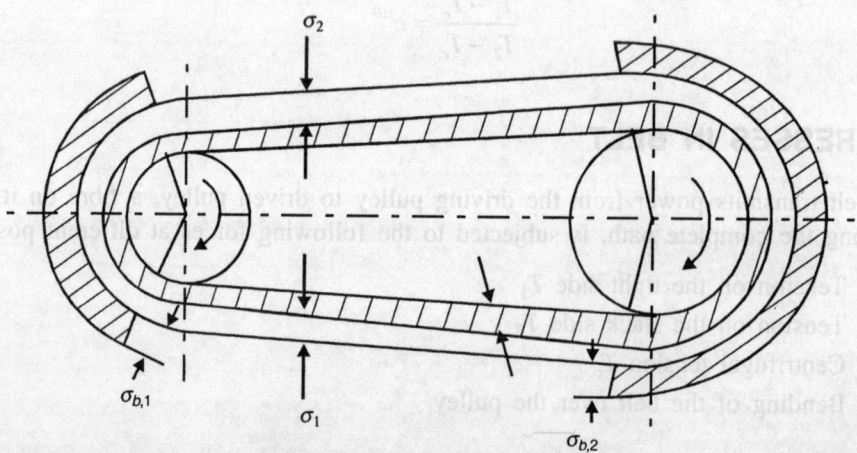

FIGURE 6.14 Stresses in a belt.

The power transmission capacity of a belt is given by the following relation:

$$P = \frac{(T - T_c)(1 - 1/e^{\mu\theta})v}{1000}$$

or

$$P = \frac{btv}{1000}\left[\sigma_d - \frac{\rho v^2}{g \times 10^6}\right]\left[\frac{e^{\mu\theta} - 1}{e^{\mu\theta}}\right] \text{ kW} \qquad (6.31)$$

where
  σ_d = design stress
and ρ = density of the belt material

6.10 ELASTIC CREEP

In a power transmitting belt, the tensions in its two branches are different. It is apparent that the tight side belt is elongated with strain e_1 whereas the slack side is contracted by strain e_2 from the original length under the action of initial tension. Therefore, in such a drive, the larger length of belt approaches to driving pulley and the shorter length leaves it (Figure 6.15). This phenomenon is called **elastic creep**. The effect of creep is to slow down the speed of belt over driving pulley, thereby reducing velocity ratio and transmission efficiency.

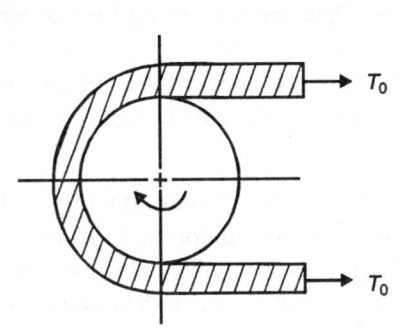

(a) Belt in idle condition   (b) Belt in normal running condition

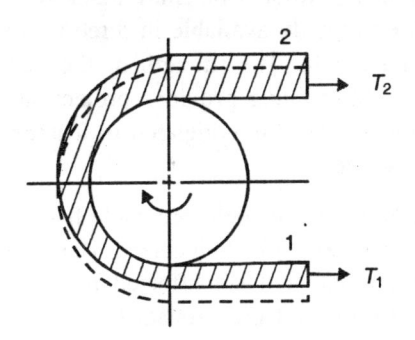

FIGURE 6.15 Elastic creep in a belt.

Velocity ratio:
$$VR = \frac{N_2}{N_1} = \frac{d_1}{d_2} \times \frac{E + \sqrt{\sigma_2}}{E + \sqrt{\sigma_1}} \qquad (6.32)$$

where
  σ_1 and σ_2 are the stresses in the belt on tight and slack sides respectively
and E is modulus of elasticity of the belt material.

If m is the mass per unit length of unstrained belt, the mass of belt passing a point per unit time is:

$$\frac{mv_1}{1 + e_1} = \frac{mv_2}{1 + e_2}$$

or
$$\frac{v_2}{v_1} = \frac{1+e_2}{1+e_1} = (1+e_2)(1+e_1)^{-1}$$
$$= 1 - e_1 + e_2$$

or
$$\frac{v_2}{v_1} = 1 - \left(\frac{T_1 - T_2}{btE}\right) \qquad (6.33)$$

6.11 BELT MATERIALS

A belt material should have a high coefficient of friction, strength, flexibility and durability. The commonly used materials are discussed further.

Leather. It is most widely used material for belts. Leather belts are available in two varieties—oak tanned and chrome tanned. Oak tanned leather is most commonly used in ordinary applications whereas for special applications involving damp environment such as chemical handling machinery and oiled surfaces, the chrome-tanned leather belts are preferred. In order to obtain a reasonable life of a belt, the layers of leather strips are cemented together and these belts are specified according to the number of layers called plys.

Rubber. Rubber belts are made from cotton ducks or canvas impregnated with rubber. These are normally available in three to ten plys. They are vulcanized for use when exposed to oil or sunlight. The strength of the belt is mainly obtained from cotton ducks or canvas whereas rubber lining provides protection against adverse working conditions. Rubber belts have short service life compared to leather belts. These belts are cheaper and most suited for outdoor service.

Balata. Balata belts are made from closely woven cotton ducks impregnated with balata gum. They need not be vulcanized. Balata does not oxidize and does not age in air or sunlight. It is waterproof, and not affected by acids, alkalies and humidity. However, it is seriously affected by mineral oils. Balata belts are 20–40 per cent stronger than rubber belts.

Fabric. Fabric is another material popularly used as belt material. Fabric belts are made from canvas or cotton ducks in which a number of layers of cotton ducks are closely woven. The woven belt is further treated in linseed oil to make it waterproof. These belts are used for temporary installations and rough service conditions where little attention is needed. The mechanical properties of fabric belts are comparable to those of rubber belts.

Plastics. Plastic strips are used as core material of the belts. Plastic-cored belts are made from nylon canvas or thin plastic sheets with a layer of rubber surrounding the whole belt structure. Plastic-cored belts can run at high speed and can be wrapped around very small pulleys. These belts possess high strength which is approximately two times that of leather belts.

EXAMPLE 6.3 A leather belt is required to transmit 10 kW with 1 m diameter pulley running at 300 rpm. The angle of contact is 160° and coefficient of friction between belt and pulley is 0.32. If the allowable strength of belt is 1.25 N/mm² and the belt thickness is 10 mm, determine the width of the belt. Density of belt material is 1000 kg/m³.

Solution

Peripheral velocity:
$$v = \frac{\pi d_1 N_1}{60} = \frac{\pi \times 1.0 \times 300}{60}$$
$$= 15.7 \text{ m/s}$$

Ratio of tensions:
$$\frac{T_1}{T_2} = e^{\mu\theta} = e^{0.32 \times \frac{\pi}{180} \times 160°}$$

or
$$\frac{T_1}{T_2} = 2.44 \qquad (i)$$

Power:
$$P = \frac{(T_1 - T_2) \times v}{1000}$$

or
$$T_1 - T_2 = \frac{10 \times 1000}{15.7} = 636.9 \text{ N} \qquad (ii)$$

Solving Eqs. (i) and (ii), we get
$$T_1 = 1079.2 \text{ N} \quad \text{and} \quad T_2 = 442.3 \text{ N}$$

Centrifugal tension:
$$T_c = mv^2 = \frac{\rho b t v^2}{10^6}$$
$$= \frac{1000 \times b \times 10 \times 15.7^2}{10^6} = 2.465\, b \text{ N}$$

$$T_{max} = bt\sigma = b \times 10 \times 1.25 = 12.5\, b \text{ N}$$

Maximum belt tension:
$$T = T_1 + T_c$$
$$= 1079.2 + 2.465\, b$$

or
$$T = T_{max}$$
$$1079.2 + 2.465\, b = 12.5\, b$$

or
Belt width: $b = \mathbf{107.54}$ **mm**   **Ans.**

EXAMPLE 6.4 An open belt running over two pulleys of diameters 200 mm and 600 mm connects two parallel shafts placed at a distance of 2.5 m. The smaller pulley rotates at 300 rpm and transmits 7.5 kW. The coefficient of friction between the belt and the pulley is 0.3. Determine:

(i) length of belt
(ii) initial tension
(iii) minimum width if the safe working tension is 12 N/mm width

Solution

(i) Approximate length of belt for open belt drive is:

$$L_o = \frac{\pi(d_1 + d_2)}{2} + \frac{(d_2 - d_1)^2}{4c} + 2c$$

$$= \frac{\pi(200 + 600)}{2} + \frac{(600 - 200)^2}{4 \times 2500} + 2 \times 2500$$

$$= 6272.64 \text{ mm} \qquad \text{Ans.}$$

(ii) We know that

$$\phi = \sin^{-1}\left(\frac{d_2 - d_1}{2c}\right) = \sin^{-1}\left(\frac{600 - 200}{2 \times 2500}\right)$$

$$= 4.59°$$

Angle of contact on smaller pulley:

$$\theta = 180° - 2\phi = 180° - 2 \times 4.59° = 170.82°$$

Ratio of tensions: $\dfrac{T_1}{T_2} = e^{\mu\theta} = e^{0.3 \times \frac{\pi}{180} \times 170.82°}$

or $\dfrac{T_1}{T_2} = 2.44$ \hfill (i)

Belt speed: $v = \dfrac{\pi d_1 N_1}{60} = \dfrac{\pi \times 0.2 \times 300}{60} = 3.14$ m/s

Difference in tensions:

$$T_1 - T_2 = \frac{P \times 10^3}{v} = \frac{7.5 \times 10^3}{3.14}$$

or $T_1 - T_2 = 2388.5$ N \hfill (ii)

Solving Eqs. (i) and (ii), we get

$$T_1 = 4047.1 \text{ N and } T_2 = 1658.68 \text{ N}$$

Initial tension: $T_0 = \dfrac{T_1 + T_2}{2} = \dfrac{4047.1 + 1658.68}{2} = 2852.89 \text{ N}$ **Ans.**

(iii) Neglecting centrifugal tension, the maximum tension on the belt:

$$T_{max} = T_1$$
$$T_{max} = 12 \times b = 4047.1$$

∴ Belt width: $b = 337.26 \text{ mm}$ **Ans.**

EXAMPLE 6.5 An open belt connects two pulleys, the smaller pulley being of 400 mm diameter. Angle of contact on the smaller pulley is 160° and coefficient of friction between belt and pulley is 0.25. Determine which of the following alternatives would be more effective in increasing the power transmission capacity:

(i) increasing the initial tension by 10 per cent
(ii) increasing the coefficient of friction by 10 per cent using suitable dressing system
(iii) increasing the angle of contact by 10 per cent with use of idler pulley

Solution

Power: $P = (T_1 - T_2)v$

or $$P = T_1 v \dfrac{(e^{\mu\theta} - 1)}{e^{\mu\theta}} \qquad (i)$$

Initial tension: $T_0 = \dfrac{T_1 + T_2}{2}$

or $T_1 = 2T_0 - T_2$

$$= 2T_0 - \dfrac{T_1}{e^{\mu\theta}}$$

or $e^{\mu\theta} T_1 = 2T_0 e^{\mu\theta} - T_1$

or $$T_1 = \dfrac{2T_0 e^{\mu\theta}}{1 + e^{\mu\theta}} \qquad (ii)$$

From Eqs. (i) and (ii), we get

$$P = \dfrac{2T_0 e^{\mu\theta}}{1 + e^{\mu\theta}} \times \dfrac{e^{\mu\theta} - 1}{e^{\mu\theta}} \times v$$

or $$P = 2T_0 v \left(\frac{e^{\mu\theta} - 1}{e^{\mu\theta} + 1} \right) \qquad \text{(iii)}$$

(i) For the same values of coefficient of friction μ, belt velocity v and contact angle θ, the power transmitted is directly proportional to initial tension [refer Eq. (iii)]. Therefore, by increasing initial tension by 10 per cent, power transmitted will also increase by 10 per cent.

(ii) If the coefficient of friction is increased to μ', keeping initial tension same, the ratio of powers is:

$$\frac{P'}{P} = \frac{e^{\mu'\theta} - 1}{e^{\mu'\theta} + 1} \times \frac{e^{\mu\theta} + 1}{e^{\mu\theta} - 1}$$

where $e^{\mu\theta} = e^{0.25 \times \frac{\pi}{180} \times 160°} = 2.01$

and $e^{\mu'\theta} = e^{0.25 \times 1.1 \times \frac{\pi}{180} \times 160°} = 2.155$

$$\therefore \quad \frac{P'}{P} = \frac{2.155 - 1}{2.155 + 1} \times \frac{2.01 + 1}{2.01 - 1} = 1.091$$

Therefore, increase in power is 9.1 per cent.

(iii) If the angle of contact is increased by 10 per cent,

$$\theta' = 1.1 \times \theta = 1.1 \times 160 = 176°,$$

$$e^{\mu\theta} = 2.01 \quad \text{and} \quad e^{\mu\theta'} = 2.155$$

Ratio of powers: $\dfrac{P'}{P} = \dfrac{2.155 - 1}{2.155 + 1} \times \dfrac{2.01 + 1}{2.01 - 1} = 1.091$

or increase in power is 9.1 per cent.

Comparing these three options, we find that the maximum power is obtained in option (i), i.e. when initial tension is increased by 10 per cent.

EXAMPLE 6.6 A belt embraces the shorter pulley by an angle of 165° and runs at a speed of 1700 m/min. Dimensions of the belt are:

Width $b = 200$ mm and thickness $t = 8$ mm

Density of belt material is 1000 kg/m^3. Determine the maximum power that can be transmitted if the maximum permissible stress in the belt is not to exceed 2.5 N/mm^2 and coefficient of friction μ is 0.25.

Solution Maximum permissible tension in the belt:

$$T_{max} = bt\sigma = 200 \times 8 \times 2.5 = 4000 \text{ N}$$

Mass of belt/m length:
$$m = \rho bt$$
$$= 1000 \times \frac{200 \times 8}{10^6} = 1.6 \text{ kg/m}$$

Centrifugal tension:
$$T_c = mv^2 = 1.6 \times \left(\frac{1700}{60}\right)^2$$
$$= 1285 \text{ N}$$

From the condition of maximum power transmission:

$$T_c = \frac{1}{3}T_{max} = \frac{4000}{3} = 1333.34 \text{ N}$$

However, a centrifugal force of this order (1333.34 N) can be obtained only if the velocity of the belt is greater than 28.34 m/s. Since the problem is very specific about the velocity of the belt so let us take $T_c = 1285$ N

Since $$T_{max} = T_1 + T_c$$
or $$T_1 = T_{max} - T_c = 4000 - 1285 = 2715 \text{ N}$$

Ratio of tensions:
$$\frac{T_1}{T_2} = e^{\mu\theta} = e^{0.25 \times 165° \times \frac{\pi}{180}}$$
$$= 2.054$$

or $$T_2 = \frac{T_1}{e^{\mu\theta}} = \frac{2715}{2.054} = 1321.8 \text{ N}$$

Power:
$$P = \frac{(T_1 - T_2)v}{1000} = \frac{(2715 - 1321.8)}{1000} \times 28.34$$
$$= 39.48 \text{ kW} \qquad \text{Ans.}$$

EXAMPLE 6.7 A shaft rotating at 250 rpm is driven by another shaft at 500 rpm and transmits 5 kW through a belt. The belt is 100 mm wide and 10 mm thick. The distance between the shaft is 3 m. The diameter of smaller pulley is 300 mm. Calculate the stress in the belt for open-belt and cross-belt drives. Coefficient of friction is 0.3.

Solution

Belt velocity:
$$v = \frac{\pi d_1 N_1}{60} = \frac{\pi \times 0.3 \times 500}{60}$$
$$= 7.85 \text{ m/s}$$

$$T_1 - T_2 = \frac{P \times 10^3}{v} = \frac{5 \times 10^3}{7.85} \text{ or } T_1 - T_2 = 636.94 \text{ N} \quad \text{(i)}$$

Assuming that the effect of slip on the peripheral velocity of belt is negligible, the velocity ratio is:
$$VR = \frac{N_2}{N_1} = \frac{d_1}{d_2}$$

or
$$d_2 = \frac{N_1}{N_2} \times d_1 = \frac{500}{250} \times 300 = 600 \text{ mm}$$

Case I Open-belt drive

Angle, $\phi = \sin^{-1}\left(\frac{d_2 - d_1}{2c}\right) = \sin^{-1}\left(\frac{600 - 300}{2 \times 3000}\right) = 2.86°$

or contact angle, $\theta = 180° - 2\phi = 180° - 2 \times 2.86°$
$$= 174.28°$$

Ratio of tensions:
$$\frac{T_1}{T_2} = e^{\mu\theta} = e^{0.3 \times \frac{\pi}{180} \times 174.28°}$$

or
$$\frac{T_1}{T_2} = 2.49 \quad \text{(ii)}$$

Solving Eqs. (i) and (ii), we get
$$T_1 = 1064.2 \text{ N} \quad \text{and} \quad T_2 = 427.4 \text{ N}$$

Neglecting the effect of centrifugal tension, the maximum stress in the belt:
$$\sigma = \frac{T_1}{bt} = \frac{1064.2}{100 \times 10} = 1.0642 \text{ N/mm}^2 \qquad \text{Ans.}$$

Case II Cross-belt drive

Angle of contact, $\theta = 180° + 2\phi$

where $\phi = \sin^{-1}\left(\frac{d_1 + d_2}{2c}\right) = \sin^{-1}\left(\frac{600 + 300}{2 \times 3000}\right) = 8.62°$

or
$$\theta = 180 + 2 \times 8.62 = 197.24°$$

Ratio of tensions, $\dfrac{T_1}{T_2} = e^{\mu\theta} = e^{0.3 \times \frac{\pi}{180} \times 197.24°} = 2.8$ \hfill (iii)

Solving Eqs. (i) and (iii), we get
$$T_1 = 990.78 \text{ N} \quad \text{and} \quad T_2 = 353.85 \text{ N}$$

Neglecting the effect of centrifugal tension, the maximum stress in the belt:

$$\sigma = \frac{T_1}{bt} = \frac{990.78}{100 \times 10} = 0.99 \text{ N/mm}^2 \qquad \text{Ans.}$$

EXAMPLE 6.8 A flat belt of 200 × 12 mm² cross section runs between two pulleys. The allowable strength of belt material is 2.5 N/mm². Determine the maximum power that can be transmitted by it if the ratio of tension is 2 and the density of the material of the belt is 1000 kg/m³.

Solution Let m be the mass of the belt per metre length.

$$m = \rho b t = \frac{1000 \times 200 \times 12}{10^6} = 2.4 \text{ kg/m}$$

Maximum permissible tension on the belt:

$$T_{max} = bt\sigma = 200 \times 12 \times 2.5 = 6000 \text{ N}$$

The limiting velocity of the belt for maximum power transmission condition is:

$$v = \sqrt{\frac{T_{max}}{3m}} = \sqrt{\frac{6000}{3 \times 2.4}} = 28.86 \text{ m/s}$$

Further, centrifugal tension T_c should be equal to one-third of maximum tension, i.e.

$$T_c = \frac{T_{max}}{3}$$

∴
$$T_1 = T_{max} - T_c = \frac{2}{3} T_{max} = \frac{2}{3} \times 6000 = 4000 \text{ N}$$

$$T_2 = 2000 \text{ N} \quad \left(\text{as } \frac{T_1}{T_2} = 2\right)$$

Power, $P = \dfrac{(T_1 - T_2)v}{1000} = \dfrac{(4000 - 2000) \times 28.86}{1000}$

$$= 57.72 \text{ kW} \qquad \text{Ans.}$$

EXAMPLE 6.9 A C-type V-belt having face width b equal to 22 mm and nominal thickness t equal to 14 mm is used to transmit power with V groove angle 40°. If the mass of the belt is 0.4 kg/m and maximum allowable stress is 1.5 N/mm², determine the maximum power that can be transmitted. Angle of contact is 155° and coefficient of friction is 0.2.

Solution The lower face width of belt is (see Figure 6.16):

$$b_1 = b - 2t \tan \alpha$$
$$= 22 - 2 \times 14 \times \tan 20°$$
$$= 11.8 \text{ mm}$$

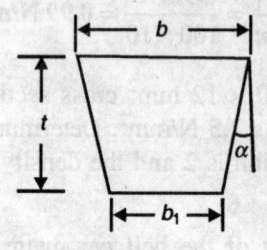

FIGURE 6.16

Area of cross section:
$$A = \frac{1}{2}(b + b_1) \times t$$
$$= \frac{1}{2}(22 + 11.8) \times 14 = 236.6 \text{ mm}^2$$

Maximum permissible tension on the belt:
$$T_{max} = \sigma \times A = 1.5 \times 236.6 = 354.9 \text{ N}$$

For maximum power, we have
$$T_c = \frac{T_{max}}{3} = \frac{354.9}{3} = 118.3 \text{ N}$$

Belt velocity:
$$v = \sqrt{\frac{T_c}{m}} = \sqrt{\frac{118.3}{0.4}} = 17.19 \text{ m/s}$$

Tension on tight side:
$$T_1 = T_{max} - T_c$$
$$= 354.9 - 118.3 = 236.6 \text{ N}$$

Ratio of tensions:
$$\frac{T_1}{T_2} = e^{\mu \theta / \sin \alpha}$$
$$= e^{0.2 \times 155 \times \frac{\pi}{180} / \sin 20°}$$
$$= 4.86$$

$$T_2 = \frac{T_1}{4.86} = \frac{236.6}{4.86} = 48.68$$

Power: $$P = \frac{(T_1 - T_2)v}{1000} = \frac{(236.6 - 48.68) \times 17.19}{1000} = 3.23 \text{ kW}$$ Ans.

EXAMPLE 6.10 A belt drive fitted with a gravity idler pulley is shown in Figure 6.17. The driving pulley rotates at 400 rpm in counterclockwise direction and coefficient of friction is 0.3. Determine the initial tension in the belt and power transmission capacity.

Solution Refer to the free body diagram of idler pulley as shown in Figure 6.18. Let P be the normal reaction at point O between the idler pulley and the lever and R be the resultant of initial belt tensions at idler pulley ($= \sqrt{2}\, T_0$).

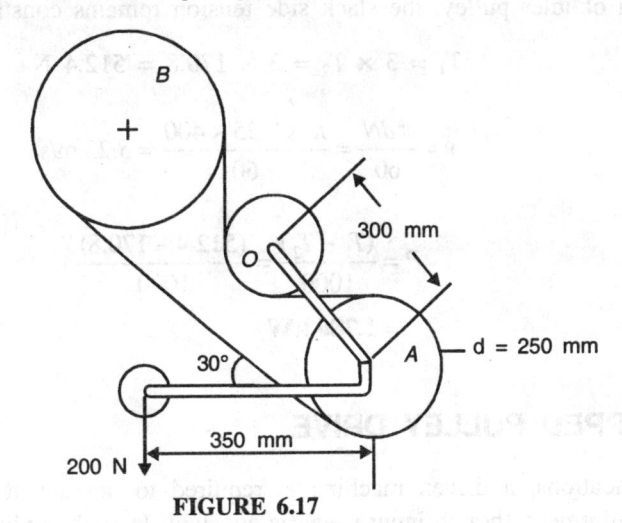

FIGURE 6.17

FIGURE 6.18

Taking moment about lever fulcrum A,
$$P \times 300 = 200 \times 350$$
or
$$P = 233.34 \text{ N}$$

Reaction:
$$R = \frac{P}{\cos 15°} = \frac{233.34}{\cos 15°} = 241.56 \text{ N}$$

Initial tension:
$$T_0 = \frac{241.56}{\sqrt{2}} = 170.8 \text{ N} \quad \text{Ans.}$$

Angle of contact on driving pulley A:
$$\theta = 180° + 30° = 210°$$

Ratio of tensions:
$$\frac{T_1}{T_2} = e^{\mu\theta} = e^{0.3 \times \frac{\pi}{180} \times 210°} = 3.0$$

Due to action of idler pulley, the slack side tension remains constant at T_0, i.e. $T_2 = T_0$

Therefore, $T_1 = 3 \times T_2 = 3 \times 170.8 = 512.4$ N

Belt velocity:
$$v = \frac{\pi d N}{60} = \frac{\pi \times 0.25 \times 400}{60} = 5.23 \text{ m/s}$$

Power:
$$P = \frac{(T_1 - T_2)}{1000} = \frac{(512.4 - 170.8)}{1000}$$
$$= 1.786 \text{ kW} \quad \text{Ans.}$$

6.12 STEPPED PULLEY DRIVE

In some applications, a driven machine is required to operate at different speeds as per functional requirement, though input speed is constant. In such applications, a pair of stepped pulley, also known as **cone pulley** is used (Figure 6.19).

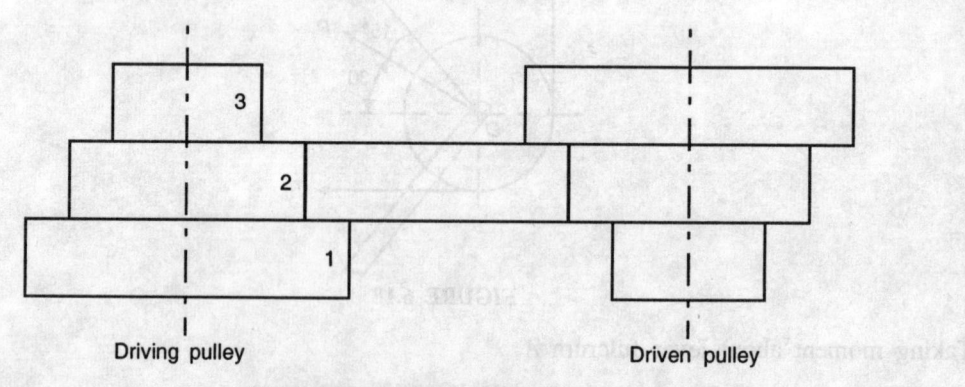

FIGURE 6.19 A stepped pulley drive.

Let
    n = speed of the driving shaft
    N_i = speed of driven shaft when belt is shifted to ith step
    d_i = diameter of driving pulley at ith step
and D_i = diameter of driven pulley at ith step

The velocity ratio of the first step:

$$\frac{N_1}{n} = \frac{d_1}{D_1} \qquad (6.33a)$$

Velocity ratio of ith step:

$$\frac{N_i}{n} = \frac{d_i}{D_i} \qquad (6.33b)$$

In multi-speed driven shaft, it is usual practice to have speed of the driven shaft in geometrical progression.

Let K be the common ratio of speed progression and $n = N_1$. Then we get

$$\frac{N_2}{N_1} = \frac{N_3}{N_2} = \cdots = \frac{N_i}{N_{i-1}} = K$$

or
$$N_i = K^{i-1} N_1 \qquad (6.34)$$

From Eqs. (6.33a) and (6.34), the speed of ith step:

$$N_i = K^{i-1} \times n \frac{d_1}{D_1}$$

or
$$\frac{N_i}{n} = K^{i-1} \times \frac{d_1}{D_1}$$

or
$$\frac{d_i}{D_i} = K^{i-1} \times \frac{d_1}{D_1} \qquad (6.35)$$

Further, in stepped pulley drive, same belt is used to obtain different speeds; thus the length of belt for all pair of steps should be the same.

or
$$l_1 = l_2 = l_3 = \cdots = l_i \qquad (6.36)$$

In the design of open-belt type stepped drive, having decided the ratio d_i/D_i, the diameters of pulleys at each step can be found from the belt length formula. In cross-belt drive, the relation for the length of belt is satisfied if the sum of radii of different pairs of the steps is constant, i.e.

$$R_1 + r_1 = R_2 + r_2 = \cdots = R_n + r_n \qquad (6.37)$$

EXAMPLE 6.11 A shaft, which rotates at a constant speed of 160 rpm, is connected to a parallel shaft at 720 mm apart by means of flat belt. The parallel shaft runs at 60, 80 and 100 rpm. If the diameter of the smallest pulley on the driving shaft is 80 mm, determine the remaining diameters of the stepped pulleys for (a) cross-belt and (b) open-belt drives.

Solution Let d_1, d_2 and d_3 be the diameters of step pulley on the driving shaft and D_1, D_2 and D_3 be corresponding diameters of step pulley on the driven shaft. (See Figure 6.20.) When the belt is on the step 3 having diameters d_3 and D_3, the speed of driven shaft will be minimum, i.e. 60 rpm.

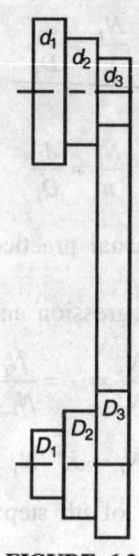

FIGURE 6.20

Therefore,
$$\text{Velocity ratio} = \frac{N_3}{n} = \frac{d_3}{D_3}$$

or
$$D_3 = \frac{n}{N_3} \times d_3 = \frac{160}{60} \times 80 = 213.34 \text{ mm}$$

(a) For cross-belt drive. In case of cross-belt drive, the length of the belt will be constant if the sum of diameters of corresponding steps is equal.

$$d_1 + D_1 = d_2 + D_2 = d_3 + D_3 = 80 + 213.34 = 293.34 \qquad (i)$$

First step

Velocity ratio:
$$\frac{d_1}{D_1} = \frac{N_1}{n}$$

or
$$\frac{d_1}{D_1} = \frac{100}{160} = 0.625 \qquad (ii)$$

Solving Eqs. (i) and (ii), we get

$$D_1 = 180.51 \text{ mm} \quad \text{and} \quad d_1 = 112.82 \text{ mm} \qquad \text{Ans.}$$

Second step

Velocity ratio: $\dfrac{d_2}{D_2} = \dfrac{N_2}{n} = \dfrac{80}{160} = 0.5$ (iii)

Solving Eqs. (i) and (iii), we get

$$D_2 = 195.56 \text{ mm} \quad \text{and} \quad d_2 = 97.78 \text{ mm} \qquad \text{Ans.}$$

(b) Open-belt drive. Length of open belt connecting step 3, i.e. d_3 and D_3:

$$L = \dfrac{\pi(d_3 + D_3)}{2} + \dfrac{(D_3 - d_3)^2}{4C} + 2C$$

$$= \dfrac{\pi(80 + 213.34)}{2} + \dfrac{(213.34 - 80)^2}{4 \times 720} + 2 \times 720$$

$$= 1906.95 \text{ mm}$$

First step

The length of the belt for step 1 should be equal to that for step 3.

$$\dfrac{\pi(d_1 + D_1)}{2} + \dfrac{(D_1 - d_1)^2}{4C} + 2C = 1906.95 \qquad \text{(iv)}$$

and $\qquad VR = \dfrac{d_1}{D_1} = \dfrac{N_1}{n} = \dfrac{100}{160} = 0.625$

or $\qquad d_1 = 0.625 \, D_1$ (v)

Solving Eqs. (iv) and (v), we get

$$D_1 = 184.32 \text{ mm} \quad \text{and} \quad d_1 = 115.2 \text{ mm} \qquad \text{Ans.}$$

Second step

$$\dfrac{\pi(d_2 + D_2)}{2} + \dfrac{(D_2 - d_2)^2}{4C} + 2C = 1906.95 \qquad \text{(vi)}$$

and $\qquad VR = \dfrac{d_2}{D_2} = \dfrac{N_2}{n} = \dfrac{80}{160} = 0.5$

or $\qquad d_2 = 0.5 D_2$ (vii)

Solving Eqs. (vi) and (vii), we get

$$D_2 = 197.75 \text{ mm} \quad \text{and} \quad d_2 = 98.88 \text{ mm} \qquad \text{Ans.}$$

6.13 ROPE DRIVE

Ropes are mainly used in elevators, mine hoist, cranes, oil well drilling, areal conveyor and haulage devices. There are two types of rope available for hoist and haulage purposes—fibre ropes and metallic wire ropes. Fibre ropes are generally made out of fibre of coir, hemp, cotton, manila, jute, nylon and polypropylene. Fibre ropes due to their low strength, high weight to strength ratio and high abrasion wear rate are suitable only for hand-operated hoisting machinery.

Metallic wire ropes are made out of cold drawn steel, wrought iron, phosphor bronze and stainless steel. These wire ropes have high strength to weight ratio, offer silent operation and greater resistance to shock load and are more reliable.

6.13.1 Wire Rope Construction

A wide varieties of wire ropes are constructed to suit the requirements of various applications and operating conditions. In a wire rope, certain number of wires, usually 7, 9, 21, 25 or 37, are first twisted together to form a strand and then a certain number of strands, usually 6, 8, 17 or 34, are twisted together around a core to form a rope. The core of wire rope may be made of fibre, steel, plastic or asbestos. The fibre or plastic core acts as an elastic support. These wires can carry lubricant to reduce wear of wire due to contact stresses. Ropes with steel core have high strength though at the cost of flexibility. Asbestos-cored rope can operate at elevated temperature. The construction details of typical wire ropes are shown in Figure 6.21.

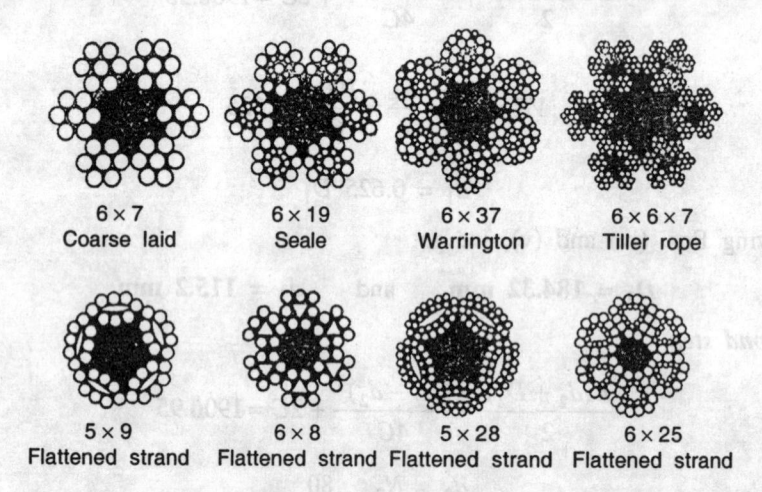

FIGURE 6.21 Construction details of a wire rope.

In a wire rope construction, besides number of wires in a strand and number of strands in a rope, the manner in which these wires and strands are twisted is important because the flexibility and life of rope depend on it. According to the hand of helix, ropes are classified into two categories—right hand or Z-lay and left hand or S-lay. If wires in a strand are coiled in the same direction as that of strand, then such rope is called **Lang's lay**. On the other hand, if wires in a strand are coiled in a direction opposite to that of the strand, then it is called

regular or **crosslay rope** (Figure 6.22). The Lang's lay ropes have high resistance to wear, high flexibility and increased fatigue strength. However, these ropes may untwist under the load, whereas crossed lay ropes offer higher crushing strength and are less likely to untwist.

Lang's lay Right lay

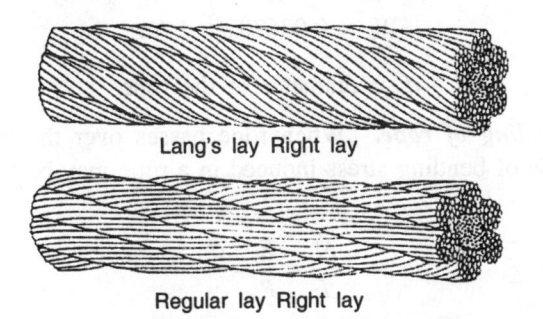

Regular lay Right lay

FIGURE 6.22 Types of wire rope.

According to BIS under code IS: 3973–1965, a code of practice for selection, installation and maintenance of wire ropes, the wire rope is designated by two numbers. The first number indicates the number of strands in a wire and the second number indicates number of wires per strand. For example, a wire rope of 6 × 19 construction means a rope has 6 strands and each strand is composed of 19 wires. Some typical standard wire ropes and their applications are given in Table 6.1.

Table 6.1 Applications of wire ropes

Type of rope	Applications
6 × 7	A coarse lay rope, useful for haulage and all those applications where resistance to abrasion is required
6 × 19	Flexible rope, useful for cranes and general purpose engineering machines
6 × 37	Extra flexible, useful for high speed machines, it can be subjected to reverse bending

6.13.2 Forces on a Wire Rope

A wire rope used in elevator and hoisting applications is generally subjected to the following types of forces:

Direct force. It is sum of weight to be lifted and weight of the rope itself, i.e.

$$F_d = W + W_r \qquad (6.38)$$

where
   W = weight to be lifted
and W_r = weight of the rope

The weight of rope is generally given in N/m length, and it can be found either from manufacture's catalogue or from the BIS under code IS: 2266–1963. In the absence of above data, the following empirical relation can be used:

$$W_r = 0.03678\, d_r^2 \text{ N/m}$$

where d_r = diameter of the rope

Force due to bending of rope. When rope passes over the drum, it is subjected to bending. The magnitude of bending stress induced in a rope may be found from elastic curve equation:

$$\frac{M}{I} = \frac{E}{R} = \frac{\sigma}{y}$$

or

$$\text{Bending stress: } \sigma_b = E \times \frac{d_w}{d_s} \tag{6.39}$$

where
- d_s = diameter of sheave (= 40 d_r)
- d_w = diameter of wire (= 0.063 d_r)
- E = elastic modulus of rope = 0.375 × elastic modulus of rope material

or

$$\text{Bending load: } F_b = \sigma_b \cdot A = \frac{d_w}{d_s} EA \tag{6.40}$$

where A is the area of cross-section of the wire rope (= 0.38 d_r^2), d_r being diameter of the rope

Force due to acceleration/retardation. When ropes are used in an elevator or haulage applications, the change in its speed causes acceleration or retardation of wire which induces additional force.

Acceleration force:

$$F_a = F_d \times \frac{a}{g} \tag{6.41}$$

where
F_d = direct force

and a = acceleration [= $(V_2 - V_1)/t$], $V_2 - V_1$ being change in the speed in time t seconds

Starting force. While starting an elevator, the rope and supported load has to be accelerated by a force transmitted through the rope. If there is a slack of amount h in the rope before starting, there will be considerable impact load on the rope. This impact load may be determined by the following relation:

$$F_i = F_d \left[1 + \sqrt{\frac{2ahE}{lg\sigma}} \right] \tag{6.42}$$

where
σ = static stress in the rope

and l = length of the rope

Total effective force on the rope at different operating conditions is given as:

(i) During normal running:
$$F_e = F_d + F_b \tag{6.43a}$$

(ii) During acceleration:
$$F_e = F_d + F_b + F_a \tag{6.43b}$$

(iii) During starting/stopping:
$$F_e = F_d + F_i \tag{6.43c}$$

6.13.3 Breaking Strength

Breaking strength of a wire rope is the maximum amount of force that can be applied to a rope without its failure. It depends upon size of the rope, size and number of the wires, the wire material and the type of core. Generally, strength of a rope is about 80 per cent of the combined strength of all wires of the rope. The nominal breaking strength of a general purpose wire rope is listed in BIS under code IS: 2266–1963. However, in the absence of this code, the following approximate relation can be used.

$$\text{Breaking strength: } F = A \times \sigma_{ut} \tag{6.44}$$

where

σ_{ut} = ultimate strength of the wire material (= 1550–1850 N/mm²)

and A = area of cross-section of the rope (= $0.38 d_r^2$),

d_r being diameter of the rope

The kinematics of a rope used in the power transmission is same as that of V-belt. Thus the same formula can be used.

EXAMPLE 6.12 A shaft running at 500 rpm carries a pulley of 500 mm diameter which drives another pulley in the same direction with a speed reduction of 2:1 by means of ropes. The drive transmits 100 kW. The angle of groove is 40°, the distance between the centres is 1.5 m. The mass of the rope is 0.12 kg/m and allowable strength is 1.8 N/mm². It is recommended that initial tension in the rope should not exceed 700 N. Find the number of ropes required and diameter of the rope. Assume $\mu = 0.2$.

Solution

Velocity of rope:
$$v = \frac{\pi d_1 N_1}{60} = \frac{\pi \times 0.5 \times 500}{60}$$
$$= 13.1 \text{ m/s}$$

Centrifugal tension: $T_c = mv^2 = 0.12 \times 13.1^2 = 20.6$ N

Angle of contact: $\theta = 180° - 2\phi$

where
$$\phi = \sin^{-1}\left(\frac{d_2 - d_1}{2c}\right) \text{ with } d_2 = 2d_1 = 1000 \text{ mm}$$

$$= \sin^{-1}\left(\frac{1000 - 500}{2 \times 1500}\right) = 9.6°$$

$\therefore \qquad \theta = 180° - 2 \times 9.6 = 160.8°$

Ratio of tension:
$$\frac{T_1}{T_2} = e^{\mu\theta/\sin\alpha}$$

$$= e^{0.2 \times 160.8 \times \frac{\pi}{180} \times \frac{1}{\sin 20°}}$$

or $\qquad \dfrac{T_1}{T_2} = 5.16$

We know that

Initial tension:
$$T_o = \frac{T_1 + T_2 + 2T_c}{2}$$

or $\qquad 700 = \dfrac{T_1 + T_2 + 2 \times 20.6}{2}$

or $\qquad T_1 + T_2 = 1358.8 \text{ N} \qquad\qquad\qquad$ (ii)

Solving Eqs. (i) and (ii), we get

$\qquad T_1 = 1138.2 \text{ N}$ and $T_2 = 220.6 \text{ N}$

Power capacity per rope $= \dfrac{(T_1 - T_2)v}{1000} = \dfrac{(1138.2 - 220.6) \times 13.1}{1000} = 12.0 \text{ kW}$

Number of ropes $= \dfrac{\text{Total power}}{\text{Capacity/rope}} = \dfrac{100}{12} = 8.34$, say **9 ropes** **Ans.**

Maximum tension:
$$T_{\max} = T_1 + T_c$$
$$= 1138.2 + 20.6$$
$$= 1158.8 \text{ N}$$

Rope strength:
$$\frac{\pi}{4}d^2 \times \sigma = 1158.8$$
or

Rope diameter: $\qquad d = \sqrt{\dfrac{4 \times 1158.8}{\pi \times 1.8}} = \textbf{28.6 mm} \qquad\qquad$ **Ans.**

EXAMPLE 6.13 Determine the number of turns a hauling rope must be wound round a rotating capstan in order to haul 30,000 kg mass up a gradient of 1:25. The rolling resistance is 50 N per 1000 kg and the pull on the free end of the rope is 150 N. Coefficient of friction is 0.4.

Solution Force on tight side of the rope = Rolling resistance
+ Effort component of the weight

or
$$T_1 = 30,000 \times \frac{50}{1000} + 30,000 \times 9.81 \times \frac{1}{25} = 13272 \text{ N}$$

Slack side tension: $T_2 = 150$ N (given)

Ratio of tension: $\dfrac{T_1}{T_2} = e^{\mu\theta}$

or
$$\theta = \frac{\log_e (T_1/T_2)}{\mu} = \frac{\log_e \left(\dfrac{13,272}{150}\right)}{0.4} = 11.2 \text{ rad}$$

Number of turns:
$$N = \frac{\theta}{2\pi} = \frac{11.2}{2\pi} = 1.78, \text{ say } \mathbf{2 \text{ turns}} \qquad \text{Ans.}$$

EXAMPLE 6.14 A wire rope of 28 mm diameter is used to lift the debris from a 60 m deep well. The weight is lifted at speed of 150 m/min, which is attained in one second and initial slack is 200 mm. Assuming that load-carrying capacity of the rope is about one-fifth of breaking strength, determine the weight of the debris which can be lifted. The ultimate strength of rope material is 1600 N/mm².

Solution The maximum force on the rope which will occur at the time of starting is:
$$F_{max} = F_d + F_i$$
where
$$F_d = W + W_r$$

W being the weight of the debris and
W_r that of the rope

We have
$$W_r = 0.03678 d_r^2 \times \text{length of the lift}$$
$$= \frac{0.03678 \times 28^2 \times 60}{1000} = 1.73 \text{ kN}$$

$$F_d = W + 1.73 \text{ kN}$$

F_i = Starting slack force
$$= F_d \left[1 + \sqrt{\frac{2ahE}{\sigma l g}}\right]$$

Acceleration:
$$a = \frac{v}{60 \times t} = \frac{150}{60 \times 1} = 2.5 \text{ m/s}^2$$

$$F_i = F_d \left[1 + \sqrt{\frac{2 \times 2.5 \times 0.2 \times 0.75 \times 10^5}{1600 \times 60 \times 9.81}} \right] = 1.282 \, F_d$$

Breaking strength:
$$F_b = A \times \sigma_{ut}$$
$$= 0.38 \, d_r^2 \times 1600$$
$$= \frac{0.38 \times 28^2 \times 1600}{1000} = 476.67 \text{ kN}$$

Effective breaking strength $= \dfrac{F_b}{5} = \dfrac{476.67}{5} = 95.3 \text{ kN}$

For safe working, effective breaking strength should be greater than the maximum force, i.e.

$$F_b \geq F_d + F_i$$

$$95.3 = F_d + 1.282 \, F_d$$

or
$$F_d = 41.76 \text{ kN}$$

Weight of the debris: $W = F_d - W_r$

$= 41.76 - 1.73 =$ **40.03 kN** Ans.

6.14 CHAIN DRIVE

A chain drive consists of an endless chain which runs over toothed driving and driven wheels called **sprockets.** A chain consists of rigid links which are hinged together to get desired flexibility for wrapping around the sprocket. It is relatively more positive drive than belt drive. However, where precise timing is required, it can not be used. Chain drive is widely used in bicycles, motorcycles, conveyors, agriculture machinery and wood working machines. The main advantages and limitations of chain drive are discussed further.

Advantages

(i) It can be used fairly over short and long distances, up to 8 m approximately.
(ii) It is compact in size compared to flat-belt drive.
(iii) The transmission efficiency is as high as 95–98 per cent.
(iv) It puts less force on the shaft compared to belt drive.

Limitations

(i) Weight of the drive is higher compared to belt drive.
(ii) It requires precise alignment of shafts.
(iii) It requires more frequent service and maintenance compared to belt drive.
(iv) It can be used for velocity ratio up to 1:8, chain velocity up to 25 m/s and power rating up to 11 kW.

6.15 TYPES OF CHAIN

According to the type of service to be performed by chains, these can be classified into three categories:

(i) Hoisting and hauling chains
(ii) Conveyor chains
(iii) Power transmission chains

A brief description about these is given further.

Hoisting chains. The hoisting chains are generally made of plain carbon steel. Their links are either round, oval or square in shape. These chains are available for load lifting and chain pulley block purpose and should conform to specifications provided in the BIS under code IS: 2429–1970. The oval link chains are used with sprockets which have teeth to receive the links. These chains are used for low speed applications such as chain hoist and anchors for marine work. Oval link chain does not kink easily (see Figure 6.23).

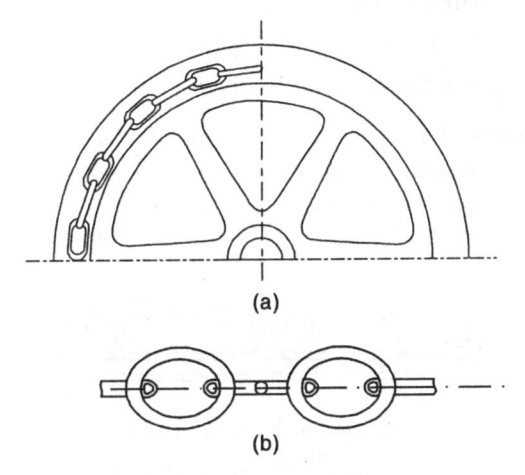

FIGURE 6.23 Hoisting chains.

Conveyor chains. The conveyor chains are meant for conveying the material continuously. They run at slow speed (up to 10 km/h). These chains are generally made of two types—hook joint or detachable type and closed joint type as shown in Figure 6.24. The sprocket teeth of these are so shaped that chain runs onto the sprocket without any interference. The conveyor chains are generally made of medium carbon steel and malleable cast iron

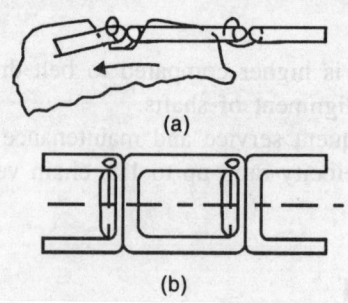

FIGURE 6.24 Conveyor chains.

having high tensile strength and abrasion wear resistance. The specification of these chains should conform to the BIS under code IS: 3748–1967.

Power transmission chains. A chain used for transmitting power from one shaft to another is called **power transmission chain.** Generally, three types of chains, namely block chain, roller chain, and inverted tooth or silent chain, are used for power transmission purpose. However, the roller chain has practically superseded other two types of chain due to the following advantages:

(a) Simple construction
(b) Due to rolling friction between the roller and the sprocket, the teeth wear rate is very small.
(c) Longer service life
(d) Lower noise compared to bush chain
(e) It can transmit more power without slippage or creep.

A brief description about these chains is given below:

Block chain. A block chain is the oldest type of chain used for power transmission purpose. Sometimes, these chains are also used as conveyer chains. When a block chain approaches to or leaves the teeth of sprocket, rubbing takes place between teeth and links. This makes operation noisy and reduces operating life of the chain (Figure 6.25).

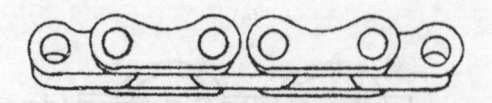

FIGURE 6.25 Block chain.

Roller chain. The roller chains are nowadays widely used in power transmission. A roller chain consists of five components—inner link, outer link, pin, bush and roller (Figure 6.26). While assembling a roller chain, bush is first passed through roller, which is free to rotate on the bush and then the bush is held between two inner links by press fitting. Outer links are placed over the inner links and a pin is passed through outer link, inner link and bush which is riveted to form a complete assembly as shown in Figure 6.26. In this chain, the bush and the pin form a swivel joint and outer link is free to swivel with respect to the

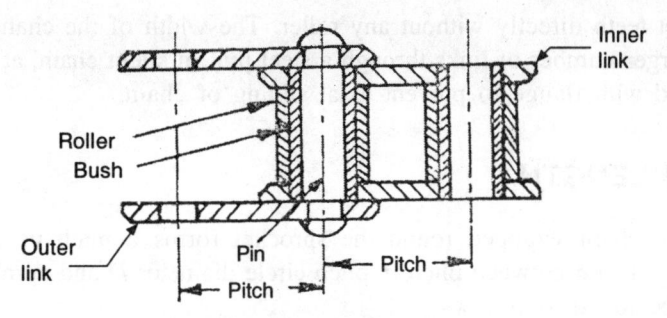

FIGURE 6.26 A roller chain.

inner link. The roller can turn freely on the bush. This type of construction results in rolling friction between the roller and sprocket teeth which reduces wear. The pin, roller and bush of a roller chain are generally made of alloy steels. Roller chains are manufactured on the basis of standard values of pitch. Depending upon the power transmission capacity, these chains are available in single strand or multiple strands called simplex, duplex or triplex chains as shown in Figure 6.27.

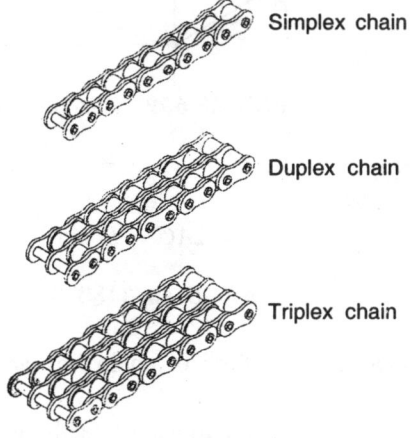

FIGURE 6.27 Multiple-strand chains.

Inverted tooth chain. The inverted tooth chain commonly known as **silent chain** was first invented by H. Renold in England. This type of chain is most widely used where maximum quietness is desired at heavier load. It consists of a series of flat links of a hooked form shape made of stamping the sheet metal (Figure 6.28). The outer faces of the teeth are ground to give an included angle of 60° or in some cases 75°. These links themselves engage

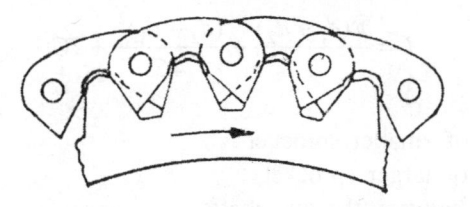

FIGURE 6.28 An inverted tooth chain.

with the sprocket teeth directly without any roller. The width of the chain can be increased by connecting larger number of links through a steel pin. In silent chain, at least one sprocket must be provided with flange to prevent axial sliding of chain.

6.16 CHAIN LENGTH

In chain drive, a chain wrapped round the sprocket forms a pitch polygon as shown in Figure 6.29. The relation between pitch p, pitch circle diameter D and number of teeth on the sprocket Z can be found as follows:

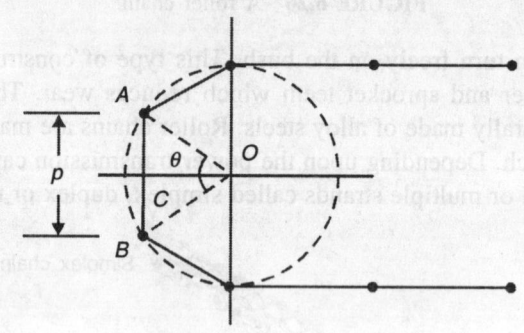

FIGURE 6.29 Pitch polygon.

Referring to Figure 6.29,

Pitch: $$p = 2AC = D \sin \frac{\theta}{2}$$

or $$p = D \sin (180°/Z)$$

where
θ = angle subtended by the chord of chain link at the centre of the sprocket ($= 360°/Z$)

For the sprocket having larger number of teeth (generally $Z > 17$), the pitch can be approximately computed by the following relation:

Pitch: $$p = \frac{\pi D}{Z} \tag{6.45}$$

The expression for approximate length of chain can be derived by considering it as an open-belt drive. The approximate length of belt is:

$$L = \frac{\pi(d_1 + d_2)}{2} + \frac{(d_2 - d_1)^2}{4C} + 2C \tag{6.46}$$

where
d_1 = pitch diameter of smaller sprocket
d_2 = pitch diameter of larger sprocket
and C = centre distance between the two shafts

Let Z_1 and Z_2 be the number of teeth on driving and driven sprockets respectively. We can rewrite the Eq. (6.46) using $\pi d = Zp$ as

$$L = p\left[0.5(Z_1 + Z_2) + \frac{p(Z_2 - Z_1)^2}{4\pi^2 C} + \frac{2C}{p}\right] \quad (6.47)$$

or
$$L = p \times l_p$$

where $l_p = \text{constant} \left[= 0.5(Z_1 + Z_2) + \frac{p(Z_2 - Z_1)^2}{4\pi^2 C} + \frac{2C}{p}\right]$

Since a chain contains integer number of links, the constant l_p should also be an integer. On account of taking l_p as an integer, the length of the chain may not fit exactly between the given centre distance. Thus a little adjustment in the centre distance is generally needed.

6.17 CHORDAL ACTION

In a chain drive, when a chain is wrapped round the sprocket, it passes over it as a series of chordal links. This action causes varying chain speed with the changing angular position of sprocket. To illustrate, let us consider a sprocket having six teeth and rotating at a constant speed N rpm as shown in Figure 6.30. In this case, the chain link is at a distance of $D/2$ from the centre of sprocket and its linear velocity is:

$$V = \omega \times OA = \pi DN.$$

Now suppose if sprocket rotates by some angle say $\theta/2$, the effective radius is reduced to OC (Figure 6.30). The chain velocity reduces to

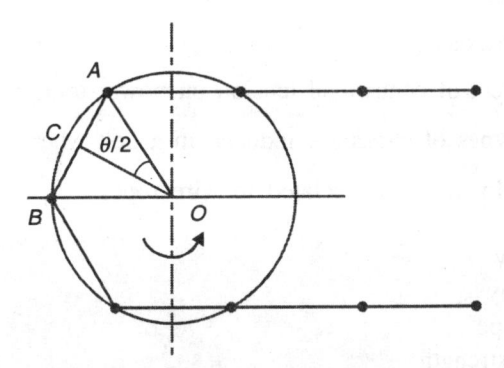

$$V' = \pi DN \cos\theta/2$$

FIGURE 6.30 Chordal action.

This action of the chain which does not transmit uniform linear velocity is called **chordal action**.

The amount of chain velocity variation is:

$$\Delta V = V - V' = \pi DN (1 - \cos\theta/2)$$

or
$$\Delta V = \pi DN \left[1 - \cos\left(\frac{180°}{Z}\right)\right] \tag{6.48}$$

This variation of velocity of the chain approaches to zero when number of teeth on a sprocket tends to infinity. It is observed that Eq. (6.48) yields 1.7 per cent variation at $Z = 17$ and 0.788 per cent at $Z = 25$. Therefore, it is good practice to use a driving sprocket with at least 17 teeth. Further, higher number of teeth results into lower level of noise.

In a chain drive, the uniform rate of wear of the chain link and sprocket can be obtained if the number of teeth on one sprocket is even while on the other sprocket, it should be odd.

EXERCISES

1. Explain the various advantages and disadvantages of belt drive.
2. Discuss the mechanics of belt drive.
3. Explain the terms slip and creep as referred to belt drive. On what factors does it depend? Explain how creep affects the power transmission capacity of the belt drive.
4. What is use of idler pulleys in the belt drive?
5. Sketch some typical arrangements for providing initial tension in the belt.
6. What will be the effect on power transmission capacity of a belt drive if
   (a) open belt is replaced with cross belt
   (b) flat belt is replaced with V-belt
   (c) idler pulley is used
   (d) belt surface is dressed
   (e) angle of lap is decreased?

   Explain with reasons.
7. What is the effect of centrifugal tension on power transmission capacity of a belt?
8. Discuss what types of stress are induced in a belt when it transmits power.
9. Explain the following terms related to wire ropes:
   (a) Strands
   (b) Lang's lay
   (c) Regular lay
   (d) 6 × 19 rope
   (e) Breaking strength
10. Explain the various advantages and limitations of a chain drive.
11. Explain the constructional details and working of roller and silent chains.

12. What do you mean by chordal action? Explain its effect on the performance of chain drive.

13. A leather belt is required to transmit 9 kW from a pulley 1.2 m in diameter running at 200 rpm. The angle embraced is 165° and coefficient of friction between the leather belt and the pulley is 0.3. If the safe working stress for the leather belt is 1.4 N/mm^2, the mass of leather is 1000 kg/m^3 and thickness of the leather belt is 10 mm, determine the width of the belt taking the centrifugal force into account.
[**Ans:** 99.7 mm]

14. Determine the maximum power that can be transmitted by a 100 × 10 mm^2 belt with angle of contact equal to 160°. The mass of the belt is 1000 kg/m^3 and coefficient of friction is 0.25. The stress in the belt should not exceed 1.5 N/mm^2.
[**Ans:** 10.89 kW]

15. A shaft rotating at 300 rpm drives another shaft at 200 rpm and transmits 6 kW through a belt drive. The belt is 100 mm wide and 10 mm thick. The distance between the centres is 4 m. The smaller pulley is 500 mm in diameter. Calculate stresses in open-belt and cross-belt. Take μ = 0.3.
[**Ans:** 1.267 N/mm^2, 1.184 N/mm^2]

16. A belt having density 1000 kg/m^3 has the maximum permissible stress of 2.5 N/mm^2. Determine the maximum power that can be transmitted by the belt of 200 mm × 12 mm, if the ratio of tensions is 2. Take belt velocity equal to 25 m/s.
[**Ans:** 56.25 kW]

17. Two pulleys of diameters 300 mm and 900 mm are on parallel shafts 1800 mm apart. Determine the necessary length of belt of an open–belt drive to connect two shafts. If the coefficient of friction is 0.2 and initial tension is same, show that the maximum powers that can be transmitted by open and cross belts are in ratio of about 1:1.34.
[**Ans:** 5.535 m]

18. An open-belt drive connects two pulleys of diameters 1.2 m and 0.5 m on parallel shafts 3.6 m apart. The belt has a mass of 0.9 kg/m length and maximum tension in it should not exceed 2 kN.
The pulley of diameter 1.2 m, which is driving, runs at 200 rpm. Due to belt slip on one of the pulleys, the speed of the driven shaft is only 450 rpm. Calculate the torque on each of the two shafts, power transmitted and power lost. What is efficiency of the drive? Take μ = 0.3.
[**Ans:** 668.94 Nm, 278.72 Nm, 13.125 kW, 0.875 kW, 93.75%]

19. A flat belt is to be used to transmit 80 kW at belt speed 20 m/s between two pulleys of diameters 250 mm and 400 mm having centre distance of 1 m. The allowable belt stress is 6 MN/m^2 and belt is available having a thickness to width ratio of 0.1 and material density of 1000 kg/m^3. If the coefficient of friction is 0.3, determine the minimum belt width required.

What would be the necessary installation force between the pulley bearings and the force between the pulley bearings when full power is being transmitted?

[Ans: 109.8 mm, 5263.9 N, 10498.16 N, 9536.8 N]

20. A compressor is driven through belt from a line shaft, the pulley on the compressor being of diameter 400 mm. The angle of lap of the belt is 160°. When the belt is moved from the loose to fast pulley, it slips for 8 seconds until the compressor attains its speed of 400 rpm. The flywheel of the compressor has a moment of inertia of 3.5 kgm^2 and the friction requires a constant torque of 3.5 Nm. If the coefficient of friction is 0.3, find the tensions (tight and slack) in the belt and also the energy lost in that time due to the belt slip.

[Ans: 192.4 N, 83.3 N, 3653.67 J]

21. A V-belt having a lap angle of 165° has a cross-sectional area of 240 mm^2 and runs in a groove of included angle 40°. The mass density of the belt is 1500 kg/m^3. The maximum stress is limited to 4 MN/m.^{2}. Find the maximum power that can be transmitted by two such belts. Take μ = 0.145. Find also the shaft speed at which power transmitted would be maximum if pulley diameter is 600 mm.

[Ans: 26.9 kW, 948.88 rpm]

22. Power is transmitted by a V-belt drive. The included angle of V-groove is 30°, the belt is 20 mm deep and maximum width is 20 mm. If the mass density of the belt material is 1200 kg/m^3 and allowable stress is 2 N/mm^2, determine the maximum power that can be transmitted. The angle of lap is 140° and coefficient of friction is 0.15.

23. A rope drive is required to transmit power from a shaft rotating at 240 rpm. On account of surging due to variation in the resisting torque, the maximum permissible tension in each rope is limited to $\left(1350 + \dfrac{8000}{v^2}\right)$ N, where v is the rope speed in m/s. The mass of the rope is 1.2 kg/m length, the angle of lap is 170° and the groove angle is 60°. The coefficient of friction between the rope and the pulley is 0.3. Taking into account centrifugal tension, determine
   (i) linear speed of the rope at which power is transmitted
   (ii) the necessary effective diameter of pulley
   (iii) the number of ropes required to transmit 150 kW.

[Ans: 19.5 m/s, 1.55 m, 7 ropes]

24. A pulley used to transmit power by means of ropes has a diameter 3.6 m and has 15 grooves of 45° angle. The angle of contact is 170° and coefficient of friction is 0.28. The maximum possible tension in the rope is 96 N and mass of the rope is 0.15 kg/m length. What is the speed of the pulley in rpm and power transmitted if the condition of maximum power prevails?

[Ans: 74.59 rpm, 11.96 kW]

25. A short vertical rope drive is required to transmit power from a pulley of 1.15 m effective diameter. The ropes have a mass of 1.2 kg/m, the groove angle is 50° and the angle of lap is 170°. The coefficient of friction is 0.3.
   (a) With initial tension of 700 N in each rope, what is the maximum power which can be transmitted per rope? What will be the load in the rope and its linear speed?
   (b) If the permissible load is 1.6 kN/rope and this is to be fully utilized for maximum power, what should be the initial tension in the rope?
   [**Ans:** 14 kW, 1.273 kN, 18.8 m/s, 1.13 kN]

MULTIPLE CHOICE QUESTIONS

1. The angular velocities of two pulleys connected by cross belt are
   (a) directly proportional to their diameters
   (b) inversely proportional to their diameters
   (c) directly proportional to square of their diameters
   (d) inversely proportional to square of their diameters

2. The power transmission capacity of a belt drive is calculated on the basis of angle of contact on the
   (a) smaller pulley   (b) larger pulley
   (c) average of smaller and larger pulley
   (d) driving pulley, irrespective of size

3. The idler pulley is fitted on the
   (a) tight side   (b) slack side
   (c) either side   (d) none of the above

4. Due to creep of belt, there is
   (a) loss of power to be transmitted
   (b) increased power transmission capacity
   (c) no change in power
   (d) none of the above

5. For stepped pulleys, if the sum of the radii of the driving pulleys is constant for all steps, then
   (a) an open belt is recommended
   (b) a cross-belt is recommended
   (c) use of an idler pulley is recommended
   (d) any one of the above is recommended

6. The crowning of pulley is done to
   (a) improve power
   (b) improve pulley strength
   (c) increase velocity ratio
   (d) prevent the belt running off the pulley

7. Considering centrifugal tension in belts, the maximum permissible velocity is
   (a) proportional to maximum tension
   (b) inversely proportional to the maximum tension
   (c) proportional to square root of the maximum tension
   (d) independent of the maximum tension

8. If the initial tension in the belt is increased, the power transmission capactiy
   (a) is increased
   (b) is decreased
   (c) remains same
   (d) is independent to initial tension

9. For constant velocity ratio with large centre distance, which one of the following drives is recommended?
   (a) flat belt
   (b) V-belt
   (c) rope
   (d) chain

10. For maximum power transmission condition, the ratio of tight side tension to the maximum tension is
    (a) $\dfrac{2}{3}$
    (b) $\dfrac{1}{3}$
    (c) $\dfrac{1}{2}$
    (d) $\dfrac{3}{4}$

11. The maximum stress in a belt
    (a) occurs at the beginning of arc of the driving pulley
    (b) occurs at the ending of arc of the driven pulley
    (c) occurs at the beginning of arc of the driving pulley
    (d) is constant all over

CHAPTER 7

Brakes and Dynamometers

7.1 INTRODUCTION

A brake is a device which is used to apply external resistance to a moving body to stop or to retard it by transforming its kinetic energy into heat energy which is ultimately dissipated by conduction and/or convection. In other words, the function of brake is to shorten the period of retardation by the application of external resistance.

Brakes are usually classified according to the means by which kinetic energy is transformed into heat energy. Thus there are three basic types of brakes—mechanical brakes, hydraulic brakes and electric brakes.

Mechanical brakes. In this type of brakes, the physical contact of two surfaces is used to create frictional resistance which transforms kinetic energy into heat energy. The main types of mechanical brakes are block brake, band brake, band and block brake and internal expanding shoe brake.

Hydraulic brakes. A hydraulic brake utilizes the fluid friction created by whirling of fluid instead of mechanical surface friction. Fluid pumps and agitators are common examples.

Electric brakes. When a moving body is transmitting high torque at high speed, mechanical brakes cause thermal distortion and failure of the braking system due to excessive heating. In such cases, electric brakes are used. In these brakes, the kinetic energy is absorbed by driving one or more generators; the output of these is either converted into useful work or dissipated through resistive heating.

In this chapter, we shall discuss in detail mechanical brakes which are useful for low to moderate speeds.

7.2 BLOCK BRAKES

A block brake is a block or shoe of wood or metal with the lining of friction material, which is pressed against a rotating drum by means of a lever as shown in Figure 7.1. The frictional force acting between the block and the drum causes transformation of kinetic energy and thereby retardation of the brake drum. In block brake, the block is either rigidly fixed or pivoted to the lever. In certain applications, if only one block is used, a side thrust on the bearing of the shaft supporting the drum will act. This can be prevented by using two blocks,

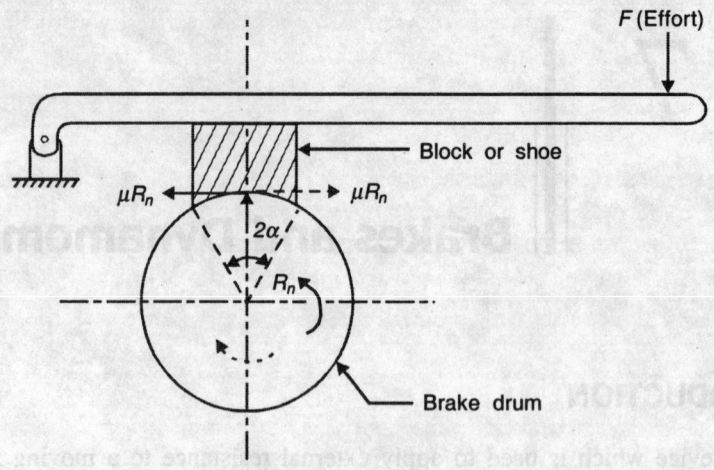

FIGURE 7.1 A block brake.

one on each side of the drum as shown in Figure 7.2. This arrangement doubles the braking capacity.

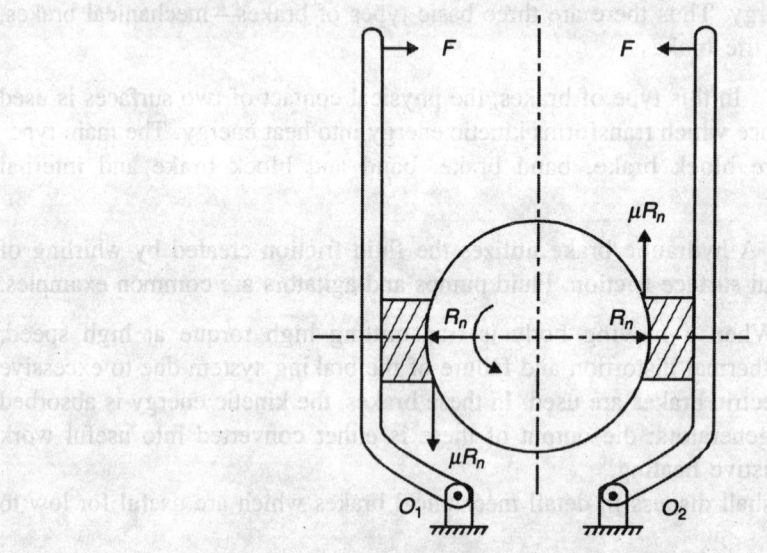

FIGURE 7.2 Double block brake.

7.2.1 Block Brakes with Small Angle of Contact

In block brakes, when the angle subtended by the block at the centre of drum is less than 45°, the distribution of pressure between the block and the drum may be assumed to be uniform. It is also fairly accurate to assume that normal reaction and the frictional force are acting at the mid-point of the block.
Suppose
$\quad\quad\quad r$ = radius of the brake drum

μ = coefficient of friction between the block and the drum
R_n = the normal reaction on the block
F = the effort, applied at the lever end
and F' = frictional force (= μR_n)

Referring to Figure 7.1, the brake shoe is in equilibrium under the action of three forces—normal reaction R_n, Frictional force μR_n and effort F. The frictional force on the drum acts in the direction opposite to the direction of motion, while on the block, it acts in the direction of rotation.

Braking torque: T = friction force × radius

or
$$T = \mu R_n \times r \tag{7.1}$$

The analysis of block brake is presented for the following three cases:

Case I *When the fulcrum point passes through the line of action of friction force.*

Let us assume that the brake drum rotates in the clockwise direction. For the brake lever to remain in equilibrium, the sum of moments of three forces about the fulcrum point O should be zero. (See Figure 7.3). Thus

$$F \times b = R_n \times a$$

or
$$R_n = F\left(\frac{b}{a}\right)$$

and braking torque:
$$T = \mu R_n r = \mu r F\left(\frac{b}{a}\right) \tag{7.2}$$

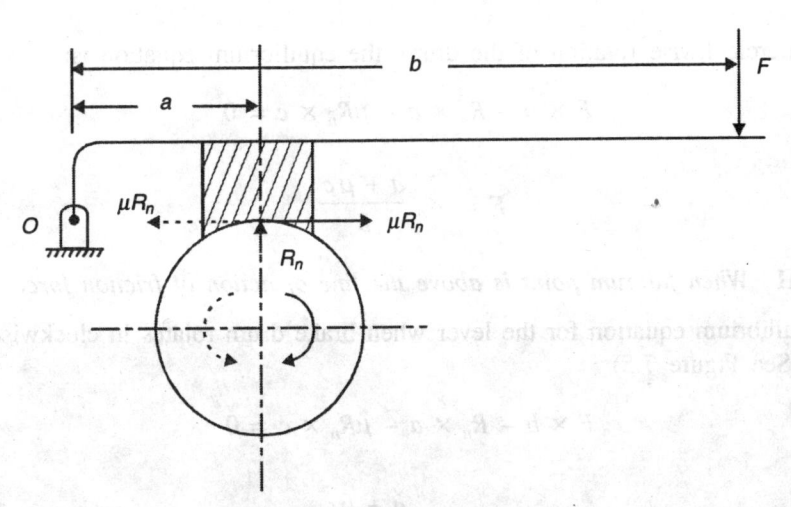

FIGURE 7.3 Block brake–Case I.

From Eq. (7.2), it is evidently clear that the brake can be applied in either direction and still braking action can be obtained by the same effort.

Case II *When the fulcrum point is below the line of action of friction force.*

The equilibrium equation of the lever for clockwise rotation of the drum can be found by summing the moments of the three forces about the fulcrum point O (see Figure 7.4).

That is
$$F \times b - R_n \times a + \mu R_n \times c = 0$$

or
$$F = \frac{R_n(a - \mu c)}{b} \qquad (7.3)$$

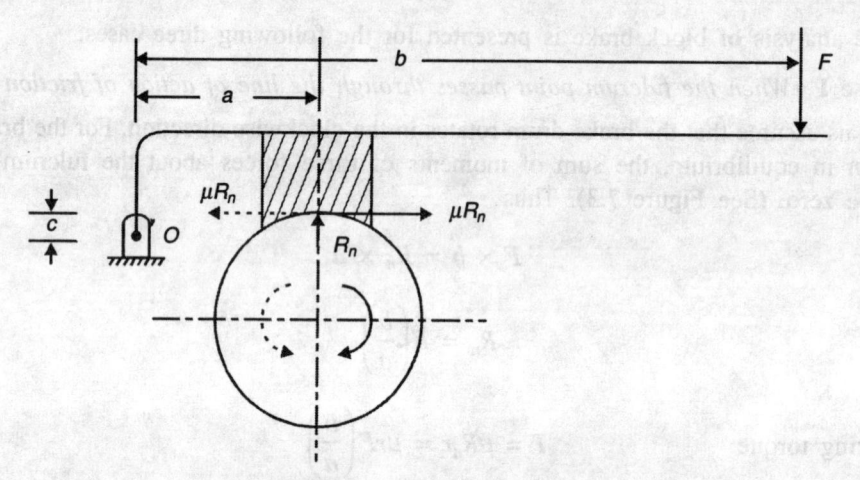

FIGURE 7.4 Block brake–Case II.

For the counterclockwise rotation of the drum, the equilibrium equation is:
$$F \times b - R_n \times a - \mu R_n \times c = 0$$

or
$$F = R_n \frac{a + \mu c}{b} \qquad (7.4)$$

Case III *When fulcrum point is above the line of action of friction force.*

The equilibrium equation for the lever when brake drum rotates in clockwise direction is given as (See Figure 7.5)
$$F \times b - R_n \times a - \mu R_n \times c = 0$$

or
$$F = R_n \frac{a + \mu c}{b} \qquad (7.5)$$

For counterclockwise rotation of the brake drum, the equilibrium equation is:

$$F \times b - R_n \times a + \mu R_n \times c = 0$$

or
$$F = R_n \frac{a - \mu c}{b} \qquad (7.6)$$

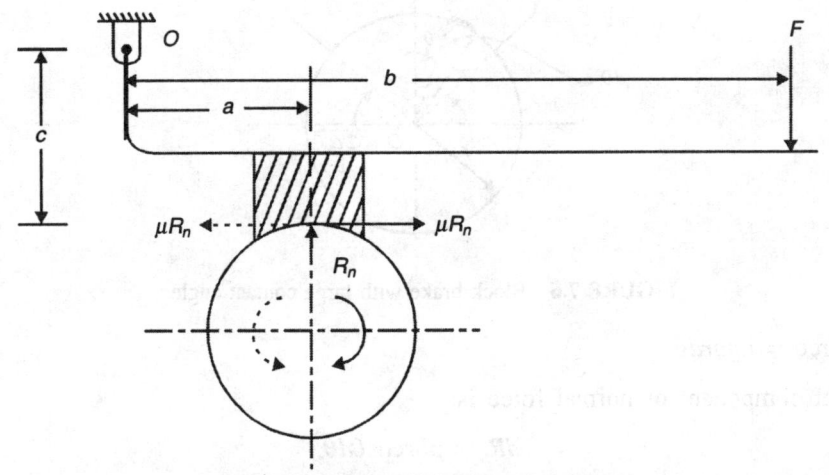

FIGURE 7.5 Block brake–Case III.

In Eqs. (7.3) and (7.6) when $a = \mu c$, the effort (F) required on the lever is zero which implies that the force needed to apply the brake is zero or in other words, once a contact is made between the block and the drum, the brake is applied by itself. Such a brake is known as **self-locking brake.** It is further observed from equilibrium equations that the moments of the friction force μR_n about the fulcrum is in the same direction as that of the effort F. Therefore, the moment due to friction force aids in applying the brake. Such a brake is known as **self-energized brake.**

7.2.2 Block Brake with Large Angle of Contact

In certain applications, where the angle subtended by the block at the centre of drum is more than 45°, the assumption that normal reaction and friction force act at the mid-point of the block is not valid. Further, the distribution of pressure does not remain constant. It is less at the ends than at the centre. Under these circumstances, it is reasonable to assume that the wear in the radial direction is uniform and the curvature of the brake shoe is unaltered.

Let
T = braking torque
R_n = resultant vertical force on the block
b = width of block
p = normal pressure
and 2α = angle of contact at the surface of the block

Consider an elemental area of contact on which the normal force is given as $pbrd\theta$ (See Figure 7.6).

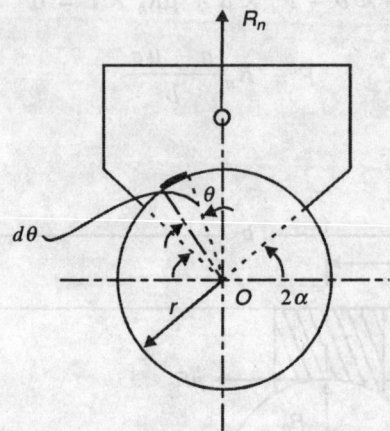

FIGURE 7.6 Block brake with large contact angle.

Friction force = $\mu pbrd\theta$

The vertical component of normal force is:

$$dR_n = pbr\cos\theta d\theta$$

or
$$R_n = \int_{-\alpha}^{+\alpha} pbr\cos\theta d\theta \qquad (7.7)$$

or Friction torque:
$$T = \int_{-\alpha}^{+\alpha} \mu pbr^2 d\theta \qquad (7.8)$$

Referring to Figure 7.7, for the assumption of uniform wear in radial direction, the normal wear W_n is proportional to the normal presure times the rubbing velocity, which is:

$$W_n \propto pv$$

or
$$W_n = k_1 pv$$

FIGURE 7.7 Block brake with uniform wear.

If the point A is displaced to new position A' after wear, the vertical displacement is given by

$$y = k_1 pv/\cos\theta$$

or

normal pressure: $p = \dfrac{y}{k_1 v}\cos\theta$

or

$$p = p_{max}\cos\theta \tag{7.9}$$

where p_{max} is the maximum normal pressure.

Substituting the value of normal pressure from Eq. (7.9) into Eq. (7.7), the resultant vertical force on the block is:

$$R_n = \int_{-\alpha}^{+\alpha} p_{max}\, br \cos^2\theta\, d\theta$$

$$= \dfrac{p_{max} br}{2}(2\alpha + \sin 2\alpha) \tag{7.10}$$

Similarly, the frictional torque:

$$T = \int_{-\alpha}^{+\alpha} \mu br^2 \times p_{max} \times \cos\theta\, d\theta$$

or

$$T = 2\mu p_{max} \times br^2 \sin\alpha \tag{7.11}$$

Substituting the value of p_{max} from Eq. (7.10) into Eq. (7.11), we get

$$T = 2\mu br^2 \sin\alpha \times \dfrac{2R_n}{br(2\alpha + \sin 2\alpha)}$$

or

$$T = \mu\left(\dfrac{4\sin\alpha}{2\alpha + \sin 2\alpha}\right) R_n r \tag{7.12}$$

$$= \mu R_n h$$

$$= \mu' R_n r$$

where μ' is the equivalent coefficient of friction

$$\left(= \mu \times \dfrac{4\sin\alpha}{2\alpha + \sin 2\alpha}\right)$$

and h is the effective brake radius $\left(= \dfrac{4\sin\alpha}{2\alpha + \sin 2\alpha} \times r\right)$ for large angle of contact.

Therefore, for the block brake with large angle of contact, the resultant of friction force and normal reaction passes through a point called **pivot point** and such block brakes are sometimes known as **pivoted shoe brakes** (see Figure 7.8).

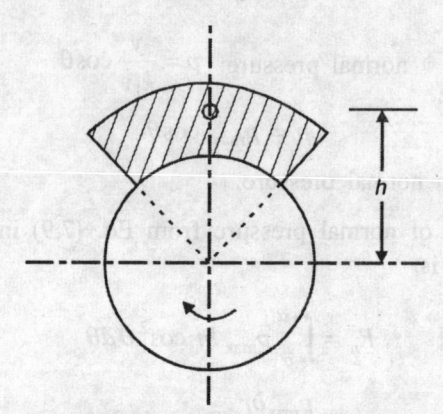

FIGURE 7.8 Pivoted shoe.

Graphically, the direction of the resultant of friction force and normal reaction can be obtained by drawing a tangent from pivot point to the friction circle of the brake as shown in Figure 7.9.

The braking effort F can be obtained by taking moment about the fulcrum point O. For clockwise rotation of the brake drum.

$$F \times b - R \times a = 0 \qquad (7.13)$$

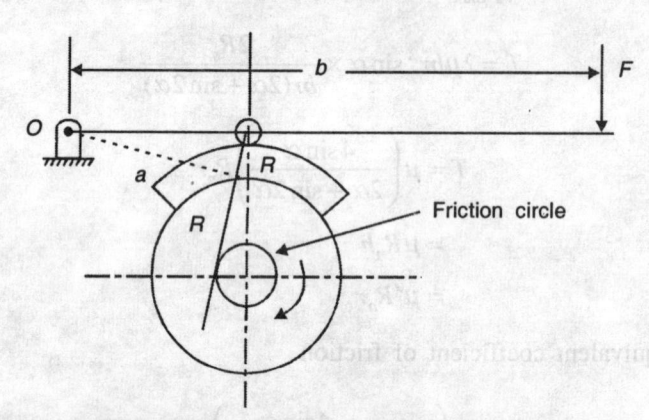

FIGURE 7.9 Pivoted shoe with friction circle.

EXAMPLE 7.1 For a block brake as shown in Figure 7.10, the diameter of the brake drum is 400 mm and angle of contact is 40°. The applied effort at the free end of lever is 2kN and coefficient of friction between brake drum and block is 0.35. Determine the maximum braking torque for the following cases:

(i) When drum rotates in clockwise direction; and

(ii) When drum rotates is counterclockwise direction. Also determine the value of dimension c for self-locking condition and heat generated if the drum rotates at 500 rpm in clockwise direction.

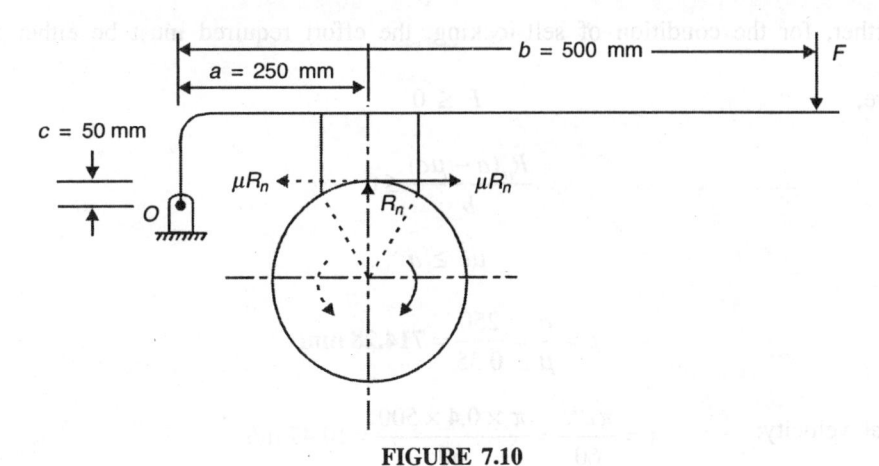

FIGURE 7.10

Solution As the given angle of contact is less than 45°, the block can be assumed to be rigidly fixed with the lever.

Case I *When brake drum rotates in clockwise direction.*

Taking moments about fulcrum O (refer Figure 7.10),

$$F \times b - R_n \times a + \mu R_n \times c = 0$$

or

$$2000 \times 500 - R_n \times 250 + 0.35 \times R_n \times 50 = 0$$

or

$$R_n = 4301 \text{ N}$$

Braking torque:

$$T = \mu R_n \times r$$

$$= 0.35 \times 4301 \times 0.2$$

$$= 301 \text{ Nm} \qquad \text{Ans.}$$

Case II *When drum rotates in counterclockwise direction.*

Taking moment about fulcrum O,

$$F \times b - R_n \times a - \mu R_n \times c = 0$$

or

$$2000 \times 500 - R_n \times 250 - 0.35 \times R_n \times 50 = 0$$

or

$$R_n = 3738.3 \text{ N}$$

Braking torque:

$$T = \mu R_n \times r$$

$$= 0.35 \times 3738.3 \times 0.2$$

$$= 261.68 \text{ Nm} \qquad \text{Ans.}$$

For clockwise rotation of brake drum, the expression for the effort is:

$$F = \frac{R_n(a - \mu c)}{b}$$

Further, for the condition of self-locking, the effort required must be either zero or negative.

Therefore,
$$F \leq 0$$

or
$$\frac{R_n(a - \mu c)}{b} \leq 0$$

or
$$\mu c \geq a$$

or
$$c = \frac{a}{\mu} = \frac{250}{0.35} = 714.28 \text{ mm} \qquad \text{Ans.}$$

Peripheral velocity:
$$v = \frac{\pi d N}{60} = \frac{\pi \times 0.4 \times 500}{60} = 10.47 \text{ m/s}$$

Heat generated:
$$H = \mu R_n v$$
$$= \frac{0.35 \times 4301 \times 10.47}{1000}$$
$$= 15.76 \text{ kW} \qquad \text{Ans.}$$

EXAMPLE 7.2 Determine the torque that a single block brake, as shown in Figure 7.11, can absorb if the diameter of brake drum is 250 mm, the angle of contact is 100° and coefficient of friction between the drum and block is 0.4. The effort on the lever is 750 N. The lever arm lengths are as follows:

$$a = 200 \text{ mm}, \quad b = 400 \text{ mm}, \quad \text{and} \quad c = 40 \text{ mm}.$$

Also determine the projected area of the block if the bearing pressure is 0.5 N/mm².

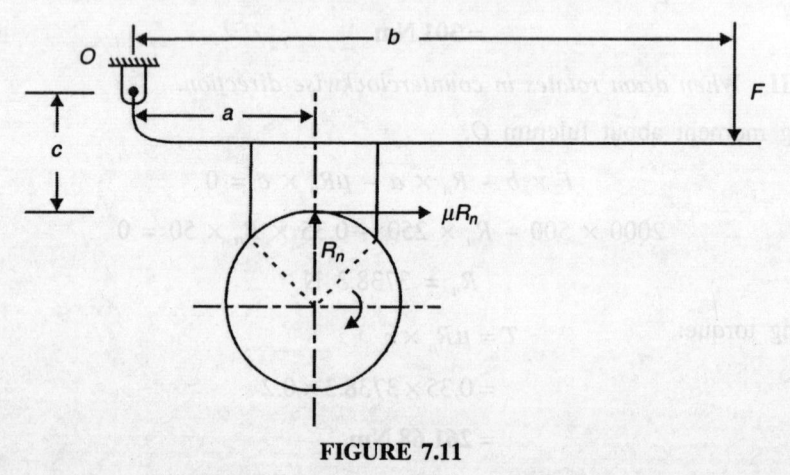

FIGURE 7.11

Solution Since the angle of contact is more than 45°, it is a case of large contact angle block brake. The equivalent coefficient of friction is:

$$\mu' = \mu\left(\frac{4\sin\alpha}{2\alpha + \sin 2\alpha}\right)$$

$$= 0.4\left(\frac{4\times \sin 50°}{\frac{\pi \times 100°}{180°} + \sin 100°}\right) = 0.449$$

Now for clockwise rotation of brake drum, taking moments about fulcrum O,

$$F \times b - R_n \times a - \mu R_n \times c = 0$$

or $\qquad 750 \times 400 - R_n \times 200 - 0.449 \times R_n \times 40 = 0$

or $\qquad R_n = 1376.4$ N

Braking torque: $\qquad T = \mu' R_n r$

$$= 0.449 \times 1376.4 \times 0.125$$

$$= 77.25 \text{ Nm} \qquad \text{Ans.}$$

Projected area: $\qquad A = \dfrac{R_n}{p_b} = \dfrac{1376.4}{0.5}$

$$= 2752.8 \text{ mm}^2 \qquad \text{Ans.}$$

EXAMPLE 7.3 A double block brake, as shown in Figure 7.12, is set by a spring that produces a force of 3.5 kN. The brake drum diameter is 400 mm and the angle of contact

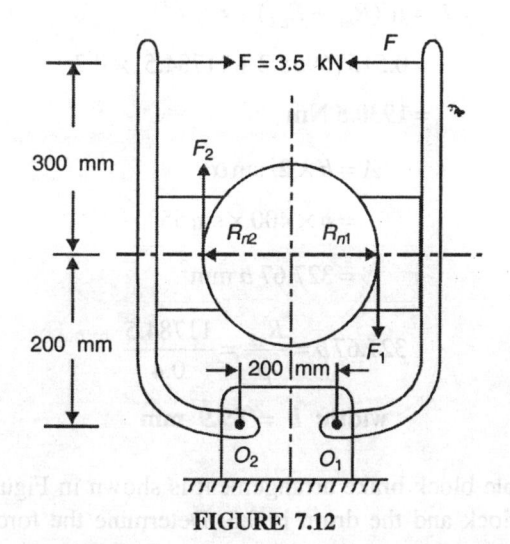

FIGURE 7.12

for each block is 110°. If the coefficient of friction between block and drum is 0.45, determine the maximum torque that can be absorbed. If the bearing pressure is not to exceed 0.4 N/mm², determine the maximum width of block brake.

Solution For a large contact angle brake, the equivalent coefficient of friction is:

$$\mu' = \mu \left(\frac{4 \sin \alpha}{2\alpha + \sin 2\alpha} \right)$$

$$= 0.45 \left(\frac{4 \times \sin 55°}{\frac{\pi \times 110°}{180°} + \sin 110°} \right) = 0.515$$

Considering the right hand lever, the moments about fulcrum O_1,

$$F \times 500 - R_{n1} \times 200 - F_1 (200 - 100) = 0$$

where
$$F_1 = \mu' R_{n1}$$

or
$$3500 \times 500 - R_{n1} \times 200 - 0.515 \times R_{n1} \times 100 = 0$$

or
$$R_{n1} = 6958.2 \text{ N}$$

Similarly, for left hand lever, taking moments about fulcrum O_2,

$$F \times 500 - R_{n2} \times 200 + F_2 \times (200 - 100) = 0$$

where
$$F_2 = \mu' R_{n2}$$

or
$$3500 \times 500 - R_{n2} \times 200 + 0.515 \times R_{n2} \times 100 = 0$$

or
$$R_{n2} = 11784.5 \text{ N}$$

Braking torque:
$$T = \mu'(R_{n1} + R_{n2}) \times r$$

$$= 0.515 (6958.2 + 11784.5) \times 0.2$$

$$= 1930.5 \text{ Nm} \qquad \text{Ans.}$$

Projected area:
$$A = b \times 2r \sin \alpha$$

$$= b \times 400 \times \sin 55°$$

$$= 327.67 \, b \text{ mm}^2$$

or
$$327.67 b = \frac{R_{n2}}{p} = \frac{11784.5}{0.4}$$

or
width: $b = 89.9$ mm Ans.

EXAMPLE 7.4 A double block brake arrangement is shown in Figure 7.13. The coefficient of friction between the block and the drum is 0.3. Determine the force required at the brake

Brakes and Dynamometers 247

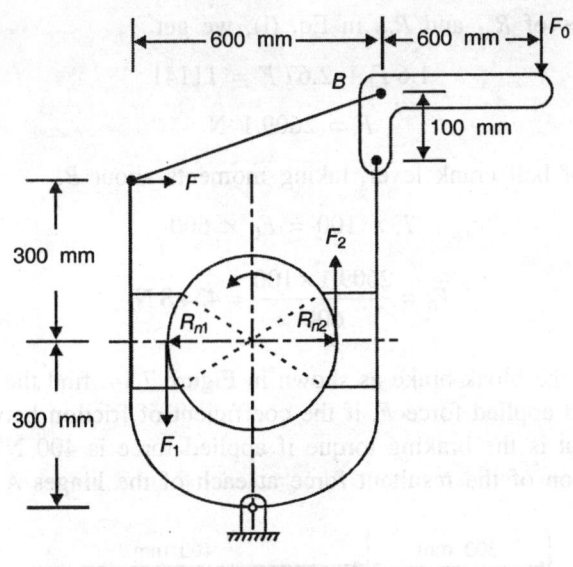

FIGURE 7.13

lever end to absorb 52.5 kW at drum speed of 600 rpm in counterclockwise direction. The diameter of brake drum is 500 mm.

Solution

Torque: $T = \dfrac{P}{\omega}$ where $\omega = \dfrac{2\pi \times 600}{60} = 62.83$ rad/s

or $$T = \dfrac{52.5 \times 10^3}{62.83} = 835.58 \text{ Nm}$$

or $$T = \mu(R_{n1} + R_{n2}) \times r$$

or $$835.58 = 0.3\,(R_{n1} + R_{n2}) \times 0.25$$

or $$R_{n1} + R_{n2} = 11141 \text{ N} \qquad (i)$$

Let F is the effort applied on the lever.

For left hand lever, taking moments about the fulcrum,

$$F \times 600 - R_{n1} \times 300 - \mu R_{n1} \times 250 = 0$$

or $$R_{n1} = \dfrac{F \times 600}{300 + 0.3 \times 250} = 1.6\,F$$

Similarly, for right hand lever, taking moments about the fulcrum,

$$F \times 600 - R_{n2} \times 300 + \mu R_{n2} \times 250 = 0$$

or $$R_{n2} = \dfrac{F \times 600}{300 - 0.3 \times 250} = 2.67\,F$$

Substituting the values of R_{n1} and R_{n2} in Eq. (i), we get

$$1.6 F + 2.67 F = 11141$$

or
$$F = 2609.1 \text{ N}$$

For the equilibrium of bell crank lever, taking moments about B,

$$F \times 100 = F_0 \times 600$$

or
$$F_0 = \frac{2609.1 \times 100}{600} = 434.8 \text{ N} \qquad \text{Ans.}$$

EXAMPLE 7.5 For the block brake as shown in Figure 7.14, find the relationship between the braking torque and applied force F, if the coefficient of friction between the brake drum and block is 0.3. What is the braking torque if applied force is 400 N? Also determine the magnitude and direction of the resultant force at each of the hinges A and B.

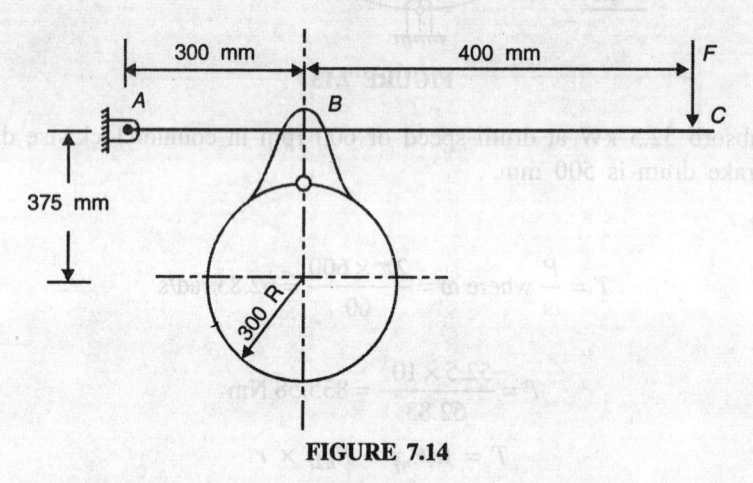

FIGURE 7.14

Solution Let the reaction R between the block and drum passes through the point B as shown in Figure 7.15 and at the point of intersection with the drum periphery D, it is inclined at the friction angle f to the radius at that point OD.

$$\phi = \tan^{-1}\mu = \tan^{-1} 0.3 = 16.7°$$

From triangle OBD,

$$\frac{300}{\sin \theta} = \frac{375}{\sin(180° - 16.7°)}$$

or
$$\theta = 13.29°$$

Taking moments about point A,

$$F \times 700 = R \times x$$

or
$$R = \frac{700 F}{300 \times \cos 13.29°}$$

Figure 7.15 (description omitted)

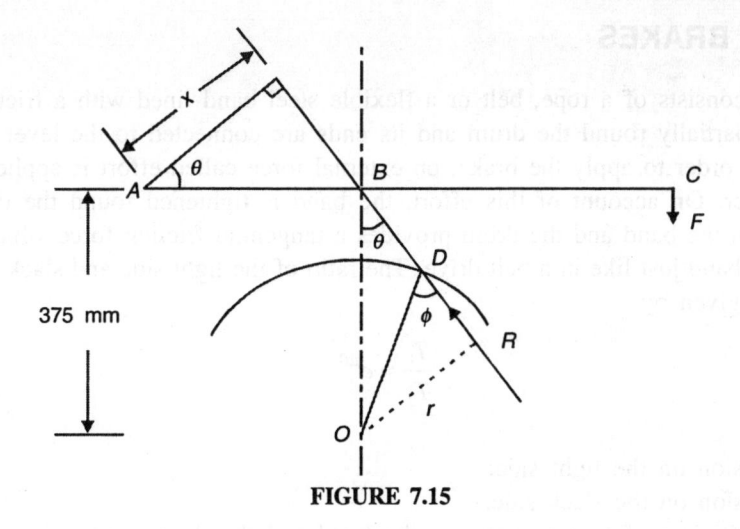

Friction torque:
$$T = R \times r$$

$$= \frac{700\,F}{300 \times \cos 13.29°} \times 375 \times \sin 13.29°$$

$$= 875\,F \times \tan 13.29°$$

$$= 206.68\,F \quad \text{Nmm} \qquad \text{Ans.}$$

When $F = 400$ N,

Torque:
$$T = \frac{206.68 \times 400}{1000} = 82.67 \text{ Nm} \qquad \text{Ans.}$$

Resultant reaction at hinge B:

$$R = \frac{700 \times 400}{300 \times \cos 13.39°} = 959.4 \text{ N} \qquad \text{Ans.}$$

Vertical component of reaction at A:

$$R_{AV} = R \cos\theta - F = 959.4 \times \cos 13.29° - 400$$

$$= 533.7 \text{ N}$$

Horizontal component of reaction at A:

$$R_{AH} = R \sin\theta = 959.4 \times \sin 13.29° = 220.5 \text{ N}$$

Resultant reaction:
$$R_A = \sqrt{R_{AV}^2 + R_{AH}^2}$$

$$= \sqrt{533.7^2 + 220.5^2}$$

$$= 577.45 \text{ N} \qquad \text{Ans.}$$

7.3 BAND BRAKES

A band brake consists of a rope, belt or a flexible steel band lined with a friction material. It is wrapped partially round the drum and its ends are connected to the lever as shown in Figure 7.16. In order to apply the brake, on external force called effort is applied at the free end of the lever. On account of this effort, the band is tightened round the drum and the friction between the band and the drum provides a tangential friction force which causes the tensions in the band just like in a belt drive. The ratio of the tight side and slack side tensions in the band is given by

$$\frac{T_1}{T_2} = e^{\mu\theta} \qquad (7.14)$$

where

T_1 = tension on the tight side;
T_2 = tension on the slack side;
μ = coefficient of friction between the band and the drum, and
θ = angle of contact.

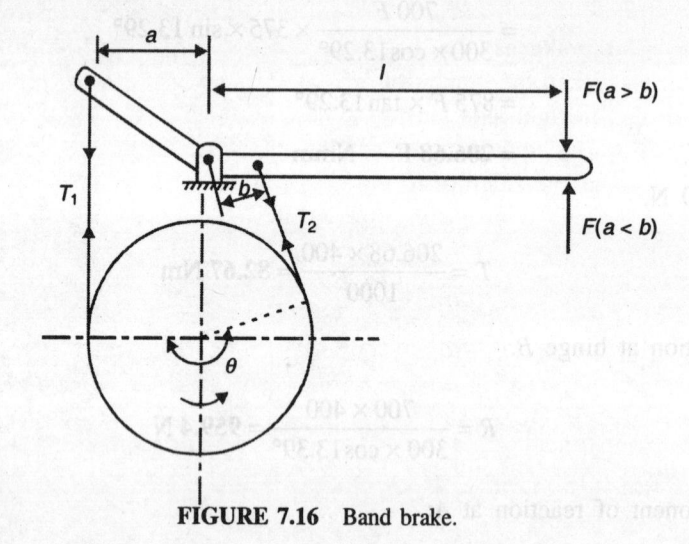

FIGURE 7.16 Band brake.

The braking torque on the brake drum is given by

$$T = (T_1 - T_2)r \qquad (7.15)$$

where r is the effective radius of the drum.

The effectiveness of the band brake depends upon the direction of rotation of the drum, the direction of effort F and the ratio of fulcrum lengths a and b. In order to prevent the loosening of the band, the effort F must act in downward direction when $a > b$ and in the upward direction when $a < b$.

Band brakes may be classified into two categories—simple band brake and differential band brake.

7.3.1 Simple Band Brake

A simple band brake is one in which one end of the band is attached to the fulcrum and other end to the brake lever. There are various arrangements of the simple band brake. One typical arrangement is shown in Figure 7.17.

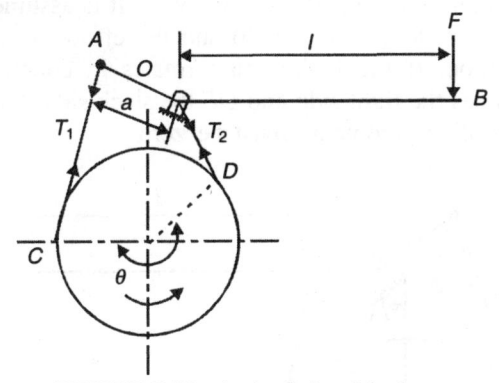

FIGURE 7.17 A simple band brake.

In simple band brake as $a > b$ (b being zero), the effort on the lever must act in downward direction. When the brake drum rotates in the counterclockwise direction, the band AC will be subjected to tight side tension T_1 and OD will be subjected to slack side tension T_2 (See Figure 7.17). The equilibrium of the lever can be found out by equating the algebraic sum of moments about the fulcrum to zero as follows:

$$F \times l - T_1 \times a = 0$$

or

$$T_1 = \frac{Fl}{a} \quad (7.16)$$

and braking torque:

$$T = (T_1 - T_2)r = T_1\left(1 - \frac{1}{e^{\mu\theta}}\right)r$$

or

$$T = \frac{Fl}{a}\left(\frac{e^{\mu\theta} - 1}{e^{\mu\theta}}\right)r \quad (7.17)$$

When the brake drum rotates in the clockwise direction, the tight and slack sides are interchanged. Then the equilibrium equation becomes

$$F \times l - T_2 \times a = 0$$

and braking torque:

$$T = \frac{Fl}{a}(e^{\mu\theta} - 1)r \quad (7.18)$$

A comparison of Eqs. (7.17) and (7.18) reveals that when the brake drum rotates in the clockwise direction, the braking torque is $e^{\mu\theta}$ times greater than that of the drum rotating in counterclockwise direction.

7.3.2 Differential Band Brake

When both the ends of the band strips are attached to the brake lever, it is called differential band brake. Figure 7.18 shows the differential band brake in which the ends of the band are attached to the lever at points A and E. Let us assume that the perpendicular distances of the band from the fulcrum point O are a and b respectively. It is assumed that the fulcrum length a is greater than the fulcrum length b ($a > b$) and the effort on the lever is applied in the vertical downward direction. If the brake drum rotates in counterclockwise direction, the portion of band CA becomes the tight side and DE the slack side. For equilibrium of the brake lever, the algebraic sum of the moments must be zero.

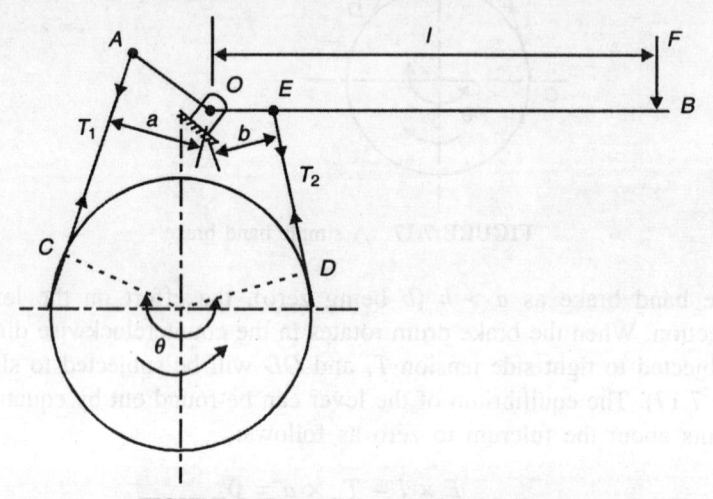

FIGURE 7.18 A differential band brake.

Referring to Figure 7.18,

$$F \times l - T_1 \times a + T_2 \times b = 0$$

or

$$F = \frac{(T_1 a - T_2 b)}{l}$$

or

$$F = \frac{T_2(ae^{\mu\theta} - b)}{l} \quad (7.19)$$

and braking torque:

$$T = (T_1 - T_2) \times r = T_2(e^{\mu\theta} - 1) \times r$$

or

$$T = \frac{Fl(e^{\mu\theta} - 1)r}{(ae^{\mu\theta} - b)} \quad (7.20)$$

In the differential band brake, when $b \geq ae^{\mu\theta}$ or $T_2 b \geq T_1 a$ [refer Eq. (7.19)], the effort required will be either zero or a negative value. This condition of brake is called **self-locking**.

When a brake drum rotates in the clockwise direction, the tight and slack side tensions are interchanged. The portion of band *CA* becomes slack while that of band *DE* gets tightened. The equilibrium equation of the lever is found by taking moment about the fulcrum as follows:

$$F \times l - T_2 \times a + T_1 \times b = 0$$

or

$$F = \frac{T_2 a - T_1 b}{l} = \frac{T_2(a - be^{\mu\theta})}{l} \tag{7.21}$$

and braking torque:

$$T = (T_1 - T_2)r = T_2(e^{\mu\theta} - 1) \times r$$

or

$$T = \frac{Fl(e^{\mu\theta} - 1)r}{(a - be^{\mu\theta})} \tag{7.22}$$

Comparison of Eqs. (7.20) and (7.22) reveals that the denominator of Eq. (7.22) has smaller value than that of Eq. (7.20). It means that when brake drum rotates in the clockwise direction, it is more effective.

EXAMPLE 7.6 A simple band brake has a drum of 300 mm diameter. The one end of the band is fixed to the fulcrum while the other end is fixed to the lever as shown in Figure 7.19. The angle of contact is 225° and the coefficient of friction is 0.25. If an effort of 300 N is applied to the lever at a distance of 500 mm, determine braking torque.

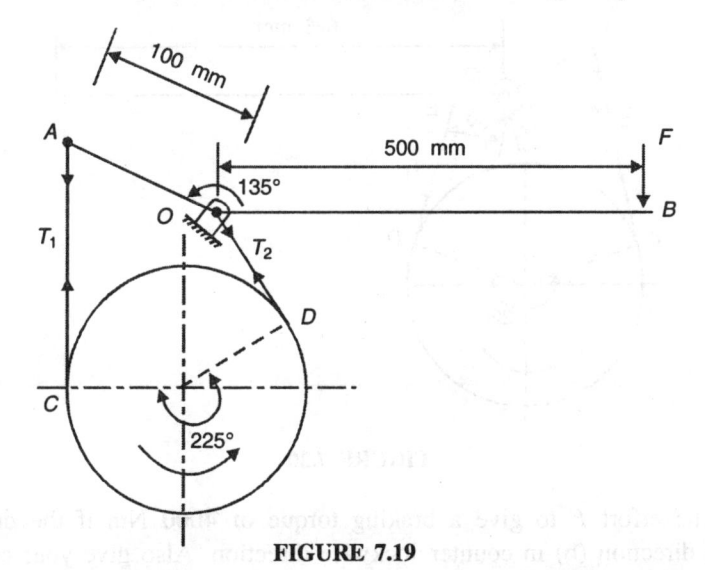

FIGURE 7.19

Solution

Ratio of tensions:

$$\frac{T_1}{T_2} = e^{\mu\theta} = e^{0.25 \times \frac{\pi}{180} \times 225°}$$

$$= 2.67$$

The perpendicular distance from fulcrum to band AC:

$$a = OA \cos 45° = 100 \times \cos 45° = 70.7 \text{ mm}$$

Taking moments about fulcrum O,

$$F \times 500 - T_1 \times a = 0$$

or

$$T_1 = \frac{300 \times 500}{70.7} = 2121.6 \text{ N}$$

or

$$T_2 = \frac{T_1}{2.67} = \frac{2121.6}{2.67} = 794.6 \text{ N}$$

Braking torque:

$$T = (T_1 - T_2) \times r$$

$$= (2121.6 - 794.6) \times 0.15$$

$$= \mathbf{199.05 \text{ Nm}}$$ **Ans.**

EXAMPLE 7.7 The Figure 7.20 shows the layout of a band brake applied to the brake drum of a hoist where the braking effort F is applied at one end of a lever which is pivoted on a fixed fulcrum O. The drum diameter is 1 m, the angle of contact is 225° and coefficient of frictions is 0.3.

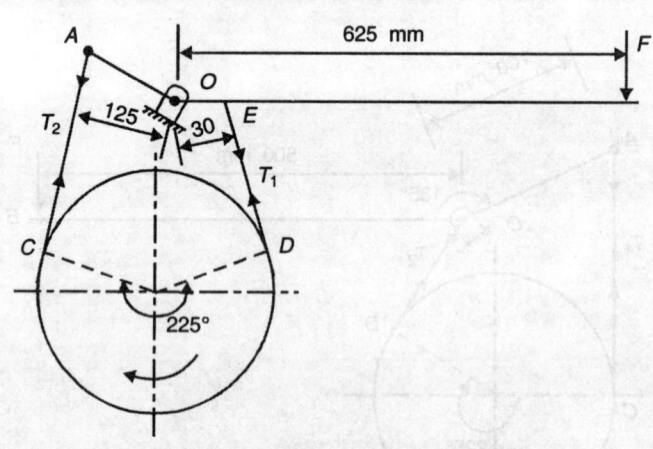

FIGURE 7.20

Calculate the effort F to give a braking torque of 4000 Nm if the drum is rotating (a) in clockwise direction (b) in counterclockwise direction. Also give your comment on the answer.

Solution

Braking torque:

$$T = (T_1 - T_2)r$$

or

$$4000 = (T_1 - T_2) \times 0.5$$

or

$$T_1 - T_2 = 8000 \text{ N} \qquad (i)$$

Ratio of tensions:
$$\frac{T_1}{T_2} = e^{\mu\theta} = e^{0.3 \times \frac{\pi}{180} \times 225°}$$

or
$$\frac{T_1}{T_2} = 3.248 \qquad (ii)$$

Solving Eqs. (i) and (ii), we get
$$T_1 = 11558.6 \text{ N and } T_2 = 3558.7 \text{ N}$$

Case I *When the brake drum rotates in clockwise direction.*
Taking moments about fulcrum O,
$$F \times l + T_1 \times b - T_2 \times a = 0$$
or
$$F \times 625 + 11558.6 \times 30 - 3558.7 \times 125 = 0$$
or
$$F = \mathbf{156.9 \text{ N}} \qquad \text{Ans.}$$

Case II *When the brake drum rotates in counterclockwise direction.*
In this case, the tight and slack sides are interchanged. Taking moments about fulcrum O,
$$F \times l + T_2 \times b - T_1 \times a = 0$$
$$F \times 625 + 3558.7 \times 30 - 11558.6 \times 125 = 0$$
or
$$F = \mathbf{2140.9 \text{ N}} \qquad \text{Ans.}$$

Comment: The effort required for clockwise rotation of brake drum is less compared to that required for counterclockwise rotation because the couple due to T_1 acts in the direction of the effort. In other words, brake becomes self-energized.

EXAMPLE 7.8 A differential band brake is used to retard a shaft carrying a flywheel weighing 3 kN with radius of gyration 350 mm. If the brake drum diameter is 250 mm and coefficient of friction is 0.15, determine the following:

  (i) the braking torque when an effort of 150 N is applied at the end of the lever
  (ii) number of revolutions of the flywheel before it comes to rest
  (iii) time taken by the flywheel to come to rest

The drum rotates at 500 rpm in clockwise direction and the layout of the brake is as shown in Figure 7.21.

Solution
The angle of contact = 180° + 40° = 220°

(i) Ratio of tensions: $\frac{T_1}{T_2} = e^{\mu\theta} = e^{0.15 \times \frac{\pi}{180} \times 220°} = 1.78$

Fulcrum lengths: $\qquad a = OA = 90 \text{ mm}$
$$b = OC \cos 50° = 40 \times \cos 50° = 25.7 \text{ mm}$$

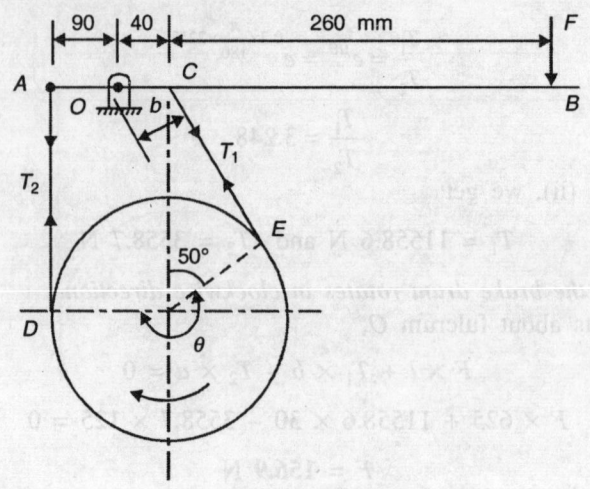

FIGURE 7.21

Taking moments about fulcrum O,

$$F \times 300 + T_1 \times b - T_2 \times a = 0$$

or $\quad 150 \times 300 + T_1 \times 25.7 - T_2 \times 90 = 0$

or $\quad 150 \times 300 + 1.78 T_2 \times 25.7 - T_2 \times 90 = 0$

$\quad T_2 = 1016.85$ N

$\therefore \quad T_1 = 1.78 \times 1016.85 = 1810$ N

Braking torque: $T = (T_1 - T_2) r$

$\qquad = (1810 - 1016.85) \times 0.125$

$\qquad = \mathbf{99.14\ Nm}$ **Ans.**

(ii) For equilibrium, the kinetic energy of flywheel should be equal to work done by the braking torque. That is

K.E. of flywheel = $T \times$ angular displacement

or $\qquad \dfrac{1}{2} I \omega^2 = T \times 2\pi n$

or $\qquad \dfrac{1}{2} m k^2 \times \left(\dfrac{2\pi N}{60}\right)^2 = T \times 2\pi \times n$

or $\qquad \dfrac{1}{2} \times \dfrac{3000}{9.81} \times 0.35^2 \times \left(\dfrac{2\pi \times 500}{60}\right)^2 = 99.14 \times 2\pi \times n$

Therefore, the number of revolutions of the flywheel before it comes to rest:

$$n = 82.43 \text{ revolution} \qquad \text{Ans.}$$

(iii) Time taken: $\qquad t = \dfrac{n}{N} = \dfrac{82.43}{500} \times 60 = 9.89 \text{ s} \qquad$ Ans.

EXAMPLE 7.9 A band brake of a crane is actuated by a lever, the free end of which is pulled downward in order to apply the brake (Figure 7.22). The tight side end of the band is attached to the fulcrum of the lever and slack end to the end of the lever 50 mm away from the fulcrum. The diameter of the brake drum is 1 m and that of the barrel is 500 mm. If an effort of 250 N is applied at the end of the lever to hold a load of 10 kN, what lever length would be required? Take μ as 0.3. The angle of contact is 270°. Also calculate the size of section of the band, if the stress in the material is not to exceed 60 N/mm².

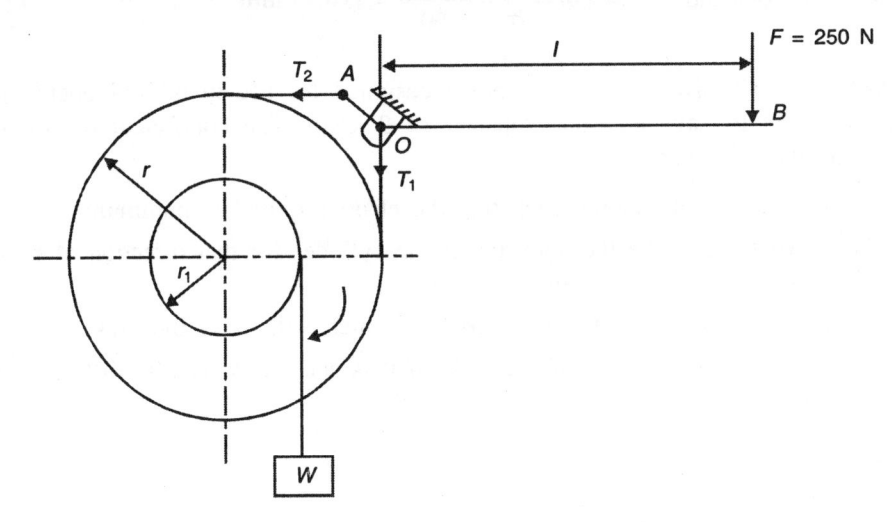

FIGURE 7.22

Solution For the equilibrium of drum and barrel,

$$\text{Braking torque} = \text{Work done by the load}$$

or $\qquad (T_1 - T_2) \times r = W \times r_1$

or $\qquad T_1 - T_2 = \dfrac{10 \times 10^3 \times 0.25}{0.5}$

or $\qquad T_1 - T_2 = 5000 \text{ N} \qquad$ (i)

Ratio of tensions: $\qquad \dfrac{T_1}{T_2} = e^{\mu\theta} = e^{0.3 \times \frac{\pi}{180} \times 270°}$

or $$\frac{T_1}{T_2} = 4.11 \qquad (ii)$$

Solving Eqs. (i) and (ii), we get

$$T_1 = 6607.7 \text{ N} \quad \text{and} \quad T_2 = 1607.7 \text{ N}$$

Taking moments about fulcrum O,

$$F \times l - T_2 \times a = 0$$

or $$l = \frac{1607.7 \times 50}{250} = 321.54 \text{ mm} \qquad \text{Ans.}$$

Cross-section of band: $$bt = \frac{T_1}{\sigma} = \frac{6607.7}{60} = 110.13 \text{ mm}^2 \qquad \text{Ans.}$$

EXAMPLE 7.10 The air-operated brake designed for use in oilfield operation is shown in Figure 7.23 to have a torque capacity of 9 kNm. The coefficient of friction is 0.35. Determine the following:

(i) Direction of rotation such that the effort required is minimum.

(ii) What should be the diameter of the cylinder for that direction of rotation, if the air pressure is 0.8 N/mm².

(iii) The value of coefficient of friction which will make the brake self-locking.

(iv) The torque capacity of the brake if it is accidently used in the direction opposite to that determined in (i).

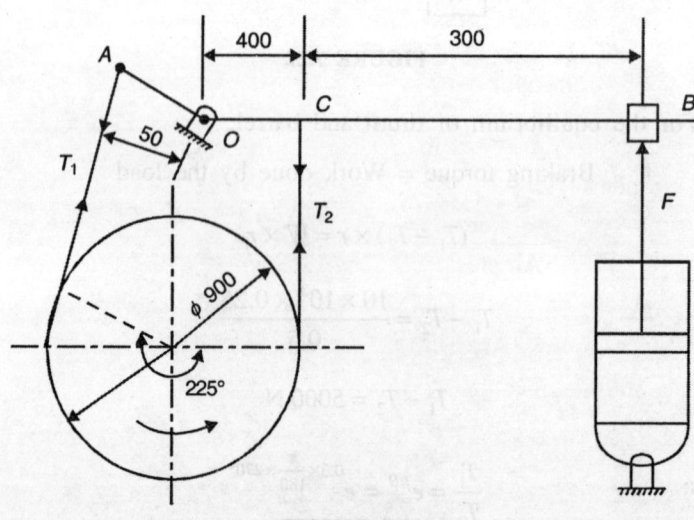

FIGURE 7.23

Solution In the given layout of brake, the lever arm $b > a$. So the effort through air cylinder should act in upward direction. Further, for minimum effort, the brake should be self-energizing, i.e. moment due to tight side tension should act in the same sense as that of effort which is possible only if the drum rotates in counterclockwise direction.

Ratio of tensions:
$$\frac{T_1}{T_2} = e^{\mu\theta} = e^{0.35 \times \frac{\pi}{180} \times 225°}$$

or
$$\frac{T_1}{T_2} = 3.95 \qquad (i)$$

Braking torque: $T = (T_1 - T_2)\,r$

or $\qquad 9 \times 10^3 = (T_1 - T_2) \times 0.45$

or $\qquad T_1 - T_2 = 20 \times 10^3 \qquad (ii)$

Solving Eqs. (i) and (ii), we get

$$T_1 = 26779.4 \text{ N} \quad \text{and} \quad T_2 = 6779.6 \text{ N}$$

(i) (a) For counterclockwise direction, taking moments about fulcrum O,

$$F \times l + T_1 \times a - T_2 \times b = 0$$

or $\qquad F \times 700 + 26779.4 \times 50 - 6779.6 \times 400 = 0$

or $\qquad F = 1961.2$ N

(b) Similarly, for clockwise rotation, taking moments about fulcrum O,

$$F \times l + T_2 \times a - T_1 \times b = 0$$

or $\qquad F \times 700 + 6779.6 \times 50 - 26779.4 \times 400 = 0$

or $\qquad F = 14818.25$ N

Therefore, for minimum effort, the drum should rotate in **counterclockwise direction.** **Ans.**

(ii) Area of cylinder:
$$A = \frac{\text{Force}}{\text{Air pressure}} = \frac{1961.2}{0.8} = 2451.5 \text{ mm}^2$$

or $\qquad$ Area: $A = \dfrac{\pi}{4}d^2 = 2451.5$

or $\qquad$ Cylinder Diameter: $d = $ **55.8 mm** $\qquad$ **Ans.**

(iii) For counterclockwise rotation,

$$F = \frac{T_2 b - T_1 a}{l} = \frac{T_2}{l}(b - ae^{\mu\theta})$$

For the condition of self-locking, substituting $F = 0$, we get

$$F = \frac{T_2}{l}(b - ae^{\mu\theta}) = 0$$

or

$$e^{\mu\theta} = \frac{b}{a}$$

Taking log on both sides,

$$\mu\theta = \log_e (b/a)$$

or

$$\mu = \frac{\log_e(b/a)}{\theta} = \frac{\log_e(400/50)}{\frac{\pi}{180} \times 225°} = 0.53 \qquad \text{Ans.}$$

(iv) Torque capacity if the drum rotates in clockwise direction. Taking moments about fulcrum O,

$$F \times l + T_2 \times a - T_1 \times b = 0$$

or

$$1961.2 \times 700 + T_2 \times 50 - T_1 \times 400 = 0$$

or

$$8T_1 - T_2 = 27456.8 \text{ N} \qquad \text{(iii)}$$

Solving Eqs. (i) and (iii), we get

$$T_1 = 3544.3 \text{ N} \quad \text{and} \quad T_2 = 897.3 \text{ N}$$

Therefore,

Braking torque:
$$T = (T_1 - T_2) \times r$$
$$= (3544.3 - 897.3) \times 0.45$$
$$= 1191.1 \text{ Nm} \qquad \text{Ans.}$$

EXAMPLE 7.11 A simple band brake operates on a drum of diameter 600 mm that is running at a speed of 200 rpm (Figure 7.24). The coefficient of friction is 0.3. The brake band has an angle of contact of 270°. One end of it is fastened to a fixed pin and the other end to the brake arm 125 mm from the fixed pin. The straight arm is 750 mm long and placed perpendicular to the diameter that bisects the angle of contact.

(a) What is the effort necessary at the end of brake arm to stop the wheel if 30 kW power is absorbed? What is the direction of rotation for the minimum pull?
(b) What width of steel band of 2.5 mm is required for this brake if the maximum tensile stress is not to exceed 50 N/mm^2?

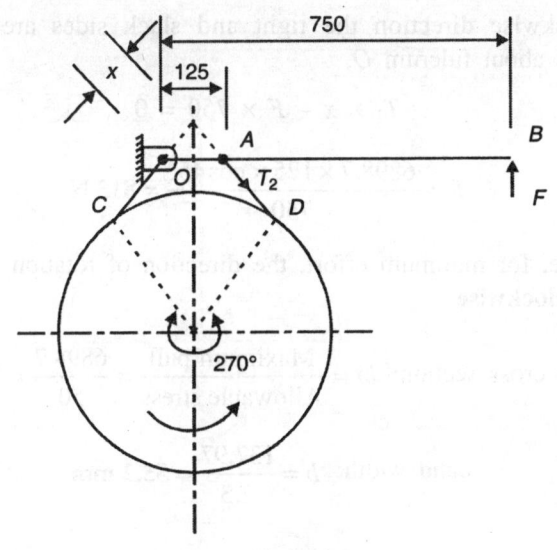

FIGURE 7.24

Solution

(a) Torque: $$T = \frac{P}{\omega} = \frac{60P}{2\pi N} = \frac{60 \times 30 \times 10^3}{2\pi \times 200} = 1432.4 \text{ Nm}$$

or $$T = (T_1 - T_2) \times r$$

or $$T_1 - T_2 = \frac{T}{r} = \frac{1432.4}{0.3}$$

or $$T_1 - T_2 = 4774.67 \text{ N} \qquad \text{(i)}$$

Ratio of tensions: $$\frac{T_1}{T_2} = e^{\mu\theta} = e^{0.3 \times \frac{\pi}{180} \times 225°}$$

or $$\frac{T_1}{T_2} = 3.248 \qquad \text{(ii)}$$

Solving Eqs. (i) and (ii), we get

$T_1 = 6898.7 \text{ N}$ and $T_2 = 2124 \text{ N}$

(i) For counterclockwise rotation of drum, band OC is tight side and AD is slack side. Taking moments about fulcrum O,

$$T_2 \times x - F \times 750 = 0$$

or $$F = \frac{2124 \times 125 \times \cos 45°}{750} = 250.3 \text{ N} \qquad \textbf{Ans.}$$

(ii) For clockwise direction the tight and slack sides are interchanged. Taking moments about fulcrum O,

$$T_1 \times x - F \times 750 = 0$$

or $$F = \frac{6898.7 \times 125 \times \cos 45°}{750} = 813 \text{ N} \qquad \text{Ans.}$$

Therefore, for minimum effort, the direction of rotation of the drum should be **counterclockwise** Ans.

(b) Band cross section: $bt = \dfrac{\text{Maximum pull}}{\text{Allowable stress}} = \dfrac{6898.7}{50} = 137.97 \text{ mm}^2$

or band width: $b = \dfrac{137.97}{2.5} = \textbf{55.2 mm} \qquad \text{Ans.}$

7.4 BAND AND BLOCK BRAKE

A band and block brake consists of a number of brake shoes fixed to the inside surface of the flexible steel band as shown in Figure 7.25. When effort is applied at the free end of the lever, the shoes are pressed against the drum. Due to the friction force between the shoes and the drum, the two sides of the bands CA and DE become tight and slack respectively. Since the block shoes have higher coefficient of friction and can easily be replaced when worn out, they increase the effectiveness of the braking system.

(a)

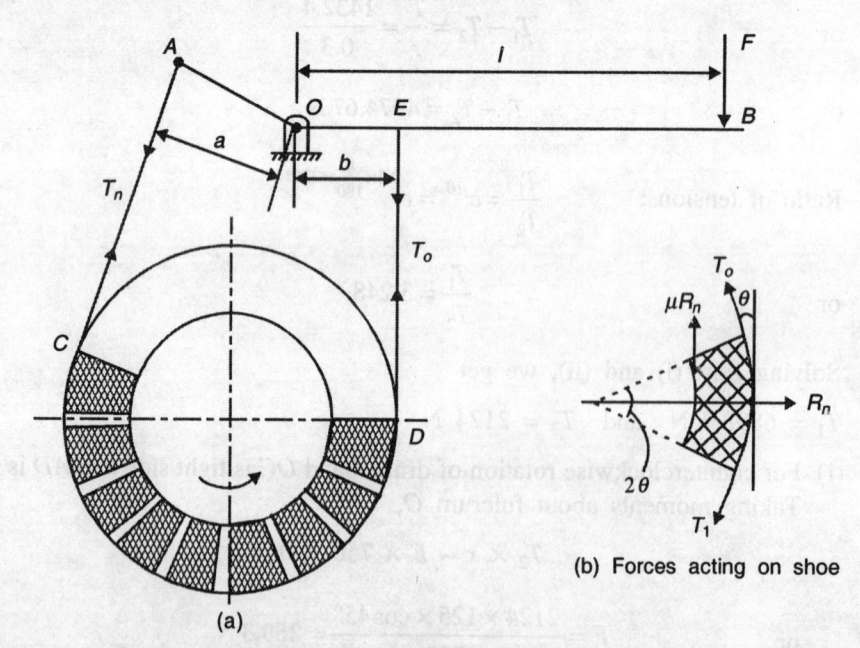

(b) Forces acting on shoe

FIGURE 7.25 Band and block brake.

Let

T_n = tension in the band on tight side.
T_o = tension in the band on slack side.
T_1 and T_2 = tensions in the band between the first and second block, between the second and third block respectively.
n = number of blocks.
2θ = angle subtended by the each block.
R_n = normal reaction on the block.

Referring to Figure 7.25(b), the equilibrium of the first block under the action of four forces—tight side tension T_1, slack side tension T_0, normal reaction R_n and frictional force μR_n—can be obtained by resolving forces in radial and tangential directions,

Resolving forces in radial direction,

$$(T_1 + T_0)\sin\theta = R_n$$

Resolving forces in tangential direction,

$$(T_1 - T_0)\cos\theta = \mu R_n$$

or

$$\frac{T_1 - T_0}{T_1 + T_0} = \mu\tan\theta \qquad (7.23)$$

Adding 1 to both sides of Eq. (7.23), we get

$$\frac{(T_1 + T_0) + (T_1 - T_0)}{T_1 + T_0} = 1 + \mu\tan\theta \qquad (7.24)$$

Similarly, subtracting 1 from both sides of Eq. (7.23), we get

$$\frac{(T_1 - T_0) - (T_1 + T_0)}{T_1 + T_0} = \mu\tan\theta - 1$$

or

$$\frac{(T_1 + T_0) - (T_1 - T_0)}{T_1 + T_0} = 1 - \mu\tan\theta \qquad (7.25)$$

Dividing Eq. (7.24) by Eq. (7.25), we get

$$\frac{(T_1 + T_0) + (T_1 - T_0)}{(T_1 + T_0) - (T_1 - T_0)} = \frac{1 + \mu\tan\theta}{1 - \mu\tan\theta}$$

or

$$\frac{T_1}{T_0} = \frac{1 + \mu\tan\theta}{1 - \mu\tan\theta} \qquad (7.26)$$

Similarly, for second block, the ratio of tensions is:

$$\frac{T_2}{T_1} = \frac{1 + \mu\tan\theta}{1 - \mu\tan\theta}$$

and for the nth block

$$\frac{T_n}{T_{n-1}} = \frac{1 + \mu \tan\theta}{1 - \mu \tan\theta}$$

Therefore,

$$\frac{T_n}{T_0} = \frac{T_n}{T_{n-1}} \times \frac{T_{n-1}}{T_{n-2}} \times \cdots \times \frac{T_1}{T_0}$$

or

$$\frac{T_n}{T_0} = \left(\frac{1 + \mu \tan\theta}{1 - \mu \tan\theta}\right)^n \qquad (7.27)$$

Using Eq. (7.27) and different arrangements of band brakes such as simple or differential band, the braking torque can be determined similar to what is discussed in Section 7.3.

EXAMPLE 7.12 A band and block brake having 10 blocks, each of which subtends an angle of 14° at the centre, is applied to a drum of effective diameter of 1 m. The brake drum and winding drum of crane mounted on the same shaft weighs 20 kN and have combined radius of gyration of 300 mm. The two ends of the band are attached to the pin on opposite side of the fulcrum at distances of 40 mm and 125 mm from the fulcrum. If an effort of 300 N is applied at distance of 800 mm from the fulcrum, find:

  (i) the braking torque
  (ii) the angular retardation of the brake
  (iii) the time taken by the system to come to rest from the rated speed of 400 rpm

Take coefficient of friction between the drum and the block as 0.25.

Solution The layout of braking system is as shown in Figure 7.26.

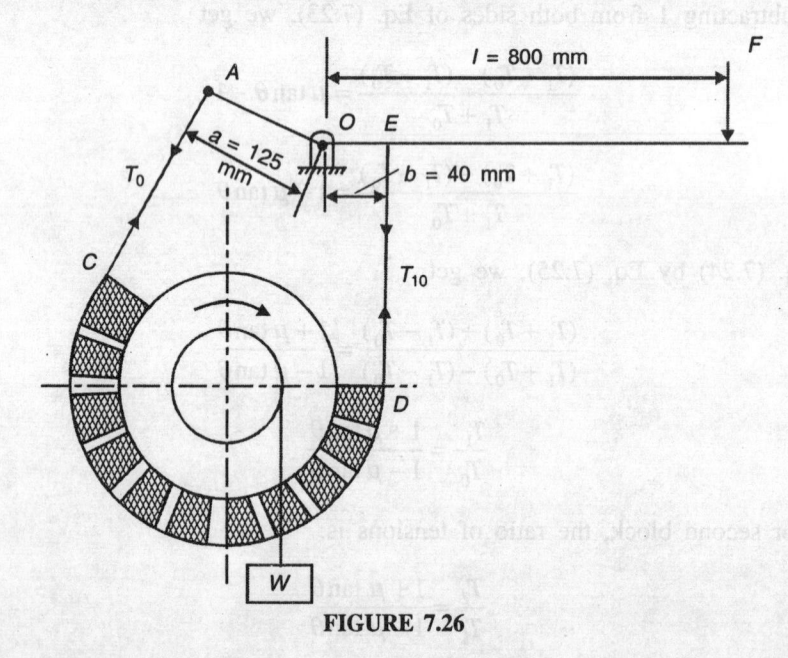

FIGURE 7.26

(i) For maximum braking torque, the drum must rotate in the clockwise direction. The side *DE* would be tight and the side *CA* would be slack. Further, when $a > b$, the effort must act in downward direction.

Ratio of tensions: $\dfrac{T_{10}}{T_0} = \left(\dfrac{1 + \mu \tan\theta}{1 - \mu \tan\theta}\right)^n = \left(\dfrac{1 + 0.25 \times \tan 7°}{1 - 0.25 \times \tan 7°}\right)^{10} = 1.848$

Taking moments about fulcrum *O*,

$$Fl + T_{10} \times b - T_0 \times a = 0$$

or

$$300 \times 800 + T_{10} \times 40 - T_0 \times 125 = 0$$

or

$$300 \times 800 + 1.848 \times T_0 \times 40 - T_0 \times 125 = 0$$

or

$$T_0 = 4698.5 \text{ N} \quad \text{and} \quad T_{10} = 1.848 \times 4698.5 = 8682.8 \text{ N}$$

Braking forque:

$$T = (T_{10} - T_0) \times r = (8682.8 - 4698.5) \times 0.5 = \textbf{1992.15 Nm Ans.}$$

(ii) Torque:

$$T = I \times \alpha = mk^2 \times \alpha$$

where α is angular retardation of the winding drum

$$\dfrac{20 \times 10^3}{9.81} \times 0.3^2 \times \alpha = 1992.15$$

or $\quad\quad\quad\quad\quad\quad\quad\quad \alpha = \textbf{10.85 rad/s}^2 \quad\quad$ **Ans.**

(iii) Let ω_0 be the initial angular speed and ω be the angular speed.

$$\omega = \omega_0 - \alpha t$$

$$0 = \dfrac{2\pi \times 400}{60} - 10.85 \times t$$

or $\quad\quad\quad\quad\quad\quad\quad$ Time: $t = \textbf{3.86 Ans.}$

7.5 INTERNAL EXPANDING SHOE BRAKE

Figure 7.27 shows a typical internal expanding shoe brake used in automobiles. It consists of two semi-circular shoes which are lined with friction material of high coefficient of friction and good wearing properties. The shoes are normally held in the inactive position by the light spring. To apply the brakes, either the cam is rotated or fluid pressure is applied through a cylinder-plunger assembly. This forces the shoes against the inner surface of the brake drum which is integral to wheel casing.

266 Theory of Mechanisms and Machines

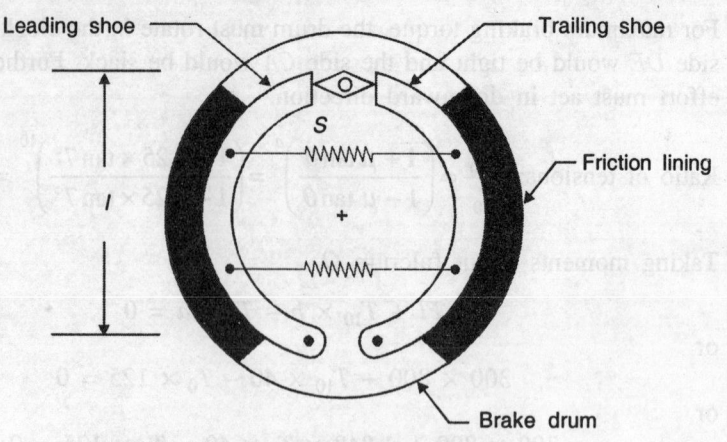

FIGURE 7.27 Internal expanding shoe brake.

The analysis of the internal expanding shoe brake is based on the assumption that each shoe is rigid enough and when compressed obeys the Hooke's law. Consider a small element AB of the shoe subtending an angle $d\theta$ at the centre of the brake drum. (See Figure 7.28). The point A makes an angle θ with the line OO_1. The pressure distribution on the shoe is assumed to be nearly uniform. However, the shoe wears out more at the free end. The rate of wear lining varies directly as O_1C. If p_{max} is the maximum intensity of the pressure on the leading shoe, then the normal pressure at the point A is:

$$p_n = p_{max} \sin\theta \qquad (7.28)$$

The element of lining is subjected to normal reaction force dR_n and friction force dF. The normal reaction force:

$$dR_n = p_n \times br\,d\theta$$

or

$$dR_n = p_{max}\, br \sin\theta\, d\theta \qquad (7.29)$$

where b is the width of brake shoe.

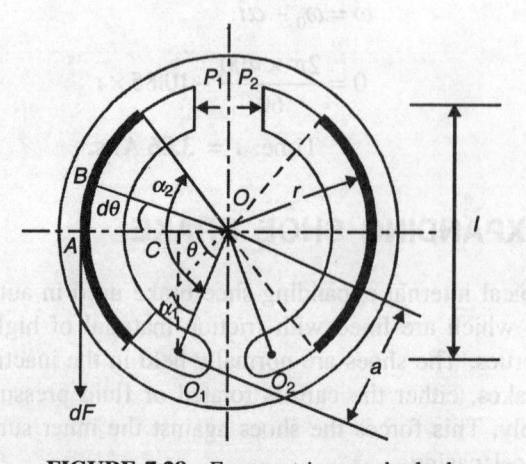

FIGURE 7.28 Forces acting on the brake.

(a) The friction force on the element:
$$dF = \mu\, p_{max}\, br\, \sin\theta\, d\theta$$

and braking torque: $\quad dT = dF \times r$
$$= \mu p_{max} \times br^2 \sin\theta\, d\theta$$

The total braking torque about centre O for whole of one shoe:
$$T = \mu p_{max} br^2 \int_{\alpha_1}^{\alpha_2} \sin\theta\, d\theta$$

or $\quad T = \mu p_{max} br^2 (\cos\alpha_1 - \cos\alpha_2) \quad\quad$ (7.30)

(b) Moment of normal reaction about fulcrum of leading shoe:
$$dM_n = dR_n \times O_1C$$
$$= p_{max} br \sin\theta\, d\theta \times OO_1 \sin\theta$$

Let OO_1 = distance a

$\therefore \quad dM_n = p_{max} br\, a\, \sin^2\theta\, d\theta$

Total moment: $\quad M_n = p_{max} br\, a \int_{\alpha_1}^{\alpha_2} \sin^2\theta\, d\theta$
$$= p_{max} br\, a \int_{\alpha_1}^{\alpha_2} \frac{1 - \cos 2\theta}{2} d\theta$$

or $\quad M_n = \dfrac{p_{max} br\, a}{4}\left[(2\alpha_2 - 2\alpha_1) - \sin 2\alpha_2 + \sin 2\alpha_1\right] \quad\quad$ (7.31)

(c) Moment of friction force about fulcrum of leading shoe:
$$dM_f = dF \times AC = dF(r - OC)$$

$\therefore \quad M_f = \mu p_{max} br \int_{\alpha_1}^{\alpha_2} \sin\theta (r - a\cos\theta)\, d\theta$
$$= \mu p_{max} br \int_{\alpha_1}^{\alpha_2} \left(r\sin\theta - \frac{a}{2}\sin 2\theta\right) d\theta$$

or $\quad M_f = \dfrac{\mu p_{max} br}{4}\left[4r(\cos\alpha_1 - \cos\alpha_2) + a(\cos 2\alpha_2 - \cos 2\alpha_1)\right] \quad\quad$ (7.32)

For the equilibrium of the leading shoe, the algebraic sum of moments about fulcrum O_1 must be zero. That is
$$P_1 \times l - M_n + M_f = 0 \quad\quad (7.33)$$

Similarly, for trailing shoe
$$P_2 \times l - M_n - M_f = 0 \qquad (7.34)$$
where P_1 and P_2 are force exerted by cylinder plunger assembly on leading and trailing shoes respectively.

For the leading shoe, if the moment of friction force M_f is greater than moment of normal force M_n, then brake will become self-locking.

The braking torque required for an internal expanding shoe brake can be approximately determined by graphical method as given in the following steps (See Figure 7.29).

1. Draw configuration of the internal expanding shoe brake including the location of the pivots of leading and trailing shoes.
2. Bisect the contact angle ($\alpha_2 - \alpha_1$) and locate the points A and B on the leading and trailing shoes respectively.
3. Draw a small circle called friction circle at the centre of the shoe. The radius of the friction circle is given by

$$r_f = r \sin\phi = \mu r$$

where
$$\phi = \tan^{-1}\mu$$

4. Draw a line tangent to the friction circle from point A and a similar tangent from the point B.

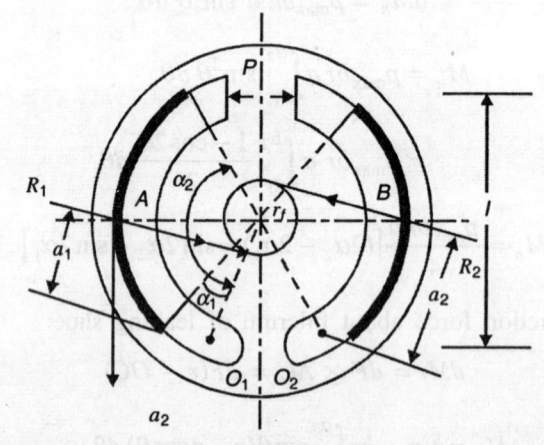

FIGURE 7.29 Internal expanding shoe with friction circle.

Let R_1 and R_2 represent the thrusts for counterclockwise rotation of the drum and let a_1 and a_2 be the perpendicular distances of their lines of action from the respective shoe pivots. Further, assuming that forces exerted by hydraulic plunger on both shoes are equal ($P_1 = P_2 = P$) and act in one line, the equilibrium equations of the shoe are:

$$P \times l = R_2 \times a_2$$

and
$$P \times l = R_1 \times a_1$$

Therefore, Braking torque: $T = \mu r(R_1 + R_2)$

$$= \mu r \left[\frac{Pl}{a_1} + \frac{Pl}{a_2} \right]$$

or
$$T = \mu r Pl \left[\frac{1}{a_1} + \frac{1}{a_2} \right] \quad (7.35)$$

EXAMPLE 7.13 An internal expanding shoe brake as shown in Figure 7.30 has width of brake lining 40 mm and intensity of pressure at point A is 3.5 sinα bar. Where α is measured as shown from either pivot. If the coefficient of friction is 0.35, determine the following:

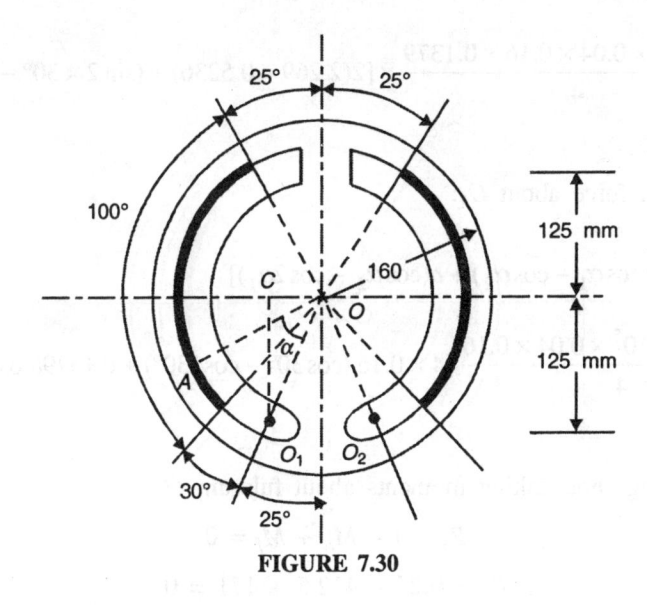

FIGURE 7.30

(i) Braking torque
(ii) Magnitude of forces P_1 and P_2

Solution Intensity of pressure: $p_n = 3.5$ sinα bar. So maximum pressure: $p_{max} = 3.5$ bar

Let $\alpha_1 = 30°$ and $\alpha_2 = 100 + 30 = 130°$

Braking torque on both shoes:

$$T = 2\mu p_{max} br^2 (\cos\alpha_1 - \cos\alpha_2)$$

$$= 2 \times 0.35 \times 3.5 \times 10^5 \times \frac{40}{1000} \times 0.16^2 (\cos 30° - \cos 130°)$$

$$= 378.53 \text{ Nm} \qquad \text{Ans.}$$

Angles: $\alpha_1 = 30° = 30° \times \dfrac{\pi}{180} = 0.5236$ rad

$$\alpha_2 = 130° = 130° \times \frac{\pi}{180} = 2.269 \text{ rad}$$

From the geometry of Figure 7.31, we get

$$OO_1 = a = \frac{O_1 C}{\cos 25°} = \frac{125}{\cos 25°} = 137.9 \text{ mm}$$

Total moment of the normal force about the fulcrum point O_1 is:

$$M_n = \frac{p_{max} b r a}{4}\left[2(\alpha_2 - \alpha_1) + (\sin 2\alpha_1 - \sin 2\alpha_2)\right]$$

$$= \frac{3.5 \times 10^5 \times 0.04 \times 0.16 \times 0.1379}{4} \times [2(2.269 - 0.5236) + (\sin 2 \times 30° - \sin 2 \times 130°)]$$

$$= 412.5 \text{ Nm}$$

Moment of friction force about O_1:

$$M_f = \frac{\mu p_{max} b r}{4}[4r(\cos\alpha_1 - \cos\alpha_2) + a(\cos\alpha_2 - \cos 2\alpha_1)]$$

$$= \frac{0.35 \times 3.5 \times 10^5 \times 0.04 \times 0.16}{4}[4 \times 0.16(\cos 30° - \cos 130°) + 0.1379(\cos 260° - \cos 60°)]$$

$$= 171.0 \text{ Nm}$$

Now for the leading shoe, taking moments about fulcrum O_1,

$$P_1 \times l - M_n + M_f = 0$$

or
$$P_1 \times 0.25 - 412.5 + 171 = 0$$

or
$$P_1 = 966 \text{ N} \qquad \text{Ans.}$$

Similarly, for the trailing shoe,

$$P_2 = \frac{M_n + F_f}{l} = \frac{412.5 + 171}{0.25} = 2334 \text{ N} \qquad \text{Ans.}$$

7.6 EFFECT OF BRAKING ON VEHICLE

In a vehicle, the quality of brake is judged by two criteria—the time required for the vehicle to come to rest and the distance traversed by the vehicle before it comes to rest. In this section, the effect of braking on vehicle when brakes are applied to front wheels, rear wheels or to all four wheels are discussed when a vehicle moves up an inclined plane, level road and down the plane.

Case I Brakes applied to front wheel

1. When vehicle moves up an inclined plane.

Figure 7.31 shows a vehicle weighing mg moving up a plane inclined at an angle of θ. The centre of gavity C of the vehicle is at a distance x, parallel to the road, from the centre of rear wheel B and a perpendicular distance y from the road. Let R_a and R_b are reactions between road and wheels A and B respectively.

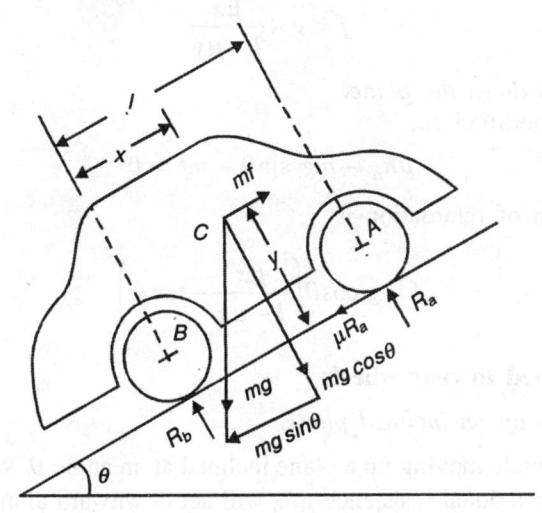

FIGURE 7.31 Brakes applied to front wheel.

When brakes are applied to the front wheels, the frictional resistance (μR_a) between the front wheel and the road opposes the motion and acts downward along the road. For equilibrium of the vehicle, algebraic sum of the forces along and perpendicular to the road and the moments about centre of gravity (c.g.) of the vehicle must be equal to zero.
Forces along the inclined plane:

$$\mu R_a + mg \sin\theta - mf = 0 \tag{7.36}$$

where f is retardation
Forces perpendicular to the inclined plane:

$$R_a + R_b - mg \cos\theta = 0 \tag{7.37}$$

Taking moments about c.g. C,

$$R_b \times x + \mu R_a \times y - R_a (l - x) = 0 \tag{7.38}$$

From Eq. (7.37), substituting the value of R_b in Eq. (7.38), we get

$$(mg \cos\theta - R_a) x + \mu R_a y - R_a(l - x) = 0$$

or

$$R_a = \frac{mgx \cos\theta}{l - \mu y}$$

Substituting the value of R_a in Eq. (7.36), we get the expression for retardation as

$$f = g\cos\theta\left(\frac{\mu x}{l - \mu y} + \tan\theta\right) \tag{7.39}$$

2. *When vehicle moves on a level road.*
   $\theta = 0°$ Therefore,

$$f = g \times \frac{\mu x}{l - \mu y} \tag{7.40}$$

3. *When vehicle moves down the plane.*
   The Eq. (7.36) is modified as

$$\mu R_a - mg\sin\theta - mf = 0$$

Therefore, the equation of retardation is:

$$f = g\cos\theta\left(\frac{\mu x}{l - \mu x} - \tan\theta\right) \tag{7.41}$$

Case II Brakes applied to rear wheels

1. *When vehicle moves up on inclined plane.*

Figure 7.32 shows a vehicle moving up a plane inclined at an angle θ. When brakes are applied to the rear wheels, the frictional resistance μR_b will act downward along the road between the rear wheel and the road.

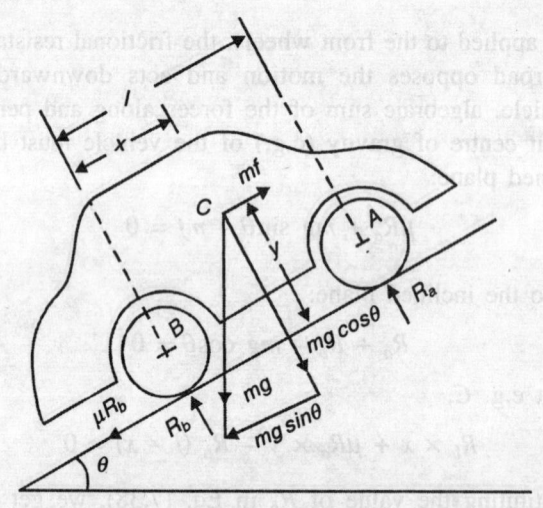

FIGURE 7.32 Brakes applied to rear wheel.

The equations for equilibrium of the vehicle are as follows:
Forces along the inclined plane:

$$\mu R_b + mg\sin\theta - mf = 0 \tag{7.42}$$

Forces perpendicular to the inclined plane:

$$R_a + R_b - mg \cos\theta = 0 \tag{7.43}$$

Taking moments about the c.g. C,

$$R_b \times x + \mu R_b \times y - R_a (l - x) = 0 \tag{7.44}$$

Solving Eqs. (7.42), (7.43) and (7.44) for unknown variable f, we get retardation:

$$f = g\cos\theta \left[\frac{\mu(l-x)}{l+\mu y} + \tan\theta \right] \tag{7.45}$$

2. When vehicle moves on level road

$$\theta = 0°$$

Therefore, retardation: $$f = g \left[\frac{\mu(l-x)}{l+\mu y} \right] \tag{7.46}$$

3. When vehicle moves down the plane

Equation (7.42) is modified as

$$\mu R_b - mg \sin\theta - mf = 0$$

Therefore, the expression for retardation of the vehicle would be modified as

$$f = g\cos\theta \left[\frac{\mu(l-x)}{l+\mu y} - \tan\theta \right] \tag{7.47}$$

Case III Brakes applied to all four wheels

1. When vehicle moves up on inclined plane.

When brakes are applied to both front and rear wheels, the friction resistance would act between road and wheels as shown in Figure 7.33. The necessary equations for equilibrium of the vehicle are as follows:

Forces along the inclined plane:

$$\mu R_a + \mu R_b + mg \sin\theta - mf = 0 \tag{7.48}$$

Forces perpendicular to the inclined plane:

$$R_a + R_b - mg \cos\theta = 0 \tag{7.49}$$

Substituting the value of $(R_a + R_b)$ from Eq. (7.49) into Eq. (7.48), we get

$$\mu(mg \cos\theta) + mg \sin\theta - mf = 0$$

or retardation: $$f = g \cos\theta \, (\mu + \tan\theta) \tag{7.50}$$

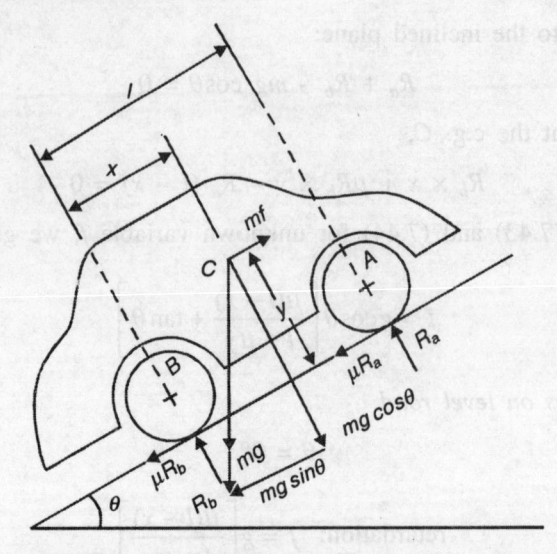

FIGURE 7.33 Brakes applied to all four wheels.

2. *When vehicle moves on level road*

$$\text{Inclination angle: } \theta = 0°$$

Therefore, retardation: $f = \mu g$ \hfill (7.51)

3. *When vehicle moves down the plane.*

The Equation (7.48) is modified as

$$\mu R_a + \mu R_b - mg \sin \theta - mf = 0$$

Therefore, retardation: $f = g \cos \theta \, (\mu - \tan \theta)$ \hfill (7.52)

7.7 DYNAMOMETERS

Dynamometer is a device by means of which energy or work done by a prime mover can be measured. A mechanical dynamometer is essentially a brake with additional device to measure frictional resistance by allowing the prime mover to run at the rated speed.

Dynamometers are mainly classified into two types—*absorption dynamometers* and *transmission dynamometers*.

7.7.1 Absorption Dynamometers

In an absorption dynamometer, the work done or energy of the prime mover is converted into heat usually against friction while being measured. In other words, it is a braking system in which some provision is made for measuring friction torque on the drum. Examples are prony brake, rope brake, hydraulic dynamometers. These dynamometers can be used to measure moderate power produced by various types of prime movers. A brief description of various absorption dynamometers is given further.

Prony brake dynamometer. Prony brake dynamometer is the simplest type of absorption dynamometer which consists of two blocks of wood (pronys). Each of which embraces pulley for less than half of the pulley rim. One of the block is connected to a lever. At one end of the lever arm, a force is applied through dead weight (Figure 7.34). The frictional torque on the pulley is varied by screwing up the nut–bolt till it balances the applied torque and the prime mover runs at the required speed. The friction between the blocks and pulley tends to rotate the brake lever in the direction of rotation. However, the lever is brought back to horizontal position by suspending dead weight at the lever end.

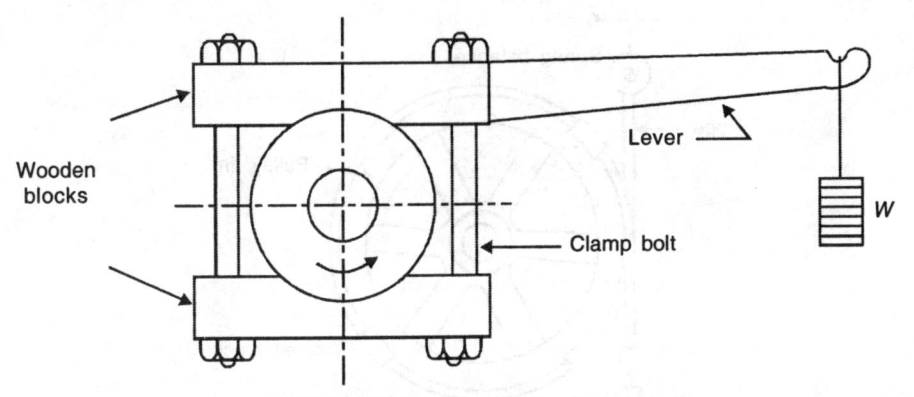

FIGURE 7.34 Prony brake dynamometer.

Frictional torque:
$$T = W \times l \tag{7.53}$$

Power:
$$P = \frac{2\pi NT}{60} \tag{7.54}$$

where
  l is the length of the lever
  N is the measured speed of the prime mover.

The main limitation of prony brake dynamometer is wear of the block and variation in the coefficient of friction between block and pulley. It necessitates continuous adjustment of bolts to maintain required grip of blocks over the pulley. Therefore, it is not suitable for absorption of large power and for long continued period.

Rope brake dynamometer. A rope brake dynamometer is most widely used to test the power of IC engines. It is inexpensive, reasonably accurate, easy to manufacture and requires little or no maintenance. Such a dynamometer consists of a hollow channelled rim wheel over which two or three turns of rope are wound around the rim. The wheel is keyed to the engine shaft. One end of the rope which is slack side is connected to the spring balance and the tight side of the rope is connected to the weight hanger as shown in Figure 7.35. The spacing of the rope is done by four U-shaped blocks of wood or Ferodo which prevent rope from slipping off the pulley.

The frictional resistance offered by the rope is obtained by measuring the spring force and dead weight. The braking torque is computed as

Power:
$$T = (W - S)\,r$$
$$P = (W - S) \times r \times \frac{2\pi N}{60} \qquad (7.55)$$

where
- W = dead weight in the weight hanger
- S = reading of spring balance
- r = effective radius of the brake wheel

and N = speed of the brake wheel.

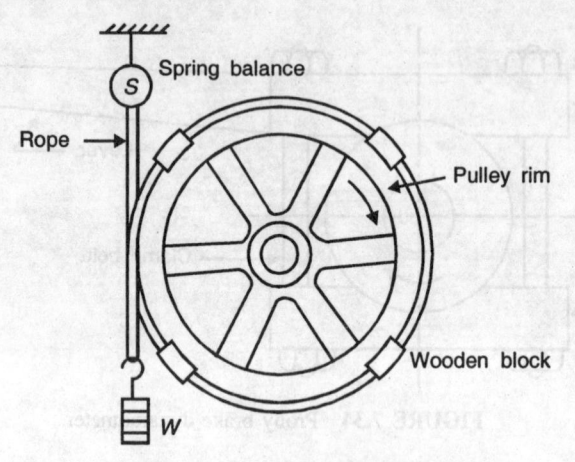

FIGURE 7.35 Rope brake dynamometer.

In rope brake dynamometer, a cooling arrangement is necessitated to carry away the heat generated due to friction between the rope and the wheel. For this purpose, cooling water is circulated in the hollow channel of the rim. A flat end extraction scoop is fitted to take out hot water.

Hydraulic dynamometer. Froude hydraulic dynamometer is a popular power measuring device which can measure the wide range of power. It is generally non-reversible and can be directly coupled to the prime mover. It works on the principle of absorption type wherein power is converted into frictional heat energy due to water friction between the rotor and the stator. The layout of dynamometer is shown in Figure 7.36. In this dynamometer, water supply and discharge are made through top of the casing so that casing is always full of water when dynamometer is in operation. The casing of dynamometer is constrained against dead weight and load spring such that by adjusting dead weight, the casing can be maintained in horizontal position.

The rotor of the dynamometer is keyed to the main shaft to which input power is supplied by the prime mover. The stator surrounds the rotor and is rigidly fixed to the casing. The main shaft is supported on ball bearings in the outer casing and the outer casing which can swing is supported on the bed plate through ball bearings. In each face of the rotor and in the adjacent face of the stator, there are semi-elliptical channels. Each channel is further divided into a number of cells by semi-circular diaphragms. These diaphragms are set at an angle of 45° to the plane of rotation such that straight edge of the diaphragm coincides with

Brakes and Dynamometers 277

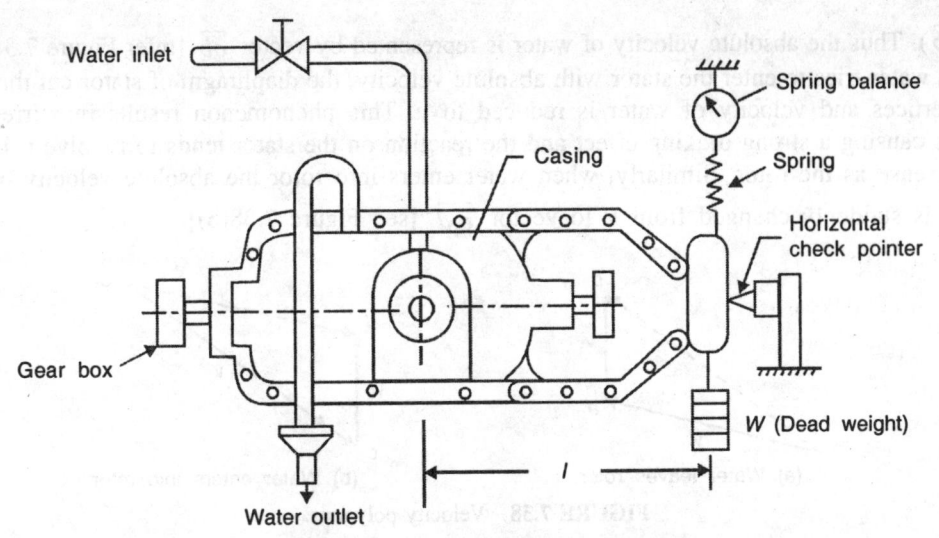

FIGURE 7.36 Froude hydraulic dynamometer.

the major axis of semi-elliptical channel (Figure 7.37). Before starting the dynamometer, water is fed to these cells through passage P. When the main shaft starts rotating, the water in the rotor cell flows outwardly whereas in the cells of the stator, it flows inwardly. The speed of circulation in the vortex depends on the rotational speed of the main shaft.

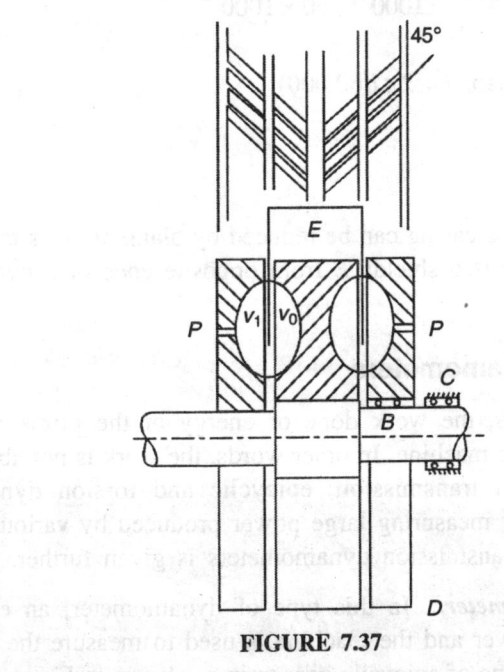

FIGURE 7.37

Considering the vortex to consist of a large number of small rings in which speed of circulation is fairly constant. The velocity of water at a point where the ring moves from rotor to stator is vector sum of the velocity of circulation v ($\overrightarrow{oa}$) and tangential velocity of the rotor

v_1 ($\overrightarrow{ab}$). Thus the absolute velocity of water is represented by vector $\overrightarrow{ob}$ [refer Figure 7.38(a)]. When water tries to enter the stator with absolute velocity, the diaphragm of stator cut through the vertices and velocity of water is reduced to v. This phenomenon results into irregular eddies causing a strong braking effect and the reaction on the stator tends to revolve it in the same sense as the rotor. Similarly, when water enters into rotor the absolute velocity of the water is suddenly changed from V to vector $\overrightarrow{o_1 d}$ [see Figure 7.38(b)].

(a) Water leaves rotor (b) Water enters into rotor

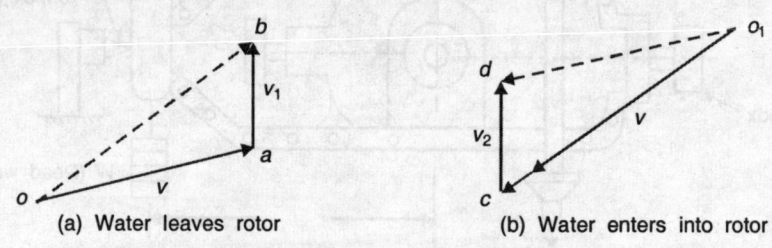

FIGURE 7.38 Velocity polygons.

Again, the reaction on the stator tends to revolve it in the same sense as the rotor. The reaction torque on the casing is measured by dead weight attached to an arm fixed to the casing.

$$\text{Power} = \frac{T \times \omega}{1000} = \frac{Wl \times 2\pi N}{60 \times 1000} = \frac{WN}{k} \text{ kW} \qquad (7.56)$$

where
k = dynamometer constant (= $2\pi l/60{,}000$)
W = dead weight
l = lever arm length
and N = speed of the rotor

The reaction torque on the casing can be reduced by blanking off some of the cells. This is usually achieved by sliding two shields E from opposite ends of a dynamometer towards the main shaft.

7.7.2 Transmission Dynamometers

In transmission dynamometers, the work done or energy of the prime mover is measured before it is utilized to drive the machine. In other words, the work is not absorbed while being measured. Examples are belt transmission, epicyclic and torsion dynamometers. These dynamometers are suitable for measuring large power produced by various prime movers. A brief description of various transmission dynamometers is given further.

Epicyclic train dynamometer. In this type of dynamometer, an epicyclic gear train placed between the prime mover and the machine is used to measure the power transmitted. Consider the sun and planet type of epicyclic gear train as shown in Figure 7.39. The sun gear B is keyed to the prime mover shaft and rotates in one direction, say counterclockwise direction. The internal gear D is keyed to the driven machine shaft and rotates in clockwise

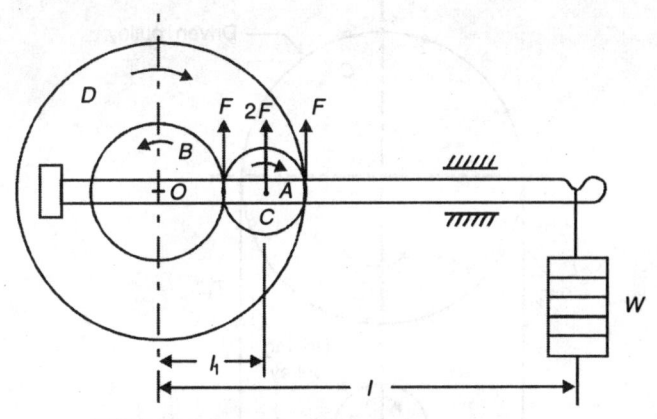

FIGURE 7.39 Epicyclic train dynamometer.

direction. The power is transmitted from gear B to D through the intermediate gear C which revolves freely on a pin fixed to the arm A. While transmitting power, the tangential force is exerted by gear B on the gear C and the reaction of the internal gear D on the gear C. If the friction of the pin on which gear C revolve is neglected, these two tangential forces act in upward direction and are equal in magnitude. Therefore, the total upward force on the lever arm A is equal to $2F$. The moment due to force $2F$ about the centre of gear B causes the lever arm A to rotate in counterclockwise direction. This moment is balanced by suspending a dead weight W from the arm, which causes the arm to float between the stops. (Refer Figure 7.39.)

Taking moments about fulcrum O,

$$W \times l = 2F \times l_1$$

or

$$F = \frac{Wl}{2l_1} \qquad (7.57)$$

Braking torque: $\qquad T = F \times r_b$

where $\qquad r_b$ is radius of the gear B

Power:

$$P = \frac{2\pi NT}{60} \qquad (7.58)$$

where N is the speed of gear B in rpm.

Belt transmission dynamometers. The belt transmission dynamometer is most widely used among transmission dynamometers. It works on the principle that when belt transmits power, there exists a difference in tension between two sides of the belt. A belt transmission dynamometer measures this difference in tensions and let us know the amount of power being transmitted. Depending upon the kinematic arrangement, there are two types of belt transmission dynamometers. These are discussed further.

Tatham dynamometer. In a Tatham dynamometer, an endless belt passes over the driving and driven pulleys through two intermediate pulleys as shown in Figure 7.40. The

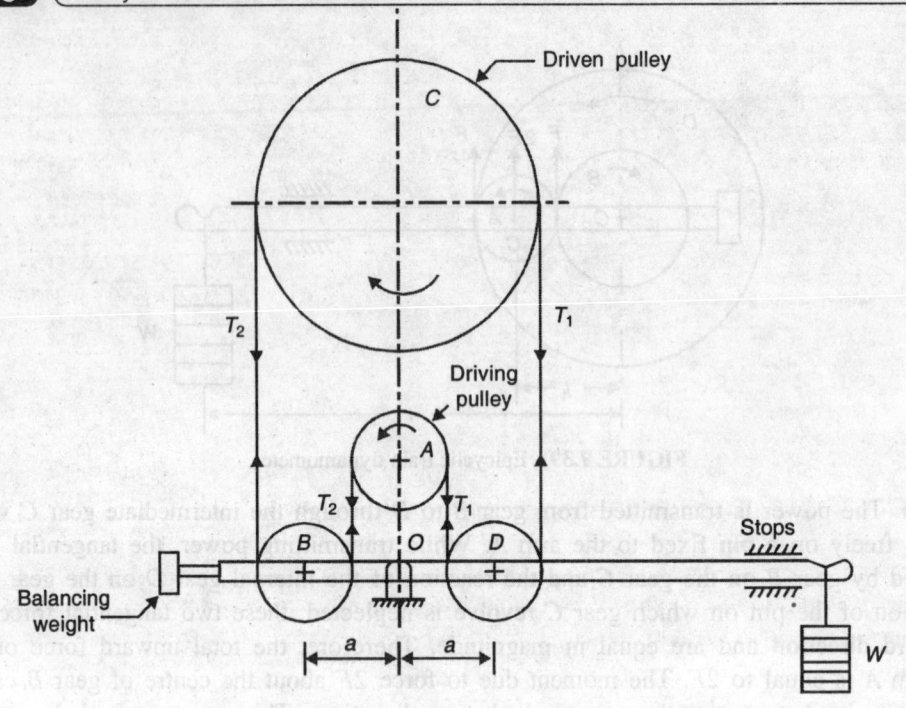

FIGURE 7.40 Tatham belt transmission dynamometer.

intermediate pulleys B and D revolve about pins mounted on the lever. The lever is pivoted about the fulcrum O on the fixed frame. When driving pulley A rotates in counterclockwise direction, the tight and slack sides of the belt are shown in Figure 7.40. The total upward forces acting on the pins of intermediate pulleys B and D are $2T_2$ and $2T_1$ respectively. The moment caused by these forces about the fulcrum O is balanced by suspending a weight W at a distance l from the fulcrum.

Taking moments about the fulcrum O, we get

$$W \times l = 2T_1 \times a - 2T_2 \times a$$

or

$$T_1 - T_2 = \frac{Wl}{2a} \tag{7.59}$$

Power:

$$P = \frac{2\pi N(T_1 - T_2) \times r}{60} \tag{7.60}$$

where
   r = radius of driving pulley A
and N = speed of driving pulley A in rpm.

Von-Hefner transmission dynamometer. The Von-Hefner belt transmission dynamometer is generally used for a horizontal belt drive. In this dynamometer, a continuous belt runs over driving and driven pulleys through two idler pulleys mounted on a triangular shaped frame as shown in Figure 7.41. The frame is free to turn about a fixed axis through point O on the line joining the centres of driving and driven pulleys. The triangular frame is connected to lever arm which is pivoted at fulcrum O_1.

Brakes and Dynamometers

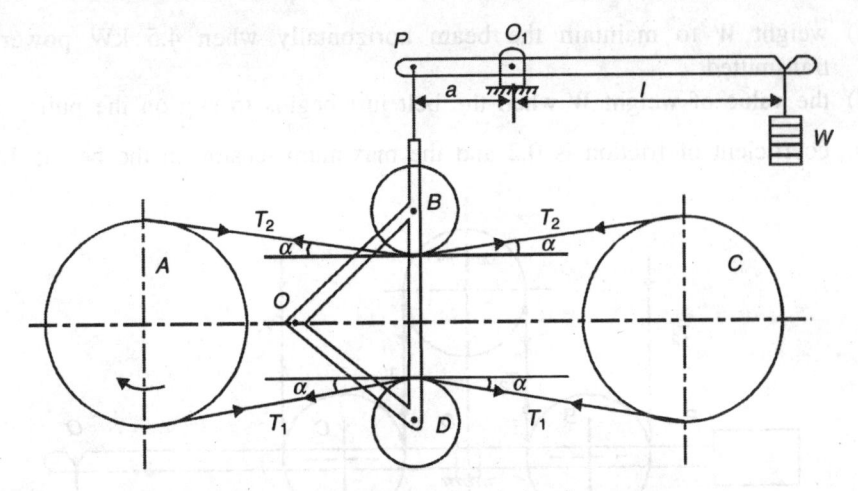

FIGURE 7.41 Von-Hefner transmission dynamometer.

The net downward force on the idler pulley caused by the difference between the belt tensions is transmitted to one end of the lever. This lever is balanced through balance weight suspending at the other end of the lever. The dynamometer is so adjusted that lever floats horizontally. Generally, the diameters of driving and driven pulleys are of equal size so that four straight portions of the belt are equally inclined at the angle α to the line joining the centres of driving and driven pulleys.

The downward force on the pulley D due to tight side tension T_1 is $2T_1 \sin \alpha$. Similarly, the upward force on the pulley B due to slack side tension is $2T_2 \sin\alpha$.

$$\text{Net force: } F = 2(T_1 - T_2) \sin \alpha \tag{7.61}$$

This net force F is acting downward on the lever arm at point P. For the equilibrium of the lever arm, taking moments about fulcrum O_1, we get

$$2(T_1 - T_2) \sin \alpha \times a = Wl \tag{7.62}$$

or

$$T_1 - T_2 = \frac{Wl}{2a \sin \alpha}$$

Power:

$$P = \frac{2\pi N(T_1 - T_2) \times r}{60}$$

where
r = radius of the driving pulley
and N = speed of the driving pulley

EXAMPLE 7.14 The essential features of a belt transmission dynamometer are shown in Figure 7.42. A is the driving pulley which runs at 600 rpm. B and C are jockey pulley mounted on a horizontal beam PQ pivoted at point D about which the complete beam is balanced when at rest. E is the driven pulley (not shown in the figure). Pulleys A, B and C are each of diameter 300 mm and the portions of the belt between the pulleys are vertical. Find the value of

(i) weight W to maintain the beam horizontally when 4.5 kW power is being transmitted

(ii) the value of weight W when the belt just begins to slip on the pulley A

The coefficient of friction is 0.2 and the maximum tension in the belt is 1.5 kN.

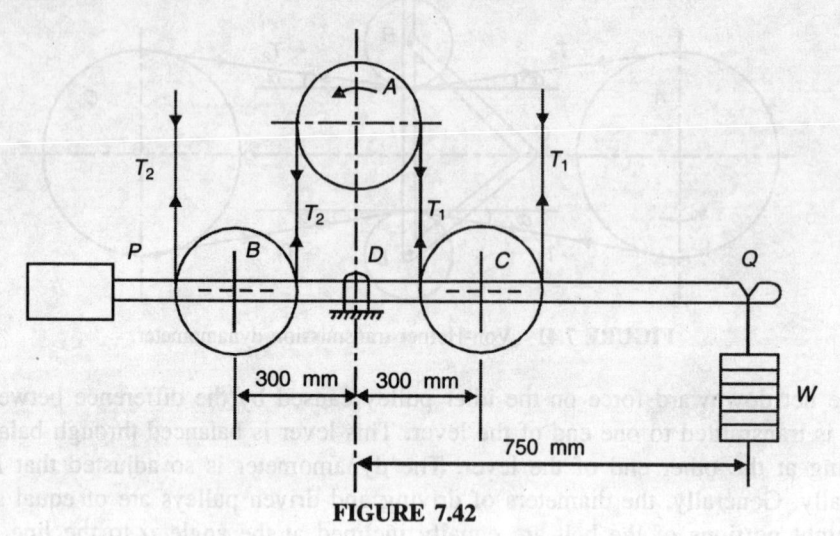

FIGURE 7.42

Solution

(i) Torque:
$$T = \frac{60P}{2\pi N} = \frac{60 \times 4.5 \times 10^3}{2\pi \times 600} = 71.62 \text{ Nm}$$

Let the forces $2T_2$ and $2T_1$ be acting at points B and C respectively.
Taking moments about fulcrum point D,

$$W \times 750 = 2T_1 \times 300 - 2T_2 \times 300$$

or $T_1 - T_2 = 1.25\ W$

Torque: $T = (T_1 - T_2) \times r = 1.25\ W \times 0.15$

or $71.62 = 1.25\ W \times 0.15$

or Weight: $W = \mathbf{381.97\ N}$ **Ans.**

(ii) Ratio of tensions:
$$\frac{T_1}{T_2} = e^{\mu\theta} = e^{0.2 \times \pi} = 1.874$$

Maximum tension: $T_1 = 1.5$ kN (given)

$$\therefore \quad T_2 = \frac{1.5 \times 10^3}{1.874} = 800 \text{ N}$$

Taking moments about fulcrum D,

$$W \times 750 = 2 \times 1500 \times 300 - 2 \times 800 \times 300$$

or Weight: $W = \mathbf{560\ N}$ **Ans.**

EXAMPLE 7.15 In the transmission dynamometer shown in Figure 7.43, the driving pulley A and driven pulley B each have a diameter of 600 mm, their centres being equidistant from

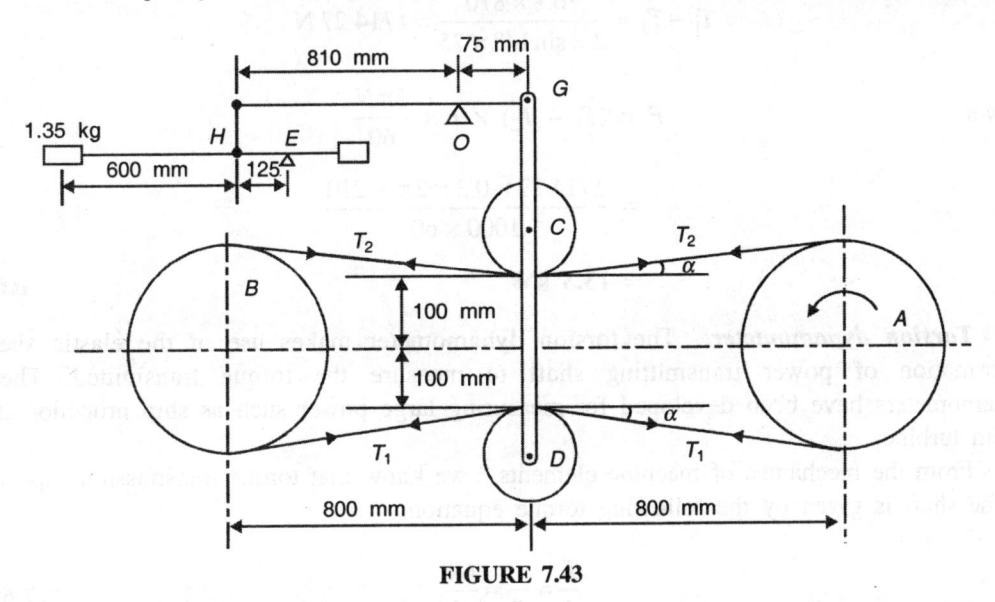

FIGURE 7.43

the jockey pulley centre line and 1.6 m apart. The jockey pulleys C and D are each of diameter 200 mm and their spindles are 400 mm apart. The masses of the jockey pulleys and levers are counterbalanced and friction at the jockey pulley is negligible. Calculate the power transmitted if 1.35 kg mass maintains balance when moved 600 mm from the fulcrum E and the speed of the pulleys A and B is 210 rpm.

Solution Since the layout of dynamometer is symmetric, all angles of belt inclination are equal.

$$\tan \alpha = \frac{300 - 100}{800}$$

or
$$\alpha = 14°$$

The driving pulley A rotates in counter clockwise direction. So lower side belt is under the tight side and upper side belt is under the slack.
Force acting downward = $2T_1 \sin\alpha$
Force acting upward = $2T_2 \sin\alpha$
Net force acting at $G = 2(T_1 - T_2) \sin\alpha$
Let F be the force applied at point H.
Taking moments about fulcrum E,

$$1.35 \times 9.81 \times 725 = F \times 125$$

or
$$F = 76.8 \text{ N}$$

Now taking moments about fulcrum O,

$$76.8 \times 810 = 2(T_1 - T_2) \sin \alpha \times 75$$

or
$$T_1 - T_2 = \frac{76.8 \times 810}{2 \times \sin 14° \times 75} = 1714.27 \text{ N}$$

Power:
$$P = (T_1 - T_2) \times r \times \frac{2\pi N}{60}$$

$$= \frac{1714.27 \times 0.3 \times 2\pi \times 210}{1000 \times 60}$$

$$= 13.3 \text{ kW} \qquad \text{Ans.}$$

Torsion dynamometer. The torsion dynamometer makes use of the elastic shear deformation of power transmitting shaft to measure the torque transmitted. These dynamometers have been developed for measuring large power such as ship propellor and steam turbine.

From the mechanics of machine elements,* we know that torque transmission capacity of the shaft is given by the following torque equation:

$$\frac{T}{J} = \frac{\tau}{R} = \frac{G\theta}{l} \qquad (7.63)$$

or
$$T = \theta \times \frac{GJ}{l} = \theta \times k$$

where
  k = constant (= GJ/l)
  G = modulus of rigidity
  J = polar moment of inertia of the shaft cross section $\left(= \frac{\pi}{32} d^4 \right)$
  l = length of the shaft
and θ = angle of twist

Hence for a given shaft, the torque capacity is directly proportional to the angle of twist. If by some means the angle of twist is measured, the corresponding torque may be calculated. One method to measure angle of twist has been suggested by Bevis–Gibson using flash–light technique. In this dynamometer, two similar discs A and B having narrow radial slot are fixed to the shaft at a certain distance from each other. Behind one disc (say disc A), a powerful electric lamp is fixed to the bearing of the shaft. This lamp is masked so that it can throw a narrow ray of light parallel to the axis of the shaft. Behind the disc B, a torque finder consisting of a vernier eyepiece is mounted on the bracket which can be moved along the circumference of the shaft (see Figure 7.44). When shaft is at rest, the eyepiece is so adjusted that it receives a ray of light passing through slots in the disc. When a shaft revolves without

---

Design of Machine Elements by C.S. Sharma and Kamlesh Purohit, 1st ed., Prentice-Hall of India, New Delhi.

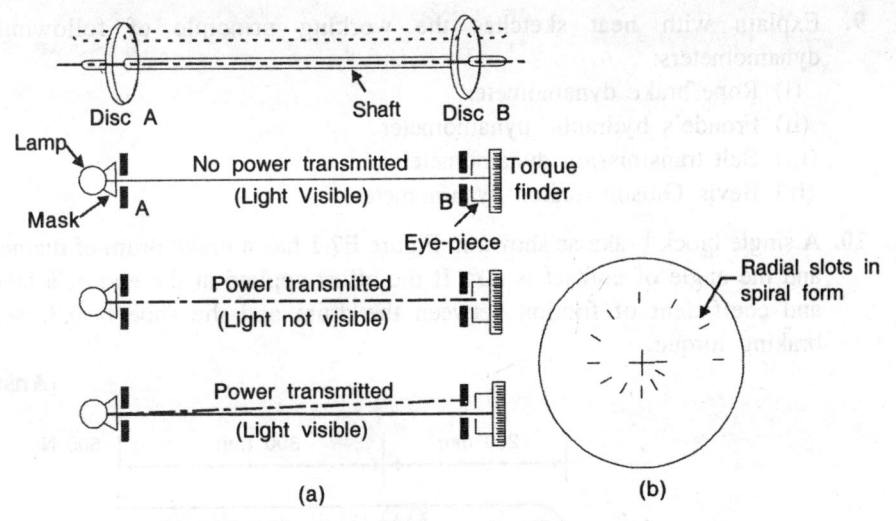

FIGURE 7.44 Bevis–Gibson torsion dynamometer.

transmitting torque, a flash of light ray will be visible at eyepiece once per revolution. When torque is transmitted, the shaft twists and the slot in the disc B changes its position. The ray of light in this position is not visible through eyepiece. However, if the eyepiece is rotated by an amount equal to the angular deformation of the shaft, the ray of light will again start flashing once in each revolution of the shaft and therefore, angle of twist of the shaft is measured.

In certain applications, where the torque is varied during each revolution of the shaft, it becomes necessary to measure the angle of twist at different positions of the shaft. In order to do this, the discs are perforated with radial slots arranged in spiral form as shown in Figure 7.44(b). The masked lamp and eyepiece are moved radially in such a way that they are brought in line with the corresponding pair of slots in discs A and B.

EXERCISES

1. What do you mean by a self-energizing brake and self–locking brake?
2. Classify the types of brake and explain each type with the help of neat sketch.
3. What is difference between the block brake with small angle of contact and that with large angle of contact?
4. Explain the principle on which a band brake works.
5. What is difference between simple band brake and differential band brake? Which is more suitable for large torque capacity?
6. Derive the formula for ratio of tensions in a band and block brake if the brake consists of n number of blocks.
7. Explain the working of an internal expanding shoe brake.
8. What do you mean by a dynamometer? How are dynamometers classified?

9. Explain with neat sketches the working principle of following types of dynamometers:
   (i) Rope brake dynamometer
   (ii) Froude's hydraulic dynamometer
   (iii) Belt transmission dynamometer
   (iv) Bevis–Gibson torsion dynamometer.

10. A single block brake as shown in Figure E7.1 has a brake drum of diameter 300 mm and the angle of contact is 60°. If the effort applied at the end of a lever is 500 N and coefficient of friction between the drum and the shoe is 0.3, determine the braking torque.

[Ans: 64.6 Nm]

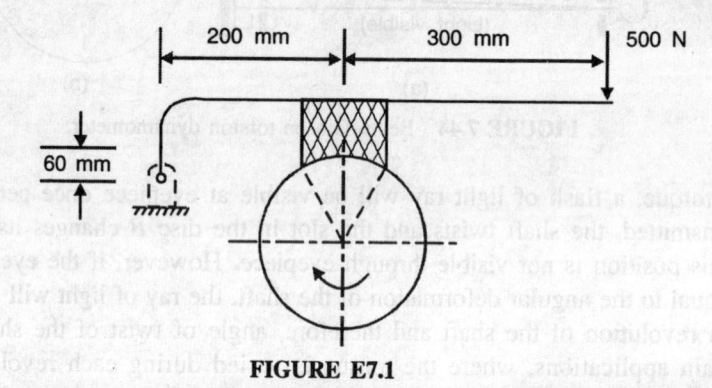

FIGURE E7.1

11. A brake shoe applied to a drum by a lever AB is pivoted at a fixed point A. The shoe is rigidly fixed to the lever. The coefficient of friction between the brake lining and the drum is 0.35. The drum rotates in clockwise direction. Find the braking torque on the drum due to an effort of 750 N applied at point B. (See Figure E7.2).

[Ans: 10.8 Nm]

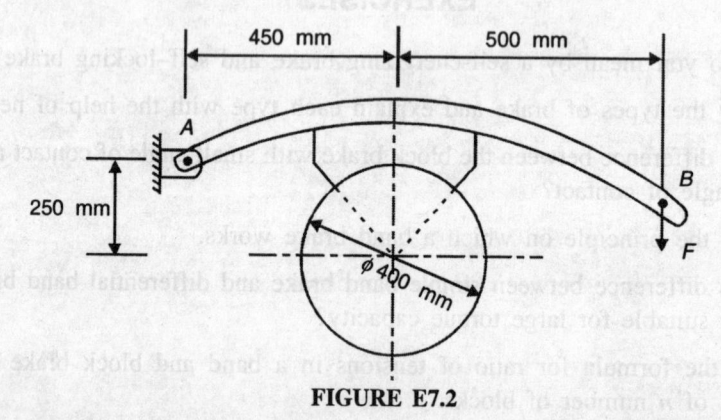

FIGURE E7.2

12. The layout and dimension of a double blocks brake are shown in Figure E7.3. The coefficient of friction is 0.4 and the angle of contact for each block is 90°.

FIGURE E7.3

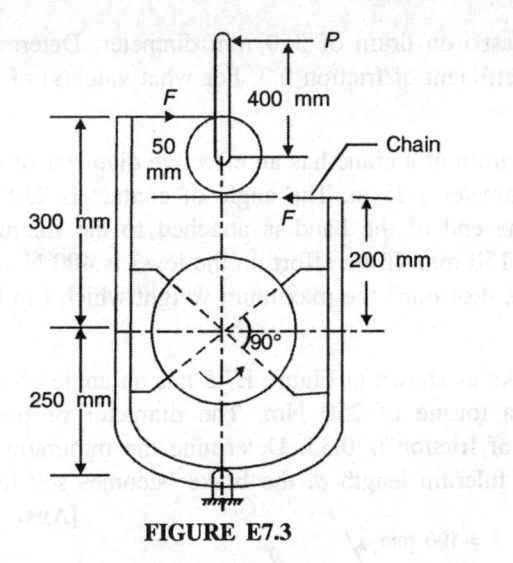

Determine the force P on the operating arm required to set the brake of counterclockwise rotation of the brake drum. The diameter of brake drum is 300 mm and the braking torque is 500 Nm.

[Ans: $P = 212.46$ N]

13. A bicycle rider weighing 1000 N is travelling at the speed of 10 km/h on the level road. A brake is applied to the rear wheel which is 800 mm in diameter. If the coefficient of friction is 0.1 and the normal reaction on the brake is 100 N, determine how far will the bicycle travel and how many turns its wheel will make before coming to rest. Neglect all other types of frictional resistance.

[Ans: 39.32m, 15.64 turns]

14. A band brake as shown in Figure E7.4 has an angle of contact 225° and is required to sustain a torque of 350 Nm. The band has a compressed woven lining and bears

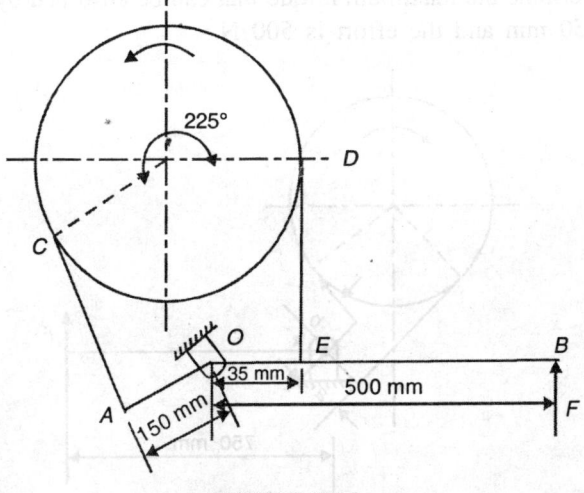

FIGURE E7.4

against a cast iron drum of 350 mm diameter. Determine the necessary effort F. Assume coefficient of friction 0.3. For what value(s) of b, the brake is self-locking.

[Ans: 190N, 114 mm]

15. A winding drum of a crane has an effective diameter of 450 mm and carries a brake drum of diameter 1.25 m. The angle of contact is 250° and coefficient of friction is 0.35. One end of the band is attached to the fulcrum while other end is at a distance of 150 mm. If the effort on the lever is 400 N acting at 800 mm away from the fulcrum, determine the maximum weight which can be raised by winding drum.

[Ans: 21451.8 N]

16. A band brake as shown in Figure E7.5 has an angle of contact 225° and is required to absorb a torque of 350 Nm. The diameter of brake drum is 400 mm and coefficient of friction is 0.35. Determine the minimum necessary effort. For what value(s) of fulcrum length a, the brake becomes self-locking?

[Ans: 546.2 N, a = 10.12 mm]

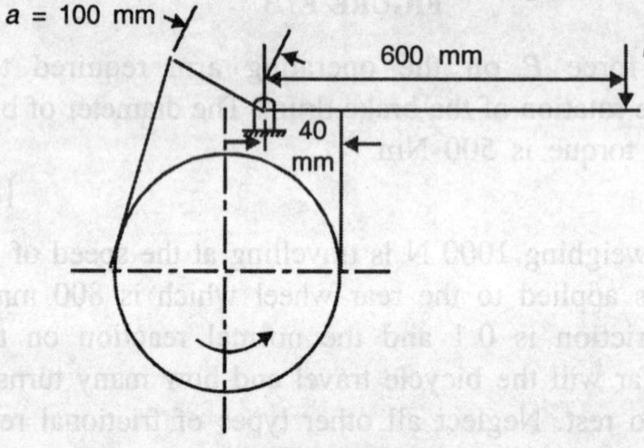

FIGURE E7.5

17. A two-way band brake as shown in Figure E7.6 has a drum diameter 500 mm and lever arm length 100 mm each. The contact angle is 260° and coefficient of friction is 0.25. Determine the maximum torque that can be absorbed by the brake. The lever length is 750 mm and the effort is 500 N.

[Ans: 481.27 Nm]

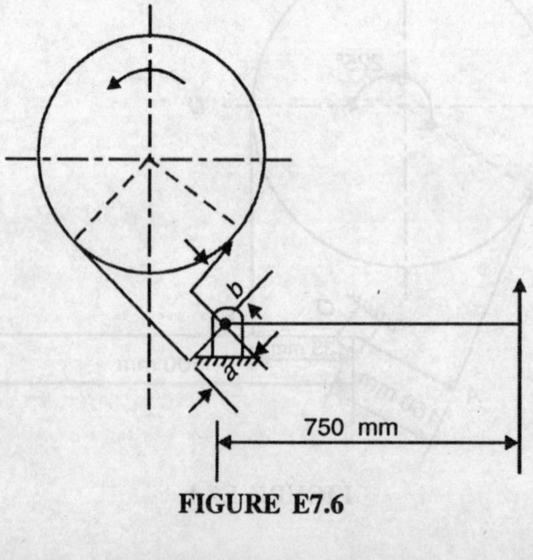

FIGURE E7.6

18. Figure E7.7 shows a band brake consisting of a drum of diameter 450 mm. The two ends of the band are attached to the brake lever at points A and B. The lever can move around the fixed point O. Determine the load to be applied to the lever 600 mm from the point O in order to apply a torque of 6.75 kNm to the drum, the rotation of the drum being clockwise and $\mu = 0.1$. The band may be considered to be inextensible so that the linkage does not depart appreciable from the position shown when under load. Length $OC = 375$ mm, $OA = 150$ mm and $OB = 300$ mm.

[Ans: 500 N]

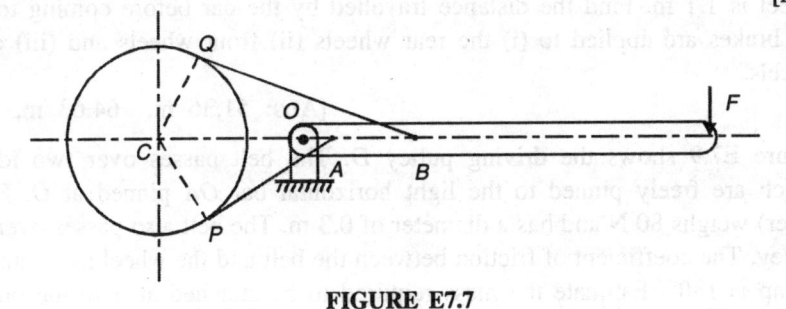

FIGURE E7.7

19. In a band and block brake, the band is lined with 14 blocks each of which subtends an angle of 15° at the centre of the drum. One end of the band is attached to the fulcrum of the brake lever and the other to a pin 125 mm away from the fulcrum. If the torque absorbed by the brake is 1500 Nm, what minimum effort must be applied to the brake at 600 mm away from the fulcrum? The diameter of the drum is 350 mm and rotates in clockwise direction. Take $\mu = 0.25$.

[Ans: 1182.6 N]

20. Figure E7.8 shows the arrangement of an internal expanding brake in which the brake shoe is pivoted at a fixed point C. The distance of this point from the centre of rotation is 75 mm. The internal radius of the drum is 100 mm. The friction lining extends over an arc AB such that the $\angle AOC$ is 135° and $\angle BOC$ is 45°. The brake is applied by means of a force at point Q perpendicular to CQ, the distance CQ being

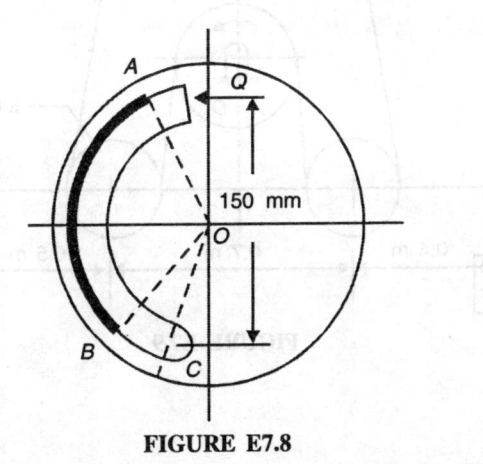

FIGURE E7.8

150 mm. The local rate of wear on the lining is proportional to the normal pressure and taking the coefficient of friction 0.4. Calculate the force required at Q to produce a braking torque of 25 Nm when the drum rotates in (a) clockwise direction and (b) counterclockwise direction.

[Ans: 450.5 N, 117.2 N]

21. A car moving on a level road at a speed of 60 km/h has a wheel base of 2.76 m. The distance of the centre of gravity from ground level is 500 mm and from rear wheel is 1.1 m. Find the distance travelled by the car before coming to rest when the brakes are applied to (i) the rear wheels (ii) front wheels and (iii) all the four wheels.

[Ans: 51.36 m, 64.63 m, 28.33 m]

22. Figure E7.9 shows the driving pulley D. The belt passes over two idler pulleys which are freely pinned to the light horizontal bar OA pinned at O. Each pulley (idler) weighs 80 N and has a diameter of 0.3 m. The belt also passes over the driven pulley. The coefficient of friction between the belt and the wheel is 0.3 and the angle of lap is 180°. Estimate the mass required to be attached at A to the bar such that 2.25 kW power is transmitted with belt speed of 15 m/s.

[Ans: 34.68 kg]

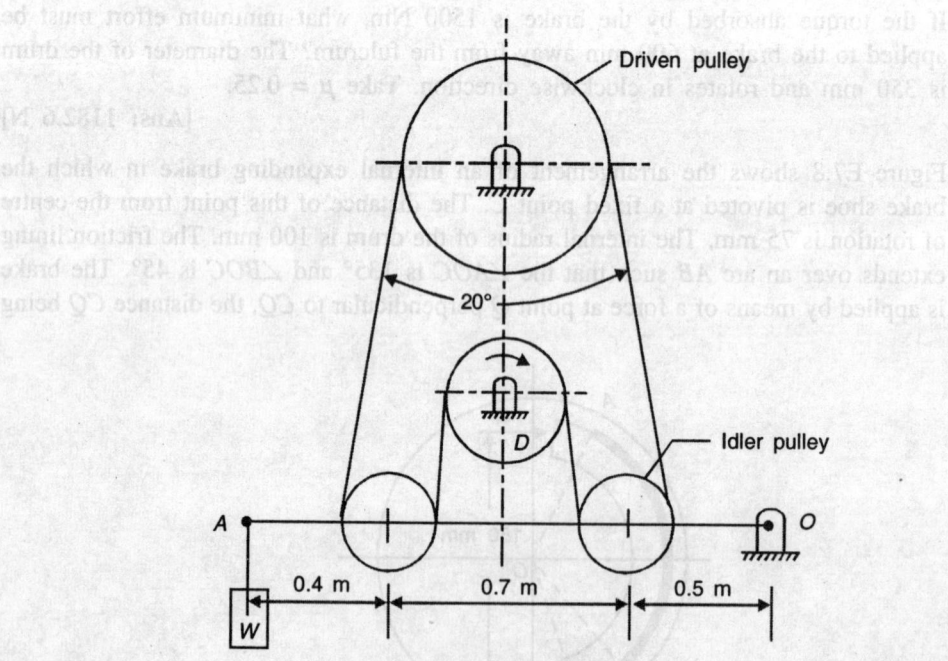

FIGURE E7.9

MULTIPLE CHOICE QUESTIONS

1. Self-locking is not possible in the case of
   (a) simple band brake
   (b) differential band brake
   (c) simple block brake
   (d) internal expanding brake

2. A short shoe brake has angle of contact less than
   (a) 60°
   (b) 120°
   (c) 45°
   (d) 90°

3. In block brake, the normal reaction and frictional force act at the mid-point of the block. This is true only if the angle of contact is
   (a) less than 45°
   (b) less than 60°
   (c) greater than 45°
   (d) greater than 60°

4. In a block brake, when the fulcrum point is below the line of action of friction point, the ratio of effort required for clockwise and counterclockwise rotation is
   (a) equal to 1
   (b) greater than 1
   (c) less than 1
   (d) unpredictable

5. When moment due to friction force aids in applying the brake, it is called
   (a) simple block brake
   (b) pivoted block brake
   (c) self-locking brake
   (d) self-energized brake

6. In a simple band brake, when direction of rotation of drum is changed from counterclockwise direction to clockwise direction, the braking torque
   (a) remains unchanged
   (b) is increased
   (c) is decreased
   (d) is unpredictable

7. In a differential band brake, the effort required for clockwise rotation of drum compared to counterclockwise rotation is
   (a) greater
   (b) equal
   (c) less
   (d) unpredictable

CHAPTER 8

Governors

8.1 INTRODUCTION

The speed of an engine varies in two ways. Firstly, due to variation in the turning moment of an engine, which is known as **cyclic variation** and can be controlled by installing a flywheel of proper size, and secondly, is due to the load-speed characteristics of the engine. As load on the engine increases, the engine has to do more work. However, due to limited supply of input fuel, speed of the engine starts decreasing. On the other hand, when the load on an engine is decreased, the speed of engine shoots up sharply. Thus a feedback control system is needed which can regulate the supply of fuel as per requirement so that speed of the engine is maintained within specified limits. The governor is a mechanically-controlled feedback device which maintains the speed of an engine within permissible range whenever there is a variation of load.

A governor has no influence over a cyclic speed fluctuation but it controls the mean speed over a long period whereas a flywheel does not exercise any control over the mean speed of the engine.

Governors are broadly classified into two types on the basis of their operating principles—centrifugal governor and inertia governor.

In a centrifugal governor, the change in centrifugal forces of the rotating masses due to change in the speed of the engine is utilized for actuation of mechanism to control the speed. In inertia governor, the angular acceleration of engine crank shaft caused by change in speed is utilized for the movement of revolving masses attached to it. The movement of revolving masses is decided by the rate of change of speed rather than the change in speed itself. Therefore these governors are more sensitive compared to centrifugal governors.

8.2 WATT GOVERNOR

One of the simplest forms of centrifugal governor is known as **Watt's governor.** It consists of a pair of balls attached to the rotating spindle with the help of arm and sleeve. In this governor, as the speed of spindle fluctuates, the centrifugal force tries to flyout or flyin the balls which results into up and down movement of the sleeve. This movement of the sleeve is transferred to the throttle valve of an engine through a suitable mechanism to regulate the speed of engine (Figure 8.1).

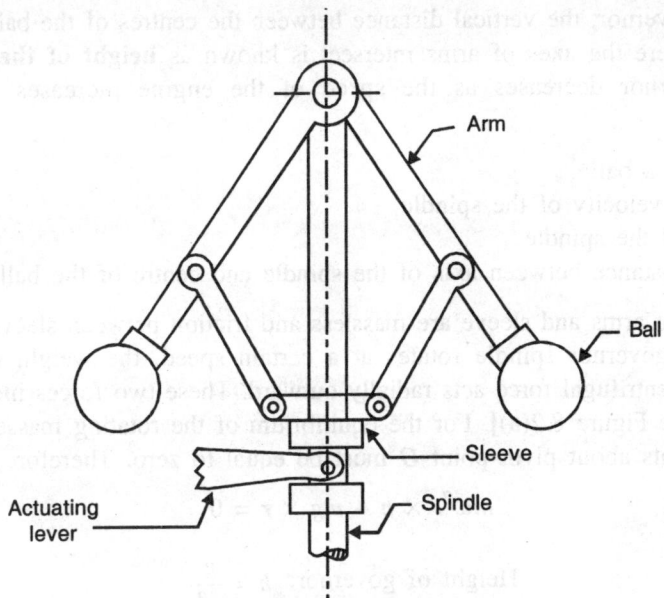

FIGURE 8.1 Watt governor.

A Watt governor is classified into three types depending upon the position of upper arms. When upper arms are joined at the intersection of the arms and the spindle axis, it is known as **pinned type governor** [Figure 8.2(a)]. When upper arms are joined by a horizontal link, it is known as **open-arm type Watt governor** [Figure 8.2(b)]. A **crossed-arm type Watt governor** is one in which both the upper arms intersect at a point but ends of both the arms are joined through a fixed horizontal link as shown in Figure 8.2(c).

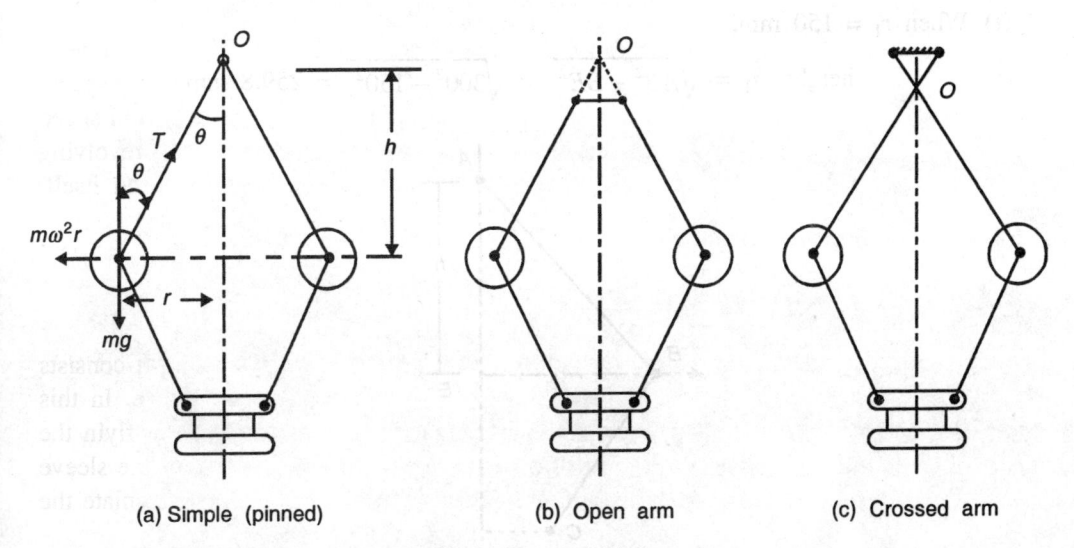

(a) Simple (pinned) (b) Open arm (c) Crossed arm

FIGURE 8.2 Types of Watt governor.

In a Watt governor, the vertical distance between the centres of the balls and the point on the spindle where the axes of arms intersect is known as **height of the governor**. The height of a governor decreases as the speed of the engine increases and vice-versa. Suppose

m = mass of a ball
ω = angular velocity of the spindle
N = speed of the spindle

and r = radial distance between axis of the spindle and centre of the ball

Assuming that arms and sleeve are massless and friction between sleeve and spindle is negligible. When governor spindle rotates at a certain speed, the weight of the ball acts downward and a centrifugal force acts radially outward. These two forces induce tensions in the upper arm [See Figure 8.2(a)]. For the equilibrium of the rotating masses, the algebraic sum of the moments about pivot point O must be equal to zero. Therefore,

$$m\omega^2 r \times h - mg \times r = 0$$

or
$$\text{Height of governor: } h = \frac{g}{\omega^2} \qquad (8.1)$$

Equation (8.1) reveals that height of governor is inversely proportional to the square of the angular velocity. This means that the governor is highly sensitive at low speeds and becomes less sensitive at higher speed.

EXAMPLE 8.1 In a simple Watt governor, the length of upper arms is each 300 mm. If the two extreme radii of rotation of the governor balls are 150 mm and 165 mm, what will be the equilibrium speeds?

Solution Refer to Figure 8.3.

(i) When $r_1 = 150$ mm,

$$\text{height: } h_1 = \sqrt{AB^2 - BE^2} = \sqrt{300^2 - 150^2} = 259.8 \text{ mm}$$

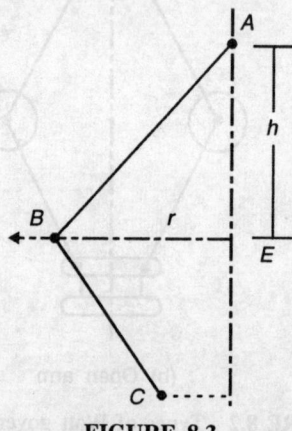

FIGURE 8.3

For Watt governor, the operating speed:

$$\omega_1 = \sqrt{\frac{g}{h}} = \sqrt{\frac{9.81}{0.2598}} = 6.145 \text{ rad/s}$$

or

the lowest speed: $N_1 = \dfrac{60 \times \omega_1}{2\pi} = \dfrac{60 \times 6.145}{2\pi} = $ **58.68 rpm**   **Ans.**

(ii) When $r_2 = 165$ mm,

$$h_2 = \sqrt{300^2 - 165^2} = 250.55 \text{ mm}$$

$$\omega_2 = \sqrt{\frac{g}{h_2}} = \sqrt{\frac{9.81}{0.25055}} = 6.257 \text{ rad/s}$$

or

the highest speed: $N_2 = \dfrac{60 \times 6.257}{2\pi} = $ **59.75 rpm**   **Ans.**

Speed range $= N_2 - N_1 = 59.75 - 58.68 = $ **1.07 rpm**   **Ans.**

8.3 PORTER GOVERNOR

The main limitation of a Watt governor is that it becomes insensitive when speed is increased which results into a smaller lift. This limitation can be rectified by introducing a heavy weight at the sleeve of a Watt governor. A Watt governor loaded with heavy weight at the sleeve is known as **Porter governor** (Figure 8.4).

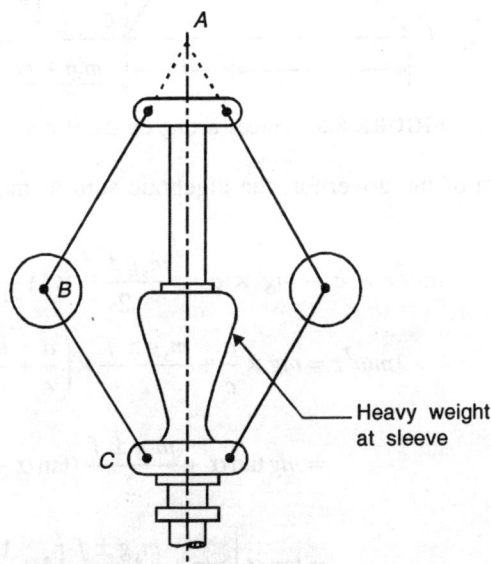

FIGURE 8.4 Porter governor.

Let

m_s = mass of the sleeve

f = friction force acting between sleeve and spindle

and r = distance between the centre of rotation of ball from the axis of rotation

In a Porter governor, due to heavy weight of the sleeve ($m_s g$), the friction between sleeve and spindle becomes important. When balls flyout and sleeve moves up, the friction force acts in downward direction which is added to the weight of the spindle ($m_s g + f$). Similarly, when the sleeve moves down, the friction force acts in upward direction and resulting weight of the sleeve will be $m_s g - f$.

Forces acting on the ball, arm and sleeve are shown in Figure 8.5. Kinematically, the links AB and BC can be treated as slider-crank mechanism in which point C is considered as slider. Therefore, the instantaneous centre of rotation of the arm BC is point I, which can be obtained as point of intersection of extended link AB and a line drawn perpendicular to the axis of rotation passing through point C.

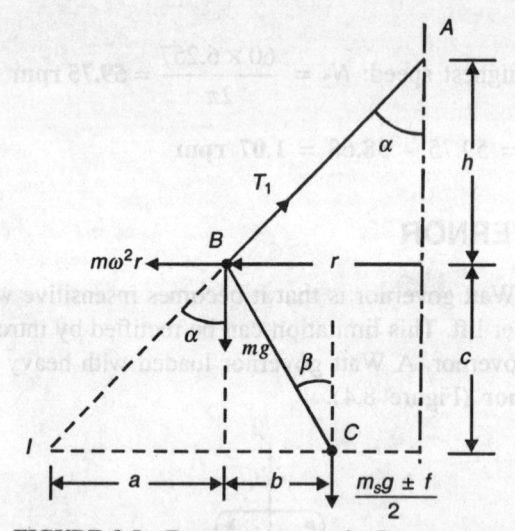

FIGURE 8.5 Forces acting on the Porter governor.

For equilibrium of the governor, the algebraic sum of moments about point I should be equal to zero.

Hence
$$m\omega^2 r \times c - mg \times a - \frac{m_s g \pm f}{2} \times (a+b) = 0 \qquad (8.2)$$

or
$$m\omega^2 r = mg \times \frac{a}{c} + \frac{m_s g \pm f}{2} \times \left(\frac{a}{c} + \frac{b}{c}\right)$$

$$= mg \tan \alpha + \frac{m_s g \pm f}{2} (\tan \alpha + \tan \beta)$$

$$= \tan \alpha \left[mg + \frac{m_s g \pm f}{2} \left(1 + \frac{\tan \beta}{\tan \alpha}\right) \right]$$

$$= \frac{r}{h}\left[mg + \frac{m_s g \pm f}{2}\left(1 + \frac{\tan\beta}{\tan\alpha}\right)\right]$$

The operating speed of the governor is given by

$$\omega^2 = \frac{1}{mh}\left[mg + \frac{m_s g \pm f}{2}\left(1 + \frac{\tan\beta}{\tan\alpha}\right)\right] \qquad (8.3)$$

and the height of the governor is expressed as

$$h = \frac{1}{m\omega^2}\left[mg + \frac{m_s g \pm f}{2}\left(1 + \frac{\tan\beta}{\tan\alpha}\right)\right] \qquad (8.4)$$

The Eq. (8.3) would provide two values of angular velocity—one with $m_s g - f$ when the sleeve has just moved down and another with force $m_s g + f$ when the sleeve has just tended to move up. The sleeve will not be able to move up until sufficient increase in the speed takes place which can overcome this frictional resistance. Thus for a given value of height of the governor, the Porter governor is insensitive between these two values of angular velocity.

EXAMPLE 8.2 A Porter governor has all the four arms 300 mm long. The upper arms are pivoted on the axis of rotation and the lower arms are attached to the sleeve at a distance of 35 mm from the axis. The mass of each ball and the sleeve are 7 kg and 54 kg respectively. If the extreme radii of rotation of the balls are 200 mm and 250 mm, determine the range of speed of the governor.

Solution Refer to Figure 8.6.

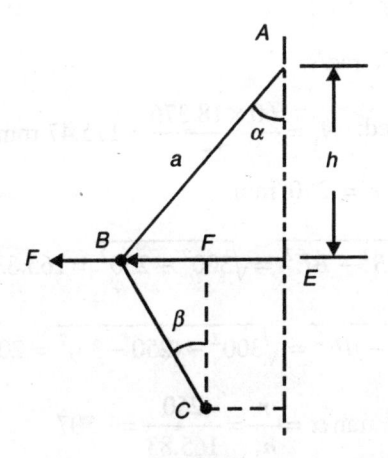

FIGURE 8.6

(i) At minimum radius, $r = 200$ mm:

$$AE = h_1 = \sqrt{AB^2 - BE^2}$$

$$= \sqrt{300^2 - 200^2}$$

$$= 223.6 \text{ mm}$$

Similarly, $CF = \sqrt{BC^2 - BF^2}$

where $BF = r - EF$

$= 200 - 35 = 165$ mm

$CF = \sqrt{300^2 - 165^2} = 250.55$ mm

$$\tan \alpha = \frac{r}{h_1} = \frac{200}{223.6} = 0.8944$$

$$\tan \beta = \frac{BF}{CF} = \frac{165}{250.55} = 0.6585$$

We know that for a Porter governor, the lowest speed is given by

$$\omega_1^2 = \frac{1}{mh_1}\left[mg + \frac{m_s g \pm f}{2}\left(1 + \frac{\tan \beta}{\tan \alpha}\right)\right]$$

Neglecting friction force f,

$$\omega_1^2 = \frac{1}{7 \times 0.2236}\left[7 \times 9.81 + \frac{54 \times 9.81}{2}\left(1 + \frac{0.6585}{0.8944}\right)\right]$$

$\omega_1 = 18.376$ rad/s

or Speed: $N_1 = \dfrac{60 \times 18.376}{2\pi} = 175.47$ rpm

(ii) At maximum radius, $r = 250$ mm:

$$h_2 = \sqrt{AB^2 - BE^2} = \sqrt{300^2 - 250^2} = 165.83 \text{ mm}$$

$$CF = \sqrt{BC^2 - BF^2} = \sqrt{300^2 - (250-35)^2} = 209.22 \text{ mm}$$

$$\tan \alpha = \frac{r}{h_2} = \frac{250}{165.83} = 1.507$$

$$\tan \beta = \frac{BF}{CF} = \frac{250 - 35}{209.22} = 1.0276$$

Operating speed: $\omega_2^2 = \dfrac{1}{mh_2}\left[mg + \dfrac{m_s g}{2}\left(1 + \dfrac{\tan\beta}{\tan\alpha}\right)\right]$

$$\omega_2^2 = \dfrac{1}{7\times 0.16583}\left[7\times 9.81 + \dfrac{54\times 9.81}{2}\left(1 + \dfrac{1.0276}{1.507}\right)\right]$$

or $\qquad\qquad\qquad\omega_2 = 21.04$ rad/s

or $\qquad\qquad$ Speed: $N_2 = \dfrac{60\times 21.04}{2\pi} = 200.9$ rpm

Range of speed: $N_2 - N_1 = 200.9 - 175.47$

$\qquad\qquad\qquad = 25.43$ rpm $\qquad\qquad\qquad\qquad\qquad$ **Ans.**

EXAMPLE 8.3 A Porter governor has equal links of 250 mm long pivoted at the axis. The weight of each ball is 3N and the weight of sleeve is 14 N. The radius of the ball is 150 mm when governor begins to lift and 200 mm at the maximum speed. If the friction at the sleeve is equivalent to 1.5 N, determine the maximum and minimum speeds.

Solution Refer to Figure 8.7.

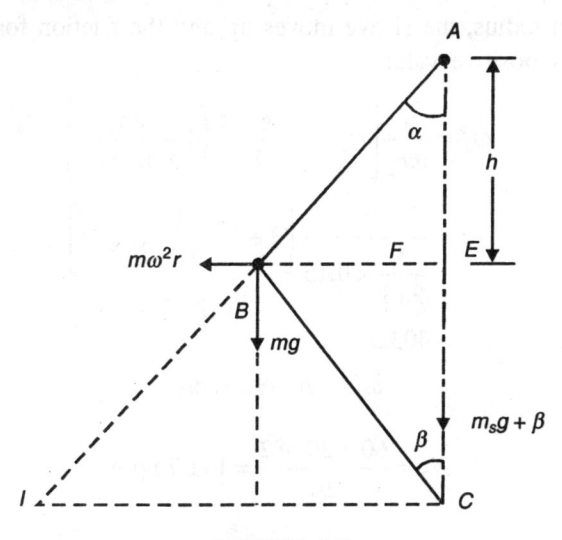

FIGURE 8.7

(i) At minimum radius, $r = 150$ mm:

$$\dfrac{\tan\beta}{\tan\alpha} = 1$$

as angle $\alpha = \beta$

Height: $\qquad h_1 = \sqrt{AB^2 - BE^2} = \sqrt{250^2 - 150^2} = 200$ mm

At minimum speed, the radius of the ball decreases, the sleeve moves down and the force of friction acts upward, so it is taken as negative.
The operating speed:

$$\omega_1^2 = \frac{1}{mh_1}\left[mg + \frac{m_s g - f}{2}\left(1 + \frac{\tan \beta}{\tan \alpha}\right)\right]$$

$$= \frac{1}{\frac{3}{9.81} \times 0.2}\left[3 + \frac{14 - 1.5}{2}(1 + 1)\right]$$

$$= 253.42$$

or $\qquad \omega_1 = 15.92$ rad/s

or $\qquad N_1 = \dfrac{60 \times 15.92}{2\pi} = \mathbf{152.0\ rpm}$ **Ans.**

(ii) At maximum radius, $r = 200$ mm:

$$h_2 = \sqrt{AB^2 - BE^2} = \sqrt{250^2 - 200^2} = 150 \text{ mm}$$

At maximum radius, the sleeve moves up and the friction force acts downward, so it is taken as positive value.

$$\omega_2^2 = \frac{1}{mh_2}\left[mg + \frac{m_s g + f}{2}\left(1 + \frac{\tan \beta}{\tan \alpha}\right)\right]$$

$$= \frac{1}{\frac{3}{9.81} \times 0.15}\left[3 + \frac{14 + 1.5}{2}(1 + 1)\right]$$

$$= 403.3$$

or $\qquad \omega_2 = 20.082$ rad/s

or $\qquad N_2 = \dfrac{60 \times 20.082}{2\pi} = \mathbf{191.7\ rpm}$ **Ans.**

8.4 PROELL GOVERNOR

In Proell governor, the rotating balls are mounted on an extension of the lower arm of the Porter governor as shown in Figure 8.8. The lower arm *CBF* is a one piece arm bent at *B*. Under the normal conditions, the extended portion of arm *BF* remains vertical; however it changes position with variation in speed. The forces acting on the Proell governor are shown in Figure 8.9. As before, assuming point *I* as instantaneous centre of rotation of the arm *CBF*.

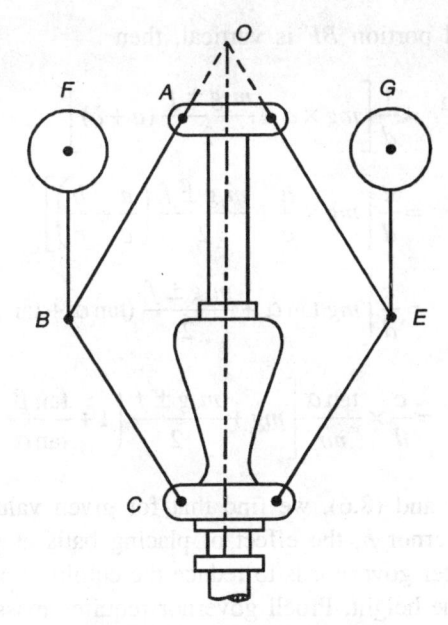

FIGURE 8.8 Proell governor.

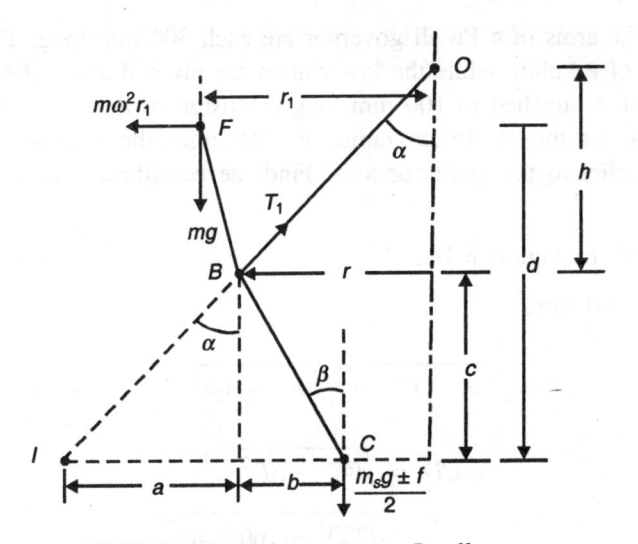

FIGURE 8.9 Forces acting on Proell governor.

For equilibrium of governor, taking moments about point I,

$$m\omega^2 r_1 \times d - mg(a + r - r_1) - \frac{m_s g \pm f}{2}(a + b) = 0 \qquad (8.5)$$

where dimensions a, b, c and d are shown in Figure 8.8.

If the position of extended portion BF is vertical, then

$$m\omega^2 r = \frac{1}{d}\left[mg \times a + \frac{m_s g \pm f}{2}(a+b)\right]$$

$$= \frac{c}{d}\left[mg \times \frac{a}{c} + \frac{m_s g \pm f}{2}\left(\frac{a}{c} + \frac{b}{c}\right)\right]$$

$$= \frac{c}{d}\left[mg \tan\alpha + \frac{m_s g \pm f}{2}(\tan\alpha + \tan\beta)\right]$$

or
$$\omega^2 = \frac{c}{d} \times \frac{\tan\alpha}{mr}\left[mg + \frac{m_s g \pm f}{2}\left(1 + \frac{\tan\beta}{\tan\alpha}\right)\right] \qquad (8.6)$$

Comparing Eqs. (8.3) and (8.6), we find that for given values of ball mass m, sleeve mass m_s and height of governor h, the effect of placing balls at point F in Proell governor instead of at point B in Porter governor is to reduce the equilibrium speed of Proell governor. In other words, for the same height, Proell governor requires mass of smaller size compared to Porter governor.

EXAMPLE 8.4 The arms of a Proell governor are each 300 mm long. The upper arms are pivoted on the axis of rotation, while the lower arms are pivoted at a radius of 40 mm. Each ball weighs 5 N and is attached to 100 mm long extension of the lower arm, the weight of the sleeve is 60 N. At the minimum radius of 160 mm, the extensions to which balls are attached are parallel to the governor axis. Find the equilibrium speed corresponding to radius 160 mm.

Solution Refer to Figure 8.10.

At radius $r = 160$ mm,

$$h = OE = \sqrt{OB^2 - BE^2} = \sqrt{300^2 - 160^2} = 253.77 \text{ mm}$$

$$c = CD = \sqrt{BC^2 - BD^2}$$
$$= \sqrt{300^2 - (160-40)^2}$$
$$= 275 \text{ mm}$$

$$\tan\alpha = \frac{BE}{h} = \frac{160}{253.77} = 0.6305$$

$$\tan\beta = \frac{BD}{CD} = \frac{120}{275} = 0.4363$$

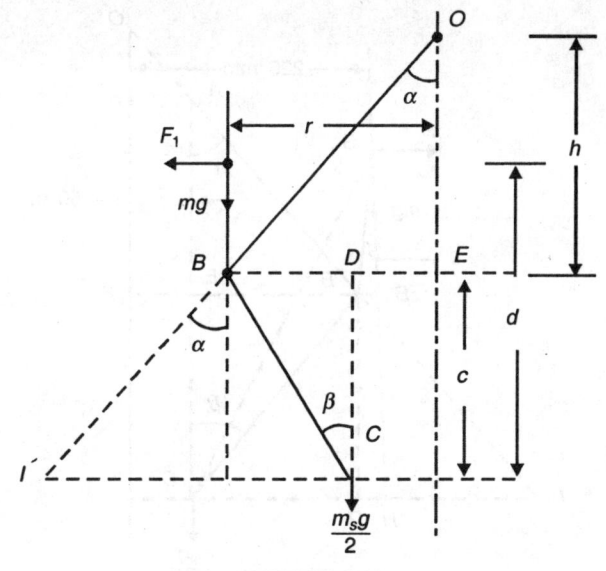

FIGURE 8.10

Neglecting obliquity, the operating speed:

$$\omega^2 = \frac{c}{d} \times \frac{\tan\alpha}{mr}\left[mg + \frac{m_s g}{2}\left(1 + \frac{\tan\beta}{\tan\alpha}\right)\right]$$

$$= \frac{275}{375} \times \frac{0.6305}{\frac{5}{9.81} \times 0.16}\left[5 + \frac{60}{2}\left(1 + \frac{0.4363}{0.6305}\right)\right]$$

or
$$\omega = 17.78 \text{ rad/s}$$

or
Speed: $N = \dfrac{60 \times 17.78}{2\pi} = \mathbf{169.78\ rpm}$ **Ans.**

EXAMPLE 8.5 The weight of each ball of a Proell governor is 9 N, the sleeve weight is 150 N and the arms are 250 mm long. The arms are pivoted at a distance of 50 mm from the axis of rotation. The extension of the lower arm to which each ball is attached is 125 mm long and the radius of rotation of the balls is 225 mm. When the arms are inclined at 40° to the axis of rotation find (i) the equilibrium speed for the above configuration and (ii) coefficient of insensitiveness, if the friction of the governor is equivalent to 5 N force at the sleeve.

Solution

(i) Refer to Figure 8.11.

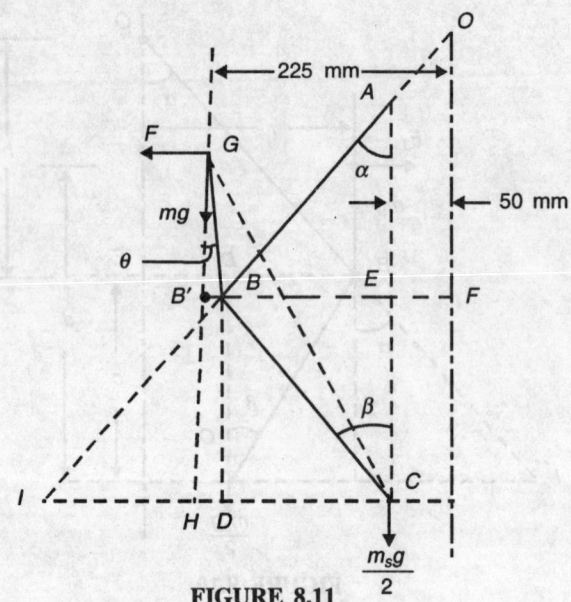

FIGURE 8.11

$$BE = AB \sin \alpha$$
$$= 250 \sin 40° = 160.7 \text{ mm}$$
$$BB' = B'F - (BE + EF)$$
$$= 225 - (160.7 + 50) = 14.3 \text{ mm}$$
$$IC = 2BE = 321.4 \text{ mm}$$
$$HC = B'F - EF$$
$$= 225 - 50 = 175 \text{ mm}$$
$$IH = IC - HC = 321.4 - 175 = 146.4 \text{ mm}$$
$$GB = 125 \text{ mm}$$
$$\alpha = \beta = 40°$$

Let $\angle B'GB = \theta$

$$\sin \theta = \frac{BB'}{BG} = \frac{14.3}{125} = 0.1144$$

or
$$\theta = 6.57°$$
$$GB' = GB \cos\theta = 125 \times \cos 6.57° = 124.18 \text{ mm}$$
$$CE = B'H = \sqrt{250^2 - 160.7^2} = 191.5 \text{ mm}$$
$$GH = B'H + B'G = 191.5 + 124.18 = 315.68 \text{ mm}$$

Taking moments about point I,

$$F \times GH = mg \times IH + \frac{m_s g}{2} \times IC$$

or $\quad F \times 0.31568 = 9 \times 0.1464 + \dfrac{150}{2} \times 0.3214$

or $\quad F = 80.53$ N

But $\quad F = m\omega^2 r$

or $\quad 80.53 = \dfrac{9}{9.81} \times \omega^2 \times 0.225$

or $\quad \omega = 19.75$ rad/s or $N = $ **188.6 rpm** Ans.

(ii) When friction force is included,

$$F \times GH = mg \times IH + \dfrac{m_s g \pm f}{2} \times IC$$

or $\quad F \times 0.31568 = 9 \times 0.1464 + \left(\dfrac{150 \pm 5}{2}\right) \times 0.3214$

with $\quad f = +5$ N, $\quad F_1 = 83.08$ N

$\quad f = -5$ N, $\quad F_2 = 78$ N

Therefore, $\quad \omega_1 = \sqrt{\dfrac{F_1}{mr}} = \sqrt{\dfrac{83.08 \times 9.81}{9 \times 0.225}} = 20.06$ rad/s

$$\omega_2 = \sqrt{\dfrac{F_2}{mr}} = \sqrt{\dfrac{78 \times 9.81}{9 \times 0.225}} = 19.44 \text{ rad/s}$$

Coefficient of insensitiveness $= \dfrac{\omega_1 - \omega_2}{\omega}$

$= \dfrac{20.06 - 19.44}{19.75} \times 100$

$= $ **3.14%** Ans.

8.5 HARTNELL GOVERNOR

A Hartnell governor is a spring loaded governor in which the balls are controlled by a spring. It consists of a casing in which a pre-compressed spring is housed so as to apply the force to the sleeve. Two bell crank levers, each carrying a ball at one end and the roller at another end, are fitted on the frame of casing (Figure 8.12). The casing alongwith frame and spring rotates about the axis of governor. When speed of governor is increased, the balls flyout away from the governor axis, the bell crank lever moves on pivot and its roller end lifts the sleeve against the spring force. This movement of sleeve is transferred to the throttle of an engine through suitable intermediate links (not shown in Figure 8.12). The spring force can be adjusted with the help of a nut.

306 Theory of Mechanisms and Machines

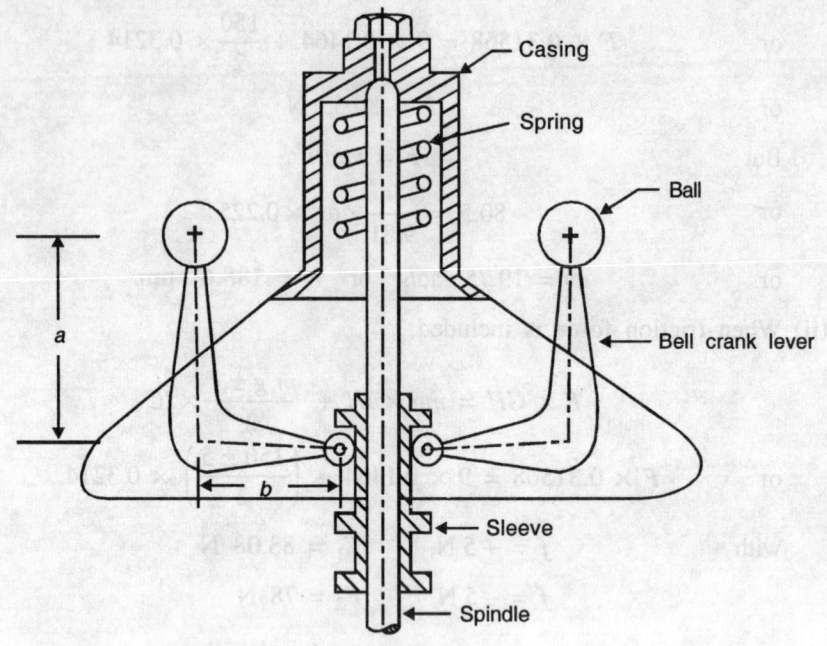

FIGURE 8.12 Hartnell governor.

The relation between the dimensions of the Hartnell governor, the equilibrium speed and the spring force can be found by considering the equilibrium of one of the bell crank levers. Figure. 8.13 shows the forces acting on the bell crank lever in two positions.

(a) When sleeve moves up   (b) When sleeve moves down

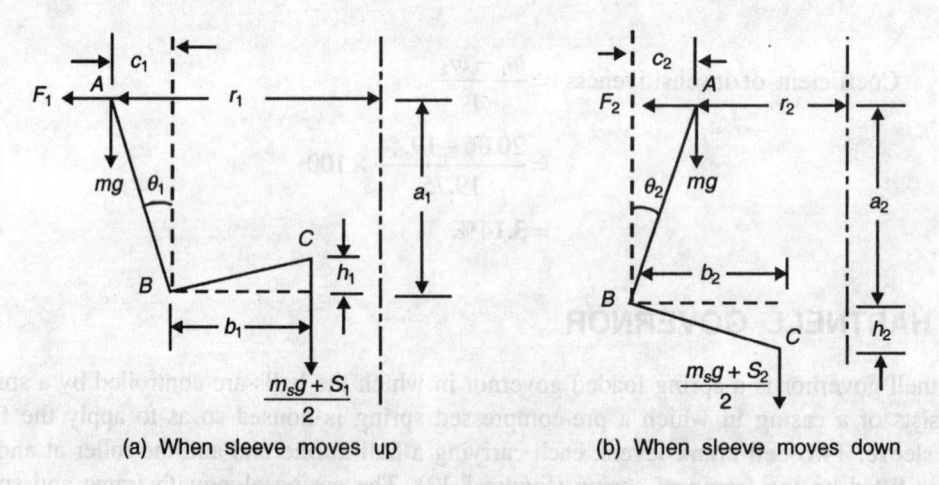

FIGURE 8.13 Forces acting on the bell crank lever.

Let F_1 and F_2 be the centrifugal force at the two extreme angular speeds and S_1 and S_2 be the spring force acting on the sleeve. Neglecting the effect of friction between sleeve and spindle for the equilibrium of bell crank lever, taking moments about fulcrum point B,

(i) When sleeve moves up [refer Figure 8.13(a)]

$$F_1 \times a_1 + mg \times c_1 - \frac{m_s g + S_1}{2} \times b_1 = 0 \qquad (8.7)$$

(ii) When sleeve moves down [refer Figure 8.13(b)]

$$F_2 \times a_2 - mg \times c_2 - \frac{m_s g + S_2}{2} \times b_2 = 0 \qquad (8.8)$$

In the working range of speeds, the obliquity angle θ is usually small so the effect of it on the moment arm length is negligible. Therefore,

$$a_1 \simeq a_2 = a, \quad b_1 \simeq b_2 = b \quad \text{and} \quad c_1 = c_2 = 0$$

Equations (8.7) and (8.8) reduce to

$$F_1 \times a - \frac{m_s g + S_1}{2} \times b = 0 \qquad (8.9)$$

and

$$F_2 \times a - \frac{m_s g + S_2}{2} \times b = 0 \qquad (8.10)$$

From Eqs. (8.9) and (8.10), we get

$$(F_1 - F_2)a = \frac{1}{2}(S_1 - S_2) \times b$$

or

$$S_1 - S_2 = \frac{2a}{b} \times (F_1 - F_2) \qquad (8.11)$$

Suppose:
 k = stiffness of the spring
and h = movement of the sleeve (= $h_1 + h_2$).

Then from similar triangles as shown in Figure 8.14, we get

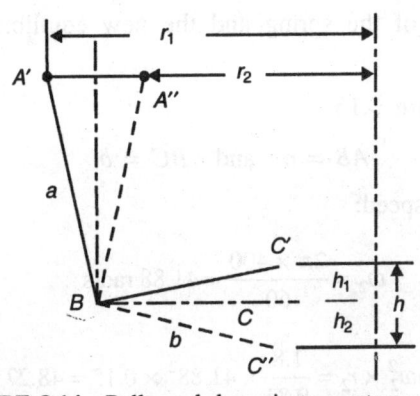

FIGURE 8.14 Bell crank lever in two extreme positions.

$$\frac{h}{b} = \frac{r_1 - r_2}{a}$$

or
$$h = \frac{b}{a}(r_1 - r_2) \tag{8.12}$$

Spring force: $S_1 - S_2 = k \times h$

or
$$S_1 - S_2 = k \times \frac{b}{a}(r_1 - r_2) \tag{8.13}$$

From Eqs. (8.11) and (8.13), we get

$$\text{Spring stiffness: } k = 2\left(\frac{a}{b}\right)^2 \left(\frac{F_1 - F_2}{r_1 - r_2}\right) \tag{8.14}$$

For any intermediate position, say at radius r, the centrifugal force F can be found as

$$\text{Stiffness: } k = 2\left(\frac{a}{b}\right)^2 \left(\frac{F - F_2}{r - r_2}\right) = 2\left(\frac{a}{b}\right)^2 \left(\frac{F_1 - F_2}{r_1 - r_2}\right)$$

or
$$\frac{F - F_2}{r - r_2} = \frac{F_1 - F_2}{r_1 - r_2}$$

or
$$F = F_2 + (F_1 - F_2) \times \left(\frac{r - r_2}{r_1 - r_2}\right) \tag{8.15}$$

EXAMPLE 8.6 A Hartnell governor has the following data:

Ball weight is 1.8 N; sleeve weight is 6 N; ball and sleeve arm of the bell crank lever are 150 mm and 120 mm respectively. Equilibrium speed and radius of rotation for the lowest position of the sleeve are 400 rpm and 150 mm respectively. Sleeve lift is 10 mm. Change in speed for full sleeve lift is 5 percent. During an overhaul, the spring was compressed 2 mm more than the correct compression for the initial setting.

Determine the stiffness of the spring and the new equilibrium speed for the lowest position of the sleeve.

Solution Refer to Figure 8.15.

Let $AB = a$ and $BC = b$

Case 1 Lowest equilibrium speed:

$$\omega_2 = \frac{2\pi \times 400}{60} = 41.88 \text{ rad/s}$$

$$F_2 = m\omega_2^2 \times r_2 = \frac{1.8}{9.81} \times 41.88^2 \times 0.15 = 48.27 \text{ N}$$

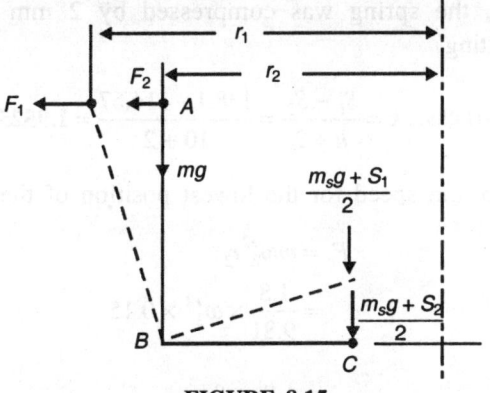

FIGURE 8.15

Taking moments about fulcrum B neglecting obliquity,

$$F_2 \times a = \frac{m_s g + S_2}{2} \times b$$

or

$$48.27 \times 150 = \frac{6 + S_2}{2} \times 120$$

or

$$S_2 = 114.67 \text{ N}$$

Case 2 Highest equilibrium speed:
Referring to the Figure 8.15 (dashed lines), we know that

$$\frac{h}{b} = \frac{r_1 - r_2}{a}$$

or

$$r_1 = r_2 + h \times \frac{a}{b} = 150 + 10 \times \frac{150}{120} = 162.5 \text{ mm}$$

$\omega_1 = 1.05 \times \omega_2$ (given that change in speed is 5 percent)

$$= 1.05 \times 41.88 = 43.97 \text{ rad/s}$$

$$F_1 = m\omega_1^2 r_1 = \frac{1.8}{9.81} \times 43.97^2 \times 0.1625 = 57.64 \text{ N}$$

Taking moments about fulcrum B (neglecting obliquity),

$$F_1 \times a = \frac{m_s g + S_1}{2} \times b$$

or

$$57.64 \times 150 = \frac{6 + S_1}{2} \times 120$$

or

$$S_1 = 138.1 \text{ N}$$

During an overhaul, the spring was compressed by 2 mm more than the correct compression for initial setting.

$$\text{Spring stiffness: } k = \frac{S_1 - S_2}{h+2} = \frac{138.1 - 114.67}{10+2} = 1.9525 \text{ N/mm} \qquad \textbf{Ans.}$$

Let ω_2' be the new equilibrium speed for the lowest position of the sleeve.

$$F_r = m\omega_2'^2 r_2$$

$$= \frac{1.8}{9.81} \times \omega_2'^2 \times 0.15$$

Mean spring force:
$$S = \frac{S_1 + S_2}{2}$$

or
$$S = \frac{138.1 + 114.67}{2} = 126.4 \text{ N}$$

Taking moments about fulcrum B,

$$F_r \times a = \frac{m_s g + S}{2} \times b$$

or
$$\frac{1.8}{9.81} \times \omega_2'^2 \times 0.15 \times 150 = \frac{6 + 126.4}{2} \times 120$$

or
$$\omega_2' = 43.86 \text{ rad/s}$$

or
$$\text{Speed: } N_2' = \frac{60 \times 43.86}{2\pi} = \textbf{418.8 rpm}$$

EXAMPLE 8.7 A Hartnell governor with a central sleeve spring and two rigid right angled bell crank levers rotates between 280 rpm and 320 rpm for a sleeve lift of 30 mm. The sleeve arms and the ball arms are 100 mm and 140 mm respectively. The levers are pivoted at 120 mm from the governor axis and the mass of each ball is 3 kg. The maximum radius of rotation of the fly ball is not to exceed 150 mm due to space restriction. Calculate

(i) load on the spring at the lowest and highest equilibrium speeds
(ii) stiffness of spring.

Solution: Highest angular speed: $\omega_1 = \dfrac{2\pi N_1}{60}$

or
$$\omega_1 = \frac{2\pi \times 320}{60} = 33.51 \text{ rad/s}$$

Lowest angular speed:
$$\omega_2 = \frac{2\pi \times 280}{60} = 29.32 \text{ rad/s}$$

Let r_2 be the radius of rotation at the lowest angular speed.

Lift: $h = (r_1 - r_2) \times \dfrac{b}{a}$

or
$$r_2 = r_1 - h \times \dfrac{a}{b}$$
$$= 150 - 30 \times \dfrac{140}{100} = 108 \text{ mm}$$

The positions of ball arm and sleeve arm at the highest and lowest equilibrium speeds are shown in Figure 8.16.

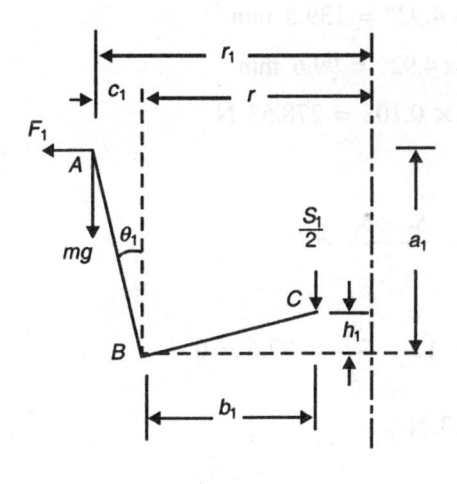

(a) Lever at highest speed (b) Lever at lowest speed

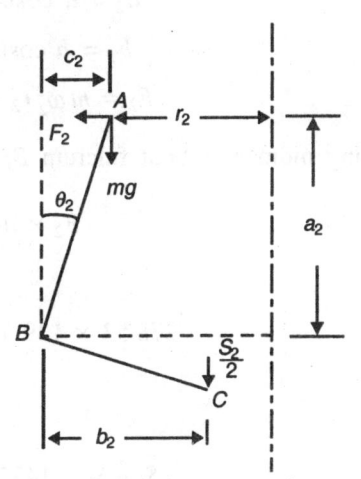

FIGURE 8.16

Referring to Figure 8.16(a),

$$c_1 = r_1 - r = 150 - 120 = 30 \text{ mm}$$

$$\theta_1 = \sin^{-1}\left(\dfrac{c_1}{a}\right) = \sin^{-1}\left(\dfrac{30}{140}\right) = 12.37°$$

$$a_1 = a\cos\theta_1 = 140 \times \cos 12.37° = 136.7 \text{ mm}$$
$$b_1 = b\cos\theta_1 = 100 \times \cos 12.37° = 97.68 \text{ mm}$$
$$F_1 = m\omega_1^2 r_1 = 3 \times 33.512^2 \times 0.15 = 505.37 \text{ N}$$

Taking moments about fulcrum B,

$$F_1 \times a_1 + mg \times c_1 - \dfrac{S_1}{2} \times b_1 = 0$$

or $\qquad 505.37 \times 136.7 + 3 \times 9.81 \times 30 - \dfrac{S_1}{2} \times 97.68 = 0$

or $\qquad\qquad\qquad S_1 = 1432.4 \text{ N}$ **Ans.**

Now referring to Figure 8.16(b),

$$c_2 = r - r_2 = 120 - 108 = 12 \text{ mm}$$

or $\qquad \theta_2 = \sin^{-1}\left(\dfrac{c_2}{a}\right) = \sin^{-1}\left(\dfrac{12}{140}\right) = 4.92°$

$$a_2 = a\cos\theta_2 = 140 \times \cos 4.92° = 139.5 \text{ mm}$$
$$b_2 = b\cos\theta_2 = 100 \times \cos 4.92° = 99.6 \text{ mm}$$
$$F_2 = m\omega_2^2 r_2 = 3 \times 29.32^2 \times 0.108 = 278.53 \text{ N}$$

Taking moments about fulcrum B,

$$F_2 \times a_2 - mg \times c_2 - \dfrac{S_2 \times b_2}{2} = 0$$

or $\qquad 278.53 \times 139.5 - 3 \times 9.81 \times 12 - \dfrac{S_2}{2} \times 99.6 = 0$

or $\qquad\qquad\qquad S_2 = 773.13 \text{ N}$ **Ans.**

Spring stiffness: $k = \dfrac{S_1 - S_2}{h} = \dfrac{1432.4 - 773.13}{30} = \mathbf{21.97 \text{ N/mm}}$ **Ans.**

8.6 WILSON-HARTNELL GOVERNOR

A Wilson-Hartnell governor is also a spring loaded type governor. In this governor, the balls, which are mounted on vertical arms of bell crank lever are connected by two radial springs as shown in Figure 8.17. These springs control the centrifugal force of the balls. In this governor, the initial tensions in the radial springs are not adjustable, so an adjustable auxiliary spring is used which is connected to sleeve through a lever. When the speed of engine increases, balls try to flyout and sleeve moves upward causing tensions in radial springs. The auxiliary spring tries to bring the sleeve down.

Let

k = stiffness of the each radial springs

k_a = stiffness of the auxiliary springs and

S_1, S_2 = spring forces in the radial spring at two different positions of the bell crank lever (refer Figure 8.18). S_{a1}, S_{a2} = Spring forces in the auxiliary spring at two different positions.

Governors 313

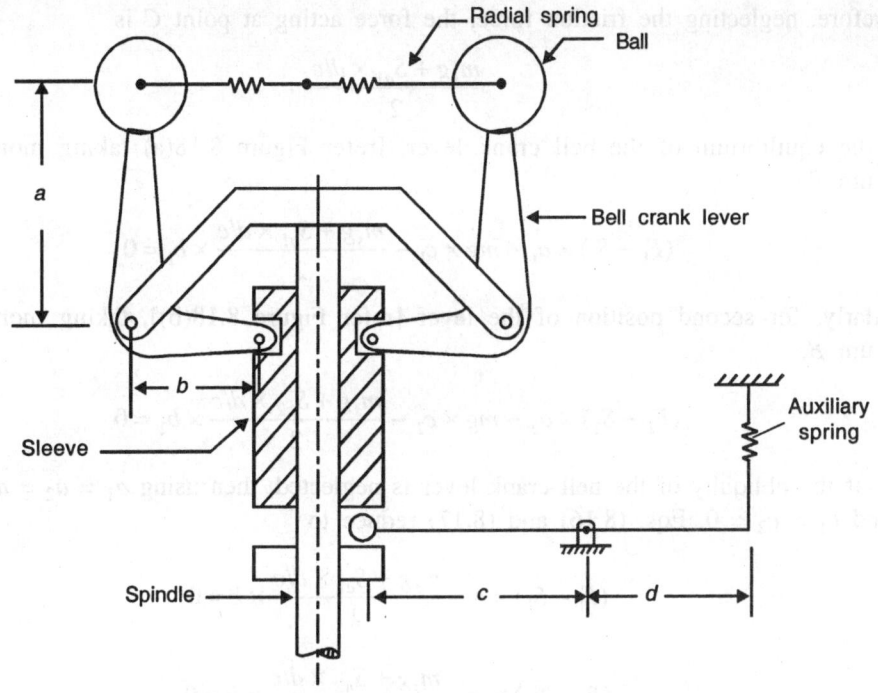

FIGURE 8.17 Wilson-Hartnell governor.

(a) At high speed

$$\frac{m_s g + S_{a1} \times d/c}{2}$$

(b) At low speed

$$\frac{m_s g + S_{a2} \times d/c}{2}$$

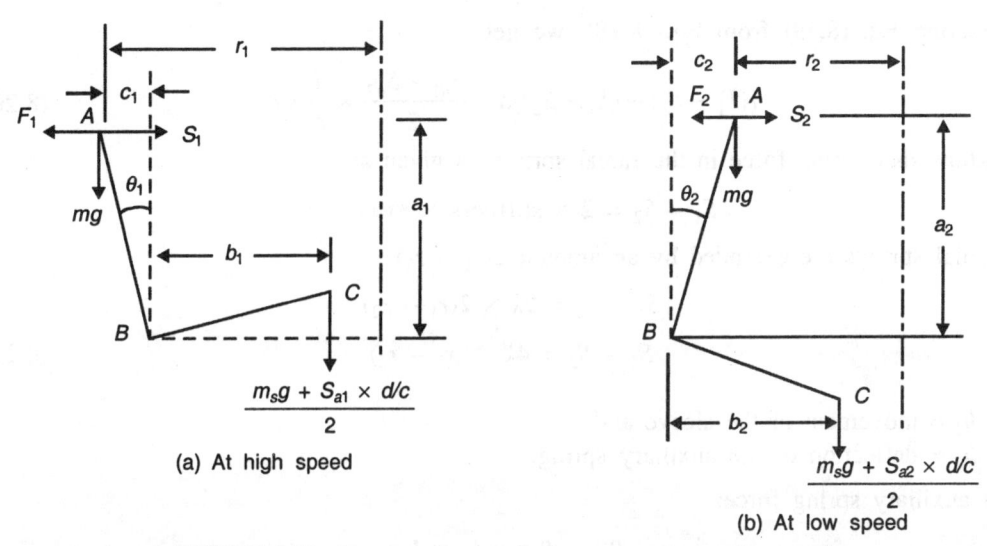

FIGURE 8.18 Forces acting on the lever at two operating speeds.

Suppose the spindle of the governor is rotating at an angular velocity of ω_1. The force at the sleeve due to auxiliary spring is given as

$$S_{a1} \times \frac{d}{c}$$

Therefore, neglecting the friction force, the force acting at point C is

$$\frac{m_s g + S_{a1} \times d/c}{2}$$

For the equilibrium of the bell crank lever, [refer Figure 8.18(a) taking moments about fulcrum B,

$$(F_1 - S_1) \times a_1 + mg \times c_1 - \frac{m_s g + S_{a1} \times d/c}{2} \times b_1 = 0 \qquad (8.16)$$

Similarly, for second position of the lever [refer Figure 8.18(b)], taking moments about fulcrum B,

$$(F_2 - S_2) \times a_2 - mg \times c_2 - \frac{m_s g + S_{a2} \times d/c}{2} \times b_2 = 0 \qquad (8.17)$$

If the obliquity of the bell crank lever is neglected, then using $a_1 \simeq a_2 \simeq a$; $b_1 \simeq b_2 \simeq b$; and $c_1 \simeq c_2 \simeq 0$, Eqs. (8.16) and (8.17) reduce to

$$(F_1 - S_1) \times a - \frac{m_s g + S_{a1} \times d/c}{2} \times b = 0 \qquad (8.18)$$

and

$$(F_2 - S_2) \times a - \frac{m_s g + S_{a2} \times d/c}{2} \times b = 0 \qquad (8.19)$$

Subtracting Eq. (8.19) from Eq. (8.18), we get

$$[(F_1 - F_2) - (S_1 - S_2)]a = \frac{S_{a1} - S_{a2}}{2} \times \frac{d}{c} \times b \qquad (8.20)$$

We know that spring force in the radial springs is given as

$$S_1 - S_2 = 2 \times \text{stiffness} \times \text{extension}$$

As radial springs are extended by an amount $2(r_1 - r_2)$,

$$S_1 - S_2 = 2k \times 2(r_1 - r_2)$$

or

$$S_1 - S_2 = 4k \times (r_1 - r_2) \qquad (8.21)$$

Let
h_1 = movement of the sleeve and
h_2 = deflection of the auxiliary spring.

Then auxiliary spring force:

$$S_{a1} - S_{a2} = k_a \times h_2$$

where

$$h_2 = h_1 \times \frac{d}{c} = (r_1 - r_2) \times \frac{b}{a} \times \frac{d}{c}$$

or

$$S_{a1} - S_{a2} = k_a \times (r_1 - r_2) \times \frac{b \times d}{a \times c} \qquad (8.22)$$

Substituting the value of radial spring force and auxiliary spring force in Eq. (8.20), we get

$$(F_1 - F_2)a - 4k(r_1 - r_2)a = k_a(r_1 - r_2)\frac{b \times d}{a \times c} \times \frac{d}{c} \times \frac{b}{2}$$

or
$$F_1 - F_2 = 4k(r_1 - r_2) + \frac{k_a(r_1 - r_2)}{2} \times \left(\frac{b \times d}{a \times c}\right)^2 \qquad (8.23)$$

In order to determine the stiffness of one spring (say radial spring) using Eq. (8.23), the stiffness of other spring (read auxiliary spring) may be fixed in advance.

EXAMPLE 8.8 In a Wilson-Hartnell type of governor, the two springs are attached to balls. Each has a stiffness of 0.75 N/mm and free length of 120 mm. The mass of each ball is 4 kg, the length of the ball arm of each bell crank lever is 80 mm and that of the sleeve arm is 70 mm; the auxiliary spring lever is pivoted at its mid-point. When the radius of rotation of the ball is 100 mm, the equilibrium speed is 300 rpm. If the sleeve is to lift by 8 mm for 5 percent increase of speed, determine the required stiffness of the auxiliary spring.

Solution Refer to Figure 8.18.
At radius $r_2 = 100$ mm,

$$F_2 = m\omega_2^2 r_2$$

$$= 4 \times \left(\frac{2\pi \times 300}{60}\right)^2 \times 0.1$$

$$= 394.8 \text{ N}$$

When the sleeve has moved through a distance of 8 mm, the new radius of rotation is:

$$r_1 = r_2 + 8 \times \frac{a}{b}$$

$$= 100 + 8 \times \frac{80}{70}$$

$$= 109.14 \text{ mm}$$

Speed: $N_1 = 1.05 \times N_2 = 1.05 \times 300 = 315$ rpm

Force: $F_1 = m\omega_1^2 \times r_1 = 4 \times \left(\frac{2\pi \times 315}{60}\right)^2 \times 0.10914 = 475$ N

Now referring to Eq. (8.23),

$$F_1 - F_2 = 4k(r_1 - r_2) + \frac{k_a(r_1 - r_2)}{2} \times \left(\frac{b}{a} \times \frac{d}{c}\right)^2$$

where $d/c = 1$ as auxiliary spring lever is pivoted at mid-span.

or $\quad 475 - 394.8 = 4 \times 0.75 (109.14 - 100) + k_a \dfrac{(109.14 - 100)}{2} \times \left(\dfrac{70}{80}\right)^2$

or $\quad\quad\quad\quad$ Stiffness of auxiliary spring: $k_a = \mathbf{15.08\ N/mm}$ $\quad\quad\quad\quad\quad\quad$ **Ans.**

8.7 SPRING-CONTROLLED GRAVITY GOVERNOR

Figure 8.19 shows a spring-controlled gravity governor in which two bell crank levers are pivoted on the moving sleeve. A spring housed in the sleeve is compressed by cap C through rollers mounted on horizontal arm of bell crank levers. The vertical arms carry rotating mass. When speed of governor spindle is increased, the balls fly out and the spring is compressed.

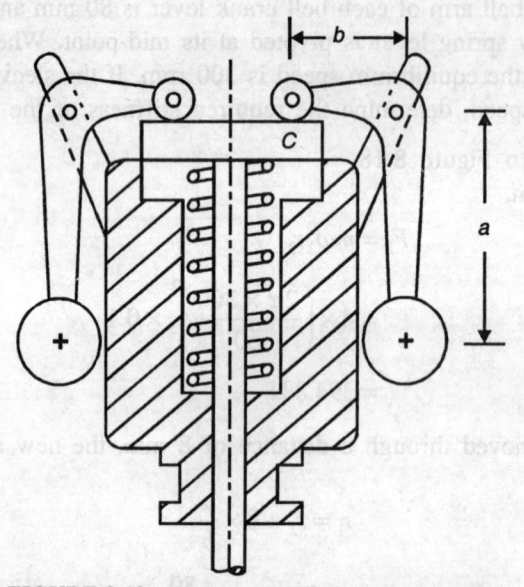

FIGURE 8.19 Spring-controlled gravity governor.

The line diagrams for two positions of the bell crank lever are shown in Figure 8.20. Various forces acting on the governor can be analysed by taking moments about the instantaneous centre I, which is intersection of the perpendicular to the fulcrum B and the path of the roller end of the bell crank lever; the path of fulcrum being parallel to the axis and that of the roller end is at right angle to the axis.

Taking moments about point I [refer Figure 8.20(a)],

$$F_1 \times a = mg \times b + \dfrac{m_s g + S_1}{2} \times b \quad\quad\quad\quad (8.24)$$

Similarly, taking moments about point I for maximum position [refer Figure 8.20(b)],

$$F_2 \times CD = mg \times ID + \dfrac{m_s g + S_2}{2} \times IA \quad\quad\quad\quad (8.25)$$

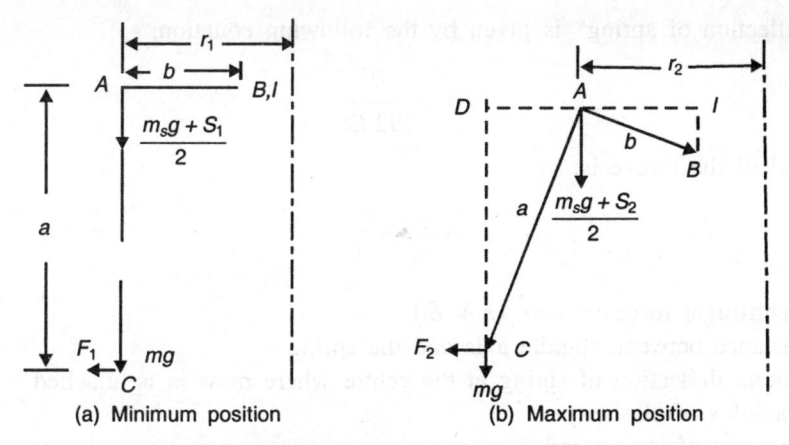

FIGURE 8.20 Positions of bell crank lever.

where $F_1 = m\omega_1^2 r_1$ and $F_2 = m\omega_2^2 r_2$ where symbols have their usual meanings.
The stiffness of the spring k can be calculated with the help of the following relation:

$$S_2 - S_1 = k \times h \tag{8.26}$$

where $\quad h =$ lift of the sleeve
and $\quad S_2 - S_1 =$ spring force.

8.8 PICKERING GOVERNOR

A pickering governor consists of a spindle and three straight leaf springs placed symmetrically around the spindle. The upper end of each spring is fixed to spindle with a nut and the lower end is fixed to the sleeve (Figure 8.21). A mass is attached at the centre of each spring. When governor operates, this mass causes centrifugal force which in turn lifts the sleeve up. Generally a stopper is provided to limit the travel of the sleeve.

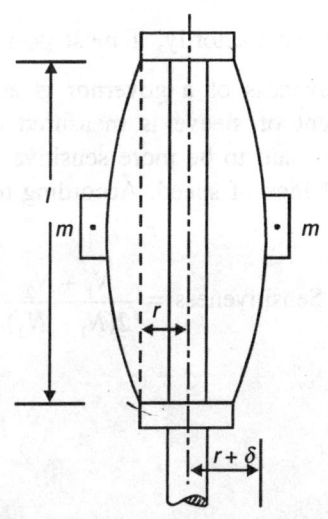

FIGURE 8.21 Pickering governor.

The deflection of spring* is given by the following equation:

$$\delta = \frac{Fl^3}{192\,EI} \tag{8.27}$$

and the travel of the sleeve is:

$$x = 2.4\frac{\delta^2}{l} \tag{8.28}$$

where
- F = centrifugal force [$= m\omega^2 (r + \delta)$]
- r = distance between spindle axis and the spring
- δ = lateral deflection of spring at the centre where mass m is attached
- E = modulus of elasticity
- I = moment of inertia and
- l = length of spring.

8.9 CHARACTERISTICS OF A GOVERNOR

A governor should have the following qualities for satisfactory performance:

(i) When its sleeve reaches to the lowest position, the engine should develop maximum power.
(ii) Its sleeve should reach the topmost position at once when the load on the engine is suddenly removed.
(iii) Its sleeve should float at some intermediate position under normal operating conditions.
(iv) Its response to change in speed should be fast.
(v) It should have sufficient power so that it can exert the required force at the sleeve to operate the control mechanism.

For a governor to function satisfactorily, it must possess the following characteristics:

Sensitiveness. The sensitiveness of a governor is a characteristic of a governor by virtue of which the displacement of sleeve is measured corresponding to change in the equilibrium speed. A governor is said to be more sensitive if the larger displacement of the sleeve takes place for a given change of speed. According to this definition of sensitiveness, it is expressed as

$$\text{Sensitiveness} = \frac{N_1 + N_2}{2(N_1 - N_2)} \tag{8.29}$$

where
- N_1 = maximum speed and
- N_2 = minimum speed

---

*Maximum deflection formula is based upon the assumption that spring acts as a fixed beam.

The above definition of sensitiveness is quite satisfactory when the governor is considered as an independent system. However, when a governor is fitted to the engine, the practical requirement is that the change of equilibrium speed from full load to no load position of the sleeve should be as small a fraction as possible of the mean equilibrium speed. The actual displacement of sleeve is immaterial provided that it is sufficient to change the fuel supply to the engine. For this reason, sensitiveness is more correctly defined as the ratio of the difference between the maximum and minimum equilibrium speeds to the mean equilibrium speed.

Stability. A governor is said to be stable when for each speed, within the working range, there is only one radius of rotation of the governor balls at which the governor is in equilibrium. In other words, a governor is called stable if it brings the speed of the engine to the required value without much fluctuation.

Isochronism. Isochronism is a property by virtue of which a governor keeps constant equilibrium speed for all radii of rotation of the balls within the working range. It is desirable when approximately constant speed is desired to be kept for all loads. In other words, a governor is said to be isochronous governor if the range of speed $(N_1 - N_2)$ is zero or its sensitiveness is infinity.

Let us find the condition of isochronism to a Porter governor. The operating speeds of Porter governor are given as (refer Eq. 8.3)

$$\omega_1^2 = \frac{1}{mh_1}\left[mg + \frac{m_s g \pm f}{2}\left(1 + \frac{\tan\beta}{\tan\alpha}\right)\right] \tag{8.30}$$

and

$$\omega_2^2 = \frac{1}{mh_2}\left[mg + \frac{m_s g \pm f}{2}\left(1 + \frac{\tan\beta}{\tan\alpha}\right)\right] \tag{8.31}$$

For the condition of isochronism, $\omega_1 = \omega_2$ and thus $h_1 = h_2$. However, from the configuration of the governor (refer Figure 8.4), it can be judged that it is impossible to have two positions of the balls at the same speed. Thus a Porter governor cannot be isochronous.

Now consider a case of Hartnell governor neglecting obliquity of bell crank lever. The equations of equilibrium for two different operating speeds are given as [refer Eqs. (8.9) and (8.10)]

$$m\omega_1^2 r_1 \times a - \frac{m_s g + S_1}{2} \times b = 0 \tag{8.32}$$

and

$$m\omega_2^2 r_2 \times a - \frac{m_s g + S_2}{2} \times b = 0 \tag{8.33}$$

For isochronism condition $\omega_1 = \omega_2$, we get

$$\frac{m_s g + S_1}{m_s g + S_2} = \frac{r_1}{r_2} \tag{8.34}$$

which is the required condition for isochronism.

Hunting. A governor is said to hunt if the speed of the engine fluctuates continuously above and below the mean speed. This is due to high sensitivity of the governor which changes the fuel supply by a large amount when a small change in the speed of rotation takes place. For example, if the load on the engine increases, the engine speed decreases and if the governor is very sensitive, the governor sleeve immediately falls to its lowest position. This will result in the opening of control valve wide and supply excess fuel to the engine so that engine speed increases rapidly and the governor rises to its highest position. Due to this movement of the sleeve, the control valve will cut off the fuel supply to the engine and the engine speed begins to fall. This cycle is repeated and is known as **hunting**.

Coefficient of insensitiveness. If there is friction force f between the sleeve and the spindle, the effective sleeve load varies between $(m_s g + f)$ and $(m_s g - f)$ when governor sleeve is rising and falling. For any sleeve position, there is thus a range of speed over which the governor is insensitive. If, for any position, the maximum and minimum speeds before sleeve movement occurs are ω' and ω'' respectively and the speed in the absence of friction is ω, the coefficient of insensitiveness is defined as the ratio $(\omega' - \omega'')/\omega$.

8.10 EFFORT OF A GOVERNOR

The effort of a governor is the force which a governor can exert on the sleeve to raise or lower it for a given change of speed, when the speed of governor is constant, the effort is zero. When the speed of the governor is either increased or decreased, a force is required on the sleeve which tends to move it. This force gradually diminishes to zero as the sleeve moves to equilibrium position corresponding to the new speed. Therefore, the mean force exerted during the given change of speed is termed the effort. For comparing different types of governor, it is defined as the force required to be applied for one percent change of speed.

To illustrate the method to be used for calculating the effort of a governor, let us take an example of Porter governor in which the effect of friction force is neglected. The height of a governor is given by following equation (refer Eq. 8.4):

$$h = \frac{1}{\omega^2}\left[g + \frac{m_s g}{2m}(1+k)\right] \qquad (8.35)$$

where k is a constant ($= \tan\beta/\tan\alpha$)

Let governor spindle speed is increased from ω rad/s to $c\omega$ rad/s. The force applied on the sleeve to prevent of from moving is F. Then the total force acting on the sleeve is algebraic sum of weight $m_s g$ and force F. The height of the governor is:

$$h = \frac{1}{c^2\omega^2}\left[g + \frac{(m_s g + F)}{2m}(1+k)\right] \qquad (8.36)$$

Dividing Eq. (8.36) by Eq. (8.35), we get

$$\frac{2mg + (m_s g + F)(1+k)}{2mg + m_s g(1+k)} = c^2$$

or
$$\frac{F(1+k)}{2mg + m_s g(1+k)} = c^2 - 1$$

or
$$F = \frac{c^2 - 1}{1+k}[2mg + m_s g(1+k)]$$

or
$$\text{Effort} = \frac{F}{2} = \frac{c^2 - 1}{2(1+k)}[2mg + m_s g(1+k)] \quad (8.37)$$

8.11 POWER OF A GOVERNOR

The power of a governor is defined as the work done at the sleeve for a given percentage change of speed. In other words, it is the product of the governor effort and the displacement of the sleeve. The power required purely depends on the type of controlling mechanism which the governor is called upon to operate. For a Porter governor, if the operating speed is changed from ω rad/s to $c\omega$ rad/s, the height of the governor changes from h to h_1.

Height of the governor when the operating speed is ω rad/s is given by

$$h = \frac{2mg + m_s g(1+k)}{2m\omega^2} \quad (8.38)$$

and height of the governor when the operating speed is $c\omega$ rad/s is given by

$$h_1 = \frac{2mg + m_s g(1+k)}{2mc^2 \omega^2} \quad (8.39)$$

From Eqs. 8.38 and 8.39,

$$\frac{h_1}{h} = \frac{1}{c^2}$$

Displacement of sleeve $= 2(h - h_1)$

$$= 2h\left(1 - \frac{h_1}{h}\right)$$

$$= 2h\left(1 - \frac{1}{c^2}\right)$$

$$= 2h\left(\frac{c^2 - 1}{c^2}\right)$$

Power $=$ Effort $\times$ Displacement of sleeve

$$= \frac{c^2 - 1}{2(1+k)}[2mg + m_s g(1+k)] \times 2h\left(\frac{c^2 - 1}{c^2}\right)$$

$$= \left(\frac{c^2 - 1}{c}\right)^2 \times \frac{h}{1+k}[2mg + m_s g(1+k)]$$

If constant $k = 1$, i.e. $\tan \beta = \tan \alpha$, then power:

$$P = \left(\frac{c^2 - 1}{c}\right)^2 (m + m_s)gh \qquad (8.40)$$

8.12 CONTROLLING FORCE

When a centrifugal governor is rotating at a uniform speed, the ball masses tend to flyout radially outward due to centrifugal force. In order to keep ball at equilibrium, an equal and opposite centripetal force acts radially inward. This inward force is called the **controlling force**. This force is resultant of all the external forces such as mass of the balls, mass of the sleeve in Porter governor or spring force in Hartnell governor. Controlling force is a function of a single variable, i.e. radius of rotation r.

Figure 8.22 shows a typical controlling force characteristic curve AB. When governor ball rotates at angular speed ω rad/s, the centripetal force for maintaining radius of rotation r is given by $m\omega^2 r$. The plot of this force against different radii of rotation for constant angular speed ω is a straight line OC with slope equal to $\phi \ (= \tan^{-1} m\omega^2)$. Thus the equilibrium radius for the angular speed ω can be determined by the intersection of controlling force curve AB and straight line OC.

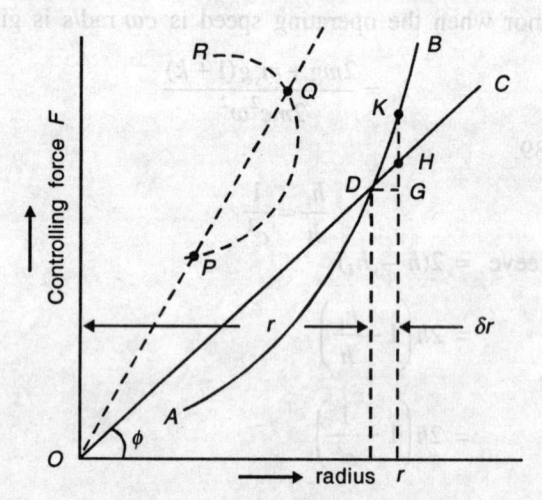

FIGURE 8.22 Controlling force curve.

The line OC, which is a constant speed line, is expressed as

$$F = m\omega^2 r$$

Let the speed of the governor be ω at equilibrium position given by point D. If the speed remains constant and radius changes from r to $(r + \delta r)$, the increment in the controlling force is:

$$KG = \left(\frac{dF}{dr}\right) \times \delta r \qquad (8.41)$$

Corresponding increase in the centripetal force is given by
$$GH = m\omega^2 \times \delta r$$

Thus the restoring force on the balls:
$$KH = KG - GH$$

or
$$KH = \left(\frac{dF}{dr} - m\omega^2\right)\delta r \tag{8.42}$$

When the value of *KH* is greater than zero, the restoring force enables the governor to rejoin equilibrium position assuring stability. But when *KH* = 0, we have

$$\frac{dF}{dr} = m\omega^2 \tag{8.43}$$

and this gives isochronous governor. It follows that in isochronos governor, the controlling force curve coincides with the centripetal force line.

Further, the term *dF/dr* in Eq. (8.42) represents the slope of the controlling force curve while the term $m\omega^2$ represents the slope of centripetal force line *OC*. Clearly when slope of one controlling force curve is larger than the slope of the centripetal force line, *KH* is positive and extra restoring force is available. Thus for a stable governor, the slope of the controlling force curve should be greater than that of the centripetal force line.

When the slope of controlling force curve is smaller than the slope of centripetal force line and centripetal force line cuts the controlling force curve *PQR* at more than one point, at each equilibrium speed, a unique radius of rotation is not possible and the governor becomes unstable.

EXAMPLE 8.9 The controlling force curve of a spring controlled governor is straight line. The weight of each governor ball is 40 N and the extreme radii of rotation are 120 mm and 180 mm. If the values of the controlling force at the above radii be respectively 200 N and 360 N and the friction of the mechanism is equivalent to 2 N at each ball, find:

(a) the extreme equilibrium speeds of the governor
(b) equilibrium speed
(c) the coefficient of insensitiveness at radius of 150 mm.

Solution At radius r_1 = 120 mm, Force:

$$F_1 = m\omega_1^2 r_1$$

or
$$200 = \frac{40}{9.81} \times \omega_1^2 \times 0.12$$

or
$$\omega_1 = 20.21 \text{ rad/s}$$

or
$$N_1 = \textbf{193 rpm} \qquad \textbf{Ans.}$$

At another radius $r_2 = 180$ mm, Force F_2:

$$F_2 = m\omega_2^2 r_2$$

or $$360 = \frac{40}{9.81} \times \omega_2^2 \times 0.18$$

or $$\omega_2 = 22.14 \text{ rad/s}$$

or $$N_2 = \mathbf{211.4 \text{ rpm}}$$ **Ans.**

Let controlling force curve equation is straight line

$$F = br + c$$

or $\qquad 200 = b \times 0.12 + c \qquad$ (i)

and $\qquad 360 = b \times 0.18 + c \qquad$ (ii)

Solving Eqs. (i) and (ii), we get

$$b = 2667 \quad \text{and} \quad c = -120$$

Therefore, $\qquad F = 2667 \times r - 120$

At radius $r = 0.15$ m,

$$F = 2667 \times 0.15 - 120 = 280.05 \text{ N}$$

However, the speed of governor at radius, $r = 150$ mm is given by

$$F = m\omega^2 r$$

or $$280.05 = \frac{40}{9.81} \times \omega^2 \times 0.15$$

or $\qquad \omega = 21.4 \text{ rad/s} \quad \text{or} \quad N = \mathbf{204.35 \text{ rpm}}$ **Ans.**

Let ω_1' and ω_2' be the maximum and minimum angular speeds of the governor considering friction.

(i) $F + f = m(\omega_1')^2 r$

or $$280.05 + 2 = \frac{40}{9.81} \times (\omega_1')^2 \times 0.15$$

or $\qquad \omega_1' = 21.474 \text{ rad/s} \quad \text{or} \quad N_1' = 205.06 \text{ rpm}$

(ii) $F - f = m(\omega_2')^2 \times r$

or $$280.05 - 2 = \frac{40}{9.81} \times (\omega_2')^2 \times 0.15$$

or $\qquad \omega_2' = 21.321 \text{ rad/s} \quad \text{or} \quad N_2' = 203.6 \text{ rpm}$

Coefficient of insensitiveness $= \dfrac{2(\omega_1' - \omega_2')}{\omega}$

$= \dfrac{21.474 - 21.321}{21.4} \times 100$

$= 0.715$ per cent **Ans.**

EXAMPLE 8.10 In a spring controlled governor, the controlling force is 60 N when the radius of rotation is 150 mm and 22.5 N when the radius is 75 mm. Each ball weighs 7.5 N. Find the equilibrium speed when the radius of rotation is 100 mm. What change is required to make the governor isochronous? Also determine isochronous speed.

Solution Assuming that the controlling force curve F varies linearly with the radius of rotation r. We have

$$F = br + c$$

when
$$r = 0.075 \text{ m,}$$
$$F = 22.5 \text{ N}$$

or $\qquad 22.5 = 0.075 \times b + c \qquad$ (i)

when
$$r = 0.15 \text{ m,}$$
$$F = 60 \text{ N}$$

or $\qquad 60 = 0.15 \times b + c \qquad$ (ii)

Solving Eqs. (i) and (ii), we have

$$b = 500 \quad \text{and} \quad c = -15$$

Hence, controlling force curve equation is:

$$F = 500 \times r - 15$$

Now at $r = 0.1$ m,

Controlling force: $\qquad F = 500 \times 0.1 - 15 = 35$ N

We know that
$$F = m\omega^2 r$$

or $\qquad 35 = \dfrac{7.5}{9.81} \times \omega^2 \times 0.1$

or $\qquad \omega = 21.39$ rad/s

or $\qquad N = 204.26$ rpm **Ans.**

Presently the governor is stable because constant b is positive and constant c is negative.

For isochronous governor,

$$F = b \times r \quad \text{where } c = 0$$

or

$$\frac{F}{r} = b = \text{constant} = 500$$

Therefore, to make the governor isochronous, the ratio of $\frac{F}{r}$ should be made constant. For isochronous speed:

$$F = m\omega^2 r = br + c$$

or

$$\frac{7.5}{9.81} \times \omega^2 \times 0.1 = 500 \times 0.1 + 0$$

or

$$\omega = 25.57 \text{ rad/s}$$

or

$$\text{Speed: } N = \frac{60 \times 25.57}{2\pi} = \mathbf{244.17 \text{ rpm}} \qquad \text{Ans.}$$

EXERCISES

1. What is the function of a governor? How is it different from that of a flywheel?
2. What is the influence of friction at the sleeve on the performance of a governor?
3. What are the limitations of a Watt governor? How are these rectified in the Porter governor?
4. Explain why a Proell governor requires balls of smaller size compared to Porter governor.
5. Define the following terms related to governor:
   (i) Height of governor
   (ii) Equilibrium speed
6. What is stability of a governor? Sketch the controlling force curve versus radius diagram for a stable, unstable and isochronous governor.
7. What are the criteria that are taken into consideration to judge the quality of a governor?
8. What is meant by the term stability of a governor? Derive the necessary condition of stability for a centrifugal governor.
9. What is meant by the controlling force curve of a governor?
10. Explain the following terms with reference to a governor:
    (i) Sensitiveness
    (ii) Hunting
    (iii) Isochronism
    (iv) Stability
    (v) Coefficient of insensitiveness

11. In an open arm type Watt governor, the lengths of upper and lower arms are each 300 mm and they are connected to horizontal link at 50 mm from the axis of the governor. If the two extreme radii of rotation of the governor balls are 160 mm and 180 mm, determine the equilibrium speeds and speed range of the governor.

[**Ans:** 51.8 rpm, 53.28 rpm, 1.48]

12. The arms of a Porter governor are each 250 mm long and pivoted on the governor axis. Weight of each ball is 5 N and the weight of central sleeve is 30 N. The radius of rotation of the balls is 150 mm when the sleeve begins to rise and reaches a value of 200 mm for maximum speed. Determine the speed range of the governor. If the friction force of the sleeve is equivalent to 2 N at the sleeve, determine how the speed range is modified.

[**Ans:** 27.32 rpm, 38.19 rpm]

13. The lengths of upper and lower arms of a Porter governor are 200 mm and 250 mm respectively. Both the arms are pivoted on the axis of rotation. The sleeve weight is 150 N, the weight of each ball is 20 N and the friction resistance is equivalent to a force of 30 N at the sleeve. If the limiting inclination of the upper arms to the vertical are 30° and 40°, determine the range of speed of the governor.

[**Ans:** 180 rpm, 225.6 rpm, 45.6 rpm]

14. The lengths of all arms of a Proell governor are 250 mm. The distance of pivot of arms from the axis of rotation is 30 mm. Length of extension of lower arm to which each ball is attached is 80 mm; mass of each ball is 5 kg and the mass of the central sleeve is 150 kg. If the radius of rotation of the ball is 200 mm when the arms are inclined to an angle of 30° to the spindle axis, calculate the equilibrium speed at which the spindle is rotating.

[**Ans:** 190.46 rpm]

15. The radius of rotation of the balls of a Hartnell governor is 80 mm at the minimum speed of 300 rpm. Neglecting gravity effect, determine the speed after the sleeve has lifted by 60 mm. Also determine the initial compression of the spring, the governor effort and the power. The particulars of the governor are given below:
    Length of ball arm = 150 mm
    Length of sleeve arm = 100 mm
    Weight of each ball = 4 N
    Stiffness of the spring = 2.5 N/mm

[**Ans:** 329 rpm, 38.64 mm, 75 N, 450 N mm]

16. In a spring loaded Hartnell type governor, the extreme radii of rotation of the balls are 80 mm and 120 mm. The ball arm and the sleeve arm of the bell crank lever are equal in length. Weight of the ball is 2 N. If speeds at the two extreme positions are 400 rpm and 420 rpm, find the (i) spring constant and (ii) initial compression of the spring.

[**Ans:** 93.5 N/m 61.2 mm]

17. In a Hartnell governor, the mass of each ball is 5 kg and the lift of the sleeve is 50 mm. The speed at which the governor begins to float is 240 rpm and at this speed, the radius of ball path is 110 mm. The mean working speed of the governor is 20 times the range of speed when friction is neglected. If the length of ball and roller arm of the bell crank lever is 120 mm and 100 mm respectively and if the distance between the centre of pivot of bell crank and axis of the governor is 140 mm, determine the initial compression of the spring, taking into account obliquity of the arm.

    If friction is equivalent to a force of 30 N at the sleeve, find the total alteration in speed before the sleeve begins to move from mid position.

    [Ans: 65 mm,   270 rpm,   263.9 rpm,   6.1 rpm]

18. A spring loaded governor is shown in Figure E8.1. The two balls each of mass 5.5 kg are connected by two springs—A. An auxiliary spring B provides an additional force at the sleeve through the medium of a lever which pivots about a fixed centre at its left hand end. In the mean position, the radius of the governor balls is 150 mm and the speed is 600 rpm. The tension in each spring A is then 1100 N. Find the tension in the spring B for this position.

    If, when sleeve moves up 20 mm, the speed is to be 630 rpm, find the necessary stiffness of the auxiliary spring B. The stiffness of each spring A is 8 kN/m.

    [Ans: 3.165 kN,   39.4 kN/m]

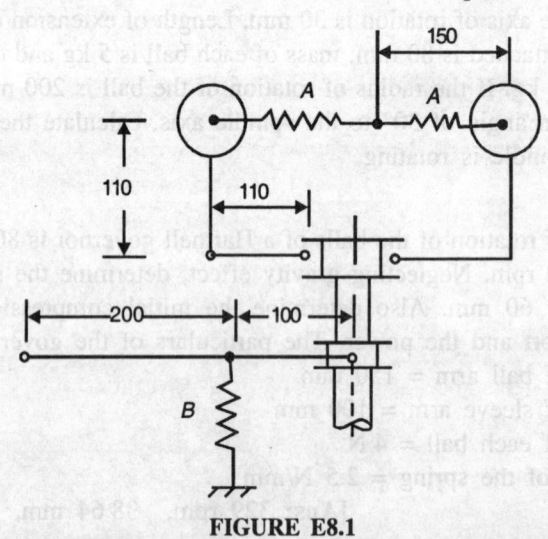

FIGURE E8.1

19. In a spring controlled governor, the radial force acting on the balls was 450 N when the centre of the balls was 200 mm from the axis and 750 N when at 300 mm. Assuming that the force varies directly as the radius, find the radius of ball path when the governor runs at 270 rpm. Also find what alteration in spring load is required in order to make the governor isochronous and speed at which it would then run. The weight of each ball is 30 N.

    [Ans: 270 mm,   initial tension must be increased by 150 N, N = 299 rpm]

20. A spring loaded shaft governor of the form is shown in Figure E8.2. The stiffness of the spring is k_s and has the spring tension T_s when the value of the radius r is zero. If the mass of the ball is m kg and the effect of gravity on the balls and the links are neglected, show that the operating speed is:

$$\omega = \left[\frac{1}{m} \left(\frac{T_s + 2k_s \left(1 - \sqrt{l^2 - r^2}\right)}{\sqrt{l^2 - r^2}} \right) \right]^{0.5}$$

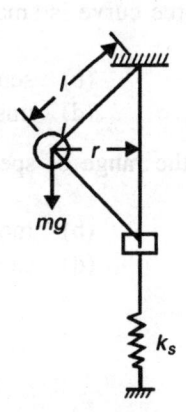

FIGURE E8.2

MULTIPLE CHOICE QUESTIONS

1. The main function of a governor is to control speed due to
   (a) cyclic variation
   (b) variation due to load-speed characteristics
   (c) random variation
   (d) not applicable

2. Inertia governors, compared to centrifugal governor are
   (a) more sensitive
   (b) less sensitive
   (c) equally sensitive
   (d) unpredictable

3. The frictional resistance at the sleeve of a governor
   (a) increases sensitivity
   (b) decreases sensitivity
   (c) does not affect sensitivity
   (d) causes isochronism

4. Isochronous governor is one which is
   (a) more stable
   (b) less stable
   (c) less sensitive
   (d) more sensitire

5. A governor which is hunting is
   (a) more stable
   (b) less sensitive
   (c) more sensitive
   (d) none of the above

6. If the balls of a governor have same speed for all radii of rotation, it is said to be
   (a) sensitive
   (b) isochronous
   (c) hunting
   (d) stable

7. Which of the following governor cannot be isochronous?
   (a) Watt (b) Hartnell
   (c) Proel (d) Porter

8. The effort of a governor is defined as the force required to be applied for what percentage change of speed.
   (a) 1 percent (b) 5 percent
   (c) 10 percent (d) any percent

9. When slope of controlling force curve is smaller than the slope of centripetal force line, governor is said to be
   (a) isochroncus (b) sensitive
   (c) stable (d) unstable

10. For the same lift of sleeve, the range of speed of Proell governor as compared to Porter governor is
    (a) same (b) more
    (c) less (d) cannot be predicted

CHAPTER 9

Cams

9.1 INTRODUCTION

A cam is a mechanical element which is used to drive another element called the follower. The cam may have rotating, reciprocating or oscillating motion whereas the follower may have reciprocating or oscillating motion. In a cam and follower mechanism, the follower usually has line contact with the cam, thus they constitute a higher pair mechanism. Cam and follower mechanisms are simple, inexpensive and have a few moving parts. The follower motions having almost any desired characteristics can be obtained. On account of these reasons, cam-follower mechanisms are most widely used in various mechanical machinery such as automatic machines, machine tools, IC engines, printing machines and so forth.

9.2 CLASSIFICATION OF CAMS AND FOLLOWERS

Cams are classified according to their basic shapes, types of follower movement and the manner of constraints of the follower. The most important types of cam are discussed further.

Wedge cam. A wedge cam has a wedge of specified contour. The translation motion of the wedge is imparted to the follower which either reciprocates or oscillates. Generally, a spring is used to maintain contact between the follower and the cam [Figure 9.1(a)].

Plate cam. A cam made out of a plate in such a way that follower moves radially from the centre of rotation is known as **plate cam.** These cams are also known as **disc cams** or **radial cams** because the surface of the cam is so shaped that the follower reciprocates or oscillates in a plane at right angle to the axis of the cam. By far, the most common is the plate cam. For this reason, we shall restrict our discussion to plate cams, although the concept presented pertains universality [Figure 9.1(b)].

Cylindrical cam. In a cylindrical cam, a circumferential contour is cut on the surface of a cylinder which rotates about its axis. The follower may either oscillate or reciprocate as shown in Figures 9.1(c) and (d). These cams are also known as **drum** or **barrel** cams.

Spiral cam. A circular plate in which spiral groove is cut and a pin gear follower meshes with the teeth cut on spiral groove, as shown in Figure 9.1(e), is called **spiral cam.** It is also known as **face cam.** The main limitation of such cam is that it has to reverse the direction to reset the position of the follower.

331

Globoidal cam. A globoidal cam may have either concave or convex surface and a circumferential contour is cut on the surface. The follower in these cams has an oscillatory motion [Figures 9.1(f) and (g)].

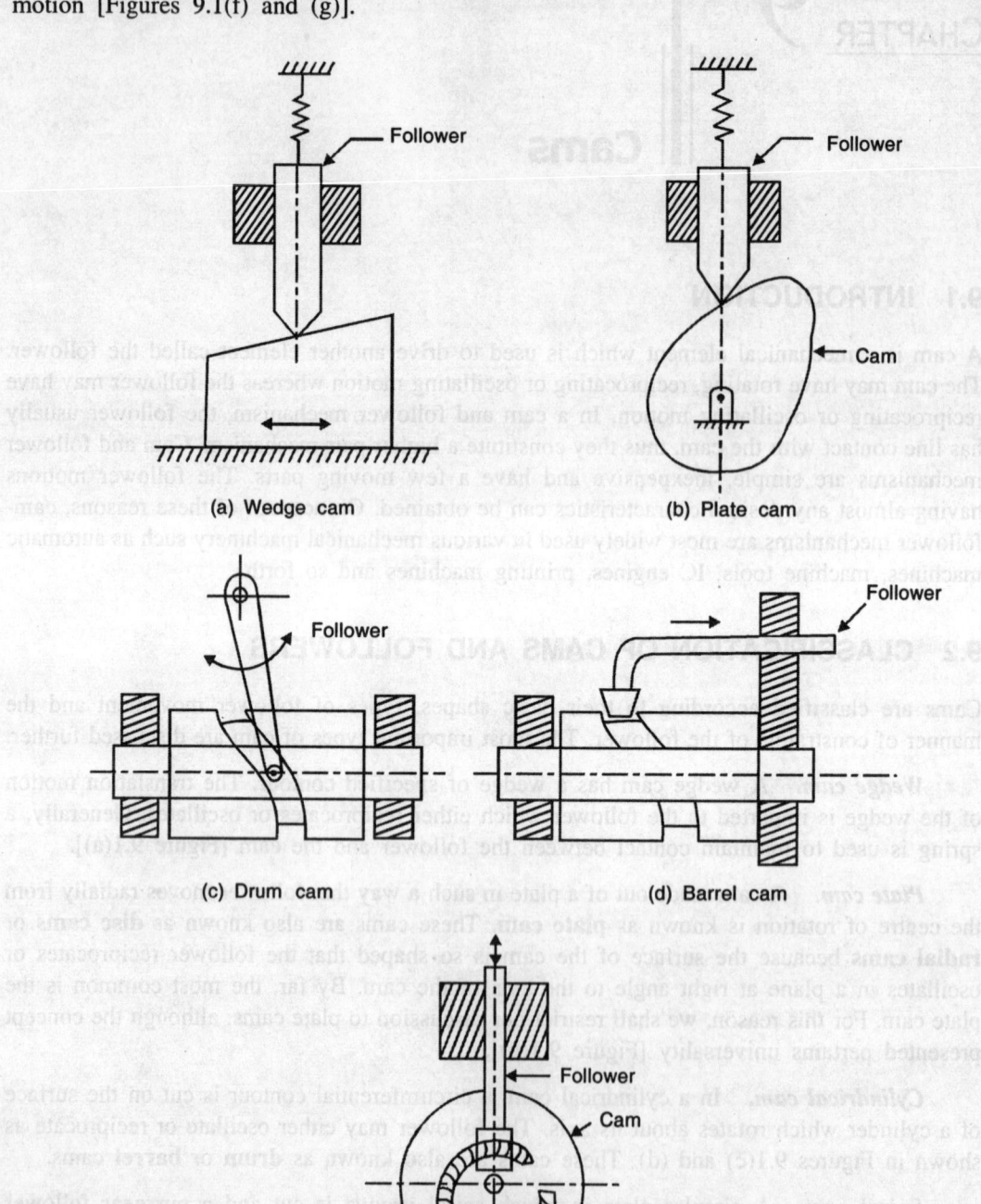

(a) Wedge cam

(b) Plate cam

(c) Drum cam

(d) Barrel cam

(e) Spiral cam

```
                            ┌── Convex surface          └── Concave surface
      (f) Convex globoidal cam              (g) Concave globoidal cam
```

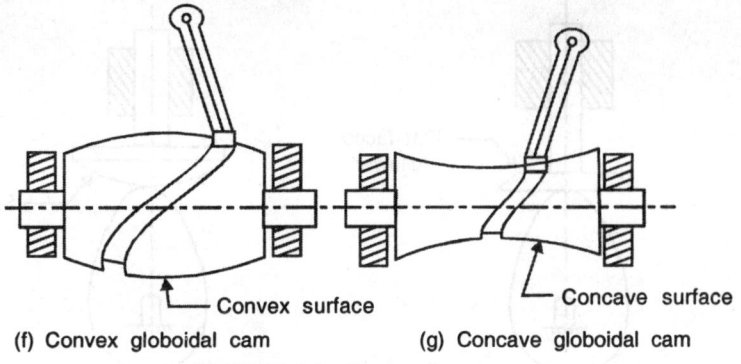

FIGURE 9.1 Types of cam.

Cam follower is guided by the movement of the cam. However, followers are classified in the following four types:

Knife edge follower. A knife edge follower is usually of the sharp pointed pencil shape. Its construction is simple but its use is limited as its edge wears out rapidly, affecting the motion imparted to the follower [Figure 9.2(a)].

Roller follower. A roller follower consists of a cylindrical roller which rolls on the cam surface. The rate of wear is considerably reduced as compared to knife edge follower. At times of steep rise, a roller follower may jam the cam; therefore it is less preferred. [Figure 9.2(b)].

Flat faced follower. The contacting surface of this type of follower is flat and it does not pose the problem of jamming the cam as found in the roller follower. However, on account of high contact stresses, the wear rate is higher. This type of follower is also called **mushroom follower** [Figure 9.2(c)].

Spherical faced follower. A mushroom follower with spherical contacting surface is called **spherical faced follower.** They are used where the space is limited. These followers are preferred to flat faced as there is less surface stress and wear of follower [Figure 9.2(d)].

According to the location of the axis of moment, a follower may be classified into two types—radial and offset followers.

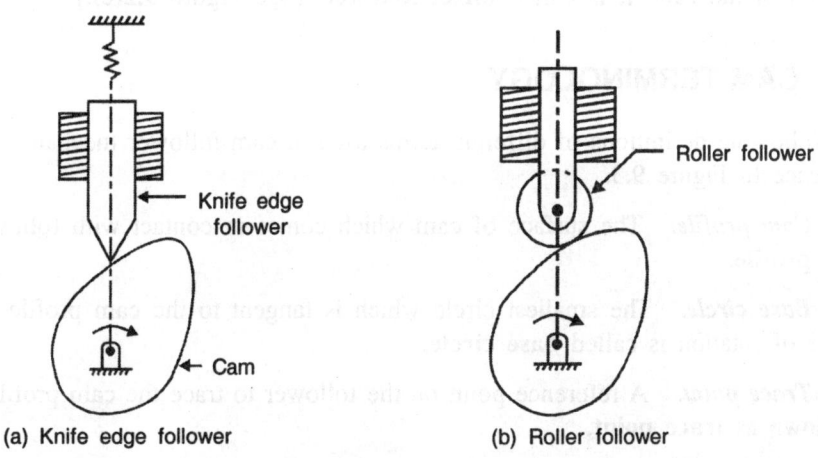

(a) Knife edge follower   (b) Roller follower

334 *Theory of Mechanisms and Machines*

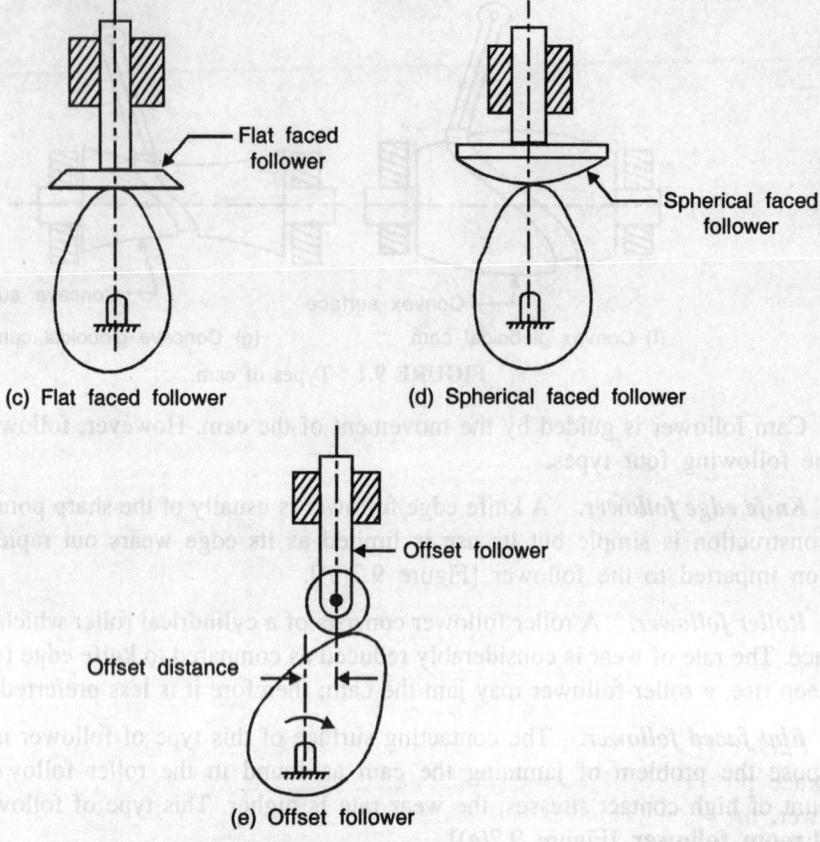

FIGURE 9.2 Types of follower.

(c) Flat faced follower
(d) Spherical faced follower
(e) Offset follower

Radial follower. If the axis of movement of the follower passes through the centre of rotation of the cam, it is called **radial follower**. [See Figure 9.2(a).]

Offset follower. If the axis of movement of the follower is away from the centre of rotation of the cam, it is called **offset follower**. [See Figure 9.2(e).]

9.3 CAM TERMINOLOGY

The following definitions of different terms used in cam-follower mechanism are given with reference to Figure 9.3.

Cam profile. The surface of cam which comes in contact with follower is known as **cam profile**.

Base circle. The smallest circle which is tangent to the cam profile drawn from the centre of rotation is called **base circle**.

Trace point. A reference point on the follower to trace the cam profile or pitch curve is known as **trace point**.

FIGURE 9.3 Cam terminology.

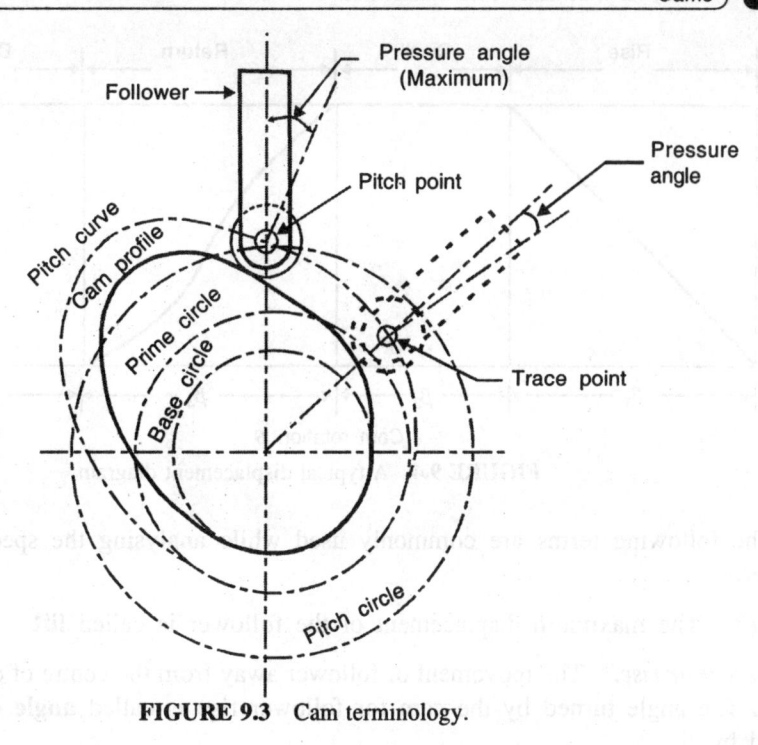

Pitch curve. It is a curve traced by the trace point assuming that the cam is fixed and the trace point of the follower rotates around the cam. In case of cam with knife edge follower, the traced curve is called **cam profile**.

Pressure angle. It is an angle between the normal to the pitch curve and line of motion of the follower. Its magnitude varies at different positions of the follower. The maximum value of pressure angle should not exceed 30°, otherwise it may jam the follower.

Pitch point. A point on the pitch curve at which the pressure angle is maximum, is called **pitch point**.

Pitch circle. A circle which passes through pitch point and is concentric to base circle is known as **pitch circle**.

Prime circle. A smallest circle that can be drawn tangent to the pitch curve and is concentric to the base circle is called **prime circle**.

9.4 ANALYSIS OF FOLLOWER MOTIONS

A cam rotating at constant angular velocity imparts a specific motion to the follower which is repeated after each revolution of the cam. The graphical representation of follower displacement y, i.e. movement of the trace point against the rotation of cam through an angle θ is called **displacement diagram.** Figure 9.4 shows a typical displacement diagram.

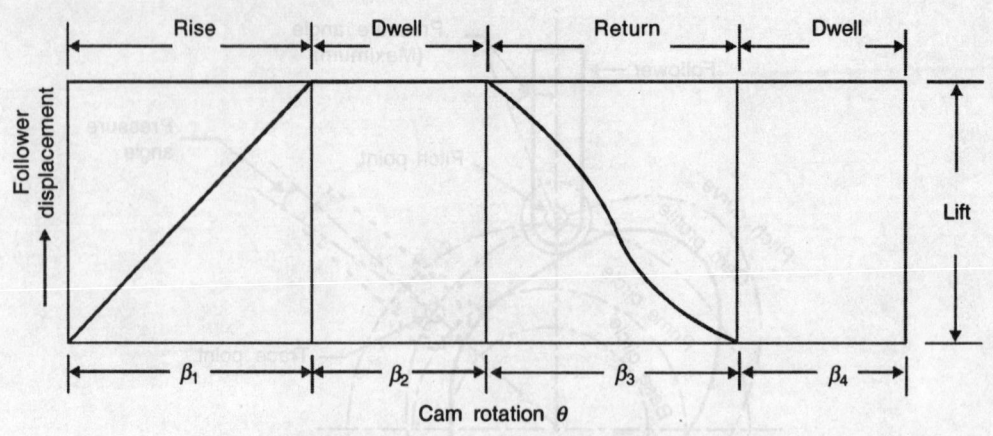

FIGURE 9.4 A typical displacement diagram.

The following terms are commonly used while analysing the specific motion of the follower.

Lift. The maximum displacement of the follower is called **lift**.

Ascent or rise. The movement of follower away from the centre of cam is called **ascent** or **rise**. The angle turned by the cam for follower rise is called **angle of ascent**, which is denoted by β_1.

Dwell. It is a period when there is no movement of the follower. The angle turned by the cam during this period is called **angle of dwell**, which is denoted by β_2.

Descent or return. The movement of the cam follower towards the centre of cam is called **descent** or **return**. The angle turned by the cam for follower return is known as **angle of descent**, which is denoted by β_3.

The rise and return of the follower can take place in many different ways. In this section, we shall discuss some of the basic follower motions.

9.4.1 Constant Velocity Motion

In a cam–follower mechanism, the constant velocity of the follower implies that the displacement of follower is directly proportional to angular rotation of the cam and displacement diagram shows a diagonal line having constant slope [Figure 9.5(a)].

Analytically the displacement can be represented as

$$y = c\theta \tag{9.1}$$

where y = displacement of the follower at an instant

θ = angle of rotation of the cam at that instant

and c = constant

The value of constant c can be determined from the following boundary conditions:

when angle $\theta = 0°$, displacement $y = 0$

angle $\theta = \beta°$, displacement $y = h$,

(a) Displacement diagram

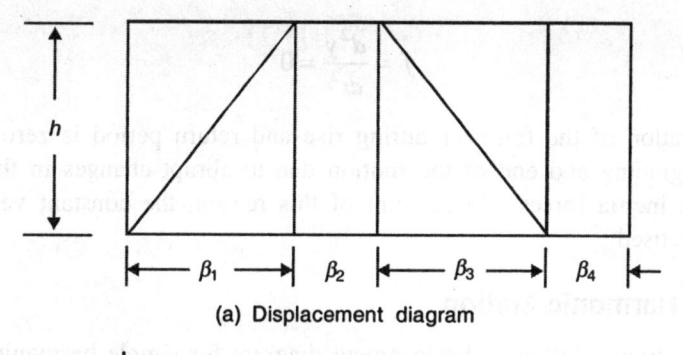

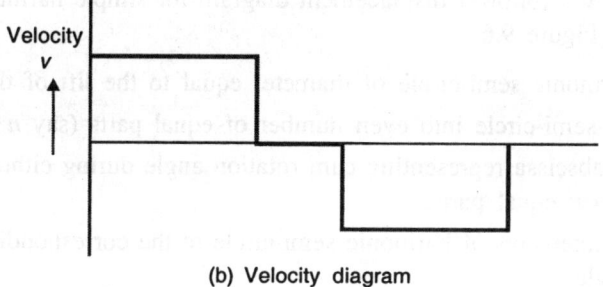

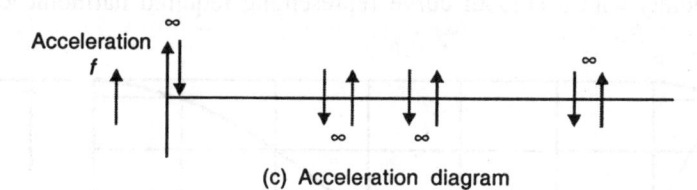

(b) Velocity diagram

(c) Acceleration diagram

FIGURE 9.5 Follower response at constant velocity.

where β is cam angle, i.e. angle of ascent or angle of descent and h is the lift of the follower. Therefore,

Follower displacement:
$$y = \frac{h\theta}{\beta} \qquad (9.2)$$

The velocity and acceleration of the follower can be obtained by differentiating Eq. (9.2) with respect to time t.

Let ω be the angular velocity of the cam. Therefore, $\theta = \omega t$

Velocity:
$$v = \frac{dy}{dt} = \frac{d}{dt}\left(\frac{h\theta}{\beta}\right) = \frac{d}{dt}\left(\frac{h\omega t}{\beta}\right)$$

or
$$v = \frac{h\omega}{\beta} \qquad (9.3)$$

338 *Theory of Mechanisms and Machines*

Acceleration:
$$f = \frac{d^2y}{dt^2} = 0 \tag{9.4}$$

The acceleration of the follower during rise and return period is zero. However, it is infinite at the beginning and end of the motion due to abrupt changes in the velocity. This results in infinite inertia forces. On account of this reason, the constant velocity motion is generally not practised.

9.4.2 Simple Harmonic Motion

The procedure to draw a follower displacement diagram for simple harmonic motion (SHM) is given below (See Figure 9.6):

1. Draw a harmonic semi-circle of diameter equal to the lift of the follower.
2. Divide this semi-circle into even number of equal parts (say n parts).
3. Divide the abscissa representing cam rotation angle during either ascent or descent or both into n equal parts.
4. Project the intercepts of harmonic semi-circle to the corresponding division of cam rotation angle.
5. Join the points with a smooth curve representing required harmonic curve.

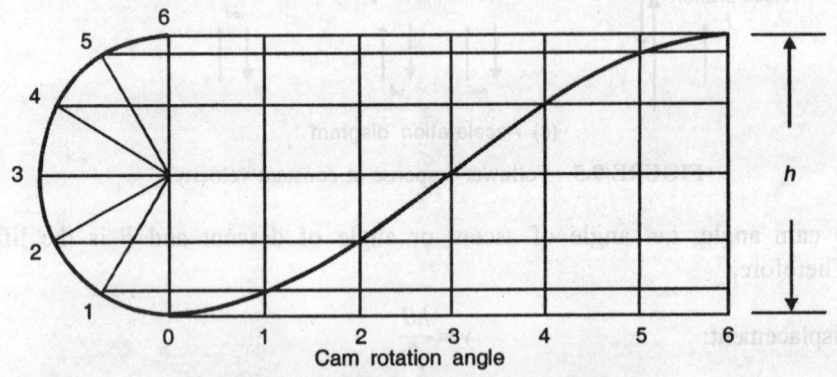

FIGURE 9.6 Displacement diagram of SHM.

In simple harmonic motion, the displacement of follower at any instant when cam has moved $\theta°$ is given as

$$y = \frac{h}{2}\left[1 - \cos\left(\frac{\pi\theta}{\beta}\right)\right] \tag{9.5}$$

Let ω be the angular velocity of the cam ($\theta = \omega t$). Then the follower displacement becomes:

$$y = \frac{h}{2}\left[1 - \cos\left(\frac{\pi\omega t}{\beta}\right)\right] \tag{9.6}$$

Follower velocity:
$$v = \frac{dy}{dt} = \frac{d}{dt}\left[\frac{h}{2}\left(1 - \cos\frac{\pi\omega t}{\beta}\right)\right]$$

or
$$v = \frac{h}{2} \times \frac{\pi\omega}{\beta} \times \sin\left(\frac{\pi\omega t}{\beta}\right) \quad (9.7)$$

$$= \frac{h}{2} \times \frac{\pi\omega}{\beta} \times \sin\left(\frac{\pi\theta}{\beta}\right)$$

The maximum velocity of the follower occurs at $\theta = \beta/2$. Therefore,

$$v_{max} = \frac{h}{2} \times \frac{\pi\omega}{\beta} \quad (9.8)$$

Acceleration:
$$f = \frac{dv}{dt} = \frac{d}{dt}\left[\frac{h}{2} \times \frac{\pi\omega}{\beta} \times \sin\left(\frac{\pi\omega t}{\beta}\right)\right]$$

or
$$f = \frac{h}{2} \times \left(\frac{\pi\omega}{\beta}\right)^2 \times \cos\left(\frac{\pi\omega t}{\beta}\right) \quad (9.9)$$

$$= \frac{h}{2} \times \left(\frac{\pi\omega}{\beta}\right)^2 \times \cos\left(\frac{\pi\theta}{\beta}\right)$$

Maximum acceleration occurs at $\theta = 0°$

or
$$f_{max} = \frac{h}{2} \times \left(\frac{\pi\omega}{\beta}\right)^2 \quad (9.10)$$

The follower response for SHM is shown in Figure 9.7. It is found from the figure that at the beginning of rise, acceleration of the follower suddenly rises from zero to maximum value and at the end of rise, it jumps from negative maximum to zero. These sudden changes in acceleration induce infinite jerk, thereby inducing vibration and noise in the cam-follower mechanism. Looking to these limitations, a follower with SHM is suggested for moderate to low speed applications.

9.4.3 Constant Acceleration and Deceleration

In this type of motion, the displacement of the follower during first half of the rise takes place at constant acceleration and during the later half period, it is at a constant deceleration of the same magnitude. The method of constructing displacement diagram is given below (See Figure 9.8):

1. Divide each half of the cam displacement interval (cam rotation angle) into n equal parts and mark them on the abscissa.
2. Divide half of the follower rise at the central ordinate of the cam displacement into same n equal parts.

340 Theory of Mechanisms and Machines

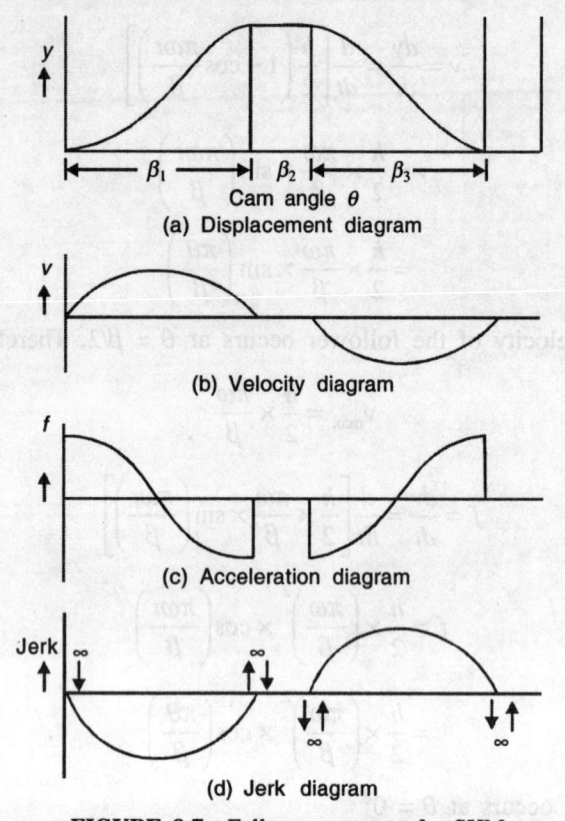

FIGURE 9.7 Follower response for SHM.
(a) Displacement diagram
(b) Velocity diagram
(c) Acceleration diagram
(d) Jerk diagram

3. Join the zero point with first point on the ordinate by a straight line 0–1. Subsequently, join all the points on the ordinate with zero point as shown in Figure 9.8. Similarly, repeat the exercise for later half portion of the follower rise at constant deceleration.
4. Pick the intersecting point and join them with a smooth curve to obtain the required displacement diagram for constant acceleration and deceleration motion.

Let f be the magnitude of constant acceleration or deceleration.

Thus
$$\frac{dv}{dt} = f$$

By integration, we get
$$v = ft + c_1 \qquad (9.11)$$

at the beginning of rise, i.e. at $t = 0$, $v = 0$ and constant $c_1 = 0$. Therefore,

Velocity:
$$v = \frac{dy}{dt} = ft \qquad (9.12)$$

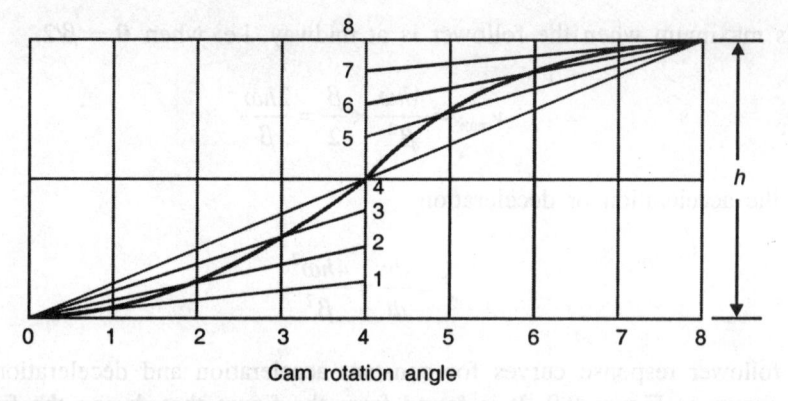

FIGURE 9.8 Displacement diagram for constant acceleration and deceleration.

Further, integrating Eq. (9.12), we get

$$y = \frac{1}{2}ft^2 + c_2$$

at $t = 0$, $y = 0$ and constant $c_2 = 0$.

Therefore,
$$y = \frac{1}{2}ft^2$$

or
$$y = \frac{1}{2}f \times \left(\frac{\theta}{\omega}\right)^2 = c\theta^2 \qquad (9.13)$$

where c is a constant.

Equation (9.13) is an equation of parabola, thus a constant acceleration and deceleration motion is also known as **parabolic motion.** This parabolic motion is most suitable for high speed cam as it minimizes the maximum inertia force. The boundary conditions for displacement equations are:

At $\qquad\qquad\qquad\qquad \theta = 0°, \quad y = 0$

and $\qquad$ at $\theta = \frac{\beta}{2}$ (i.e. at midway of follower rise), $y = \frac{h}{2}$

Using these boundary conditions in Eq. (9.13), we get

Constant:
$$c = \frac{2h}{\beta^2}$$

or follower displacement:
$$y = 2h\left(\frac{\theta}{\beta}\right)^2 \qquad (9.14)$$

Velocity:
$$v = \frac{dy}{dt} = \frac{4h\omega\theta}{\beta^2} \quad (\text{as } \theta = \omega t) \qquad (9.15)$$

Velocity is maximum when the follower is at midway, i.e. when $\theta = \beta/2$.

Therefore,
$$v_{max} = \frac{4h\omega}{\beta^2} \times \frac{\beta}{2} = \frac{2h\omega}{\beta} \tag{9.16}$$

Similarly, the acceleration or deceleration

$$f = \frac{dv}{dt} = \frac{4h\omega^2}{\beta^2} \tag{9.17}$$

The follower response curves for constant acceleration and deceleration or parabolic motion is shown in Figure 9.9. It is found from the figure that during the first half of the follower rise velocity increases at constant rate and in the second half, it decreases at constant rate to zero. Further, at the beginning of the follower rise, midway and at the end of the follower motion, there are sudden jumps in acceleration which causes infinite jerks at these points.

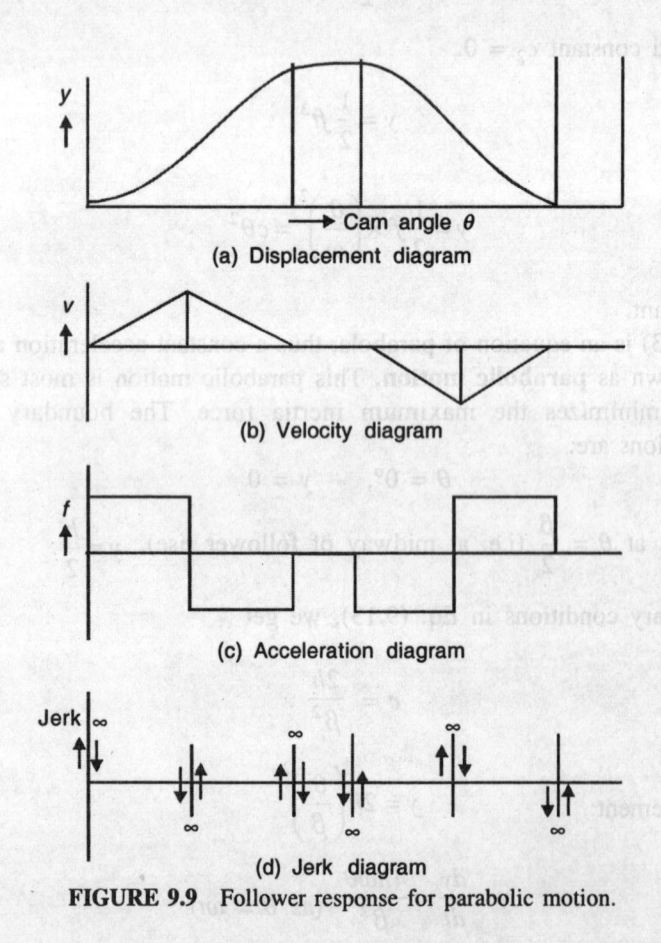

FIGURE 9.9 Follower response for parabolic motion.

9.4.4 Cycloidal Motion

A **cycloid** is the locus of a point on a circle which rolls on a straight line. Therefore, the cycloidal motion is obtained by rolling a circle of radius equal to $h/2\pi$ on the ordinate of the displacement diagram. The method of constructing displacement diagram is given below (see Figure 9.10):

1. Divide the cam displacement interval (cam rotation angle) into n equal parts.
2. For given maximum lift of the follower, draw a diagonal line.
3. Draw a circle of radius $h/2\pi$ at centre O and divide it into n equal parts.
4. Project these points on the circle to the vertical line OC.
5. Draw lines parallel to the diagonal line from these projected points to intersect the respective cam displacement angle.
6. Join these points with a smooth curve to produce follower displacement diagram.

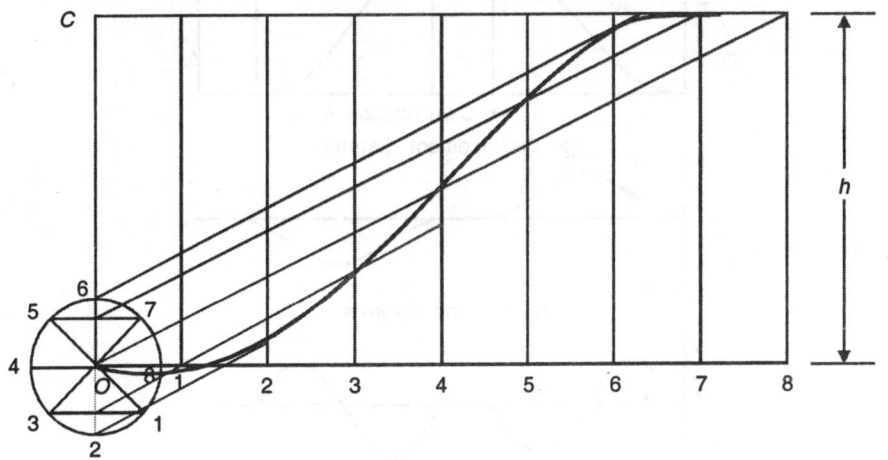

FIGURE 9.10 Displacement diagram for cycloidal motion.

Mathematically, the equation of cycloidal motion is given by

Displacement:
$$y = h\left[\frac{\theta}{\beta} - \frac{1}{2\pi} \times \sin\left(\frac{2\pi\theta}{\beta}\right)\right] \quad (9.18)$$

Velocity:
$$v = \frac{dy}{dt} = \frac{h\omega}{\beta}\left[1 - \cos\left(\frac{2\pi\theta}{\beta}\right)\right] \quad (9.19)$$

The maximum velocity occurs at $\theta = \beta/2$, which is:

$$v_{max} = \frac{2h\omega}{\beta} \quad (9.20)$$

Acceleration:
$$f = \frac{dv}{dt} = \frac{2\pi h\omega^2}{\beta^2} \times \sin\left(\frac{2\pi\theta}{\beta}\right) \quad (9.21)$$

and maximum acceleration which occurs at $\theta = \beta/4$ is:

$$f_{max} = \frac{2\pi h\omega^2}{\beta^2} \qquad (9.22)$$

Jerk:
$$J = \frac{df}{dt} = \frac{4\pi^2 h\omega^3}{\beta^3} \times \cos\left(\frac{2\pi\theta}{\beta}\right) \qquad (9.23)$$

The follower response curves for cycloidal motion is shown in Figure 9.11. It is found from the figure that there is no sudden changes in the velocity and acceleration at any instant. Thus for high speed applications, the cycloidal motion is the best of all the motions described earlier, if the accuracy of the profile is maintained.

(a) Displacement diagram

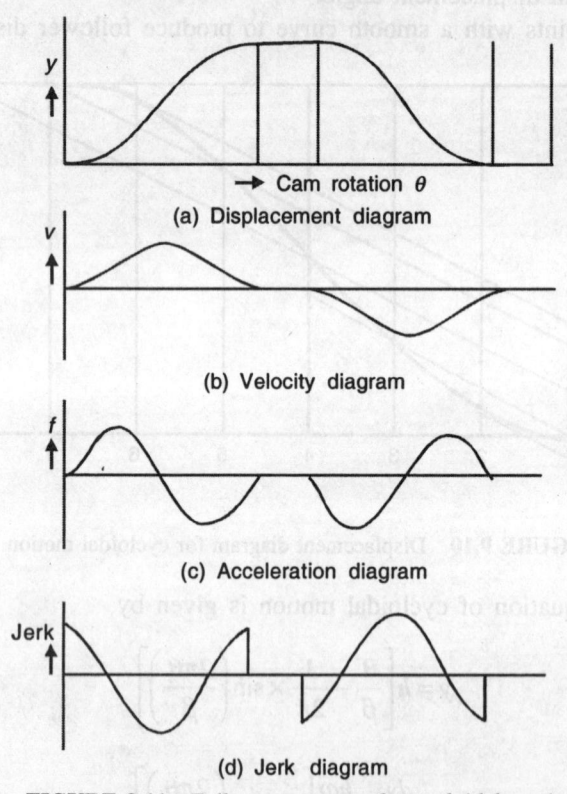

(b) Velocity diagram

(c) Acceleration diagram

(d) Jerk diagram

FIGURE 9.11 Follower response for cycloidal motion.

9.4.5 Polynomial Motion

From previous discussion about various types of follower motion, we found that for smooth functioning of cam-follower mechanism, there should be continuity in the displacement and velocity curves of the follower motion. Further, for high speed applications, the acceleration curve must be continuous and should have the lowest maximum value. The derivative of acceleration, called jerk, should have a finite value. Prof. Stoddart could achieve these characteristics by using polynomial equation for follower motion.

Consider a 3–4–5 polynomial full rise motion cam defined by a fifth degree polynomial equation:

$$y = c_0 + c_1\theta + c_2\theta^2 + c_3\theta^3 + c_4\theta^4 + c_5\theta^5 \qquad (9.24)$$

The above polynomial motion fulfill the following boundary conditions:

At $\theta = 0$, $y = 0$, $v = 0$ and $f = 0$

At $\theta = \beta$, $y = h$, $v = 0$ and $f = 0$ $\qquad (9.25)$

Substituting these boundary conditions in Eq. (9.24) and its derivatives and solving six simultaneous equations, we get the values of six constants which are as under:

$$c_0 = 0; \quad c_1 = 0; \quad c_2 = 0$$

$$c_3 = \frac{10h}{\beta^3}; \quad c_4 = -\frac{15h}{\beta^4}; \quad c_5 = \frac{6h}{\beta^5} \qquad (9.26)$$

Hence, the equation of displacement curve for 3–4–5 polynomial motion* is:

$$y = \frac{10h}{\beta^3}\theta^3 - \frac{15h}{\beta^4}\theta^4 + \frac{6h}{\beta^5}\theta^5 \qquad (9.27)$$

The derivatives of displacement equation with respect to time are:

Velocity:
$$v = \frac{dy}{dt} = \frac{30h\omega}{\beta^3}\theta^2 - \frac{60h\omega}{\beta^4}\theta^3 + \frac{30h\omega}{\beta^5}\theta^4 \qquad (9.28)$$

Acceleration:
$$f = \frac{dv}{dt}$$

or
$$f = \frac{60h\omega^2}{\beta^3}\theta - \frac{180h\omega^2}{\beta^4}\theta^2 + \frac{120h\omega^2}{\beta^5}\theta^3 \qquad (9.29)$$

Jerk:
$$J = \frac{df}{dt}$$

or
$$J = \frac{60h\omega^3}{\beta^3} - \frac{360h\omega^3}{\beta^4}\theta + \frac{360h\omega^3}{\beta^5}\theta^2 \qquad (9.30)$$

The graphical representation of displacement, velocity, acceleration and jerk for the period of rise of 3–4–5 cam follower is shown in Figure 9.12. It is found from the figure that acceleration and jerks are finite and hence suitable for high speed applications.

---

* Since the three terms of polynomial equation have powers of 3, 4 and 5, it is commonly referred to as 3–4–5 polynomial motion equation.

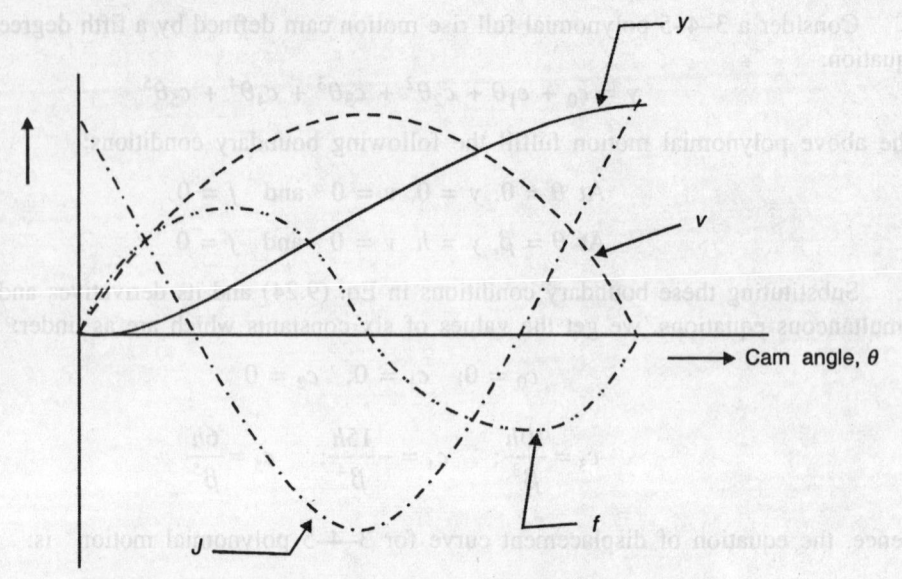

FIGURE 9.12 Follower response for 3–4–5 polynomial motion.

9.5 GRAPHICAL SYNTHESIS OF CAM PROFILE

The following general procedure is adopted for graphical synthesis of cam profile.

1. Draw the displacement diagram for the given type of follower motion as discussed in previous section.
2. Assume that cam remains stationary and the follower moves round it in the direction opposite to the direction of rotation of the cam.
3. In case of knife edge and flat faced followers, draw a base circle of given minimum radius. While in the case of roller follower, draw a pitch circle of radius equal to sum of base circle radius and roller radius.
4. Divide its circumference into a number of parts equal to the divisions used in displacement diagram.
5. Draw various positions of the follower corresponding to angular position of the cam.
6. Draw a smooth curve tangential to the contact surface in different positions of the follower.

The above procedure is the most general one. The detailed procedure is illustrated through some worked examples further.

EXAMPLE 9.1 Draw the profile of a cam operating a knife-edge follower from the following data:

(i) Follower move to rise through 40 mm during 90° rotation of the cam.
(ii) Follower to dwell for next 45° rotation of the cam.
(iii) Follower to return to its original position during next 120° rotation.
(iv) Follower to dwell for the remaining period.

The displacement of the follower is to take place with simple harmonic motion during both rise and return strokes. The least radius of the cam is 50 mm, If the cam rotates at 300 rpm, determine the maximum velocity and acceleration of the follower during rise and return stroke.

Solution Draw the displacement diagram as per the following procedure:

(i) Draw a semi-circle of 40 mm diameter and divide it into any number of equal parts, say in six parts by radial lines.
(ii) On the abscissa, take cam displacement for 90° rise, next 45° for dwell and 120° for return.
(iii) Through each division points on semi-circle, say 1, 2, ... 6, draw horizontal lines.
(iv) Next, divide the distance on abscissa 0–6 and 6′–0′ into six equal parts.
(v) Intercept points a, b, c, d, e, f and f', e', d', c', b' and a' for rise and return stroke.
(vi) Join these points with smooth curve to produce displacement diagram. See Figure 9.13(a).

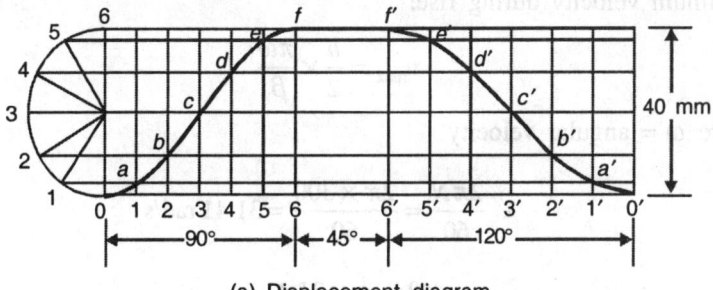

(a) Displacement diagram

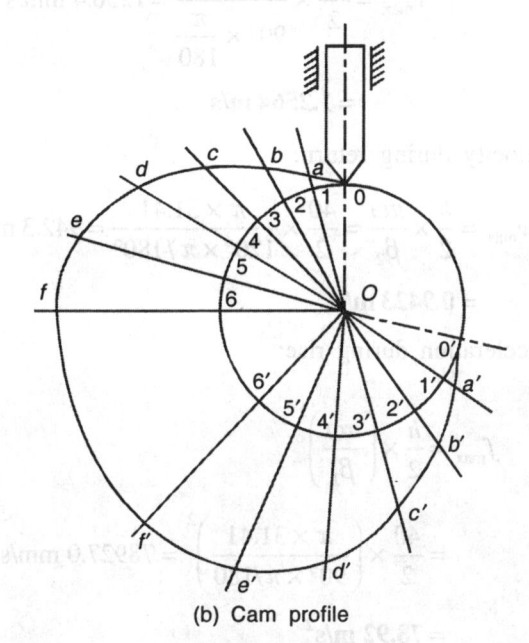

(b) Cam profile

FIGURE 9.13

Procedure to draw cam profile

(i) With O as centre and radius equal to 50 mm, draw circle with same scale chosen for displacement diagram.
(ii) With reference to vertical axis, mark the points on circle at 90°, 45° and 120° in order to represent rise, dwell and return strokes.
(iii) Divide the angle of rise into six equal parts (Same as that used in displacement diagram). Draw radial lines as shown in Figure 9.13(b).
(iv) Similarly, divide angle of return into six parts and draw radial lines.
(v) Above the least radius circle, mark the distances $1 - a$, $2 - b$, ... $6 - f$ corresponding to lift of the follower (taken from displacement diagram) on the radial lines drawn for angle of rise.
(vi) Similarly, mark distances $6' - f'$, $5' - e'$, ..., $1' - a'$ on radial lines of angle of return.
(vii) Join these points with smooth curve to generate a cam profile.

(i) Maximum velocity during rise:

$$v_{max} = \frac{h}{2} \times \frac{\pi \omega}{\beta_1}$$

where ω = angular velocity

$$= \frac{2\pi N}{60} = \frac{2\pi \times 300}{60} = 31.41 \text{ rad/s}$$

$\therefore$
$$v_{max} = \frac{40}{2} \times \frac{\pi \times 31.41}{90° \times \frac{\pi}{180}} = 1256.4 \text{ mm/s}$$

$$= 1.2564 \text{ m/s} \qquad \text{Ans.}$$

(ii) Maximum velocity during return:

$$v_{max} = \frac{h}{2} \times \frac{\pi \omega}{\beta_2} = \frac{40}{2} \times \frac{\pi \times 31.41}{120° \times \pi/180°} = 942.3 \text{ mm/s}$$

$$= 0.9423 \text{ m/s} \qquad \text{Ans.}$$

(iii) Maximum acceleration during rise:

$$f_{max} = \frac{h}{2} \times \left(\frac{\pi \omega}{\beta_1}\right)^2$$

$$= \frac{40}{2} \times \left(\frac{\pi \times 31.41}{90° \times \pi/180}\right)^2 = 78927.0 \text{ mm/s}^2$$

$$= 78.92 \text{ m/s}^2 \qquad \text{Ans}$$

(iv) Maximum acceleration during return:

$$f_{max} = \frac{h}{2} \times \left(\frac{\pi \omega}{\beta_2}\right)^2$$

$$= \frac{40}{2} \times \left(\frac{\pi \times 31.41}{120° \times \pi/180}\right)^2 = 44396.4 \text{ mm/s}^2$$

$$= 4439 \text{ m/s}^2 \qquad \text{Ans.}$$

EXAMPLE 9.2 A cam with 30 mm minimum radius is rotating clockwise at 1200 rpm to give the follower motion to a roller follower of 20 mm diameter.

  (i) Lift = 25 mm
  (ii) Follower rises during 120° cam rotation with simple harmonic motion
  (iii) Follower to dwell for 60° cam rotation
  (iv) Follower to return during 90° cam rotation with uniform acceleration and deceleration
  (v) Follower to dwell for remaining period.

Draw the profile of the cam and determine the maximum velocity and acceleration during rise and return stroke.

Solution Construct the displacement diagram as described in the following procedure. See Figure 9.14(a).

  (i) On the ordinate, draw a semi-circle of diameter equal to lift, i.e. 25 mm and divide it into any number of equal parts (say six) by radial lines.
  (ii) On the abscissa, take cam displacement 120° for rise, next 60° for dwell and next 90° for return.
  (iii) Divide the angle of rise into six equal parts.
  (iv) Through each division of semi-circle, draw horizontal lines up to angle of rise (ascent) period.
  (v) Mark points a, b, c, d, e and f as shown in the Figure 9.14(a) and join them with smooth curve.
  (vi) Now divide the angle of return (descent) into six equal parts and mark points on abscissa (say 6′, 5′, ..., 0′)
  (vii) Pick point 3′ on abscissa and divide vertical line on it into six equal parts.
  (viii) Join points 1, 2, 3 to point 0′ and 3, 4, 5 to point f'.
  (ix) Pick intersecting points marked as f', e', d', c', b', a' and join them by smooth curve to produce displacement diagram for angle of return.

Procedure to draw a cam profile

  (i) Draw a circle equal to radius of prime circle, i.e. sum of radii of the base circle and the roller circle (= 30 + 10 = 40 mm).
  (ii) With reference to vertical axis, mark the points on circle at 120°, 60° and 90° intervals to represent angle of rise, dwell and return strokes respectively.

(iii) Divide the angle of rise and return into six equal parts by drawing radial lines.
(iv) Above the prime circle, mark the distances 1 – a, 2 – b, 3 – c, 4 – d, 5 – e and 6 – f corresponding to lift of the follower, on the radial lines drawn for angle of rise.
(v) Similarly, mark the distance 6' – f', 5' – e', 4' – d', 3' – c', 2' – b', 1' – a' corresponding to position of the follower during the return stroke, on the radial lines drawn for angle of return.
(vi) Join these points to generate pitch curve of the cam profile. [Figure 9.14(b)]
(vii) From points a, b, c, ..., f and f', e', ..., a', draw arcs of radius equal to the radius of roller.
(viii) Draw a smooth curve which is tangent to these arcs to produce required cam profile.

(a) Displacement diagram

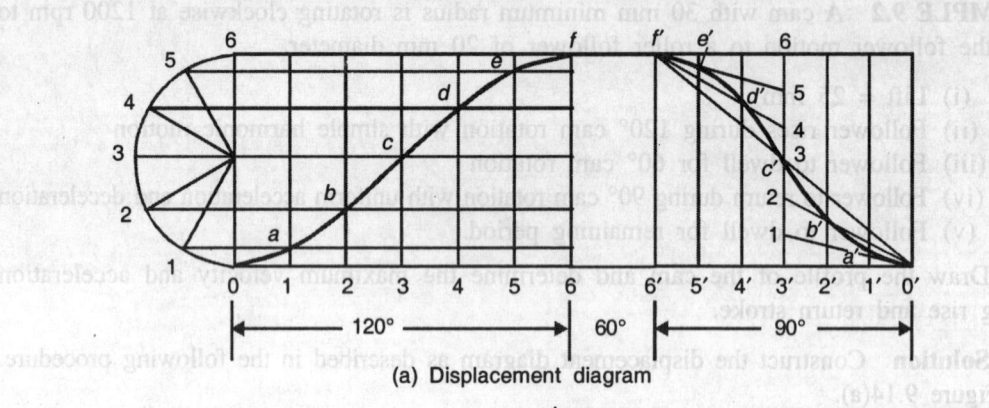

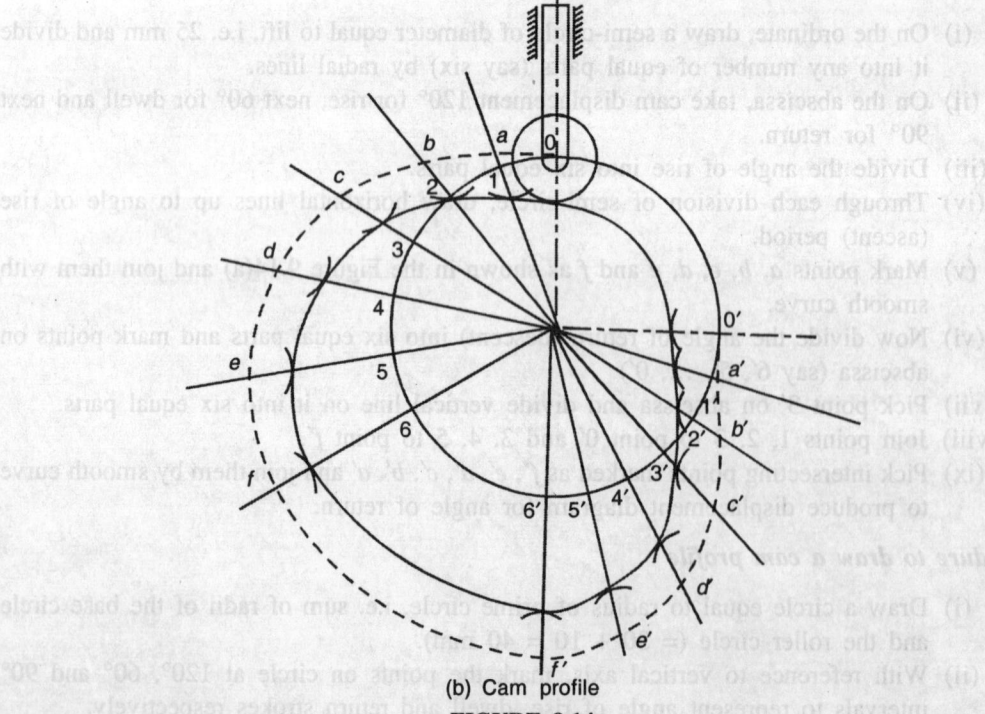

(b) Cam profile

FIGURE 9.14

Let ω = angular velocity

$$= \frac{2\pi N}{60} = \frac{2\pi \times 1200}{60} = 125.66 \text{ rad/s}$$

(a) Maximum velocity during the follower rise with SHM:

$$v_{max} = \frac{h}{2} \times \frac{\pi \omega}{\beta_1}$$

$$= \frac{25}{2 \times 1000} \times \frac{\pi \times 125.66}{120° \times \pi/180} = 2.35 \text{ m/s} \qquad \text{Ans.}$$

(b) Maximum velocity during the follower return with uniform acceleration and deceleration:

$$v_{max} = \frac{2h \times \omega}{\beta_2} = \frac{2 \times 25 \times 125.66}{1000 \times 90° \times \pi/180}$$

$$= 4.0 \text{ m/s} \qquad \text{Ans.}$$

(c) Maximum acceleration during the follower rise:

$$f_{max} = \frac{h}{2} \times \left(\frac{\pi \omega}{\beta_1}\right)^2 = \frac{25}{2 \times 1000} \times \left(\frac{\pi \times 125.66}{120° \times \pi/180}\right)^2$$

$$= 444.1 \text{ m/s}^2 \qquad \text{Ans.}$$

(d) Maximum acceleration during the follower return:

$$f_{max} = \frac{4h\omega^2}{\beta_2^2} = \frac{4 \times 25}{1000} \times \left(\frac{125.66}{90° \times \pi/180}\right)^2$$

$$= 639.9 \text{ m/s}^2 \qquad \text{Ans.}$$

EXAMPLE 9.3 A cam rotating clockwise with a uniform speed of 300 rpm is to give the roller follower of 15 mm diameter the following motion:

(i) Follower to rise through a distance of 45 mm during 120° cam rotation.
(ii) Follower to dwell for 60° of cam rotation.
(iii) Follower to return to its initial position during next 120° of cam rotation.
(iv) Follower to dwell for remaining period of 60° of cam rotation.

The minimum radius of the cam is 30 mm and the line of stroke of the follower is offset by 15 mm from the axis of the cam and the displacement of the follower is to take place with cycloidal motion on both rise and return strokes. Draw the cam profile and calculate maximum velocity and acceleration during the rise stroke.

Solution Construct the displacement diagram as described in the following procedure:

(i) Divide the cam displacement interval for rise and return into eight equal parts.
(ii) For a given maximum lift, draw a diagonal line for follower rise and return portion as shown in Figure 9.15(a).
(iii) Draw circle of radius $\dfrac{h}{2\pi}\left(=\dfrac{45}{2\pi}=7.16\text{ mm}\right)$ at point O for rise stroke.
(iv) Divide this circle into eight equal parts.
(v) Project these points to the vertical line OC.
(vi) Draw lines parallel to the diagonal line from these projection points to intersect the respective cam displacement angle lines.
(vii) Join these points with a smooth curve to produce follower displacement curve.
(viii) Repeat this procedure for follower stroke.

Procedure to draw a cam profile

(i) Draw a circle equal to radius of prime circle, i.e. sum of radii of base circle and roller circle (= 30 + 7.5 = 37.5 mm).
(ii) Draw another concentric circle of radius equal to offset distance.
(iii) Join OO_1 and divide the prime circle into four parts equal to angle of rise, dwell, angle of return and dwell.
(iv) Divide angle of rise and return into eight equal parts.
(v) Draw tangents to the offset circle from point 1, 2, 3, etc.
(vi) On the extension of tangent lines, mark distances $1 - a$, $1 - b$, ..., $8 - h$ from displacement diagram.
(vii) With a, b, c, etc. as centres, draw a number of arcs of radius equal to the roller radius.
(viii) Draw a smooth curve tangential to all the arcs.
(ix) Repeat this procedure for angle of return to obtain the required cam profile. [Figure 9.15(b)].

$$\text{Let } \omega = \text{angular velocity} = \frac{2\pi N}{60}$$

$$= \frac{2\pi \times 300}{60} = 31.4 \text{ rad/s}$$

(a) Maximum velocity of the follower during rise stroke:

$$v_{max} = \frac{2h\omega}{\beta_1}$$

$$= \frac{2 \times 45 \times 31.4}{1000 \times 120 \times \pi/180}$$

$$= 1.35 \text{ m/s} \hspace{2cm} \textbf{Ans.}$$

(a) Displacement diagram

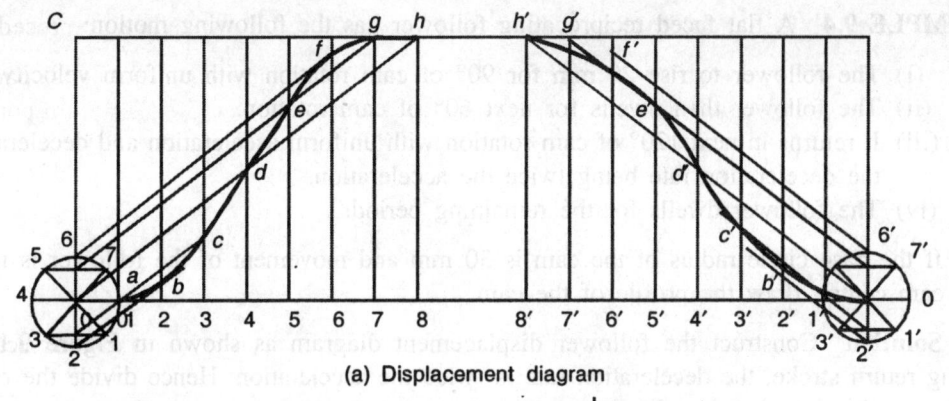

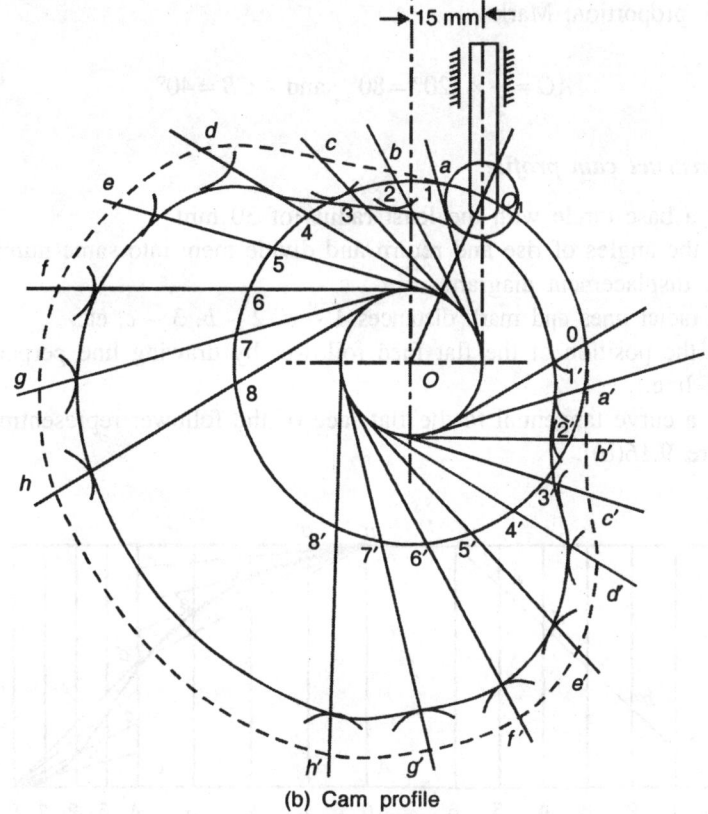

(b) Cam profile

FIGURE 9.15

(b) Maximum acceleration of the follower during rise stroke:

$$f_{max} = \frac{2\pi h \omega^2}{\beta_1^2}$$

$$= \frac{2\pi \times 45 \times 31.4^2}{1000 \times (120° \times \pi/180)^2}$$

$$= 63.55 \text{ m/s}^2 \qquad \text{Ans.}$$

EXAMPLE 9.4 A flat faced reciprocating follower has the following motion:
   (i) The follower to rise 20 mm for 90° of cam rotation with uniform velocity.
   (ii) The follower then dwells for next 60° of cam rotation.
   (iii) It returns in next 120° of cam rotation with uniform acceleration and deceleration, the deceleration rate being twice the acceleration.
   (iv) The follower dwells for the remaining period.

If the base circle radius of the cam is 30 mm and movement of the follower is inline with cam centre, draw the profile of the cam.

Solution Construct the follower displacement diagram as shown in Figure 9.16(a). During return stroke, the deceleration rate is twice the acceleration. Hence divide the return stroke into 2 : 1 proportion. Mark

$$AC = \frac{2}{3} \times 120° = 80° \quad \text{and} \quad CB = 40°$$

Procedure to construct cam profile
   (i) Draw a base circle with the least radius of 30 mm.
   (ii) Mark the angles of rise and return and divide them into same number of parts as in the displacement diagram.
   (iii) Draw radial lines and mark distances $1 - a$, $2 - b$, $3 - c$, etc.
   (iv) Draw the position of the flat face follower by drawing line perpendicular to the radial line.
   (v) Draw a curve tangential to the flat face of the follower representing cam profile. [Figure 9.16(b)].

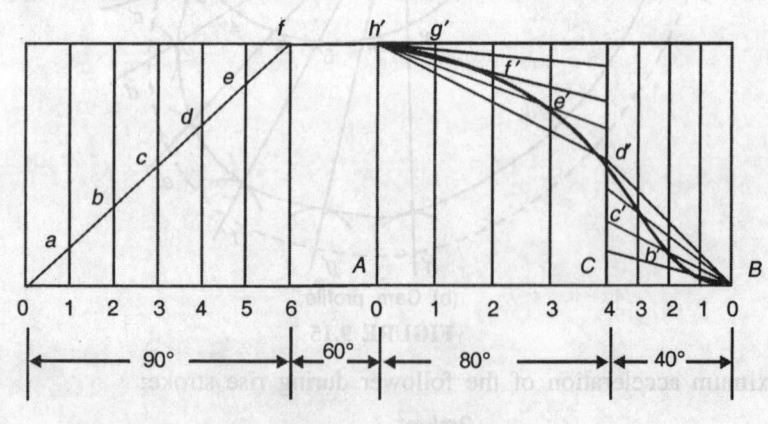

(a) Displacement diagram

(b) Cam profile

FIGURE 9.16

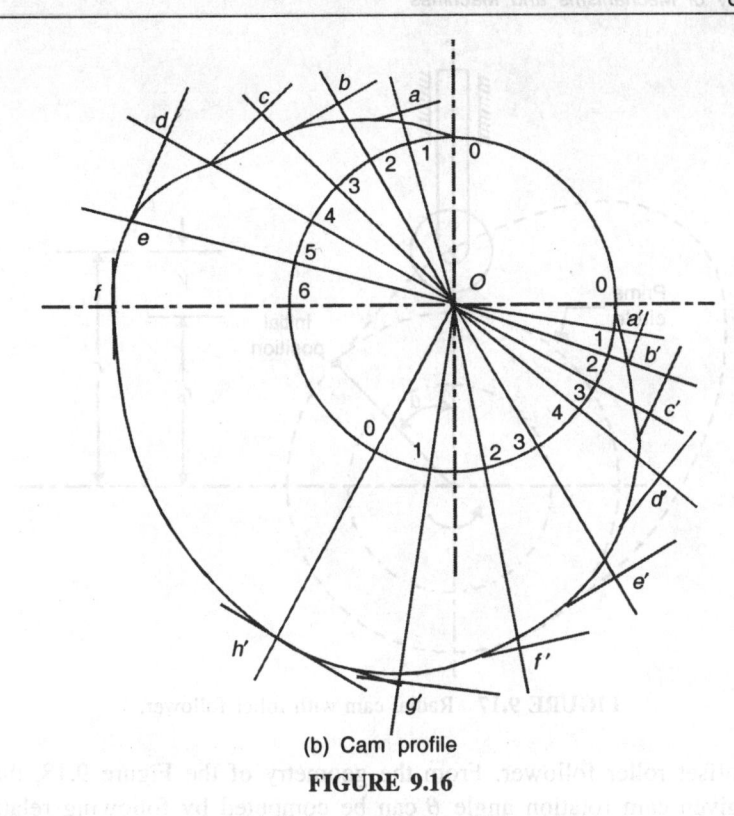

9.6 ANALYTICAL SYNTHESIS OF CAM PROFILE

If the displacement of the follower y is expressed as a function of cam rotation angle θ, i.e. $y = f(\theta)$, then the parametric equation of the cam profile can be obtained as discussed for the following cases:

Radial roller follower. Figure 9.17 shows the radial cam with a roller follower. The line of movement of the follower passes through the centre of rotation of the cam, so it is called radial roller follower. For every cam rotation angle θ the pitch curve radius r can be computed by the following equation:

$$r = r_p + y \qquad (9.31)$$

where

r_p = radius of prime circle (= $r_b + r_r$)

with r_b = base circle radius

r_r = radius of the roller

and y = follower displacement for a given type of motion at cam rotation angle θ (See Table 9.1.)

Offset roller follower. Figure 9.18 shows radial cam with an offset roller follower. The line of movement of the follower is at eccentricity e from the centre of rotation of the cam,

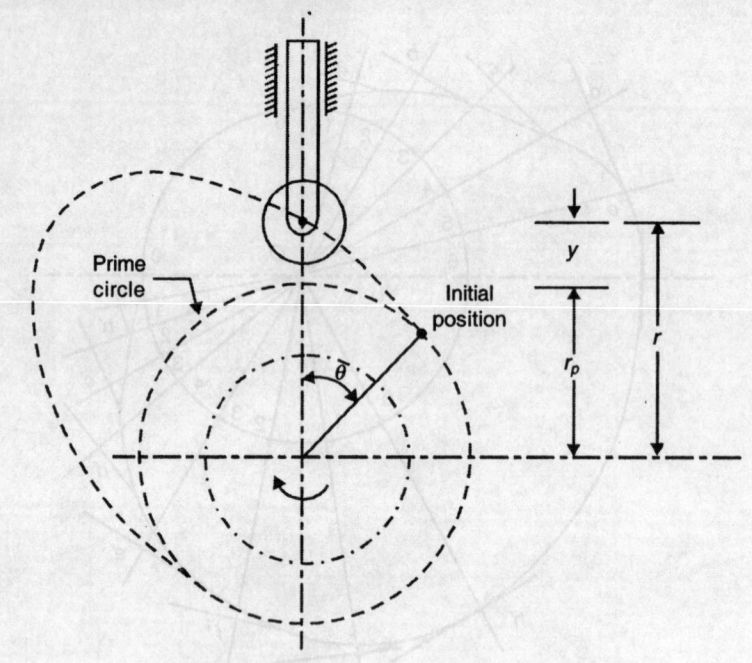

FIGURE 9.17 Radial cam with roller follower.

so it is called offset roller follower. From the geometry of the Figure 9.18, the pitch curve radius r for a given cam rotation angle θ can be computed by following relation:

$$r = [(y + y_0)^2 + e^2]^{0.5} \quad (9.32)$$

where y = follower displacement (See Table 9.1.)

and y_0 = distance labelled on the Figure 9.18 [$y_0 = (r_p^2 - e^2)^{0.5}$] (9.33)

Angles:
$$\cos(180° - \alpha) = \frac{e}{r} \quad (9.34)$$

$$\cos(180° - \beta) = \frac{e}{r_p} \quad (9.35)$$

Therefore, angle $\gamma = \beta - \alpha$ and inclination angle of pitch curve radius:

$$\phi = \theta - \gamma \quad (9.36)$$

Angles α, β, γ and θ are shown in Figure 9.18.

Using the above geometric relations, the coordinates of a point on the pitch curve of radial cam with offset follower can be determined.

Table 9.1 Characteristic equation of follower motions

Type of motion	Displacement	Velocity	Acceleration
1. SHM	$\dfrac{h}{2}\left[1-\cos\left(\dfrac{\pi\theta}{\beta}\right)\right]$	$\dfrac{h\pi\omega}{2\beta}\times\sin\left(\dfrac{\pi\theta}{\beta}\right)$	$\dfrac{h}{2}\times\left(\dfrac{\pi\omega}{\beta}\right)^2\times\cos\left(\dfrac{\pi\theta}{\beta}\right)$
2. Constant velocity	$\dfrac{h\theta}{\beta}$	$\dfrac{h\omega}{\beta}$	0
3. Constant acceleration and deceleration	$2h\left(\dfrac{\theta}{\beta}\right)^2$	$\dfrac{4h\omega\theta}{\beta^2}$	$\dfrac{4h\omega^2}{\beta^2}$
4. Cycloidal	$h\left[\dfrac{\theta}{\beta}-\dfrac{1}{2\pi}\times\sin\left(\dfrac{2\pi\theta}{\beta}\right)\right]$	$\dfrac{h\omega}{\beta}\left[1-\cos\left(\dfrac{2\pi\theta}{\beta}\right)\right]$	$\dfrac{2\pi h\omega^2}{\beta^2}\times\sin\left(\dfrac{2\pi\theta}{\beta}\right)$
5. 3–4–5 polynomial	$\dfrac{10h}{\beta^3}\theta^3-\dfrac{15h}{\beta^4}\theta^4+\dfrac{6h}{\beta^5}\theta^5$	$\dfrac{30h\omega}{\beta^3}\theta^2-\dfrac{60h\omega}{\beta^4}\theta^3+\dfrac{30h\omega}{\beta^5}\theta^4$	$\dfrac{60h\omega^2}{\beta^3}\theta-\dfrac{180h\omega^2}{\beta^4}\theta^2+\dfrac{120h\omega^2}{\beta^5}\theta^3$

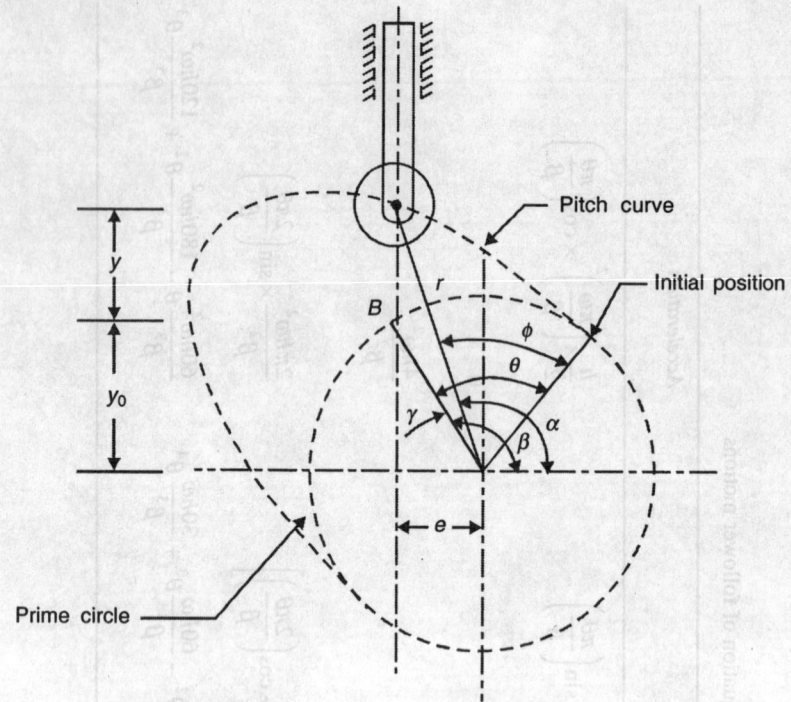

FIGURE 9.18 Radial cam with offset follower.

Flat faced follower. Figure 9.19 shows a radial cam with flat faced follower in which point O is the centre of cam, OB is the line of the follower motion and point C is the point of contact.

Let e be the eccentricity of point of contact from the cam centre. Since C is the point of contact, the component of velocity of point C treated as a point on the cam in the direction of the follower travel must be equal to the follower lift velocity.

Therefore,
$$\omega \times e = v = \frac{dy}{dt} \tag{9.37}$$

The radius of contact point r_c and respective angle ϕ represents the location of pitch curve. From the geometry of the Figure 9.19, we get

$$r_c = \left[(r_b + y)^2 + e^2 \right]^{0.5} \tag{9.38a}$$

and
$$\tan \alpha = \frac{e}{r_b + y} \tag{9.38b}$$

where r_b = radius of the base circle
and y = follower displacement (See Table 9.1)

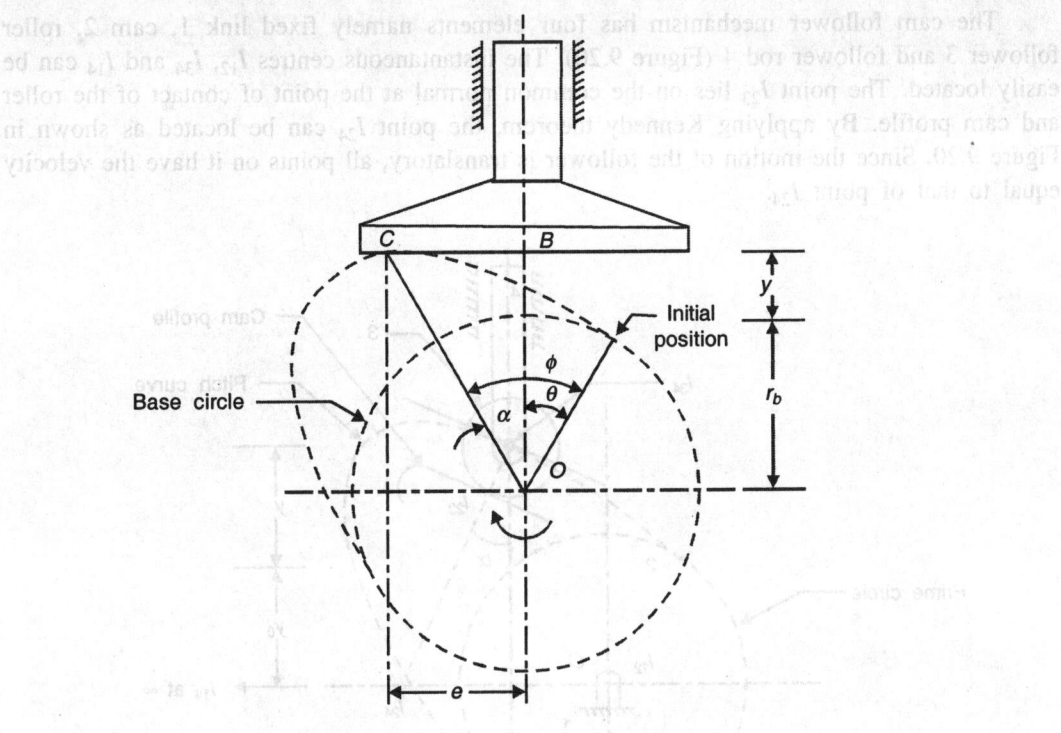

FIGURE 9.19 Radial cam with offset follower.

The inclination angle

(a) on the rise period of the follower motion:

$$\phi = \theta + \alpha \quad (9.39a)$$

(b) on the remaining period of the follower motion:

$$\phi = \theta - \alpha \quad (9.39b)$$

9.7 PRESSURE ANGLE

Pressure angle is the angle between the normal to the pitch curve at a point and the direction of follower motion. In other words, it measures the steepness of the cam profile. Pressure angle varies in magnitude at all instants of the follower motion. In an offsetted follower as shown in Figure 9.20, the pressure angle reduces during the follower rise but at the cost of increased pressure during the return stroke.

Let

α = pressure angle
r_p = prime circle radius
y = follower displacement which is function of cam rotation angle θ
ω = angular velocity

The cam follower mechanism has four elements namely fixed link 1, cam 2, roller follower 3 and follower rod 4 (Figure 9.20). The instantaneous centres I_{12}, I_{34} and I_{14} can be easily located. The point I_{23} lies on the common normal at the point of contact of the roller and cam profile. By applying Kennedy theorem, the point I_{24} can be located as shown in Figure 9.20. Since the motion of the follower is translatory, all points on it have the velocity equal to that of point I_{24}.

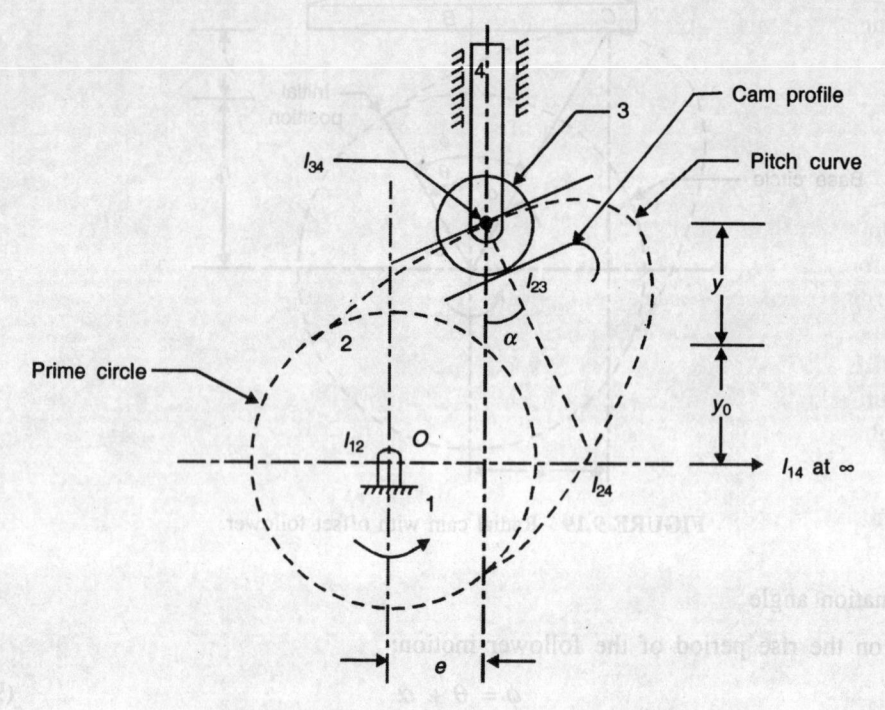

FIGURE 9.20 Determination of pressure angle.

Thus the velocity of follower.

$$\dot{y} = \frac{dy}{dx} = v = \omega \times OI_{24}$$

or
$$\dot{y} = \omega[e + (y_0 + y)\tan\alpha] \qquad (9.40)$$

where y_0 is distance labelled on Figure 9.20 ($= \sqrt{r_p^2 - e^2}$)

Therefore,
$$\tan\alpha = \frac{\dot{y}/\omega - e}{y + \sqrt{r_p^2 - e^2}} \qquad (9.41)$$

We know that
$$\dot{y} = \frac{dy}{dt} = \frac{dy}{d\theta} \cdot \frac{d\theta}{dt} = \frac{dy}{d\theta} \times \omega$$

Substituting the value of $\dot{y}$ in Eq. (9.41), we get the expression for pressure angle:

$$\tan \alpha = \frac{\dfrac{dy}{d\theta} - e}{y + \sqrt{r_p^2 - e^2}} \qquad (9.42)$$

and during the return period of follower:

$$\tan \alpha = \frac{\dfrac{dy}{d\theta} + e}{y + \sqrt{r_p^2 - e^2}} \qquad (9.43)$$

From Eqs. (9.42) and (9.43) we see that once the displacement equation y have been selected for a given follower motion, the prime circle radius r_p and eccentricity e can be selected for a suitable pressure angle.

EXAMPLE 9.5 A roller follower has cycloidal motion and its 30 mm lift is completed in 90° of cam rotation. The follower is offseted in a direction opposite to the direction of rotation by 6.5 mm and radius of the roller is 10 mm. Determine the base circle radius which would limit the pressure angle to 30°.

Solution Pressure angle is given by

$$\tan \alpha = \frac{\dfrac{dy}{d\theta} - e}{y + \sqrt{r_p^2 - e^2}}$$

where
r_p = radius of prime circle (= $r_1 + r_r$)
e = eccentricity (= 6.5 mm)
and ω = angular speed

For cycloidal motion,

$$y = h\left[\frac{\theta}{\beta_1} - \frac{1}{2\pi} \times \sin\left(\frac{2\pi\theta}{\beta_1}\right)\right]$$

or

$$\frac{dy}{d\theta} = h\left[\frac{1}{\beta_1} - \frac{1}{2\pi} \times \frac{2\pi}{\beta_1} \times \cos\left(\frac{2\pi\theta}{\beta_1}\right)\right]$$

$$= \frac{h}{\beta_1}\left[1 - \cos\left(\frac{2\pi\theta}{\beta_1}\right)\right]$$

The pressure angle α will be maximum when $dy/d\theta$ is maximum, i.e. when $\theta = \beta_1/2$ (= half of the angle of ascent).

$$\left(\frac{dy}{d\theta}\right)_{max} = \frac{h}{\beta_1}\left[1 - \cos\left(\frac{2\pi \times \beta_1}{\beta_1 \times 2}\right)\right] = \frac{2h}{\beta_1}$$

$$= \frac{2 \times 0.03}{90° \times \pi/180°} = 0.0382$$

$$y = \frac{b}{2} = \frac{30}{2} = 15 \text{ mm} \qquad (\because \theta = \beta_1/2)$$

and

$$\tan 30° = \frac{0.0382 - 0.00625}{0.015 + \sqrt{r_p^2 - 0.00625^2}}$$

or

$$r_p = 0.04 \text{ m}$$

Base circle radius:
$$r_1 = r_p - r_r$$
$$= 0.04 - 0.01$$
$$= 0.039 \text{ m} \quad \text{or} \quad \mathbf{39 \text{ mm}} \qquad \text{Ans.}$$

9.8 CAMS WITH SPECIFIED PROFILE

In Sections 9.5 and 9.6, synthesis of cam profile for a given type of follower motion has been performed by both graphical and analytical methods. However, there are some applications where a cam with specified profile is used with a follower. For these cases, one need to determine the displacement, velocity and acceleration characteristics of the follower. In this section, we shall deal with the following three types of cam profile:

(i) Tangent cam with roller follower
(ii) Circular cam with roller follower
(iii) Circular cam with flat faced follower.

9.8.1 Tangent Cam with Roller Follower

A tangent cam is symmetrical about the centre line. It has straight flanks and circular nose. Let the centre of cam is at O_1 and that of the nose at O_2. The straight flank commences from point A and continues till it meets arc of the nose radius at point E (Figure 9.21).
Let
  r = distance between the cam and nose centres
  r_1 = base circle radius or least radius
  r_2 = nose radius
  r_r = radius of roller
  l = sum of nose radius and roller radius (= $r_2 + r_r$)
and β = angle of ascent or rise.

Roller in contact with flank. Consider a tangent cam with roller follower in contact with straight flank. The centre of the roller lies at point C on the pitch curve as shown in Figure 9.21. Let angle turned by the cam from its original position, i.e. beginning of the

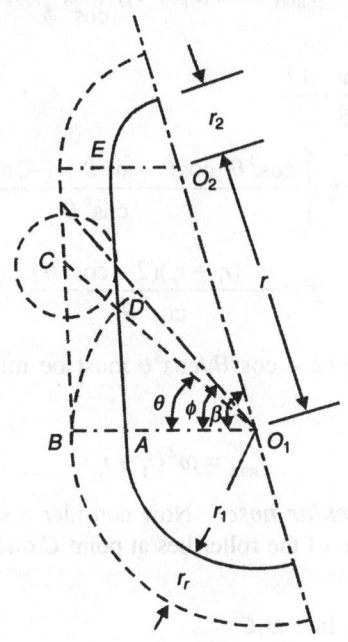

FIGURE 9.21 Tangent cam with roller follower in contact with straight flank.

follower motion is $\theta°$. The follower changes its position from B to C. Thus the lift of the follower is:

$$x = O_1C - O_1D$$

or

$$x = \frac{O_1B}{\cos\theta} - O_1B \qquad (\because\ O_1D = O_1B)$$

or

$$x = (r_1 + r_r)\left(\frac{1}{\cos\theta} - 1\right) \qquad (9.44)$$

Velocity:

$$v = \frac{dx}{dt} = \frac{dx}{d\theta} \cdot \frac{d\theta}{dt}$$

$$= (r_1 + r_r) \times \frac{\sin\theta}{\cos^2\theta} \cdot \omega$$

or

$$v = \omega(r_1 + r_r)\frac{\sin\theta}{\cos^2\theta} \qquad (9.45)$$

In Eq. (9.45) as cam rotation angle θ increases, $\sin\theta$ inreases while $\cos^2\theta$ decreases. This means velocity of the follower increases. The maximum velocity will occur at a point where roller leaves the contact with straight flank.

364 *Theory of Mechanisms and Machines*

Let ϕ be the angle turned by the cam when the roller leaves the contact with straight flank.

Therefore,
$$v_{max} = \omega(r_1 + r_r)\frac{\sin\phi}{\cos^2\phi} \tag{9.46}$$

Acceleration:
$$f = \frac{dv}{dt} = \frac{dv}{d\theta}\cdot\frac{d\theta}{dt}$$

$$= \omega^2(r_1 + r_r)\left[\frac{\cos^2\theta\cdot\cos\theta - \sin\theta\times(-2\sin\theta\cos\theta)}{\cos^4\theta}\right]$$

or
$$f = \frac{\omega^2(r_1 + r_r)(2 - \cos^2\theta)}{\cos^3\theta} \tag{9.47}$$

For minimum acceleration $(2 - \cos^2\theta)/\cos^3\theta$ must be minimum, which can be obtained if angle $\theta = 0°$. Therefore,

Minimum acceleration:
$$f_{min} = \omega^2(r_1 + r_r) \tag{9.48}$$

Roller in contact with circular nose. Now consider a second case when roller follower is in contact with nose. The centre of the roller lies at point C on the pitch curve of nose as shown in Figure 9.22.

Let O_2E = perpendicular on the line O_1C
β = angle of ascent (rise)
θ = angle of cam rotation
and ϕ = angle $(\beta - \theta)$

The displacement of the follower is given as (See Figure 9.22)

$$x = O_1C - O_1D$$
$$= (O_1E + EC) - O_1D$$
$$= \{r\cos\phi + (r_2 + r_r)\cos\gamma\} - (r_1 + r_r) \tag{9.49}$$

From the geometry of the Figure 9.22, we know that

$$r\sin\phi = (r_2 + r_r)\sin\gamma$$

$\therefore$
$$\sin\gamma = \left(\frac{r}{r_2 + r_r}\right)\sin\phi = \frac{\sin\phi}{n}$$

where
$$n = \frac{r_2 + r_r}{r}$$

$\therefore$
$$\cos\gamma = \sqrt{1 - \left(\frac{\sin\phi}{n}\right)^2}$$

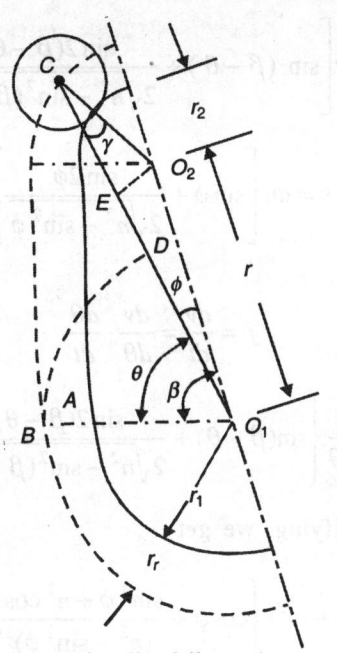

FIGURE 9.22 Tangent cam with roller follower in contact with nose of the cam.

Substituting the value of $\cos\gamma$ in Eq. (9.49), we get

$$x = r\cos\phi + (r_2 + r_r) \times \sqrt{1 - \left(\frac{\sin\phi}{n}\right)^2} - (r_1 + r_r)$$

After simplifying, the equation of displacement is:

$$x = r\left\{\cos\phi + \sqrt{n^2 - \sin^2\phi}\right\} - (r_1 + r_r) \tag{9.50}$$

Substituting $\phi = \beta - \theta$ in Eq. (9.50), we get

$$x = r\left\{\cos(\beta - \theta) + \sqrt{n^2 - \sin^2(\beta - \theta)}\right\} - (r_1 + r_r) \tag{9.51}$$

Velocity: $v = \dfrac{dx}{dt} = \dfrac{dx}{d\theta} \cdot \dfrac{d\theta}{dt}$

$$= \omega r\left[\frac{d}{d\theta}\left\{\cos(\beta - \theta) + \sqrt{n^2 - \sin^2(\beta - \theta)} - (r_1 + r_r)\right\}\right]$$

$$= \omega r\left[-\sin(\beta - \theta) \times -1 + \frac{1}{2\sqrt{n^2 - \sin^2(\beta - \theta)}} \times -2\sin(\beta - \theta) \times \cos(\beta - \theta) \times (-1)\right]$$

or
$$v = \omega r \left[\sin(\beta - \theta) + \frac{\sin 2(\beta - \theta)}{2\sqrt{n^2 - \sin^2(\beta - \theta)}} \right] \quad (9.52)$$

or
$$v = \omega r \left[\sin \phi + \frac{\sin 2\phi}{2\sqrt{n^2 - \sin^2 \phi}} \right] \quad (9.53)$$

Acceleration:

$$f = \frac{dv}{dt} = \frac{dv}{d\theta} \cdot \frac{d\theta}{dt}$$

or
$$f = \omega r \left[\frac{d}{d\theta} \left\{ \sin(\beta - \theta) + \frac{\sin 2(\beta - \theta)}{2\sqrt{n^2 - \sin^2(\beta - \theta)}} \right\} \right] \times \omega$$

After differentiating and simplifying, we get

$$f = -\omega^2 r \left\{ \cos \phi + \frac{\sin^4 \phi + n^2 \cos 2\phi}{(n^2 - \sin^2 \phi)^{3/2}} \right\} \quad (9.54)$$

The graphical representation of follower response for a tangent cam during angle of ascent is shown in Figure 9.23. It is very interesting to note from the figure that the values of maximum acceleration and maximum deceleration occurs at the point which lies on the flank as well as the nose.

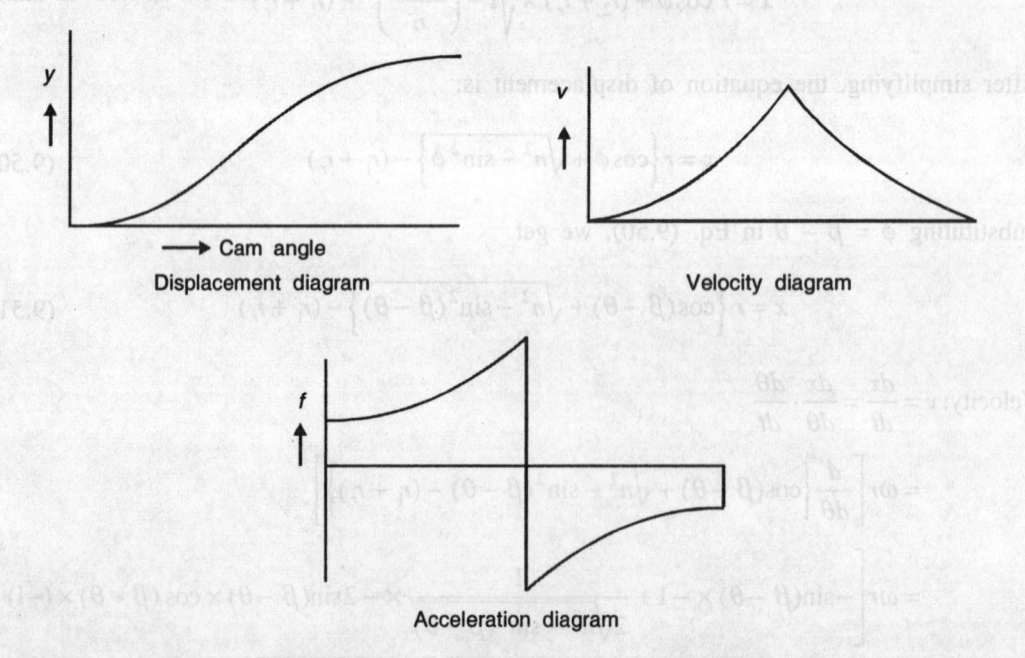

FIGURE 9.23 Follower response for tangent cam.

EXAMPLE 9.6 In a symmetrical tangent cam operating a roller follower, the least radius of the cam is 30 mm and the roller radius is 15 mm. The angle of ascent is 75° and the total lift is 20 mm. The speed of the cam is 600 rpm. Calculate

(i) the principal dimensions of the cam
(ii) the acceleration of the follower at the beginning of lift, where straight flank merges into the circular nose and at the apex of the nose.

Solution

(i) $\beta = 75°$, $r_1 = 30$ mm, $r_r = 15$ mm and lift = 20 mm. From the geometry of Figure 9.24,

$$r + r_2 = r_1 + \text{lift}$$

or
$$r = r_1 + \text{lift} - r_2$$
$$= 30 + 20 - r_2$$
$$= 50 - r_2$$

In $\triangle O_1 E O_2$,
$$\cos \beta = \frac{O_1 E}{O_1 O_2}$$

$$\cos 75° = \frac{r_1 - r_2}{50 - r_2}$$

or
$$30 - r_2 = 0.2588(50 - r_2)$$

or
nose radius: $r_2 = 23.0$ mm **Ans.**

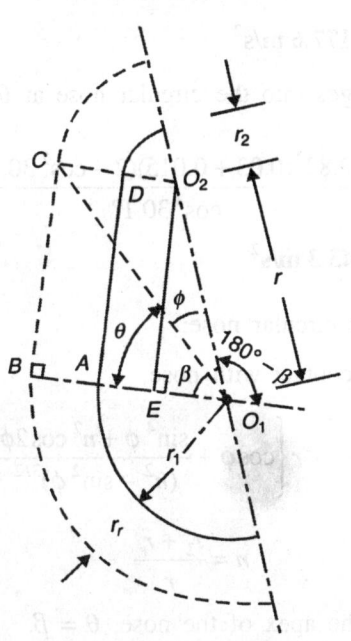

FIGURE 9.24

Distance between the centres:

$$r = 50 - r_2 = 50 - 23 = 27 \text{ mm}$$ **Ans.**

From $\Delta O_1 BC$, $\tan\theta = \dfrac{BC}{O_1 B}$

or $\tan\theta = \dfrac{r\sin\beta}{r_1 + r_r}$

$$= \dfrac{27 \times \sin 75°}{30 + 15} = 0.5795$$

or $\theta = 30.1°$ **Ans.**

(ii) Acceleration:
 (a) At the beginning of lift, $\theta = 0°$

$$f = \dfrac{\omega^2(r_1 + r_r)(2 - \cos^2\theta)}{\cos^3\theta}$$

where

$$\omega = \dfrac{2\pi N}{60} = \dfrac{2\pi \times 600}{60} = 62.83 \text{ rad/s}$$

$$f = \dfrac{62.83^2(0.03 + 0.015)(2 - \cos^2 0°)}{\cos^3 0°}$$

$$= 177.6 \text{ m/s}^2$$ **Ans.**

(b) Straight flank merges into the circular nose at $\theta = 30.1°$.

$$f = \dfrac{62.83^2(0.03 + 0.015)(2 - \cos^2 30.1°)}{\cos^3 30.1°}$$

$$= 343.3 \text{ m/s}^2$$ **Ans.**

(c) At the apex of the circular nose:

When roller is in contact with nose,

$$f = -\omega^2 r \left\{ \cos\phi + \dfrac{\sin^4\phi + n^2 \cos 2\phi}{(n^2 - \sin^2\phi)^{3/2}} \right\}$$

where $n = \dfrac{r_2 + r_r}{r}$

when roller is at the apex of the nose, $\theta = \beta$.

Therefore,
$$\phi = \beta - \theta = 0$$

$$f = -\omega^2 r\left(1 + \frac{1}{n}\right)$$

$$n = \frac{23 + 15}{27} = 1.407$$

∴ Acceleration: $f = -62.83^2 \times 0.027\left(1 + \frac{1}{1.407}\right)$

$$= -182.34 \text{ m/s}^2 \qquad \text{Ans.}$$

EXAMPLE 9.7 A symmetrical cam has a base circle 60 mm radius, arc of action 110°, straight flanks and tip is a circular arc. The line of action of the follower passes through the centre line of the cam shaft. The follower which has 40 mm diameter roller has a lift of 26 mm. Calculate the velocity and acceleration of the follower when moving outward and contact is just reaching the end of the straight flank. The cam rotates at 500 rpm.

Solution Arc of action is 110° so angle of ascent, being symmetrical cam, should equal to $\frac{110°}{2} = 55°$. Referring to Figure 9.25,

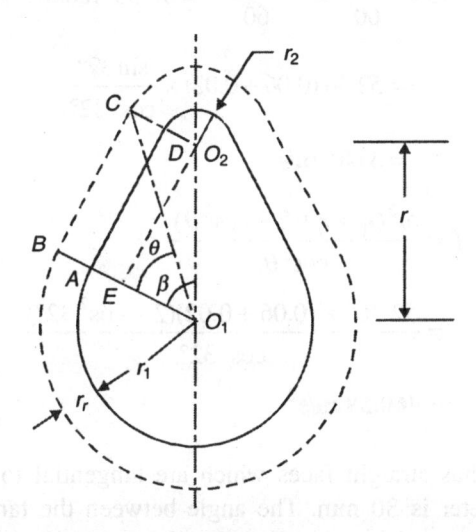

FIGURE 9.25

$$r = r_1 + \text{lift} - r_2$$
$$= 60 + 26 - r_2$$
$$= 86 - r_2$$

In $\triangle O_1EO_2$,
$$\cos\beta = \frac{O_1E}{O_1O_2}$$

$$\cos 55° = \frac{r_1 - r_2}{86 - r_2}$$

or $\quad 60 - r_2 = (86 - r_2) \times \cos 55°$

or $\quad r_2 = 25.0$ mm

Centre distance = $O_1O_2 = r = 86 - 25 = 61$ mm

When contact is just reaching at the end of straight flank, i.e. at point C, we have from $\triangle O_1BC$,

$$\tan\theta = \frac{BC}{O_1B} = \frac{r\sin\beta}{r_1 + r_r} = \frac{61 \times \sin 55°}{60 + 20} = 0.6246$$

or $\quad \theta = 32°$

(i) **Velocity:** $\quad v = \omega(r_1 + r_r)\dfrac{\sin\theta}{\cos^2\theta}$

where $\quad \omega = \dfrac{2\pi N}{60} = \dfrac{2\pi \times 500}{60} = 52.36$ rad/s

$$v = 52.36 (0.06 + 0.02) \times \frac{\sin 32°}{\cos^2 32°}$$

$= 3.086$ m/s **Ans.**

(ii) **Acceleration:** $\quad f = \dfrac{\omega^2(r_1 + r_r)(2 - \cos^2\theta)}{\cos^3\theta}$

$$= \frac{52.36^2 \times (0.06 + 0.02)(2 - \cos^2 32°)}{\cos^3 32°}$$

$= 460.59$ m/s^2 **Ans.**

EXAMPLE 9.8 A cam has straight faces which are tangential to a base circle of 70 mm diameter. The roller diameter is 30 mm. The angle between the tangential faces of the cam is 90° and these faces are joined by a nose circle of 8 mm radius. The speed of the cam is 120 rpm. Find the acceleration of the roller centre, when the roller.

(i) is just about to leave the straight face
(ii) is at the apex of the nose.

Solution Referring to Figure 9.26, angle between the faces is 90°. Therefore the angle of each face with axis = 90°/2 = 45°.

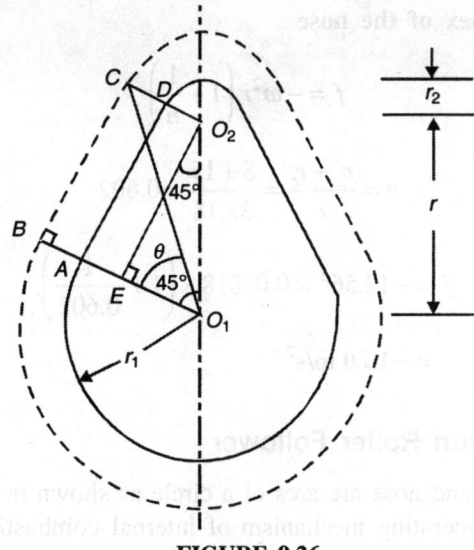

FIGURE 9.26

In $\triangle O_1EO_2$, angle of ascent:
$$\beta = 180° - 90° - 45°$$
$$= 45°$$

$$\cos 45° = \frac{O_1E}{O_1O_2} = \frac{r_1 - r_2}{r}$$

or
$$r = \frac{r_1 - r_2}{\cos 45°} = \frac{35 - 8}{\cos 45°} = 38.18 \text{ mm}$$

(i) Roller at straight face

In $\triangle O_1BC$,
$$\tan\theta = \frac{BC}{O_1B} = \frac{r \sin\beta}{r_1 + r_r} = \frac{38.18 \times \sin 45°}{35 + 15} = 0.54$$

or
$$\theta = 28.37°$$

Acceleration:
$$f = \frac{\omega^2(r_1 + r_r)(2 - \cos^2\theta)}{\cos^3\theta}$$

where
$$\omega = \frac{2\pi N}{60} = \frac{2\pi \times 120}{60} = 12.56 \text{ rad/s}$$

$\therefore$
$$f = \frac{12.56^2 \times (35 + 15)(2 - \cos^2 28.37°)}{1000 \times \cos^3 28.37°}$$

$$= 14.19 \text{ m/s}^2 \qquad \text{Ans.}$$

(ii) Roller at the apex of the nose

Acceleration: $$f = -\omega^2 r \left(1 + \frac{1}{n}\right)$$

where $$n = \frac{r_2 + r_r}{r} = \frac{8 + 15}{38.18} = 0.602$$

$\therefore$ $$f = -12.56^2 \times 0.03818 \times \left(1 + \frac{1}{0.602}\right)$$

$$= -16.0 \text{ m/s}^2 \qquad \text{Ans.}$$

9.8.2 Circular Cam with Roller Follower

In circular cams, the flank and nose are arcs of a circle as shown in Figure 9.27. These cams are widely used in valve operating mechanism of internal combustion engine.

r_f = radius of circular flank
r_1 = least radius of cam
r_2 = nose radius
r_r = radius of roller
β = angle of ascent
and θ = cam rotation angle at any instant.

Roller in contact with flank. Consider a circular cam having radius of flank r_f equal to pf with centre P. (See Figure 9.27.) The cam is in contact with flank at point F with roller centre at C, when the cam rotates by angle θ from initial position B.

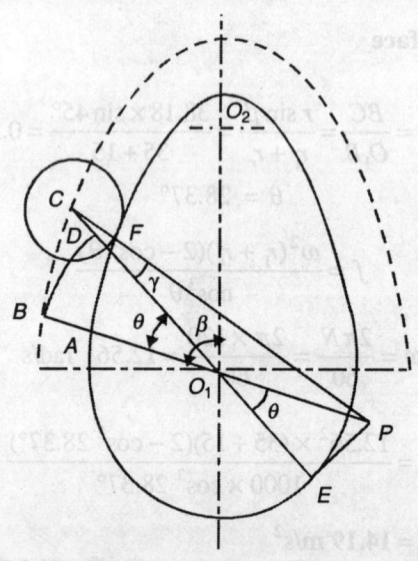

FIGURE 9.27 Circular cam with roller in contact at flank.

The lift of the follower:

$$x = O_1C - O_1D$$
$$= (EC - O_1E) - O_1B$$
$$= (CP\cos\gamma - O_1P\cos\theta) - (r_1 + r_r)$$
$$= (r_f + r_r)\cos\gamma - (AP - AO_1)\cos\theta - (r_1 + r_r)$$

or
$$x = (r_f + r_r)\cos\gamma - (r_f - r_1)\cos\theta - (r_1 + r_r) \quad (9.55)$$

where
$$\cos\gamma = \sqrt{1 - \sin^2\gamma}$$
$$= \sqrt{1 - (EP/CP)^2}$$
$$= \sqrt{1 - \left(\frac{O_1P\sin\theta}{CP}\right)^2}$$
$$= \sqrt{1 - \left[\frac{(r_f - r_1)\sin\theta}{r_f + r_r}\right]^2}$$

Let $r_f - r_1 = a$ and $r_f + r_r = b$.

$\therefore$
$$\cos\gamma = \frac{1}{b}\sqrt{b^2 - a^2\sin^2\theta}$$

Substituting the value of $\cos\gamma$ in Eq. (9.55), we get the expression for follower displacement as

$$x = \sqrt{b^2 - a^2\sin^2\theta} - a\cos\theta - (r_1 + r_r) \quad (9.56)$$

Follower velocity:
$$v = \frac{dx}{dt} = \frac{dx}{d\theta} \cdot \frac{d\theta}{dt}$$

or
$$v = \frac{d}{d\theta}\left[\sqrt{b^2 - a^2\sin^2\theta} - a\cos\theta - (r_1 + r_r)\right]\omega$$

$$= \left[-\frac{a^2 \times 2\sin\theta\cos\theta}{2\sqrt{b^2 - a^2\sin^2\theta}} + a\sin\theta\right]\omega$$

or
$$v = \omega a\left[\sin\theta - \frac{a\sin 2\theta}{2\sqrt{b^2 - a^2\sin^2\theta}}\right] \quad (9.57)$$

Acceleration:
$$f = \frac{dv}{dt} = \frac{dv}{d\theta} \cdot \frac{d\theta}{dt}$$

or
$$f = \frac{d}{d\theta}\left\{\omega a \times \left[\sin\theta - \frac{a\sin 2\theta}{2\sqrt{b^2 - a^2\sin^2\theta}}\right]\right\} \times \omega$$

or
$$f = \omega^2 a\left[\cos\theta - \frac{a\cos 2\theta}{\sqrt{b^2 - a^2\sin^2\theta}} - \frac{a^3 \sin 2\theta}{4(b^2 - a^2\sin^2\theta)^{3/2}}\right] \quad (9.58)$$

Roller in contact with nose. This case is similar to that of tangent cam when roller is in contact with the nose. The displacement, velocity and acceleration of the follower can be computed from Eqs. (9.50), (9.53) and (9.54) respectively.

9.8.3 Circular Cam with Flat Faced Follower

Figure 9.28 shows a circular cam with flat faced follower in contact at point E. Let P be the centre of circular flank. The radius of flank arc is r_f ($= PE = PA$).

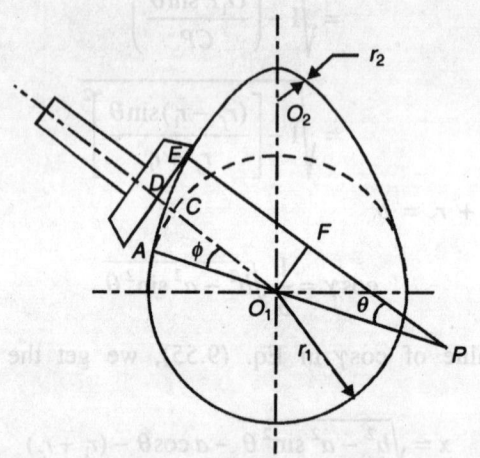

FIGURE 9.28 Circular cam with flat faced follower at the flank.

Follower at the flank. Referring to Figure 9.28, the lift of the follower is:

$$x = \text{distance } CD$$
$$= O_1 D - O_1 C$$
$$= FE - O_1 C$$
$$= (PE - PF) - O_1 C$$
$$= r_f - O_1 P \cos\theta - r_1$$
$$= r_f - (AP - AO_1)\cos\theta - r_1$$

or
$$x = r_f - (r_f - r_1)\cos\theta - r_1$$

or
$$x = (r_f - r_1)(1 - \cos\theta) \quad (9.59)$$

Velocity of the follower:
$$v = \frac{dx}{dt} = \frac{dx}{d\theta} \cdot \frac{d\theta}{dt}$$

or
$$v = [(r_f - r_1)\sin\theta] \times \omega$$

or
$$v = \omega(r_f - r_1)\sin\theta \quad (9.60)$$

The follower velocity increases with cam rotation angle θ and is maximum when angle $\theta = \beta$, i.e. when it leaves the flank.

$$v_{max} = \omega(r_f - r_1)\sin\beta \quad (9.61)$$

Acceleration:
$$f = \frac{dv}{dt} = \frac{dv}{d\theta} \cdot \frac{d\theta}{dt}$$

or
$$f = \omega^2(r_f - r_1)\cos\theta \quad (9.62)$$

The follower acceleration will be maximum at $\theta = 0°$, i.e. when the rise starts.

Follower in contact with nose. When follower is in contact with nose at point E, the lift of the follower is (see Figure 9.29):

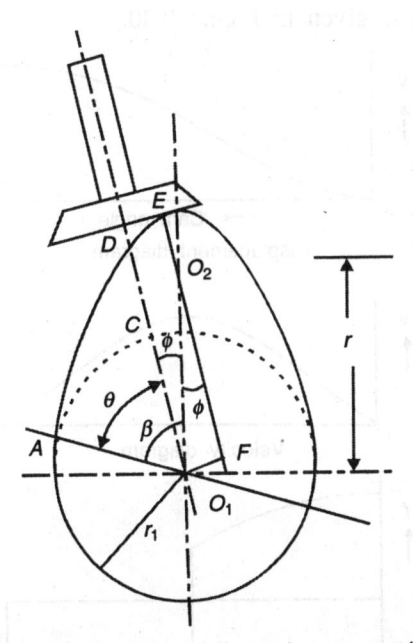

FIGURE 9.29 Flat-faced follower in contact with nose.

$$x = O_1D - O_1C$$
$$= EF - O_1C$$
$$= (EO_2 + O_2F) - O_1C$$
$$= EO_2 + r\cos(\beta - \theta) - O_1C$$
$$= r_2 + r\cos(\beta - \theta) - r_1$$

or
$$x = (r_2 - r_1) + r\cos(\beta - \theta) \tag{9.63}$$

Velocity: $$v = \frac{dx}{dt} = \frac{dx}{d\theta} \cdot \frac{d\theta}{dt}$$

or
$$v = \omega r \sin(\beta - \theta) \tag{9.64}$$

When the follower just touches the nose of the cam, velocity is maximum and when it is at the apex of the nose, velocity is minimum.

Acceleration: $$f = \frac{dv}{dt} = \frac{dv}{d\theta} \cdot \frac{d\theta}{dt}$$

or
$$f = -\omega^2 r \cos(\beta - \theta) \tag{9.65}$$

The minus sign indicates that it is retardation which is maximum when the follower is at the apex and minimum at the starting of the nose travel.

The follower response is given in Figure 9.30.

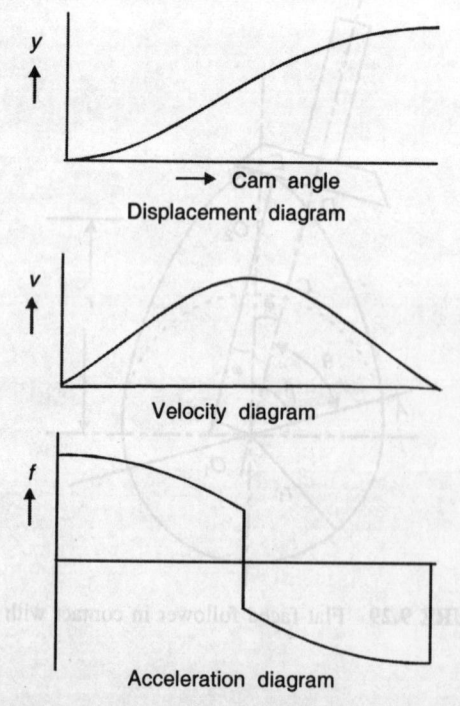

FIGURE 9.30 Follower response for the circular cam with flat-faced follower.

EXAMPLE 9.9 A symmetrical circular cam operating a flat faced follower has the following particulars:

Minimum radius	= 30 mm
Lift h	= 20 mm
Angle of ascent β	= 75°
Nose radius	= 5 mm
Speed	= 600 rpm

Find:
 (i) the principal dimensions of the cam
 (ii) the acceleration of the follower at the beginning of the lift, at the end of contact with circular flank, at the beginning of contact with nose and at the apex of the nose.

Solution

(i) From the geometry of Figure 9.31, we know that

$$r + r_2 = r_1 + h$$

or
$$r = r_1 + h - r_2$$
$$= 30 + 20 - 5$$
$$= 45 \text{ mm} \qquad \text{Ans.}$$

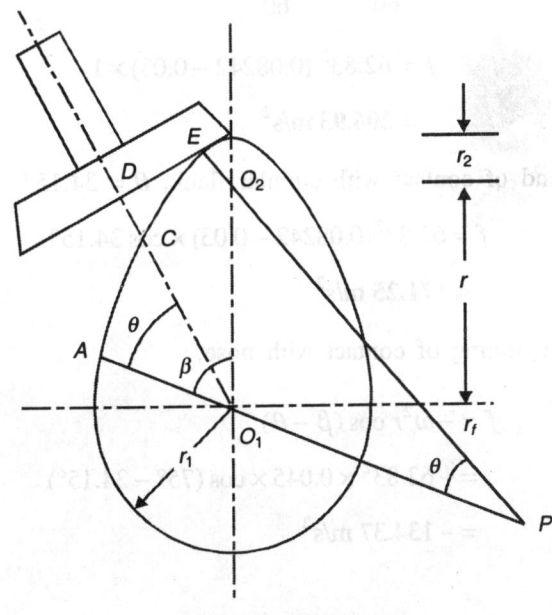

FIGURE 9.31

From $\triangle O_1O_2P$,

$$O_2P^2 = O_1P^2 + O_1O_2^2 - 2 \times O_1P \times O_1O_2 \times \cos \angle PO_1O_2$$

or $(r_f - 5)^2 = (r_f - 30)^2 + 45^2 - 2 \times (r_f - 30) \times 45 \times \cos(180° - 75°)$

or $\qquad r_f = 82.42$ mm **Ans.**

Applying sine rule to $\triangle O_1PO_2$,

$$\frac{O_1O_2}{\sin \theta} = \frac{O_2P}{\sin(180° - \beta)}$$

or $\qquad \dfrac{r}{\sin \theta} = \dfrac{r_f - r_2}{\sin(180° - \beta)}$

or $\qquad \dfrac{45}{\sin \theta} = \dfrac{82.42 - 5}{\sin 105°}$

or $\qquad \theta = 34.15°$

(ii) Acceleration

(a) At the beginning of lift, $\theta = 0°$

$$f = \omega^2(r_f - r_1) \cos \theta$$

where $\qquad \omega = \dfrac{2\pi N}{60} = \dfrac{2\pi \times 600}{60} = 62.83$ rad/s

$\therefore \qquad f = 62.83^2 (0.08242 - 0.03) \times 1$

$\qquad \qquad = 206.93$ m/s² **Ans.**

(b) At the end of contact with circular flank, $\theta = 34.15°$

$\therefore \qquad f = 62.83^2(0.08242 - 0.03) \times \cos 34.15°$

$\qquad \qquad = 171.25$ m/s² **Ans.**

(c) At the beginning of contact with nose,

$$f = -\omega^2 r \cos(\beta - \theta)$$

$\qquad \qquad = -62.83^2 \times 0.045 \times \cos(75° - 34.15°)$

$\qquad \qquad = -134.37$ m/s² **Ans.**

(d) At the apex of the nose, angle $\beta = \theta$

$$\therefore \quad f = -\omega^2 r \cos 0°$$
$$= -62.83^2 \times 0.045$$
$$= -177.64 \text{ m/s}^2 \qquad \text{Ans.}$$

EXAMPLE 9.10 A flat-faced valve tappet is operated by a symmetrical cam with circular arcs for flank and nose profiles; the straight line path of the tappet passes through the cam axis. The total angle of action is 150°, the lift is 6 mm, the base circle diameter is 30 mm and period of acceleration is half that of deceleration during the lift; the cam rotates at 1250 rpm. Determine:

(a) nose and flank radii
(b) the maximum acceleration and deceleration while lifting.

Solution From the geometry of Figure 9.32, we have

$$r = r_1 + h - r_2$$
$$= 15 + 6 - r_2$$
$$= 21 - r_2$$

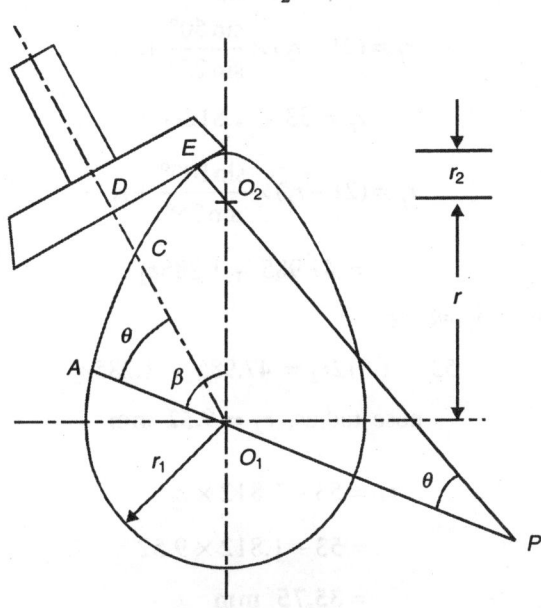

FIGURE 9.32

$$\angle \beta = \frac{\text{Angle of action}}{2}$$
$$= \frac{150}{2} = 75°$$

$$\omega = \frac{2\pi N}{60} = \frac{2\pi \times 1250}{60}$$

$$= 130.9 \text{ rad/s}$$

Further, it is given that
$$\angle O_1PO_2 = \frac{1}{2} \angle O_1O_2P \qquad \text{(i)}$$

$$\angle PO_1O_2 = 180° - 75° = 105°$$

$$\angle O_1PO_2 + \angle O_1O_2P = 180° - \angle PO_1O_2 = 180° - 105°$$

or
$$\angle O_1PO_2 + \angle O_1O_2P = 75° \qquad \text{(ii)}$$

Solving Eqs. (i) and (ii), we get

$$\angle O_1PO_2 = \theta = 25° \quad \text{and} \quad \angle O_1O_2P = \alpha = 50°$$

From ΔO_1O_2P,
$$\frac{O_1O_2}{\sin\theta} = \frac{O_1P}{\sin\alpha} = \frac{O_2P}{\sin 105°}$$

or
$$\frac{21 - r_2}{\sin 25°} = \frac{r_f - 15}{\sin 50°} = \frac{r_f - r_2}{\sin 105°}$$

or
$$r_f = (21 - r_2) \times \frac{\sin 50°}{\sin 25°} + 15$$

or
$$r_f = 53 - 1.812 r_2 \qquad \text{(iii)}$$

Similarly
$$r_f = (21 - r_2) \times \frac{\sin 105°}{\sin 25°} + r_2$$

or
$$r_f = 47.985 - 1.285 r_2 \qquad \text{(iv)}$$

Equating Eqs. (iii) and (iv), we get
$$53 - 1.812 r_2 = 47.985 - 1.285 r_2$$

or nose radius, $r_2 = $ **9.52 mm** Ans.

Radius of flank:
$$r_f = 53 - 1.812 \times r_2$$
$$= 53 - 1.812 \times 9.52$$
$$= \mathbf{35.75 \text{ mm}} \qquad \text{Ans.}$$

Centre distance:
$$r = 21 - r_2$$
$$= 21 - 9.52$$
$$= \mathbf{11.48 \text{ mm}} \qquad \text{Ans.}$$

Acceleration:

(i) Maximum acceleration when lift starts, i.e. when $\theta = 0°$:

$$f = \omega^2 (r_f - r_1) \cos\theta$$
$$= 130.9^2 \times (0.03575 - 0.015) \times 1$$
$$= 355.5 \text{ m/s}^2 \quad \text{Ans.}$$

(ii) Maximum retardation occurs when the followers is at apex, i.e. $\theta = \beta$:

$$f = -\omega^2 r \cos(\beta - \theta)$$
$$= -\omega^2 r$$
$$= -130.9^2 \times 0.01148$$
$$= -196.7 \text{ m/s}^2 \quad \text{Ans.}$$

EXAMPLE 9.11 The following data refers to a cam used to operate suction valve mechanism of a four stroke petrol engine.

Lift	= 10 mm
Least radius	= 25 mm
Nose radius	= 5 mm
Suction valve opens 6° after TDC	
Suction valve closes 40° after BDC	
Engine speed	= 2000 rpm

The cam used is circular flank type with circular nose and operates a flat faced follower. Estimate:

(i) Maximum velocity of the valve
(ii) Maximum acceleration and retardation
(iii) Minimum force to be exerted by the spring to overcome inertia of the valve parts of mass 0.25 kg.

Solution: For a four stroke engine, the cam speed is half of the engine speed, i.e.

$$N = \frac{1}{2} \times 2000 = 1000 \text{ rpm}$$

$$\therefore \quad \omega = \frac{2\pi N}{60} = \frac{2\pi \times 1000}{60} = 104.72 \text{ rad/s}$$

Angle moved by the crank when valve is opened

$$= 180° - 6° + 40° = 214°$$

Angle moved by cam = $\frac{1}{2}$ × crank rotation angle

$$= \frac{214}{2} = 107°$$

Angle of ascent: $\beta = \frac{107}{2} = 53.5°$

Referring to Figure 9.33,

$$r_1 + h = r + r_2$$

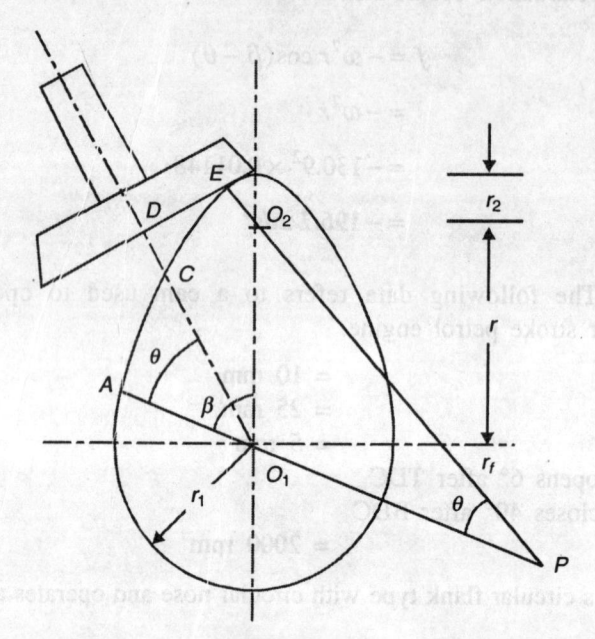

FIGURE 9.33

or
$$r = r_1 + h - r_2$$
$$= 25 + 10 - 5$$
$$= 30 \text{ mm}$$

In $\Delta O_1 O_2 P$,

$$O_2 P^2 = O_1 O_2^2 + O_1 P^2 - 2O_2 O_1 \times O_1 P \times \cos \angle O_2 O_1 P$$

or $\quad (r_f - 5)^2 = 30^2 + (r_f - 25)^2 - 2 \times 30 \times (r_f - 25) \times \cos (180° - 53.5°)$

or $\quad\quad\quad$ radius of flank: $r_f = 140.95$ mm

From $\Delta O_1 P O_2$, $\quad \dfrac{O_1 O_2}{\sin \theta} = \dfrac{O_2 P}{\sin (180° - 53.5°)}$

or
$$\frac{30}{\sin\theta} = \frac{(r_f - r_2)}{\sin 126.5°}.$$

or
$$\frac{30}{\sin\theta} = \frac{140.95 - 5}{\sin 126.5°}$$

or
$$\sin\theta = 0.1774$$

or
$$\theta = 10.22°$$

(i) Maximum velocity when follower leaves the flank, i.e. $\theta = \beta$:

$$v = \omega(r_f - r_1)\sin\beta$$

$$= 104.72 \times \frac{(140.95 - 25)}{1000} \times \sin 53.5°$$

$$= 9.76 \text{ m/s} \qquad \text{Ans.}$$

(ii) Maximum acceleration, when follower begins to rise, i.e. at $\theta = 0°$

$$f = \omega^2(r_f - r_1)\cos 0°$$

$$= 104.72^2 \times \frac{(140.95 - 25)}{1000} \times 1$$

$$= 1271.54 \text{ m/s}^2 \qquad \text{Ans.}$$

Maximum retardation occurs when follower is at apex of the nose, i.e. when $\theta = \beta$:

$$f = -\omega^2 r \cos(\beta - \theta)$$

$$= -\omega^2 r$$

$$= -104.72^2 \times \frac{30}{1000}$$

$$= -328.98 \text{ m/s}^2 \qquad \text{Ans.}$$

(iii) Spring force is needed to maintain contact during the retardation of the follower.

Minimum force = mass × retardation

= 0.25 × 328.98

= 82.24 N Ans.

EXERCISES

1. Describe various types of cams and follower commonly used. Write their relative merits and demerits.
2. Explain why a roller follower is preferred to a knife edge follower?
3. Describe the various factors which govern the choice of profile.
4. Describe the various factors which determine the size of base circle of a cam.
5. What do you mean by the pressure angle of a cam? Discuss its importance in cam design.
6. From the following data, draw the profile of a cam in which the follower moves with SHM during ascent and with uniform acceleration and deceleration when it returns to base.

Least radius of cam	= 50 mm
Angle of ascent	= 48°
Angle of dwell	= 42°
Angle of descent	= 60°
Lift	= 40 mm
Diameter of roller	= 30 mm
Distance between the line of action and follower axis	= 20 mm

   If the follower rotates at 360 rpm, find the maximum velocity and acceleration during ascent and descent.

   [**Ans:** 2.827 m/s, 399.7 m/s^2, 2.88 m/s, 207.36 m/s^2]

7. A cam drives a flat reciprocating follower in the following manner:
   During the first 90° rotation of the cam, the follower moves outward through a distance of 30 mm with SHM. The follower dwells during next 90° cam rotation. During the next 90° cam rotation, the follower moves inward with SHM. The follower then dwells for next 90° cam rotation. Draw the cam profile and calculate the maximum values of velocity and acceleration when cam rotates at 10 rad/s.

   [**Ans:** 1.885 m/s, 236.87 m/s^2]

8. Draw a cam profile to drive an oscillating roller follower to the specification given below:
   (i) Follower to move outward through an angular displacement of 20° during the first 120° rotation of the cam.
   (ii) Follower to return to its initial position during next 120° rotation of the cam.
   (iii) Follower to dwell during the next 120° cam rotation.

   The distance between the pivot centre and the roller centre = 120 mm
   Distance between the pivot centre and the cam axis = 130 mm
   Minimum radius of the cam = 40 mm
   Radius of roller = 10 mm
   Inward and outward strokes take place with SHM.

9. Construct the profile of the cam to suit the following specification:

   Least radius of cam = 25 mm
   Diameter of roller = 20 mm
   Angle of ascent = 120°
   Angle of dwell = 45°
   Angle of descent = 150°
   Lift of the follower = 40 mm

   The rise and return of the follower take place with cycloidal motion. The line of stroke of the follower is offsetted by 10 mm from the centre of cam. If cam rotate at 480 rpm. Calculate maximum velocity and acceleration during ascent and descent.
   [Ans: 1.917 m/s, 144.38 m/s^2, 1.534 m/s, 92.4 m/s^2]

10. The exhaust valve of a petrol engine opens 55° before bdc and closes 15° after tdc. A cam operates this valve. The maximum radius and lift of the cam are 40 mm and 15 mm respectively. The valve opens with constant acceleration and deceleration, the acceleration being twice the retardation. The period for closing the valve is same as that for opening it. The follower returns with SHM. Draw the profile of the cam if the roller radius is 15 mm and offset is 10 mm to the right of the cam centre. The cam rotates in clockwise direction.

11. Explain, with the help of velocity and acceleration diagram, why a cycloidal profile is preferred over SHM profile for cams used in high speed application.

12. The following data relates to a tangential cam:

    Base circle radius = 42 mm
    Lift = 14 mm
    Nose radius = 16 mm
    Roller radius = 20 mm
    Cam action angle = 150°

    If the roller follower axis passes through the centre of the cam and speed of the cam is 300 rpm, determine the velocity and acceleration of the follower at the point where straight flank merge into circular nose.
    [Ans: 1.378 m/s, 124.0 m/s^2]

13. A cam shaft of a petrol engine operates an overhead valve through a vertical tappet and push rod and horizontal rocker. A spring acting against a collar on the valve stem maintains contact between the tappet and the cam. The effective mass of the tappet is 0.455 kg, the arm length of the rocker on push rod side is 25 mm and on the valve side is 40 mm.
    The cam has a base circle of 25 mm diameter straight sides, a nose of 6 mm radius and a lift of 6 mm. The roller on the tappet is 12 mm diameter. Find the spring force necessary to keep the tappet in contact with the cam when valve has its maximum opening and the camshaft is rotating at 2000 rpm.
    [Ans: 298.2 N]

14. A cam with convex flanks, operating a flat follower whose lift is 18 mm, has a base circle radius of 36 mm and nose radius of 9.6 mm. The cam is symmetrical about the line drawn through the centre of curvature of the nose and the centre of the cam shaft. Find the radius of the flank if the total angle of cam action is 120°. Also determine the maximum velocity and the maximum acceleration and retardation when cam shaft speed is 500 rpm.

[Ans: 188 mm, 1.716 m/s, 414 m/s^2, 122 m/s^2]

15. The following data relates to a circular arc cam working with a flat faced follower:

   Minimum radius of cam = 30 mm
   Angle of action = 120°
   Radius of circular arcs = 80 mm
   Nose radius = 10 mm

   Determine the following:
   (i) Distance between the centre of nose to the centre of cam
   (ii) The angle through which the cam turns when the point of contact moves from the junction of minimum radius arc and circular arc to the junction of nose radius arc and circular arc.
   (iii) The velocity and acceleration of the follower when cam has turned through an angle of 20°. The angular velocity of the cam is 10 rad/s.

   [Ans: 30 mm, θ = 21.78°, 0.171 m/s, 4.69 m/s^2]

16. A symmetrical cam with circular arcs for flank and nose profile operates a flat ended follower. The axis of the follower passes through the cam axis. If the total angle of action is 144°, lift is 5 mm, base circle diameter is 24 mm and the period of acceleration is half of the period of ratardation during the lift, find:
   (i) Nose and flank radii
   (ii) Maximum acceleration and retardation during the lift.

   [Ans: 4.24 mm, 48.74 mm, 393.3 m/s^2, 139.7 m/s^2]

MULTIPLE CHOICE QUESTIONS

1. The minimum radius circle of a cam profile is called
   (a) base circle   (b) prime circle
   (b) pitch circle   (d) pitch curve

2. The size of a cam depends upon
   (a) pitch circle   (b) prime circle
   (c) base circle   (d) pitch curve

3. The locus of the trace point is called
   (a) pitch circle   (b) prime circle
   (c) base circle   (d) pitch curve

4. The reference point on the follower for the purpose of drawing cam profile is called
   (a) cam centre   (b) trace point
   (c) pitch point   (d) roller centre

5. The cam pitch curve and cam profile are same for
   (a) roller follower
   (b) flat faced follower
   (c) knife edge follower
   (d) all of the above

6. The point on the cam with maximum pressure angle is called
   (a) trace point
   (b) pitch point
   (c) cam centre
   (d) roller centre

7. For the same lift of the follower and same angle of action for ascent of the follower in the cams, smaller base circle diameter will give
   (a) smaller pressure angle
   (b) larger pressure angle
   (c) same pressure angle
   (d) pressure angle does not depend on these

8. The lift of a cam is the maximum distance of the follower from
   (a) base circle
   (b) prime circle
   (c) pitch circle
   (d) pitch curve

9. The term $\dfrac{df}{dt}$ in cam and follower motion represents
   (a) displacement
   (b) velocity
   (c) acceleration
   (d) jerk

10. For a follower moving with cycloidal motion, the acceleration of the follower at the time of starting the lift is
    (a) zero
    (b) infinity
    (c) moderate positive
    (c) moderate negative

11. In a tangent cam and roller follower, base circle diameter is 60 mm and roller diameter is 20 mm. Cam rotates for 60° with roller just leaves contact with the flank. The lift of the follower at this moment is
    (a) 40 mm
    (b) 20 mm
    (c) 10 mm
    (d) 80 mm

CHAPTER 10

Gears

10.1 INTRODUCTION

Gear is a machine element which, by means of progressive engagement of projections called teeth, transmits motion and power between two rotating shafts. Gear teeth, in general, have an involute profile which provides a constant pitch line velocity although some times cycloids and other profiles are used to advantage in some cases. The action of such mating gear teeth consists of a combination of rolling and sliding motion thus producing a positive drive.

Gears are classified on the following basis:

(i) Relation between axes of power transmitting shafts
    (a) Parallel shafts—Spur gear, rack and pinion, helical gears, herringbone gears, etc.
    (b) Intersecting shafts—Straight bevel gear
    (c) Non-parallel non-intersecting shafts—Worm gears
(ii) Shape of the solid on which teeth are cut
(iii) Curvature of the teeth profile—involute profile, cycloidal profile

Generally, the following types of gear are most commonly used in industry for power transmission purposes:

Spur gear. A gear having straight teeth cut on the cylindrical disc along the axis is called spur gear. These gears are used to transmit power between two parallel shafts as shown in Figure 10.1(a). Customarily, a smaller size gear is called **pinion**. A rack is a straight tooth gear which can be thought of as a segment of spur gear of infinite diameter. The rack and pinion arrangement is used to convert rotary motion into translatory motion or vice-versa.

Helical gears. These gears are also used to transmit power between two parallel shafts and teeth are cut on cylindrical disc. The tooth faces of these gears have certain degree of helix angle of opposite hand on pinion and gear as shown in Figure 10.1(b). These gears are smooth in operation and therefore can transmit power at a high pitch line velocity. Helical gears have the disadvantage of having a force component along the axis.

Herringbone gears. It is a double helical gear in which one pair having left hand helix and the other having right hand helix are secured together in such a way that left hand helix and right hand helix meet at a common point and there is no groove in between the gears. The axial thrust of two rows of helical gears cancel each other out. Therefore, these gears can run at high speed without noise.

Bevel gear. When power is transmitted between two intersecting shafts, bevel gears are used. The angle of intersection of shafts is called **shaft angle**. The gear blank is a frustum of cone on which teeth are generated. A bevel gear with straight teeth which points towards intersection of shaft axes and vary in cross-section throughout their length is called **straight bevel gear** [Figure 10.1(c)]. Gears of the same size connecting two shafts at right angle to each other are called **miter gears**. When the teeth of a bevel gear are inclined at an angle of the face of the bevel, they are known as **spiral bevel gears**. These gears are smoother in action than straight bevel gears.

Worm and worm gears. In this system of gearing, the axes of the power transmitting shafts are neither parallel nor intersecting but the plane containing the axes are generally at right angle to each other [Figure 10.1(d)].

In a gear drive, the smaller of the two gears in mesh is called the pinion and the larger gear is customarily designated as gear. In most of the applications, the pinion is the driving element whereas the gear is the driven element. There are some applications like the epicyclic gear train where the gear teeth are cut on the inside of the rim. Such gears are known as **internal gears**.

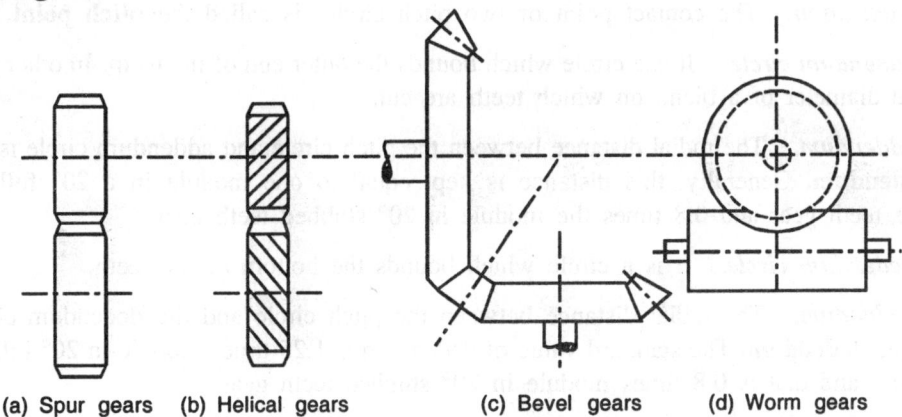

(a) Spur gears (b) Helical gears (c) Bevel gears (d) Worm gears

FIGURE 10.1 Types of gears.

10.2 GEAR TERMINOLOGY

The terminology and notations for toothed gearing are covered by Bureau of Indian Standards (BIS) in their codes IS: 2458–1965 and IS: 2467–1965. The following definitions of different terms are given with reference to these codes. (See Figure 10.2.)

Pitch cylinder. It is an imaginary friction cylinder which by pure rolling, transmit the same motion as that of a pair of gears.

Pitch surface. The surface on which teeth are cut to ensure positive drive is called **pitch surface**.

Pitch circle. The intersection of the pitch surface with a plane perpendicular to the axis of rotation is called the **pitch circle**.

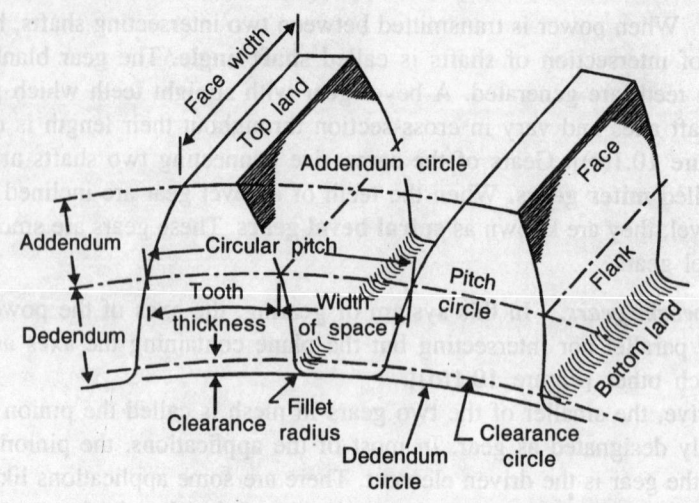

[Figure: Terminology of spur gear tooth showing labels: Face width, Top land, Addendum circle, Face, Addendum, Circular pitch, Flank, Tooth thickness, Pitch circle, Dedendum, Width of space, Bottom land, Clearance, Fillet radius, Dedendum circle, Clearance circle]

FIGURE 10.2 Terminology of spur gear tooth.

Pitch point. The contact point of two pitch circles is called the **pitch point.**

Addendum circle. It is a circle which bounds the outer end of the teeth. In other words, it is the diameter of a blank on which teeth are cut.

Addendum. The radial distance between the pitch circle and addendum circle is called the **addendum.** Generally, this distance is kept equal to one module in a 20° full depth involute teeth gear and 0.8 times the module in 20° stubbed teeth gear.

Dedendum circle. It is a circle which bounds the bottom of the teeth.

Dedendum. The radial distance between the pitch circle and the dedendum circle is called the **dedendum.** The standard value of dedendum is 1.25 times module in 20° full depth teeth gear and that is 0.8 times module in 20° stubbed teeth gear.

Total depth of teeth. The sum of addendum and dedendum is called the total depth of the **tooth.**

Clearance. The difference between dedendum and addendum of a mating gear teeth is commonly referred to as clearance.

Base circle. A circle from which the tooth profile curve is generated is known as **base circle.**

Tooth thickness. The chord length measured along the pitch circle between the two opposite faces of the same tooth is called **tooth thickness.**

Top land. It is the surface of the top of the tooth.

Bottom land. The surface at the bottom of the tooth between the adjacent fillets.

Face. Tooth surface between the pitch circle and the topland is known as **face.**

Flank. Tooth surface between the pitch circle and the bottom land is known as **flank.**

Circular pitch. The distance measured along the pitch circle from a point on one tooth to the corresponding point on the adjacent tooth is called the **circular pitch**. It is denoted by p and given by the following relation:

$$p = \frac{\pi d_1}{Z_1} = \frac{\pi d_2}{Z_2} \qquad (10.1a)$$

where d_1, d_2 = pitch circle diameters of pinion and gear respectively
and Z_1, Z_2 = number of teeth on the pinion and gear respectively

Module. The ratio of the pitch circle diameter to the number of teeth is called **module** and is denoted by m.

$$m = \frac{d_1}{Z_1} \qquad (10.1b)$$

Diametral pitch. The ratio of number of teeth in the pitch circle diameter is called the **diametral pitch**. It is reciprocal of the module.

$$DP = \frac{Z_1}{d_1} = \frac{1}{m} \qquad (10.1c)$$

Gear ratio. It is the ratio of number of teeth on the gear to that on the pinion.

$$G = \frac{Z_2}{Z_1} \qquad (10.1d)$$

Velocity ratio. It is the ratio of the angular velocity of the driven gear to the angular velocity of the driving pinion.

$$VR = \frac{\omega_2}{\omega_1} = \frac{N_2}{N_1} = \frac{d_1}{d_2} = \frac{Z_1}{Z_2} = \frac{1}{G} \qquad (10.1e)$$

10.3 LAW OF GEARING

In a gear drive, the action of tooth profile is to transmit motion at constant angular velocity ratio for which the gears must satisfy the fundamental law of gearing. Accordingly, the common normal at the point of action between two teeth must always pass through a point, called the **pitch point,** in such a way that it divides the line joining the centres of two matting gears in the inverse ratio of angular velocities.

Consider two rigid bodies 1 and 2 representing a portion of two gears in mesh rotating about fixed centres O_1 and O_2. The point A on the rigid body 1 is in contact with a point B on the rigid body 2. Let $X - X$ and $Y - Y$ represent common tangent and common normal at the point of contact (See Figure 10.3).

Let ω_1 = angular velocity of the rigid body 1
and ω_2 = angular velocity of the rigid body 2

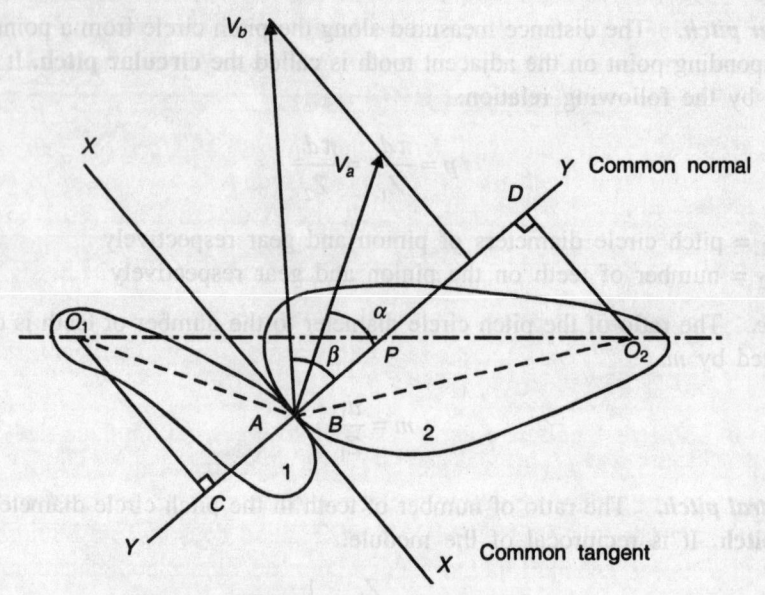

FIGURE 10.3 Law of gearing.

At a given instant, the point A on the rigid body 1 is moving in the direction perpendicular to O_1A with a velocity $V_a = \omega_1 . O_1A$. At the same time, the point B on the rigid body 2 is moving in the direction perpendicular to O_2B with a velocity $V_b = \omega_2 . O_2B$.

If the two rigid bodies are always to remain in contact, their relative velocities at any instant along the common normal must be zero, i.e. the component of velocities V_a and V_b along common normal must be equal.

Thus
$$V_a \cos \alpha = V_b \cos \beta$$

or
$$\omega_1 \times O_1A \cos \alpha = \omega_2 \times O_2B \cos \beta$$

or
$$\omega_1 \times O_1A \times \frac{O_1C}{O_1A} = \omega_2 \times O_2B \times \frac{O_2D}{O_2B}$$

or
$$\omega_1 \times O_1C = \omega_2 \times O_2D$$

or
$$\frac{\omega_1}{\omega_2} = \frac{O_2D}{O_1C} \tag{10.2}$$

Let P is a point of intersection of line joining the centres of rigid bodies and common normal. Therefore, from similar triangles O_1PC and O_2DP,

$$\frac{O_2D}{O_1C} = \frac{O_2P}{O_1P} \tag{10.3}$$

Equating Equations (10.2) and (10.3), we get

$$\frac{\omega_1}{\omega_2} = \frac{O_2P}{O_1P} \tag{10.4}$$

Thus for a constant angular velocity ratio of a gear pair in mesh the normal at the point of contact divides the line joining the centre of rotation in the inverse ratio of the angular velocities. Thus the dividing point P is called **pitch point**. This is the fundamental law of gearing which must be satisfied by the profiles adopted for the teeth of gear in mesh. This condition is fulfilled by teeth of involute form, provided that the base circle from which the profiles are generated are tangential to the common normal. Since all points of contact lie on the common normal, it is called the **line of action**.

The gear tooth action is shown in Figure 10.4 in which a line AB, which is normal to the line joining the centres of mating gear, meet at the pitch point P and another line CD, which is tangent to the base circle of gear and pinion, also passes through the pitch point and is normal to teeth in action. The angle between the lines AB and CD is called the **pressure angle** (α) and the normal force that one tooth exerts on the other passes through the pressure line. The pressure line in a gear pair can be located by rotating line AB through the pressure angle α in the direction opposite to the direction of rotation of the driving gear.

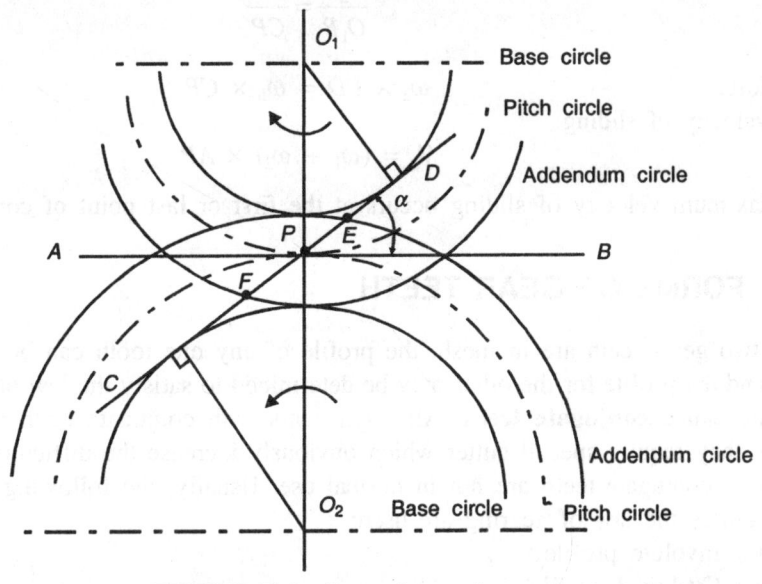

FIGURE 10.4 Gear pair in action.

10.4 VELOCITY OF SLIDING

The relative velocity of points A and B along the common tangent is known as **velocity of sliding**. In other words, it represents the sliding velocity of the surface of rigid body 2 relative to the surface of rigid body 1 at the point of contact. (See Figure 10.3.)

Velocity of sliding: $\qquad V_s = V_b \sin \beta - V_a \sin \alpha$

$$V_s = \omega_2 \times O_2B\sin\beta - \omega_1 \times O_1A\sin\alpha$$
$$= \omega_2 \times O_2B \times \frac{BD}{O_2B} - \omega_1 \times O_1A \times \frac{AC}{O_1A}$$
$$= \omega_2 \times BD - \omega_1 \times AC$$
$$= \omega_2(BP + PD) - \omega_1(CP - AP)$$
$$= (\omega_1 + \omega_2) \times AP + \omega_2 \times PD - \omega_1 \times CP \quad (\because AP = BP)$$

From Eq. (10.4), we know that
$$\frac{\omega_1}{\omega_2} = \frac{O_2P}{O_1P}$$
and from similar ΔO_1CP and ΔO_2DP,
$$\frac{O_2P}{O_1P} = \frac{DP}{CP}$$

Therefore,
$$\omega_2 \times PD = \omega_1 \times CP$$
Thus velocity of sliding:
$$V_s = (\omega_1 + \omega_2) \times AP \quad (10.5)$$

The maximum velocity of sliding occurs at the first or last point of contact.

10.5 FORMS OF GEAR TEETH

When two gears teeth are in mesh, the profile of any one tooth can be chosen of arbitrary shape and the profile for the other may be determined to satisfy the law of gearing. Such gear teeth are called **conjugate teeth.** Although gears with conjugate teeth transmit the desired motion, they require special cutter which obviously increase the difficulty in manufacturing. Therefore, conjugate teeth are not in normal use. Usually, the following geometrical curves which satisfy the law of gearing are used:
   (i) Involute profile
   (ii) Cycloidal profile
These are discussed further.

10.5.1 Involute Profile

An involute is the locus of a point on a straight line which rolls on the circumference of a circle without slippage. In other words, a point on a taut rope, when unwound from a cylinder, would trace an involute curve. To illustrate, consider a cylinder around which a cord *abc*, which held tightly, is wrapped (Figure 10.5). The point *e* on the cord represents the tracing point. When the cord is unwrapped about the cylinder, the point *e* will trace out the involute curve *d e f*. The radius of curvature of the involute is zero at point *d* and maximum at point *f*. The radius of curvature at point *e* is equal to the distance *eb* as point *e* is rotating

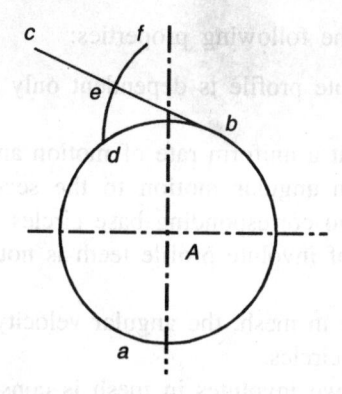

FIGURE 10.5 Generation of an involute.

about point b. Thus the generating line bc is normal to the involute at all points of intersection and is always tangent to the cylinder.

Now to see whether or not the involute profile satisfies the law of gearing, let us consider two gear blanks fixed at centres O_1 and O_2 (See Figure 10.6). Imagine that a cord is wound clockwise around the base circle of gear 1. This cord is pulled tightly between points A and B and is wound counterclockwise around the base circle of gear 2. If the base circles are rotated in different directions to keep the cord tight, a point G on the cord will trace out the involute CD on gear 1 and EF on gear 2. Since both involutes are generated simultaneously, the tracing point represents the point of contact. The portion of cord AB is called **generating line**. In this arrangement, the point of contact moves along the generating line which is always normal to the involutes at the point of contact and intersects the line joining the centres of gear at the pitch point. Thus the requirement of law of gearing is satisfied.

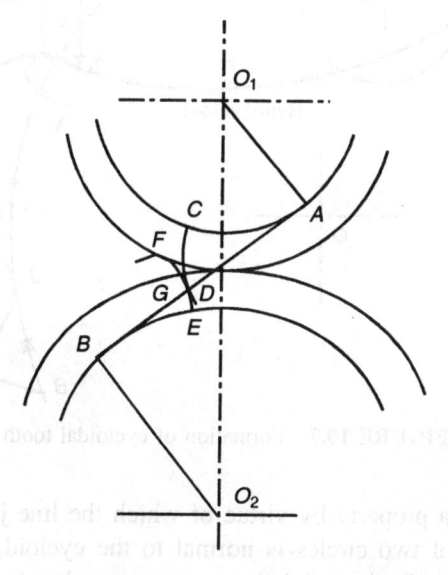

FIGURE 10.6 Involute in action.

The involute profile has the following properties:

1. The shape of the involute profile is dependent only on the dimensions of the base circle.
2. If one involute rotates at a uniform rate of motion and is in contact with another, it will transmit a uniform angular motion to the second irrespective of the centre distance between the two corresponding base circles.
3. Angular velocity ratio of involute profile teeth is not sensitive to centre distance of their base circles.
4. When two involutes are in mesh, the angular velocity ratio is inversely proportional to the size of the base circles.
5. The pressure angle of two involutes in mesh is constant.

10.5.2 Cycloidal Profile

A cycloid is the locus of a point on the circumference of a circle which rolls without slipping on a fixed straight line. It has two variants—epicycloid and hypocycloid. An **epicycloid** is the locus of a point on the circumference of a circle, which rolls without slipping on outside circumference of another circle of finite radius. Similarly, a hypocycloid is a locus of a point on the circumference of circle which rolls without slipping on inside circumference of another circle.

In a gear having cycloidal teeth, the face of the tooth is made by epicycloid curve whereas flank is made up of hypocycloid as shown in Figure 10.7.

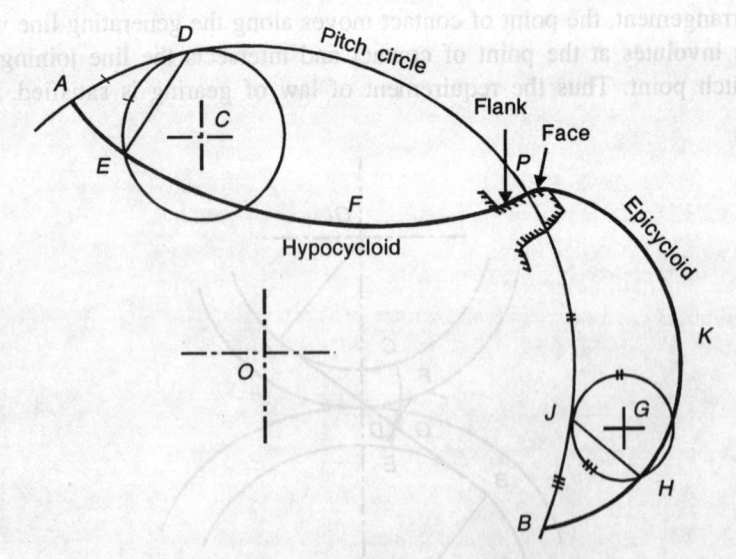

FIGURE 10.7 Formation of cycloidal tooth.

A cycloid curve has a property by virtue of which the line joining the generating point with the point of contact of two circles is normal to the cycloid. For example, when circle C touches the pitch circle at D, the point A moves to new location at E and the line joining points E and D is normal to hypocycloid AFP, (see Figure 10.7). Similarly when circle G rolls outside the pitch circle starting from point P, an epicycloid PKB is generated.

A cycloidal teeth gear-pinion in mesh is shown in Figure 10.8. Consider the pinion rotating in counterclockwise direction. The two pitch circles are tangent at the pitch point P

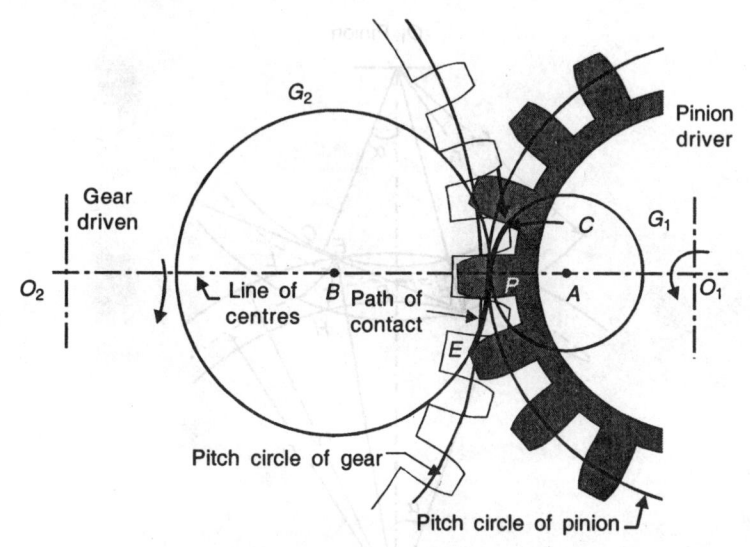

FIGURE 10.8 Cycloidal teeth in mesh.

and they roll upon each other without slipping. The two generating circles G_1 and G_2 having centres at A and B also rolls with the moving pitch circles. The initial contact between the teeth occurs at point C where the addendum circle of the driven gear cuts the generating circle. Thus the complete path of approach is the arc CP. However, at that instant, the contact is made at point E, which is at the intersection of the generating circle G_1 and the two contacting profiles. The pitch point P is the instantaneous centre of rotation of point E on the generating circle and line PE is normal to both tooth profiles. Thus cycloidal teeth satisfies the law of gearing that the normal to the tooth profile always passes through the pitch point. However, the line PE, which is pressure line, does not have a constant inclination. The varying pressure of cycloidal teeth results in noise and wear. It also results into changing bearing reactions at the support. Further, these gears must be operated at exactly the correct centre distance; otherwise the contacting portion of the profiles will not be conjugate. Further, since the cycloidal teeth are made up of two curves, it is very difficult to produce them accurately. On account of these reasons, the cycloidal teeth have become obsolete.

10.6 ARC OF CONTACT

When two gears start transmitting motion, the initial contact occurs at a point where the flank of the driving gear (pinion) tooth comes in contact with the face tip of the driven gear tooth and the contact ends when the face tip of the pinion tooth comes in contact with the flank of the driven gear tooth. In other words, the contact between the two gears starts when the addendum circle of the driven gear cuts the normal pressure line at point E and ends when the addendum circle of the pinion cuts the normal pressure line at point F. (See Figure 10.9.) The distance between these two points, EF is called **path of contact** or **length of contact**. Usually, the path of contact is divided into two parts—path of approach and path of recess.

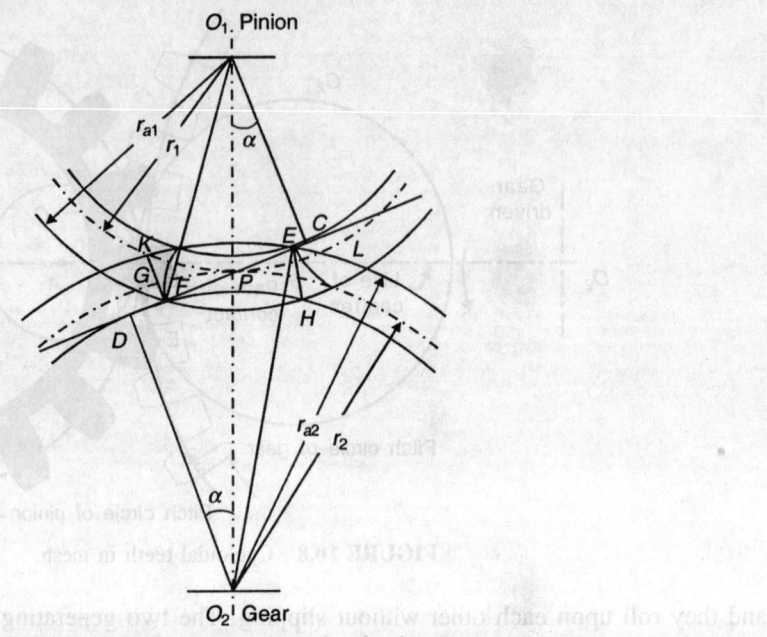

FIGURE 10.9 Representation of path of contact.

Path of approach. A portion of path of contact from the beginning of engagement to the pitch point, i.e. the length EP, is called **path of approach.**

Path of recess. The portion of path of contact from pitch point to the end of engagement, i.e. the length PF, is called **path of recess.**

Let r_1 = pitch circle radius of the pinion
 r_2 = pitch circle radius of the gear
 r_{a1} = addendum circle radius of the pinion
 r_{a2} = addendum circle radius of the gear
and α = pressure angle
Referring to Figure 10.9,
Path of contact = Path of approach + Path of recess

or $$EF = EP + PF$$

Path of approach: $$EP = ED - PD$$

or $$EP = \sqrt{O_2E^2 - O_2D^2} - PD$$

or $$EP = \sqrt{r_{a2}^2 - r_2^2 \cos^2\alpha} - r_2 \sin\alpha \qquad (10.6)$$

The maximum possible length of path of approach is:
$$CP = r_1 \sin \alpha$$

Similarly, the path of recess: $PF = FC - PC$

or
$$PF = \sqrt{O_1F^2 - O_1C^2} - PC$$

or
$$PF = \sqrt{r_{a1}^2 - r_1^2 \cos^2 \alpha} - r_1 \sin \alpha \qquad (10.7)$$

The maximum possible length of path of recess is:
$$PD = r_2 \sin \alpha$$

Therefore, the path of contact:

$$EF = \sqrt{r_{a2}^2 - r_2^2 \cos^2 \alpha} + \sqrt{r_{a1}^2 - r_1^2 \cos^2 \alpha} - (r_1 + r_2)\sin \alpha \qquad (10.8)$$

The maximum length of path of contact occurs when point E lies at point C and point F lies at point D. Thus
$$CD = (r_1 + r_2) \sin \alpha \qquad (10.9)$$

The arc of contact is the distance travelled by a point on either pitch circle of gear or a pinion during the period of contact of a pair of teeth. Referring to Figure 10.9, the contact on pitch circle of gear begins at point G and ends at point H. Similarly, on the pitch circle of pinion, the contact is between K and L. Therefore, the arc of contact is:

$$\text{arc } GH = \text{arc } KL$$

But we know from the geometry of the involute curve that

$$\text{arc of contact, } GH = \frac{\text{path of contact}}{\cos \alpha}$$

or
$$= \frac{EF}{\cos \alpha}$$

Since the arc of contact is defined as the length of the pitch circle during the mating of teeth, the number of teeth lying in between the arc of contact GH will be meshing with the teeth on the pinion. Therefore,

$$\text{Number of teeth in contact} = \frac{\text{arc } GH}{\text{Circular pitch}}$$

$$= \frac{EF}{p \cos \alpha}$$

or
$$\text{Number of teeth in contact} = \frac{EF}{\cos \alpha} \times \frac{1}{\pi m} \qquad (10.10)$$

Number of teeth in contact at any instant is also called **contact ratio**. For continuous transmission of power, atleast one pair of teeth should always remain in contact. Generally, gears are designed for contact ratio from 1 to 1.6.

EXAMPLE 10.1 A pair of gears having 20° involute teeth is required to transmit motion at a velocity ratio of 1:4. If the module of both pinion and gear is 5 mm and centre distance is 250 mm, determine the number of teeth and base circle radius of pinion and gear.

Solution:

(i) Velocity ratio:
$$VR = \frac{N_2}{N_1} = \frac{Z_1}{Z_2} = \frac{1}{4}$$

or
$$Z_2 = 4Z_1 \qquad (i)$$

Centre distance:
$$C = \frac{d_1 + d_2}{2} = \frac{m(Z_1 + Z_2)}{2}$$

or
$$250 = \frac{5(Z_1 + Z_2)}{2}$$

or
$$Z_1 + Z_2 = 100 \qquad (ii)$$

Solving Equation (i) and (ii), we get

$$Z_1 = 20 \text{ teeth} \quad \text{and} \quad Z_2 = 80 \text{ teeth} \qquad \textbf{Ans.}$$

(ii) Pitch circle radius of

(a) Pinion:
$$r_1 = \frac{mZ_1}{2} = \frac{5 \times 20}{2} = 50 \text{ mm}$$

Base circle radius: $r_{b1} = r_1 \cos \alpha = 50 \times \cos 20° = \textbf{46.98 mm}$ **Ans.**

(b) Gear:
$$r_2 = \frac{mZ_2}{2} = \frac{5 \times 80}{2} = 200 \text{ mm}$$

Base circle radius: $r_{b2} = r_2 \cos \alpha = 200 \times \cos 20° = \textbf{187.93 mm}$ **Ans.**

EXAMPLE 10.2 A pair of spur gear having 20 and 40 teeth are in mesh. The pinion being driving element rotates at 2000 rpm. Find the sliding velocity between the teeth faces (i) at the point of engagement (ii) at the pitch point and (iii) at the point of disengagement. Assume that gear teeth are of 20° involute form. Addendum is 5 mm and module is 5 mm. Find also the angle through which pinion turns while one pair of teeth are in contact.

Solution Pitch circle radius of pinion:

$$r_1 = \frac{mZ_1}{2} = \frac{5 \times 20}{2} = 50 \text{ mm}$$

Pitch circle radius of gear: $r_2 = \dfrac{mZ_2}{2} = \dfrac{5 \times 40}{2} = 100$ mm

Radius of addendum circle of pinion and gear:

$$r_{a1} = r_1 + \text{addendum} = 50 + 5 = 55 \text{ mm}$$

$$r_{a2} = r_2 + \text{addendum} = 100 + 5 = 105 \text{ mm}$$

Path of approach
$$= \sqrt{r_{a2}^2 - r_2^2 \cos^2\alpha} - r_2 \sin\alpha$$
$$= \sqrt{105^2 - 100^2 \times \cos^2 20°} - 100 \times \sin 20°$$
$$= 12.64 \text{ mm}$$

Path of recess:
$$= \sqrt{r_{a1}^2 - r_1^2 \cos^2\alpha} - r_1 \sin\alpha$$
$$= \sqrt{55^2 - 50^2 \times \cos^2 20°} - 50 \times \sin 20°$$
$$= 11.49 \text{ mm}$$

Angular velocity of pinion: $\omega_1 = \dfrac{2\pi \times 2000}{60} = 209.44$ rad/s

Angular velocity of gear: $\omega_2 = \dfrac{2\pi \times 1000}{60} = 104.72$ rad/s

(i) Sliding velocity at the point of engagement:

$$V_{s1} = (\omega_1 + \omega_2) \times \text{path of approach}$$

$$= (209.44 + 104.72) \times \dfrac{12.64}{1000} = \mathbf{3.97 \text{ m/s}} \qquad \text{Ans.}$$

(ii) Sliding velocity at the pitch point:

$$V_{s2} = \mathbf{0} \qquad \text{Ans.}$$

(iii) Sliding velocity at the point of disengagement:

$$V_{s3} = (\omega_1 + \omega_2) \times \text{path of recess}$$

$$= (209.44 + 104.72) \times \dfrac{11.49}{1000}$$

$$= \mathbf{3.61 \text{ m/s}} \qquad \text{Ans.}$$

(iv) Angle through which the pinion turns:

$$\theta = \dfrac{\text{Length of arc of contact} \times 360°}{\text{Circumference of pinion}}$$

$$\text{Arc of contact} = \frac{\text{Length of path of contact}}{\cos \alpha}$$

$$= \frac{12.64 + 11.49}{\cos 20°} = 25.67 \text{ mm}$$

$$\theta = \frac{25.67 \times 360°}{2\pi \times 50} = 29.41° \quad \text{Ans.}$$

EXAMPLE 10.3 The following data refer to a pair of spur gear in mesh having 20° involute profile teeth:

Number of teeth on pinion = 24
Number of teeth on gear = 48
Speed of pinion = 300 rpm
Module = 6 mm

If the addendum on each gear is such that the path of approach and path of recess are half of their maximum possible values find:
  (i) the addendum on gear and pinion
  (ii) the length of arc of contact, and
  (iii) the maximum velocity of sliding of gears

Solution We have

$$r_1 = \frac{mZ_1}{2} = \frac{6 \times 24}{2} = 72 \text{ mm}$$

$$r_2 = \frac{mZ_2}{2} = \frac{6 \times 48}{2} = 144 \text{ mm}$$

(i) Path of approach $= \sqrt{r_{a2}^2 - r_2^2 \cos^2 \alpha} - r_2 \sin \alpha$

The maximum possible path of approach $= r_1 \sin \alpha$

Therefore, $\sqrt{r_{a2}^2 - r_2^2 \cos^2 \alpha} - r_2 \sin \alpha = \dfrac{r_1 \sin \alpha}{2}$

$$\sqrt{r_{a2}^2 - 144^2 \times \cos^2 20°} - 144 \times \sin 20° = \frac{72 \times \sin 20°}{2}$$

or $\quad r_{a2} = 148.66$ mm

Addendum on gear $= r_{a2} - r_2$

$$= 148.66 - 144 = \textbf{4.66 mm} \quad \text{Ans.}$$

Similarly, the length of path of recess is equal to half of the maximum possible length of path of recess.

$$\sqrt{r_{a1}^2 - r_1^2 \cos^2 \alpha} - r_1 \sin \alpha = \frac{r_2 \sin \alpha}{2}$$

$$= \sqrt{r_{a1}^2 - 72^2 \times \cos^2 20°} - 72 \times \sin 20° = \frac{144 \times \sin 20°}{2}$$

or radius of addendum: $r_{a1} = 83.68$ mm

Addendum of pinion $= r_{a1} - r_1$

$\qquad = 83.68 - 72 = \mathbf{11.68\ mm}$ **Ans.**

(ii) Length of arc of contact $= \dfrac{\text{Length of path of contact}}{\cos \alpha}$

$$= \frac{\dfrac{r_1 \sin \alpha}{2} + \dfrac{r_2 \sin \alpha}{2}}{\cos \alpha}$$

$$= (r_1 + r_2) \frac{\tan \alpha}{2}$$

$$= (72 + 144) \times \frac{\tan 20°}{2} = \mathbf{39.3\ mm} \qquad \text{Ans.}$$

(iii) Maximum velocity of sliding:

$$V_s = (\omega_1 + \omega_2) \times \text{Length of path of recess}$$

$$= \frac{2\pi}{60}(300 + 150) \times \frac{144 \times \sin 20°}{2 \times 1000}$$

$$= \mathbf{1.16\ m/s} \qquad \text{Ans.}$$

EXAMPLE 10.4 The number of teeth on each of the two equal spur gears in mesh is 40. The teeth have 20° involute profile and the module is 6 mm. If the arc of contact is 1.75 times the circular pitch, find the addendum.

Solution Circular pitch:

$$p_c = \pi \times m = \pi \times 6 = 18.85\ \text{mm}$$

Pitch circle radii of each gear:

$$r_1 = r_2 = \frac{mZ_2}{2} = \frac{6 \times 40}{2} = 120\ \text{mm}$$

Length of arc of contact $= 1.75 \times p_c$

$\qquad\qquad\qquad\qquad\qquad = 1.75 \times 18.85 = 33$ mm

Length of path of contact $=$ Arc of contact $\times \cos \alpha$

$\qquad\qquad\qquad\qquad\qquad = 33 \times \cos 20° = 31$ mm

We know that path of contact:

$$31 = \sqrt{r_{a2}^2 - r_2^2 \cos^2\alpha} + \sqrt{r_{a_1}^2 - r_1^2 \cos^2\alpha} - (r_1 + r_2)\sin\alpha$$

$$= 2 \times \sqrt{r_{a2}^2 - r_2^2 \cos^2\alpha} - 2 \times r_2 \sin\alpha$$

$$= 2 \times \sqrt{r_{a2}^2 - 120^2 \times \cos^2 20°} - 2 \times 120 \times \sin 20°$$

or $\sqrt{r_{a2}^2 - 120^2 \times \cos^2 20°} = 56.54$

or $r_{a2} = 126.14$ mm

$$\text{Addendum} = r_{a2} - r_2$$
$$= 126.14 - 120.0$$
$$= 6.14 \text{ mm} \qquad \text{Ans.}$$

EXAMPLE 10.5 A pinion of 100 mm pitch circle diameter (pcd) drives a gear of 300 mm pcd. The teeth are of involute form with a pressure angle of 20°. If the addendum on each gear is 5 mm, the pinion speed is 2000 rpm and coefficient of friction between teeth is 0.1, find for the first point of contact (i) sliding velocity between the teeth and (ii) transmission efficiency.

Solution Speed of gear:

$$N_2 = N_1 \times \frac{d_1}{d_2} = 2000 \times \frac{100}{300} = 666.67 \text{ rpm}$$

Pitch circle radius of gear: $r_2 = \frac{300}{2} = 150$ mm

Addendum circle radius: $r_{a2} = r_2 + $ addendum $= 150 + 5 = 155$ mm

Path of approach $= \sqrt{r_{a2}^2 - r_2^2 \cos^2\alpha} - r_2 \sin\alpha$

$$= \sqrt{155^2 - 150^2 \times \cos^2 20°} - 150 \times \sin 20°$$

$$= 13.17 \text{ mm}$$

Angular velocity of pinion: $\omega_1 = \frac{2\pi N_1}{60} = \frac{2\pi \times 2000}{60}$

$$= 209.44 \text{ rad/s}$$

Angular velocity of gear: $\omega_2 = \frac{2\pi \times 666.67}{60} = 69.81$ rad/s

(i) Velocity of sliding: $V_s = (\omega_1 + \omega_2) \times$ path of approach

$$= (209.44 + 69.81) \times \frac{13.17}{1000} = 3.67 \text{ m/s} \qquad \text{Ans.}$$

(ii) At the first point of contact, let the pitch circles rotate through a distance δ. The angle turned through by the pinion is δ/r_1 and the time taken is $\delta/\omega_1 r_1$. Therefore, the distance of sliding in this time period

$$= \text{velocity of sliding} \times \frac{\delta}{\omega_1 r_1}$$

$$= 3.67 \times 1000 \times \frac{\delta}{209.44 \times 50}$$

$$= 0.35\,\delta \text{ mm}$$

If F_t is the tangential force between the teeth, the normal force along the line of action is $F_t/\cos\alpha$.

Therefore, friction force at the point of contact $= \dfrac{\mu F_t}{\cos\alpha}$

Work lost in friction $= \dfrac{\mu F_t}{\cos\alpha} \times 0.35\,\delta$

Instantaneous transmission efficiency

$$= \frac{F_t \times \delta - \dfrac{\mu F_t}{\cos\alpha} \times 0.35\delta}{F_t \times \delta}$$

$$= 1 - \mu \times \sec\alpha \times 0.35$$

$$= 1 - 0.1 \times \sec 20° \times 0.35$$

$$= 0.962 \quad \text{or} \quad \textbf{96.2 per cent} \qquad \textbf{Ans.}$$

10.7 INTERFERENCE

When a pair of gear transmits power, the normal force is passed through common normal to the two involutes at the point of contact, which is also a tangent to the base circles of mating gear pair. If, by any reason, any of the two surfaces is not involute, the two surfaces would not touch each other tangentially and transmission of power would not be proper. The mating of non-conjugate gear teeth will violate the fundamental law of gearing and this non-conjugate action is called **interference**. To illustrate, let us consider a pair of gear and pinion in mesh as shown in Figure 10.10, in which the path of contact is *EF*. If, now, the radius of addendum circle of the gear is increased, the point of engagement *E* shifts along *PC* towards point *C*. The limiting position of point *E* is the point *C*. Any further increase in the value of addendum circle radius will shift the point of contact *E* inside the base circle of the pinion. Since there is no involute profile below the base circle of the pinion, it will form a non-conjugate contact and resulting phenomenon is called **interference**. In other words, the condition of interference arises when contact between teeth occurs outside the points *C* and *D*, i.e. when path of contact *EF* is greater than distance *CD*.

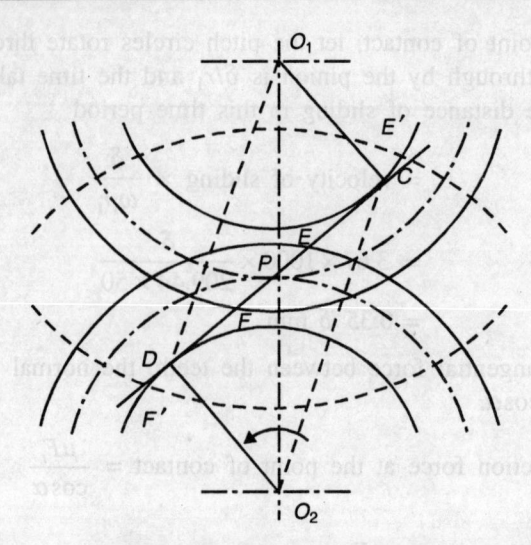

FIGURE 10.10 Gear interference.

The condition of interference with revised addendum circles marked with dashed line and path of contact $E'F'$ greater than distance CD is shown in Figure 10.10. The phenomenon of interference can be avoided by employing some preventive measures as given below:

Undercutting. When gear teeth are manufactured by a generating process, a portion of tooth flank which causes interference is cut away by the cutting tools. Thus there will be no conjugate action between teeth. However, by undercutting, the actual ratio of contact decreases which causes more noisy and rough gear action. Secondly, it also reduces the thickness of tooth which ultimately reduces the beam strength of tooth.

Stubbed tooth. When a portion of a tooth near the top is cut away, such a tooth is called **stubbed tooth**. Such a measure prevents interference but it reduces the contact ratio.

Number of teeth. Interference in a gear pair can be avoided by increasing number of teeth on the gears. This makes the gear larger in diameter and also increases the pitch line velocity. This increased pitch line velocity makes noiser gear action and also reduces the power transmission to some extent.

Pressure angle. Increasing the pressure angle decreases the base circle diameter of the gear, which means that it increases the involute portion of the tooth profile and hence eliminates the chances of interference. This also demands for smaller number of teeth on gear. However, it will increase the radial force component which may try to dislodge the gear. The contact ratio is also decreased which means it results into rough gear action.

10.8 MINIMUM NUMBER OF TEETH TO AVOID INTERFERENCE

As mentioned earlier, the interference between two mating gears can be avoided provided that the addendum circles of two gears in mesh cut the line which is tangent to the two base circles within points of tangency. Under limiting conditions, the addendum circles will pass through the point of tangency. Thus for a pair of gear there should be some minimum number of teeth which can prevent interference. This can be achieved only if the maximum value of the addendum radius of the gear to avoid interference is increased upto O_2C (see Figure 10.11).

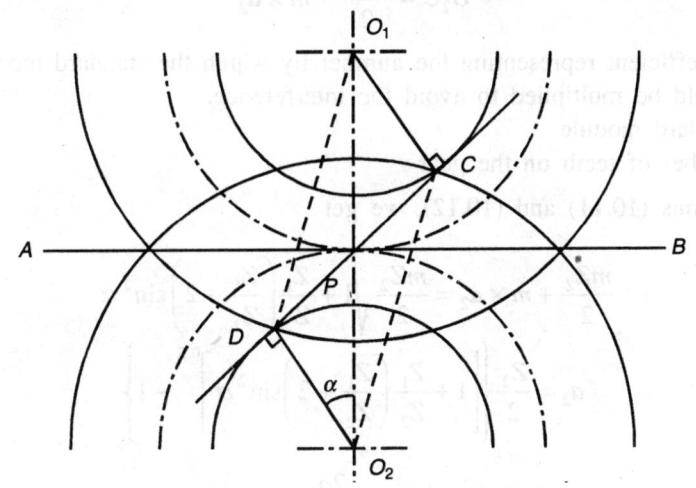

FIGURE 10.11 Limiting condition of interference.

From the geometry (refer Figure 10.11)

$$O_2C^2 = O_2D^2 + DC^2$$
$$= O_2D^2 + (DP + PC)^2$$
$$= r_2^2 \cos^2\alpha + (r_2 \sin\alpha + r_1 \sin\alpha)^2$$
$$= r_2^2 \cos^2\alpha + r_2^2 \sin^2\alpha + r_1^2 \sin^2\alpha + 2r_1r_2 \sin^2\alpha$$
$$= r_2^2 (\cos^2\alpha + \sin^2\alpha) + (r_1^2 + 2r_1r_2) \sin^2\alpha$$
$$= r_2^2 + (r_1^2 + 2r_1r_2) \sin^2\alpha$$
$$= r_2^2 \left[1 + \left(\frac{r_1^2}{r_2^2} + \frac{2r_1r_2}{r_2^2}\right) \sin^2\alpha\right]$$

or
$$O_2C = r_2\sqrt{1 + \frac{r_1}{r_2}\left(\frac{r_1}{r_2} + 2\right)\sin^2\alpha}$$

or
$$O_2C = \frac{mZ_2}{2}\sqrt{1 + \frac{Z_1}{Z_2}\left(\frac{Z_1}{Z_2} + 2\right)\sin^2\alpha} \qquad (10.11)$$

We also know that addendum circle radius:

$$O_2C = O_2P + m \times a_2$$

or
$$O_2C = \frac{mZ_2}{2} + m \times a_2 \tag{10.12}$$

where a_2 = a coefficient representing the number by which the standard module of the gear should be multiplied to avoid the interference.

m = standard module

and $Z_{2'}$ = number of teeth on the gear

Equating Equations (10.11) and (10.12), we get

$$\frac{mZ_2}{2} + m \times a_2 = \frac{mZ_2}{2}\sqrt{1 + \frac{Z_1}{Z_2}\left(\frac{Z_1}{Z_2} + 2\right)\sin^2\alpha}$$

or
$$a_2 = \frac{Z_2}{2}\left\{\left[1 + \frac{Z_1}{Z_2}\left(\frac{Z_1}{Z_2} + 2\right)\sin^2\alpha\right]^{0.5} - 1\right\}$$

or
$$Z_2 = \frac{2a_2}{\left[1 + \frac{Z_1}{Z_2}\left(\frac{Z_1}{Z_2} + 2\right)\sin^2\alpha\right]^{0.5} - 1} \tag{10.13}$$

$$= \frac{2a_2}{\left[1 + \frac{1}{G}\left(\frac{1}{G} + 2\right)\sin^2\alpha\right]^{0.5} - 1}$$

or
$$Z_2 = \frac{2a_2 G}{\left[G^2 + (1 + 2G)\sin^2\alpha\right]^{0.5} - G} \tag{10.14}$$

A similar expression for minimum number of teeth on pinion to avoid interference can be derived which is as follows:

$$Z_1 = \frac{2a_1}{\left[1 + G(G + 2)\sin^2\alpha\right]^{0.5} - 1} \tag{10.15}$$

10.9 MINIMUM NUMBER OF TEETH ON PINION WITH RACK

Rack is a segment of gear having infinite pitch circle diameter. We know that as the size of pitch circle or thereby that of base circle increases, the involute curve becomes more and more flat. Therefore, the teeth profile of a rack becomes straight line. In a rack pinion arrangement, to avoid interference, the point of contact E must lie between the points P and C. In otherwords, the limiting value of addendum should be such that the point E coincides with point C. Thus addendum of rack must be less than distance CG, as shown in Figure 10.12.

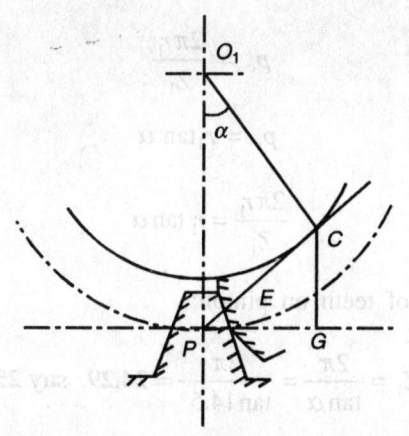

FIGURE 10.12 Interference in rack and pinion.

Maximum addendum:
$$CG = PC \sin \alpha$$
$$= (r_1 \sin \alpha) \times \sin \alpha$$
$$= r_1 \sin^2 \alpha$$

or
$$CG = \frac{mZ_1}{2} \sin^2 \alpha \qquad (10.16)$$

Let a_r = addendum coefficient
and m = module
Therefore, to avoid interference,
$$CG \geq ma_r$$

or
$$\frac{mZ_1}{2} \times \sin^2 \alpha \geq ma_r$$

or
$$Z_1 \geq \frac{2a_r}{\sin^2 \alpha} \qquad (10.17)$$

When $a_r = 1$ and pressure angle $\alpha = 20°$, the minimum number of teeth required to avoid interference is 17.1, say 18.

EXAMPLE 10.6 A pair of spur gear with involute teeth is to give a gear ratio of 4:1. The Length of arc of approach should not to be less than circular pitch and smaller gear is the driving element. The pressure angle is 14.5°. Determine:
(i) the least number of teeth that can be used on each wheel
(ii) the addendum of the gear in terms of the circular pitch

Solution To avoid interference, the path of approach should not be more than $r_1 \sin \alpha$.

$$\text{Length of arc of approach} = \frac{r_1 \sin \alpha}{\cos \alpha} = r_1 \tan \alpha$$

(i) Circular pitch:
$$p_c = \frac{2\pi r_1}{Z_1}$$

Therefore,
$$p_c = r_1 \tan \alpha$$

or
$$\frac{2\pi r_1}{Z_1} = r_1 \tan \alpha$$

Minimum number of teeth on pinion:

$$Z_1 = \frac{2\pi}{\tan \alpha} = \frac{2\pi}{\tan 14.5°} = 24.29, \text{ say } 25 \qquad \text{Ans.}$$

Number of teeth on gear: $Z_2 = Z_1 \times G$
$$= 25 \times 4 = \textbf{100 teeth} \qquad \text{Ans.}$$

(ii) Pitch circle radius of gear:
$$r_2 = \frac{Z_2 \times p_c}{2\pi} = \frac{100 \times p_c}{2\pi}$$

The limiting radius of addendum circle:
$$r_{a2} = r_2 \sqrt{1 + \frac{r_1}{r_2}\left(\frac{r_1}{r_2} + 2\right)\sin^2 \alpha}$$

Therefore, addendum on gear,
$$r_{a2} - r_2 = r_2 \left[\sqrt{1 + \frac{r_1}{r_2}\left(\frac{r_1}{r_2} + 2\right)\sin^2 \alpha} - 1\right]$$

$$= \frac{100 p_c}{2\pi}\left[\sqrt{1 + \frac{1}{4}\left(\frac{1}{4} + 2\right)\sin^2 14.5°} - 1\right]$$

$$= \textbf{0.278 } p_c \qquad \text{Ans.}$$

EXAMPLE 10.7 A pinion having 30 teeth drives a spur gear having 80 teeth. The profile of the gear is involute with 20° pressure angle, 12 mm module and addendum 10 mm. Find the (i) length of path of contact (ii) arc of contact and (iii) the contact ratio. Also show that there is no interference between the teeth.

Solution

Pitch circle radius: $r_1 = \dfrac{mZ_1}{2}$

or $$r_1 = \frac{12 \times 30}{2} = 180 \text{ mm}$$

Pitch circle radius of gear: $$r_2 = \frac{mZ_2}{2} = \frac{12 \times 80}{2} = 480 \text{ mm}$$

Radius of addendum circle of pinion:
$$r_{a1} = r_1 + \text{addendum}$$
$$= 180 + 10$$
$$= 190 \text{ mm}$$
$$r_{a2} = r_2 + \text{addendum}$$
$$= 480 + 10$$
$$= 490 \text{ mm}$$

Path of approach $= \sqrt{r_{a2}^2 - r_2^2 \cos^2 \alpha} - r_2 \sin \alpha$
$$= \sqrt{490^2 - 480^2 \times \cos^2 20°} - 480 \times \sin 20°$$
$$= 27.27 \text{ mm}$$

Path of recess $= \sqrt{r_{a1}^2 - r_1^2 \cos^2 \alpha} - r_1 \sin \alpha$
$$= \sqrt{190^2 - 180^2 \times \cos^2 20°} - 180 \times \sin 20°$$
$$= 24.98 \text{ mm}$$

(i) Path of contact = path of approach + path of recess
$$= 27.27 + 24.98$$
$$= \mathbf{52.25 \text{ mm}} \qquad \text{Ans.}$$

(ii) Length of arc of contact $= \dfrac{\text{Length of path of approach}}{\cos \alpha}$

$$= \frac{52.25}{\cos 20°} = \mathbf{55.6 \text{ mm}} \qquad \text{Ans.}$$

(iii) Contact ratio $= \dfrac{\text{Length of arc of contact}}{\text{Circular pitch}}$

Circular pitch: $p_c = \pi m$

Contact ratio $= \dfrac{55.6}{\pi \times 12} = 1.47$, say **2 pairs** $\qquad$ Ans.

In order to avoid interference between pinion and gear, the path of approach should be less than or equal to the maximum possible path of approach.

$$\text{Maximum path of approach} = r_1 \sin\alpha$$
$$= 180 \times \sin 20°$$
$$= 61.56 \text{ mm}$$

Since the actual path of approach is 27.27 mm, which is less than maximum path of approach, therefore there is no interference between a pair of gear.

EXAMPLE 10.8 A pinion of 20° involute teeth and 120 mm pitch circle diameter drives a rack. The addendum of both pinion and rack is 6 mm. What is the least pressure angle which can be used to avoid interference. Also find the length of arc of contact and contact ratio.

Solution Pitch circle radius of pinion: $r_1 = 60$ mm

Addendum circle radius: $r_{a1} = r_1 +$ Module
$$= 60 + 6 = 66 \text{ mm}$$

For no interference,

Rack addendum $\leq r_1 \sin^2\alpha$

or
$$\sin^2\alpha = \frac{6}{60}$$

or

the least angle: $\alpha = 18.43°$ **Ans.**

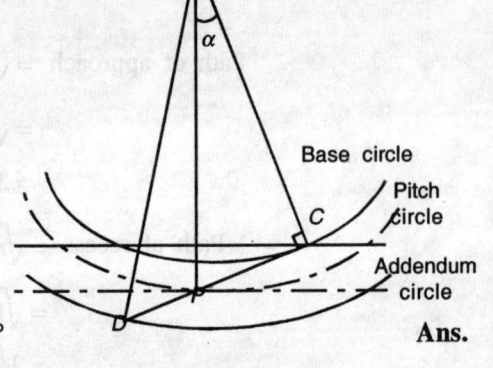

Length of path of contact $= CD = \sqrt{O_1 D^2 - O_1 C^2}$ (refer adjoining figure)

$$= \sqrt{r_{a1}^2 - r_1^2 \cos^2\alpha}$$

$$= \sqrt{66^2 - 60^2 \times \cos^2 20°}$$

$$= 34.3 \text{ mm}$$

Length of arc of contact $= \dfrac{\text{Length of path of contact}}{\cos\alpha}$

$$= \frac{34.3}{\cos 18.43°} = \mathbf{36.15 \text{ mm}} \quad \textbf{Ans.}$$

Contact ratio $= \dfrac{\text{Arc of contact}}{\text{Circular pitch}} = \dfrac{36.15}{\pi \times 6} = 1.91$, say **2 pairs** **Ans.**

EXAMPLE 10.9 A pair of spur gear has 12 and 20 teeth. The module is 12.5 mm and addendum is 12.5 mm. Show that the gears have interference, if the pressure angle is 20°, Also find the minimum number of teeth maintaining the same gear ratio to avoid interference.

Solution Pitch circle radius of pinion:

$$r_1 = \frac{mZ_1}{2} = \frac{12.5 \times 12}{2} = 75 \text{ mm}$$

Pitch circle radius of gear: $r_2 = \dfrac{mZ_2}{2} = \dfrac{12.5 \times 20}{2} = 125 \text{ mm}$

Addendum circle radius of gear:

$$r_{a2} = r_2 + \text{addendum} = 125 + 12.5 = 137.5 \text{ mm}$$

Path of approach $= \sqrt{r_{a2}^2 - r_2^2 \cos^2 \alpha} - r_2 \sin \alpha$

$$= \sqrt{137.5^2 - 125^2 \times \cos^2 20°} - 125 \times \sin 20°$$

$$= 28.72 \text{ mm}$$

To avoid interference, the path of approach should be less than or equal to maximum possible path of approach, which is

$$r_1 \sin 20° = 75 \times \sin 20° = 25.65 \text{ mm}$$

Since the actual path of approach, i.e. 28.72 mm, is greater than maximum possible path of approach, i.e. 25.65 mm there is interference between a pair of gear.

In order to avoid interference, the minimum number of teeth on gear should be

$$Z_2 \geq \frac{2a_2 G}{[G^2 + (1+2G)\sin^2 \alpha]^{0.5} - G}$$

Addendum: $\qquad a_2 m = 12.5$

or $\qquad a_2 \times 12.5 = 12.5$

or $\qquad a_2 = 1$

Gear ratio: $\qquad G = \dfrac{Z_2}{Z_1} = \dfrac{20}{12} = 1.667$

or $\qquad Z_2 \geq \dfrac{2 \times 1 \times 1.667}{[1.667^2 + (1+2\times 1.667)\times \sin^2 20°]^{0.5} - 1.667}$

$\qquad \geq 22.88$

So let us adopt number of teeth on gear:

$$Z_2 = 25 \text{ teeth} \qquad \text{Ans.}$$

Number of teeth on pinion: $Z_1 = Z_2 \times \dfrac{1}{1.667}$

$$= \dfrac{25}{1.667} = 15 \text{ teeth} \qquad \textbf{Ans.}$$

EXAMPLE 10.10 A pair of involute spur gears is in mesh having 13 and 39 teeth on pinion and gear respectively. Both pinion and gear have addendum equal to one module which is 12 mm. Determine whether the gear interfere with the pinion or not. If there is any interference, suggest suitable value of pressure angle which can avoid interference.

Solution Pitch circle radius of pinion:

$$r_1 = \dfrac{mZ_1}{2} = \dfrac{10 \times 13}{2} = 65 \text{ mm}$$

Pitch circle radius of gear: $r_2 = \dfrac{mZ_2}{2} = \dfrac{10 \times 39}{2} = 195 \text{ mm}$

Addendum circle radius of gear = r_2 + addendum

or $\qquad r_{a2} = 195 + 12 = 207 \text{ mm}$

Referring Figure (10.11), the maximum possible addendum circle radius of gear is:

$$r_{a2\,max} = \sqrt{(r_2 \cos\alpha)^2 + (r_2 \sin\alpha + r_1 \sin\alpha)^2}$$

$$= \sqrt{(195 \times \cos 20°)^2 + (195 \times \sin 20° + 65 \times \sin 20°)^2}$$

$$= \sqrt{(195 \times \cos 20°)^2 + (260 \times \sin 20°)^2}$$

$$= 203.67 \text{ mm}$$

The actual addendum circle radius is greater than the maximum possible addendum circle radius, i.e.

$$r_{a2} > r_{a2\,max}$$

Therefore, interference between gear and pinion will occur.

To avoid interference, $r_{a2\,max}$ should be equal to r_{a2}

$$r_{a2} = r_{a2\,max}$$

or $\qquad 207 = \sqrt{(195 \times \cos\alpha)^2 + (260 \times \sin\alpha)^2}$

or $\qquad \cos\alpha = 0.9148$

or $\qquad \alpha = 23.82° \qquad \textbf{Ans.}$

Thus, the pressure angle is required to be increased to 23.82° to avoid interference.

10.10 FORCE ANALYSIS OF A SPUR GEAR

A gear drive is generally specified by power to be transmitted, the speed of the driving shaft and the velocity ratio. The power is transmitted by means of a force exerted by the tooth of driving gear on the mating driven gear. According to the law of gearing, this force F_n is always normal to the tooth surface and acts along the pressure angle line. This normal force is designated by two subscripts, for example F_{12}, which means the force exerted by gear 1 against gear 2.

Figure 10.13(a) shows a pair of spur gear in mesh mounted on respective shafts. The driving gear 2 is mounted on shaft 1 and rotates in the clockwise direction. The driven gear 3 is mounted on shaft 4. The free body diagram of the forces acting upon two gears along the pressure line is shown in Figure 10.13(b). The driving gear 2 exerts a force F_{23} on the driven gear 3. Similarly, the driving gear experiences a reaction force F_{32}.

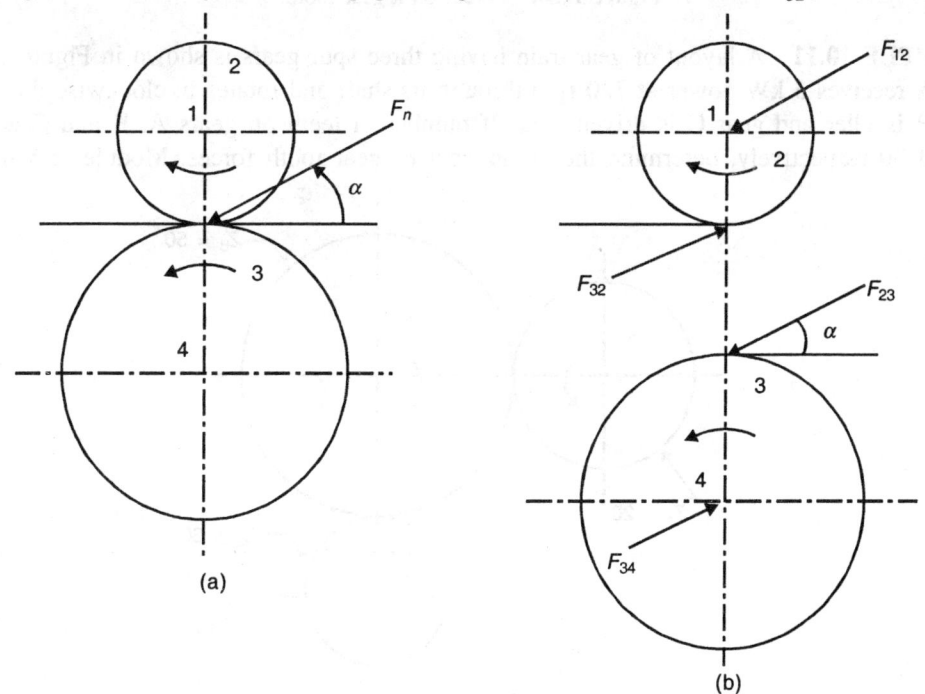

FIGURE 10.13 Forces acting on two mating spur gears.

The normal force F_n (or F_{23}), as shown in Figure 10.14, acting along the pressure line can be resolved into two components—the tangential force F_t and radial force F_r. Thus

$$F_t = F_n \cos \alpha \tag{10.18}$$

$$F_r = F_n \sin \alpha = F_t \tan \alpha \tag{10.19}$$

where α is the pressure angle.

The tangential component of force F_t is mainly responsible for transmitting torque and consequently the power. The radial force F_r is called the **separating force,** which always acts towards the centre of the gear.

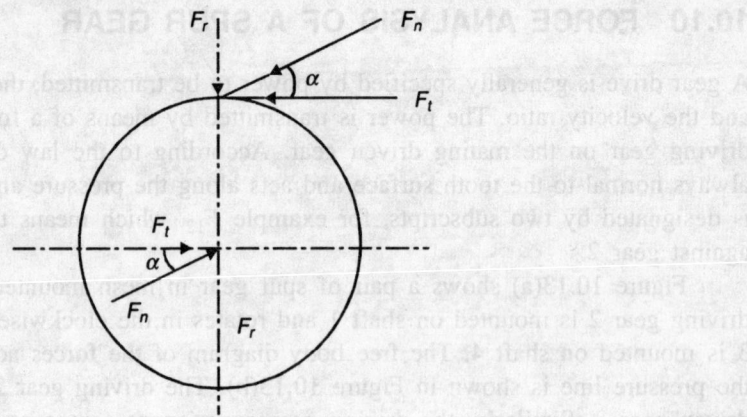

Figure 10.14 Forces on a gear tooth.

EXAMPLE 10.11 A layout of gear train having three spur gears is shown in Figure 10.15. Gear A receives 3 kW power at 720 rpm through its shaft and rotates in clockwise direction. Gear B is idler and gear C is driven gear. If number of teeth on gears A, B and C are 20, 50 and 30 respectively, determine the component of gear tooth forces. Module is 5 mm.

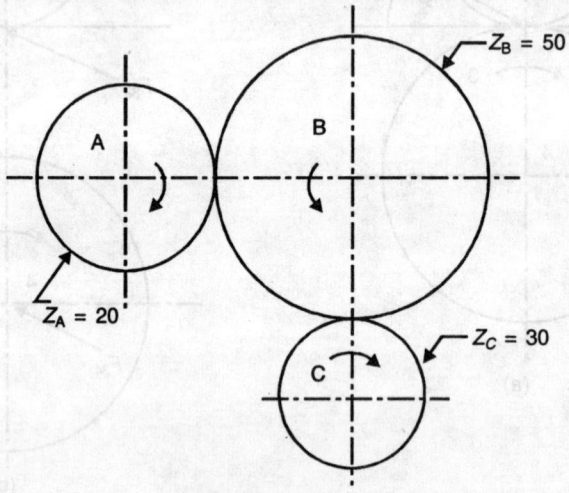

FIGURE 10.15

Solution

$$Z_A = 20, \quad Z_B = 50 \text{ and } Z_C = 30 \text{ with } m = 5 \text{ mm.}$$

Let us assume that all three gears have 20° full depth involute profile tooth.
Pitch circle diameter of gears:

$$d_A = mZ_A = 5 \times 20 = 100 \text{ mm}$$
$$d_B = mZ_B = 5 \times 50 = 250 \text{ mm}$$
$$d_C = mZ_C = 5 \times 30 = 150 \text{ mm}$$

Torque transmitted by gear A:

$$T_A = \frac{60 \times P}{2\pi N_A} = \frac{60 \times 3 \times 10^3}{2\pi \times 720} = 39.79 \text{ Nm}$$

Tangential force on gear A:

$$F_{tA} = \frac{2T_A}{d_A}$$

$$= \frac{2 \times 39.79 \times 1000}{100} = 795.8 \text{ N}$$

Radial force:

$$F_{rA} = F_{tA} \tan \alpha$$
$$= 795.8 \times \tan 20°$$
$$= 289.65 \text{ N}$$

Normal force:

$$F_{nA} = \frac{F_{tA}}{\cos \alpha}$$

$$= \frac{795.8}{\cos 20°} = \mathbf{846.87 \text{ N}}$$

Since gear B is idler, whatever torque is received from gear A is transmitted to gear C. Therefore, tangential and radial components between the gears B and C must be equal to the tangential and radial components between gears A and B.

$$F_{tB} = 795.8 \text{ N}, \quad F_{rB} = 289.69 \text{ N} \quad \text{and} \quad F_{nB} = 846.87 \text{ N}$$

The free body diagram of gear tooth forces is shown in Figure 10.16.

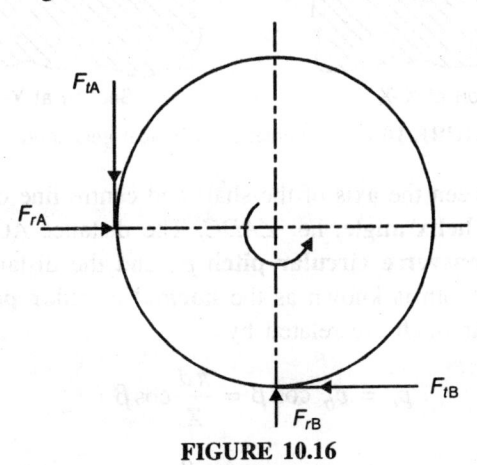

FIGURE 10.16

10.11 HELICAL GEAR

A helical gear may be considered to be composed of an infinite number of infinitesimally narrow staggered spur teeth. The result is that each tooth slants across the face so as to form a cylindrical helix. In helical gears, there is progressive engagement of the tooth and gradual pick up of the load by it. It results into a smooth engagement and quiet operation even at a high pitch line velocity. Helical gears are of two types—parallel helical gears used for parallel shafts and cross helical gears used for non-parallel shafts. In a parallel helical gear, the slope of the helix may be either right handed or left handed. A general rule to determine the hand of helix is same as that used for the screw.

Figure 10.17 shows the top view of a helical gear in which the lines AB and CD are the centre lines of two adjacent helical teeth taken on the pitch line and line AD is perpendicular

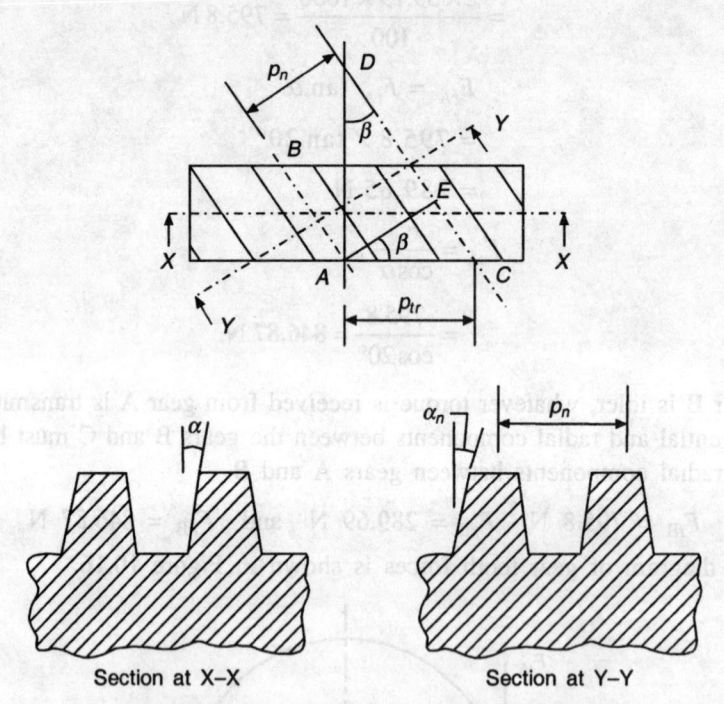

FIGURE 10.17 Kinematics of helical gear tooth.

to the edge. The angle between the axis of the shaft and centre line of the tooth taken on the pitch line is known as the **helix angle**, i.e. $\angle ADC$. The distance AC measured in the plane of rotation XX is called **transverse circular pitch** p_{tr} and the distance AE measured in the plane perpendicular to the tooth is known as the **normal circular pitch** p_n. Normal circular pitch and transverse circular pitch are related by

$$p_n = p_{tr} \cos \beta = \frac{\pi d}{Z} \cos \beta \tag{10.20}$$

and normal module:

$$m_n = m \cos \beta \tag{10.21}$$

The distance AD, called the axial pitch p_a, is related to transverse pitch.

$$p_a = \frac{p_{tr}}{\tan \beta} \tag{10.22}$$

The centre distance C between two helical gears having Z_1 and Z_2 as the number of teeth is computed by the following relation:

$$C = \frac{d_1 + d_2}{2} = \frac{m_n(Z_1 + Z_2)}{2\cos\beta} \qquad (10.23)$$

10.11.1 Formative Number of Teeth

If the pitch cylinder of a helical gear is cut by an oblique plane X–X at an helix angle β which is normal to the tooth as shown in Figure 10.18, the oblique plane cuts out an arc having a radius of curvature R. When the helix angle β is zero, the radius of curvature is equal to

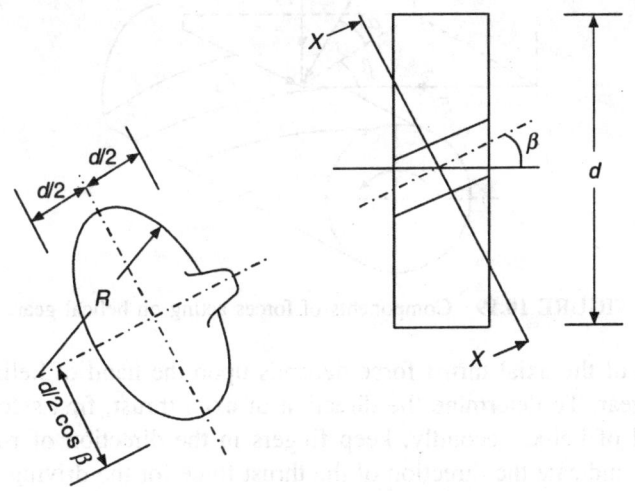

FIGURE 10.18 Formative pitch circle.

$d/2$ as in the case of spur gear. If the helix angle β is increased from 0° to 90°, the radius of curvature approaches infinity. This radius is the apparent pitch circle radius of the helical gear when viewed in the direction of the tooth element. Thus it is the radius of an equivalent spur gear having a greater number of teeth. In the helical gear terminology, it is called the **formative** or **virtual** number of teeth. It is defined as the number of teeth on an equivalent spur gear which will give the same tooth profile when measured normal to the helix. The selection of a cutter for milling a helical gear is based on the formative number of teeth. The formative number of teeth Z' is related to the actual number of teeth by the following relation:

$$Z' = \frac{Z}{\cos^3\beta} \qquad (10.24)$$

10.11.2 Forces on Helical Gear

The point of application of the forces in a helical gear is in the pitch plane and at the centre of the face as shown in Figure 10.19. The resultant force F acting on the gear tooth has three components—tangential force F_t, radial force F_r and axial force F_a. From the geometry of Figure 10.19, the three components of resultant force F are:

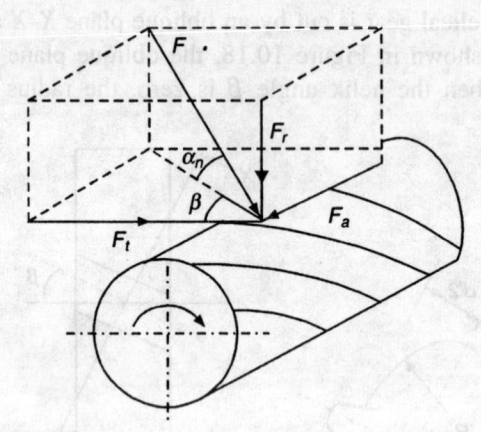

$$F_r = F \sin\alpha_n \tag{10.25a}$$

$$F_t = F \cos\alpha_n \cos\beta \tag{10.25b}$$

$$F_a = F \cos\alpha_n \sin\beta \tag{10.25c}$$

FIGURE 10.19 Components of forces acting on helical gear.

The direction of the axial thrust force depends upon the hand of helix and the direction of rotation of the gear. To determine the direction of axial thrust, first select the driving gear and mark the hand of helix. Secondly, keep fingers in the direction of rotation of the gear, the thumb will then indicate the direction of the thrust force for the driving gear. The direction of thrust for the driven gear is opposite to that of the driving gear.

10.12 SPIRAL GEAR

Helical gears used to connect non-parallel, non-intersecting shafts are known as **spiral gears** or **crossed helical gears.** In these gears, there is a point contact between two gears in mesh instead of a line contact as for helical gears. Thus these gears are suitable for light load only.

A pair of spiral gear in mesh is shown in Figure 10.20. These gears connect two shafts inclined at an angle θ, called **shaft angle.**

Let β_1 = spiral angle of teeth on gear A
and β_2 = spiral angle of teeth on gear B

The shaft angle θ is defined as the angle through which one shaft is rotated relative to other about a line joining gear centres to bring the shafts parallel and cause rotation in the opposite direction.

Therefore, for gear pair having teeth of the same hand,

$$\theta = \beta_1 + \beta_2 \tag{10.26a}$$

and for gear pair having teeth of opposite hand,

$$\theta = \beta_1 - \beta_2 \quad \text{or} \quad \beta_2 - \beta_1 \tag{10.26b}$$

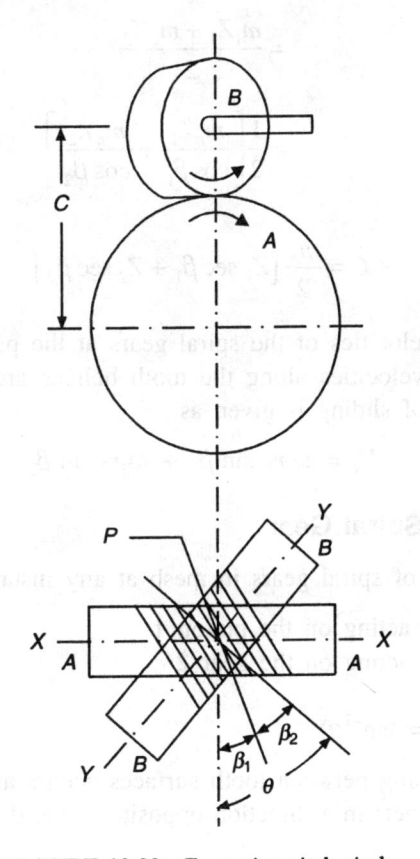

FIGURE 10.20 Formative pitch circle.

Let m_n = normal module which is same for gear and pinion

and m_1, m_2 = modules measured in the plane of revolution which depend on the spiral angle.

$$m_n = m_1 \cos \beta_1 = m_2 \cos \beta_2$$

Also
$$m_1 = \frac{d_1}{Z_1} \quad \text{and} \quad m_2 = \frac{d_2}{Z_2}$$

where d_1, d_2 are pitch circle diameters and Z_1, Z_2 are number of teeth of gears.

Gear ratio:
$$G = \frac{\omega_1}{\omega_2} = \frac{Z_2}{Z_1} \quad \text{or} \quad G = \frac{m_1 d_2}{m_2 d_1} \tag{10.27}$$

Centre distance:
$$C = \frac{1}{2}(d_1 + d_2)$$

$$= \frac{m_1 Z_1 + m_2 Z_2}{2}$$

$$= \frac{1}{2}\left[\frac{m_n Z_1}{\cos \beta_1} + \frac{m_n Z_2}{\cos \beta_2}\right]$$

or
$$C = \frac{m_n}{2}\left[Z_1 \sec \beta_1 + Z_2 \sec \beta_2\right] \qquad (10.28)$$

The circumferential velocities of the spiral gears at the pitch point are $\omega_1 r_1$ and $\omega_2 r_2$. The components of these velocities along the tooth helices are $\omega_1 r_1 \sin \beta_1$ and $\omega_2 r_2 \sin \beta_2$ respectively. The velocity of sliding is given as

$$V_s = \omega_1 r_1 \sin \beta_1 + \omega_2 r_2 \sin \beta_2 \qquad (10.29)$$

10.12.1 Efficiency of Spiral Gear

Figure 10.21 shows a pair of spiral gears in mesh at any instant.

Let F_{t1} = tangential force acting on the pinion 1
F_{t2} = tangential force acting on the gear 2
R_n = normal reaction
and ϕ = friction angle (= $\tan^{-1} \mu$)

In spiral gear, the sliding between tooth surfaces accurs along the tangent to the pitch helix and the friction force acts in a direction opposite to the direction of sliding of the gear surface.

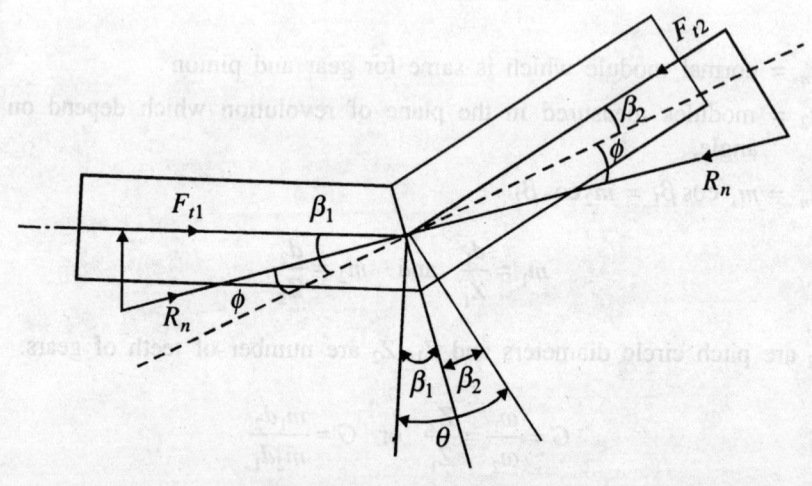

FIGURE 10.21 Spiral gears in mesh.

Referring to Figure 10.21,

Tangential force:
$$F_{t1} = R_n \cos(\beta_1 - \phi) \tag{10.30}$$

and
$$F_{t2} = R_n \cos(\beta_2 + \phi) \tag{10.31}$$

or
$$\frac{F_{t1}}{F_{t2}} = \frac{\cos(\beta_1 - \phi)}{\cos(\beta_2 + \phi)} \tag{10.32}$$

If there is no friction between the mating teeth, then tangential force required for pinion is:

$$F'_{t1} = F_{t2} \times \frac{\cos \beta_1}{\cos \beta_2}$$

Therefore, the efficiency of the spiral gear drive is given by

$$\eta = \frac{F'_{t1}}{F_{t1}} = \frac{\cos \beta_1}{\cos \beta_2} \times \frac{\cos(\beta_2 + \phi)}{\cos(\beta_1 - \phi)} \tag{10.33}$$

or
$$\eta = \frac{\cos(\beta_1 + \beta_2 + \phi) + \cos(\beta_1 - \beta_2 - \phi)}{\cos(\beta_2 + \beta_1 - \phi) + \cos(\beta_2 - \beta_1 + \phi)}$$

$$= \frac{\cos(\theta + \phi) + \cos(\beta_1 - \beta_2 - \phi)}{\cos(\theta - \phi) + \cos(\beta_2 - \beta_1 + \phi)}$$

Using trigonometrical identity $\cos\theta = \cos(-\theta)$, we get

$$\eta = \frac{\cos(\theta + \phi) + \cos(\beta_1 - \beta_2 - \phi)}{\cos(\theta - \phi) + \cos(\beta_1 - \beta_2 - \phi)} \tag{10.34}$$

For maximum efficiency,
$$\cos(\beta_1 - \beta_2 - \phi) = 1$$

or
$$\beta_1 - \beta_2 - \phi = 0 \tag{10.35}$$

Substituting the value of angles β_1 and β_2 from Eq. (10.35) and Eq. (10.26a) in Eq. (10.34), we get the expression for maximum efficiency as

$$\eta_{max} = \frac{1 + \cos(\theta + \phi)}{1 + \cos(\theta - \phi)} \tag{10.36}$$

EXAMPLE 10.12 The centre distance between two spiral gears in mesh is 150 mm and the angle between the shaft axes is 60°. The gear ratio is 2 and normal circular pitch is 10 mm. If the driven gear has a helix angle of 25°, determine

(i) the number of teeth on pinion and gear
(ii) the exact centre distance
(iii) the efficiency if the coefficient of friction is 0.07.

Solution Gear ratio

$$G = \frac{Z_2}{Z_1} \quad \text{or} \quad Z_2 = 2Z_1 \qquad (i)$$

Angle $\beta_1 = \theta - \beta_2 = 60° - 25° = 35°$

(i) Centre distance: $\quad C = \dfrac{m_n}{2}\left[Z_1 \sec \beta_1 + Z_2 \sec \beta_2\right]$

or $\quad C = \dfrac{p_n}{2\pi}\left[Z_1 \sec \beta_1 + Z_2 \sec \beta_2\right]$

or $\quad 150 = \dfrac{10}{2\pi}\left[Z_1 \times \sec 35° + Z_2 \times \sec 25°\right]$

or $\quad 150 = 1.943\, Z_1 + 2 \times Z_1 \times 1.756$

or $\quad Z_1 = 27.49$, say **28 teeth**   **Ans.**

$Z_2 = 2 \times Z_1 = 2 \times 28 = \textbf{56 teeth}$ **Ans.**

(ii) Exact centre distance:

$$C = \frac{10}{2\pi}\left[28 \times \sec 35° + 56 \times \sec 25°\right]$$

$$= \textbf{152.74 mm} \qquad \textbf{Ans.}$$

(iii) Friction angle: $\phi = \tan^{-1}\mu = \tan^{-1} 0.07 = 4°$

Efficiency: $\quad \eta = \dfrac{\cos \beta_1}{\cos \beta_2} \times \dfrac{\cos(\beta_2 + \phi)}{\cos(\beta_1 - \phi)}$

$$= \frac{\cos 35°}{\cos 25°} \times \frac{\cos(25° + 4°)}{\cos(35° - 4°)}$$

$$= 0.922 \quad \text{or} \quad \textbf{92.2 per cent} \qquad \textbf{Ans.}$$

EXAMPLE 10.13 A pair of spiral gear is required to transmit motion between shafts having non-intersecting axes as 90°. The pitch circle diameter of the gear is equal to 1.5 times that of pinion and the gear ratio is 3. If the approximate centre distance is 250 mm and normal module is 4 mm, determine (i) helix angles (ii) number of teeth on pinion and gear and (iii) exact centre distance.

Solution Gear ratio: $\quad G = \dfrac{Z_2}{Z_1} = 3 \quad \text{or} \quad Z_2 = 3Z_1$

Also $$m_1 = \frac{d_1}{Z_1} \quad \text{and} \quad m_2 = \frac{d_2}{Z_2}$$

$$m_n = \frac{d_1}{Z_1} \times \cos \beta_1 = \frac{d_2}{Z_2} \times \cos \beta_2$$

or $$\frac{d_2}{d_1} = \frac{Z_2}{Z_1} \times \frac{\cos \beta_1}{\cos \beta_2}$$

or $$\frac{1.5 d_1}{d_1} = \frac{Z_2}{Z_1} \times \frac{\cos \beta_1}{\cos \beta_2}$$

or $$\frac{\cos \beta_1}{\cos \beta_2} = 1.5 \times \frac{Z_1}{Z_2} = 1.5 \times \frac{1}{3} = 0.5$$

Since $\beta_1 + \beta_2 = 90°$, $\beta_2 = 90° - \beta_1$

$$\frac{\cos \beta_1}{\cos(90° - \beta_1)} = 0.5$$

or $\cot \beta_1 = 0.5$

∴ $\beta_1 = 63.43°$ and $\beta_2 = 90° - 63.43° = 26.57°$ **Ans.**

Centre distance: $$C = \frac{m_n}{2}(Z_1 \sec \beta_1 + Z_2 \sec \beta_2)$$

or $$250 = \frac{4}{2}(Z_1 \times \sec 63.43° + Z_2 \times \sec 26.57°)$$

or $Z_1 = 22.36$, say **23 teeth**

$Z_2 = 3 \times 23 =$ **69 teeth** **Ans.**

Exact centre distance: $$C = \frac{4}{2}(23 \times \sec 63.43° + 69 \times \sec 26.57°)$$

$= 257.13$ mm **Ans.**

EXAMPLE 10.14 Two spiral gears in mesh have the following data:
  (i) Shaft angle, $\theta = 90°$
  (ii) Normal module, $m_n = 6$ mm
  (iii) Gear ratio, $G = 2$
  (iv) Approximate centre distance $= 400$ mm
  (v) Coefficient of friction, $\mu = 0.105$

Determine the following parameters of the gear if the drive is to transmit power at the maximum efficiency:

(a) Spiral angle
(b) Number of teeth on pinion and gear
(c) Pitch diameter of pinion and gear
(d) Exact centre distance
(e) Efficiency of the drive
(f) Virtual number of teeth on gear and pinion.

Solution

(a) The condition of maximum efficiency is:
$$\beta_1 - \beta_2 - \phi = 0$$

Further, we also know that $\beta_1 + \beta_2 = \theta$

Therefore, $\beta_1 - (\theta - \beta_1) - \phi = 0$

or $$\beta_1 = \frac{\theta + \phi}{2}$$

where
$$\phi = \tan^{-1}\mu = \tan^{-1} 0.105 = 6°$$

∴ $\beta_1 = \dfrac{90 + 6}{2} = 48°$ and $\beta_2 = 90° - 48° = 42°$ **Ans.**

(b) Gear ratio: $G = \dfrac{Z_2}{Z_1} = 2$ or $Z_2 = 2Z_1$

Centre distance: $C = \dfrac{m_n}{2}\left[Z_1 \sec\beta_1 + Z_2 \sec\beta_2\right]$

or $400 = \dfrac{6}{2}\left[Z_1 \times \sec 48° + Z_2 \times \sec 42°\right]$

$= 3 \times \left[Z_1 \times 1.4944 + 2Z_1 \times 1.3456\right]$

or $Z_1 = 31.85$, say **32 teeth** **Ans.**

$Z_2 = 2 \times 32 = $ **64 teeth** **Ans.**

(c) We know that $m_n = m_1 \cos\beta_1 = m_2 \cos\beta_2$

or $m_1 = \dfrac{m_n}{\cos\beta_1} = \dfrac{6}{\cos 48°} = 8.967$ mm

and $m_2 = \dfrac{m_n}{\cos\beta_2} = \dfrac{6}{\cos 42°} = 8.073$ mm

Pitch circle diameter of pinion:
$$d_1 = m_1 Z_1 = 8.967 \times 32 = \mathbf{286.94 \text{ mm}}$$
Pitch circle diameter of gear:
$$d_2 = m_2 Z_2 = 8.073 \times 64 = \mathbf{516.67 \text{ mm}}$$

(d) Exact centre distance:
$$C = \frac{1}{2}(d_1 + d_2) = \frac{1}{2}(286.94 + 516.67)$$
$$= \mathbf{401.805 \text{ mm}} \qquad \text{Ans.}$$

(e) Efficiency of the drive:
$$\eta = \frac{\cos \beta_1}{\cos \beta_2} \times \frac{\cos(\beta_2 + \phi)}{\cos(\beta_1 - \phi)}$$
$$= \frac{\cos 48°}{\cos 42°} \times \frac{\cos(42° + 6°)}{\cos(48° - 6°)}$$
$$= 0.81 \text{ or } \mathbf{81 \text{ per cent}} \qquad \text{Ans.}$$

(f) Formative or virtual number of teeth:
$$Z_1' = \frac{Z_1}{\cos^3 \beta_1} = \frac{32}{\cos^3 48°} = 106.8, \text{ say } \mathbf{107 \text{ teeth}} \qquad \text{Ans.}$$

and
$$Z_2' = \frac{Z_2}{\cos^3 \beta_2} = \frac{64}{\cos^3 42°} = 155.94, \text{ say } \mathbf{156 \text{ teeth}} \qquad \text{Ans.}$$

10.13 WORM GEAR SET

A worm gear set is used to transmit power between two non-parallel. non-intersecting shafts. It consists of worm, which is very much similar to a threaded screw and a worm gear (Figure 10.22). It is widely used in machine tools, automobiles, material handling equipments and cement plants for rotating the kiln. The worm gear has the following important features:

  (i) Reduction of high speed in the smallest possible space.
  (ii) Tooth engagement occurs without shock, hence operation is quieter.
  (iii) The provision for self-locking can also be made.
  (iv) The transmission efficiency is very low compared to spur and helical gears and generates considerable amount of heat.

In a worm gear set, the number of threads, i.e. number of start of worm plays an important role in deciding mechanical advantage and efficiency of the drive. As the number of starts increases, the transmission efficiency improves but at the cost of mechanical advantage. Figure 10.23 shows the nomenclature of the worm and worm gear.

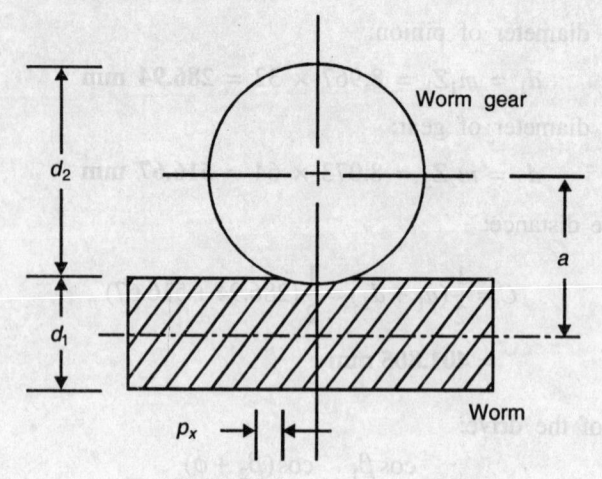

FIGURE 10.22 A worm gear set.

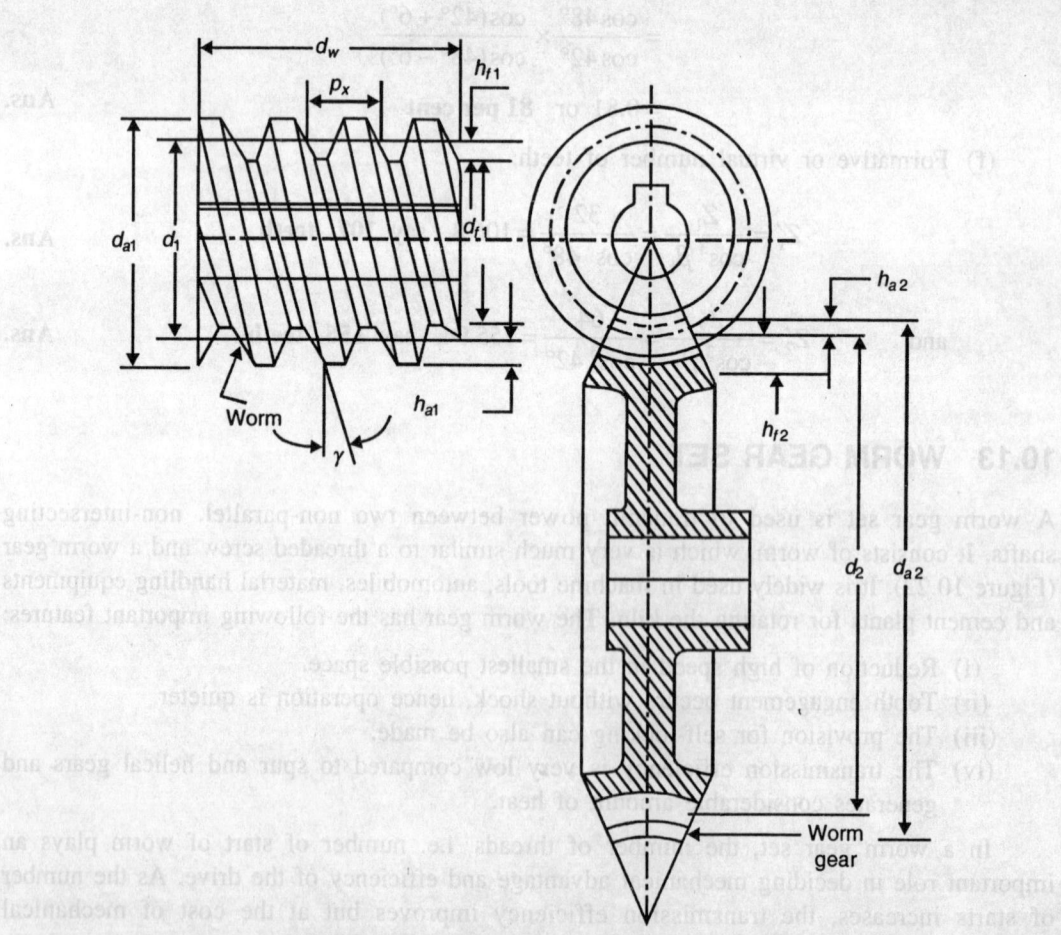

FIGURE 10.23 Nomenclature of worm gear set.

Axial pitch (p_x). It is the distance measured axially between two corresponding point on adjacent teeth. This is equal to transverse circular pitch (p_{cir}) of the mating worm gear, i.e.

$$p_x = p_{cir} = \frac{\pi d_2}{Z_2} \qquad (10.37)$$

where d_2 is the pitch diameter of the worm gear.

Lead (l). It is the axial distance by which a worm advances during its one revolution. The lead is equal to the product of the number of starts and axial pitch of the worm, i.e.

$$l = Z_1 \times p_x = \pi m Z_1 \qquad (10.38)$$

where Z_1 is the number of starts of worm.

Lead Angle (γ). It is the angle between the tangent to the helix on the pitch circle and the plane normal to the worm axis. It is given by

$$\tan \gamma = \frac{l}{\pi d_1} \qquad (10.39)$$

For a shaft with a shaft angle of 90° the lead angle:

$$\gamma = \frac{\pi}{2} - \text{helix angle} \qquad (10.40)$$

Velocity ratio. The ratio of number of teeth on worm gear and number of starts of the worm is called velocity ratio, i.e.

$$VR = \frac{Z_2}{Z_1} \qquad (10.41)$$

Centre distance. The centre distance between the worm and worm gear is given by the following relation:

$$C = \frac{d_1 + d_2}{2}$$

$$= \frac{m_n}{2}\left(\frac{Z_1}{\cos \beta_1} + \frac{Z_2}{\cos \beta_2}\right)$$

$$= \frac{m_2 \cos \beta_2}{2} \times \frac{1}{\cos \beta_2}\left(\frac{\cos \beta_2}{\cos \beta_1} Z_1 + Z_2\right)$$

as we know that $\beta_1 + \beta_2 = \theta = 90°$ and lead angle of worm is equal to helix angle of worm gear ($\gamma = \beta_2$).

$$C = \frac{m_2}{2}\left[\frac{\cos \gamma}{\cos(90° - \gamma)} Z_1 + Z_2\right]$$

or

$$C = \frac{m_2}{2}(Z_1 \cot \gamma + Z_2) \qquad (10.42)$$

10.13.1 Efficiency of Worm Gear Set

The worm and worm gear is considered as a special case of spiral gears in mesh, in which the angle between driving and driven axes is right angle.

$$\beta_1 + \beta_2 = \theta = \frac{\pi}{2} \tag{10.43}$$

We also know that the lead angle is compliment of the helix angle, i.e.

$$\gamma = \frac{\pi}{2} - \beta_1$$

or

$$\left(\frac{\pi}{2} - \gamma\right) + \beta_2 = \frac{\pi}{2}$$

or

$$\gamma = \beta_2 \tag{10.44}$$

that is lead angle of worm is equal to the helix angle of the worm gear.
Referring to Eq. (10.33), the expression for efficiency of spiral gear pair in mesh is:

$$\eta = \frac{\cos\beta_1}{\cos\beta_2} \times \frac{\cos(\beta_2 + \phi)}{\cos(\beta_1 - \phi)}$$

or

$$\eta = \frac{\cos(\pi/2 - \gamma)}{\cos\gamma} \times \frac{\cos(\gamma + \phi)}{\cos\left(\frac{\pi}{2} - \gamma - \phi\right)}$$

$$= \frac{\sin\gamma}{\cos\gamma} \times \frac{\cos(\gamma + \phi)}{\sin(\gamma + \phi)}$$

or

$$\eta = \frac{\tan\gamma}{\tan(\gamma + \phi)} \tag{10.45}$$

If the worm gear is driver, the expression for efficiency can be deduced as

$$\eta = \frac{\tan(\gamma - \phi)}{\tan\gamma} \tag{10.46}$$

The condition for maximum efficiency is:

$$\gamma = \frac{\pi}{4} - \frac{\phi}{2} \tag{10.47}$$

and the maximum efficiency is:

$$\eta_{max} = \frac{1 - \sin\phi}{1 + \sin\phi} \tag{10.48}$$

EXAMPLE 10.15 A worm gear set transmits power between two shafts which are at right angle to each other. The gear ratio is 6 and the worm has 4 teeth of normal pitch 20 mm. The pitch diameter of the worm is 50 mm. Calculate the tooth angles of the worm and worm gear and the distance between the shafts. If the efficiency of the drive is 85 per cent, the worm being the driver, find the coefficient of friction between the mating tooth surfaces.

Solution Gear ratio:

$$G = \frac{Z_2}{Z_1} = 6$$

or
$$Z_2 = 6 \times Z_1$$

or
$$Z_2 = 6 \times 4 = 24 \text{ teeth}$$

Lead angle:
$$\tan \gamma = \frac{l}{\pi d_1} = \frac{4p}{\pi d_1} = \frac{4 \times p_n}{\pi d_1 \times \cos \gamma}$$

$$= \frac{4 \times 20}{\pi \times 50 \times \cos \gamma}$$

or
$$\sin \gamma = 0.5093$$

Lead angle:
$$\gamma = 30.61°$$

Spiral angle (gear): $\beta_2 = \gamma = \mathbf{30.61°}$ **Ans.**

Spiral angle (worm): $\beta_1 = 90° - \beta_2 = 90° - 30.61°$

$= \mathbf{59.39°}$ **Ans.**

Pitch circle diameter of worm gear:

$$d_2 = Z_2 \times m = Z_2 \times \frac{p}{\pi}$$

$$= \frac{Z_2 p_n \times \sec \gamma}{\pi}$$

$$= \frac{24 \times 20 \times \sec 30.61°}{\pi}$$

$$= 177.52 \text{ mm}$$

Centre distance $= \frac{1}{2}(d_1 + d_2) = \frac{50 + 177.52}{2} = \mathbf{113.76 \text{ mm}}$ **Ans.**

Efficiency of the drive:
$$\eta = \frac{\tan \gamma}{\tan(\gamma + \phi)}$$

or
$$\frac{\tan 30.61°}{\tan(30.61° + \phi)} = 0.85$$

or
$$\tan(30.61° + \phi) = \frac{\tan 30.61°}{0.85} = 0.696$$

or
$$\text{Friction angle: } \phi = 4.23°$$

or
$$\text{Coefficient of friction: } \mu = \tan\phi = \tan 4.23° = \mathbf{0.074} \qquad \textbf{Ans.}$$

10.14 BEVEL GEARS

Bevel gears are used to transmit power between two intersecting shafts. These gears are cut in conical pitch surfaces. Two types of bevel gears—straight tooth and spiral tooth—as shown in Figure 10.24 are commonly used. The teeth on a straight tooth bevel gear are on straight line which converge to a common point called the **apex of the pitch cone,** which is also the point of intersection of the gear axes. Involute profile straight bevel gears are used for

(a) Straight tooth bevel gear   (b) Spiral bevel gear

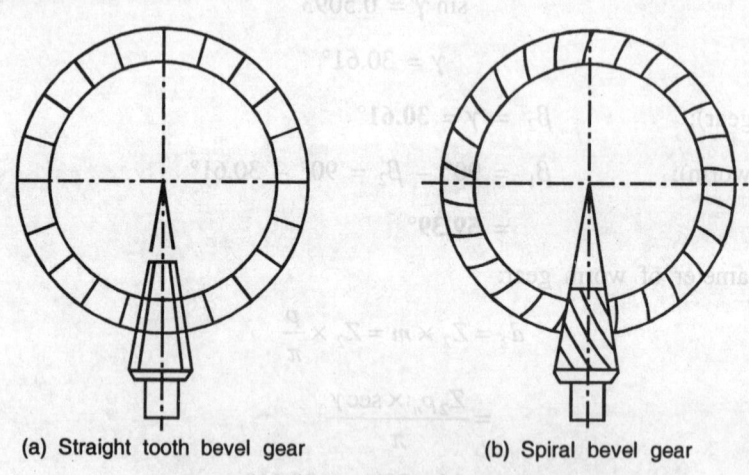

FIGURE 10.24 Types of bevel gears.

relatively low speed application with pitch line velocity upto 10 m/s. When smooth tooth engagement, quiet operation, greater strength and high pitch line velocity are the major requirements, spiral bevel gears with curved teeth are used.

Bevel gears are not interchangeable, hence these gears are designed in pair. In the majority of applications, the angle between the axes of two intersecting shaft is 90°. However, the intersecting angle may be acute or obtuse angle. These gears are manufactured either by casting, machining or generating process.

The definitions and dimensions relating to bevel gear are shown in Figure 10.25. In a bevel gear, if the pitch line distance L, called cone distance, is revolved about the axis of the

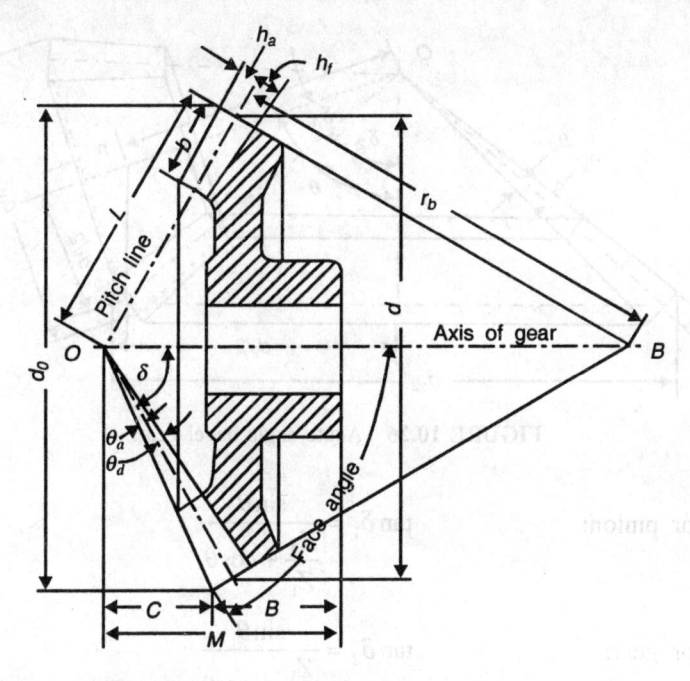

FIGURE 10.25 Dimensions of a bevel gear.

gear, it generates an imaginary pitch cone with the apex at O. The angle formed between the cone distance and the axis is called the **pitch angle** δ. The angles θ_a and θ_d are called **addendum** and **dedendum** angles, respectively. The sum of pitch and addendum angles $(\delta + \theta_a)$ is known as the **face angle**. Dimensions of the bevel gear such as addendum h_a, dedendum h_f and pitch diameter d are specified at the larger end of the tooth. A line drawn perpendicular to the pitch line intersects the axis at point B and forms a cone called the **back cone**. The length of the back cone is called the **back cone radius** r_b. The distance C is called the **crown hight**. The B is **backing distance** and M is known as **mounting distance**.

In bevel gear, for the purpose of design, an imaginary spur gear in a plane perpendicular to the tooth at the larger end having pitch circle radius r_b is considered for finding the formative or virtual number of teeth, i.e.

Formative number of teeth:
$$Z' = \frac{Z}{\cos \delta} \tag{10.49}$$

The shaft angle for any pair of bevel gears is an angle between two intersecting axes which meet at an apex. It is equal to the sum of the pitch angle of two mating gears, i.e. $\theta = \delta_1 + \delta_2$.

The relationship from the gear geometry, shown in Figure 10.26, for shaft angle θ are given below:
1. *Pitch angle* (δ)
   (a) For acute angle bevel gear $(0° < \theta < 90°)$

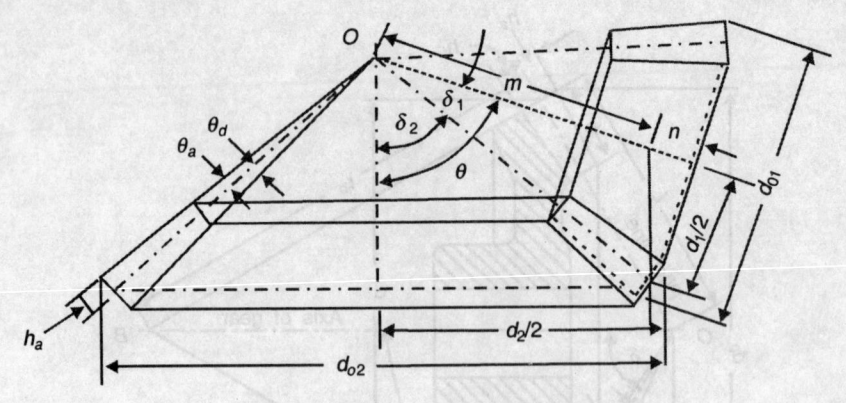

FIGURE 10.26 Acute angle bevel gear.

(i) For pinion: $$\tan \delta_1 = \frac{\sin \theta}{\frac{Z_2}{Z_1} + \cos \theta}$$

(ii) For gear: $$\tan \delta_2 = \frac{\sin \theta}{\frac{Z_1}{Z_2} + \cos \theta}$$

(b) For right angled bevel gear ($\theta = 90°$)

(i) For pinion: $$\tan \delta_1 = \frac{Z_1}{Z_2}$$

(ii) For gear: $$\tan \delta_2 = \frac{Z_2}{Z_1}$$

(c) For obtuse angle bevel gear ($90° < \theta \leq 180°$)

(i) For pinion: $$\tan \delta_1 = \frac{\sin(180° - \theta)}{\frac{Z_2}{Z_1} - \cos(180° - \theta)}$$

(ii) For gear: $$\tan \delta_2 = \frac{\sin(180° - \theta)}{\frac{Z_1}{Z_2} - \cos(180° - \theta)}$$

2. *Pitch diameter at the larger end*
   (i) For pinion: $d_1 = mZ_1$
   (ii) For gear: $d_2 = mZ_2$

3. *Outer diameter at the larger end*
   (i) For pinion: $d_{o1} = d_1 + 2h_a \cos \delta_1$
   (ii) For gear: $d_{o2} = d_2 + 2h_a \cos \delta_2$

4. *Length of the cone distance*

$$L = 0.5\left(d_1^2 + d_2^2\right)^{0.5}$$

In a bevel gear, the resultant force acting between two meshing teeth is assumed to be a concentrated force acting at the mid-point along the face width of the tooth, while the actual force acts somewhere between mid-point and the larger end of the tooth. Thus there is a small error in making this assumption. The resultant force has two components—tangential force F_t and separating force F_s as shown in Figure 10.27. The tangential force shown perpendicular to the plane of rotation is the driving force. Its magnitude can be computed by the following relation:

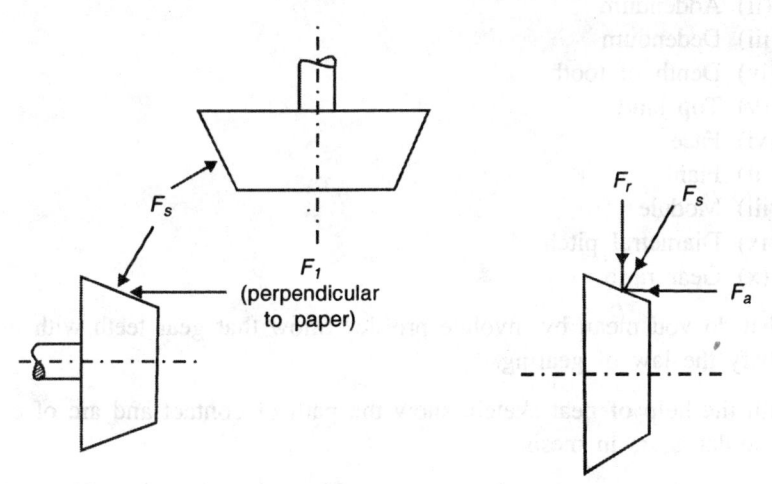

$$F_t = \frac{T}{r_{mid}} \qquad (10.50)$$

FIGURE 10.27 Components of forces on bevel gear.

where
$\quad T$ = torque transmitted by a gear
and $\quad r_{mid}$ = pitch radius of the gear under consideration at the mid-point of the tooth

$$= \frac{d_1}{2} - \frac{b\sin\delta_1}{2}$$

The separating force which acts perpendicular to the pitch line is determined by the following relation:

$$F_s = F_t \tan\alpha \qquad (10.51)$$

where α is the pressure angle

The separating force can be further resolved into two components—one along the axis of the gear, called the axial force F_a and other perpendicular to the axis of the gear, called radial force F_r.

Therefore,
$$F_a = F_s \sin \delta_1 = F_t \tan \alpha \sin \delta_1 \qquad (10.52a)$$
$$F_r = F_s \cos \delta_1 = F_t \tan \alpha \cos \delta_1 \qquad (10.52b)$$

The forces acting on the mating gears are equal in magnitude but opposite in direction.

EXERCISES

1. Enumerate main types of toothed gears with their usual field of application.

2. What do you mean by law of gearing? Define conjugate action.

3. Define the following terms related to gears:
    (i) Pitch
    (ii) Addendum
    (iii) Dedendum
    (iv) Depth of tooth
    (v) Top land
    (vi) Face
    (vii) Flank
    (viii) Module
    (ix) Diametral pitch
    (x) Gear ratio

4. What do you mean by involute profile? Show that gear teeth with involute profile satisfy the law of gearing.

5. With the help of neat sketch, show the path of contact and arc of contact for two cycloidal gears in mesh.

6. Compare the properties of involute profile and cycloidal profile teeth.

7. Define the following terms:
    (i) Path of approach
    (ii) Path of recess
    (iii) Length of path of contact
    (iv) Arc of contact
    (v) Contact ratio

8. What is significance of contact ratio in gear drive?

9. What do you mean by interference between two mating gears? State the conditions under which interference can be avoided.

10. Explain various preventive measures used to prevent interference between the gears.

11. From first principle, drive an expression for minimum number of teeth on gear to avoid interference with mating pinion.

12. Show that for a rack and pinion arrangement with 20° pressure angle, the minimum number of teeth required on pinion to avoid interference is 18.

13. What is the main limitation of a helical gear? Explain with the help of a neat sketch, how that limitation can be overcome in herringbone gear?

14. Derive an expression for centre distance for a pair of spiral gears in mesh.

15. Derive an expression for maximum efficiency of a spiral gear pair in mesh.

16. In what respect, a worm and worm gear is a special case of spiral gear drive? Hence prove that the lead angle of worm is equal to the helix angle of worm gear.

17. Two mating involute spur gears of 20° pressure angle have a gear ratio of 2. The number of teeth on the pinion is 20 and its speed is 250 rpm. The module is 12 mm. If the addendum on each wheel is such that path of approach and path of recess on each side are half the maximum possible length, find:
    (i) addendum for pinion and gear
    (ii) length of arc of contact
    (iii) maximum velocity of sliding during approach and recess.
    [Ans: 19.5 mm, 7.7 mm, 65.51 mm, 0.805 m/s 1.61 m/s]

18. A pinion having 20 teeth of involute profile, 20° pressure angle and 6 mm module drives a gear having 40 teeth of addendum equal to one module. Find:
    (i) pitch circle radii of the two gears
    (ii) the length of the approach
    (iii) path of contact
    (iv) arc of contact.
    [Ans: 60 mm, 120 mm, 15.17 mm, 28.95 mm, 30.8 mm]

19. A pinion having 30 teeth drives a gear having 80 teeth. The profile of both gears being involute with 20° pressure angle, 12 mm module and addendum 10 mm. Find the length of path of contact, arc of contact and the contact ratio.
    [Ans: 51.81 mm, 55.14 mm, 1.46 pairs]

20. Two mating gears have 20 and 40 involute teeth of module 10 mm and 20° pressure angle. If the addendum on each wheel is such that the path of contact is maximum and interference is just avoided find the path of contact, arc of contact and contact ratio. Also find the addendum for each gear.
    [Ans: 102.6 mm, 109.2 mm, 3.47, 39 mm, 14 mm]

21. A pair of involute spur gears has 16 and 18 teeth, a module 12.5 mm, an addendum 8.75 mm and pressure angle 14.5°. Prove that the gears have interference. Also determine the minimum number of teeth required on the gears to avoid interference and to maintain the original gear ratio.
    [Ans: 24 teeth, 27 teeth]

22. Two mating spur gears with module 6 mm have 20 and 47 teeth of 20° pressure angle. The addendum is equal to one module. Determine the number of pair of teeth

in contact and angle turned through by gear for one pair of teeth in contact. Also determine the ratio of sliding velocity to the rolling velocity at the following instant when
   (i) engagement commences
   (ii) engagement terminates and
   (iii) at the pitch point.

[Ans: 1.649, 12.63°, 0.3667, 0.327, 0]

23. A pinion having 20 involute teeth of 6 mm module rotates at 200 rpm and transmits 1.5 kW to a gear having 50 teeth. The addendum on both gears is one-fourth of the circular pitch and pressure angle is 20°. Find
   (i) length of path of approach and the arc of approach and
   (ii) the normal force between the teeth at an instant when there is only one pair of teeth in contact.

[Ans: 12.5 mm, 13.3 mm, 1.27 kN]

24. Two involute gears operate with a pressure angle of 20° and rotate in opposite directions. The pinion has 20 teeth of 3 mm module and runs at 200 rpm. The gear has 80 teeth. If the contact ratio is 1.4 and the arc of recess is 1.2 times as large as the arc of approach, determine:
   (i) the addenda of teeth on both the pinion and gear
   (ii) the greatest speed of sliding
   (iii) centre distance
   (iv) the working depth

[Ans: 3.4 mm, 4.89 mm, 2.95 m/s, 250 mm, 8.29 mm]

25. A pinion of 250 mm pitch circle diameter drives a rack. The addendum height for both pinion and rack is 12.5 mm and teeth of involute profile have a pressure angle 20°. Show that there is no interference between them. Also find the number of teeth on the pinion to ensure continuity of contact.

[Ans: 12 teeth]

26. A pinion of 100 mm diameter drives a rack. The teeth are of involute profile with 20° pressure angle and total depth of the teeth is 12.5 mm, the root clearance being 1 mm. Determine the length of addendum and root for both pinion and rack, the addendum of rack being as large as possible, consistent with correct tooth action. Also calculate the length of the path of approach.

[Ans: 10.65 mm, 6.85 mm, 5.85 mm, 11.65 mm, 17.1 mm]

27. A spiral gear drive has a shaft angle of 60°. The speed ratio is 9:4. The approximate centre distance is 140 mm and coefficient of friction is 0.105. The module is 4 mm. If the efficiency of the drive is maximum, determine:
   (i) number of teeth on gears
   (ii) exact centre distance
   (iii) efficiency of the drive

[Ans: 16, 36, 133.95 mm, 83.9%]

28. Two spiral gears of equal diameter are used for a drive on a machine tool. The angle between the shaft is 75° and the approximate centre distance is 115 mm, speed of pinion is 1.5 times the speed of the gear and normal pitch is 10 mm. Find:
    (i) number of teeth on each gear
    (ii) spiral angles for each gear
    (iii) rubbing speed, if the pinion rotates at 100 rpm.
    [Ans: 22, 23, 52.1°, 22.9°, 0.63 m/s]

29. A pair of spiral gears connects two shafts inclined at 80°. The velocity ratio is 2 and driver has 25 teeth of normal pitch 12 mm and spiral angle of 30°. Find the centre distance between the shafts.
    [Ans: 203.5 mm]

30. For a right angled worm gear drive, the normal module is 5 mm, the speed ratio is 30 and coefficient of friction is 0.15. The distance between the axes is 150 mm and efficiency is maximum. Find:
    (i) the lead angle of the worm
    (ii) the number of teeth
    (iii) exact centre distance
    (iv) the efficiency of the drive.

MULTIPLE CHOICE QUESTIONS

1. Which of the following gears are used to connect two non-parallel non-intersecting shafts?
   (a) spur (b) helical
   (c) bevel (d) worm gear

2. Which of the following gear have minimum axial thrust?
   (a) helical (b) bevel
   (c) herringbone (d) worm

3. The product of circular pitch p_c and diametral pitch is equal to
   (a) 2π (b) π
   (c) $\pi/2$ (d) 1

4. The common normal at the point of contact between the mating teeth of the gears must pass through the
   (a) centre of pinion (b) centre of gear
   (c) pitch point (d) centre of both pinion and gear

5. The involute profile of the teeth extends from its addendum circle to its
   (a) pitch (b) root circle
   (c) base circle (d) addendum circle

6. Pressure angle is constant for
   (a) cycloidal teeth (b) involute teeth
   (c) epicycloidal (d) none of the above

7. Undercutting occurs in the case of
   (a) involute teeth
   (b) cycloidal teeth
   (c) helical gear
   (d) none of the above

8. Which of the following profiles satisfy the law of gearing?
   (a) involute
   (b) cycloidal
   (c) conjugate profile
   (d) all of the above

9. The path of point of contact between the involute teeth profile gears is
   (a) circle
   (b) parabola
   (c) straight line
   (c) involute

10. The minimum number of teeth for involute rack and pinion arrangement for pressure angle of 20° is
    (a) 32
    (b) 34
    (c) 17
    (d) any number

11. From the point of view of strength which gear profile is better
    (a) involute
    (b) cycloidal
    (c) conjugate
    (d) none of the above

12. Teeth of a bevel gear are laid on the surface of a
    (a) pyramid
    (b) plane
    (c) cone
    (d) cylinder

13. The profile of a cycloidal gear tooth below the pitch circle is
    (a) hypocycloid
    (b) epicycloid
    (c) involute
    (d) none of the above

14. The surface of the tooth below the pitch surface is called
    (a) face
    (b) Flank
    (c) clearance
    (d) bottom land

CHAPTER 11

Gear Train

11.1 INTRODUCTION

A gear train is a mechanism which transmits power or motion from one shaft to another by means of a combination of gears. The gear train enables us to either step-up or step-down the speed of a prime mover and to obtain a different range of speeds as per the requirement of the driven machine. For example, in a lathe machine, different materials need to be cut at different cutting speeds. These speeds can be obtained by a suitable gear train. Gear trains are widely used in mechanical clocks, automotive vehicles, turbine generator sets, machine tools and so forth.

In gear train terminology, a term called **train value** is quite often used to designate the characteristics of a gear train. It is defined as the ratio of the speed of driven gear to that of driving gear. In other words, it is reciprocal of speed ratio (SR). That is,

Train value:
$$T = \frac{\text{Speed of driven gear}}{\text{Speed of driving gear}}$$

or
$$T = \frac{1}{SR} \tag{11.1}$$

Gear trains, depending upon the functional requirement of an application, are classified into one of the following three categories:
1. Simple gear train
2. Compound gear train
3. Epicyclic gear train

11.2 SIMPLE GEAR TRAIN

In a simple gear train, a series of gears are mounted on individual shafts to receive and transmit motion (Figure 11.1). In a gear train, the sense of rotation depends upon the direction of rotation of driving gear. For example, in a simple gear train, if the direction of rotation of driving gear is clockwise and it is designated positive direction, then the direction of mating gear, which is naturally anticlockwise, is termed negative direction. In other words, all odd number of gears rotate in one direction and even number of gears rotate in other direction.

Figure 11.1 shows a simple gear train consisting of four gears each having Z_1, Z_2, Z_3 and Z_4 number of teeth and N_1, N_2, N_3 and N_4 rpm respectively. Suppose gear 1 is mounted on driving shaft rotating in clockwise direction and gear 4 is mounted on driven shaft, which should rotate in anticlockwise direction. However, the pitch line velocity of two mating gears is always constant.

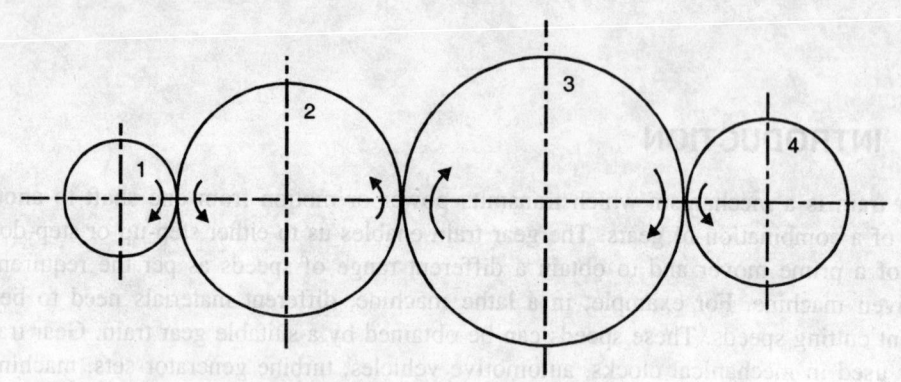

FIGURE 11.1 A simple gear train.

For gear pair 1–2, the pitch line velocity:

$$\pi d_1 N_1 = \pi d_2 N_2$$

Train Value:

$$T_{1-2} = \frac{N_2}{N_1} = \frac{d_1}{d_2} = \frac{Z_1}{Z_2}$$

Similarly, for pair 2–3 and 3–4, the train values are:

$$T_{2-3} = \frac{N_3}{N_2} = \frac{Z_2}{Z_3} \quad \text{and} \quad T_{3-4} = \frac{N_4}{N_3} = \frac{Z_3}{Z_4}$$

The train value of complete gear train can be found by multiplying the train values of individual pair as follows:

$$T = T_{1-2} \times T_{2-3} \times T_{3-4}$$

$$= \frac{N_2}{N_1} \times \frac{N_3}{N_2} \times \frac{N_4}{N_3} = \frac{N_4}{N_1}$$

$$= \frac{Z_1}{Z_2} \times \frac{Z_2}{Z_3} \times \frac{Z_3}{Z_4}$$

or

Train value:

$$T = \frac{N_4}{N_1} = \frac{Z_1}{Z_4} \tag{11.2}$$

Speed ratio:
$$SR = \frac{N_1}{N_4} = \frac{Z_4}{Z_1} \tag{11.3}$$

Thus from Eqs. (11.2) and (11.3), we observed that the intermediate gears 2 and 3 have no effect on the train value. Therefore, these gears are called **idler gears.** Idler gears are used for changing the direction of rotation of the driven shaft.

11.3 COMPOUND GEAR TRAIN

In a compound gear train, a series of gears are connected in such a way that intermediate shaft carries two gears which are fastened together rigidly. Such a gear train is called **compound gear train.** Figure 11.2 shows a compound gear train constituted of four gears in which gears 2 and 3 are compounded on the intermediate shaft.

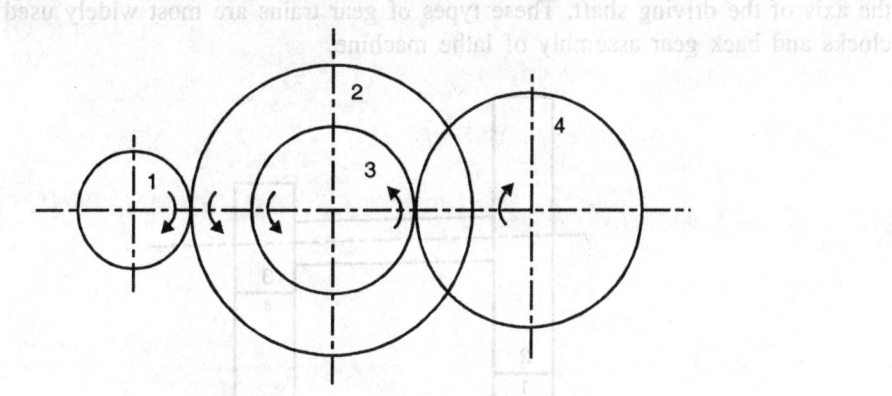

FIGURE 11.2 A compound gear train.

For a gear pair 1–2, the train value:
$$T_{1-2} = \frac{N_2}{N_1} = \frac{Z_1}{Z_2} \tag{11.4a}$$

Similarly, for pair 3–4
$$T_{3-4} = \frac{N_4}{N_3} = \frac{Z_3}{Z_4} \tag{11.4b}$$

Multiplying Eqs. [11.4(a)] and [11.4(b)], we get the train value of the compound gear train as follows:
$$T = \frac{N_2}{N_1} \times \frac{N_4}{N_3} = \frac{Z_1}{Z_2} \times \frac{Z_3}{Z_4}$$

As gears 2 and 3 are compound gears and mounted on the same shaft, their speed will be equal, i.e. $N_2 = N_3$. The train value of compound gear train:

$$T = \frac{N_4}{N_1} = \frac{Z_1}{Z_2} \times \frac{Z_3}{Z_4} \tag{11.5}$$

In other words, the train value of a compound gear train is the quotient of the product of teeth on the driver gears to that of the product of teeth on the driven gears, of each pair in mesh.

$$T = \frac{\text{Product of number of teeth on driving gears}}{\text{Product of number of teeth on driven gears}}$$

Reverted Gear Train. It is a special type of compound gear-train in which the axes of the driving and driven gear shaft coincide. Figure 11.3 shows a riveted gear train in which the driving shaft gear 1 drives gear 2 mounted on the intermediate shaft. Gears 2 and 3 are compound gears. Gear 3 drives gear 4 which is mounted on the driven shaft coinciding with the axis of the driving shaft. These types of gear trains are most widely used in mechanical clocks and back gear assembly of lathe machine.

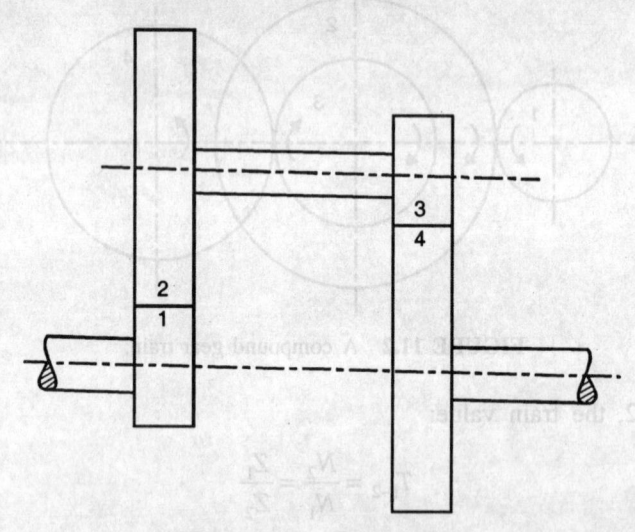

FIGURE 11.3 Reverted gear.

The distance between centres of two shafts is

$$C = r_1 + r_2 = r_3 + r_4$$

For constant module of gears in mesh,

$$C = Z_1 + Z_2 = Z_3 + Z_4 \tag{11.6}$$

and train value:

$$T = \frac{N_4}{N_1} = \frac{Z_1}{Z_2} \times \frac{Z_3}{Z_4} \tag{11.7}$$

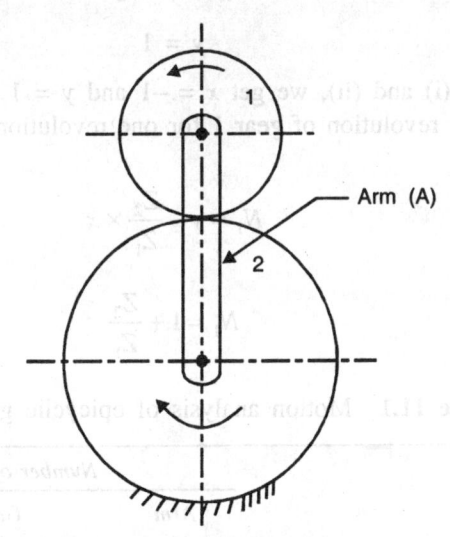

11.4 EPICYCLIC GEAR TRAIN

In the types of gear train discussed so far, the axes of the gears remain fixed and there is no relative motion between the axes. In epicyclic gear train, there exists a relative motion between two axes of the gears constituting the train. An epicyclic gear train usually consists of three elements—driving gear, driven gear and an arm which is pivoted about a fixed centre as shown in Figure 11.4. In this gear train, if the arm A is held fixed, the driving gear 1 and driven gear 2 constitute a simple gear train. However, if gear 2 is held fixed, the arm A can revolve about the centre of gear 2 and gear 1 rolls around the pitch circle circumference of the stationary gear 2. In such a gear train, the driving gear 1 rolls around the driven gear 2 and traces an epicyclic path; hence it is called **epicyclic gear train**. This motion also resembles the motion of planets around the sun. So sometimes, it is also called **planetary gear train.**

FIGURE 11.4 Epicyclic gear train.

In some gear trains, the fixed gear may be internal gear and pinion which rolls inside of internal gear traces hypocycloid. However, it has become customary to call even these gears epicyclic gears.

The motion analysis of an epicyclic gear train is complex in nature. The train value or velocity ratio of these trains can be found by the methods discussed further.

11.4.1 Tabulation Method

In tabulation method, the complex motion of the gear train is splitted into different motions of individual gear pair and their train value or speed ratio is written in tabular form. Finally these splitted motion segments are added as per their connectivity. A detailed stepwise procedure is summarised as follows (See Figure 11.4):

(i) Assume that the arm is locked and all other gears are free to rotate.
(ii) Mark any gear (say gear 2) as reference gear and rotate it through one revolution in the clockwise direction which is designated as positive direction.

(iii) Calculate the number of revolutions made by all the gears and record them in Table 11.1. This can be calculated by known values of number of teeth.
(iv) Multiply each train value by x, assuming that reference gear rotates at x rpm. Write down the corresponding number of revolutions of all the gears in second row of the Table 11.1.
(v) Now it is assumed that arm is unlocked and allowed to rotate in clockwise direction by y rpm. Thus add y to all the elements of second row and write down in the third row.
(vi) Apply the given boundary conditions and find the values of x and y. For example, if gear 2 is fixed and arm A rotates by one revolution in clockwise direction, then from Table 11.1

$$x + y = 0 \qquad \text{(i)}$$

and

$$y = 1 \qquad \text{(ii)}$$

Solving Eqs. (i) and (ii), we get $x = -1$ and $y = 1$

Therefore, the revolution of gear 1 for one revolution of arm A and gear 2 being fixed, is.

$$N_1 = y - \frac{Z_2}{Z_1} \times x$$

or

$$N_1 = 1 + \frac{Z_2}{Z_1}$$

Table 11.1 Motion analysis of epicyclic gear train

Operation	Arm (A)	Gear 2 (Z_2)	Gear 1 (Z_1)
1. Arm A is locked. Gear 2 is given one turn in clockwise direction.	0	+1	$-\frac{Z_2}{Z_1}$
2. Multiply by x.	0	x	$-\frac{Z_2}{Z_1} \times x$
3. Unlock arm and rotate it by y turns. Add y.	y	$x + y$	$y - \frac{Z_2}{Z_1} \times x$

11.4.2 Relative Velocity Method

Let the arm A and the gears 1 and 2 each be revolving at the speeds of N_a, N_1 and N_2 rpm respectively. Then the angular velocity of gear 1 is algebraic sum of angular velocity of gear

1 relative to arm A and angular velocity of arm A. Thus the speed of gear 1 relative to the arm A is given as $N_1 - N_a$ and similarly, the speed of gear 2 relative to arm is $N_2 - N_a$. The ratio of these two speeds depends upon the number of teeth on the mating gears. Thus.

$$\frac{N_1 - N_a}{N_2 - N_a} = -\frac{Z_2}{Z_1} \tag{11.8}$$

If the given conditions are applied, we can find the speed of unknown parameter. For example, if gear 2 is fixed, then $N_2 = 0$ and arm is revolved by one revolution, i.e. $N_a = 1$. Then the speed of gear 1 can be computed by substituting these values in Eq. (11.8) as follows:

$$\frac{N_1 - 1}{0 - 1} = -\frac{Z_2}{Z_1}$$

or

$$N_1 = 1 + \frac{Z_2}{Z_1} \tag{11.9}$$

11.5 TORQUE TRANSMITTED BY EPICYCLIC GEAR TRAIN

When an epicyclic gear train transmits power, torques are transmitted from one element to another.

Let T_i = input torque on the driving gear
T_o = output torque or resisting torque on the driven gear
and T_b = bracking torque or fixing torque on the fixed element

If the elements of an epicyclic gear train are all moving at uniform angular speed such that there is no angular acceleration and each gear is in equilibrium under the action of torques acting on it. The following equation holds true:

$$T_i + T_o + T_b = 0 \tag{11.10}$$

Further, if ω_i, ω_o and ω_b are uniform angular velocities of input, output and fixed gears respectively and assuming that there is no loss of power during transmission due to friction, then

Input power = Output power

or

$$\Sigma T \omega = 0$$

or

$$T_i \omega_i + T_o \omega_o + T_b \omega_b = 0 \tag{11.11}$$

While computing power through Eq. (11.11), the proper sense of torque and speed must be taken into account.

EXAMPLE 11.1 An epicyclic gear train arrangement is shown in Figure 11.5. Gear E is a fixed gear and gears C and D are compounded and mounted on one shaft. If the arm A makes 60 rpm in counterclockwise direction, determine the speed and direction of rotation of gears B and F. The number of teeth on different gears are as given below:

$$Z_b = 25, \ Z_c = 15, \ Z_d = 50, \ Z_e = 20 \text{ and } Z_f = 30.$$

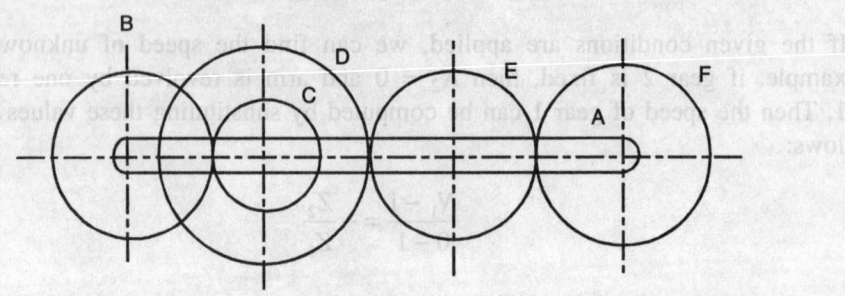

FIGURE 11.5

Solution Let us perform the motion analysis by the tabular method.

Operation	Arm A	Gear B	Compound gears C	D	Gear E	Gear F
1. Arm A is locked and gear B is rotated by one revolution in clockwise direction	0	+1	$-\dfrac{Z_b}{Z_c}$	$-\dfrac{Z_b}{Z_c}$	$\dfrac{Z_b}{Z_c} \times \dfrac{Z_d}{Z_e}$	$-\dfrac{Z_b}{Z_c} \times \dfrac{Z_d}{Z_e} \times \dfrac{Z_e}{Z_f}$
2. Multiply by x.	0	x	$-\dfrac{Z_b}{Z_c}x$	$-\dfrac{Z_b}{Z_c}x$	$\dfrac{Z_b}{Z_c} \times \dfrac{Z_d}{Z_e}x$	$-\dfrac{Z_b}{Z_c} \times \dfrac{Z_d}{Z_f} \times x$
3. Unlock arm A and rotate it by y revolution Add y.	y	$x+y$	$-\dfrac{Z_b}{Z_c}x+y$	$-\dfrac{Z_b}{Z_c}x+y$	$\dfrac{Z_b}{Z_c} \times \dfrac{Z_d}{Z_e}x+y$	$-\dfrac{Z_b}{Z_c} \times \dfrac{Z_d}{Z_f}x+y$

Now applying boundary conditions:
(a) Gear E is fixed, i.e.

$$\frac{Z_b}{Z_c} \times \frac{Z_d}{Z_e} x + y = 0$$

or

$$\frac{25}{15} \times \frac{50}{20} x + y = 0$$

or

$$\frac{25}{6} x + y = 0 \tag{i}$$

(b) Arm A makes 60 rpm in counterclockwise direction, i.e.

$$y = -60 \qquad \text{(ii)}$$

Solving Equations (i) and (ii), we get

$$x = +\frac{72}{5} \text{ and } y = -60$$

Speed of gear B: $N_b = x + y = \frac{72}{5} - 60$

$$= -45.6 \text{ rpm (counterclockwise)} \qquad \text{Ans.}$$

Speed of gear F: $N_f = -\frac{Z_b}{Z_c} \times \frac{Z_d}{Z_f} x + y$

$$= -\frac{25}{15} \times \frac{50}{30} \times \frac{72}{5} - 60$$

$$= -100 \text{ rpm} \quad \text{(counterclockwise)} \qquad \text{Ans.}$$

EXAMPLE 11.2 In an epicyclic gear train as shown in Figure 11.6, the arm A is fixed to the shaft S. The gear B having 80 teeth rotates freely on the shaft S and gear D with 120 teeth is separately driven. If the arm A runs at 100 rpm and gear D at 50 rpm in same direction, find the speed of gear B.

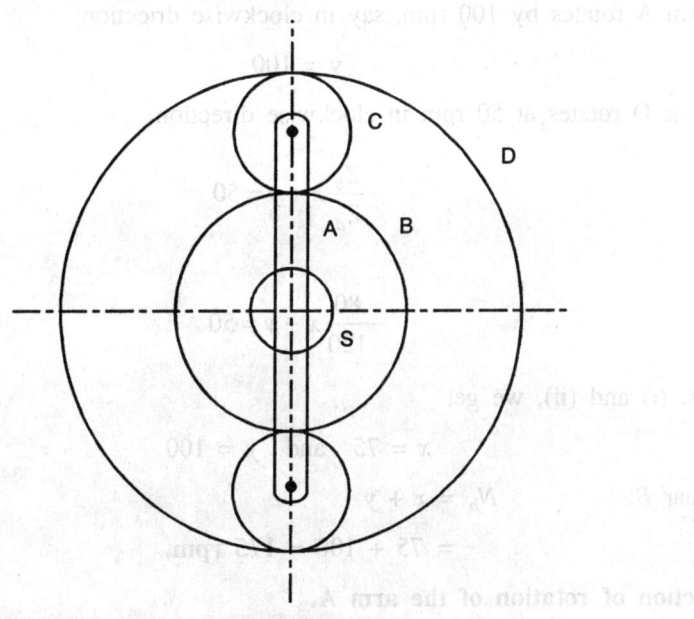

FIGURE 11.6

Solution Let $Z_B = 80$ and $Z_D = 120$.

The diameter of gear D is:
$$d_d = d_b + 2 \times d_c$$
or
$$Z_d = Z_b + 2 \times Z_c \quad \text{for the same module}$$
or
$$Z_c = \frac{Z_d - Z_b}{2} = \frac{120 - 80}{2} = 20 \text{ teeth}$$

The motion analysis of gear train is given in the following table.

Operation	Arm A	Gear B	Gear C	Annulus gear D
1. Lock arm A and rotate gear B by one revolution in clockwise direction.	0	+1	$-\dfrac{Z_b}{Z_c}$	$-\dfrac{Z_b}{Z_c} \times \dfrac{Z_c}{Z_d}$
2. Multiply by x.	0	x	$-\dfrac{Z_b}{Z_c} x$	$-\dfrac{Z_b}{Z_d} x$
3. Unlock arm A and rotate it by y revolution. Add y.	y	$x + y$	$-\dfrac{Z_b}{Z_c} x + y$	$-\dfrac{Z_b}{Z_d} x + y$

Conditions:

(a) Arm A rotates by 100 rpm, say in clockwise driection

$\therefore \qquad\qquad\qquad y = 100$ \hfill (i)

(b) Gear D rotates at 50 rpm in clockwise direction.

$\therefore \qquad\qquad\qquad -\dfrac{Z_b}{Z_d} x + y = 50$

or
$$-\frac{80}{120} x + y = 50 \qquad\qquad\qquad\qquad (ii)$$

Solving Eqs. (i) and (ii), we get

$$x = 75 \quad \text{and} \quad y = 100$$

Speed of gear B: $\quad N_b = x + y$

$\qquad\qquad\qquad = 75 + 100 = \mathbf{175\ rpm}$,

in the direction of rotation of the arm A. \hfill **Ans.**

EXAMPLE 11.3 Figure 11.7 shows an epicyclic gear train in which gear A drives the internal gear D through compound gears B and C. The number of teeth on gear A is 20 and centre distance between the centres of gears A and B is 300 mm. If the module of all gears is 10 mm and gear C has 30 teeth, find the speed of gear D. The arm rotates at 600 rpm in counterclockwise direction and gear A is fixed.

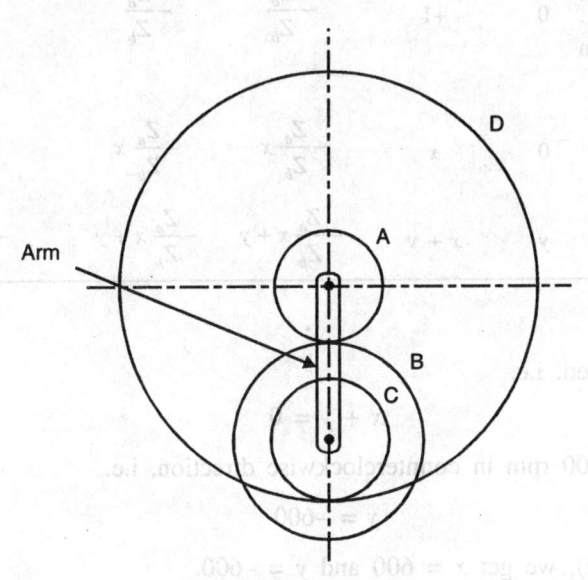

FIGURE 11.7

Solution Let C is the centre distance.

$$C = \frac{m}{2}(Z_a + Z_b)$$

or
$$300 = \frac{10}{2}(20 + Z_b)$$

or
$$Z_b = 40 \text{ teeth}$$

Pitch circle radius of gear D, $r_d = r_a + r_b + r_c$

or
$$r_d = \frac{1}{2}(20 \times 10 + 40 \times 10 + 30 \times 10) = 450 \text{ mm}$$

Number of teeth on gear D,
$$Z_d = \frac{2r_d}{m} = \frac{2 \times 450}{10}$$
$$= 90 \text{ teeth}$$

The motion analysis of the gear train is given in the following table.

Operation	Arm	Gear A	Compound gears B	Compound gears C	Gear D
1. Lock arm and rotate gear A by one revolution in clockwise direction.	0	+1	$-\dfrac{Z_a}{Z_b}$	$-\dfrac{Z_a}{Z_b}$	$-\dfrac{Z_a}{Z_b} \times \dfrac{Z_c}{Z_d}$
2. Multiply by x.	0	x	$-\dfrac{Z_a}{Z_b} x$	$-\dfrac{Z_a}{Z_b} x$	$-\dfrac{Z_a}{Z_b} \times \dfrac{Z_c}{Z_d} x$
3. Add y.	y	x + y	$-\dfrac{Z_a}{Z_b} x + y$	$-\dfrac{Z_a}{Z_b} x + y$	$-\dfrac{Z_a}{Z_b} \times \dfrac{Z_c}{Z_d} x + y$

Conditions:

(a) Gear A is fixed, i.e.

$$x + y = 0 \qquad (i)$$

(b) Arm makes 600 rpm in counterclockwise direction, i.e.

$$y = -600 \qquad (ii)$$

Solving Eqs. (i) and (ii), we get $x = 600$ and $y = -600$.

Therefore, the speed of gear D:

$$N_d = -\dfrac{Z_a}{Z_b} \times \dfrac{Z_c}{Z_d} x + y$$

$$= \dfrac{-20}{40} \times \dfrac{30}{90} \times 600 - 600$$

$$= -700 \text{ rpm, i.e. in the direction of rotation of the arm.} \qquad \text{Ans.}$$

EXAMPLE 11.4 In an epicyclic gear train shown in Figure 11.8, a disc E, which also acts as an arm, is attached to the shaft X. Gear A meshes with an internal gear D fixed to the casing and the gear B meshes with gear C attached to the shaft Y. Find the velocity ratio between the shafts X and Y. The teeth on gears A, B and C are 20, 40 and 20 respectively.

Solution From the geometry of Figure 11.8, we find that radius of internal gear D is:

$$r_d = r_a + r_b + r_c$$

We also know that number of teeth on a gear is proportional to its pitch circle diameter. Therefore,

$$Z_d = Z_a + Z_b + Z_c = 20 + 40 + 20 = 80 \text{ teeth}$$

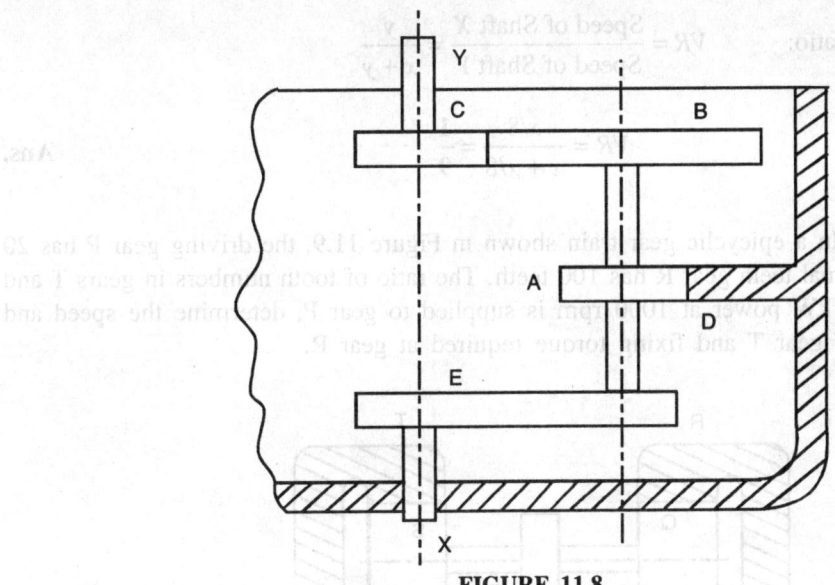

FIGURE 11.8

The motion analysis of the gear train is given in the following table.

Operation	Arm E	Gear C	Compound gears B	Compound gears A	Gear D
1. Lock arm E and rotate gear C by one revolution in clockwise direction.	0	+1	$-\dfrac{Z_c}{Z_b}$	$-\dfrac{Z_c}{Z_b}$	$-\dfrac{Z_c}{Z_b} \times \dfrac{Z_a}{Z_d}$
2. Multiply by x.	0	x	$-\dfrac{Z_c}{Z_b} x$	$-\dfrac{Z_c}{Z_b} x$	$-\dfrac{Z_c}{Z_b} \times \dfrac{Z_a}{Z_d} x$
3. Unlock arm E and rotate it by y revolution. Add y.	y	$x + y$	$-\dfrac{Z_c}{Z_b} x + y$	$-\dfrac{Z_c}{Z_b} x + y$	$-\dfrac{Z_c}{Z_b} \times \dfrac{Z_a}{Z_d} x + y$

Conditions:
(a) Gear D is fixed, i.e.

$$-\frac{Z_c}{Z_b} \times \frac{Z_a}{Z_d} x + y = 0$$

or

$$-\frac{20}{40} \times \frac{20}{80} x + y = 0$$

or

$$y = \frac{x}{8}$$

Velocity ratio: $VR = \dfrac{\text{Speed of Shaft } X}{\text{Speed of Shaft } Y} = \dfrac{y}{x+y}$

or
$$VR = \dfrac{x/8}{x + x/8} = \dfrac{1}{9}$$ **Ans.**

EXAMPLE 11.5 In a epicyclic gear train shown in Figure 11.9, the driving gear P has 20 teeth and fixed internal teeth gear R has 100 teeth. The ratio of tooth numbers in gears T and S is 98 : 30. If 10 kW power at 1000 rpm is supplied to gear P, determine the speed and sense of rotation of gear T and fixing torque required at gear R.

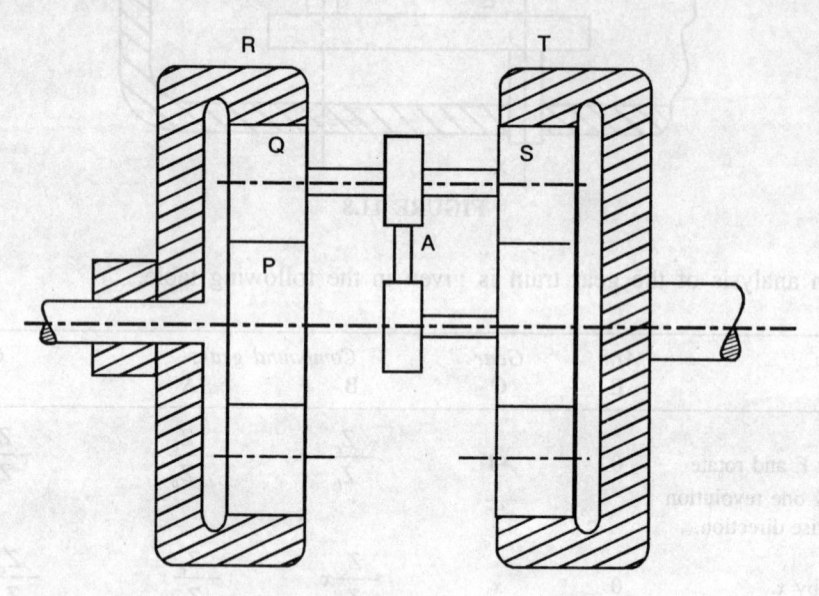

FIGURE 11.9

Solution We have

$$Z_p = 20, \; Z_r = 100 \quad \text{and} \quad N_p = 1000 \text{ rpm}$$

We know that number of teeth are proportional to the diameter of gear. Therefore,

$$Z_r = Z_p + 2Z_q$$

or
$$Z_q = \dfrac{Z_r - Z_p}{2} = \dfrac{100 - 20}{2} = 40 \text{ teeth}$$

The motion analysis of the gear train is given in the following table.

Operation	Arm A	Gear P	Compound gear Q	Compound gear S	Gear R	Gear T
1. Arm A is locked and gear P is given one revolution in clockwise direction.	0	+1	$-\dfrac{Z_p}{Z_q}$	$-\dfrac{Z_p}{Z_q}$	$-\dfrac{Z_p}{Z_q}\times\dfrac{Z_q}{Z_r}$	$-\dfrac{Z_p}{Z_q}\times\dfrac{Z_s}{Z_t}$
2. Multiply by x.	0	x	$-\dfrac{Z_p}{Z_q}x$	$-\dfrac{Z_p}{Z_q}x$	$-\dfrac{Z_p}{Z_r}x$	$-\dfrac{Z_p}{Z_q}\times\dfrac{Z_s}{Z_t}x$
3. Unlock arm A and rotate it by y revolution. Add y.	y	x+y	$-\dfrac{Z_p}{Z_q}x+y$	$-\dfrac{Z_p}{Z_q}x+y$	$-\dfrac{Z_p}{Z_r}x+y$	$-\dfrac{Z_p}{Z_q}\times\dfrac{Z_s}{Z_t}x+y$

Conditions:

(a) Gear R is fixed. Therefore,

$$y - \dfrac{Z_p}{Z_r} x = 0$$

or

$$y - \dfrac{20}{100} x = 0 \qquad (i)$$

(b) Gear P rotates at 1000 rpm, in clockwise direction. Therefore,

$$x + y = 1000 \qquad (ii)$$

Solving Eqs. (i) and (ii), we get

$$x = 833.34 \quad \text{and} \quad y = 166.66$$

Speed of gear T:

$$N_t = -\dfrac{Z_p}{Z_q} \times \dfrac{Z_s}{Z_t} x + y$$

$$= -\dfrac{20}{40} \times \dfrac{30}{98} \times 833.34 + 166.66$$

$$= \mathbf{39.1 \text{ rpm (clockwise)}} \qquad \textbf{Ans.}$$

Input torque:

$$T_i = \dfrac{60P}{2\pi N_p} = \dfrac{60 \times 10 \times 10^3}{2\pi \times 1000}$$

$$= 95.49 \text{ Nm}$$

Output torque:

$$T_o = \dfrac{60P}{2\pi N_t} = \dfrac{60 \times 10 \times 10^3}{2\pi \times 39.1}$$

$$= 2442.27 \text{ Nm}$$

Therefore, fixing or braking torque at gear R is:

$$T_b = T_o - T_i$$

$$= 2442.27 - 95.49$$

$$= \mathbf{2346.78 \text{ Nm}} \qquad \textbf{Ans.}$$

EXAMPLE 11.6* In an epicyclic gear train shown in Figure 11.10, the gear A, fixed to the shaft S_1, has 30 teeth and rotates at 500 rpm. Gear B, which meshes with gear A, is compounded with gear C. Gears B and C both are free to rotate on shaft S_2. The gears B, C and D have 50, 70 and 90 teeth respectively. If the gear D rotates at 80 rpm in the direction opposite to that of the gear A, find the speed of the shaft S_2.

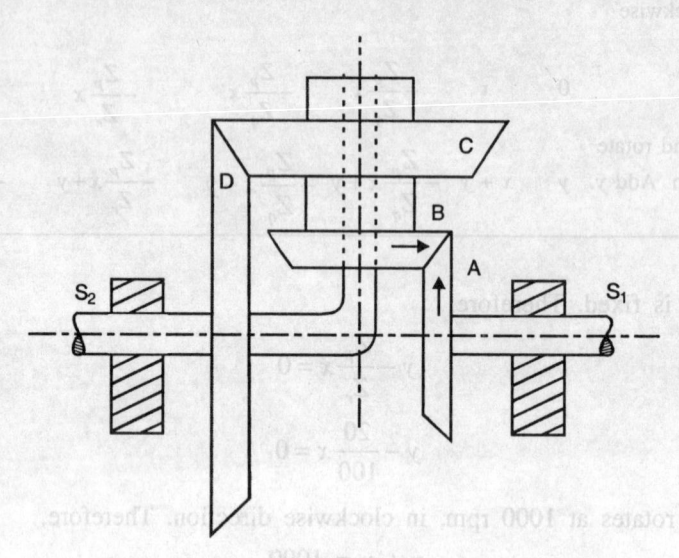

FIGURE 11.10

Solution Let Z_a, Z_b, Z_c and Z_d be the number of teeth on gears A, B, C and D respectively. The motion analysis of the gear train is given below:

Operation	Arm or Shaft	Gear A	Compound gears B	Compound gears C	Gear D
1. Lock the shaft S_2 which acts as arm and rotate gear A by one revolution.	0	+1	$\dfrac{Z_a}{Z_b}$	$\dfrac{Z_a}{Z_b}$	$-\dfrac{Z_a}{Z_b} \times \dfrac{Z_c}{Z_d}$
2. Multiply by x.	0	x	$\dfrac{Z_a}{Z_b} x$	$\dfrac{Z_a}{Z_b} x$	$-\dfrac{Z_a}{Z_b} \times \dfrac{Z_c}{Z_d} x$
3. Add y.	y	$x + y$	$\dfrac{Z_a}{Z_b} x + y$	$\dfrac{Z_a}{Z_b} x + y$	$-\dfrac{Z_a}{Z_b} \times \dfrac{Z_c}{Z_d} x + y$

---

* In bevel epicyclic gear train for gears whose axes are inclined to the main axis, the term clockwise and counter-clockwise directions of rotation are not applicable. So in the tubular method of solution, plus or minus sign is omitted. Further, the addition of y is not convenient to make as the rotation is about different axes. However, it is used in the context of spur gear.

Conditions:

(a) Gear A rotates at 500 rpm clockwise. Therefore,

$$x + y = 500 \qquad (i)$$

(b) Gear D rotates at 80 rpm in the direction opposite to that of gear A. Therefore,

$$-\frac{Z_a}{Z_b} \times \frac{Z_c}{Z_d} \times x + y = -80$$

or

$$-\frac{30}{50} \times \frac{70}{90} \times x + y = -80$$

or

$$\frac{7}{15} x - y = 80 \qquad (ii)$$

Solving Eqs. (i) and (ii), we get

$$x = 395.45 \quad \text{and} \quad y = 104.55$$

The speed of shaft S_2:

$$N_{s2} = 104.55 \text{ rpm (clockwise)} \qquad \text{Ans.}$$

EXAMPLE 11.7 Two bevel gears P and Q having 40 and 30 teeth respectively are mounted on two co-axial shafts A and B. A bevel gear R meshes with gears P and Q and rotates freely on one end of the arm. The other end of the arm is welded to a sleeve which rides freely on the axes of the shafts A and B. If the shaft A rotates at 100 rpm clockwise and the arm rotates at 100 rpm counterclockwise, find the speed of shaft B. The number of teeth on gear R is 50. (See Figure 11.11.)

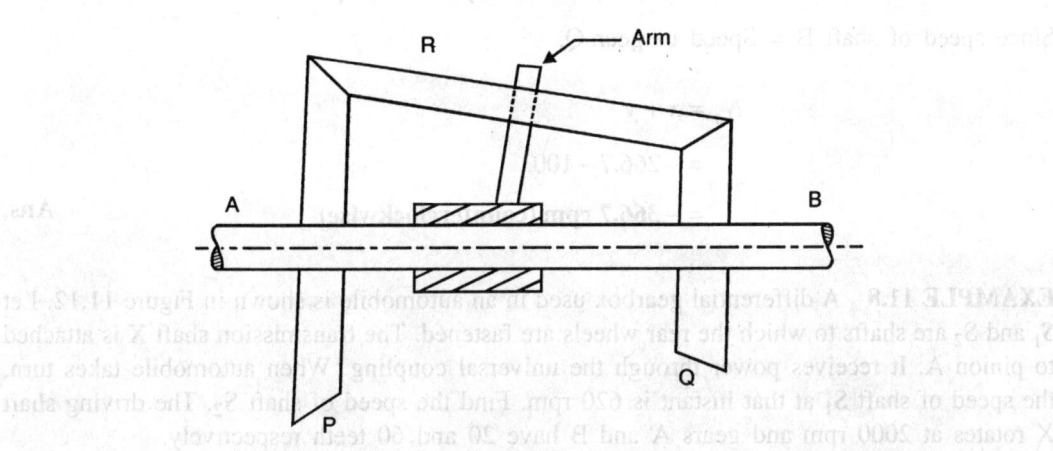

FIGURE 11.11

Solution Let Z_p, Z_q and Z_r denote number of teeth on gears P, Q and R respectively. The motion analysis of the gear train is given in the following table.

Operation	Arm	Gear Q	Gear R	Gear P
1. Lock arm and rotate gear Q by one revolution in clockwise direction.	0	+1	$\dfrac{Z_q}{Z_r}$	$-\dfrac{Z_q}{Z_r} \times \dfrac{Z_r}{Z_p}$
2. Multiply by x.	0	x	$\dfrac{Z_q}{Z_r} x$	$-\dfrac{Z_q}{Z_p} x$
3. Add y.	y	$x + y$	$\dfrac{Z_q}{Z_r} x + y$	$-\dfrac{Z_q}{Z_p} x + y$

Conditions:
(a) Shaft A carrying gear P rotates at 100 rpm clockwise. Therefore,

$$N_p = -\frac{Z_q}{Z_p} x + y = 100$$

or

$$-\frac{30}{40} x + y = 100 \quad \text{(i)}$$

(b) Arm rotates at 100 rpm in counterclockwise. Therefore,

$$y = -100$$

Solving Eqs. (i) and (ii), we get

$$x = -266.7 \text{ rpm} \quad \text{and} \quad y = -100 \text{ rpm}$$

Since speed of shaft B = Speed of gear Q,

$$N_q = x + y$$
$$= -266.7 - 100$$
$$= -366.7 \text{ rpm (counterclockwise)} \quad \text{Ans.}$$

EXAMPLE 11.8 A differential gearbox used in an automobile is shown in Figure 11.12. Let S_1 and S_2 are shafts to which the rear wheels are fastened. The transmission shaft X is attached to pinion A. It receives power through the universal coupling. When automobile takes turn, the speed of shaft S_1 at that instant is 620 rpm. Find the speed of shaft S_2. The driving shaft X rotates at 2000 rpm and gears A and B have 20 and 60 teeth respectively.

Gear Train **459**

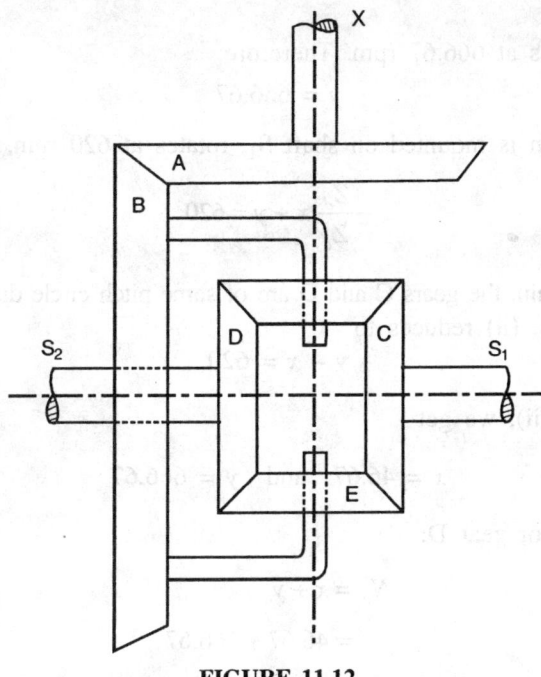

FIGURE 11.12

Solution The gearing between gears A and B is external to the epicyclic gear train. The speed of gear B:

$$N_b = \frac{Z_a}{Z_b} \times N_a = \frac{20}{60} \times 200 = 666.67 \text{ rpm}$$

Let the gear B acts as arm of epicyclic gear train consisting of gears C, D and E. The motion analysis is given in the following table.

Operation	Arm Gear B	Gear D	Gear E	Gear C
1. Lock arm gear B and rotate gear D by one revolution in clockwise direction.	0	+1	$\frac{Z_d}{Z_e}$	$-\frac{Z_d}{Z_e} \times \frac{Z_e}{Z_c}$
2. Multiply by x.	0	x	$\frac{Z_d}{Z_e} \times x$	$-\frac{Z_d}{Z_c} \times x$
3. Add y.	y	$x + y$	$\frac{Z_d}{Z_e} x + y$	$-\frac{Z_d}{Z_c} \times x + y$

Conditions:

(a) Gear B rotates at 666.67 rpm. Therefore,
$$y = 666.67 \tag{i}$$

(b) Gear C, which is mounted on shaft S_1, rotates at 620 rpm. Therefore,
$$-\frac{Z_d}{Z_c} x + y = 620 \tag{ii}$$

In this gear train, the gears C and D are of same pitch circle diameter. Hence $Z_c = Z_d$. Therefore, Eq. (ii) reduces to
$$y - x = 620 \tag{iii}$$

Solving Eqs. (i) and (iii), we get
$$x = 46.67 \quad \text{and} \quad y = 666.67$$

The speed of shaft S_2 or gear D:
$$N_d = x + y$$
$$= 46.67 + 666.67$$
$$= 713.34 \text{ rpm} \qquad \text{Ans.}$$

EXAMPLE 11.9 An epicyclic gear train, known as Ferguson's Paradox, is shown in Figure 11.13. Gear B is fixed to the frame. The arm A and gears C and D are free to rotate on the shaft. Gears B, C and D have 80, 81 and 79 teeth respectively. Pitch circle diameters

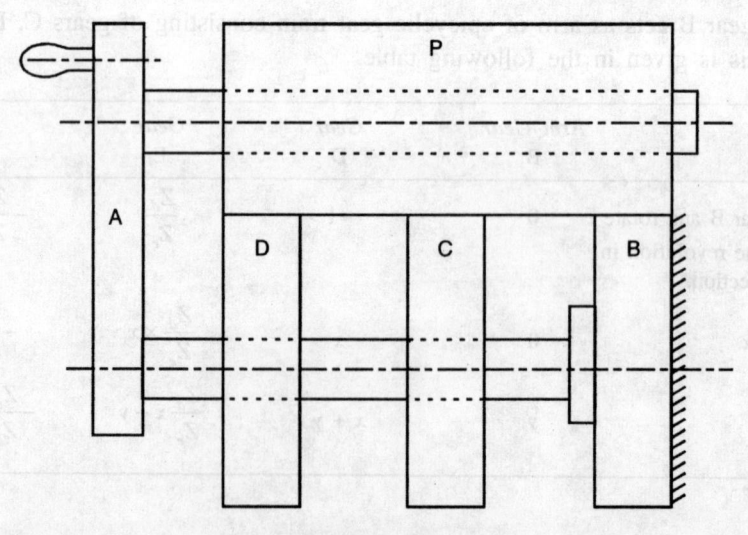

FIGURE 11.13

of all the gears are same so that the planet gear P having 20 teeth meshes with all of them. Determine the speed of gears C and D for one revolution of the arm A.

Solution Let Z_b, Z_c and Z_d are number of teeth on gears B, C and D. The motion analysis of the gear train is given in the following table.

Operation	Arm A	Gear B	Gear P	Gear C	Gear D
1. Lock arm A and rotate gear B by one revolution in clockwise direction.	0	+1	$-\dfrac{Z_b}{Z_p}$	$\dfrac{Z_b}{Z_p} \times \dfrac{Z_p}{Z_c}$	$\dfrac{Z_b}{Z_p} \times \dfrac{Z_p}{Z_d}$
2. Multiply by x.	0	x	$-\dfrac{Z_b}{Z_p} x$	$\dfrac{Z_b}{Z_c} x$	$\dfrac{Z_b}{Z_d} x$
3. Add y.	y	$x+y$	$-\dfrac{Z_b}{Z_p} x + y$	$\dfrac{Z_b}{Z_c} x + y$	$\dfrac{Z_b}{Z_d} x + y$

Conditions:

(a) Gear B is fixed. Therefore,

$$x + y = 0 \qquad \text{(i)}$$

(b) Arm A makes 1 revolution in clockwise direction, Therefore,

$$y = +1 \qquad \text{(ii)}$$

Solving Eqs. (i) and (ii), we get

$$x = -1 \quad \text{and} \quad y = 1$$

Speed of gear C:
$$N_c = \dfrac{Z_b}{Z_c} x + y$$
$$= \dfrac{80}{81} \times -1 + 1$$
$$= \dfrac{1}{81} \text{ revolution in clockwise direction} \qquad \text{Ans.}$$

Speed of gear D:
$$N_d = \dfrac{Z_b}{Z_d} x + y$$
$$= -\dfrac{80}{79} + 1$$
$$= -\dfrac{1}{79} \text{ revolution in counterclockwise direction} \qquad \text{Ans.}$$

EXAMPLE 11.10 In an epicyclic gear train shown in Figure 11.14, the gears B and G are integral with the driving shaft A. The internal gear D is fixed and the gear C rotates about

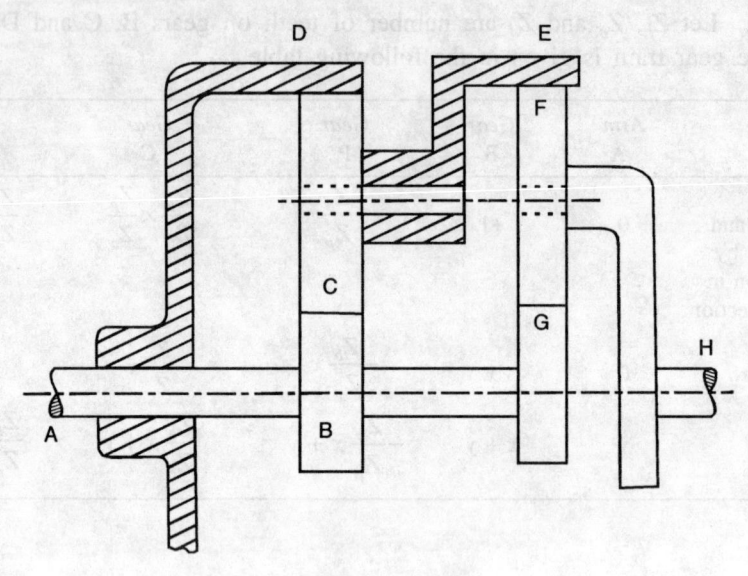

FIGURE 11.14

a pin carried by internal gear E. The gear F is carried by an arm which is keyed to output shaft H. The number of teeth on each gear are:

$$Z_b = 20, \ Z_d = 80, \ Z_g = 24 \ \text{and} \ Z_e = 80$$

Determine the speed of the output shaft and direction in which it rotates when shaft A rotates at a speed of 1000 rpm in clockwise direction and the input power is 11 kW. Also determine the output torque and torque required to keep the gear D fixed.

Solution A close look at the gear train suggests that it is a compound epicyclic gear train consisting of two different arms, one keyed to shaft H and other carried by E. Let us first consider gear train B–C–D–E.

Operation	Arm A	Gear B	Gear C	Gear D
1. Lock arm E and rotate gear B by one turn in clockwise direction.	0	+1	$-\dfrac{Z_b}{Z_c}$	$-\dfrac{Z_b}{Z_c} \times \dfrac{Z_c}{Z_d}$
2. Multiply by x.	0	x	$-\dfrac{Z_b}{Z_c}x$	$-\dfrac{Z_b}{Z_d}x$
3. Add y.	y	$x + y$	$-\dfrac{Z_b}{Z_c}x + y$	$-\dfrac{Z_b}{Z_d}x + y$

Gear Train 463

Conditions:
(a) Gear D is fixed. Therefore,

$$-\frac{Z_b}{Z_d}x + y = 0$$

or

$$-\frac{20}{80}x + y = 0 \qquad (i)$$

(b) Gear B rotates at 1000 rpm in clockwise direction. Therefore,

$$x + y = 1000 \qquad (ii)$$

Solving Eqs. (i) and (ii), we get

$$x = 800 \quad \text{and} \quad y = 200$$

Speed of arm E: $N_e = y = 200$ rpm in clockwise direction.

Now let us consider the second gear train B–G–E–F–H. The velocity analysis is given in the following table.

Operation	Arm H	Gear G	Gear F	Gear E
1. Lock arm H and rotate gear G by one revolution in clockwise direction.	0	+1	$-\dfrac{Z_g}{Z_f}$	$-\dfrac{Z_g}{Z_f} \times \dfrac{Z_f}{Z_e}$
2. Multiply by x.	0	x	$-\dfrac{Z_g}{Z_f}x$	$-\dfrac{Z_g}{Z_e}x$
3. Add y.	y	$x + y$	$-\dfrac{Z_g}{Z_f}x + y$	$-\dfrac{Z_g}{Z_e}x + y$

Conditions:
(a) Speed of gear E from previous analysis is 200 rpm.

∴

$$-\frac{Z_g}{Z_e}x + y = 200$$

or

$$-\frac{24}{80}x + y = 200 \qquad (iii)$$

(b) Gear G, which is compounded with gear B, rotates at 1000 rpm, clockwise.
Therefore,

$$x + y = 1000 \qquad (iv)$$

Solving Eqs. (iii) and (iv), we get

$$x = \frac{8000}{13} \quad \text{and} \quad y = \frac{5000}{13}$$

Speed of output shaft H: $N_h = y = \dfrac{5000}{13} =$ **384.6 rpm, clockwise** Ans.

Input torque: $$T_i = \dfrac{60P}{2\pi N} = \dfrac{60 \times 11 \times 10^3}{2\pi \times 1000}$$
$$= 105.04 \text{ Nm}$$

Output torque: $$T_0 = \dfrac{60 \times 11 \times 10^3}{2\pi \times 384.6} = 273.1 \text{ Nm}$$

Therefore, fixing or braking torque at gear D is:
$$T_b = T_o - T_i$$
$$= 273.1 - 105.04$$
$$= \mathbf{168.06 \text{ Nm}}$$ Ans.

EXERCISES

1. Name the different types of gear train and give the examples where each of them is used.
2. What do you mean by the term train value? How is it related to velocity ratio?
3. What is an epicyclic gear train? In what manner, does it differ from a simple or compound gear train?
4. List the advantages offered by epicyclic gear train compared to simple and compound gear train.
5. What do you mean by reverted gear train? Explain with suitable sketch.
6. Discuss how the fixing torque in case of fixed wheel in epicyclic train can be evaluated?
7. In an epicyclic gear train, the pinion A has 15 teeth and is rigidly fixed to the motor shaft. The wheel B has 20 teeth and gears with A and also with annular fixed gear D. Pinion C has 15 teeth and is integral with B, the gears B and C being a compound gear. Gear C meshes with annular wheel E, which is keyed to the machine shaft. The arm rotates about the same shaft on which gear A is fixed and carries the compound wheels B and C. If the motor runs at 1440 rpm, find the speed of machine shaft. Also determine the torque exerted on the machine shaft if the motor develops a torque of 100 Nm.

[**Ans:** 308.6 rpm, 566.62 Nm]

8. In an epicyclic gear train of the sun and planet type shown in Figure E11.1, the pitch circle diameter of the internally toothed ring is to be as nearly as possible, 216 mm and the module 4 mm. When the ring is stationary, the spider which carries three planet wheels of equal size, is to make one revolution in same sense as the sun wheel for every five revolutions of the driving spindle carrying the sun wheel. Determine suitable number of teeth for all the wheels. Also determine the torque necessary to keep the ring stationary, if a torque of 20 Nm is applied to the spindle carrying the sun wheel.

[Ans: $Z_b = 14$, $Z_c = 21$, $Z_d = 56$, $D_d = 224$ mm, $T_b = 80$ Nm]

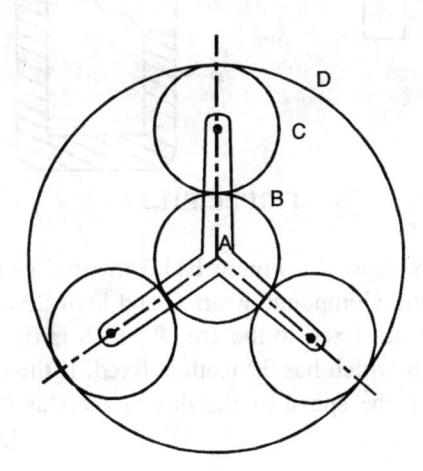

FIGURE E11.1

9. An epicyclic gear train is composed of a fixed annular wheel A having 150 teeth. Meshing with A is a wheel B, which drives wheel D through an idler wheel C, D being concentric with A. Wheels B and C are carried on an arm which revolves clockwise at 100 rpm about the axis of A and D. If the wheel B and D have 25 teeth and 40 teeth respectively, find the number of teeth of gear C and the speed and sense of rotation of gear C.

[Ans: $Z_c = 30$, $N_c = 600$ rpm, clockwise direction]

10. Two shafts A and B are co-axial (Figure E11.2). A gear C having 50 teeth is rigidly mounted on shaft A. A compound gear D–E mesh with gear C and internal gear G. Gear D has 30 teeth and meshes with C and gear E having 35 teeth gears with internal gear G. Gear G is fixed and is concentric with shaft axis. The compound gear D–E is mounted on a pin which projects from an arm keyed to the shaft B. Determine:

(a) The number of teeth on internal gear G, assuming that all gears have same module.
(b) If shaft A rotates at 110 rpm, determine the speed of shaft B.

[Ans: $Z_g = 105$, 50 rpm]

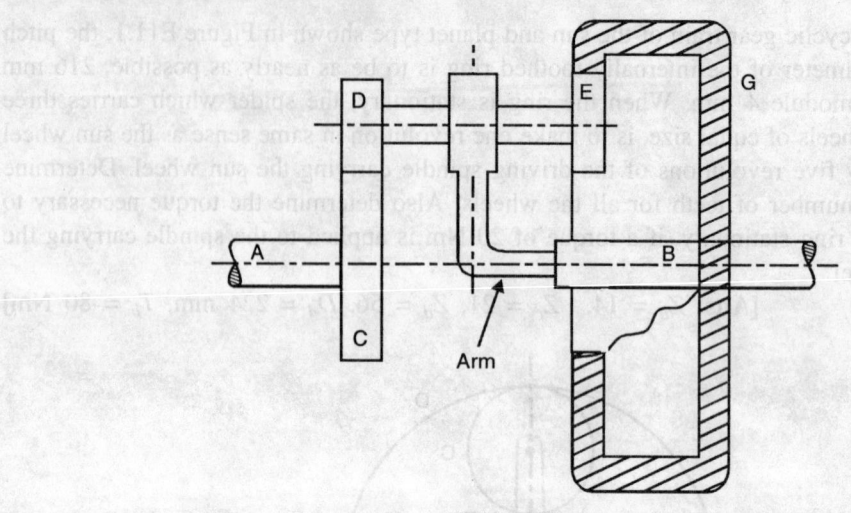

FIGURE E11.2

11. An epicyclic train is shown in Figure E11.3 internal gear A is keyed to the driving shaft and has 30 teeth. Compound gears C and D of 20 and 22 teeth respectively are free to rotate on the pin fixed to the arm P which is rigidly connected to the driven shaft. Internal gear B which has 32 teeth is fixed. If the driving shaft runs at 60 rpm clockwise, determine the speed of the driven shaft and its sense of rotation.

[Ans: 1980 rpm, clockwise]

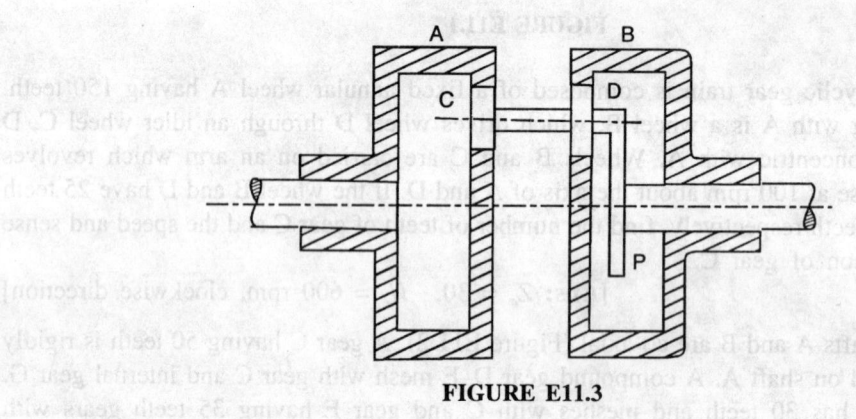

FIGURE E11.3

12. Figure E11.4 shows a gear train in which bevel gear A is fixed. All the bevel gears of the train are identical. The input shaft P carrying the gear B on the arm as shown in the Figure rotates at 1000 rpm in counterclockwise direction. Determine the speed and direction of shaft Q.

[Ans: 2000 rpm, counterclockwise]

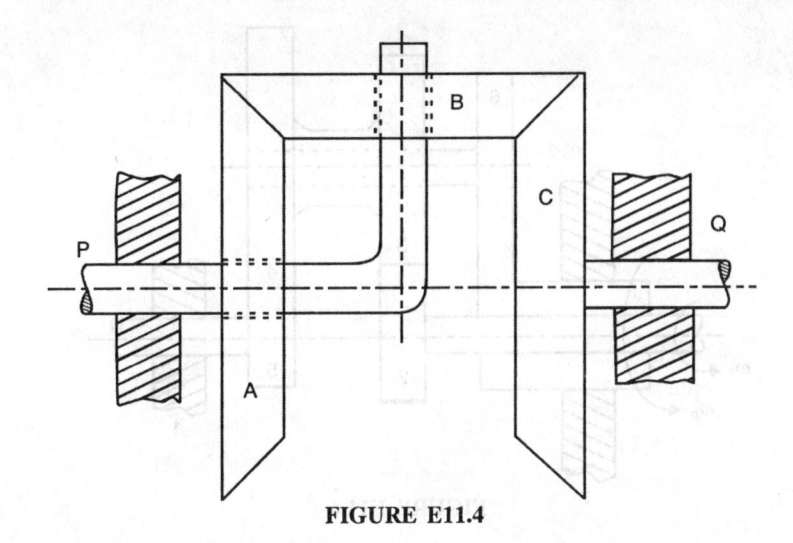

FIGURE E11.4

13. If the shaft A of a gear system shown in Figure E11.5 turns at 100 rpm, obtain the speed of shaft B and its direction of rotation.

[Ans: 269.4 rpm, counterclockwise]

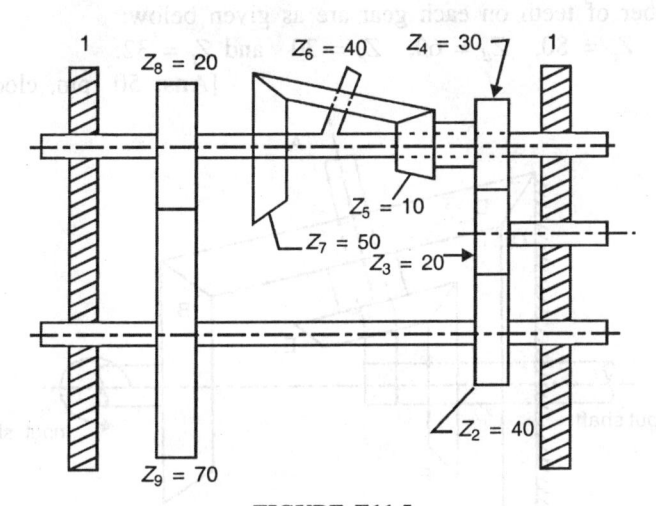

$Z_8 = 20$, $Z_6 = 40$, $Z_4 = 30$, $Z_5 = 10$, $Z_7 = 50$, $Z_3 = 20$, $Z_2 = 40$, $Z_9 = 70$

FIGURE E11.5

14. Figure E11.6 shows a planetary gear train that has two inputs—sun gear 2, which rotates at 500 rpm and arm 6, which rotates at 750 rpm both clockwise as viewed from left. Determine the speed and direction of rotation of gear 5, where $Z_2 = 18$, $Z_3 = 22$, $Z_4 = 25$ and $Z_5 = 15$.

[Ans: $N_5 = 409.1$ rpm, clockwise]

FIGURE E11.6

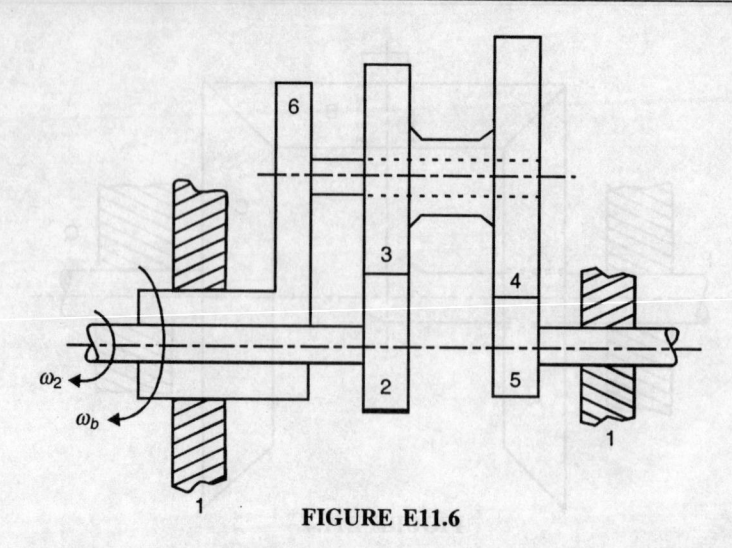

15. In a gear train shown in Figure E11.7 wheel C is fixed. The gear B is connected to the input shaft and gear F is connected to the output shaft. Arm A carrying the compound wheel D and turn freely on the output shaft. If the input speed is 1000 rpm clockwise when seen from the right, determine the speed of the output shaft. The number of teeth on each gear are as given below:

   $Z_b = 20$, $Z_c = 80$, $Z_d = 60$, $Z_e = 30$ and $Z_f = 32$.

   [Ans: 50 rpm, clockwise direction]

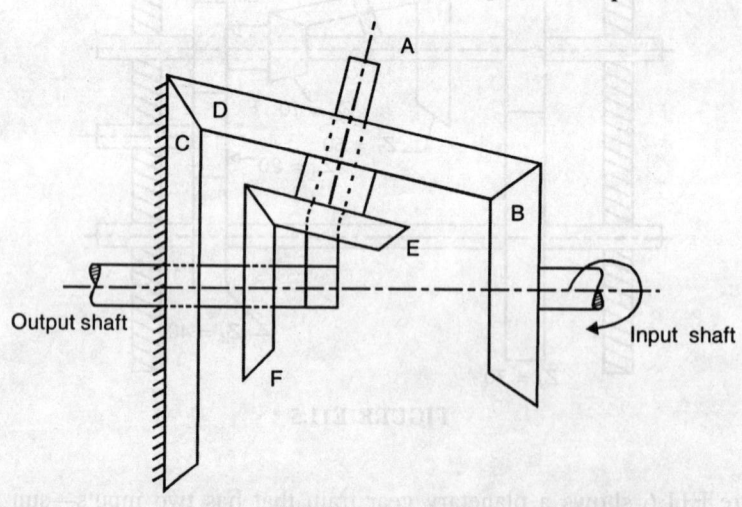

FIGURE E11.7

16. In a epicyclic gear train as shown in Figure E11.8 given below, shafts I and II rotate at 2000 rpm and 350 rpm respectively in opposite directions. Determine the speed of rotation of the shaft III and the direction of rotation. The number of teeth on various gears are as follows:

   $Z_1 = 20$, $Z_2 = 40$, $Z_3 = 30$, $Z_4 = 64$, $Z_5 = 24$ and $Z_6 = 78$.

   [Ans: 1084.3 rpm, opposite to shaft I]

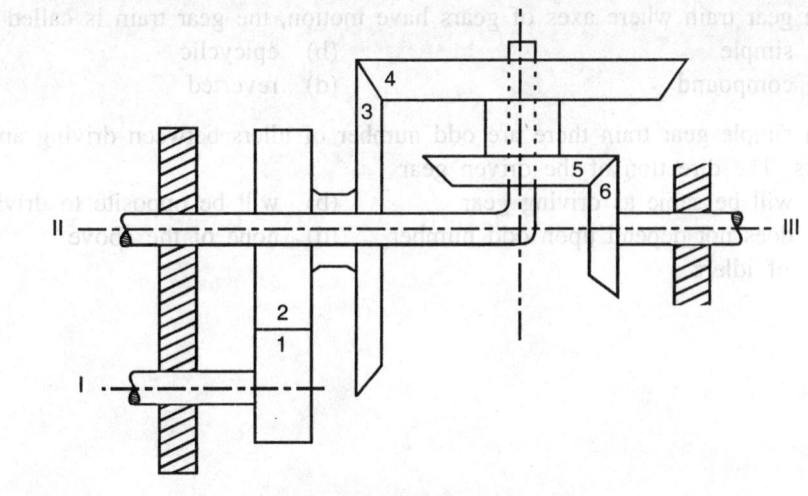

FIGURE E11.8

MULTIPLE CHOICE QUESTIONS

1. The train value of a gear train is
   - (a) equal to speed ratio
   - (b) reciprocal of speed ratio
   - (c) always greater than unity
   - (d) always less than unity

2. In a simple gear train having three gears with 20, 40 and 60 teeth respectively. Find the speed ratio of third to first gear.
   - (a) $\dfrac{1}{3}$
   - (b) 3
   - (c) $\dfrac{1}{2}$
   - (d) $\dfrac{2}{3}$

3. An idler gear helps to
   - (a) increase the speed
   - (b) decrease the speed
   - (c) change the direction of rotation
   - (d) both increase the speed and change the direction

4. When axes of the first and last gear of a compound gear train are co-axial, the gear train is called
   - (a) simple gear train
   - (b) compound gear train
   - (c) epicyclic train
   - (d) reverted gear train

5. In a simple gear train of N gears, the train value is defined as
   - (a) $\dfrac{Z_n}{Z_1}$
   - (b) $\dfrac{N_1}{N_n}$
   - (c) $N_1 \times N_n$
   - (d) $\dfrac{N_n}{N_1}$

6. In a gear train where axes of gears have motion, the gear train is called
   (a) simple
   (b) epicyclic
   (c) compound
   (d) reverted

7. In a simple gear train there are odd number of idlers between driving and driven gears. The direction of the driven gear
   (a) will be same as driving gear
   (b) will be opposite to driving gear
   (c) does not depend upon odd number of idlers
   (d) none of the above

CHAPTER 12

Force Analysis

12.1 INTRODUCTION

The kinematic synthesis of a mechanism or a machine is not sufficient for its proper functioning. A machine should be able to resist external forces without any undue deformation or stresses. Thus it is imperative to calculate the magnitude and direction of forces in various elements of a mechanism.

The forces acting on machine members may arise from different sources such as weight of components, combustion of fuel, change in temperature, external force, spring force, inertia force and many more. These forces may be classified into two categories—static forces and dynamic forces. The static forces are those forces acting on the members whose magnitude doesnot depend on acceleration and mass of the component whereas forces produced due to dynamic action of machine element or inertia are called **dynamic forces.** In this chapter, a complete force analysis (static and dynamic) is presented.

12.2 STATIC FORCE ANALYSIS

While analyzing the mechanism, if the inertia force produced due to accelerating masses are smaller than externally applied force, they can be neglected. Such an analysis is known as **static force analysis.** In static force analysis, for a known value of force acting on one element of a mechanism, the forces acting in other elements are determined without any consideration for motion of other elements. For example, in an IC engine mechanism, the gas force acting on piston is generally known or can be computed from thermodynamic cycle. The determination of forces in connecting rod, crank shaft and frame is termed as **static force analysis.**

12.2.1 Static Equilibrium

A rigid body is said to be in static equilibrium, if it retains its state of affair, i.e. if it is in rest position, tends to remain at rest or is in motion, then it tends to keep itself in motion. The state of equilibrium can be changed by application of external forces or moments. For a rigid body to be in static equilibrium, the vector sum of all the forces and moments about any reference point is zero. In other words, the force and moment (couple) polygon must be closed. Mathematically, the condition of static equilibrium can be written as

472 Theory of Mechanisms and Machines

$$\left.\begin{array}{r}\Sigma F = 0 \\ \Sigma M = 0\end{array}\right\} \quad (12.1)$$

In a two-dimensional planer system, the forces can be described by two-dimensional vector. Therefore,

$$\left.\begin{array}{r}\Sigma F_x = 0 \\ \Sigma F_y = 0 \\ \Sigma M_z = 0\end{array}\right\} \quad (12.2)$$

A member can achieve the state of equilibrium if it satisfies any of the following conditions:

(i) A member under the action of two forces will be in equilibrium if the forces are of the same magnitude and act along the same line in opposite direction (Figure 12.1).

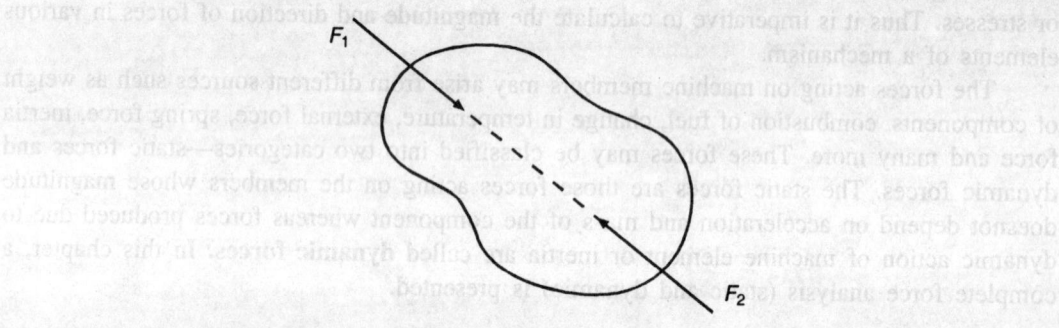

FIGURE 12.1 Member in equilibrium under the action of two forces.

(ii) A member under the action of three forces is said to be in equilibrium, if the resultant of forces is zero and lines of action of the forces intersects at a point called **point of concurrency** (Figure 12.2).

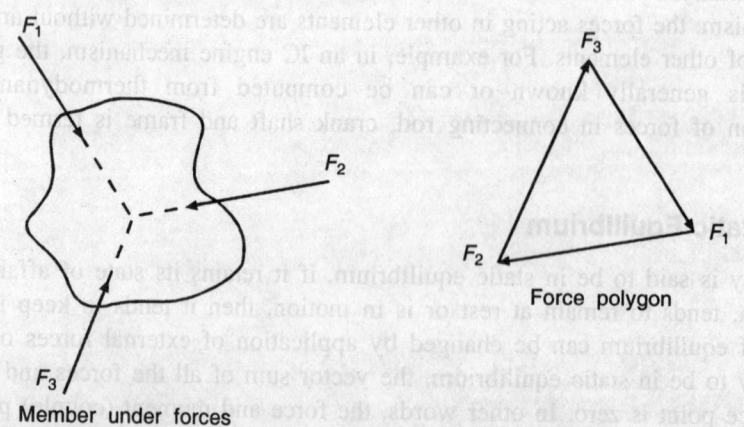

Member under forces

Force polygon

FIGURE 12.2 Member in equilibrium under the action of three forces.

(iii) A member under the action of four forces is said to be in equilibrium if the vector sum of all forces is zero in such a way that resultant of first two forces, say F_1 and F_2 and remaining two forces F_3 and F_4 are collinear as shown in Figure 12.3.

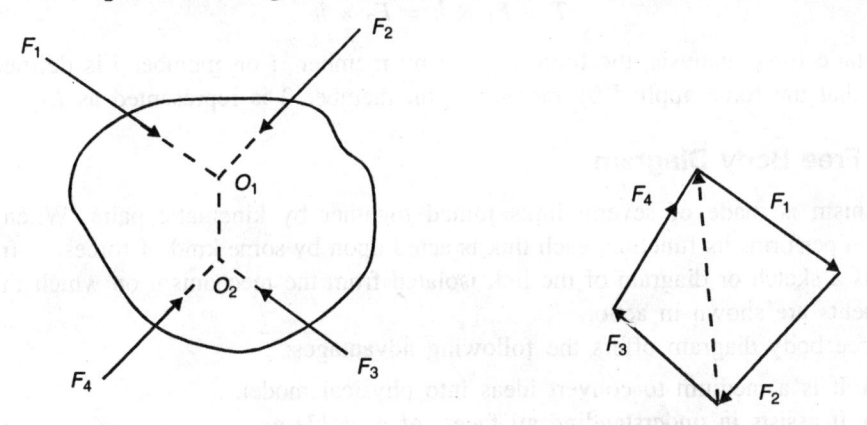

FIGURE 12.3 Member in equilibrium under the action of four forces.

(iv) A member under the action of three or more parallel forces is said to be in equilibrium if the algebraic sum of forces and moments is zero (Figure 12.4).

$$\Sigma F = F_1 - F_2 + F_3 = 0$$

$$\Sigma M = F_3 \times l - F_2 \times l_1 = 0$$

FIGURE 12.4 Member in equilibrium under parallel forces.

(v) A member under the action of two forces and an applied couple will be in equilibrium if the forces are (a) equal in magnitude (b) parallel in direction and in opposite sense and (c) the couple formed by them should be equal in magnitude and should act opposite to the applied torque (Figure 12.5).

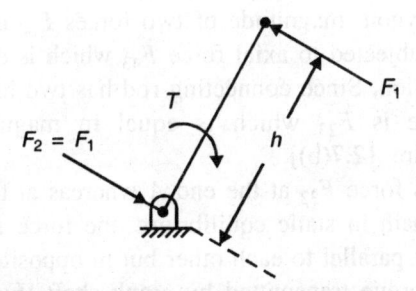

FIGURE 12.5 Member with two forces and a couple.

Equilibrium conditions are:

$$F_1 = F_2$$

and Couple:

$$T = F_1 \times h = F_2 \times h \tag{12.3}$$

In static force analysis, the force applied by member i on member j is defined as F_{ij}. It means that the force applied by member 1 on member 2 is represented as F_{12}.

12.2.2 Free Body Diagram

A mechanism is made of several links joined together by kinematic pairs. When such a mechanism performs its function, each link is acted upon by some kind of forces. A free body diagram is a sketch or diagram of the link isolated from the mechanism on which the forces and moments are shown in action.

A free body diagram offers the following advantages:

(i) It is a medium to convert ideas into physical model.
(ii) It assists in understanding all facets of a problem.
(iii) It helps to build up a suitable mathematical relation between various forces acted upon a link.

Figure 12.6 shows a slider-crank mechanism used in IC engine. The piston of engine is subjected to gas force F, generated due to burning of fuel. This gas force is ultimately transmitted to crank shaft which delivers power. The free body diagrams of individual links are shown in Figure 12.7.

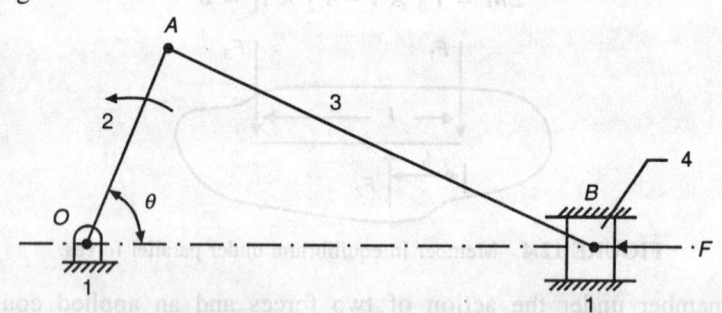

FIGURE 12.6 A Slider–crank mechanism.

The free body diagram of piston shows that three forces act at the point of concurrency. For static equilibrium of piston, the force polygon should be closed as shown in Figure 12.7(a). From this polygon, magnitude of two forces F_{34} and F_{14} can be determined.

The connecting rod is subjected to axial force F_{43} which is equal in magnitude to force F_{34} but acts in opposite direction. Since connecting rod has two hinge ends, it is a system of two forces. The other force is F_{23} which is equal in magnitude to F_{32}. Thus force $F_{34} = F_{43} = F_{32} = F_{23}$ [Figure 12.7(b)].

The crank shaft receives force F_{32} at the end A whereas at the end O, it receives force F_{12}. For a crank shaft to remain in static equilibrium, the force F_{32} must be equal to force F_{12}. However, these forces act parallel to each other but in opposite direction. Thus they form a couple which is equal to torque transmitted by crank shaft [Figure 12.7(c)].

Force Analysis 475

(a) Forces on piston (link 4)

(b) Forces on connecting rod (link 3)

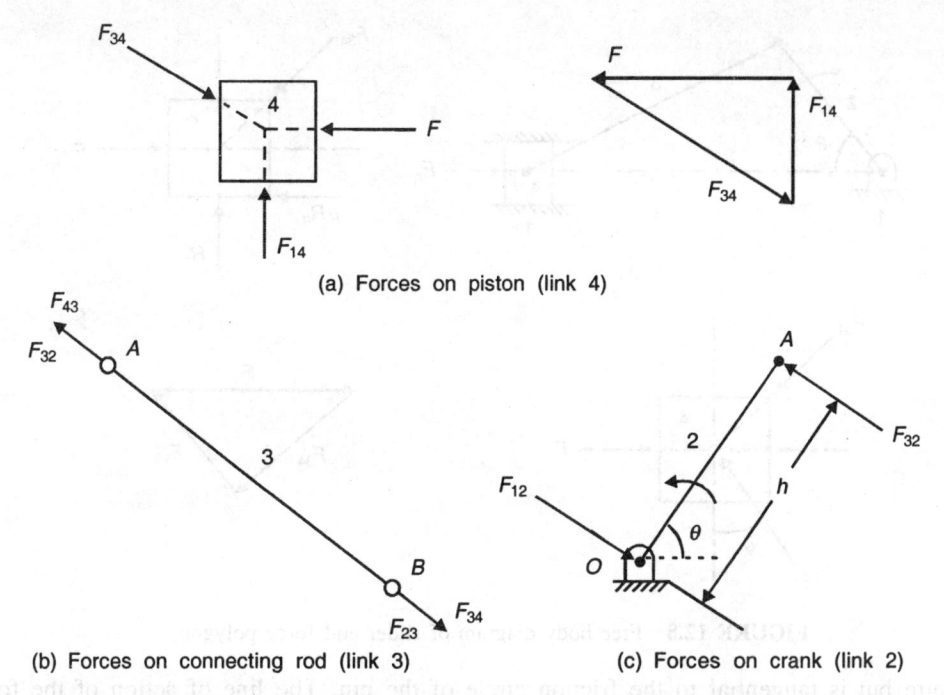

(c) Forces on crank (link 2)

FIGURE 12.7 Free body diagrams of various links of a slider-crank mechanism.

Torque: $$T = F_{32} \times h \tag{12.4}$$

In a machine, if a number of forces act on different elements, the net effect of forces can be found by principle of superposition. Accordingly, the free body diagrams of individual elements are drawn by considering one force at a time and the forces and/or moments thus generated on individual link are super positioned to get the overall effect.

12.2.3 Effect of Friction

When two links of a machine, joined as kinematic pair, move relative to each other, some power is lost due to frictional contact. Generally, two types of contacts occur between mating links—sliding contact and turning contact.

Accordingly, the friction in a machine may be classified as—sliding friction and turning friction.

Sliding Friction. In a machine, the friction generated due to sliding of one link over another is called **sliding friction** such as friction between piston and cylinder of a slider–crank mechanism. In sliding friction the resultant force is inclined by an angle ϕ, called **friction angle.** The free body diagram of a slider with sliding friction is shown in Figure 12.8.

Turning Friction. The turning pair is most commonly used kinematic pair in which a circular pin supported in a bearing allows turning or revolving motion between the links. The friction between the pin and link for revolving motion is called **turning friction**. When a link of a mechanism is joined to another link by such a pair, the force does not pass through the

476 *Theory of Mechanisms and Machines*

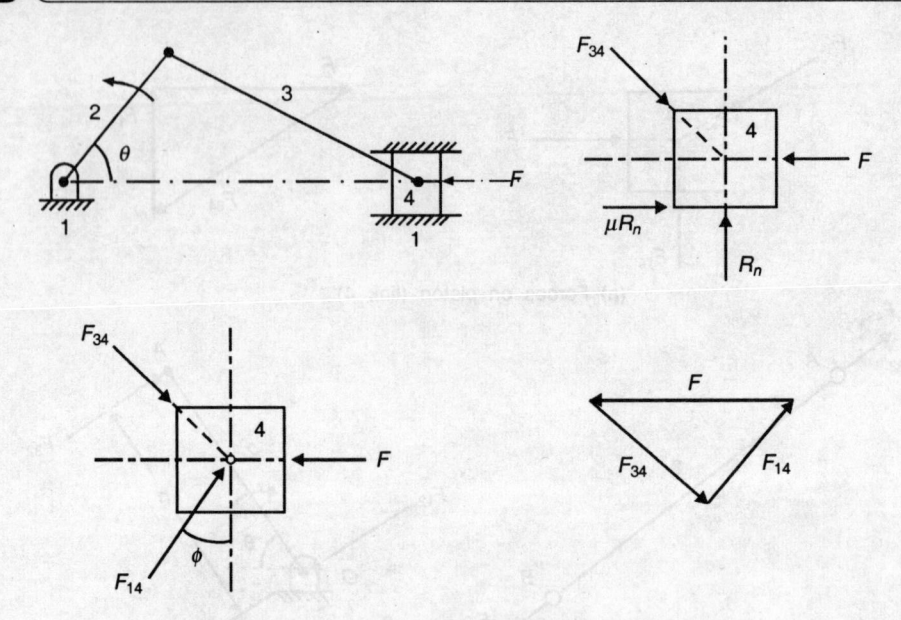

FIGURE 12.8 Free body diagram of slider and force polygon.

pin centre but is tangential to the friction circle of the pin. The line of action of the force is a common tangent to the friction circles of two pins. It is called **friction axis.** For detailed discussion on the subject, reader is advised to refer to Chapter 5.

12.2.4 Principle of Virtual Work

When a mechanism is subjected to number of forces and the entire mechanism is to be investigated for force analysis, the principle of virtual work can be applied. The principle of virtual work states that if a physical system is in equilibrium and undergoes on infinitesimal displacement consistent with the system constraints and if this displacement is imagined to take place without any lapse of time, the net work done by the forces during the process is zero. In other words, the workdone during virtual displacement of various elements from the equilibrium is equal to zero.

For example, consider a slider–crank mechanism as shown in Figure 12.9. If the piston is subjected to force F causing displacement dx, the crank rotates through angular displacement $d\theta$ producing torque T.

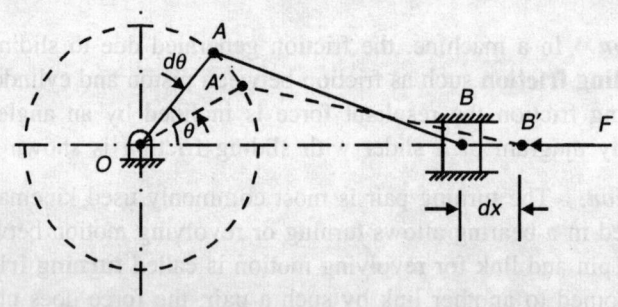

FIGURE 12.9 Slider–crank mechanism.

Force Analysis 477

According to principle of virtual work,

$$Td\theta + Fdx = 0 \tag{12.5}$$

As the forces acting in connecting rod and crank shaft bearing do not perform any work.

$$T \cdot \frac{d\theta}{dt} + \frac{Fdx}{dt} = 0$$

or

$$T\omega + F \times v = 0 \tag{12.6}$$

where ω = angular velocity of the crank
and v = linear velocity of the piston

EXAMPLE 12.1 Determine the torque required to be applied at the crank shaft of a slider–crank mechanism to bring it in equilibrium. The slider is subjected to a horizontal force of 5000 N and a force of magnitude 1000 N is applied on the connecting rod as shown in Figure 12.10. The dimensions of various links are as under:

OA = 250 mm, AB = 750 mm   and AC = 250 mm, $\angle BOA$ = 40°

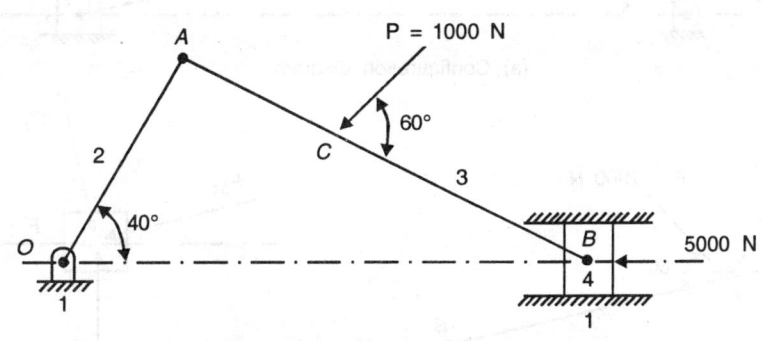

FIGURE 12.10 Slider–crank mechanism.

Solution The graphical solution procedure is given as under:

1. Draw configuration diagram of the mechanism as shown in Figure 12.11(a).
2. Draw free body diagram of link 3 as shown in Figure 12.11(b). The force F_{43} is broken into two components—one along the link F_{43}^t and another normal to the link F_{43}^n.
3. Taking moments about point A,

$$P \times x = F_{43}^n \times AB$$

where $x = AC \cos 30° = 250 \times \cos 30° = 216.5$ mm

Force: $$F_{43}^n = \frac{1000 \times 216.5}{750} = 288.67 \text{ N}$$

4. Out of forces acting on link 4, forces F and F_{43}^n are known completely [Figure 12.11(c)]. Force F_{14} is perpendicular to the sliding surface and force F_{34}^t is along the link 3. Construct a force polygon as shown in Figure 12.11(d) with suitable scale (1 mm = 100 N). By measurement, the magnitude of force F_{34} = 5200 N.
5. Draw the force polygon of link 3 with forces F_{43}, P and F_{23}. By measurement, force F_{23} = 5700 N. [See Figure 12.11(e)].
6. Draw the free body diagram of link 2. Measure distance h and calculate torque [Figure 12.11(f)].

$$T_2 = F_{32} \times h$$
$$= 5700 \times 0.175$$
$$= 997.5 \text{ Nm clockwise} \quad \text{Ans.}$$

(a) Configuration diagram

(b) Equilibrium of link 3

(c) Forces on link 4

(d) Force polygon of link 4

(e) Force polygon of link 3

(f) Free body diagram of link 2

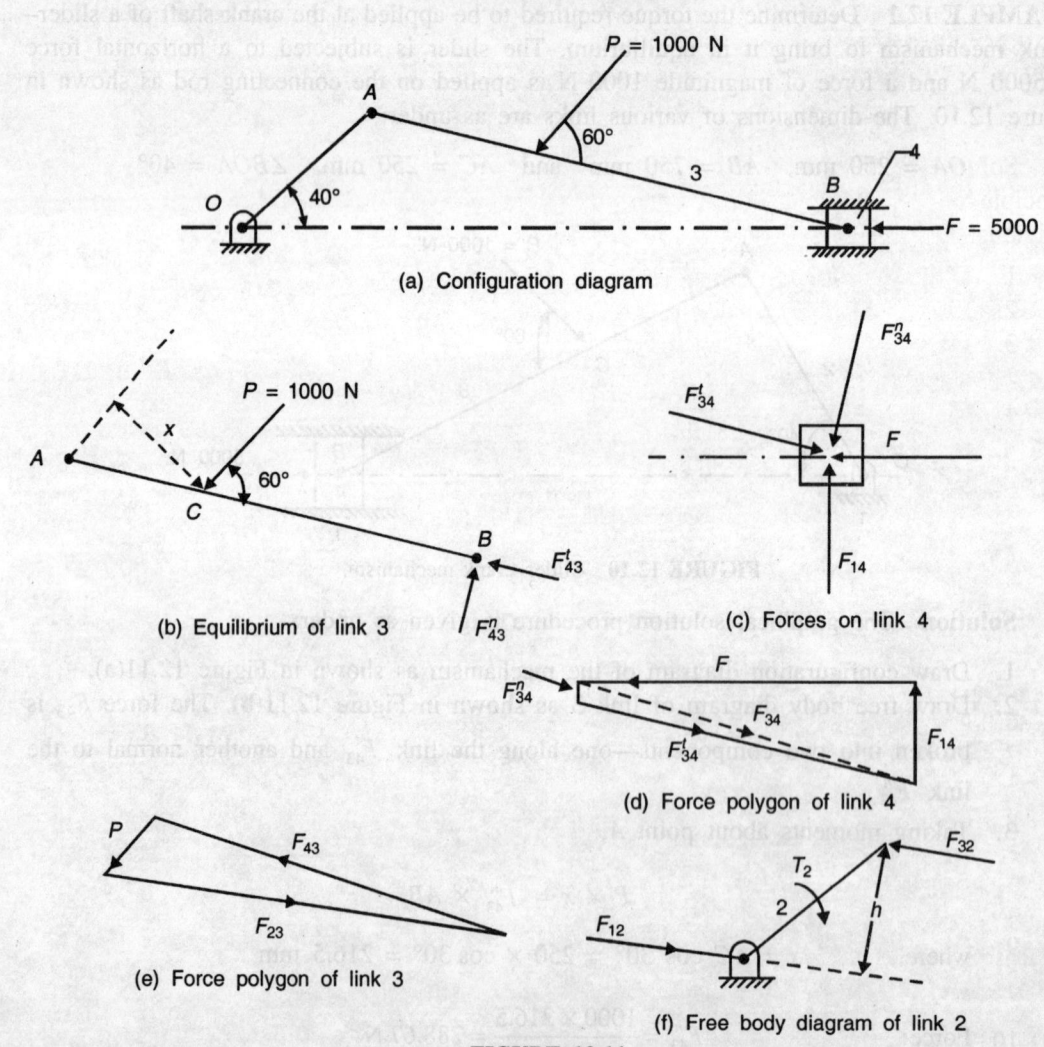

FIGURE 12.11

Force Analysis 479

EXAMPLE 12.2 A four bar mechanism as shown in Figure 12.12 is subjected to two forces, $F_3 = 1000$ N at 60° from horizontal and $F_4 = 2000$ N at 45° from link 4. The dimensions of links are as under:

$AB = 300$ mm, $BC = 400$ mm, $CD = 450$ mm and $AD = 600$ mm

Perform static force analysis and determine resisting torque on link 2.

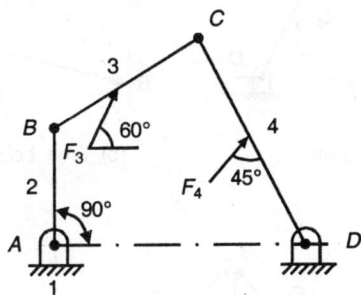

FIGURE 12.12

Solution The problem of a number of loads acting on a mechanism can be solved by principle of superposition in which net effect is equal to superposition of the effect of individual loads taken one at a time.

1. Draw the configuration diagram of mechanism as shown in Figure 12.13(a).
2. Let us first consider the effect of force F_3, neglecting force F_4.
3. Draw the free body diagram of link 3 and find the line of action (loa) of forces F_{23} and F_{43} [Figure 12.13(b)].
4. Construct force polygon [Figure 12.13(c)] with suitable scale (250 N = 10 mm). Measure the magnitude of forces F_{43} and F_{23}.
   By measurement,
   $$F_{43} = 450 \text{ N} \quad \text{and} \quad F_{23} = 750 \text{ N}$$
5. Draw free body diagram of link 2 [Figure 12.13(d)]. Measure distance $h_1 = 200$ mm. Torque due to force F_3,
   $$T_{23} = F_{23} \times h_1 = 750 \times \frac{200}{1000} = 150 \text{ Nm (counterclockwise)}$$
6. Now consider the effect of force F_4 neglecting the effect of force F_3.
7. Draw the free body diagram of link 4 and find the loa of forces F_{34} and F_{14} [Figure 12.13(e)].
8. Construct force polygon for link 4 [Figure 12.13(f)] with suitable scale (500 N = 10 mm). Measure the magnitude of forces F_{14} and F_{34}.
   $$F_{14} = 1750 \quad \text{and} \quad F_{34} = 750 \text{ N}$$
9. Draw the free body diagram of link 3 [Figure 12.13(g)].
   $$F_{34} = F_{43} = F_{23} = F_{32}$$
10. Draw the free body diagram of link 2 [Figure 12.13(h)]. Measure the distance $h_2 = 280$ mm.

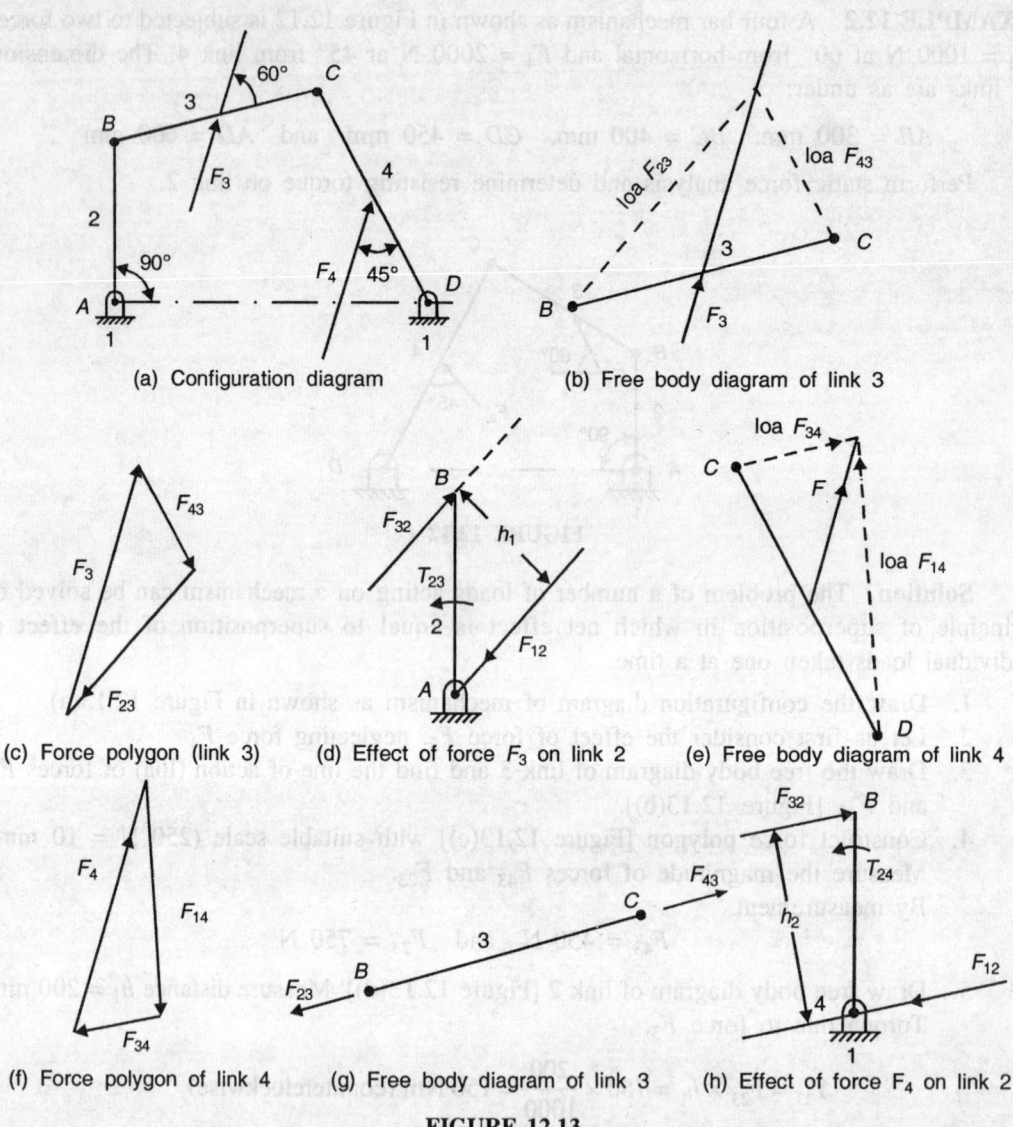

FIGURE 12.13

Torque: $$T_{24} = F_{32} \times h_2$$

$$= 750 \times \frac{280}{1000} = 210 \text{ Nm (counterclockwise)}$$

Total resisting torque: $$T_2 = T_{23} + T_{24}$$

$$= 150 + 210$$

$$= \mathbf{360 \text{ Nm (counterclockwise)}} \qquad \text{Ans.}$$

Force Analysis 481

EXAMPLE 12.3 Determine the torque required to be applied for static equilibrium of a press mechanism as shown in Figure 12.14. The force acting on the slider is 5000 N and various dimensions are as given below:

$$OA = 240 \text{ mm}, \quad AB = 1000 \text{ mm}, \quad BC = 620 \text{ mm}$$
$$CD = 400 \text{ mm}, \quad DE = 600 \text{ mm}$$

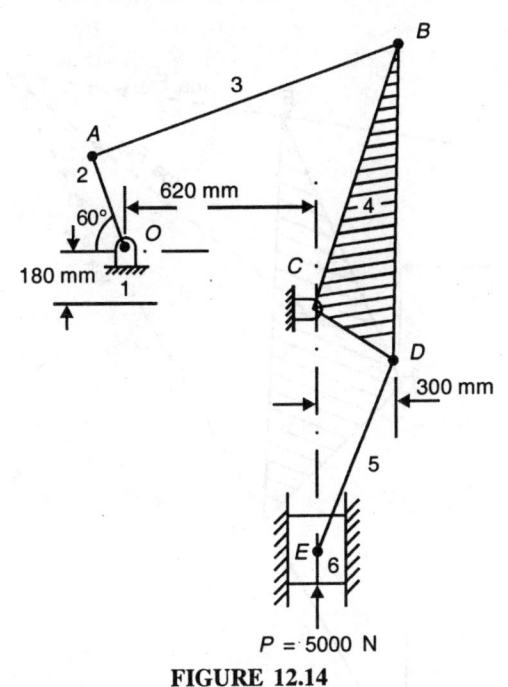

FIGURE 12.14

Solution The graphical solution procedure is given as under:
1. Draw the configuration diagram of the mechanism as shown in Figure 12.15(a).
2. The free body diagram of slider with forces is shown in Figure 12.15(a).
3. Draw the force polygon of slider link 6 with suitable scale (1000 N = 10 mm). Find the magnitude of forces F_{56} and F_{16} by measurement [Figure 12.15(b)].

$$F_{56} = 5700 \text{ N} \quad \text{and} \quad F_{16} = 2800 \text{ N}$$

4. Draw the line of action (loa) of forces F_{34}, F_{41} and F_{54} as shown by dashed line [see Figure 12.15(a)].
5. Draw the force polygon of link 4 as shown in Figure 12.15(c). By measurement,

$$F_{41} = 9200 \text{ N} \quad \text{and} \quad F_{34} = 4600 \text{ N}$$

6. Draw the free body diagram of link 3 [Figure 12.15(d)].

$$F_{34} = F_{43} = F_{32} = F_{23}$$

7. Draw the free body diagram of link 2 [Figure 12.15(e)]. Measure the distance $h = 230$ mm.

482 Theory of Mechanisms and Machines

Torque:
$$T_2 = F_{32} \times h$$
$$= 4600 \times \frac{230}{1000}$$
$$= 1058 \text{ Nm (clockwise)} \qquad \text{Ans.}$$

(a) Configuration diagram
(b) Force polygon of link 6
(c) Force polygon of link 4
(d) Free body diagram of link 3
(e) Free body diagram of link 2

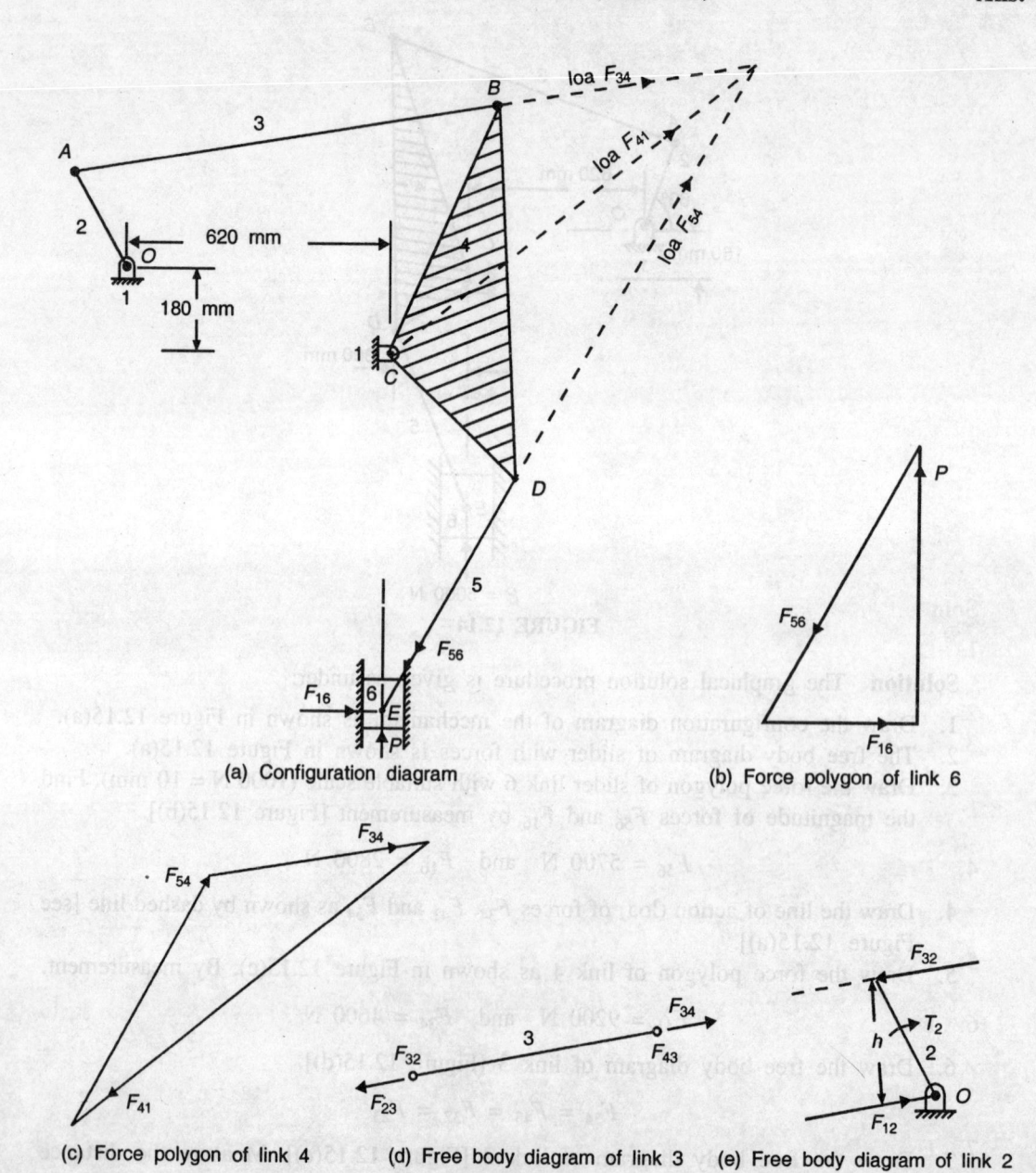

FIGURE 12.15

EXAMPLE 12.4 For the static equilibrium of a quick-return motion mechanism shown in Figure 12.16, determine the required input torque for a force of 4000 N.

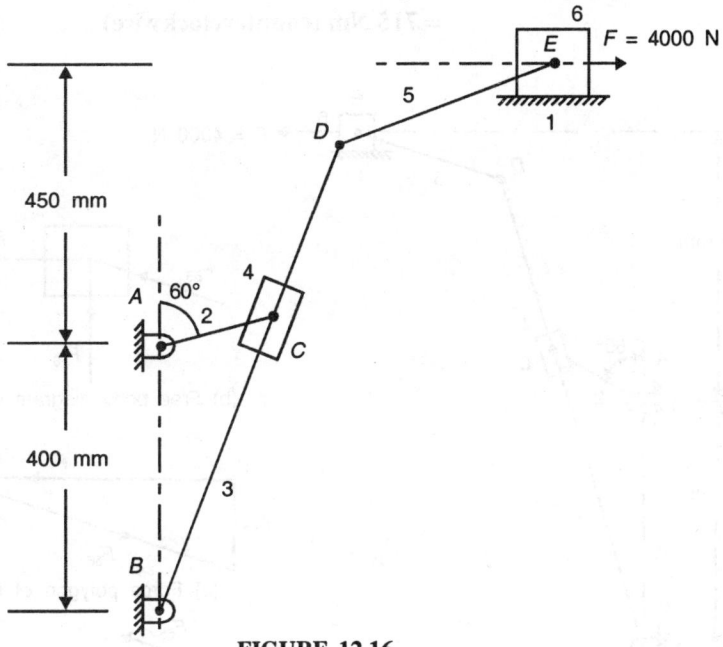

$CD = 300$ mm, $DE = 250$ mm and $AC = 175$ mm

FIGURE 12.16

Solution The graphical solution procedure is given below:
1. Draw the configuration diagram of a quick-return motion mechanism as shown in Figure 12.17(a).
2. Various forces acting on the slider link 6 are shown in Figure 12.17(b).
3. Draw the force polygon for link 6 and measure forces F_{56} and F_{16} (Scale 1000 N = 10 mm) [Figure 12.17(c)].

$$F_{56} = 4200 \text{ N} \quad \text{and} \quad F_{16} = 1200 \text{ N}$$

4. Draw the free body diagram of link 5 [Figure 12.17(d)].

$$F_{65} = F_{56} = F_{35} = F_{53}$$

5. Draw the force diagram of link 3 to determine the line of action (loa) of forces F_{35}, F_{43} and F_{13} [Figure 12.17(e)].
6. Draw the force polygon of link 3 with suitable scale and measure the forces [Figure 12.17(f)].

$$F_{13} = 3200 \text{ N} \quad \text{and} \quad F_{43} = 5500 \text{ N}$$

7. Draw the free body diagram of link 2 [Figures 12.17(g) and (h)].

$$F_{42} = F_{24} = F_{34} = F_{43}$$

484 *Theory of Mechanisms and Machines*

Torque:
$$T_2 = F_{42} \times h$$
$$= 5500 \times \frac{130}{1000}$$
$$= 715 \text{ Nm (counterclockwise)} \qquad \text{Ans.}$$

FIGURE 12.17

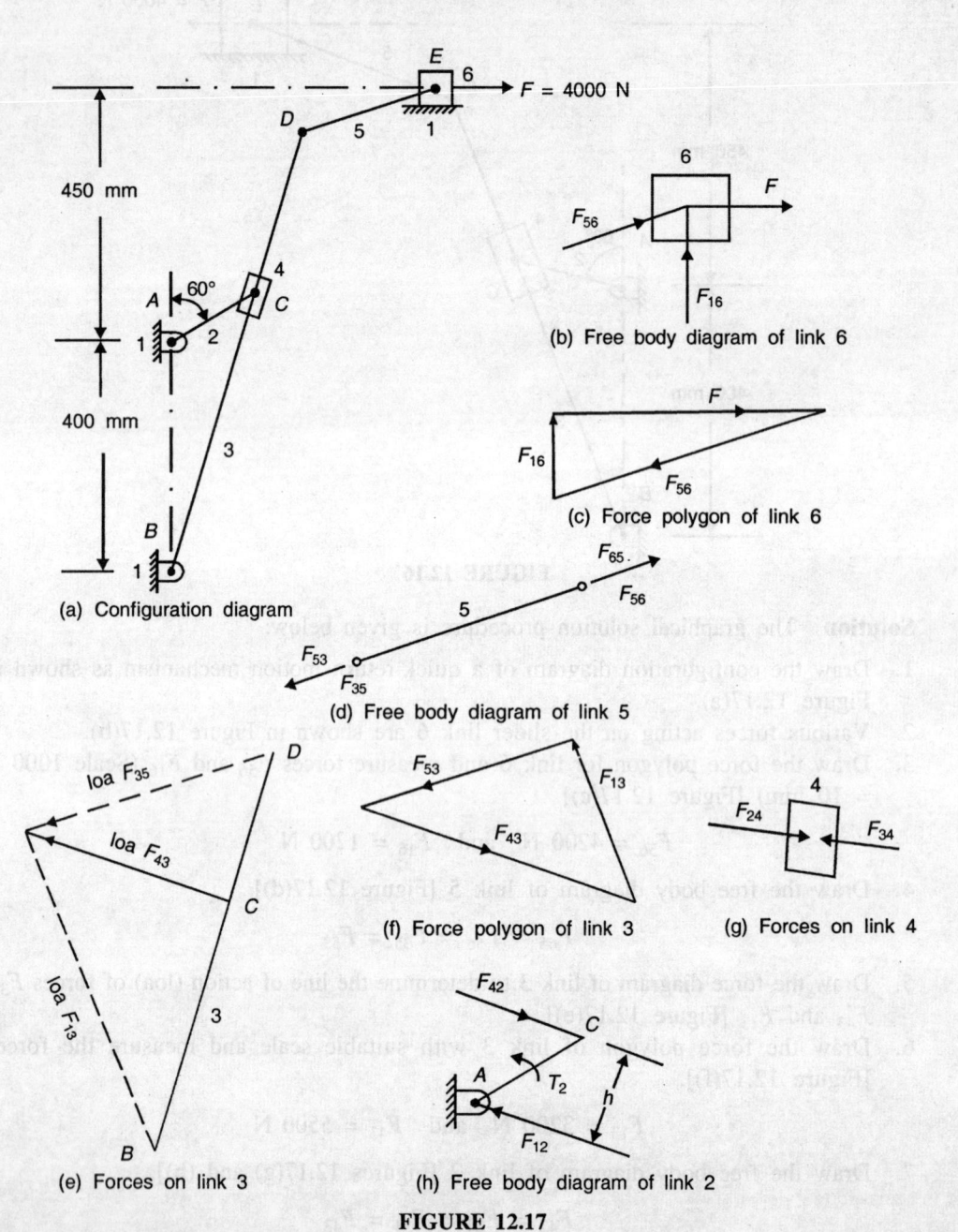

Force Analysis   **485**

EXAMPLE 12.5 A Slider-crank mechanism is shown in Figure 12.18 given below. The force acting on slider is 5000 N and coefficient of friction* between all the links is 0.25. Calculate the driving torque if the pin diameters at joints O, A and B are 80 mm, 80 mm and 40 mm respectively. The dimensions of links are:

$$OA = 300 \text{ mm}, \quad AB = 1050 \text{ mm} \quad \text{and} \quad \angle BOA = 60°$$

FIGURE 12.18

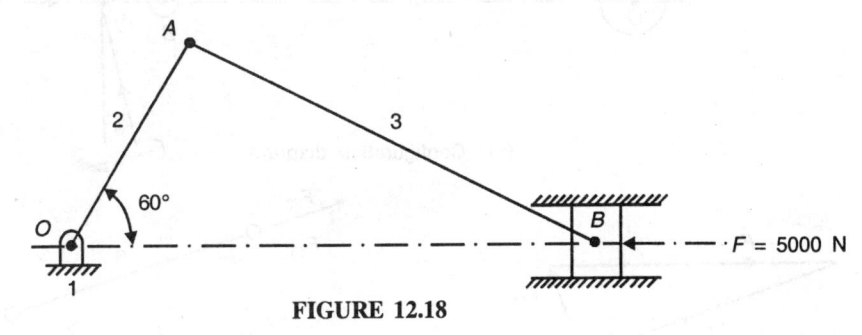

Solution Friction circle radius:

$$\text{at point } O = \mu r_1 = 0.25 \times \frac{80}{2} = 10 \text{ mm}$$

$$\text{at point } A = \mu r_2 = 0.25 \times \frac{80}{2} = 10 \text{ mm}$$

$$\text{at point } B = \mu r_3 = 0.25 \times \frac{40}{2} = 5 \text{ mm}$$

The graphical procedure is given below:

1. Draw configuration diagram of slider-crank mechanism as shown in Figure 12.19(a).
2. Draw friction circle of different radius at points O, A and B.
3. Draw tangents and decide friction axis of links 2 and 3.
4. Calculate friction angle $\phi = \tan^{-1}(0.25) = 14°$ and draw the resultant side thrust force F_{14} inclined at friction angle.
5. Consider various forces at the slider and draw the force polygon with suitable scale (1000 N = 10 mm) as shown in Figure 12.19(b). Find the forces F_{34} and F_{14} by measurement.

$$F_{34} = 4800 \text{ N} \quad \text{and} \quad F_{14} = 1150 \text{ N}$$

6. Draw the free body diagram of link 3 as shown in Figure 12.19(c) with

$$F_{34} = F_{43} = F_{32} = F_{23}$$

7. Mark Forces F_{32} and F_{12} at the crank as shown in Figure 12.19(a).

---

* In actual slider-crank mechanism, the coefficient of friction is very low. In this problem, a fictitious high value has been taken to facilitate drawing of friction circle.

(a) Configuration diagram

(b) Force polygon of slider link 4

(c) Free body diagram of link 3

FIGURE 12.19

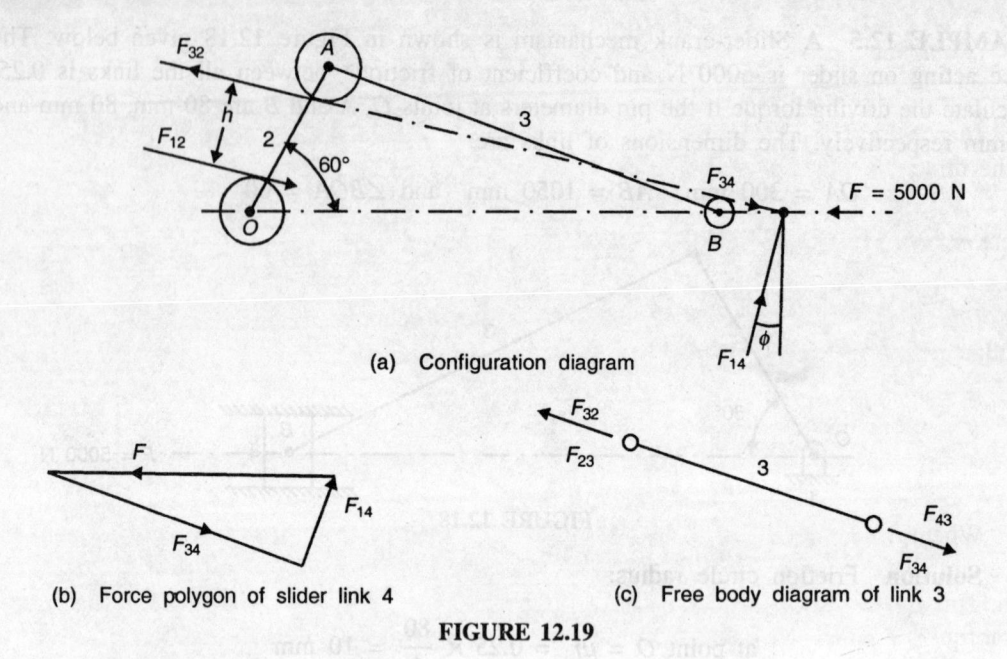

8. Measure distance $h = 165$ mm.

   Torque: $\quad T_2 = F_{32} \times h$

   $$= 4800 \times \frac{165}{1000}$$

   $$= 792 \text{ Nm (counterclockwise)} \qquad \text{Ans.}$$

12.3 DYNAMIC FORCE ANALYSIS

In the mechanism or a machine, when a link moves with zero acceleration, from Newton's second law, it is said that the resultant of forces acting on the link is zero and the link is in the state of static equilibrium. However, when a link moves with certain acceleration, the moving mass of the link experiences on additional force called **inertia force** or **dynamic force**. Inertia is a property of matter by virtue of which a rigid body resists any change in velocity. In a force analysis, if the magnitude of these forces are large compared to externally applied forces, they are also accounted. This type of force analysis is called **dynamic force analysis**.

The problem of dynamic force analysis can be converted into static force analysis using D'Alembert's principle. According to it, the vector sum of all external forces and inertia force acting on a rigid body is zero and that the vector sum of all external torques/moments and inertia couple acting on a rigid body is also separately equal to zero. A rigid body following above principle is said to be in dynamic equilibrium. Thus we can write the equation of equilibrium just similar to that of static equilibrium as follows:

$$\Sigma F + F_i = 0 \qquad (12.7a)$$
$$\Sigma T + C_i = 0 \qquad (12.7b)$$

Inertia force. The inertia force F_i is the product of mass and linear acceleration. It acts in the direction opposite to that of acceleration.

$$\therefore \qquad F_i = -mf \qquad (12.8)$$

where m = mass of the rigid body
and f = acceleration of the mass of the body

Inertia couple. The inertia couple C_i is the product of mass moment of inertia and angular acceleration of the rigid body.

Let I = mass moment of inertia ($= mk^2$)
k = radius of gyration
and α = angular acceleration about an axis passing through centre of mass and perpendicular to the plane of motion

When dynamic force analysis of a link is performed by force polygon method, the inertia force and inertia couple are replaced by an equivalent offset inertia force which accounts for both. This is done by displacing the line of action of acceleration by distance h in such a way that torque produced is equal to the inertia couple acting on the link. This inertia couple should oppose the direction of angular acceleration.

Therefore, $\qquad T = C_i = F_i \times h \qquad (12.9)$

or $$h = \frac{C_i}{F_i} = \frac{I\alpha}{mf} = \frac{mk^2\alpha}{mf} \qquad (12.10)$$

$$= \frac{k^2\alpha}{f}$$

For example, consider a link which is subjected to two forces F_{43} and F_{23} as shown in Figure 12.20(a). The resultant force R (vector sum of F_{43} and F_{23}) produces an acceleration f_g of the mass centre of the link and angular acceleration α because the line of action of the resultant force does not pass through the mass centre. We choose two forces $\pm mf_g$ at a distance h such that the couple formed by these forces opposes the direction of angular acceleration [Figure 12.20(b)].

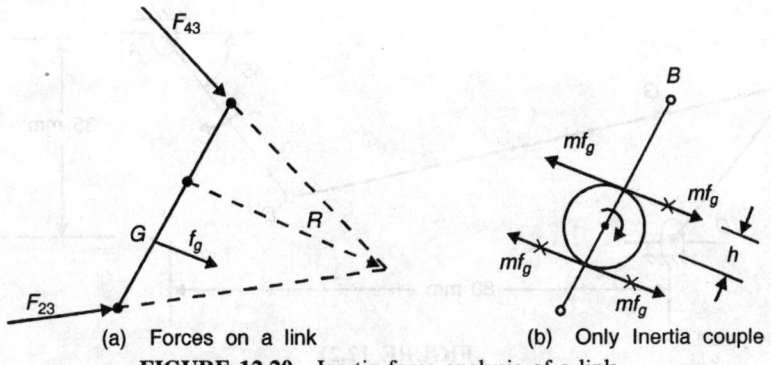

(a) Forces on a link (b) Only Inertia couple
FIGURE 12.20 Inertia force analysis of a link.

12.4 INERTIA FORCE ANALYSIS

The inertia force analysis of a mechanism can be performed just similar to the static force analysis except that inertia force and couple are included as suggested by D'Alembert's principle. The following procedure may be adopted:

1. Draw the configuration diagram of a given mechanism.
2. Draw the velocity and acceleration diagram of the mechanism as discussed in Chapter 2.
3. Determine the linear acceleration of centre of mass of a given link and also its angular acceleration.
4. Calculate the inertia force and inertia couple from Equations (12.8) and (12.9) respectively.
5. In order to replace the inertia force and torque by a single offset inertia force, Calculate the offset distance h from Eq. (12.10).
6. Draw a circle of radius h with centre of gravity of link as centre.
7. Draw two tangent lines parallel to the direction of acceleration and mark arrows in two opposite directions as shown in Figure 12.20(b).
8. Out of these four directions, two directions which are along the direction of acceleration are dropped as the inertia force opposes the motion.
9. Out of remaining two directions, choose one which produces a couple about the mass centre in the direction opposite to the direction of angular acceleration.

EXAMPLE 12.6 Figure 12.21 given below shows a rotating crank AB driving an oscillating link CD through the coupler link BC. The mass of link BC is 0.75 kg and its radius of gyration about an axis through point G is 40 mm. The distance CG is 26 mm. For the above configuration of the mechanism, when link AB rotates in an anticlockwise direction at constant speed of 700 rpm find:

(i) the acceleration of point G
(ii) the angular acceleration of the link BC
(iii) the inertia force on link BC and its point of application.

$AB = 30$ mm, $BC = 90$ mm, $CD = 40$ mm and $CG = 26$ mm

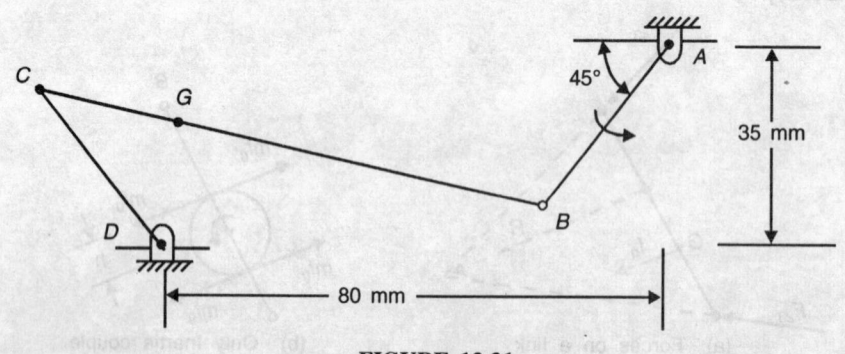

FIGURE 12.21

Solution Draw the configuration diagram as shown in the Figure 12.22(a).

Velocity of point B: $\quad v_b = \omega \times AB$

$$= \frac{2\pi \times 700}{60} \times \frac{30}{1000} = 2.2 \text{ m/s}$$

Construct velocity polygon [Figure 12.22(b)] by the method outlined in Chapter 2. Measure the velocities of links BC and CD. The construct a table of centripetal acceleration as given below:

Link	Length (mm)	Velocity m/s	Centripetal acceleration $\frac{v^2}{r}$ (m/s^2)
AB	30	ab = 2.2	161.34
BC	90	bc = 3.9	169
CD	40	cd = 3.2	256

(a) Configuration diagram

(b) Velocity polygon

(c) Acceleration diagram

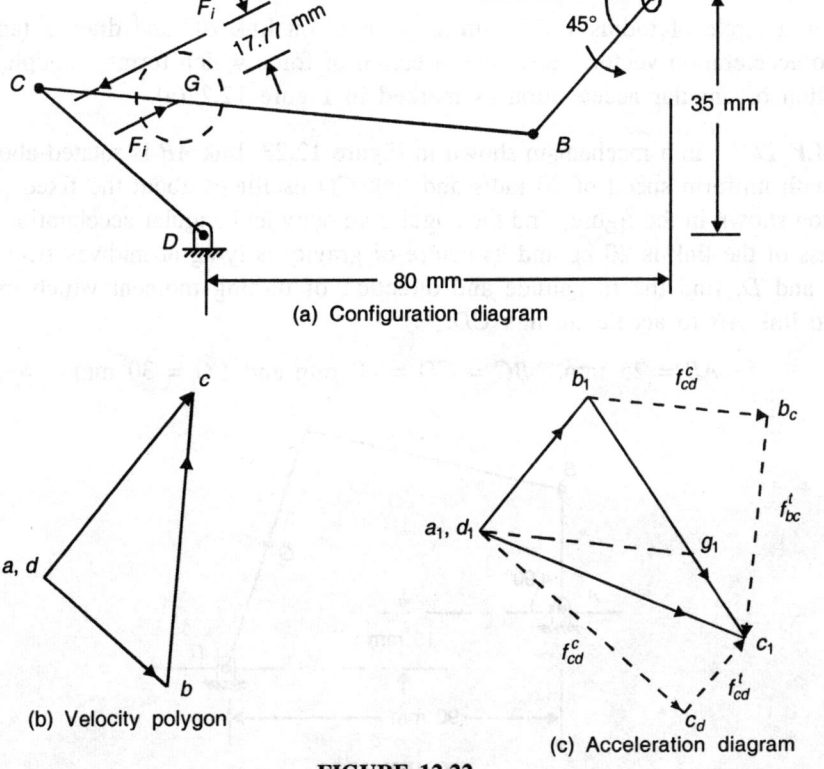

FIGURE 12.22

Construct the acceleration polygon as shown in Figure 12.22(c) by the method outlined in Chapter 2. The $a_1 b_1 c_1 d_1$ is the total acceleration diagram. Vectors $\overrightarrow{b_1 b_c}$ and $\overrightarrow{b_c b_1}$ represent centripetal and tengential acceleration of the link BC respectively.

The point g_1 corresponding to point G on the link BC is determined by dividing $b_1 c_1$ at g_1 such that

$$\frac{BG}{BC} = \frac{b_1 g_1}{b_1 c_1}$$

(i) Join the line $a_1 g_1$ [Figure 12.22(c)] and measure the distance. Acceleration of point G: $f_g = \overrightarrow{a_1 g_1} = 205$ m/s² Ans.

(ii) Angular acceleration of link BC:

$$\alpha_{BC} = \frac{b_c c_1}{BC} = \frac{205}{0.09} = 2277.78 \text{ m/s}^2$$ Ans.

(iii) (a) Inertia force: $F_i = m f_g = 0.75 \times 205 = \mathbf{153.75}$ **N** Ans.

(b) Inertia couple: $C_i = I \alpha_{BC} = m k^2 \alpha_{BC} = 0.75 \times 0.04 \times 2277.78 = 2.733$ Nm

$$\therefore \quad h = \frac{C_i}{F_i} = \frac{2.733 \times 1000}{153.75} = 17.77 \text{ mm}$$

Draw a circle of radius 17.77 mm at point G on link BC and draw a tangential line parallel to acceleration vector. Select the direction of force which forms a couple opposite to the direction of angular acceleration as marked in Figure 12.22(a).

EXAMPLE 12.7 In a mechanism shown in Figure 12.23, link AB is rotated about the fixed point A with uniform speed of 20 rad/s and link CD oscillates about the fixed point D. For the position shown in the figure, find the angular velocity and angular acceleration of link CD. If the mass of the link is 20 kg and its centre of gravity is lying at midway between the two points C and D, find the magnitude and direction of turning moment which must then be applied to link AB to accelerate link CD.

$AB = 25$ mm, $BC = CD = 60$ mm and $CG = 30$ mm

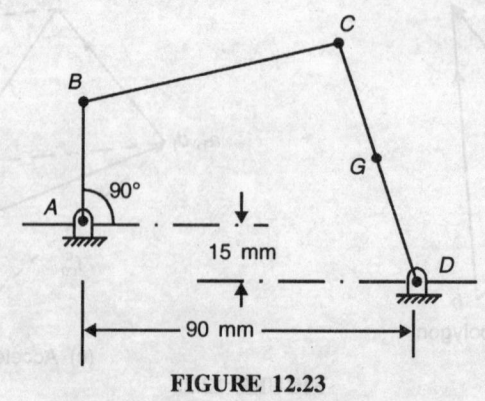

FIGURE 12.23

Force Analysis 491

Solution Draw the configuration diagram as shown in the Figure 12.24(a).

Velocity of point B: $\qquad v_b = \omega \times AB = 20 \times \dfrac{25}{1000} = 0.5$ m/s

Construct the velocity polygon [Figure 12.24(b)] by the method as outlined in the Chapter 2. Measure the velocities of links BC and CD. Then construct the table of centripetal acceleration as given below:

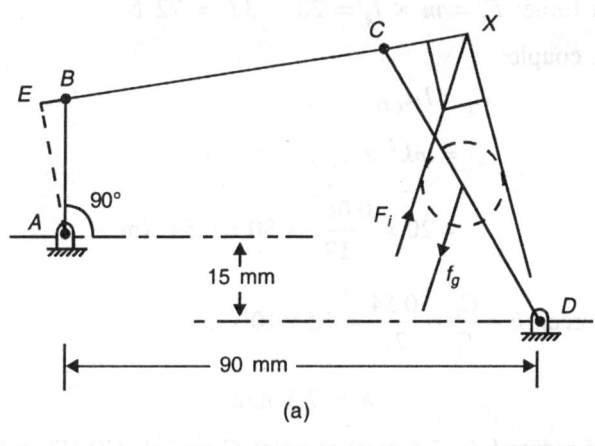

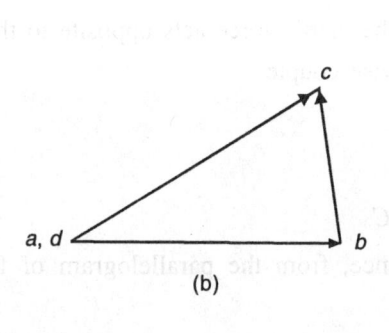

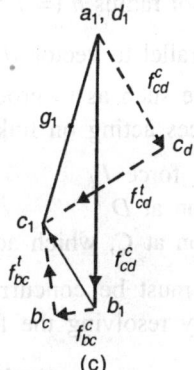

Link	Length (mm)	Velocity (m/s)	Centripetal acceleration $\dfrac{v^2}{r}$ (m/s^2)
AB	25	$ab = 0.5$	10
BC	60	$bc = 0.29$	1.4
CD	60	$cd = 0.54$	4.86

FIGURE 12.24

Construct the acceleration polygon as shown in Figure 12.24(c) by the method outlined in Chapter 2. Locate the point g_1 corresponding to point G on the link CD such that it satisfies the following condition:

$$\frac{CG}{CD} = \frac{c_1 g_1}{c_1 d_1}$$

(i) Angular velocity of link CD:

$$\omega_{CD} = \frac{v_{cd}}{CD} = \frac{0.54}{0.06} = 9 \text{ rad/s}$$

(ii) Angular acceleration of link CD:

$$\alpha_{CD} = \frac{f_{cd}^t}{CD} = \frac{\overline{c_1 c_d}}{CD} = \frac{5.4}{0.06} = 90 \text{ rad/s}^2$$

(iii) (a) Acceleration of point G:

$$f_g = \overrightarrow{d_1 g_1} = 3.6 \text{ m/s}^2$$

(b) Inertia force: $F_i = m \times f_g = 20 \times 3.6 = 72$ N

(c) Inertia couple:

$$C_i = I\alpha_{CD}$$
$$= mk^2 \times \alpha_{CD}$$
$$= 20 \times \frac{0.06^2}{12} \times 90 = 0.54 \text{ Nm}$$

Location of inertia force: $h = \dfrac{C_i}{F_i} = \dfrac{0.54}{72} = 7.5 \times 10^{-3}$ m

or
$$h = 7.5 \text{ mm}$$

Draw a circle of radius h (= 7.5 mm) at point G on link CD [Figure 12.24(c)] and draw a tangential line parallel to vector $\overrightarrow{d_1 g_1}$ or f_g. The inertia force acts opposite to the direction of acceleration force such as to produce clockwise couple.

The three forces acting on link CD are:

(a) the inertia force F_i
(b) the reaction at D
(c) the reaction at C, which acts along BC

These forces must be concurrent and hence, from the parallelogram of forces at X [Figure 12.24(a)] by resolving the force F_i,

reaction at point C = 41.4 N

Therefore,

Torque at link AB: $T = 41.4 \times AE$

$$= 41.4 \times \frac{26}{1000} = 10.76 \text{ Nm (counterclockwise)} \qquad \text{Ans.}$$

EXAMPLE 12.8 In the mechanism shown in the Figure 12.25, crank OA rotates about fixed centre in clockwise direction with an angular velocity of 7.5 rad/s. The connecting rod AB moves the slider B. The mass of connecting rod is 6.3 kg. Determine the external torque required to overcome the inertia of connecting rod. Given:

$OA = 75$ mm, $AB = 175$ mm, $OC = 137.5$ mm, $\angle OCB = 110°$ and $\angle COA = 65°$

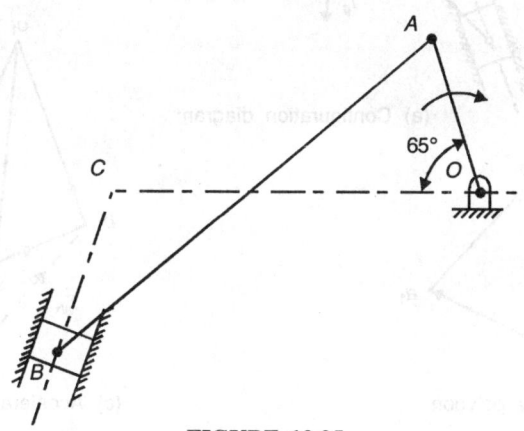

FIGURE 12.25

Solution Draw the configuration diagram as shown in Figure 12.26(a).

Velocity of point A: $v_a = \omega \times OA = 7.5 \times \dfrac{75}{1000} = 0.5625$ m/s

Construct the velocity polygon [Figure 12.26(b)] by the method as outlined in Chapter 2. Measure the velocities of links OA and AB and slider B.

Construct the acceleration polygon $o_1 a_1 b_1$ and mark the location of point g_1 corresponding to point G on link AB.

Acceleration of point G: $f_g = o_1 g_1 = 3.2$ m/s²

Angular acceleration of link AB:

$$\alpha_{AB} = \frac{f^t_{ab}}{AB} = \frac{b_a b_1}{AB} = \frac{2.8}{0.175} = 16 \text{ rad/s}^2$$

(i) Inertia force: $F_i = m f_g = 6.3 \times 3.2 = 20.16$ N

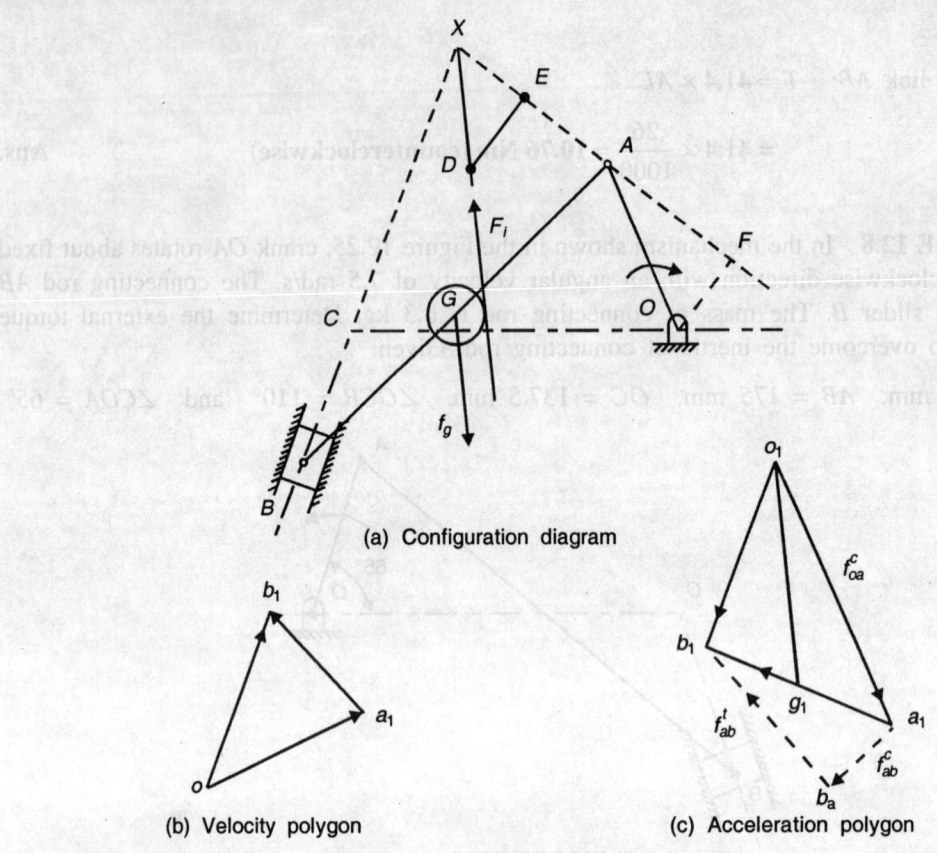

494 Theory of Mechanisms and Machines

FIGURE 12.26
(a) Configuration diagram
(b) Velocity polygon
(c) Acceleration polygon

(ii) Inertia couple: $C_i = I\, \alpha_{AB} = mk^2\, \alpha_{AB}$

$$= 6.3 \times \frac{0.175^2}{12} \times 16 = 0.2572 \text{ Nm}$$

Location of inertia force: $h = \dfrac{C_i}{F_i} = \dfrac{0.2572}{20.16} = 0.0128$ m

or $h = 12.8$ mm

Draw a circle of radius h (= 12.8 mm) at point G on link AB and draw a tangential line parallel to vector $\overrightarrow{o_1 g_1}$. The inertia force acts opposite to the direction of acceleration force.

The connecting rod is subjected to three froces—inertia force F_i, force at slider and force at point A. The inertia force F_i is resolved along the line XAF. The force component along XE represents force on the crank pin.

Force on the crank pin = 14 N (by measurement)

∴ Torque at crank OA: $T = 14 \times OF = 14 \times \dfrac{35}{1000} = $ **0.49 Nm**   **Ans.**

Force Analysis

12.5 DYNAMICS OF SLIDER–CRANK MECHANISM

In a slider-crank mechanism, a slider, also known as piston, performs to and fro reciprocating motion. This motion is possible when at the instant of changing direction of motion, velocity of the slider is zero. In other words, at that instant, an element of mechanism is subjected to varying acceleration. This results into a member subjected to verying kinetic force from instant to instant. Besides this force, members are also subjected to variation of fluid pressure on account of its compression and expansion. The dynamics of slider-crank mechanism deals with the study of effect of variation of forces on its elements.

12.5.1 Displacement, Velocity and Acceleration of Piston

Consider a slider-crank mechanism in which crank OB rotates in counterclockwise direction. When crank OB is rotated by an angle θ from initial position OA, the piston is displaced from position P_1 to P_2. Thus the displacement of piston for $\theta°$ crank rotation is P_1P_2 (Figure 12.27).

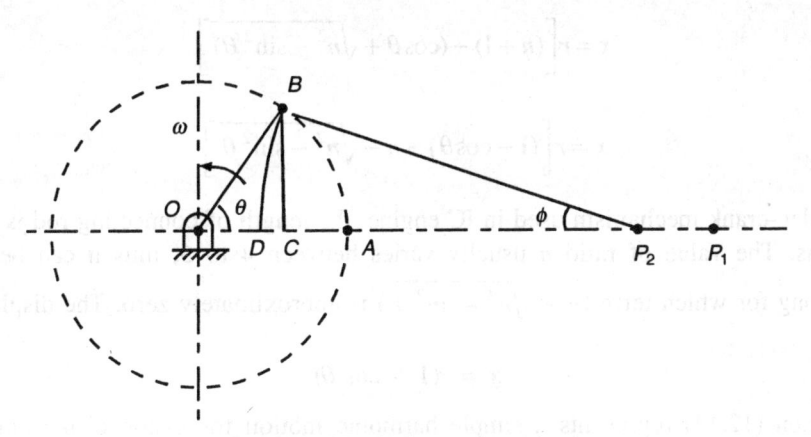

FIGURE 12.27 Slider-crank mechanism.

Let x = displacement of piston (= distance P_1P_2)
 r = radius of crank
 l = length of connecting rod
 n = ratio of length of connecting rod to crank radius (= l/r)
 θ = crank rotation angle
and ϕ = obliquity angle

Displacement. The displacement of piston is given as (see Figure 12.27):

$$\begin{aligned} x &= \text{distance } P_1P_2 \\ &= OP_1 - OP_2 \\ &= OP_1 - (OC + CP_2) \\ &= (l + r) - (r \cos \theta + l \cos \phi) \\ &= (nr + r) - (r \cos \theta + nr \cos \phi) \\ &= r[(n + 1) - (\cos \theta + n \cos \phi)] \end{aligned} \qquad (12.11)$$

We know that $\cos\phi = \sqrt{1-\sin^2\phi}$

where $\sin\phi = \dfrac{BC}{BP_2}$ (refer Figure 12.27)

$= \dfrac{r\sin\theta}{l} = \dfrac{\sin\theta}{n}$

or
$$\cos\phi = \sqrt{1-\left(\dfrac{r\sin\theta}{l}\right)^2}$$

$$= \dfrac{1}{n}\sqrt{n^2 - \sin^2\theta}$$

Substituting the value of $\cos\phi$ in Eq. (12.11), we get

$$x = r\left[(n+1) - (\cos\theta + \sqrt{n^2 - \sin^2\theta})\right]$$

or
$$x = r\left[(1-\cos\theta) + n - \sqrt{n^2 - \sin^2\theta}\right] \quad (12.12)$$

In slider–crank mechanism used in IC engine, the length of connecting rod is larger than crank radius. The value of ratio n usually varies between 4 to 5, thus it can be treated as infinitely long for which term $(n - \sqrt{n^2 - \sin^2\theta})$ is approximately zero. The displacement of piston is:

$$x = r(1 - \cos\theta) \quad (12.13)$$

Equation (12.13) represents a simple harmonic motion for piston displacement.

Velocity. The rate of change of piston displacement is known as **piston velocity.**

$$v = \dfrac{dx}{dt} = \dfrac{dx}{d\theta}\cdot\dfrac{d\theta}{dt}$$

$$= r\dfrac{d}{d\theta}\left[(1-\cos\theta) + n - \sqrt{n^2 - \sin^2\theta}\right]\omega$$

or
$$v = \omega r\left(\sin\theta + \dfrac{\sin 2\theta}{2\sqrt{n^2 - \sin^2\theta}}\right) \quad (12.14)$$

For large value of ratio $n(n > 4)$, the term $\sqrt{n^2 - \sin^2\theta}$ is approximately equal to n. Thus the piston velocity:

$$v = \omega r\left(\sin\theta + \dfrac{\sin 2\theta}{2n}\right) \quad (12.15)$$

Acceleration. The acceleration of piston:

$$f = \frac{dv}{dt} = \frac{dv}{d\theta} \cdot \frac{d\theta}{dt}$$

$$= \omega^2 r \frac{d}{d\theta}\left(\sin\theta + \frac{\sin 2\theta}{2n}\right)$$

or
$$f = \omega^2 r \left(\cos\theta + \frac{\cos 2\theta}{n}\right) \qquad (12.16)$$

12.5.2 Velocity and Acceleration of Connecting Rod

Let the angle made by connecting rod with the line of stroke be ϕ when crank has rotated through an angle θ.

Referring to Figure 12.27, in $\triangle OBP_2$, we know that

$$\frac{OB}{\sin\phi} = \frac{BP_2}{\sin\theta}$$

or
$$\frac{r}{\sin\phi} = \frac{l}{\sin\theta}$$

or
$$\sin\phi = \frac{\sin\theta}{n} \qquad (12.17)$$

Differentiating Eq. (12.17) with respect to time,

$$\cos\phi \cdot \frac{d\phi}{dt} = \frac{\cos\theta}{n} \cdot \frac{d\theta}{dt}$$

or angular velocity of connecting rod:

$$\omega_c = \frac{d\phi}{dt} = \frac{\omega \times \cos\theta}{n\cos\phi}$$

We know that

$$\cos\phi = \frac{1}{n}\sqrt{n^2 - \sin^2\theta}$$

or
$$\omega_c = \frac{\omega \times \cos\theta}{n \times \frac{1}{n}\sqrt{n^2 - \sin^2\theta}}$$

or
$$\omega_c = \omega\left(\frac{\cos\theta}{\sqrt{n^2 - \sin^2\theta}}\right) \qquad (12.18)$$

Angular acceleration of connecting rod:

$$\alpha_c = \frac{d\omega_c}{dt}$$

$$= \frac{d\omega_c}{d\theta} \cdot \frac{d\theta}{dt}$$

$$= \omega^2 \frac{d}{d\theta}\left(\frac{\cos\theta}{\sqrt{n^2 - \sin^2\theta}}\right)$$

or
$$\alpha_c = -\omega^2 \sin\theta \left[\frac{n^2 - 1}{(n^2 - \sin^2\theta)^{3/2}}\right] \quad (12.19)$$

The negative sign indicates that the direction of angular acceleration of the connecting rod is such that it tends to reduce the obliquity angle ϕ.

12.5.3 Crank Effort

The driving force acting on the piston is termed as **piston effort**. In a vertical cylinder IC engine, following three types of forces act:

Gas Force. The force due to variation of working fluid pressure is known as **gas force**.

Gas force:
$$F_g = \frac{\pi}{4} D^2 \times p \quad (12.20)$$

where
D = diameter of the piston
and p = gas pressure

Inertia force. In IC Engine, during the first half of the stroke, the reciprocating mass accelerates and the inertia force tends to resist the motion. Thus the net force on the piston is decreased. However, during the second half of the stroke, the reciprocating mass decelerate and inertia force opposes this deceleration. Thus it increases the effective force on the piston.

Inertia force:
$$F_i = \mp mf \quad (12.21)$$

Weight of reciprocating mass. The weight of reciprocating mass assists the piston during its movement towards bottom dead centre (*BDC*). Therefore, piston effort is increased by an amount equal to the weight of the piston. However, when the piston moves towards top dead centre (*TDC*), the piston effort is decreased by the same amount.

Net piston effort:
$$P = F_g \mp F_i \pm W \quad (12.22)$$

In IC engine mechanism, the gudgen pin, which connects piston and connecting rod is in equilibrium under the action of the following three forces:

(i) piston effort P

(ii) axial force in the connecting rod F_c
(iii) reaction thrust on cylinder surface F_r

Geometrically, the axial force in connecting rod and the thrust on cylinder surface can be expressed in terms of piston effort P and obliquity angle ϕ as given below:

Axial force
$$F_c = \frac{P}{\cos \phi} \tag{12.23}$$

and
reaction thrust: $F_r = P \tan \phi$ (12.24)

Referring to Figure 12.28, a force F_c equal and opposite to axial force F_c in connecting rod acts at crank pin B. This force can be resolved into two components:

(i) A force acting along the crank, called radial force F_{cr}.
$$F_{cr} = F_c \cos(\theta + \phi) \tag{12.25}$$

(ii) A force acting perpendicular to the crank OB. This force constitutes a driving torque which is called **crank effort**.

Driving force:
$$F_{ct} = F_c \sin(\theta + \phi) \tag{12.26}$$

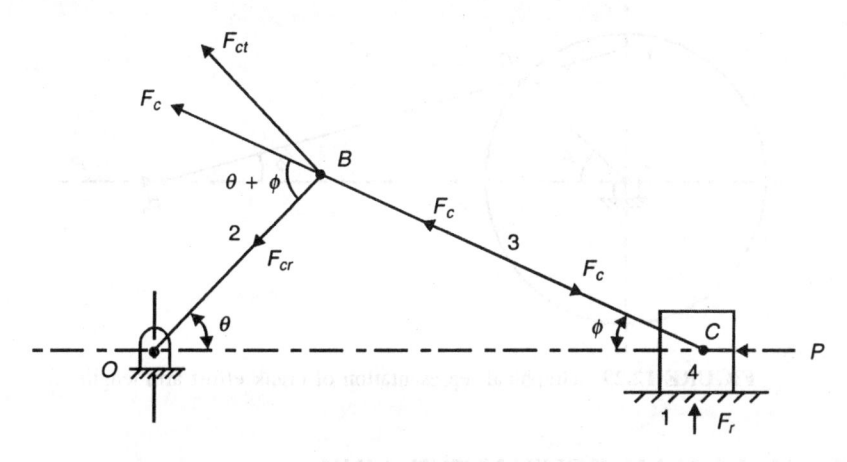

FIGURE 12.28 Forces acting on slider-crank mechanism.

Crank effort:
$$T = F_{ct} \times r$$

or
$$T = rF_c \sin(\theta + \phi) \tag{12.27}$$

or
$$T = \frac{P \times r}{\cos \phi}(\sin \theta \cdot \cos \phi + \cos \theta \cdot \sin \phi)$$

$$= Pr(\sin \theta + \cos \theta \times \tan \phi)$$

We know that $\sin\phi = \dfrac{\sin\theta}{n}$ and $\cos\phi = \dfrac{1}{n}\sqrt{n^2 - \sin^2\theta}$.

Therefore,
$$\tan\phi = \dfrac{\sin\theta}{\sqrt{n^2 - \sin^2\theta}}$$

or
$$T = Pr\left(\sin\theta + \cos\theta \times \dfrac{\sin\theta}{\sqrt{n^2 - \sin^2\theta}}\right)$$

or
$$T = Pr\left(\sin\theta + \dfrac{\sin 2\theta}{2\sqrt{n^2 - \sin^2\theta}}\right) \tag{12.28}$$

Graphically, the crank effort or torque can be expressed as

$$T = P \times \text{distance } OY \tag{12.29}$$

Where OY is the distance measured between centre of crank and a point of intersection of Y axis and extension of connecting rod P_2B (see Figure 12.29).

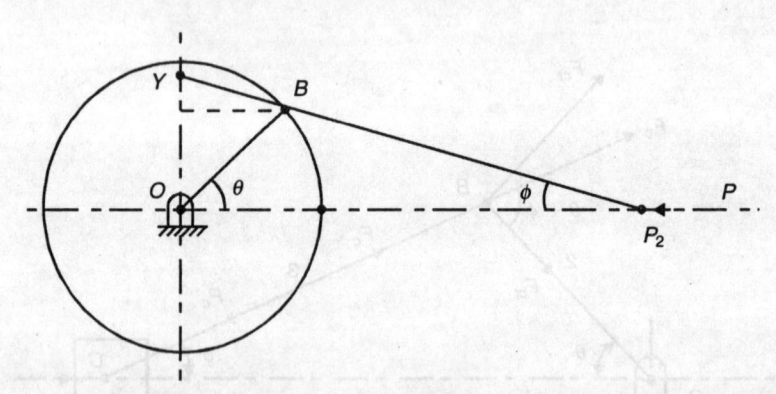

FIGURE 12.29 Graphical representation of crank effort arm length.

12.6 DYNAMICALLY EQUIVALENT LINK

In Slider-crank mechanism, the motion of connecting rod is non-linear. Thus its inertia can be found. However, the same can be computed only if the connecting rod is replaced by a dynamically equivalent link that satisfies the following conditions, yet gives same dynamical properties.

    (i) Mass of connecting rod is divided into two concentrated masses acting at both ends.
   (ii) The centre of mass of equivalent link coincides with that of connecting rod.
  (iii) The moment of inertia of both links is same.

Consider a connecting rod AB of slider-crank mechanism as shown in Figure 12.30.

FIGURE 12.30 Dynamically equivalent link.

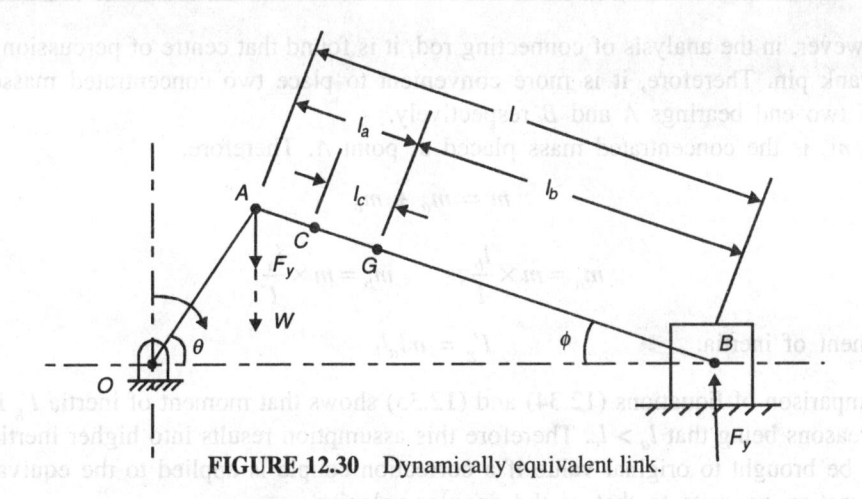

Let m = mass of connecting rod
 m_b = concentrated mass placed at point B
and m_c = concentrated mass placed at point C.

The mass of connecting rod is assumed to be concentrated at the centre of gravity G. For dynamic equivalent link, we divide this mass into two parts; the mass m_b is placed at the gudgen pin end B and the mass m_c is placed at the centre of percussion for oscillation of the rod about gudgen pin B. This disposition of the mass of the connecting rod is dynamic equivalent to the original connecting rod if the condition of dynamic equivalence are satisfied.

Therefore,
$$m = m_b + m_c \qquad (12.30)$$
$$m_b l_b = m_c l_c \qquad (12.31)$$

and
$$I_g = m_b l_b^2 + m_c l_c^2 \qquad (12.32)$$

Solving Eqs. (12.30) and (12.31), we find the portion of mass to be placed at each end, i.e. at points B and C.

$$m_b = m \times \frac{l_c}{l_b + l_c} \qquad (12.33a)$$

$$m_c = m \times \frac{l_b}{l_b + l_c} \qquad (12.33b)$$

Substituting the values of these masses in Eq. (12.32), we get the value of moment of inertia as

$$I_g = m \times \frac{l_c}{l_b + l_c} \times l_b^2 + m \times \frac{l_b}{l_b + l_c} \times l_c^2$$

or
$$I_g = m l_b l_c \qquad (12.34)$$

However, in the analysis of connecting rod, it is found that centre of percussion is closer to the crank pin. Therefore, it is more convenient to place two concentrated masses at the centre of two end bearings A and B respectively.

Let m_a is the concentrated mass placed at point A. Therefore,

$$m = m_a + m_b$$

and
$$m_a = m \times \frac{l_b}{l}, \qquad m_b = m \times \frac{l_a}{l}$$

and moment of inertia:
$$I'_g = m l_a l_b \qquad (12.35)$$

Comparison of Equations (12.34) and (12.35) shows that moment of inertia I'_g is greater than I_g, reasons being that $l_a > l_c$. Therefore this assumption results into higher inertia torque. This can be brought to original value if a correction couple is applied to the equivalent link in the direction opposite to that of the angular velocity.

Correction couple:
$$T_c = m(l_a l_b - l_b l_c)\alpha_c$$

or
$$T_c = m\alpha_c l_b (l_a - l_c) \qquad (12.36)$$

where α_c is angular acceleration of connecting rod.

The correction couple in connecting rod is produced by two equal parallel and opposite forces F_y acting at the crank pin and gudgen pin ends perpendicular to the line of stroke (see Figure 12.30).

$$T_c = F_y \times l \cos\phi \qquad (12.37)$$

Inertia torque on the crank shaft. Analytically, the resultant inertia torque on the crank shaft can be found by algebraic sum of the following torques:
  (i) torque due to mass of reciprocating parts as derived in Eq. (12.28)
  (ii) the dynamic force F_y, induced due to correction couple T_c applied on dynamic equivalent link, also produces torque on the crank shaft (see Figure 12.30)

$$T_2 = F_y \times r \cos\theta$$

or
$$T_2 = \frac{T_c}{l \cos\phi} \times r \cos\theta \quad \text{[refer Eq. (12.37)]}$$

We know that $\cos\phi = \dfrac{1}{n}\sqrt{n^2 - \sin^2\theta}$

Therefore, torque on the crank shaft due to correction couple:

$$T_2 = T_c \times \frac{\cos\theta}{\sqrt{n^2 - \sin^2\theta}}$$

$$= m\alpha_c l_b (l_a - l_c) \times \frac{\cos\theta}{\sqrt{n^2 - \sin^2\theta}} \qquad (12.38)$$

(iii) In dynamic equivalent link, some mass m_a is placed at crank pin. This mass also causes additional torque.
$$T_3 = m_a \times g \times r\cos\theta \qquad (12.39)$$
(iv) The mass of connecting rod offsetted at gudgen pin reciprocates along with mass of reciprocating parts. The expression for torque is similar to Eq. (12.28), as given below:

$$T_4 = m_b \times g \times r\left[\sin\theta + \frac{\sin 2\theta}{2\sqrt{n^2 - \sin^2\theta}}\right] \qquad (12.40)$$

EXAMPLE 12.9 In a slider–crank mechanism, the length of stroke is 120 mm and connecting rod is 250 mm long. When the crank rotates at 2000 rpm and has travelled 60° from dead centre, find;

(i) displacement of piston
(ii) velocity and acceleration of the piston
(iii) angular velocity and acceleration of connecting rod.

Solution Crank radius:
$$r = \frac{\text{stroke length}}{2} = \frac{120}{2} = 60 \text{ mm}$$

Angular speed: $\omega = \dfrac{2\pi N}{60} = \dfrac{2\pi \times 2000}{60} = 209.44$ rad/s

Ratio: $n = \dfrac{l}{r} = \dfrac{250}{60} = 4.167$

(i) Displacement of piston:
$$x = r\left[(1 - \cos\theta) + (n - \sqrt{n^2 - \sin^2\theta})\right]$$
$$= 60\left[(1 - \cos 60°) + (4.167 - \sqrt{4.167^2 - \sin^2 60°})\right]$$
$$= 35.46 \text{ mm} \qquad \text{Ans.}$$

(ii) Velocity:
$$v = \omega r\left[\sin\theta + \frac{\sin 2\theta}{2\sqrt{n^2 - \sin^2\theta}}\right]$$
$$= 209.44 \times \frac{60}{1000}\left[\sin 60° + \frac{\sin 2 \times 60°}{2\sqrt{4.167^2 - \sin^2 60°}}\right]$$
$$= 12.21 \text{ m/s} \qquad \text{Ans.}$$

Acceleration: $f = \omega^2 r \left(\cos\theta + \dfrac{\cos 2\theta}{n} \right)$

or $f = 209.44^2 \times \dfrac{60}{1000} \left(\cos 60° + \dfrac{\cos 120°}{4.167} \right)$

$= 1000.15$ m/s² **Ans.**

(iii) Angular velocity of connecting rod:

$$\omega_c = \omega \left[\dfrac{\cos\theta}{\sqrt{n^2 - \sin^2\theta}} \right]$$

$$= 209.44 \left[\dfrac{\cos 60°}{\sqrt{4.167^2 - \sin^2 60°}} \right]$$

$= 25.69$ rad/s **Ans.**

Angular acceleration:

$$\alpha_c = -\omega^2 \sin\theta \left[\dfrac{n^2 - 1}{(n^2 - \sin^2\theta)^{3/2}} \right]$$

$$= -209.44^2 \times \sin 60° \left[\dfrac{4.167^2 - 1}{(4.167^2 - \sin^2 60°)^{3/2}} \right]$$

$= -9179.72$ rad/s² **Ans.**

EXAMPLE 12.10 A vertical cylinder petrol engine has a bore of 100 mm and stroke 120 mm. The length of connecting rod between centres is 250 mm. The mass of the piston is 1.1 kg. The speed of engine is 1500 rpm. In the expansion stroke with a crank at 30° from TDC, the gas pressure is 700 kN/m². Determine:

(i) net force on the piston
(ii) force on the connecting rod
(iii) thrust on the cylinder wall
(iv) crank effort
(v) speed above which the gudgen pin force would reverse in direction.

Solution Crank radius:

$$r = \dfrac{\text{stroke length}}{2} = \dfrac{120}{2} = 60 \text{ mm}$$

Ratio: $n = \dfrac{l}{r} = \dfrac{250}{60} = 4.167$

Angular speed: $$\omega = \frac{2\pi N}{60} = \frac{2\pi \times 1500}{60} = 157.08 \text{ rad/s}$$

(i) Piston effort:

(a) Gas force: $F_g = \frac{\pi}{4}d^2 \times p$

$$= \frac{\pi}{4} \times 0.1^2 \times 700 = 5.4977 \text{ kN}$$

(b) Inertia force: $F_i = m\omega^2 r \left(\cos\theta + \frac{\cos 2\theta}{n} \right)$

$$F_i = 1.1 \times 157.08^2 \times 0.06 \left(\cos 30° + \frac{\cos 60°}{4.167} \right)$$

$$= 1605.7 \text{ N}$$

(c) Weight of reciprocating parts:

$$W_r = m \times g = 1.1 \times 9.81 = 10.79 \text{ N}$$

Piston effort: $P = F_g - F_i + W_r$

$$= 5497.7 - 1605.7 + 10.79$$

$$= 3902.79 \text{ N} \qquad \text{Ans.}$$

(ii) Force on connecting rod:

$$F_c = \frac{P}{\cos\phi}$$

Where ϕ is obliquity angle, $\sin\phi = \frac{\sin\theta}{n}$

or $$\sin\phi = \frac{\sin 30°}{4.167} = 0.1199$$

or $$\phi = 6.89°$$

Force: $$F_c = \frac{3902.79}{\cos 6.89°} = 3931.18 \text{ N} \qquad \text{Ans.}$$

(iii) Thrust on the cylinder wall:

$$F_r = P \tan\phi$$

$$= 3902.79 \times \tan 6.89°$$

$$= 471.6 \text{ N} \qquad \text{Ans.}$$

(iv) Crank effort:

$$T = Pr\left(\sin\theta + \frac{\sin 2\theta}{2\sqrt{n^2 - \sin^2\theta}}\right)$$

$$= 3902.79 \times 0.06\left(\sin 30° + \frac{\sin 60°}{2\sqrt{4.167^2 - \sin^2 30°}}\right)$$

$$= 141.59 \text{ Nm} \qquad \text{Ans.}$$

(v) Speed: The direction of the force would be reversed if $F_i > F_g + W_r$. Let N_1 is the speed at which this condition is obtained.

$$m\omega_1^2 r\left(\cos\theta + \frac{\cos 2\theta}{n}\right) \geq 5497.7 + 10.79$$

or $\qquad 1.1 \times \omega_1^2 \times 0.06\left(\cos 30° + \dfrac{\cos 60°}{4.167}\right) = 5508.49$

or $\qquad \omega_1 = 290.94$ rad/s

or $\qquad$ Speed $N_1 = \dfrac{60 \times \omega_1}{2\pi} = \dfrac{60 \times 290.94}{2\pi}$

$$= 2778.27 \text{ rpm} \qquad \text{Ans.}$$

EXAMPLE 12.11 The following data relate to a connecting rod of a petrol engine:

Mass	= 60 kg
Distance between bearings	= 850 mm
Diameter of small end bearing	= 75 mm
Diameter of big end bearing	= 100 mm
Time of oscillation when the connecting rod is suspended from small end	= 1.83 s
Time of oscillation when the connecting rod is suspended from big end	= 1.68 s

Determine the following:

(i) radius of gyration of connecting rod
(ii) moment of inertia of connecting rod
(iii) masses of dynamic equivalent link

Solution: Refer to Figure 12.31.

FIGURE 12.31

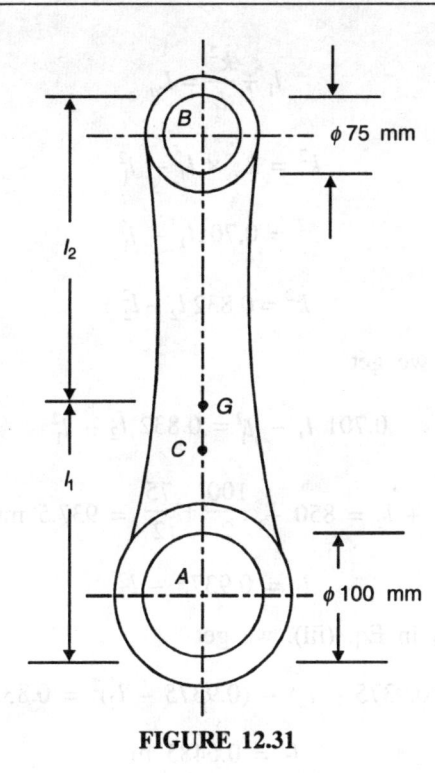

Let L_a = length of equivalent simple pendulum when it is suspended from the top of big end bearing
L_b = length of equivalent simple pendulum when it is suspended from the top of small end bearing
l_1 = distance between centre of gravity point G and top of the big end bearing
and l_2 = distance between centre of gravity point G and top of the small end bearing

We know from theory of simple pendulum that the time period of oscillation is given as

$$t_a = 2\pi \sqrt{\frac{L_a}{g}}$$

or $$L_a = \frac{t_a^2 g}{4\pi^2}$$

or $$L_a = \frac{1.68^2 \times 9.81}{4\pi^2} = 0.701 \text{ m}$$

Similarly, $$L_b = \frac{t_b^2 g}{4\pi^2} = \frac{1.83^2 \times 9.81}{4\pi^2} = 0.832 \text{ m}$$

For simple pendulum, $l_1 + \dfrac{k^2}{l_1} = L_a$

or $k^2 = L_a \times l_1 - l_1^2$

$k^2 = 0.701 l_1 - l_1^2$ ...(i)

Similarly, $k^2 = 0.832 \, l_2 - l_2^2$ ...(ii)

Equating Eqs. (i) and (ii), we get

$0.701 \, l_1 - l_1^2 = 0.832 \, l_2 - l_1^2$ ...(iii)

However, $l_1 + l_2 = 850 + \dfrac{100}{2} + \dfrac{75}{2} = 937.5$ mm

or $l_1 = 0.9375 - l_2$

Substituting the value of l_1 in Eq. (iii), we get

$0.701 \times (0.9375 - l_2) - (0.9375 - l_2)^2 = 0.832 \, l_2 - l_2^2$

or $l_2 = 0.6485$ m

$l_1 = 0.9375 - 0.6485$

or $= 0.289$ m

(i) Radius of gyration: $k = \sqrt{0.701 \, l_1 - l_1^2}$

or $k = \sqrt{0.701 \times 0.289 - 0.289^2}$

$= \mathbf{0.345}$ m **Ans.**

(ii) Moment of inertia: $I = mk^2$

$= 60 \times 0.345^2$

$= \mathbf{7.14 \; kgm^2}$ **Ans.**

(iii) Masses of dynamic equivalent link:

Distance of CG from centre of small end: $l_b = l_2 - \dfrac{d_2}{2}$

$= 0.6485 - \dfrac{0.075}{2}$

$= 0.611$ m

We know that $l_c l_b = k^2$

where l_c is the distance between CG and point C where the mass m_c is placed.

or $$l_c = \frac{0.345^2}{0.611} = 0.1948 \text{ m}$$

Mass at small end B: $$m_b = m \times \frac{l_c}{l_b + l_c} = 60 \times \frac{0.1948}{0.611 + 0.1948}$$

$$= 14.5 \text{ kg} \qquad \text{Ans.}$$

Mass at point C: $$m_c = m - m_b$$

$$= 60 - 14.5$$

$$= 45.5 \text{ kg} \qquad \text{Ans.}$$

EXAMPLE 12.12 The following data relate to a vertical cylinder four stroke IC engine:

(i) Mass of reciprocating parts $\qquad m_r = 8.0$ kg
(ii) Radius of crank $\qquad r = 100$ mm
(iii) Mass of connecting rod $\qquad m_c = 12$ kg
(iv) Length of connecting rod $\qquad l = 500$ mm
(v) Engine speed $\qquad = 300$ rpm
(vi) The distance between CG of connecting rod and big end of the bearing $\qquad = 200$ mm

When connecting rod is suspended as pendulum from the gudgen pin, it takes 11 seconds to complete 8 oscillations. Calculate:

(i) the radius of gyration of the connecting rod.
(ii) torque on the crank shaft when crank has turned 60° and piston moves down.

Solution Let L_b = length of pendulum when it is suspended from the top of the small end bearing and l_b = distance between CG and small end bearing (= $l - 200$)

Time period of oscillation: $$t_b = 2\pi \sqrt{\frac{L_b}{g}}$$

or $$L_b = \frac{t_b^2 \times g}{4\pi^2} = \frac{(11/8)^2 \times 9.81}{4\pi^2} = 0.47 \text{ m}$$

For simple pendulum, $$l_b + \frac{k^2}{l_b} = L_b$$

where $l_b = l - 200 = 500 - 200 = 300$ mm or 0.3 m

or
$$0.3 + \frac{k^2}{0.3} = 0.47$$

or   radius of gyration: $k = 0.226$ m   Ans.

Refer to Figure 12.32.

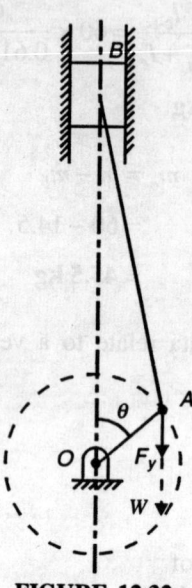

FIGURE 12.32

Mass at the crank pin: $m_a = \dfrac{m_c l_b}{l_a + l_b} = 12 \times \dfrac{0.3}{0.5} = 7.2$ kg

Mass at gudgen pin: $m_b = m_c - m_a$

$= 12 - 7.2 = 4.8$ kg

Total mass at gudgen pin $= m_r + m_b = 8 + 4.8 = 12.8$ kg

Inertia force: $F_i = m\omega^2 r \left(\cos\theta + \dfrac{\cos 2\theta}{n} \right)$

where
$$\omega = \frac{2\pi N}{60} = \frac{2\pi \times 300}{60} = 31.4 \text{ rad/s}$$

$$n = \frac{l}{r} = \frac{500}{100} = 5$$

$$\therefore \quad F_i = 12.8 \times 31.4^2 \times 0.1 \times \left(\cos 60° + \frac{\cos 120°}{5}\right)$$

$$= 504.8 \text{ N}$$

(i) Torque:

$$T_1 = F_i r \left[\sin\theta + \frac{\sin 2\theta}{2\sqrt{n^2 - \sin^2\theta}}\right]$$

$$= 504.8 \times 0.1 \left[\sin 60° + \frac{\sin 120°}{2\sqrt{5^2 - \sin^2 60°}}\right]$$

$$= 48.15 \text{ Nm (counterclockwise)}$$

(ii) Angular acceleration:

$$\alpha_c = -\omega^2 \sin\theta \left[\frac{n^2 - 1}{(n^2 - \sin^2\theta)^{3/2}}\right]$$

$$= -31.4^2 \times \sin 60° \left[\frac{5^2 - 1}{(5^2 - \sin^2 60°)^{3/2}}\right]$$

$$= -171.6 \text{ rad/s}^2$$

Correction couple:
$$T_c = m_c \alpha_c L_b (l - l_b)$$

$$= -12 \times 171.6 \times 0.3 \times (0.5 - 0.47)$$

$$= -18.53 \text{ Nm}$$

Torque on crank shaft due to correction couple:

$$T_2 = T_c \times \frac{\cos\theta}{\sqrt{n^2 - \sin^2\theta}}$$

$$= -18.53 \times \frac{\cos 60°}{\sqrt{5^2 - \sin^2 60°}}$$

$$= -1.88 \text{ Nm (clockwise)}$$

(iii) Torque due to mass at crank:

$$T_3 = m_a g \times r \sin\theta$$

$$= 7.2 \times 9.81 \times 0.1 \times \sin 60°$$

$$= 6.117 \text{ Nm (clockwise)}$$

Total inertia torque: $T = T_1 - T_2 - T_3$

$ = 48.15 - (-1.88) - 6.117$

$ = 43.913$ Nm (counterclockwise) **Ans.**

12.7 INERTIA FORCE IN RECIPROCATING ENGINE (GRAPHICAL METHOD)

Besides analytical method of inertia force analysis of reciprocating engine, graphical method is quite popular. The detailed procedure is as follows:

1. Draw a configuration diagram OAB, (Figure 12.33) wherein OA and AB represent crank and connecting rod respectively. The centre of gravity of connecting rod is at point G.

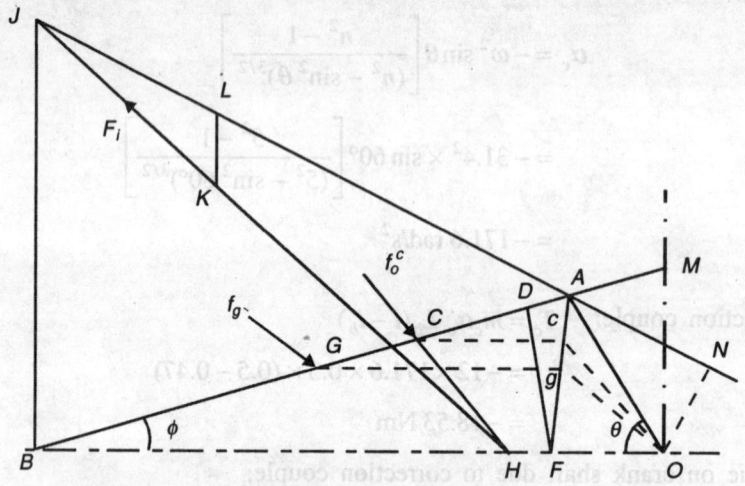

FIGURE 12.33 Graphical method of inertia force analysis.

2. The position of one of the dynamically equivalent mass is assumed to be at point B, the gudgen pin centre. The position of another dynamically equivalent mass situated at point C is determined by the method explained is Section 12.6.
3. With the help of Klien's construction method, explained in the Chapter 2, construct the acceleration polygon $OADF$.
4. Draw lines Gg and Cc parallel to the line of stroke and cutting the acceleration image AF at point g and c respectively. The Og and Oc then represent the magnitude and direction of acceleration of G and C respectively.
5. The inertia force due to mass at B acts along BO and that due to mass at point C acts through C parallel to cO, so draw a line CH parallel to Oc which cuts line of stroke at point H. The resultant inertia force of the rod must pass through the intersecting point H and must also be parallel to the direction of acceleration of point G given by Og.

6. The three forces acting on the connecting rod are:
   (i) the inertia force mf_g
   (ii) the side thrust reaction on the piston F_r
   (iii) the force at the crank pin A.

   The line of action of these forces must be concurrent so that the force at the crank pin A must pass through the intersection of the inertia force and vertical side thrust at J.

7. Resolving the inertia force F_i into components parallel to AJ and BJ, JL represents the force at crank pin and KL represents the side rection thrust. The crank shaft torque is then product of the crank pin force and the perpendicular distance ON of its line of action from O. This torque is in addition to that due to gas force and inertia force of the reciprocating part.

EXAMPLE 12.13 A single cylinder horizontal steam engine has a stroke of 0.75 m and a connecting rod 1.8 m long. The mass of reciprocating part is 520 kg and that of the connecting rod is 230 kg. The centre of gravity of the connecting rod is 0.8 m away from the crankpin. The moment of inertia about an axis through the C.G. perpendicular to the plane of motion is 100 kg m². For an engine speed of 90 rpm and crank position of 45°, determine the torque on the crankshaft and the forces on the crank shaft bearing due to the inertia of these parts.

Solution Draw a configuration diagram OAB with suitable scale (Figure 12.34).

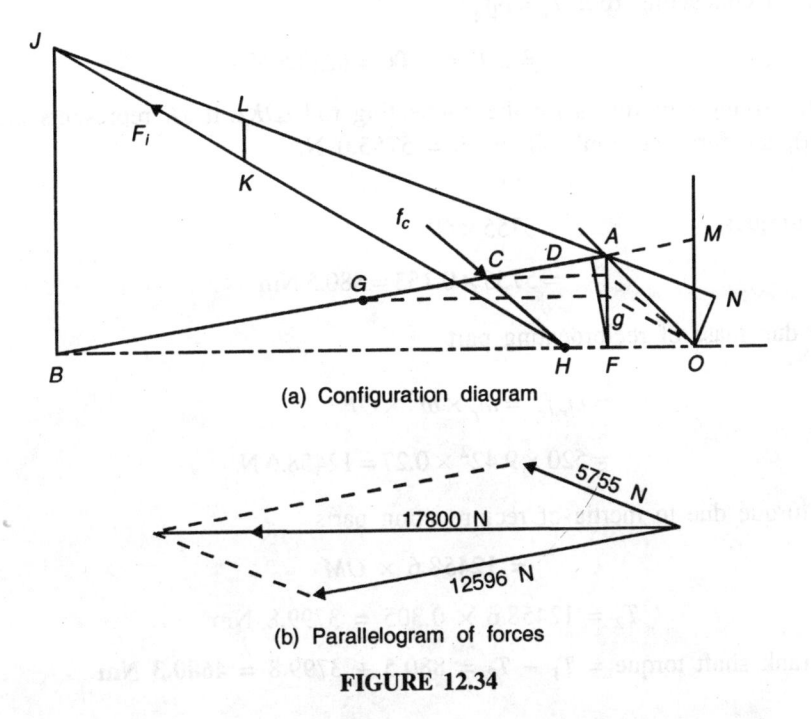

FIGURE 12.34

Crank radius: $OA = \dfrac{0.75}{2} = 0.375$ m

The position of one of the dynamically equivalent mass is assumed to be at B. The position of another dynamically equivalent mass situated at point C is such that

$$BG \times GC = k^2 = \dfrac{I}{m} = \dfrac{100}{230} = 0.4347 \text{ m}^2$$

But $\qquad BG = 1$ m (given)

Therefore, $\qquad GC = \dfrac{0.4347}{1} = 0.4347$ m

Now construct acceleration diagram by Klein's construction method. $OADF$ is the acceleration diagram. Draw Gg and Cc parallel to the line of stroke and gO and cO then represents the magnitude and direction of acceleration of points G and C respectively.

Angular velocity: $\qquad \omega = \dfrac{2\pi N}{60} = \dfrac{2\pi \times 90}{60} = 9.42$ rad/s

Acceleration of point G: $\qquad f_g = \omega^2 \times gO$

$\qquad\qquad\qquad\qquad\qquad = 9.42^2 \times 0.305 = 27.06$ m/s^2

Inertia force on connecting rod: $F_i = mf_g$

$\qquad\qquad\qquad\qquad\qquad = 230 \times 27.06 = 6223.8$ N

From the triangle of forces for the connecting rod ΔJKL if JK represents inertia force (= 6223.8 N), the force at crank pin = JL = 5755.0 N.
Therefore,

Crank shaft torque: $\qquad T_i = 5755 \times ON$

$\qquad\qquad\qquad\qquad = 5755 \times 0.153 = 880.5$ Nm

Inertia force due mass of reciprocating part

$$= m_r f_r^2 = m_r \times \omega^2 \times OF$$

$$= 520 \times 9.42^2 \times 0.27 = 12458.6 \text{ N}$$

Crank shaft torque due to inertia of reciprocating parts

$$= 12458.6 \times OM$$

or $\qquad T_2 = 12458.6 \times 0.305 = 3799.8$ Nm

Total crank shaft torque = $T_1 + T_2$ = 880.5 + 3799.8 = **4680.3 Nm** $\qquad$ **Ans.**

The forces acting on crank pin are:
(i) component of inertia force on connecting rod along JA acting in the direction A to J = 5755 N
(ii) forces along the connecting rod due to inertia of reciprocating parts

$$= \frac{12458.6}{\cos\phi}$$

where $\sin\phi = \frac{\sin\theta}{n}$, $n = \frac{l}{r} = \frac{1.8}{0.375} = 4.8$

$$= \frac{\sin 45°}{4.8} = 0.1473$$

or $\phi = 8.47°$

Therefore, the force along the connecting rod

$$= \frac{12458.6}{\cos 8.47°} = 12596 \text{ N}$$

acting in the direction A to B.

The resultant force acting on crank pin is obtained from the parallelogram of forces [Figure 12.34b)].

Resultant force = 17800 N

The reaction at the crank shaft bearing is equal and opposite to this force. **Ans.**

12.8 TURNING MOMENT DIAGRAM

In earlier section, we obtained an expression for crank effort which is found to be function of piston effort P and crank rotation angle θ. Further, the piston effort is also a function of crank angle θ. The diagram showing the crank effort or torque as a function of crank rotation angle θ for any reciprocating engine is called crank-effort diagram or turning moment diagram. The turning moment diagram of any engine can be plotted if the gas pressure p is known for all positions of the crank. The value of gas pressure can be found from a given pressure-volume (P-V) diagram (Figure 12.35). Using these pressure values, gas forces can be computed and plotted as shown in Figure 12.36. Further, the variation of inertia force due to mass of reciprocating parts can be plotted as shown in Figure 12.36 with dashed line.

The graphical addition of these two forces results into piston effort* at different crank positions (Figure 12.37).

Referring to Figure 12.37 and Eq. (12.28), we can calculate crank effort or turning moment at different position of crank.

---

*In case of vertical cylinder engine, the weight of reciprocating part, i.e. piston, is added algebraically.

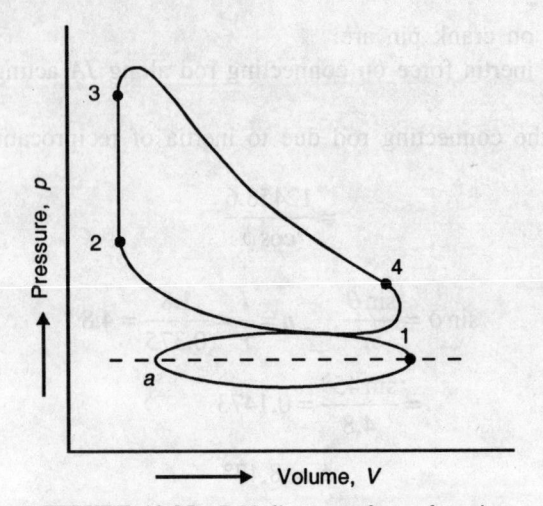

FIGURE 12.35 P-V diagram of petrol engine.

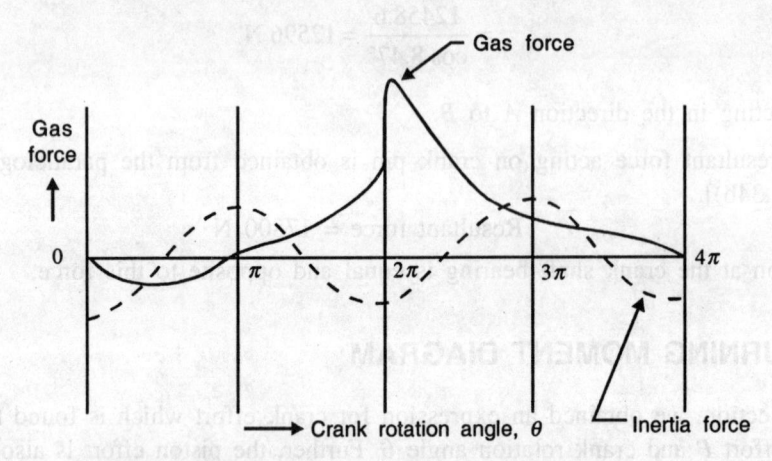

FIGURE 12.36 Variation of gas force and inertia force.

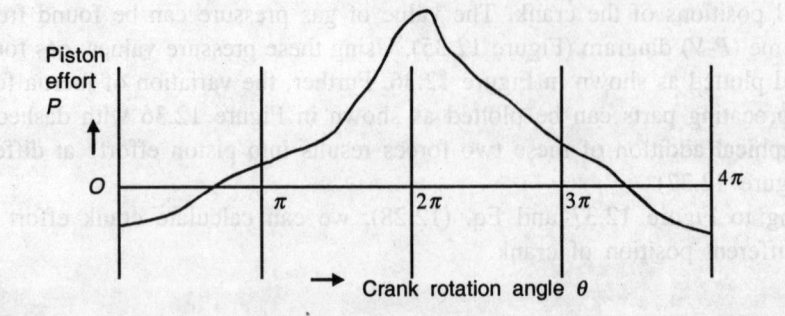

FIGURE 12.37 Variation of piston effort.

$$T = \text{piston effort} \times OY \tag{12.41}$$

where OY is the crank effort arm length.

The variation in crank effort arm length for different crank positions is shown in Figure 12.38. Finally the turning moment diagram is shown in Figure 12.39.

A close look at the turning moment diagram (Figure 12.39) shows that torque T is entirely positive in expansion stroke of engine whereas in suction, compression and exhaust strokes, it is negative. This indicates that in these strokes, power is consumed. Thus there is a large variation of torque which may cause fluctuation of speed.

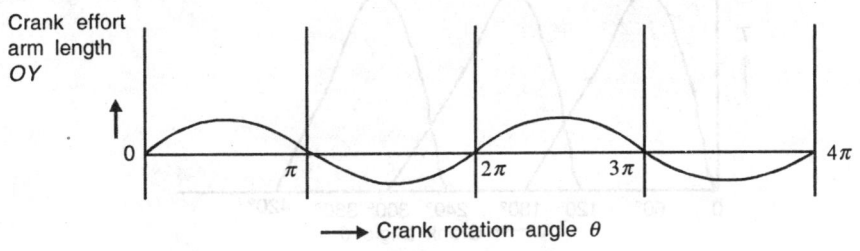

FIGURE 12.38 Variation of crank effort arm length.

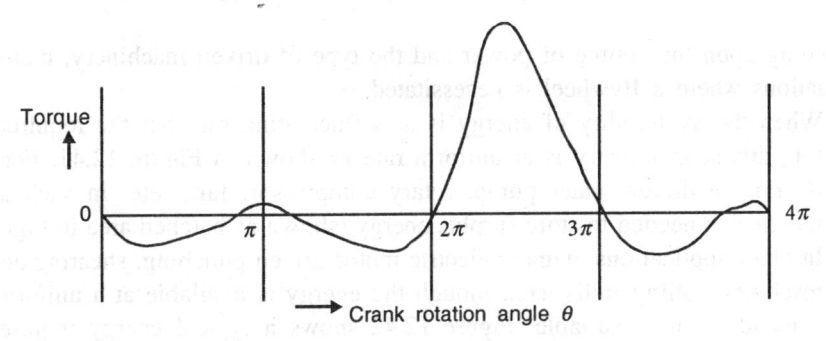

FIGURE 12.39 Turning moment diagram.

In multicylinder engine, the turning moment diagram of each cylinder is obtained separately and they are superimposed over each other with starting point shifted to phase difference of angle between respective crank positions. A typical turning moment diagram of multicylinder engine is shown in Figure 12.40.

12.9 FLYWHEEL

In an IC engine, the amount of fluctuation of kinetic energy of the crank shaft depends on the nature of the turning moment diagram. This fluctuation results into fluctuation of crank shaft speed. However, the maximum permissible variation in the speed of crank shaft depends upon the purpose for which the engine is to be used. In an engine, to keep the maximum fluctuation of speed within a permissible limit, a flywheel is attached to the crank shaft. Flywheel is a heavy rotating mass which causes additional inertia to the crank shaft and acts as a reservoir of energy. It stores up energy when availability is more than demand and delivers back when there is a lean period.

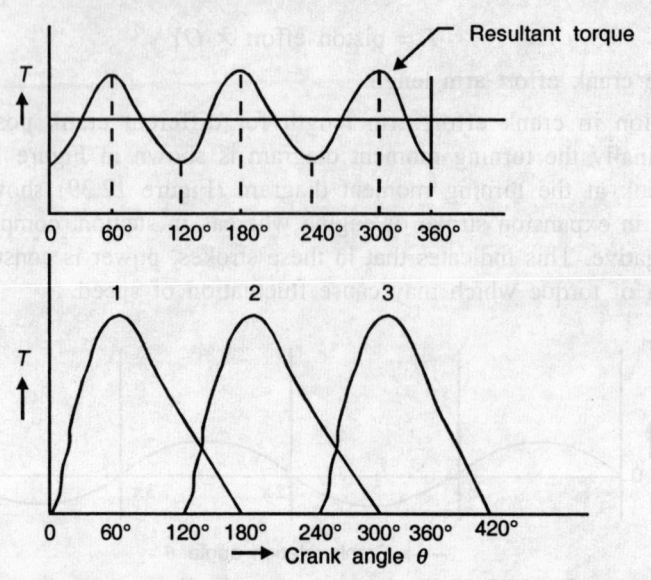

FIGURE 12.40 Turning moment diagram for a multicylinder engine.*

Depending upon the source of power and the type of driven machinery, there are three distinct situations where a flywheel is necessitated.

(i) When the availability of energy is at a fluctuating rate but the requirement of it for driven machinery is at uniform rate as shown in Figure 12.41. For example, IC Engine driven water pump, rotary compressor, fans, etc. In such a situation flywheel is needed to store surplus energy (shown as hatched area in Figure 12.41).

(ii) In other applications, namely electric motor driven punching, shearing and riveting machines, rolling mills, etc., though the energy is available at a uniform rate, the demand for it is variable. Figure 12.42 shows a typical energy requirement and availability curve for an electric motor driven rolling mill which shows that for a small fraction of the cycle period, there is huge requirement of energy. Thus again a flywheel is needed. The variation of angular speed during the cycle period is shown in Figure 12.43.

(iii) In the third situation, both the requirement and availability of energy represent a variable rate, e.g. IC engine driven reciprocating air compressor or pump.

Generally, three types of flywheel—disc type, web type and arm type are most commonly used (Figure 12.44).

A flywheel stores up energy in the form of kinetic energy of a rotating mass, which is given by

$$E = \frac{1}{2} I \omega^2 \qquad (12.42)$$

---

*Diagram is plotted for 3 cylinders IC engine. Only expansion stroke is shown in the figure with assumption that work consumed in other strokes is negligible.

Force Analysis 519

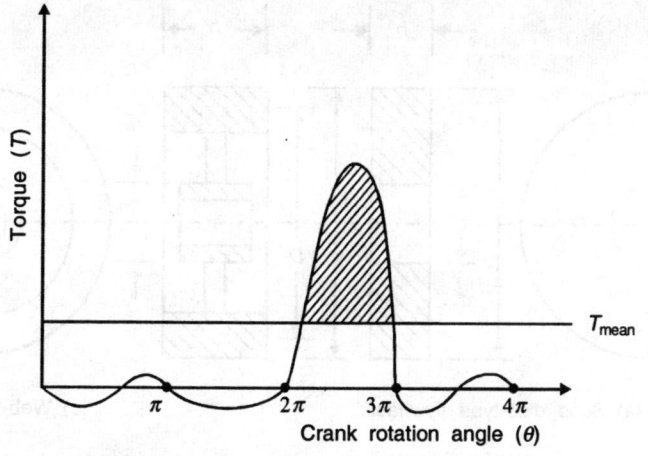

FIGURE 12.41 $T-\theta$ diagram of IC Engine.

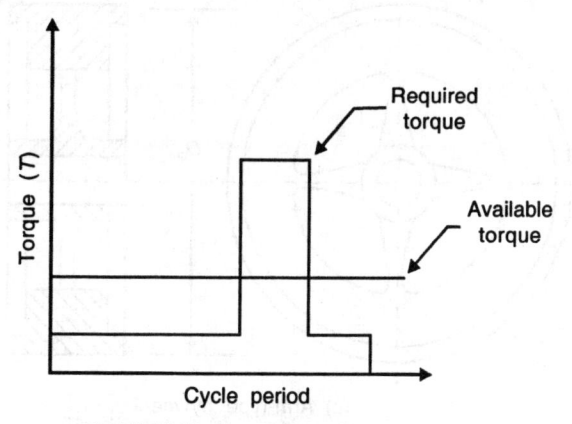

FIGURE 12.42 Torque diagram.

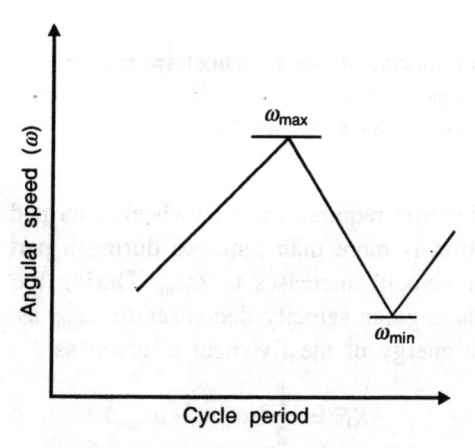

FIGURE 12.43 Angular speed variation.

(a) Solid disc-type flywheel

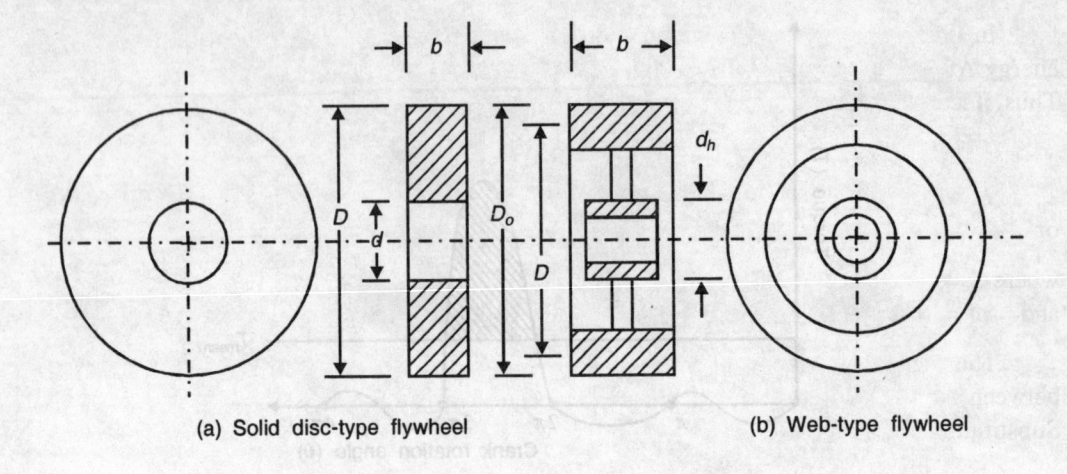

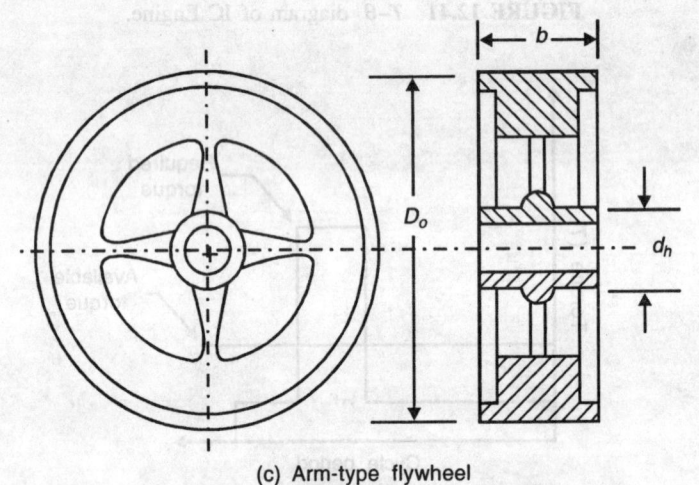

(b) Web-type flywheel

(c) Arm-type flywheel

FIGURE 12.44 Types of flywheels.

where I = mass moment of inertia of the flywheel ($= mk^2$)
  m = mass of the flywheel
  k = radius of gyration of the flywheel
and ω = angular speed.

The mass moment of inertia required for a flywheel is termed as **flywheel effect.** When input power to an application is more than required during a part of cycle, the flywheel is accelerated and its angular velocity increases to ω_{max}. During the power deficit period, the flywheel is retarded and its angular velocity decreases to ω_{min} as shown in Figure 12.43. Thus the change in kinetic energy of the flywheel is given as

$$\Delta KE = \frac{1}{2} I(\omega_{max}^2 - \omega_{min}^2) \qquad (12.43)$$

Force Analysis

In order to keep the variation of speed within the permissible range, the fluctuation of energy ΔE of the combined driver/driven system should be equal to change in kinetic energy. Thus fluctuation of energy:

$$\Delta E = \frac{1}{2} I(\omega_{max}^2 - \omega_{min}^2) \qquad (12.44)$$

or

$$\Delta E = I\omega^2 C_s$$

where C_s = coefficient of fluctuation of speed [= $(\omega_{max} - \omega_{min})/\omega$]
and ω = mean angular speed [= $(\omega_{max} + \omega_{min})/2$]

The value of coefficient of fluctuation of speed depends on the permissible variation between the highest and the lowest speeds during the operating cycle of the driven machine. Substituting $I = mk^2$ in Eq. (12.44), we get

Mass of flywheel:
$$m = \frac{\Delta E}{k^2 \omega^2 C_s} \qquad (12.45)$$

The amount of fluctuation of energy can be computed either graphically or analytically from known characteristics of the power source and driven machinery. However, in IC engine, the fluctuation of energy can be determined by the use of coefficient of fluctuation of energy C_E which is defined as the ratio of change of energy ΔE to mean energy E per revolution or per cycle.

$$C_E = \frac{E_{max} - E_{min}}{E} = \frac{\Delta E}{E} \qquad (12.46)$$

The values of coefficient of fluctuation of energy C_E for a single cylinder, four stroke engine range between 2.3–2.5.

Based upon the amount of fluctuation of energy, the different dimensions of flywheel such as mean radius, cross section of rim and details of arm section can be decided. The detailed design of a flywheel is beyond the scope of the subject. Reader is advised to refer to a book on Design of Machine Elements*.

EXAMPLE 12.14 A single cylinder four stroke petrol engine develops 20 kW at 600 rpm. The mechanical efficiency of the engine is 80 per cent and the work done by the gases during the expansion stroke is three times the work consumed in compression stroke. If the total fluctuation of speed is not to exceed ±1.5 per cent of the mean speed and turning moment diagram during expansion stroke is assumed to be triangular. Determine the mass of the flywheel if the diameter of it is 1 m. Also determine the cross section of flywheel rim. The width to thickness ratio is 1.5 and mass density of flywheel material is 7100 kg/m³.

Solution Permissible speed variation = ±1.5 per cent or coefficient of speed fluctuation, $C_s = 0.03$

Design of Machine Elements by C.S. Sharma and Kamlesh Purohit, Prentice-Hall of India, New Delhi.

Angular speed: $$\omega = \frac{2\pi N}{60} = \frac{2\pi \times 600}{60} = 62.83 \text{ rad/s}$$

Indicated power: $$ip = \frac{bp}{\eta_{mech}} = \frac{20}{0.8} = 25 \text{ kW}$$

ip = Indicated work done/cycle × Number of explosions/sec

or $$25 \times 10^3 = WD \times \frac{600}{2 \times 60}$$

or work done/cycle: $WD = 5000$ Nm (i)

Since the work during suction and exhaust strokes is neglected, the net work done during the cycle

$$WD = W_{exp} - W_{comp}$$
$$= W_{exp} - \frac{1}{3}W_{exp}$$

or $$WD = \frac{2}{3}W_{exp}$$ (ii)

Equating Eqs. (i) and (ii), we get

$$WD = 5000 = \frac{2}{3}W_{exp}$$

or $$W_{exp} = \frac{3}{2} \times 5000 = 7500 \text{ Nm}$$

Refer to turning moment diagram of engine as shown in Figure 12.45.

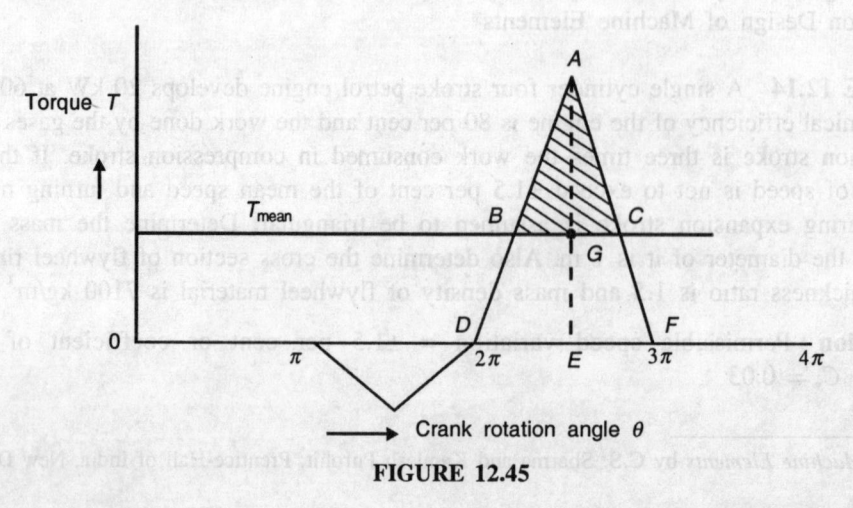

FIGURE 12.45

$$W_{exp} = 7500 \text{ Nm} = \text{area of } \triangle ADF$$

$$= \frac{1}{2} \times DF \times AE$$

or

$$AE = T_{max} = \frac{2W_{exp}}{DF} = \frac{2 \times 7500}{\pi} = 4774.6 \text{ Nm}$$

Mean torque $= \dfrac{\text{Work done/cycle}}{4\pi} = \dfrac{5000}{4\pi} = 397.88$ Nm

Excess torque:

$$AG = AE - GE$$
$$= 4774.6 - 397.88$$
$$= 4376.72 \text{ Nm}$$

From similar triangles ABC and ADF,

$$\frac{BC}{DF} = \frac{AG}{AE}$$

or

$$BC = DF \times \frac{AG}{AE}$$

or

$$BC = \pi \times \frac{4376.72}{4774.6} = 0.9167\pi$$

Therefore,

Fluctuation of energy: $\Delta E = $ area of $\triangle ABC$

or

$$\Delta E = \frac{1}{2} \times 0.9167\pi \times 4376.72$$
$$= 6302.2 \text{ Nm}$$

Mass of flywheel:

$$m = \frac{\Delta E}{k^2 \omega^2 C_s}$$

where k is radius of gyration (= radius of flywheel)

∴

$$m = \frac{6302.2}{0.5^2 \times 62.83^2 \times 0.03} = \textbf{212.86 kg} \qquad \textbf{Ans.}$$

Assuming that 90 per cent of mass is contributed by rim and remaining 10 per cent by hub, arms, etc.,

$$0.9 \, m = \pi d \times b \times h \times \rho$$

Rim cross section: $b \times h = \dfrac{0.9 \times 212.86}{\pi \times 1.0 \times 7100} = 0.0086 \text{ m}^2$

or $\qquad b \times h = 0.0086$

or $\qquad 1.5 h^2 = 0.0086 \qquad \because b = 1.5\,h$

Rim height: $\qquad h = 0.0757$ m or **75.7 mm** **Ans.**

Rim width: $\qquad b = 1.5 \times h = 1.5 \times 75.7 =$ **113.55 mm** **Ans.**

EXAMPLE 12.15 A machine that requires a torque of $T_M = 1500 - 600\cos\theta$ Nm for its drive is coupled to an engine that develops an effective torque of $T_E = 1500 + 800\sin 2\theta$ Nm at 200 rpm. A flywheel is fitted on the crank shaft to limit the fluctuation of speed to 2 per cent of the mean speed. Determine:

(i) power of the engine
(ii) minimum mass moment of inertia of flywheel
(iii) maximum and minimum acceleration of the flywheel and corresponding angular position of the engine shaft.

Solution Change in torque: $\Delta T = T_E - T_M$

or $\qquad \Delta T = (1500 + 800\sin 2\theta) - (1500 - 600\cos\theta)$

$\qquad\qquad = 800\sin 2\theta + 600\cos\theta$

ΔT will be zero when

$\qquad 800\sin 2\theta + 600\cos\theta = 0$

or $\qquad 800 \times 2 \sin\theta \cos\theta + 600\cos\theta = 0$

or $\qquad \cos\theta\,(1600\sin\theta + 600) = 0$

Therefore, if $\cos\theta = 0$, then $\theta = 90°$ or $270°$

and $\qquad 1600\sin\theta + 600 = 0$

or $\qquad \sin\theta = -\dfrac{600}{1600} = -0.375$

or $\qquad \theta = 202°$ or $338°$

Figure 12.46 shows the variation of torque against crank rotation angle θ. The largest area representing fluctuation of energy lies between $90°$ and $202°$. Therefore, fluctuation of energy:

$$\Delta E = \int_{90°}^{202°} (800\sin 2\theta + 600\cos\theta)\,d\theta$$

$$= \int_{90°}^{202°} 800\sin 2\theta\, d\theta + \int_{90°}^{202°} 600\cos\theta\, d\theta$$

$$= -800 \times \left.\dfrac{\cos 2\theta}{2}\right|_{90°}^{202°} + 600 \times \sin\theta\,\Big|_{90°}^{202°}$$

$$= -1512.5 \text{ Nm}$$

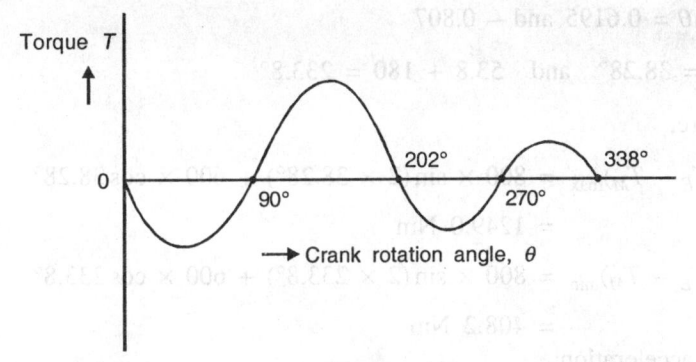

FIGURE 12.46

Since T–θ diagram for engine is function of 2θ, the cycle will repeat after every crank rotation of $180°$.

$$T_{mean} = \frac{1}{\pi}\int_0^\pi (1500 + 800\sin 2\theta)d\theta$$

$$= \frac{1}{\pi}\left[1500 \times \theta - \frac{800 \times \cos 2\theta}{2}\right]_0^\pi$$

$$= 1500 \text{ Nm}$$

(i) Power of engine: $bp = \dfrac{2\pi NT}{60 \times 10^3} = \dfrac{2\pi \times 200 \times 1500}{60 \times 10^3}$

$$= 31.41 \text{ kW} \qquad \text{Ans.}$$

(ii) Fluctuation of energy: $\Delta E = I\omega^2 C_s$

where $\omega = \dfrac{2\pi N}{60} = \dfrac{2\pi \times 200}{60} = 20.94$ rad/s

Mass moment of Inertia:

$$I = \frac{\Delta E}{\omega^2 C_s} = \frac{1512.5}{20.94^2 \times 0.02} = 172.47 \text{ kg m}^2 \qquad \text{Ans.}$$

(iii) For maximum acceleration, change in torque should be maximum. Therefore,

$$\frac{d}{d\theta}(800\sin 2\theta + 60\cos\theta) = 0$$

or $\quad 2 \times 800\cos 2\theta - 600\sin\theta = 0$

or $\quad 1600(1 - 2\sin^2\theta) - 600\sin\theta = 0$

or $\quad 3200\sin^2\theta + 600\sin\theta - 1600 = 0$

or $\sin\theta = 0.6195$ and -0.807

or $\theta = 38.28°$ and $53.8 + 180 = 233.8°$

Therefore,

$$(T_E - T_M)_{max} = 800 \times \sin(2 \times 38.28°) + 600 \times \cos 38.28°$$
$$= 1249.0 \text{ Nm}$$
$$(T_E - T_M)_{min} = 800 \times \sin(2 \times 233.8°) + 600 \times \cos 233.8°$$
$$= 408.2 \text{ Nm}$$

Maximum angular acceleration:

$$\alpha_{max} = \frac{(T_E - T_M)_{max}}{I}$$

$$= \frac{1249}{172.47} = 7.24 \text{ rad/s}^2 \qquad \text{Ans.}$$

Minimum angular acceleration:

$$\alpha_{min} = \frac{(T_E - T_M)_{min}}{I}$$

$$= \frac{408.2}{172.47} = 2.36 \text{ rad/s}^2 \qquad \text{Ans.}$$

EXAMPLE 12.16 A machine requires a torque $T_M = 5000 + 600 \sin\theta$ Nm for its drive, is coupled to a three cylinder engine that develops on effective torque $T_E = 5000 + 1500 \sin 3\theta$ at crank shaft rotating at 300 rpm. Determine the cross-sectional area of flywheel rim if the mean diameter of flywheel is 1.5 m and permissilbe fluctuation of speed is 1 per cent. Also calculate the power developed by engine.

Solution Work done by engine per revolution:

$$WD = \int_0^{2\pi} (5000 + 1500 \sin 3\theta) \, d\theta$$

$$= \left[5000\theta + \frac{1500 \times \cos 3\theta}{3} \right]_0^{2\pi}$$

$$= 10000\pi$$

Mean torque $= \dfrac{WD/\text{revol}}{2\pi} = \dfrac{10000\pi}{2\pi} = 5000$ Nm

Power $= \dfrac{2\pi NT}{60 \times 1000} = \dfrac{2\pi \times 300 \times 5000}{60 \times 1000} = \mathbf{157.08}$ **kW** \qquad Ans.

The turning moment diagram of machine and engine both are shown in Figure 12.47. It is found from this diagram that there are two points A and B where the torque developed by engine is equal to the forque required by machine. Therefore, at these points,

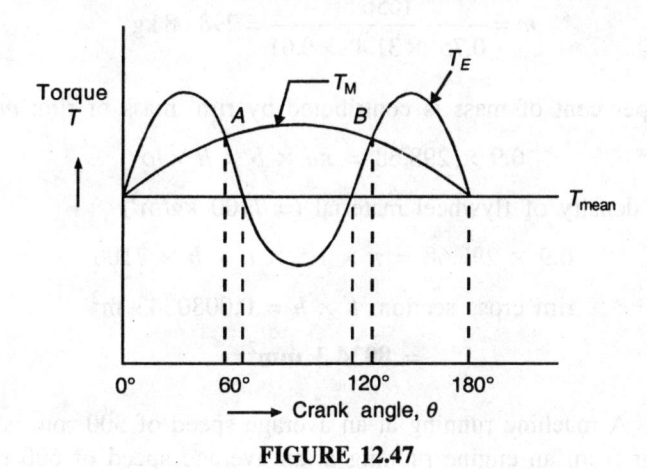

FIGURE 12.47

$$5000 + 1500 \sin 3\theta = 5000 + 600 \sin\theta$$

or $\qquad 2.5(3\sin\theta - 4\sin^3\theta) = \sin\theta$

or $\qquad 7.5 - 10 \times \sin^2\theta = 1$

or $\qquad \sin\theta = 0.806$

or $\qquad \theta_a = 53.7°$ and $\theta_b = 126.3°$

Thus the maximum fluctuation of energy:

$$\Delta E = \int_{53.7°}^{126.3°} (T_E - T_M)\, d\theta$$

$$= \int_{53.7°}^{126.3°} (1500 \times \sin 3\theta - 600 \times \sin\theta)\, d\theta$$

$$= \left[-\frac{1500 \times \cos 3\theta}{3} + 600 \times \cos\theta \right]_{53.7°}^{126.3°}$$

$$= -1656.5 \text{ Nm}$$

Note: Negative sign indicates deficit of energy and that has to be supplied by flywheel.

Mass of flywheel: $\qquad m = \dfrac{\Delta E}{k^2 \omega^2 C_s}$

where $\qquad \omega = \dfrac{2\pi N}{60} = \dfrac{2\pi \times 300}{60} = 31.4$ rad/s

Radius of gyration: $k = \dfrac{d}{2} = \dfrac{1.5}{2} = 0.75$ m

Therefore, $$m = \dfrac{1656.5}{0.75^2 \times 31.4^2 \times 0.01} = 298.68 \text{ kg}$$

Assuming that 90 per cent of mass is contributed by rim, mass of rim: $m_r = 0.9\, m$

or $$0.9 \times 298.68 = \pi d \times b \times h \times \rho$$

Let ρ be the mass density of flywheel material (= 7100 kg/m^3)

or $$0.9 \times 298.68 = \pi \times 1.5 \times b \times h \times 7100$$

or rim cross section: $b \times h = 0.0080343$ m^2

$$= 8034.3 \text{ mm}^2 \qquad \text{Ans.}$$

EXAMPLE 12.17 A machine running at an average speed of 300 rpm is driven through a single reducing gear from an engine running at an average speed of 600 rpm. The moment of inertia of the rotating parts on the machine shaft is equivalent to 110 kg at a radius of 0.3 m and that of rotating parts on the engine shaft is 18 kg at a radius of 0.3 m.

The torque transmitted to the machine from the engine is $2500 + 675 \sin 2\theta$ Nm, where θ is angle of rotation of the machine from some datum. The torque required to drive the machine is $2500 + 270 \sin\theta$ Nm.

Draw T–θ diagram for above arrangement and find the coefficient of fluctuation of speed.

Solution The T–q diagram is as Figure 12.48.

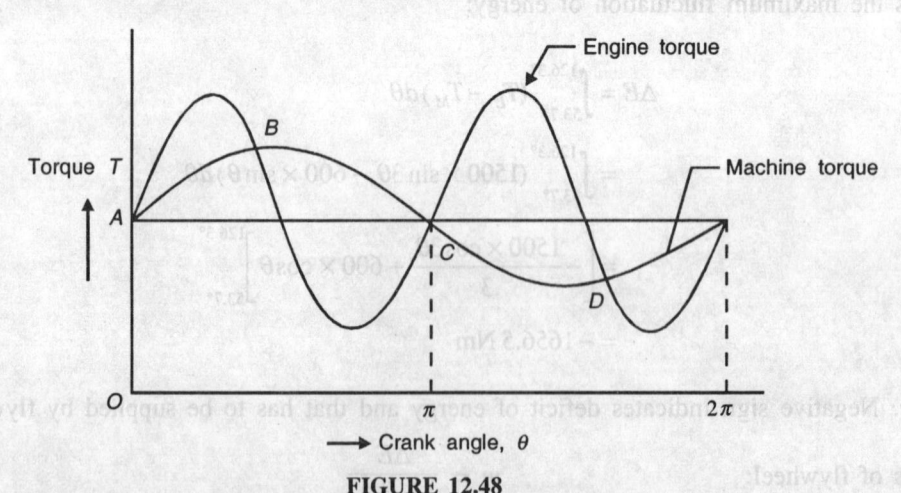

FIGURE 12.48

The engine and resisting torque of machine are equal when

$$2500 + 675 \sin 2\theta = 2500 + 270 \sin\theta$$

or $\quad\quad\quad\quad\quad\quad 2.5 \times 2\sin\theta \cos\theta = \sin\theta$

or $\quad\quad\quad\quad\quad\quad \cos\theta = 0.2 \quad \text{and} \quad \sin\theta = 0$

Therefore, $\quad\quad\quad\quad \theta = 78.46° \quad \text{and} \quad 281.54°$

or $\quad\quad\quad\quad\quad\quad \theta = 0, \pi, 2\pi$

The greatest fluctuation of energy occurs between points B and C or between points C and D. Therefore,

Fluctuation of energy: $\quad \Delta E = \int_{78.46°}^{180°} (270\sin\theta - 675\sin 2\theta)d\theta$

$$= 270 \times -\cos\theta \Big|_{78.46°}^{180°} + \frac{675 \times \cos 2\theta}{2}\Big|_{78.46°}^{180°}$$

$$= 972 \text{ Nm}$$

Equivalent moment of inertia of machine shaft

$$= 110 \times 0.3^2 + 18 \times 0.3^2 \times \left(\frac{600}{300}\right)^2$$

$$= 16.38 \text{ kg m}^2$$

Fluctuation of energy: $\Delta E = I\omega^2 C_s$

where $\quad\quad\quad\quad \omega = \frac{2\pi N}{60} = \frac{2\pi \times 300}{60} = 31.4 \text{ rad/s}$

Coefficient of fluctuation of speed:

$$C_s = \frac{\Delta E}{I\omega^2} = \frac{972}{16.38 \times 31.4^2} = 0.06$$

$$= 6 \text{ per cent} \quad\quad\quad\quad\quad\quad\quad\quad\quad\quad \text{Ans.}$$

EXAMPLE 12.18 A three cylinder single acting engine has its cranks set equally at 120° and runs at 700 rpm. The turning moment diagram for each cylinder is a triangle and maximum torque is 80 Nm at 60° from dead centre of the corresponding crank. The torque on the return stroke is zero.

Determine:

(i) power developed
(ii) coefficient of fluctuation of speed if the flywheel has mass of 10 kg and radius of gyration is 100 mm
(iii) coefficient of fluctuation of energy
(iv) the maximum angular acceleration of the flywheel

Solution The turning moment diagram of three cylinder engine is shown in Figure 12.49.

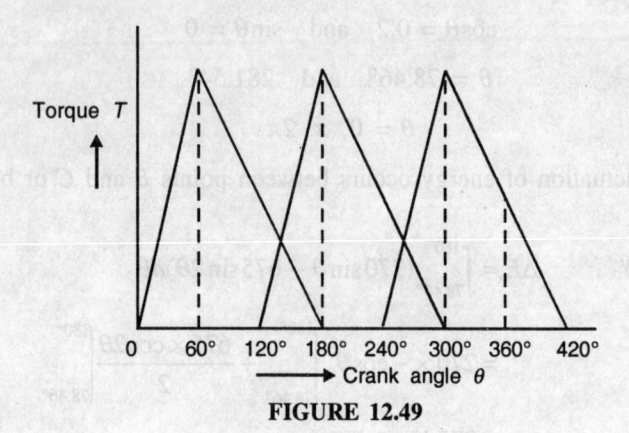

FIGURE 12.49

Workdone/cycle = area of three triangles

$$= 3 \times \frac{1}{2} \times \pi \times 80$$

$$= 120\pi \text{ Nm}$$

Mean torque: $T_m = \dfrac{\text{WD/cycle}}{2\pi} = \dfrac{120\pi}{2\pi} = 60$ Nm

Angular speed: $\omega = \dfrac{2\pi N}{60} = \dfrac{2\pi \times 700}{60} = 73.3$ rad/s

(i) Power: $P = \dfrac{2\pi NT}{60 \times 1000} = \dfrac{2\pi \times 700 \times 60}{60 \times 1000} = \mathbf{4.4\ kW}$ **Ans.**

(ii) When turning moment diagram of three cylinder is superimposed a combined turning moment diagram as shown in Figure 12.50 is obtained.

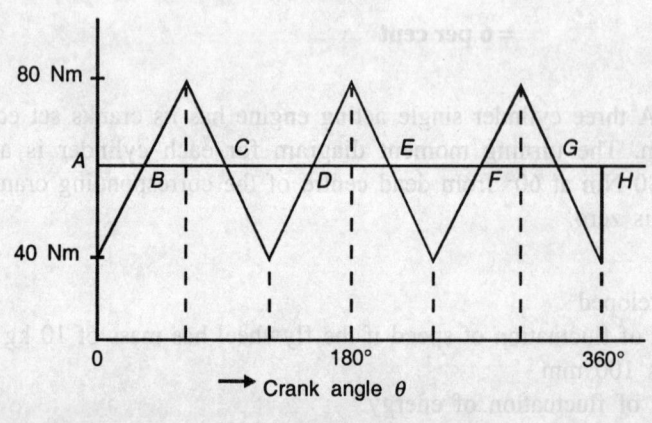

FIGURE 12.50

Let energy at point $A = e$

Energy at point $B = e - \dfrac{1}{2} \times \dfrac{\pi}{6} \times 20 = e - \dfrac{10\pi}{6}$

at point $C = e - \dfrac{10\pi}{6} + \dfrac{1}{2} \times \dfrac{\pi}{3} \times 20 = e + \dfrac{10\pi}{6}$

at point $D = e + \dfrac{10\pi}{6} - \dfrac{10\pi}{3} = e - \dfrac{10\pi}{6}$

at point $E = e - \dfrac{10\pi}{6} + \dfrac{10\pi}{3} = e + \dfrac{10\pi}{6}$

at point $F = e + \dfrac{10\pi}{6} - \dfrac{10\pi}{3} = e - \dfrac{10\pi}{6}$

at point $G = e - \dfrac{10\pi}{6} + \dfrac{10\pi}{3} = e + \dfrac{10\pi}{6}$

at point $H = e + \dfrac{10\pi}{6} - \dfrac{10\pi}{6} = e$

Therefore, maximum fluctuation of energy:

$$\Delta E = \left(e + \dfrac{10\pi}{6}\right) - \left(e - \dfrac{10\pi}{6}\right) = \dfrac{10\pi}{3} = 10.472 \text{ Nm}$$

Coefficient of fluctuation of speed:

$$C_s = \dfrac{\Delta E}{I\omega^2} = \dfrac{10.472}{10 \times 0.1^2 \times 73.3^2} = 0.0195$$

or $C_s = \mathbf{1.95 \text{ per cent}}$ **Ans.**

(iii) Coefficient of fluctuation of energy:

$$C_E = \dfrac{\text{Maximum fluctuation of energy}}{\text{WD/cycle}}$$

$$= \dfrac{10.472}{120\pi} = 0.0278$$

or $= \mathbf{2.78 \text{ per cent}}$ **Ans.**

(iv) Angular acceleration:

$$\alpha = \dfrac{T_{max} - T_{min}}{I} = \dfrac{80 - 60}{10 \times 0.1^2} = \mathbf{200 \text{ rad/s}^2}$$ **Ans.**

EXAMPLE 12.19 A bicycle driven single acting air compressor is being used for spray painting applications. The data on capability of human being indicate that a man can provide energy in the following order:

    1000W for 5 seconds
    150W for next 5 seconds
    100W for next 30 minutes
    75W for next 10 minutes
    750W for next 5 seconds

None for the next 19 minutes and 45 seconds.

The cycle is then repeated. Assuming a 90 per cent mechanical efficiency, Calculate the average power out put at 100 rpm. If a speed variation of 15 per cent is allowed, determine the moment of inertia of flywheel.

Solution Total work output

$= 1000 \times 5 + 150 \times 5 + 100 \times 30 \times 60 + 75 \times 10 \times 60 + 750 \times 5$

$= 2.345 \times 10^5$ Nm

Cycle time $= \dfrac{5}{60} + \dfrac{5}{60} + 30 + 10 + \dfrac{5}{16} + 19 + \dfrac{45}{60} = 60$ minutes

$= 1$h

Angular speed: $\omega = \dfrac{2\pi N}{60} = \dfrac{2\pi \times 100}{60} = 10.47$ rad/s

Average power $= \dfrac{\text{Work output}}{\text{cycle time}} \times \eta_{mech}$

$= \dfrac{2.345 \times 10^5}{3600} \times 0.9 =$ **58.625 W**   Ans.

Fluctuation of energy = mean output × idle time

$= 58.625 \times (19 \times 60 + 45)$

$= 69470.625$ Nm

$\Delta E = I\omega^2 C_s$

or mass moment of inertia: $I = \dfrac{\Delta E}{\omega^2 C_s}$

or $I = \dfrac{69470.625}{(10.47)^2 \times 0.15} = 4224.9$ kg m²   Ans.

EXAMPLE 12.20 A punching machine has a capacity of producing 30 holes of 20 mm diameter per minute in a steel plate of 16 mm thickness. The material of plate has a ultimate shear strength of 360 N/mm². The actual punching operation lasts for a period of 36° rotation of the crank shaft. This crank shaft is powered by a flywheel through a reduction gear having a ratio of 1:8.

(i) Mechanical efficiency = 80 per cent
(ii) Speed fluctuation = 10 per cent
(iii) Mean diameter of flywheel = 0.75 m

Determine the following:
(a) Power requied
(b) Fluctuation of energy
(c) Cross section of rim if the width to thickness ratio is 1.5.

Solution Speed of flywheel shaft = hole capacity × gear ratio
$$= 30 \times 8 = 240 \text{ rpm}$$

Force required = hole shear area × shear strength
$$= \pi dt \times \tau$$
$$= \pi \times 20 \times 16 \times 360$$
$$= 361.9 \text{ kN}$$

Work required = mean force × distance
$$= \frac{0 + 361.9}{2} \times \frac{16}{1000} = 2.8952 \text{ kNm}$$

Work required at flywheel $= \dfrac{2.8952}{\eta_{mech}} = \dfrac{2.8952}{0.8} = 3.619 \text{ kNm}$

(a) Power: $P = \dfrac{\text{Work required} \times \text{number of holes/ min}}{60} = \dfrac{3.619 \times 30}{60} = 1.81 \text{ kW}$ **Ans.**

Punching period $= \dfrac{36°}{360°} = 0.1$

(b) Energy supplied to flywheel or fluctuation of energy:
$$\Delta E = 0.9 \times 3.619 = 3.257 \text{ kNm} \qquad \text{Ans.}$$

Mass of flywheel: $m = \dfrac{\Delta E}{k^2 \omega^2 C_s}$

where $\omega = \dfrac{2\pi N}{60} = \dfrac{2\pi \times 240}{60} = 25.13 \text{ rad/s}$

$$m = \frac{3.257 \times 10^3}{0.375^2 \times 25.13^2 \times 0.1} = 366.75 \text{ kg}$$

Assume mass of rim: m_r = 0.9 × mass of flywheel
$$= 0.9 \times 366.75$$
$$= 330.0 \text{ kg}$$
$$m_r = \pi d \times b \times h \times \rho$$

Let ρ be the mass density of flywheel material (= 7100 kg/m)
$$330 = \pi \times 0.75 \times b \times h \times 7100$$
or $$b \times h = 0.01972 \text{ m}^2$$
Since $$b/h = 1.5$$

(c) Width: b = 172 mm   Ans.
    Thickness: h = 114.66 mm Ans.

EXERCISES

1. Explain the conditions for static equilibrium of a link.
2. What do you mean by free body diagram? Explain with suitable example.
3. Explain the principle of virtual work.
4. Explain the procedure to perform static force analysis of a shaper mechanism with suitable free body diagram of various links.
5. Define the terms inertia force and inertia couple. How is their magnitude and direction found?
6. What do you mean by dynamical equivalent system? State the important role played by such system for determining the line of action of inertia force.
7. Explain how is the inertia force in a connecting rod of reciprocating engine is determined.
8. From the first principle, derive expressions for displacement, velocity and acceleration of piston in an IC engine.
9. Derive an expression for crank effort for any given crank position.
10. Explain procedure to construct turning moment diagram of a four stroke IC engine.
11. Define the following terms:
    (a) Flywheel effect
    (b) Coefficient of fluctuation of speed
    (c) Coefficient of fluctuation of energy
12. Explain the function of a flywheel with reference to turning moment diagram.
13. What are the different types of flywheels? state how the size of a flywheel is calculated.

14. Determine the torque required to be applied at the crank shaft of a slider–crank mechanism to bring it in equilibrium. The slider is subjected to a horizontal force of 4000 N and connecting rod and crank shaft are subjected to forces of 500 N and 2000 N respectively as shown in Figure E12.1. The dimensions of various links are as given below:

$AB = 30$ mm, $BC = 45$ mm, $BE = 17$ mm and $AD = 15$ mm.

Crank AB makes an angle of 45° from line of stroke.

[Ans: 1220 Nm]

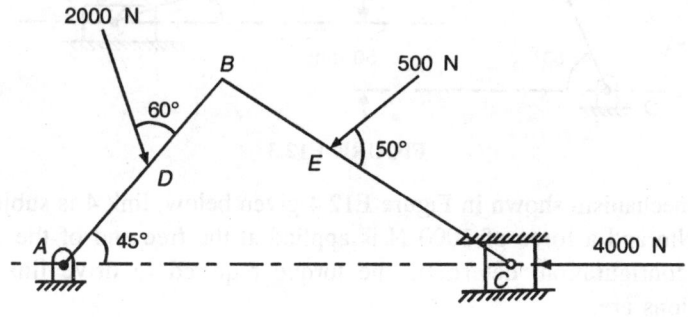

FIGURE E12.1

15. In a four bar mechanism shown in Figure E12.2 torques T_3 and T_4 of 30 Nm and 20 Nm are applied to links 3 and 4 respectively. The lengths of links are:

$AB = 300$ mm, $BC = 700$ mm, $CD = 400$ mm and $AD = 800$ mm.

For the static equilibrium of the mechanism at a given instant, determine the input torque T_2 required to drive link 2.

[Ans: 16.2 Nm, counterclockwise]

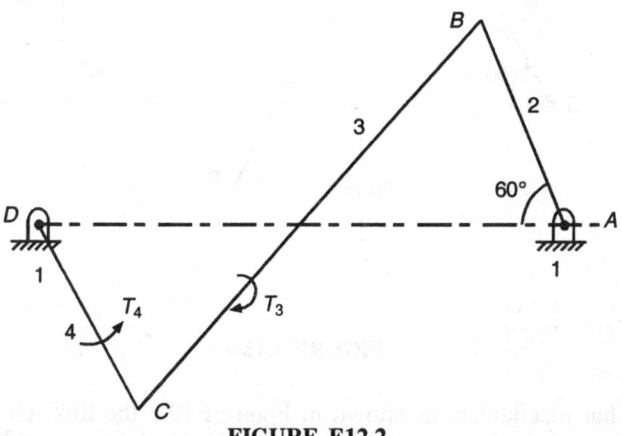

FIGURE E12.2

16. For the offset slider mechanism shown in Figure E12.3, determine the required input torque for the condition of static equilibrium. The lengths of links OA and AB are 250 mm and 650 mm respectively.

[Ans: 68 Nm, clockwise]

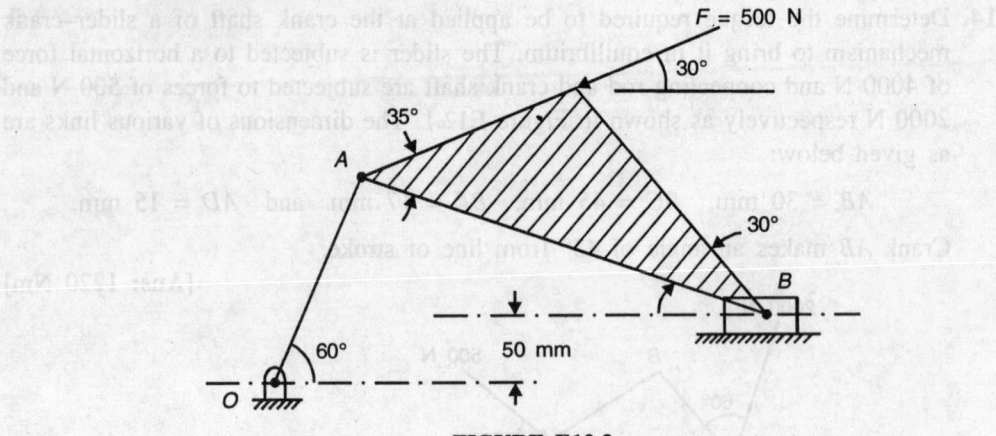

FIGURE E12.3

17. In the mechanism shown in Figure E12.4 given below, link 4 is subjected to a torque of 56 Nm and a force of 1000 N is applied at the free end of the link BD. For the above configuration, determine the torque required to drive link 2. The various dimensions are:

$AB = 140$ mm, $BC = 170$ mm, $BD = 300$ mm,   and $CE = 110$ mm
$\angle ABD = 50°$

[Ans: 100 Nm, counterclockwise]

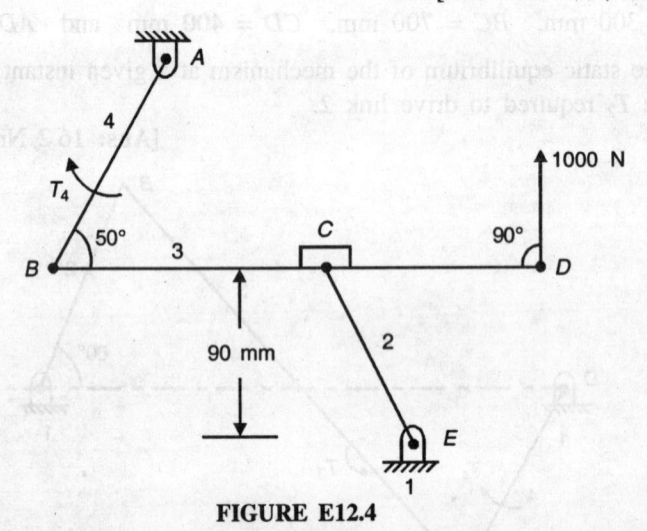

FIGURE E12.4

18. In a four bar mechanism as shown in Figure E12.5 the link AB rotates with angular velocity of 20 rad/s and angular acceleration of 100 rad/s^2 both in clockwise direction when it makes an angle of 45° with link AD, which is fixed. The dimensions of various links are:

$AB = CD = 400$ mm, $BC = 500$ mm   and $AD = 750$ mm.

Neglecting the gravitational effect and friction determine the torque required to overcome inertia forces. The mass of link is 5 kg per metre length.

[Ans: 60 Nm, counter clockwise]

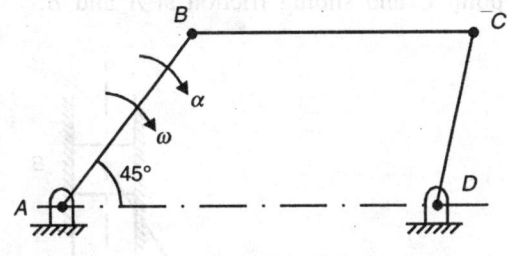

FIGURE E12.5

19. Describe the method to find the torque which would overcome the inertia forces of a shaper mechanism.

20. Figure E12.6 shows a mechanism in which crank AB rotates anticlockwise about A at 70 rpm. The link CD swings about D and is connected to the link BC. The lengths are:

$$AB = 0.25 \text{ m}, \quad BC = 1.0 \text{ m}, \quad CD = 0.75 \text{ m}$$

The connecting link BC has a mass of 14 kg. Its centre of gravity G is at the centre of its length and its radius of gyration about a transverse axis through G is 0.3 m. The joints B, C and D are friction less and link CD is of negligible mass.

For the given position in which link AB is at 45° from the horizontal, find the forces exerted at the joints B and C resulting from the inertia of the rod BC.

[Ans: 150.5 N, 29.4 N]

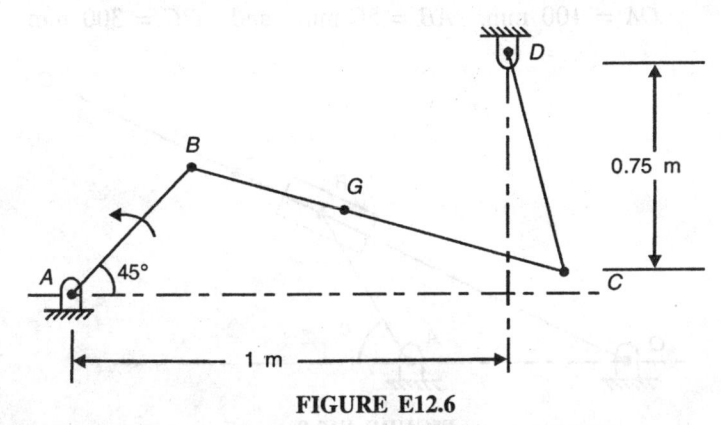

FIGURE E12.6

21. A slider A slides between horizontal guides as shown in Figure E12.7. It is connected by a light rod AB, 200 mm long, to another slider B which slides between vertical guides. A concentrated body C, of mass 3 kg, is fixed to the rod at a point 125 mm from A. The coefficient of friction between each slider and the guides is 0.1 and friction at turning pairs may be reglected.

At the instant when angle $BAO = 60°$, slider A has a velocity to the right of 1.5 m/s and acceleration to the right of 12 m/s². Determine the horizontal force required under these conditions taking into account the weight and inertia of the mass placed at point C and sliding friction at A and B.

[Ans: 8.9 N]

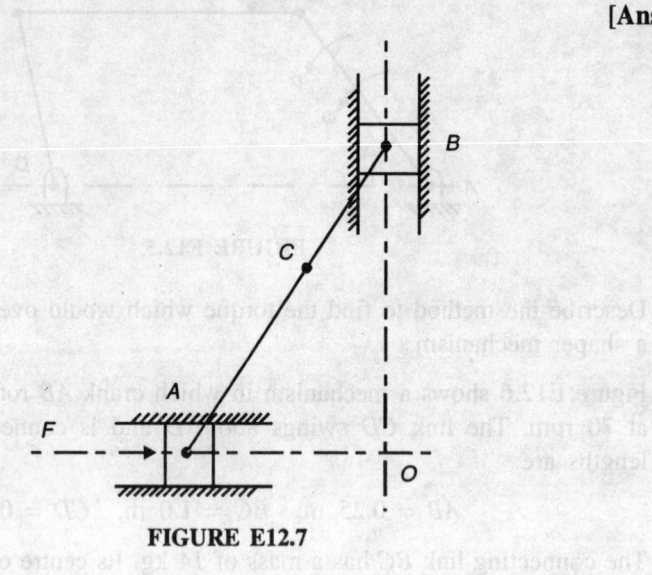

FIGURE E12.7

22. In the mechanism shown in Figure E12.8 a small block B can slide along the link OC which rotates about the fixed point O. The crank AB rotates about the fixed point A and is pinned to the block at B. The link OC has a mass of 5 kg/m length. The dimensions of various links are as given below:

$OA = 100$ mm, $AB = 50$ mm and $OC = 300$ mm

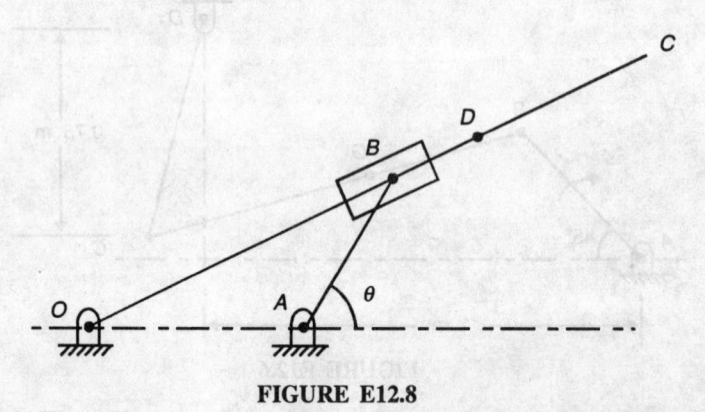

FIGURE E12.8

At the instant when the angle θ is 90°, the crank AB has an angular velocity of 20 rad/s in anticlockwise direction and angular acceleration of 400 rad/s² in clockwise direction. For this configuration, determine:

(i) angular acceleration of link *OC*
(ii) force between the link *OC* and turning pair at *B*
(iii) the bending moment at point *D* if *OD* = 175 mm.

23. A crank and connecting rod of a steam engine are 300 mm and 1500 mm in length respectively. The crank rotates at 180 rpm clockwise. Determine the velocity and acceleration of the piston when the crank is at 40° from the *IDC* position. Also determine the position of the crank for zero acceleration of the piston.

[**Ans:** 4.19 m/s, 85.26 m/s^2, 79.27°]

24. The crank and connecting rod of a petrol engine running at 1800 rpm are 50 mm and 200 mm in length respectively. The diameter of the piston is 80 mm and the mass of reciprocating parts is 1 kg. At a point during the power stroke, the pressure on the piston is 0.7 N/mm^2 when it has moved 10 mm from *IDC* and corresponding crank displacement being 33°.
Determine:
   (i) thrust in the connecting rod
   (ii) reaction between the piston and cylinder
   (iii) turning moment on the crank shaft

[**Ans:** 5240 N, 719.3 N, 147.3 Nm]

25. In a slider–crank mechanism *ABC*, the crank *AB* is of 40 mm length. The connecting rod *BC* is of 120 mm and piston mass is 0.1 kg. The crank at a given position makes an angle of 45° from *IDC* and rotates at 1500 rpm. Determine the torque required at the crank to accelerate the piston.

[**Ans:** 1.97 Nm]

26. The length of a connecting rod of an engine is 500 mm measured between the centres and its mass is 18 kg. The centre of gravity is 125 mm from the crank pin centre. Determine the dynamically equivalent systems keeping one mass at the small end. The frequency of oscillation of the rod when suspended from the centre of the small end is 43 vibrations per minute.

[**Ans:** 4.1 kg, 13.9 kg, l_1 = 375 mm, l_2 = 110 mm]

27. The connecting rod of vertical cylinder IC engine is 600 mm long between centres and has a mass of 3 kg. The mass centre of connecting rod is 200 mm way from the big end. When connecting rod is suspended as pendulum from the gudgen pin, it makes 45 complete oscillations in 30 seconds. The crank radius is 125 mm and the mass of piston is 1.2 kg. Determine the inertia torque on the crank shaft when crank makes an angle of 140° from TDC and crank shaft speed is 1500 rpm.

[**Ans:** 361.9 Nm]

28. The turning moment curve for an engine is represented by equation T = 2000 + 950 sin 2θ – 570 cos 2θ Nm, where θ is the angle moved by the crank from the IDC. If the resisting torque is constant of 2000 Nm find:
   (i) power developed by the engine

(ii) moment of inertia of the flywheel if fluctuation of speed is 0.01 and mean speed is 180 rpm.

(iii) angular acceleration of the flywheel when crank has turned 45° from IDC

[Ans: 37.7 kW, 311.8 kgm^2, 3.047 rad/s^2]

29. The areas of the turning moment diagrams for one revolution of multicylinder engine with reference to mean torque below and above the line (in mm^2) are −32, 408, −267, 333, −310, 226, −374, 260 and −244. The scale for abscissa and ordinate are 1 mm = 2.4° and 1 mm = 650 Nm respectively. The mean speed is 300 rpm with percentage speed fluctuation of ± 1.5%. Determine the mass of the flywheel if the maximum speed is 25 m/s.

[Ans: 688.3 kg]

30. The turning moment diagram for an engine is given by $T = 2100 \sin\theta + 900 \sin 2\theta$ Nm for values of θ, the crank rotation angle, between 0 and π and by $T = 375 \sin\theta$ for values of θ between π and 2π. This is repeated for every revolution of the engine.

The resisting torque is constant and the speed is 850 rpm. The total moment of inertia of the rotating parts of the engine is 270 kgm^2. Determine:

(i) power
(ii) fluctuation in speed
(iii) the maximum instantaneous angular acceleration of the engine and the value of θ at which it occurs.

[Ans: 307 kW, 1.103 rpm, 7.6 rad/s^2]

31. A four stroke engine develops 18.5 kW at 250 rpm. The turning moment diagram is rectangular for the expansion and compression strokes. The turning moment diagram during the expansion stroke is 2.8 times of that during the compression stroke. Assuming constant load, determine the moment of inertia of the flywheel to keep total speed fluctuation within ±1 per cent of the average speed.

[Ans: 2855 kg m^2]

32. A punching machine operates at the rate of 600 holes/hr. It does 45 Nm of work per sq mm of shear, cutting 25 mm diameter hole in a 3 mm thick plate. The machine is operated by a constant torque motor. The speed of machine fluctuates between 250 and 230 rpm. The frictional losses are 20% of the work done during punching and actual punching time per hold is 2 seconds. Sketch the turning moment diagram for one complete cycle. Determine:

(i) motor power
(ii) maximum fluctuation of energy
(iii) mass of the flywheel if radius of gyration is 500 mm.

MULTIPLE CHOICE QUESTIONS

1. A rigid body is said to be in equilibrium if
   (a) $\Sigma F_x = 0$
   (b) $\Sigma F_y = 0$
   (c) $\Sigma M_z = 0$
   (d) all of the above

2. A rigid body subjected to three forces can be brought to equilibrium condition when
   (a) the forces meet at point of concurrency
   (b) the force polygon is open
   (c) the force polygon is closed
   (d) either (a) or (c)

3. In a reciprocating horizontal engine, the inertia force due to reciprocating mass helps the piston effort at
   (a) $\theta = 45°$
   (b) $\theta = 120°$
   (c) $\theta = 30°$
   (d) $\theta = 180°$

4. If P is the piston effort and ϕ is the obliquity angle of connecting rod, the expression for thrust in connecting rod is
   (a) $P/\cos\phi$
   (b) $P \times \cos\phi$
   (c) $P \tan\phi$
   (d) $P/\sin\phi$

5. For a link to be dynamically equivalent, which condition needs to be satisfied
   (a) $k^2 = l_a/l_b$
   (b) $k^2 = l_b/l_a$
   (c) $k^2 = l_a l_b$
   (d) $k^2 = l_a + l_b$

6. The displacement of the piston is given by
   (a) $r(1 - \cos\theta)$
   (b) $\omega r\left(\sin\theta + \dfrac{\sin 2\theta}{n}\right)$
   (c) $\omega^2 r\left(\cos\theta + \dfrac{\cos 2\theta}{n}\right)$
   (d) none of the above

CHAPTER 13

Balancing

13.1 INTRODUCTION

In high speed machinery such as internal combustion engines, steam and gas turbines, centrifugal pumps, fans and blowers, the distribution of mass along the axis of rotation depends upon the configuration of component. It is designer's responsibility to design a machine or it components in such a way that a line joining all mass centres will be a straight line coinciding with the axis of rotation. However, due to limitation of manufacturing processes and cost thereof, the perfect manufacturing of components and their assembly is seldom attained. Consequently, a line joining all mass centres may coincide with the axis of rotation occasionally at some locations. Therefore, such eccentricity between the mass centre and axis of rotation may cause inertia force which becomes main source of vibration. These vibrations at times may reach dangerous amplitudes and lead to fatigue failure of components.

Balancing is a technique of eliminating or at least reducing these unwanted inertia forces and moments to some acceptable limits. Thus determining the unbalanced forces and the application of method of correction is principal task of balancing.

13.2 STATIC BALANCING OF ROTATING MASSES

When a rotor rotates with mass centre away from the axis of rotation, it experiences a centripetal acceleration which generates a radially inward force. An equal and opposite force called **centrifugal force** acts radially outward which is the main source of unbalancing of the rotor system. The magnitude of this unbalanced force remains constant but its direction changes with the rotation (Figure 13.1).

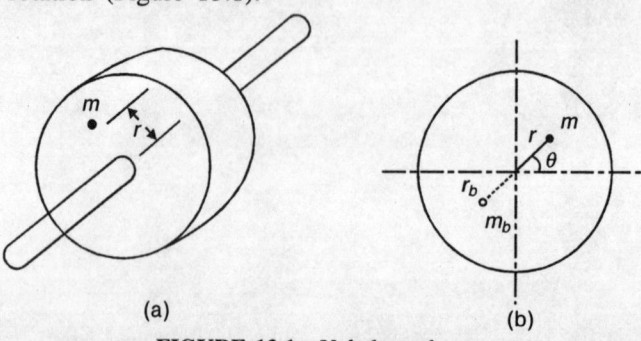

FIGURE 13.1 Unbalanced rotor.

The magnitude of unbalanced force or centrifugal force:

$$F_c = m\omega^2 r \qquad (13.1)$$

where m = mass of the rotor
ω = angular velocity
and r = eccentricity, i.e. the distance between axis of mass centre and axis of rotation

To counteract the effect of centrifugal force (inertia force), a balancing mass can be introduced in the plane of rotation of unbalance mass, such that the centrifugal forces produced by these two masses are equal and opposite as shown in Figure 5.1.

Let m_b = balancing mass
r_b = distance between the centre of gravity of the balancing mass and the axis of rotation

Then for balanced rotor system,

$$m\omega^2 r = m_b \omega^2 r_b$$

or
$$mr = m_b r_b \qquad (13.2)$$

Therefore, in a single rotor system by satisfying Eq. (13.2), we bring the mass centre of rotor to the axis of rotation. Such balancing is called **static balancing.**

In a system of multiple rotating masses, if the combined mass centre of the system lies on the axis of rotation, then it is called a **statically balanced system.** In other words, the vector sum of various unbalanced forces acting in a system must be equal to zero.

Consider a rigid rotor system consisting of four masses revolving with a constant angular velocity ω rad/s (Figure 13.2). If the masses m_1, m_2, m_3 and m_4 revolve at radii r_1, r_2, r_3 and r_4 respectively in the same plane, then each mass produces a centrifugal force acting radially outward. For a rotor to be statically balanced, the vector sum of these forces must be equal to zero. However, if the rotor is unbalanced, i.e. the vector sum of forces is not equal to zero, then a balancing mass m_b acting at radius r_b is required to be introduced in such a way that condition of statically balancing is satisfied.

(a) Unbalanced masses   (b) Force polygon

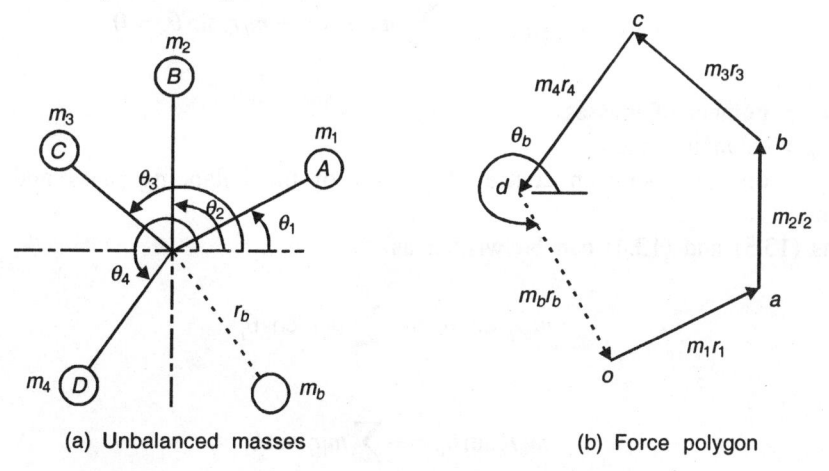

FIGURE 13.2 Unbalanced masses rotating in the same plane.

The problem of balancing several unbalanced masses rotating in the same plane can be solved either by graphical or by analytical method as discussed further.

Graphical Method. Figure 13.2 shows a system of unbalanced masses revolving in the same plane at constant angular velocity. Therefore, the magnitude of unbalanced centrifugal force is proportional to the product of unbalanced mass and its eccentricity. The following procedure may be used to determine the magnitude and location of balancing mass.

1. Select a suitable scale to represent graphically the product of unbalanced mass and eccentricity (mr).
2. Draw a line oa representing force $m_1 r_1$ (ω^2 being common to all is assumed to be unity) along the direction of unbalanced mass m_1 as shown in Figure 13.2(b).
3. From point a, draw another line ab parallel to the mass m_2 and equal to force $m_2 r_2$.
4. Similarly, draw a line bc from point b, parallel to the mass at C representing force $m_3 r_3$.
5. From point c, draw a line cd parallel and equal to the force $m_4 r_4$.
6. If this force polygon is not closed then the system of revolving masses is unbalanced. Therefore, join the points d and o by a straight line. The vector $\overrightarrow{do}$ represents the direction and magnitude of the balancing force ($m_b r_b$) [See Figure 13.2(b)]. For known value of eccentricity r_b, the balancing mass m_b can be determined.

Analytical Method. In analytical method, the problem of balancing can be solved by resolving each force horizontally and vertically. For a statically balanced system, the vector sum of these horizontal and vertical force components must be separately equal to zero. Therefore, resolving horizontally,

$$\sum_{i=1}^{n} m_i r_i \cos\theta_i + m_b r_b \cos\theta_b = 0 \qquad (13.3)$$

and resolving vertically, $\sum_{i=1}^{n} m_i r_i \sin\theta_i + m_b r_b \sin\theta_b = 0 \qquad (13.4)$

where n = number of masses
m_b = balancing mass
r_b = distance between centre of gravity of the balancing mass and the axis of rotation

Equations (13.3) and (13.4) can be written as

$$m_b r_b \cos\theta_b = -\sum_{i=1}^{n} m_i r_i \cos\theta_i$$

$$m_b r_b \sin\theta_b = -\sum_{i=1}^{n} m_i r_i \sin\theta_i$$

The magnitude of balancing force is:

$$m_b r_b = \sqrt{\left(\sum_{i=1}^{n} m_i r_i \cos\theta_i\right)^2 + \left(\sum_{i=1}^{n} m_i r_i \sin\theta_i\right)^2} \qquad (13.5)$$

and its angular position is given by

$$\tan\theta_b = \frac{-\sum_{i=1}^{n} m_i r_i \sin\theta_i}{-\sum_{i=1}^{n} m_i r_i \cos\theta_i} \qquad (13.6)$$

The quadrant of the angle θ_b is identified by sign of numerator and denominator of Eq. (13.6).

EXAMPLE 13.1 A rotor has three unbalanced masses acting in the same plane as shown in Figure 13.3. The three masses m_1, m_2 and m_3 are 3 kg, 4 kg and 5 kg respectively. The corresponding radii of rotations are 60 mm, 40 mm and 20 mm and angles θ_1, θ_2 and θ_3 are 30°, 120° and 270° from the horizontal axis. Determine the magnitude and position of the balance mass if the radius of rotation is 30 mm.

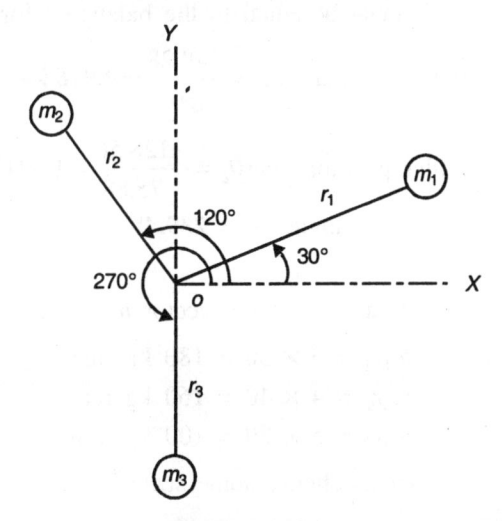

FIGURE 13.3

Solution Since all masses act on the rotor is one plane, so it is a problem of static balancing which can be solved either by analytical or graphical method. Let us solve this problem first by analytical method and later we should check it through graphical method.
Analytical method

$$\text{Centrifugal force} = m\omega^2 r$$

As all masses are rotating at the same angular speed,

$$\text{Centrifugal force} = mr \times \text{constant}$$

or mr can be taken as centrifugal force.

Resolving all forces horizontally,

$$F_H = m_1 r_1 \cos\theta_1 + m_2 r_2 \cos\theta_2 + m_3 r_3 \cos\theta_3$$
$$= 3 \times 60 \times \cos 30° + 4 \times 40 \times \cos 120° + 5 \times 20 \times \cos 270°$$
$$= 75.88 \text{ kg mm}$$

Similarly, resolving all forces vertically,

$$F_V = 3 \times 60 \times \sin 30° + 4 \times 40 \times \sin 120° + 5 \times 20 \times \sin 270°$$
$$= 128.56 \text{ kg mm}$$

Let m_b = balanced mass
and r_b = radius at which the mass m_b will act.

Resultant unbalanced force $= \sqrt{F_H^2 + F_V^2}$

$$= \sqrt{75.88^2 + 128.56^2} = 149.28 \text{ kg mm}$$

The resultant unbalanced force must be equal to the balancing force $m_b r_b$.

or balancing mas: $m_b = \dfrac{149.28}{30} = \mathbf{4.976 \text{ kg}}$ **Ans.**

Angular position: $\tan\theta_b = \dfrac{-128.56}{-75.88} = 1.6942$

or angle: $\theta_b = \mathbf{239.4°}$ **Ans.**

Graphical method

In graphical method, let us first calculate force vector mr for all three masses.

$$m_1 r_1 = 3 \times 60 = 180 \text{ kg mm}$$
$$m_2 r_2 = 4 \times 40 = 160 \text{ kg mm}$$
and $$m_3 r_3 = 5 \times 20 = 100 \text{ kg mm}$$

To construct a force polygon, let us choose some scale.

Let 25 kg mm = 10 mm

(i) Draw a vector $\overrightarrow{oa}$ parallel to mass m_1 representing 180 kg mm [refer Figure 13.4 (b)].

(ii) At point a, draw another vector $\overrightarrow{ab}$ parallel to mass m_2 of magnitude 160 kg mm.

(iii) Draw a vector $\overrightarrow{bc}$ parallel to mass m_3, representing force of magnitude 100 kg mm.

(iv) Join points c and o. The length of vector $\overrightarrow{co}$ represents resultant unbalanced force which is equal to $m_b r_b$ to calculate mass of balance m_b.

(a) Configuration diagram

(b) Force polygon

FIGURE 13.4

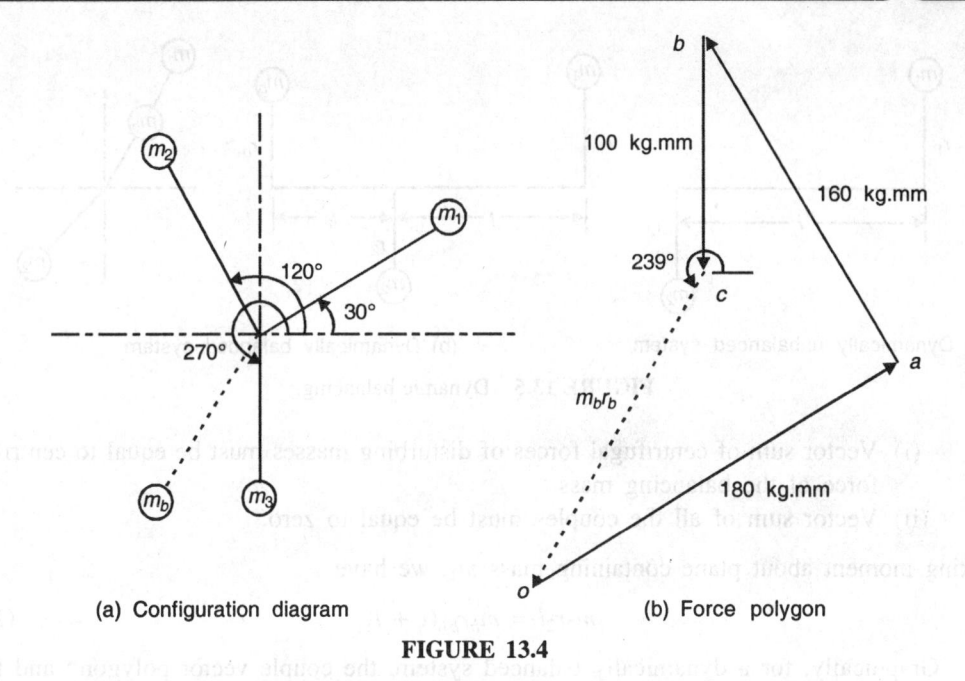

(v) The orientation of vector $\overrightarrow{co}$ gives the angular position θ_b.

$$m_b r_b = \text{vector } \overrightarrow{co} = 6 \times 25$$

or $\qquad$ balancing mass: $m_b = \dfrac{150}{30} = 5 \text{ kg}$ $\qquad$ **Ans.**

Angle: $\theta_b = 239°$ $\qquad$ **Ans.**

13.3 DYNAMIC BALANCING

When several masses rotate in a single plane, the system is said to be statically balanced if the vector sum of all forces is zero. However, when several masses rotate in different planes, the system is subjected to two types of unbalance—centrifugal force and couple. Thus a system having neither any unbalanced centrifugal force nor any resultant couple is termed as balanced system. Such type of a balancing is often called **dynamic balancing.** In other words, a system consisting of several rotating masses in different planes is said to be dynamically balanced, if the vector sums of centrifugal forces and couples are zero.

Consider two systems of masses as shown in Figure 13.5. The masses m_1 and m_2 revolves diametrically opposite to each other in different planes. The system can be brought to statically balance if $m_1 r_1 = m_2 r_2$. However, in this system an unbalance couple of magnitude $m_1 r_1 l$ is introduced additionally. Therefore, the system is statically balanced but dynamically unbalanced. In order to balance the above system, let us introduce another mass m_b revolving at radius r_b in such a way that [see Figure 13.5(b)]:

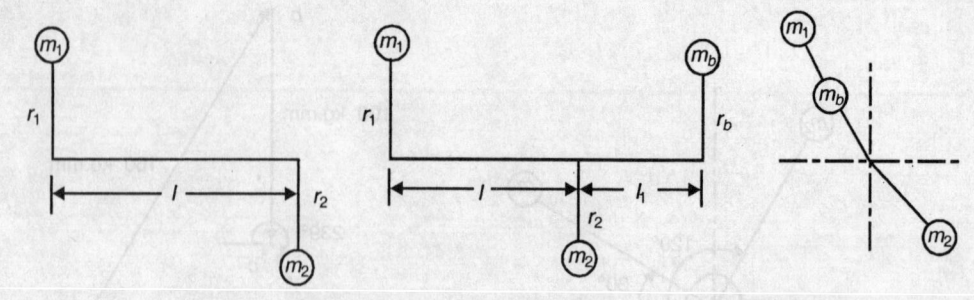

(a) Dynamically unbalanced system (b) Dynamically balanced system

FIGURE 13.5 Dynamic balancing.

(i) Vector sum of centrifugal forces of disturbing masses must be equal to centrifugal force of the balancing mass.
(ii) Vector sum of all the couples must be equal to zero.

Taking moment about plane containing mass m_1, we have

$$m_2 r_2 l = m_b r_b (l + l_1) \tag{13.7}$$

Graphically, for a dynamically balanced system, the couple vector polygon* and force vector polygon must be closed.

13.4 TWO PLANES BALANCING

Let us consider a rotor rotating with a uniform angular velocity ω rad/s. On this rotor, several unbalanced masses (say three masses m_1, m_2 and m_3) are attached at radii r_1, r_2 and r_3 respectively revolving in different planes as shown in Figure 13.6. The above problem of balancing can be solved by the two plane balancing method in which the effect of rotating masses, i.e. centrifugal forces and couples are transferred to the two arbitrarily chosen reference planes, say XX and YY.

Let m_x = balancing mass put up on the plane XX at a radius r_x
and m_y = balancing mass put up on the plane YY at a radius r_y

The rotor system with these reference planes becomes dynamically balanced when it satisfies the following necessary and sufficient conditions:

(i) The vector sum of all the centrifugal forces is equal to zero.
(ii) The vector sum of all the couples is zero.

The above conditions can be verified by both the graphical and analytical methods.

Graphical method. Refer to Figure 13.6(a) which shows a system of unbalanced

---

*The original direction of couple vector is perpendicular to that of the force vector. However, in balancing problem it has become practice to draw couple polygon by turning it through 90°, i.e. vectors are drawn parallel to the force vectors. The same practice is adopted in this chapter.

Balancing **549**

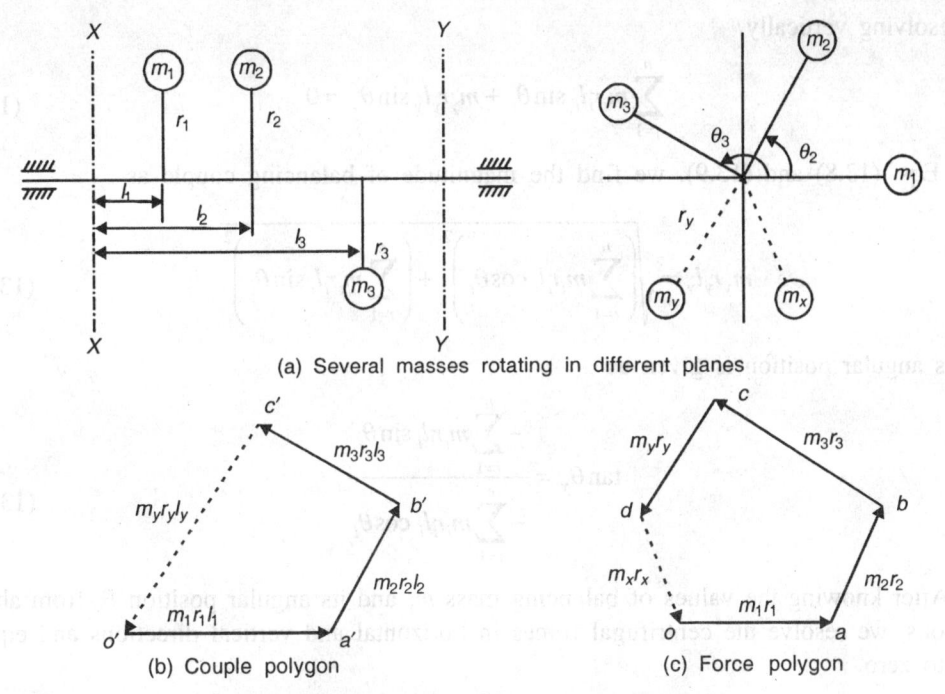

FIGURE 13.6 Two plane balancing.

(a) Several masses rotating in different planes
(b) Couple polygon
(c) Force polygon

masses revolving in different planes at constant angular velocity. The following procedure may be used to determine the magnitude and angular position of balancing masses.

1. Choose a reference plane (say *XX*) and find the distances of all the unbalanced masses from it.
2. Compute the magnitude of couple (*mrl*) of all the unbalanced masses omitting ω^2.
3. Draw a couple polygon for the known values of couples as shown in Figure 13.6(b). The closing couple vector, gives the value of the unbalanced couple, i.e. $m_y r_y l_y$.
4. Calculate the magnitude of the balancing mass m_y for the known radius r_y. The angular position of the mass m_y can be decided on the basis of the direction of the couple vector.
5. With the known value of mass m_y and centrifugal force $m_y r_y$, draw a force polygon as shown in Figure 13.6(c). The closing force vector gives the magnitude of $m_x r_x$. For the known value of radius r_x, the magnitude of the balancing mass m_x required at the plane *XX* can be determined.

Analytical method. The problem of two plane balancing can be solved easily by the analytical method, in which the unbalanced centrifugal forces and couples are resolved in horizontal and vertical directions.

First, resolve the couple vector in horizontal and vertical directions and equate them to zero to get the necessary condition for balancing. Thus resolving horizontally,

$$\sum_{i=1}^{n} m_i r_i l_i \cos\theta_i + m_y r_y l_y \cos\theta_y = 0 \qquad (13.8)$$

and resolving vertically

$$\sum_{i=1}^{n} m_i r_i l_i \sin\theta_i + m_y r_y l_y \sin\theta_y = 0 \tag{13.9}$$

From Eqs. (13.8) and (13.9), we find the magnitude of balancing couple as

$$m_y r_y l_y = \sqrt{\left(\sum_{i=1}^{n} m_i r_i l_i \cos\theta_i\right)^2 + \left(\sum_{i=1}^{n} m_i r_i l_i \sin\theta_i\right)^2} \tag{13.10}$$

and its angular position is given as

$$\tan\theta_y = \frac{-\sum_{i=1}^{n} m_i r_i l_i \sin\theta_i}{-\sum_{i=1}^{n} m_i r_i l_i \cos\theta_i} \tag{13.11}$$

After knowing the values of balancing mass m_y and its angular position θ_y from above equations, we resolve the centrifugal forces in horizontal and vertical directions and equate them to zero.

$$\sum_{i=1}^{n} m_i r_i \cos\theta_i + m_x r_x \cos\theta_x + m_y r_y \cos\theta_y = 0 \tag{13.12}$$

and

$$\sum_{i=1}^{n} m_i r_i \sin\theta_i + m_x r_x \sin\theta_x + m_y r_y \sin\theta_y = 0 \tag{13.13}$$

or the magnitude of unbalanced force:

$$m_x r_x = \sqrt{\left(\sum_{i=1}^{n} m_i r_i \cos\theta_i + m_y r_y \cos\theta_y\right)^2 + \left(\sum_{i=1}^{n} m_i r_i \sin\theta_i + m_y r_y \sin\theta_y\right)^2} \tag{13.14}$$

and the angular position of mass m_x is given as

$$\tan\theta_x = \frac{-\left(\sum_{i=1}^{n} m_i r_i \sin\theta_i + m_y r_y \sin\theta_y\right)}{-\left(\sum_{i=1}^{n} m_i r_i \cos\theta_i + m_y r_y \cos\theta_y\right)} \tag{13.15}$$

The quadrant of the angular position is identified by sign of numerator and denominator of Eq. (13.15).

Balancing

EXAMPLE 13.2 A shaft carries three rotating masses in different planes having the following properties:

$$m_1 = 5 \text{ kg} \quad r_1 = 60 \text{ mm} \quad \theta_1 = 0°$$
$$m_2 = 4 \text{ kg} \quad r_2 = 70 \text{ mm} \quad \theta_2 = 90°$$
$$m_3 = 3 \text{ kg} \quad r_3 = 80 \text{ mm} \quad \theta_3 = 135°$$

The length of the shaft is 700 mm and planes containing masses are 150 mm apart. The balancing planes XX and YY are located 100 mm and 600 mm away from the left hand bearing. Determine the magnitude of balancing masses to be placed at planes XX and YY and their angular position. The radius at which these masses act are 50 mm and 75 mm respectively.

Solution The layout of the system is as shown in Figure 13.7.

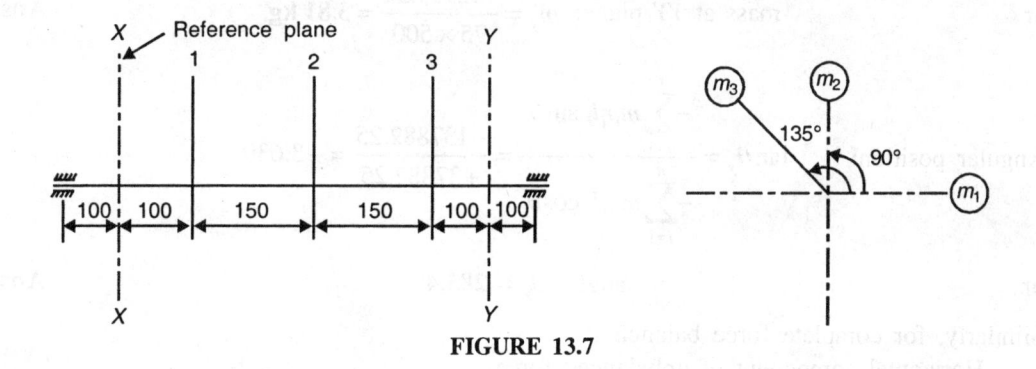

FIGURE 13.7

Let m_x be the balancing mass to be placed at the plane XX and m_y be the balancing mass to be placed at the plane YY. Assume the plane XX as reference plane. For complete balancing of couple,

$$m_y r_y l_y = \left[\left(\sum_{i=1}^{n} m_i r_i l_i \cos\theta_i\right)^2 + \left(\sum_{i=1}^{n} m_i r_i l_i \sin\theta_i\right)^2\right]^{0.5}$$

where n is the number of unbalanced masses and l_1, l_2, l_3, l_y are distances of respective planes from the reference plane.

The horizontal component of unbalanced couple:

$$\sum_{i=1}^{n} m_i r_i l_i \cos\theta_i = 5 \times 60 \times 100 \times \cos 0° + 4 \times 70 \times 250 \times \cos 90°$$
$$+ 3 \times 80 \times 400 \times \cos 135°$$
$$= -37882.25 \text{ kg mm}^2$$

The vertical component of unbalanced couple:

$$\sum_{i=1}^{n} m_i r_i l_i \sin\theta_i = 5 \times 60 \times 100 \times \sin 0° + 4 \times 70 \times 250 \times \sin 90° +$$
$$3 \times 80 \times 400 \times \sin 135°$$
$$= 137882.25 \text{ kg mm}^2$$

Therefore, resultant unbalanced couple

$$= \sqrt{(-37882.25)^2 + (137882.25)^2}$$
$$= 142991.54 \text{ kg mm}^2$$

For balancing, $\quad m_y r_y l_y = 142991.54$

or $\quad$ mass at YY plane: $m_y = \dfrac{142991.54}{75 \times 500} = $ **3.81 kg** $\quad$ **Ans.**

Angular position: $\quad \tan\theta_y = \dfrac{-\sum_{i=1}^{n} m_i r_i l_i \sin\theta_i}{-\sum_{i=1}^{n} m_i r_i l_i \cos\theta_i} = \dfrac{-137882.25}{+37882.25} = -3.639$

or $\quad$ angle: $\theta_y = $ **285.4°** $\quad$ **Ans.**

Similarly, for complete force balancing:
Horizontal component of unbalanced force

$$= \sum_{i=1}^{n} m_i r_i \cos\theta_i + m_y r_y \cos\theta_y$$
$$= 5 \times 60 \times \cos 0° + 4 \times 70 \times \cos 90° + 3 \times 80 \times \cos 135°$$
$$+ 3.81 \times 75 \times \cos 285.4°$$
$$= 206.17 \text{ kg mm}$$

Vertical component of unbalanced force

$$= \sum_{i=1}^{n} m_i r_i \sin\theta_i + m_y r_y \sin\theta_y$$
$$= 5 \times 60 \times \sin 0° + 4 \times 70 \times \sin 90° + 3 \times 80 \times \sin 135°$$
$$+ 3.81 \times 75 \times \sin 285.4°$$
$$= 174.16 \text{ kg mm}$$

Resultant unbalanced force

$$= \sqrt{206.17^2 + 174.16^2} = 269.88 \text{ kg mm}$$

For balancing, $m_x r_x = 269.88$

or $m_x = \dfrac{269.74}{50} = 5.397$ kg   Ans.

Angular position: $\tan \theta_x = \dfrac{-174.16}{-206.17} = 0.8445$

or angle $\theta_x = 220.2°$ Ans.

EXAMPLE 13.3 A shaft carries four masses A, B, C and D which are placed in parallel planes perpendicular to the longitudinal axis. The unbalanced masses at planes B and C are 3.6 kg and 2.5 kg respectively and both are assumed to be concentrated at a radius of 25 mm while the masses in planes A and D are both at radius of 40 mm. The angle between the planes B and C is $100°$ and that between B and A is $190°$, both angles being measured in counterclockwise direction from the plane B. The planes containing A and B are 250 mm apart and those containing B and C are 500 mm. If the shaft is to be completely balanced, determine:

  (i) masses at the planes A and D
  (ii) the distance between the planes C and D
  (iii) the angular position of the mass D

Solution The procedure of graphical method of solution is as follows:

1. Draw the given configuration as shown in Figure 13.8(a). Consider plane A as the reference plane.
2. Prepare the following table. Let x be the distance between planes C and D.

Plane	Mass (kg)	Radius r (mm)	Centrifugal force $\div \omega^2$ mr(kg mm)	Distance from reference plane A l (mm)	Couple $\div \omega^2$ mrl
A	m_a	40	$40\, m_a$	0	0
B	3.6	25	90	250	22500
C	2.5	25	62.5	750	46875
D	m_d	40	$40\, m_d$	$(750 + x)$	$40 m_d (750 + x)$

3. Assume suitable scale to represent couple vector, say 1000 units = 1 mm. Draw the couple polygon obc as shown in the Figure 13.8(b). From this couple polygon, measure vector $\overrightarrow{co}$ which represents the couple produced by mass D.
Couple due to mass at $D = 48 \times 1000 = 48000$ kg mm^2 and angular position measured with reference to mass B is $253°$.

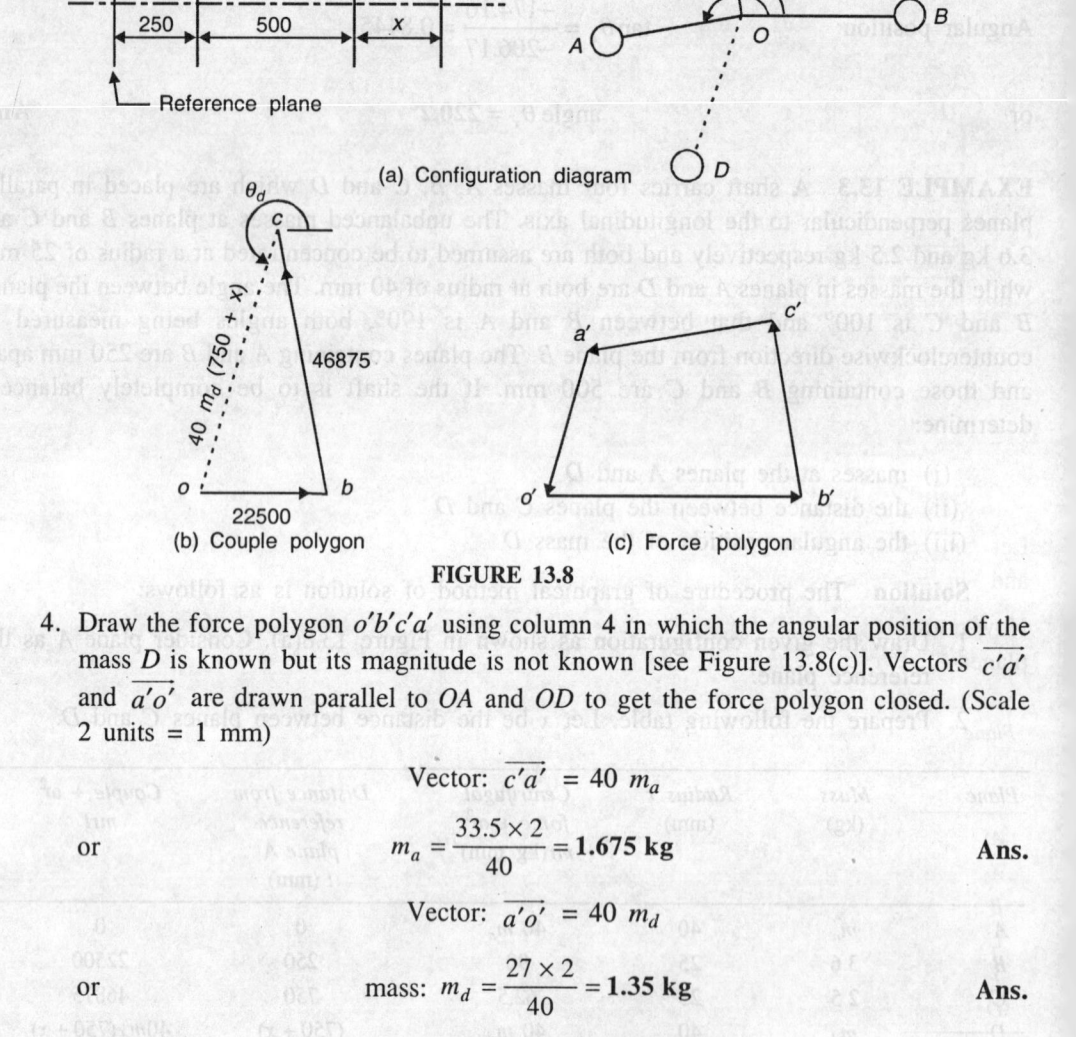

FIGURE 13.8

(a) Configuration diagram
(b) Couple polygon
(c) Force polygon

4. Draw the force polygon $o'b'c'a'$ using column 4 in which the angular position of the mass D is known but its magnitude is not known [see Figure 13.8(c)]. Vectors $\overrightarrow{c'a'}$ and $\overrightarrow{a'o'}$ are drawn parallel to OA and OD to get the force polygon closed. (Scale 2 units = 1 mm)

$$\text{Vector: } \overrightarrow{c'a'} = 40\, m_a$$

or
$$m_a = \frac{33.5 \times 2}{40} = 1.675 \text{ kg} \qquad \text{Ans.}$$

$$\text{Vector: } \overrightarrow{a'o'} = 40\, m_d$$

or
$$\text{mass: } m_d = \frac{27 \times 2}{40} = 1.35 \text{ kg} \qquad \text{Ans.}$$

Vector $\overrightarrow{co}$ in couple polygon equals to $40 m_d(750 + x)$

or $40 m_d(750 + x) = 48000$

or $40 \times 1.35(750 + x) = 48000$

or distance: $x = 138.89$ mm Ans.

EXAMPLE 13.4 Determine the masses to be added at planes X and Y at radius 600 mm if the system shown in the Figure 13.9 is to be dynamically balanced. The unbalanced masses and eccentricities are given in the table:

Plane	Mass (kg)	Eccentricities (m)
A	400	0.22
B	600	0.18
C	480	0.25
D	520	0.3

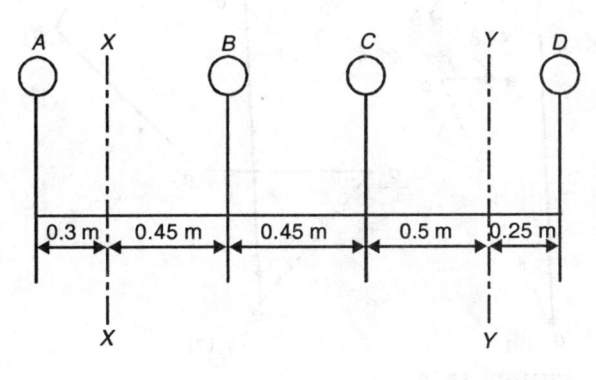

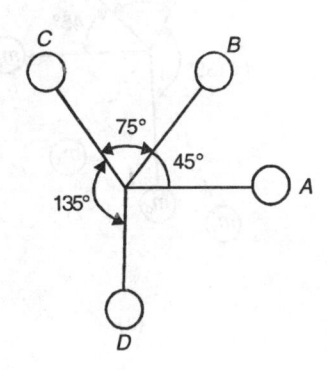

FIGURE 13.9

Solution Let us draw a configuration diagram [Figure 13.10(a)].

Let m_x = balancing mass to be placed at the plane X
and m_y = balancing mass to be placed at the plane Y

Now construct the following table to compute forces and couples with plane X as reference plane:

Plane	Mass m (kg)	Radius r (m)	Force ÷ ω^2 mr (kg mm)	Distance from reference plane (l) (mm)	Couple ÷ ω^2 mrl kg mm²
A	400	0.22	88	–0.3	–26.4
X	m_x	0.6	0.6 m_x	0	0
B	600	0.18	108	0.45	48.6
C	480	0.25	120	0.90	108
Y	m_y	0.6	0.6 m_y	1.4	0.84 m_y
D	520	0.3	156	1.65	257.4

Draw a couple polygon as shown in Figure 13.10(b) with scale 25 units = 10 mm. Vector $\overrightarrow{oa}$ is drawn in opposite direction as its value carries negative sign. Vector $\overrightarrow{od}$ represents couple of mass m_y, i.e.

$$\overrightarrow{od} = \frac{56}{10} \times 25 = 0.84 \, m_y$$

or mass at plane Y: m_y = **166.67 kg at 249° from the plane A** Ans.

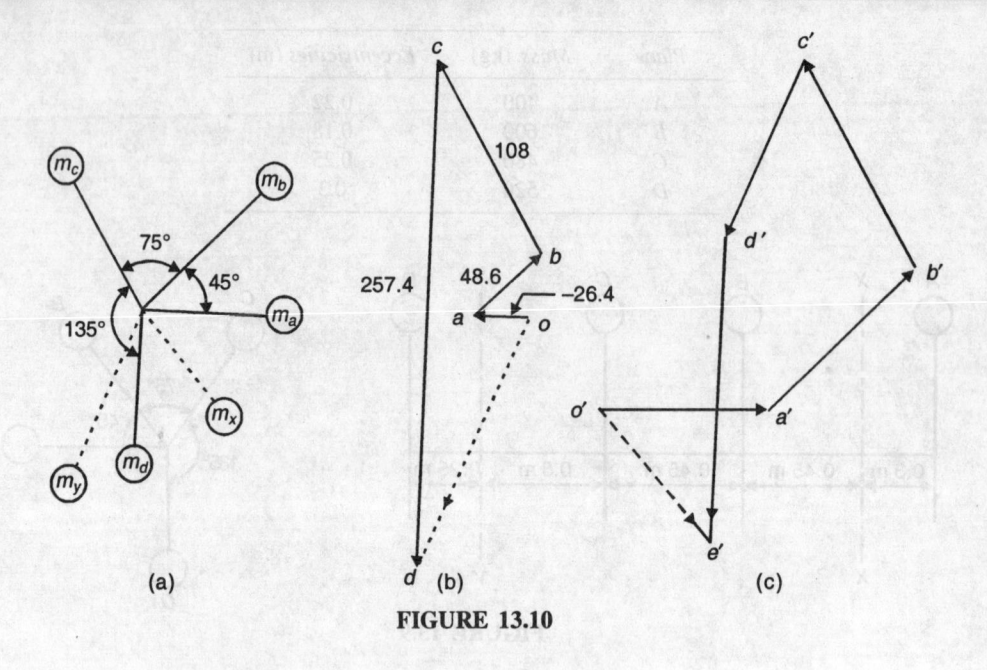

FIGURE 13.10

Draw a force polygon $o'a'b'c'd'e'o'$ [(Figure 13.10(c))]. The closing vector $\overrightarrow{o'e'}$ is equal to the balancing force $0.6\, m_x$, i.e.

$$0.6\, m_x = \frac{38}{10} \times 25$$

or   mass at plane X: m_x = **158.33 kg at 313° from the plane A**   **Ans.**

EXAMPLE 13.5 A shaft rotating at a uniform speed carries two discs A and B of mass 5 kg and 4 kg respectively. The centre of gravity of each disc is 2.7 mm away from the axis of rotation and the angle between them is 90°. The shaft has bearings C and D, between A and B, such that AC = 250 mm, AD = 500 mm and AB = 1000 mm. It is desired to make the dynamic forces on the bearings equal and opposite and to have a minimum value for a given speed by means of a mass in plane E at a radius of 20 mm.

Determine:

  (i) the magnitude of the mass to be attached at E and its angular position with respect to that of A

  (ii) the distance of plane E from the plane A

Solution The configuration of the rotating shaft is shown in Figure 13.11(a). Let the balancing mass in the plane E be m_e and the plane E is situated at a distance x from bearing C. Considering the plane C as reference plane, we make a table as under:

Plane	Mass m (kg)	Radius r (m)	Force ÷ ω^2 mr (kg mm)	Distance from reference plane l (mm)	Couple ÷ ω^2 mrl kg mm^2
A	5	2.7	13.5	−250	−3375
B	4	2.7	10.8	750	8100
C	−	−	−	0	0
D	−	−	−	250	$F_d \times 250/\omega^2 g$
E	m_e	20	20 m_e	x	20 $m_e x$

where F_d is force acting on the bearing D.

Since the dynamic forces of reactions on bearings C and D are to be equal and opposite in force polygon, they mutually cancel out. This requires that the remaining forces in column 4 of the above table should be in equilibrium. Draw a force polygon with scale 2 kg mm = 10 mm as shown in Figure 13.11(b).

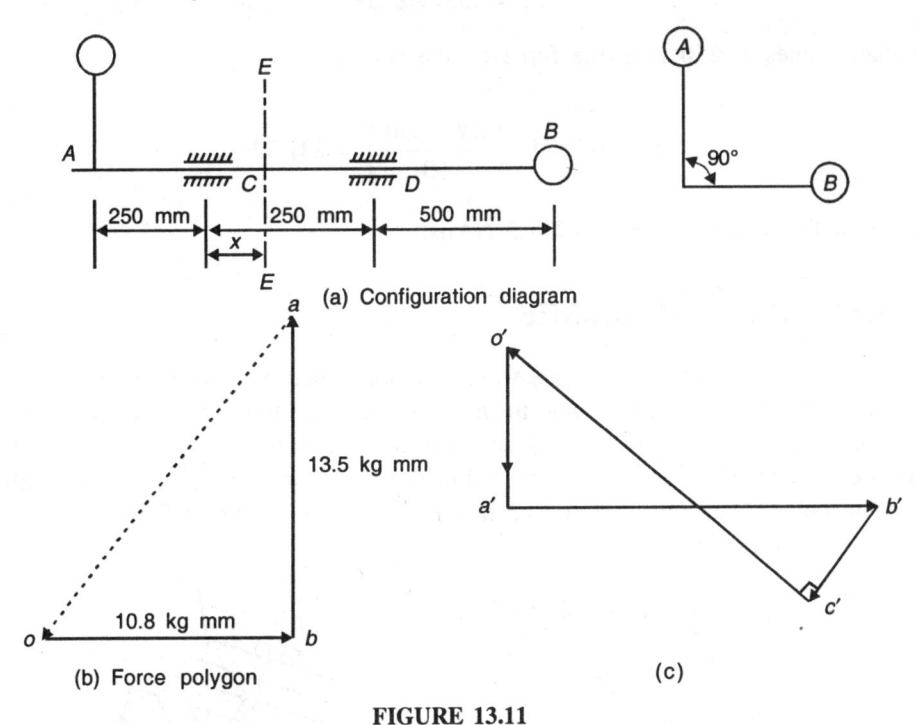

FIGURE 13.11

From the force polygon,

Vector: $\vec{ao}$ = 8.7 × 2 = 17.4 kg mm

or 20 m_e = 17.4

or   mass at plane E: m_e = **0.87 kg at 231° in counterclockwise from plane B**   Ans.

For force F_d to be minimum, the couple $F_d \times 250/g\omega^2$ should be minimum.

Draw a couple polygon with scale 1000 units = 10 mm and $\overrightarrow{o'a'}$ = 3375 kg mm², $\overrightarrow{a'b'}$ = 8100 kg mm as shown in Figure 13.11(c). From point b', draw a line parallel to the direction of mass m_e and from point o', we drop a perpendicular to $\overrightarrow{b'c'}$.

Therefore, $\overrightarrow{o'c'}$ is the minimum value of couple $F_d \times 250/\omega^2 g$

and vector: $\overrightarrow{bc}$ = 2.1 × 1000 = 20 $m_e x$

or $\qquad$ distance: $x = \dfrac{2100}{20 \times 0.87} = 120.7$ mm

or $\qquad$ distance from the plane A = 250 + 120.7 = **370.7 mm** $\qquad$ Ans.

Now $\qquad \dfrac{F_d \times 250}{g \times \omega^2 \times 1000} = \text{Vector } \overrightarrow{c'o'} = \dfrac{8600}{1000 \times 1000}$ kg m²

or $\qquad F_d = 0.3374\, \omega^2$ N

If the shaft rotates at 250 rpm, the force on the bearing D,

$$F_d = 0.3374 \times \left(\dfrac{2\pi \times 250}{60}\right)^2 = 231.3 \text{ N}$$

and force on the bearing C: F_c = –231.3 N**Ans.**

13.5 BALANCING MACHINES

A balancing machine is used to indicate whether a component is in balance or not. If it is out of balance, then machine must be able to measure the magnitude and location of unbalance. Mechanical components whose axial dimensions are small such as gears, pulley, fans and impeller require static balancing. Such balancing is often done through a single plane balancing machine. One typical static balancing machine is shown in Figure 13.12. When a

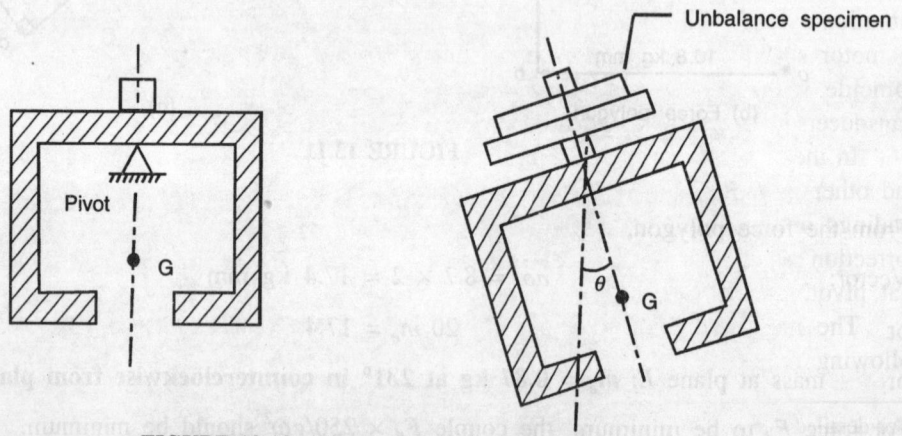

FIGURE 13.12 Single plane balancing machine.

unbalanced component is mounted on the plateform, the pendulum tilts. The direction of tilt gives the location of unbalance and the angle θ indicates the magnitude of unbalance.

In the case of axially longer components, such as turbine, rotor, armature, etc., the unbalanced centrifugal forces result in couples whose effect is to cause the rotor turn over the end. Thus the purpose of balancing is to measure the unbalanced couple and to add a new couple of same magnitude in the opposite direction. Therefore, balancing of these components require both static and dynamic balancing. The most common types of balancing machines are discussed below:

13.5.1 Pivoted-Cradle Balancing Machine

In a pivoted-creadle balancing machine, the rotor is mounted on half-bearings attached to the cradle and is connected to the electric motor as shown in Figure 13.13.

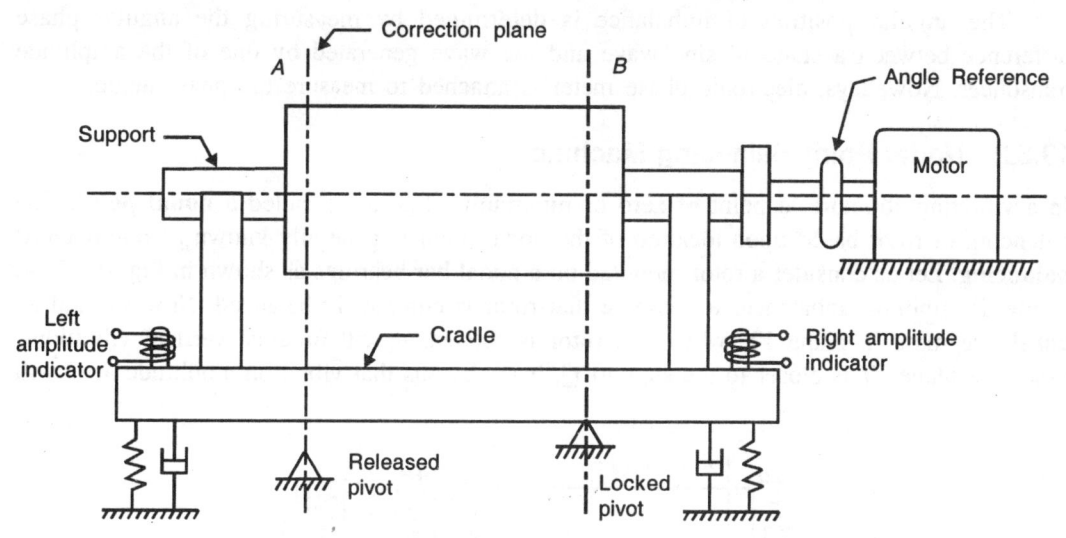

FIGURE 13.13 Pivoted-cradle balancing machine.

The cradle is mounted on spring–dashpots to provide a single degree of freedom vibration system. Often they are made adjustable so that the natural frequency can be tuned to motor speed. Further, the cradle is pivoted about two points which can be adjusted to coincide with the plane of correction. The amplitude of vibration is measured through transducers mounted at each end.

In the test run, the pivots are positioned at the plane of correction. One pivot is locked and other is kept free. When the rotor rotates, its amplitude of vibration is measured. The readings obtained will be completely independent of the measurement taken at the other correction plane because an unbalance in the plane of locked pivot will have no moment about that pivot.

The measured amplitude of vibration and amount of unbalance are related by the following relation*.

---

*For detailed derivation, refer to Chapter 14 on Mechanical Vibrations.

$$X = \frac{m_0 e \left(\dfrac{\omega}{\omega_n}\right)^2}{m\sqrt{\left[1-\left(\dfrac{\omega}{\omega_n}\right)^2\right]^2 + \left(2\xi\left(\dfrac{\omega}{\omega_n}\right)\right)^2}}$$

or
$$X = \frac{m_0 e r^2}{m\sqrt{(1-r^2)^2 + (2\xi r)^2}} \tag{13.16}$$

where $m_0 e$ = magnitude of unbalance
and r = ratio of forcing frequency to the natural frequency (ω/ω_n)

The angular position of unbalance is determined by measuring the angular phase difference between a standard sine wave and the wave generated by one of the amplitude transducer. Nowadays, electronic phase meter is attached to measure the phase angle.

13.5.2 Nodal Point Balancing Machine

In a vibrating structure, a point of zero or minimum vibration is called a **nodal point.** The balancing of rotor based upon location of the nodal point is generally known as **nodal point balancing.** Let us consider a rotor mounted on a nodal bar bearings as shown in Figure 13.14. While determining unbalance, we assume that rotor is completely balanced. Now we add an unbalance mass in plane *YY*, when this rotor is rotated, it will produce vertical vibrations. Since the plane *YY* is closer to the support *Q*, it is obvious that vibration amplitude at *Q* will

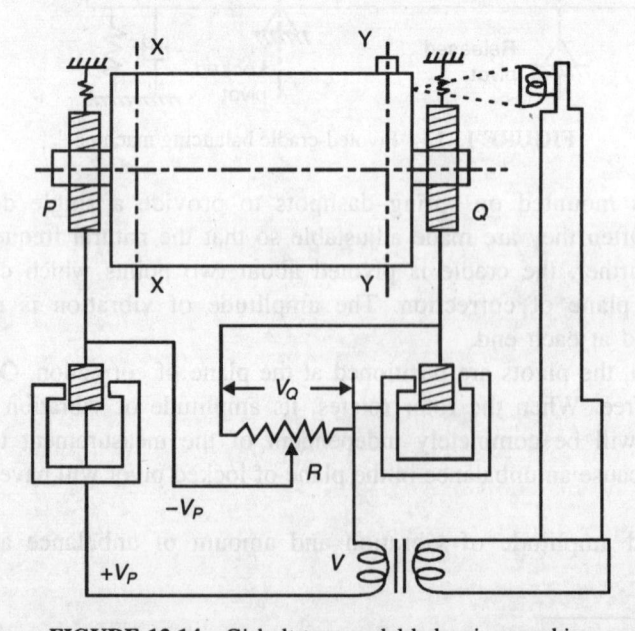

FIGURE 13.14 Gisholt type nodal balancing machine.

be more than that at the support P. However, the ratio of amplitude of vibration is independent of the magnitude of unbalance.

Let V_P and V_Q are voltages, which corresponds to vibration amplitude at bearings P and Q respectively. The voltage divider at R is set so that a portion of voltage V_Q, equal to V_P, is applied opposite to the voltage V_P, the resultant voltage V thus being reduced to zero. As a result, the electric circuit becomes insensitive to any unbalance in the plane YY.

In actual test run of an unbalanced rotor, the voltage V corresponds to the magnitude of unbalance in the plane XX. The angular position of unbalance is measured by applying the resultant voltage V to a stroboscope which flashes for microseconds when amplitude is maximum, i.e. in the direction of unbalance. The amount and direction of unbalance in plane YY can be determined in the similar way by making circuit insensitive to any unbalance present in plane XX.

13.5.3 Field Balancing

The rotors of turbines and generators are very large in size and they can not be put on the balancing machine. Therefore, balancing of such rotors has to be done on their bearings. This type of balancing is known as **field balancing.** In this method of balancing, a vibration pickup is used to measure the amplitude of vibration. Further, in order to measure the phase angle a sine wave generator is connected to the rotor and its output is fed to dual channel oscilloscope, as shown in Figure 13.15. This sine wave acts as a reference for measuring the phase angle.

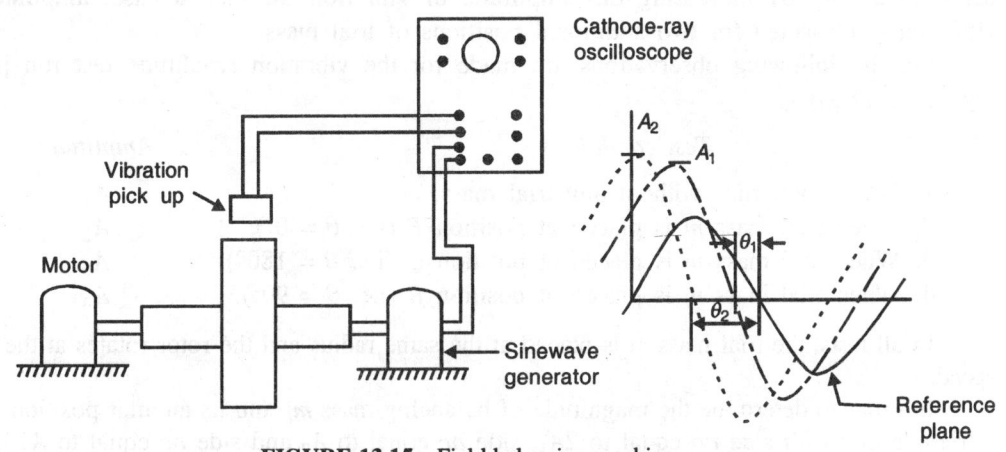

FIGURE 13.15 Field balancing machine.

In actual balancing, the rotor is rotated at a constant speed sufficient to produce a measurable amplitude of vibration. Let A_1 be the amplitude and θ_1 be its phase difference with reference to the generated sine wave. Graphically, this amplitude is represented by a vector $\overrightarrow{oa}$ as shown in Figure 13.16(a). Next, a trial mass m is attached to the rotor at a known angular position and at a known radius. The rotor is again rotated at the speed of first run. Let the amplitude of vibration at this time be A_2 and the phase difference be angle θ_2. This is represented by vector $\overrightarrow{ob}$. The vector difference $\overrightarrow{ab}$ represents the effect of trial mass. Thus

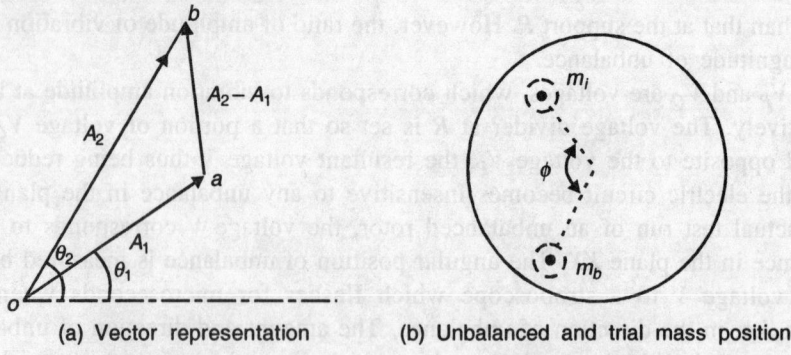

(a) Vector representation (b) Unbalanced and trial mass position

FIGURE 13.16 Vector representation of unbalanced mass.

to balance the rotor a balancing mass m_b can be placed at the same radius of trial mass m such that the balancing mass:

$$m_b = m \times \frac{oa}{ob} \qquad (13.17)$$

and this balancing mass should be located at an angle ϕ in counterclockwise direction from the position of trial mass as shown in Figure 13.16(b).

In the field, sometimes it may be difficult to couple sine wave generator with the rotor assembly. Under such circumstances, the magnitude and location of balancing mass can be determined only by measuring the amplitude of vibration. In such a case, amplitude of vibration is measured for two additional positions of trial mass.

Let the following observations are made for the vibration amplitude test run [Refer Figure 13.17(a)]:

Test conditions	Amplitude
1. When rotor runs without any trial mass.	A_1
2. When trial mass m is placed at position P (i.e. $\theta = 0°$).	A_2
3. When trial mass m is placed at position Q (i.e. $\theta = 180°$).	A_3
4. When trial mass m is placed at position R (i.e. $\theta = 90°$).	A_4

In all tests, the trial mass m is placed at the same radius and the rotor rotates at the same speed.

In order to determine the magnitude of balancing mass m_b and its angular position, draw a triangle abc with side ab equal to $2A_1$, side ac equal to A_2 and side bc equal to A_3 [Refer Figure 13.17(b)]. Let d be the mid–point of line ab. In this Figure, the vector $\overrightarrow{dc}$ represents the effect of only the trial mass at the position P. Hence, the magnitude of the required balancing mass to be placed at the same radius is obtained as*

$$m_b = m \times \frac{ad}{dc} \qquad (13.18)$$

---

*For detailed proof, refer *Theory of Mechanism and Machines* by Ghosh and Mallik, Affiliated East-West Press, New Delhi, 1990.

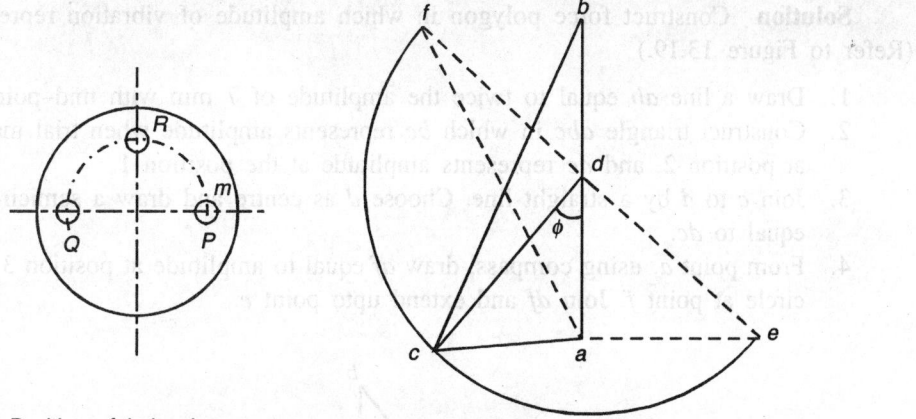

(a) Position of balancing masses (b) Vector diagram of vibration amplitude

FIGURE 13.17 Field balancing with three mass locations.

The angular position of balancing mass, angle ϕ, is measured from the reference plane. The angle should be measured in counterclockwise direction if the amplitude A_4 is represented by vector $\overrightarrow{ae}$ or it should be in clockwise direction when it is represented by $\overrightarrow{af}$. The points e and f can be obtained by rotating vector $\overrightarrow{dc}$ about point d through 90° in either direction.

EXAMPLE 13.6 During the field balancing of a blower fan, the following measurements are taken:

  (i) When fan is run at 2000 rpm, the amplitude of vibration due to original imbalance is 7 mm.
  (ii) When a trial mass of 200 g is fastened at position 1, shown in Figure 13.18 the vibration amplitude is 10.6 mm.
  (iii) When the same trial mass is fastened at position 2, the vibration amplitude is 5 mm and
  (iv) When the same trial mass is fastened at position 3, the vibration amplitude is 10.8 mm.

Determine the magnitude of balancing mass and its angular position with respect to position 1.

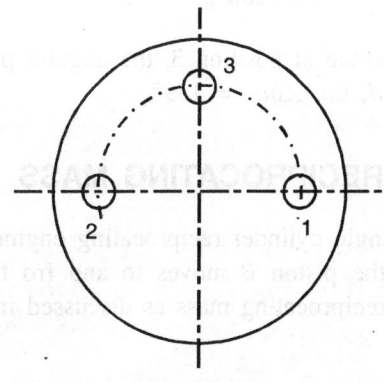

FIGURE 13.18

Solution Construct force polygon in which amplitude of vibration represents forces. (Refer to Figure 13.19.)

1. Draw a line *ab* equal to twice the amplitude of 7 mm with mid–point *d*.
2. Construct triangle *abc* in which *bc* represents amplitude when trial mass is placed at position 2, and *ac* represents amplitude at the position 1.
3. Join *c* to *d* by a straight line. Choose *d* as centre and draw a semicircle of radius equal to *dc*.
4. From point *a*, using compass, draw *af* equal to amplitude at position 3 to cut semicircle at point *f*. Join *df* and extend upto point *e*.

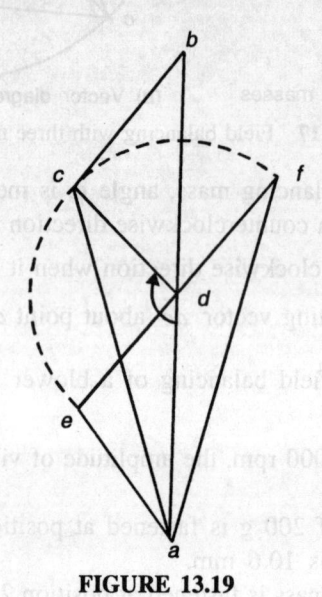

FIGURE 13.19

Magnitude of balancing mass: $m_b = m \times \dfrac{ad}{dc} = 200 \times \dfrac{7}{7.4}$ **Ans.**

$= 318.2$ g

Since *af* is equal to amplitude at position 3, the angular position of balancing mass is measured in clockwise from *ad*, i.e. $\angle adc = 139°$. **Ans.**

13.6 BALANCING OF RECIPROCATING MASS

Let us consider a horizontal single cylinder reciprocating engine as shown in Figure 13.20. The reciprocating mass *m* at the piston *B* moves to and fro from its mean position. The expression for acceleration of reciprocating mass as discussed in Chapter 12, can be written as

$$f = \omega^2 r \left(\cos\theta + \frac{\cos 2\theta}{n} \right) \qquad (13.19)$$

where ω = angular velocity of the crank
  r = radius of the crank
  θ = angle turned by the crank
and n = ratio of length of connecting rod to the radius of the crank

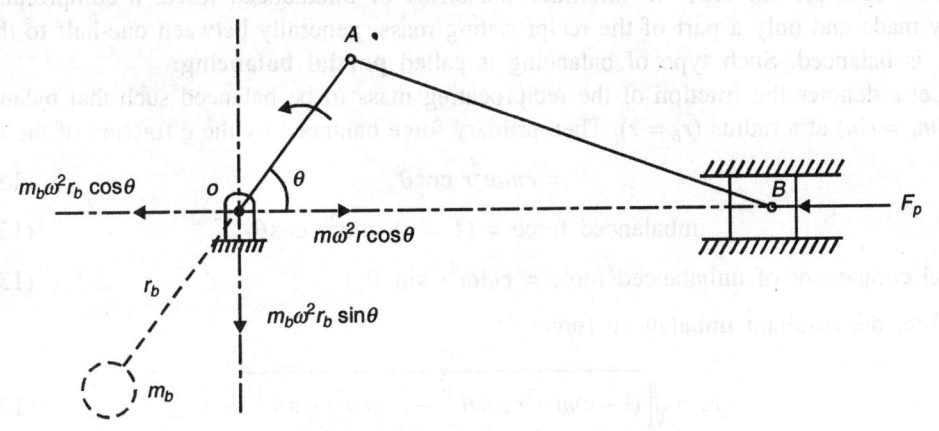

FIGURE 13.20 Balancing of reciprocating mass.

Therefore, the force required to accelerate the reciprocating mass m is:

$$F = m\omega^2 r \left(\cos\theta + \frac{\cos 2\theta}{n} \right)$$

or
$$F = m\omega^2 r \cos\theta + m\omega^2 r \times \frac{\cos 2\theta}{n} \qquad (13.20)$$

This accelerating force is composed of the following two components:

(i) Primary acceleration force
$$F_p = m\omega^2 r \cos\theta \qquad (13.21)$$

(ii) Secondary acceleration force
$$F_s = m\omega^2 r \frac{\cos 2\theta}{n} \qquad (13.22)$$

In slow speed engines with large value of ratio n, the effect of secondary accelerating force is much smaller than primary force; hence it can be neglected.

In an engine mechanism when piston moves towards inner dead centre, i.e. towards the BO direction, an accelerating force F_p is acted upon. The reaction of this force tends to move the frame in the direction opposite to that of piston. To prevent this, an attempt has to be made to balance this reaction by placing a balancing mass m_b at radius r_b directly opposite to the crank as shown in Figure 13.20. This balancing mass m_b produces centrifugal force $m_b\omega^2 r$, having two components:

(i) the horizontal component of balancing force $m_b\omega^2 r_b \cos\theta$ which neutralises the unbalance force due to reciprocating mass ($m\omega^2 r \cos\theta$)
(ii) the vertical component of balancing force $m_b\omega^2 r_b \sin\theta$ acts perpendicular to the line of stroke which remains unbalanced.

Thus in this type of balancing, the unbalance force is shifted at right angle to the original direction of unbalance and the mechanism tends to jump up and down instead of slide to and fro on its structure. In order to minimize the effect of unbalanced force, a compromise is usually made and only a part of the reciprocating mass, generally between one-half to three-fourth, is balanced. Such type of balancing is called **partial balancing**.

Let c denotes the fraction of the reciprocating mass to be balanced such that balancing mass ($m_b = cm$) at a radius ($r_b = r$). Then primary force balanced by the c fraction of the mass

$$= cm\omega^2 r \cos\theta \qquad (13.23)$$

and
$$\text{unbalanced force} = (1-c)m\omega^2 r \cos\theta \qquad (13.24)$$

Vertical component of unbalanced force $= cm\omega^2 r \sin\theta \qquad (13.25)$

Therefore, the resultant unbalanced force:

$$F_r = \sqrt{\left[(1-c)m\omega^2 r \cos\theta\right]^2 + \left[cm\omega^2 r \sin\theta\right]^2} \qquad (13.26)$$

The resultant unbalanced force F_r is minimum when fraction c is 0.5.

The engines operating at higher speed, the second term of accelerating force, called **secondary force**, cannot be neglected. Thus secondary force:

$$F_s = m\omega^2 r \frac{\cos 2\theta}{n}$$

or
$$F_s = m(2\omega)^2 \frac{r}{4n}\cos 2\theta \qquad (13.27)$$

If we compare the expressions of secondary force [Eq. (13.27)] and primary force [Eq. (13.21)], we can say that the effect of secondary force is equivalent to an imaginary crank of length $r/4n$ rotating at twice the speed of engine and it makes an angle two times the angle of original crank. This imaginary crank coincides with the original crank at the inner dead centre, i.e. at an angle $\theta = 0°$.

Thus the secondary unbalanced force can also be balanced just similar to the primary force balancing. For complete balancing of the reciprocating masses of multicylinder engine, the following conditions must be satisfied:

(i) Primary and secondary forces must be balanced individually and
(ii) Primary and secondary couples must be balanced individually.

EXAMPLE 13.7 A single cylinder reciprocating engine has speed 240 rpm, stroke 300 mm, mass of reciprocating parts 5 kg and mass of revolving parts at 150 mm radius is 3.7 kg. If two-third of the reciprocating parts and all the revolving parts are to be balanced, find:

(i) the balancing mass required at a radius of 400 mm
(ii) the residual unbalanced force when the crank has rotated 60° from top dead centre.

Solution Let r_b be the radius at which the balancing mass m_b is placed and m_r be the mass of revolving parts.

We know that
$$m_b \times r_b = (m_r + cm) \times r$$

or
$$m_b \times 400 = \left(3.7 + \frac{2}{3} \times 5\right) \times 150$$

or balancing mass: $m_b = \mathbf{2.638\ kg}$ **Ans.**

Residual unbalanced force:
$$F_r = m\omega^2 r \left[(1-c)^2 \cos^2\theta + c^2 \sin^2\theta\right]^{0.5}$$

where
$$\omega = \frac{2\pi N}{60} = \frac{2\pi \times 240}{60} = 25.13 \text{ rad/s}$$

∴
$$F_r = 5 \times (25.13)^2 \times 0.15 \left[(1-0.667)^2 \times \cos^2 60° + 0.667^2 \times \sin^2 60°\right]^{0.5}$$

$$= \mathbf{284.7\ N} \qquad \textbf{Ans.}$$

13.7 BALANCING OF LOCOMOTIVE

Locomotive engines, depending upon the position of cylinders, are classified in two types inside cylinder engine, and outside cylinder engine. They are further classified as coupled and uncoupled engines. If two or more pairs of wheels are coupled together, it is called coupled locomotive otherwise it is an uncoupled locomotive. While balancing an uncoupled locomotive, one has to consider four planes—two for the cylinders and other two for driving wheels where balancing masses are to be positioned. Whereas in coupled locomotive two additional planes for coupling rods are required to be considered.

Locomotive engine operates at low speed and the ratio of length of connecting rod to radius of crank is generally large enough to neglect the effect of secondary force.

Consider a two cylinder locomotive having their cranks at 90° to each other as shown in Figure 13.21. The distance between cylinder centre lines is l. Both the cylinders have crank radius r and also the mass of reciprocating parts of both cylinders is m. The partial balancing of reciprocating masses of the said locomotive, similar to internal combustion engine mechanism, results into some unbalanced force along the line of stroke and some unbalanced force perpendicular to the line of stroke. The effect of these unbalanced forces is to produce three effects, namely

(i) the variation of tractive force
(ii) the swaying couple and
(iii) the hammer blow

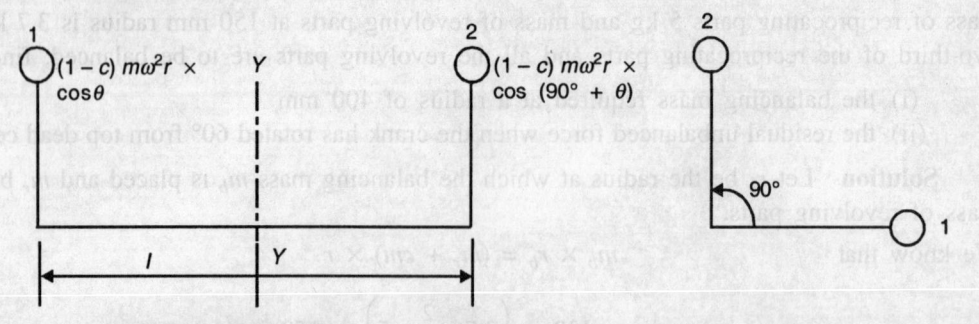

FIGURE 13.21 Effect of partial balancing of two cylinder locomotive.

1. Variation of tractive force. The unbalanced portion of the primary force which acts along the line of stroke causes the variation in the tractive force or effort. Let c is the fraction of reciprocating mass that is balanced. The unbalanced primary forces along the line of stroke

$$= (1-c)m\omega^2 r \cos\theta + (1-c)m\omega^2 r \cos(90°+\theta)$$
$$= (1-c)m\omega^2 r (\cos\theta - \sin\theta) \tag{13.28}$$

This unbalanced force is maximum when

$$\frac{\partial}{\partial \theta}\left[(1-c)m\omega^2 r (\cos\theta - \sin\theta)\right] = 0$$

or $\theta = 135°$ and $315°$

Therefore, maximum variation of tractive force is:

$$\pm\sqrt{2}(1-c)m\omega^2 r \tag{13.29}$$

2. Swaying couple. The unbalanced portions of forces acting along the line of stroke are distance l apart. These forces constitutes a couple about YY axis, which tends to make the leading wheels sway from side to side. This couple is known as **swaying couple**.

$$\text{Swaying couple} = (1-c)m\omega^2 r \cos\theta \times \frac{l}{2} - (1-c)m\omega^2 r \cos(90°+\theta) \times \frac{l}{2}$$

or $$\text{swaying couple} = (1-c)\frac{m\omega^2 rl}{2}(\cos\theta + \sin\theta) \tag{13.30}$$

For maximum magnitude of couple,

$$\frac{\partial}{\partial \theta}\left[(1-c)m\omega^2 r \frac{l}{2}(\cos\theta + \sin\theta)\right] = 0$$

or $\theta = 45°$ or $225°$

Thus $$\text{maximum swaying couple} = \pm \frac{1}{\sqrt{2}}(1-c)m\omega^2 rl \tag{13.31}$$

Balancing

3. Hammer blow. The maximum magnitude of unbalanced force acting perpendicular to the line of stroke is called **hammer blow**. At high speed, this unbalanced force tries to lift the wheels from the rail.

Let W be the dead weight of the wheel. The maximum value of unbalanced force is equal to $m\omega^2 r$ (when $\theta = 90°$ or $270°$). Thus the threshold condition for lifting of wheel can be written as

$$m\omega^2 r = W$$

or limiting angular velocity of the crank:

$$\omega = \sqrt{\frac{W}{mr}} \text{ rad/s} \qquad (13.32)$$

EXAMPLE 13.8 The following data refer to a two cylinder locomotive with cranks at 90°.

(i) mass of reciprocating parts/cylinder = 300 kg
(ii) crank radius = 300 mm
(iii) diameter of driving wheel = 1800 mm
(iv) distance between cylinder axis = 600 mm
(v) distance between the driving wheel = 1600 mm

Determine:

(a) The fraction of the reciprocating masses to be balanced if the hammer blow is not to exceed 45 kN at 95 kmph
(b) variation in tractive effort
(c) maximum swaying couple

Solution Let us construct a configuration diagram of a plane of locomotive in which the planes 1 and 4 represent central planes of driving wheel and the planes 2 and 3 represent planes of cylinder centre line [Figure 13.22(a)]. Assume plane 1 as reference plane. Construct the following table with fraction of reciprocating mass c.

Plane	Mass m (kg)	Crank radius r (mm)	Primary force ÷ ω^2 mr	Distance from plane 1 (l)	Couple ÷ ω^2 (mrl)
1	m_1	r_1	$m_1 r_1$	0	0
2	300 c	0.3	90 c	0.5	45 c
3	300 c	0.3	90 c	1.1	99 c
4	m_4	r_4	$m_4 r_4$	1.6	1.6 $m_4 r_4$

Let us assume $m_1 = m_4$ and $r_1 = r_4$.

Construct a couple polygon as shown in Figure 13.22(b). From the couple diagram, the closing vector $\overrightarrow{bo} = 5.4 \times 20 = 108 c$

or

$$1.6\, m_4 r_4 = 108\, c$$

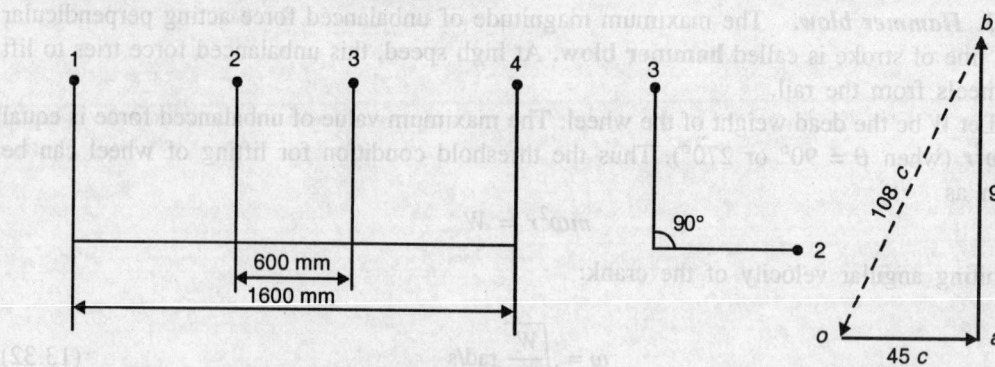

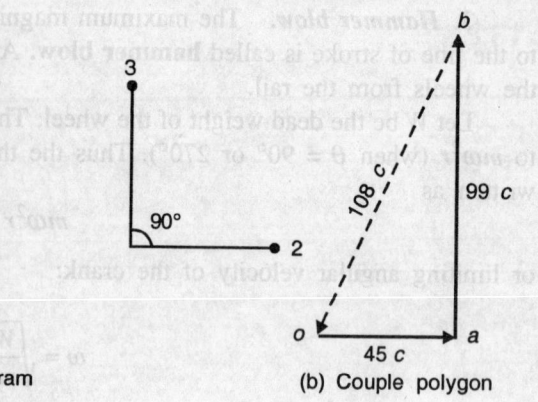

(a) Configuration diagram
(b) Couple polygon

FIGURE 13.22

or mass: $m_4 = \dfrac{108\,c}{1.6 \times r_4} = \dfrac{67.5c}{r_4}$

(a) Maximum hammer blow

Linear speed: $v = \dfrac{95 \times 10^3}{3600} = 26.389$ m/s

Angular velocity: $\omega = \dfrac{v}{\text{Radius of wheels}} = \dfrac{26.389}{1.8/2}$

$= 29.32$ rad/s

Hammer blow: $m_4 \omega^2 r_4 = 45 \times 10^3$

or $\dfrac{67.5 \times c}{r_4} \times 29.32^2 \times r_4 = 45 \times 10^3$

or Fraction: $c = 0.775$ **Ans.**

(b) Maximum variation in tractive force

$= \pm \sqrt{2}\,(1-c)\,m\omega^2 r$

$= \pm \sqrt{2}\,(1 - 0.775) \times 300 \times 29.32^2 \times 0.3$

$= 24618.86$ N **Ans.**

(c) Maximum swaying couple

$c = \pm \dfrac{(1-c)}{\sqrt{2}} m\omega^2 r \times l$

$= \pm \dfrac{(1 - 0.775)}{\sqrt{2}} \times 300 \times 29.32^2 \times 0.3 \times 0.6$

$= 7385.6$ Nm **Ans.**

Balancing

13.8 BALANCING OF IN-LINE ENGINE

The in-line engine is a multicylinder engine in which the line of stroke of all reciprocating parts are placed parallel to each other. The connecting rods of all cylinder–piston assembly are connected to a common crank shaft. The layout and angular positions of cranks generally depend upon number of cylinders. Figure 13.23 shows the crank position layout of a four stroke, four cylinders in-line, vertical engine. The reciprocating masses for all the cylinders

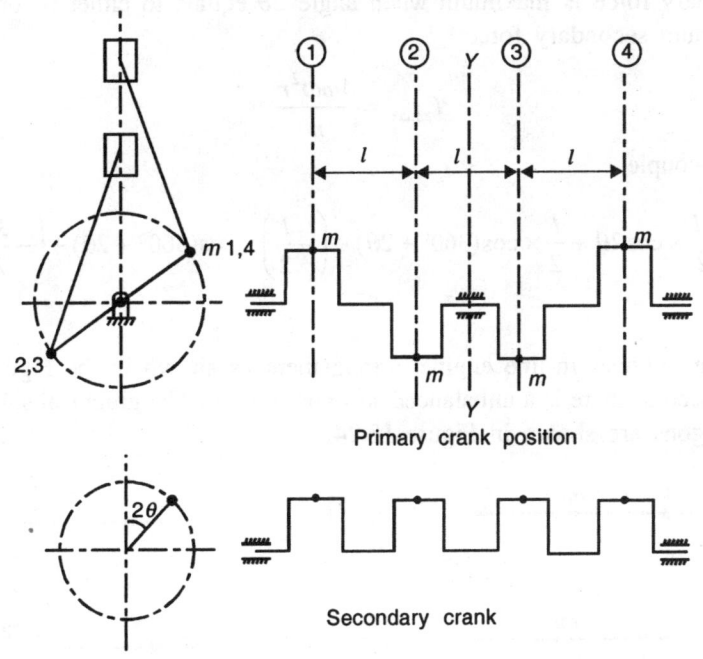

FIGURE 13.23 Balancing of four cylinders in-line engine.

are equal. In this layout, the angular positions of cranks are θ, $180° + \theta$, $180° + \theta$ and θ respectively. For complete balancing of such engine, primary and secondary force polygons as well as primary and secondary couple polygons should be closed. For finding unbalance couple, any plane can be taken as a reference plane, say for the layout shown in Figure 13.23, the middle bearing is chosen as reference plane, as the whole arrangement is symmetrical about the reference plane. Therefore,

Primary force

$$= m\omega^2 r [\cos\theta + \cos(180° + \theta) + \cos(180° + \theta) + \cos\theta]$$
$$= 0$$

Primary couple

$$= m\omega^2 r \left[\frac{3l}{2}\cos\theta + \frac{l}{2}\cos(180° + \theta) + \left(-\frac{l}{2}\right)\cos(180° + \theta) + \left(-\frac{3l}{2}\right)\cos\theta\right]$$
$$= 0$$

Secondary force

$$= \frac{m\omega^2 r}{n}\left[\cos 2\theta + \cos(360° + 2\theta) + \cos(360° + 2\theta) + \cos 2\theta\right]$$

$$= \frac{4m\omega^2 r}{n}\cos 2\theta \qquad (13.33)$$

The secondary force is maximum when angle 2θ equals to either 0° or 180° or 360°, therefore, maximum secondary force:

$$F_{s\max} = \frac{4m\omega^2 r}{n} \qquad (13.34)$$

Secondary couple

$$= \frac{m\omega^2 r}{n}\left[\frac{3l}{2}\times\cos 2\theta + \frac{l}{2}\times\cos(360°+2\theta) + \left(-\frac{l}{2}\right)\times\cos(360°+2\theta) + \left(-\frac{3l}{2}\right)\times\cos 2\theta\right]$$

$$= 0$$

Thus a four cylinder in-line engine, arrangement as shown in the Figure 13.23 is not completely balanced as there is a unbalanced secondary force. The graphical solution, i.e. force and couple polygons are shown in Figure 13.24.

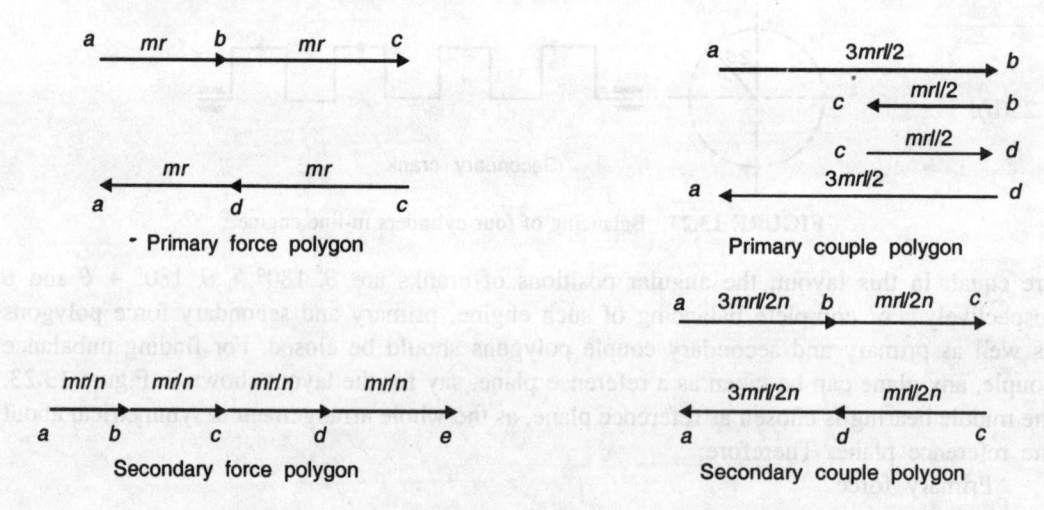

FIGURE 13.24 Graphical method for balancing of a four cylinder in-line engine.

EXAMPLE 13.9 A four cylinder in-line engine has two outer cranks set at 120° to each other and their reciprocating masses are 400 kg. The distance between the planes of rotation of adjacent cranks are 450 mm, 750 mm and 600 respectively. If the engine is to be in complete primary balance, find the magnitude of reciprocating masses and their relative angular positions for each of the inner cranks.

Balancing 573

If the length of each crank is 300 mm, the length of each connecting rod is 1200 mm and the speed of rotation is 240 rpm what is the secondary unbalanced force?

Solution: Angular velocity:
$$\omega = \frac{2\pi N}{60} = \frac{2\pi \times 240}{60} = 25.13 \text{ rad/s}$$

Also ratio:
$$n = \frac{l}{r} = \frac{1200}{300} = 4$$

The position of the planes of cranks and their angular positions are shown in configuration diagram [Figure 13.25(a)]. Choose the plane of crank 2 as the reference plane. Prepare the following table for primary forces and couples. Let m_2 and m_3 be the masses of reciprocating parts of cranks 2 and 3.

Plane	Mass m kg	Radius r (m)	force ÷ ω^2 mr kg.mm	Distance from reference plane l (m)	Couple ÷ ω^2 mrl kg mm^2
1	400	0.3	120	−0.45	−54.0
2	m_2	0.3	0.3 m_2	0	0
3	m_3	0.3	0.3 m_3	0.75	0.225 m_3
4	400	0.3	120	1.35	162

Construct a primary couple polygon *oab* with scale 25 units = 10 mm as shown in Figure 13.25(b). The closing vector $\overrightarrow{bo}$ represents couple 0.225 m_3. Therefore,

$$\overrightarrow{bo} = 7.8 \times 25 = 0.225 \ m_3$$

or mass at the crank 3:

$$m_3 = \textbf{866.67 kg at 314°} \text{ from the position of crank 1.} \qquad \textbf{Ans.}$$

Construct primary force polygon $o'a'b'c'$ as shown in Figure 13.25(c). The closing vector $\overrightarrow{o'c'}$ represents primary force 0.3 m_2. Therefore,

$$\overrightarrow{o'c'} = 4.3 \times 60 = 0.3 \ m_2$$

or mass at the crank 2: $m_2 = \textbf{860 kg at 161°}$ from the position of crank 1 **Ans.**

Draw secondary crank position by rotating all cranks by the angle twice the original inclination [Figure 13.25(d)]. Draw secondary force polygon $o''a'' b'' c'' d''$. The closing vector $\overrightarrow{o''d''}$ represents secondary unbalance [Figure 13.25(e)].

$$\overrightarrow{o''d''} = mr = 9.6 \times 60 = 576$$

Secondary force: $$F_s = m(2\omega)^2 \frac{r}{4n} = mr \frac{\omega^2}{n}$$

574 Theory of Mechanisms and Machines

(a) Configuration diagram

(b) Primary couple Polygon

(c) Primary force Polygon

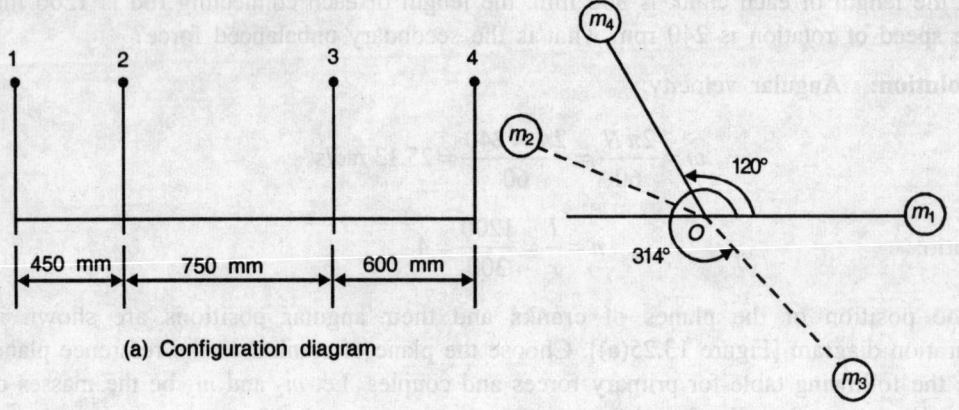

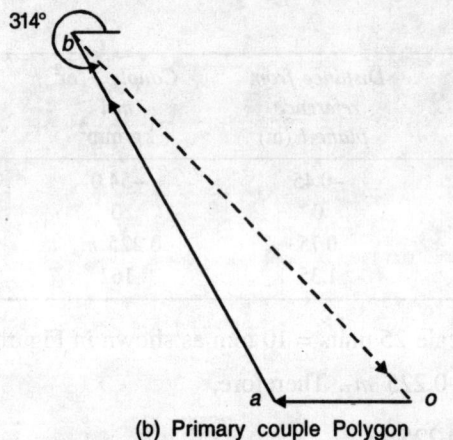

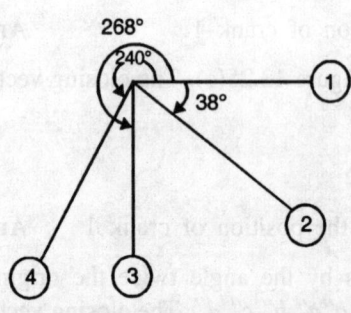

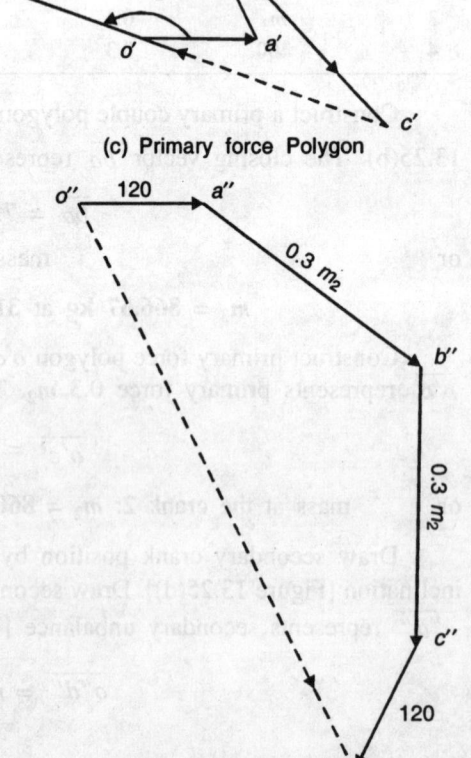

(d) Position of secondary crank

(e) Secondary force Polygon

FIGURE 13.25

where

$$\omega = \frac{2\pi N}{60} = \frac{2\pi \times 240}{60} = 25.1 \text{ rad/s}$$

$$\therefore \quad F_s = 576 \times \frac{25.1^2}{4} = 90721.4 \text{ N} \qquad \text{Ans.}$$

EXAMPLE 13.10 In a four cylinder in-line IC engine, the mass of reciprocating parts of cylinder number 1 and 4 are 100 kg and that of cylinder number 2 and 3 are 173 kg. If the crank radius is 150 mm, length of connecting rod is 450 mm and engine speed is 1200 rpm, determine the primary and secondary forces and couples. The cylinders are placed 600 mm apart as shown in Figure 13.26.

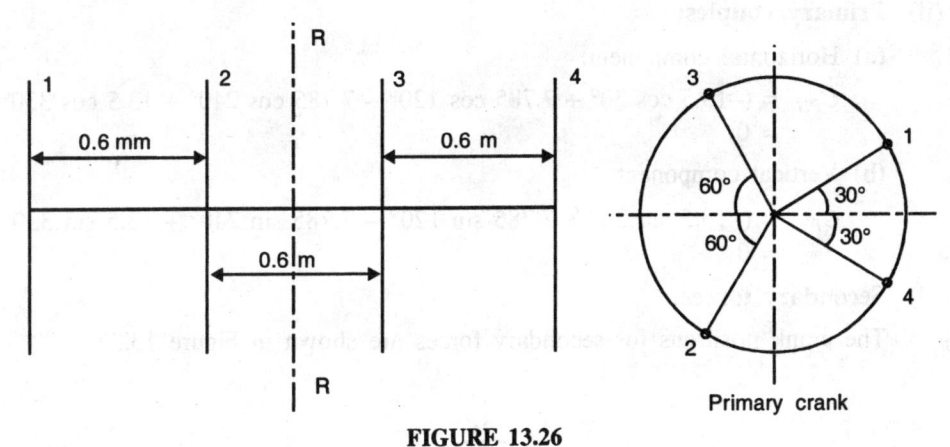

FIGURE 13.26

Solution: Angular velocity:

$$\omega = \frac{2\pi N}{60} = \frac{2\pi \times 1200}{60} = 125.66 \text{ rad/s}$$

Ratio:
$$n = \frac{l}{r} = \frac{450}{150} = 3$$

Assuming central plane R–R as reference plane, prepare the table as given below:

Plane	Mass m (kg)	Radius r (m)	Primary force mr kg mm	Secondary force mr/n kg mm	Distance from reference plane l (m)	Primary couple mrl kg mm²	Secondary couple mrl/n kg mm²
1	100	0.15	15	5	−0.9	−13.5	−4.5
2	173	0.15	25.95	8.65	−0.3	−7.785	−2.595
3	173	0.15	25.95	8.65	+0.3	7.785	+2.595
4	100	0.15	15	5	+0.9	13.5	+4.5

(i) **Primary forces:**

(a) Horizontal component:
$$F_{PH} = (15 \cos 30° + 25.95 \cos 120° + 25.95 \cos 240° + 15 \cos 330°)\, \omega^2$$
$$= 0$$

(b) Vertical component:
$$F_{PV} = (15 \sin 30° + 25.95 \sin 120° + 25.95 \sin 240° + 15 \sin 330°)\, \omega^2$$
$$= 0$$

Thus, the primary forces are completely balanced.

(ii) **Primary couples:**

(a) Horizontal component:
$$C_{PH} = (-13.5 \cos 30° + 7.785 \cos 120° - 7.785 \cos 240° + 13.5 \cos 330°)\, \omega^2$$
$$= 0$$

(b) Vertical component:
$$C_{PV} = (-13.5 \sin 30° + 7.785 \sin 120° - 7.785 \sin 240° + 13.5 \sin 330°)\, \omega^2$$
$$= 0$$

(iii) **Secondary forces:**

The crank positions for secondary forces are shown in Figure 13.27.

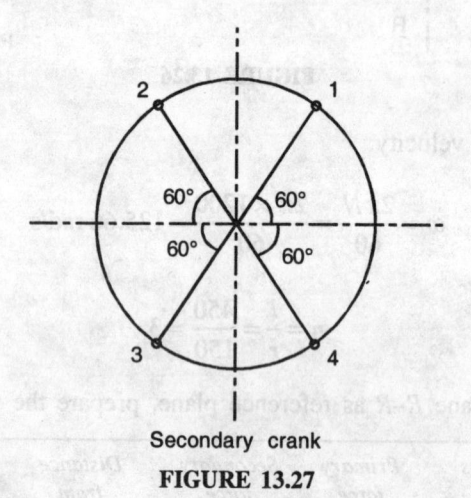

Secondary crank

FIGURE 13.27

(a) Horizontal component:
$$F_{SH} = [5 \cos 60° + 8.65 \cos 240° + 8.65 \cos 480° + 5 \cos 660°]\, \omega^2$$
$$= -3.65\, \omega^2$$

(b) Vertical component:
$$F_{SV} = [5 \sin 60° + 8.65 \sin 240° + 8.65 \sin 480° + 5 \sin 660°]\, \omega^2$$
$$= 0$$

So the resultant secondary unbalanced force:

$$F_S = 3.65\ \omega^2 = 3.65 \times 125.66^2 = \mathbf{57635\ N} \qquad \text{Ans.}$$

(iv) **Secondary couple:**

(a) Horizontal component:

$$C_{SH} = [-4.5 \cos 60° - 2.595 \cos 480° + 2.595 \cos 240° + 4.5 \cos 660°]\ \omega^2$$
$$= 0$$

(b) Vertical component:

$$C_{SV} = [-4.5 \sin 60° - 2.595 \sin 480° + 2.595 \sin 240° + 4.5 \sin 660°]\ \omega^2$$
$$= -12.2889\ \omega^2$$
$$= -12.2889 \times 125.66^2$$
$$= \mathbf{194047.0\ Nm} \qquad \text{Ans.}$$

13.9 BALANCING OF V-ENGINES

In a V-engine, the cylinders are arranged along radial lines and the lines of strokes of the cylinders form a letter V. These engines are also called **radial engines.** In a V-engine, all cylinders have a common crank which revolves in one plane; hence the primary and secondary couples are not produced. The analytical method for balancing of V-engine is discussed below (See Figure 12.28):

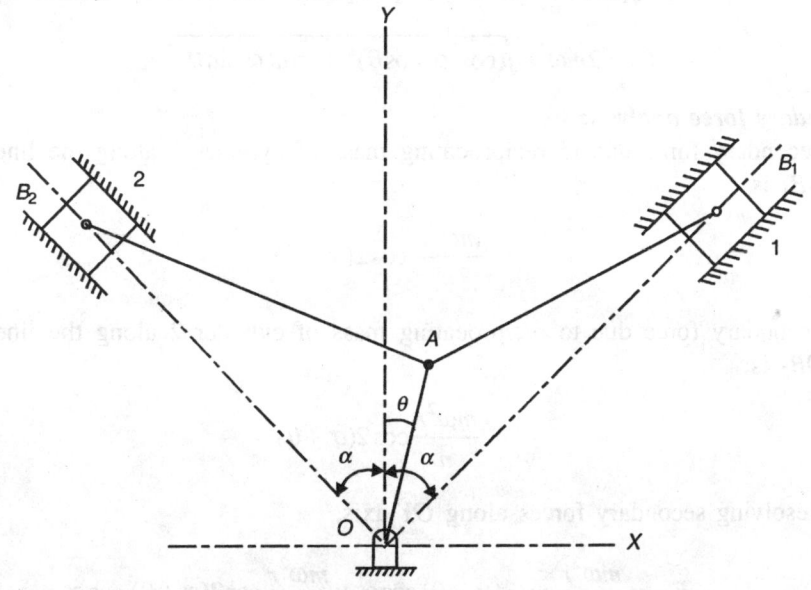

FIGURE 13.28 Balancing of V-engine.

Let us consider a two cylinder symmetrically arranged V-engine as shown in Figure 13.28. The line of stroke of each engine is inclined at an angle α from OY axis. If the common crank of this engine, at any instant, is rotated by angle θ from OY axis, then the inertia forces (both primary and secondary) due to reciprocating parts are calculated as given below:

Primary force analysis:

(a) Primary force due to cylinder 1 along the line of stroke OB_1 is:
$$m\omega^2 r \cos(\alpha - \theta) \qquad (13.35a)$$

(b) Primary force due to cylinder 2 along the line of stroke OB_2 is:
$$m\omega^2 r \cos(\alpha + \theta) \qquad (13.35b)$$

Resolving primary forces along vertical axis OY,
$$F_{PV} = m\omega^2 r \cos(\alpha - \theta) \cos\alpha + m\omega^2 r \cos(\alpha + \theta) \cos\alpha$$
or
$$F_{PV} = 2m\omega^2 r \cos^2\alpha \cos\theta \qquad (13.36)$$

Similarly, resolving primary forces along horizontal axis OX,
$$F_{PH} = m\omega^2 r \cos(\alpha - \theta) \sin\alpha - m\omega^2 r \cos(\alpha + \theta) \sin\alpha$$
or
$$F_{PH} = 2m\omega^2 r \sin^2\alpha \sin\theta \qquad (13.37)$$

Therefore, resultant primary forces:
$$F_P = \sqrt{F_{PV}^2 + F_{PH}^2}$$
$$= \sqrt{[2m\omega^2 r \cos^2\alpha \cos\theta]^2 + [2m\omega^2 r \sin^2\alpha \sin\theta]^2}$$
or
$$F_P = 2m\omega^2 r \sqrt{(\cos^2\alpha \cos\theta)^2 + (\sin^2\alpha \sin\theta)^2} \qquad (13.38)$$

Secondary force analysis:

(a) Secondary force due to reciprocating mass of cylinder 1 along the line of stroke OB_1 is
$$\frac{m\omega^2 r}{n} \cos 2(\alpha - \theta) \qquad (13.39a)$$

(b) Secondary force due to reciprocating mass of cylinder 2 along the line of stroke OB_2 is:
$$\frac{m\omega^2 r}{n} \cos 2(\alpha + \theta) \qquad (13.39b)$$

Resolving secondary forces along OY axis,
$$F_{SV} = \frac{m\omega^2 r}{n} \cos 2(\alpha - \theta) \cos\alpha + \frac{m\omega^2 r}{n} \cos 2(\alpha + \theta) \cos\alpha$$
or
$$F_{SV} = \frac{2m\omega^2 r}{n} \cos 2\alpha \cdot \cos\alpha \cdot \cos 2\theta \qquad (13.40)$$

Similarly, resolving secondary forces along OX axis,

$$F_{SH} = \frac{m\omega^2 r}{n}\cos 2(\alpha - \theta)\sin\alpha - \frac{m\omega^2 r}{n}\cos 2(\alpha + \theta)\sin\alpha$$

or

$$F_{SH} = \frac{2m\omega^2 r}{n}\sin 2\alpha \cdot \sin\alpha \cdot \sin 2\theta \qquad (13.41)$$

Therefore, resultant secondary forces:

$$F_S = \sqrt{F_{SV}^2 + F_{SH}^2}$$

$$= \frac{2m\omega^2 r}{n}\sqrt{(\cos 2\alpha \cdot \cos\alpha \cdot \cos 2\theta)^2 + (\sin 2\alpha \cdot \sin\alpha \sin 2\theta)^2} \qquad (1342)$$

EXAMPLE 13.11 A two cylinder engine has the cylinder axes at right angle and the connecting rods operate a common crank as shown in Figure 13.29. The reciprocating mass

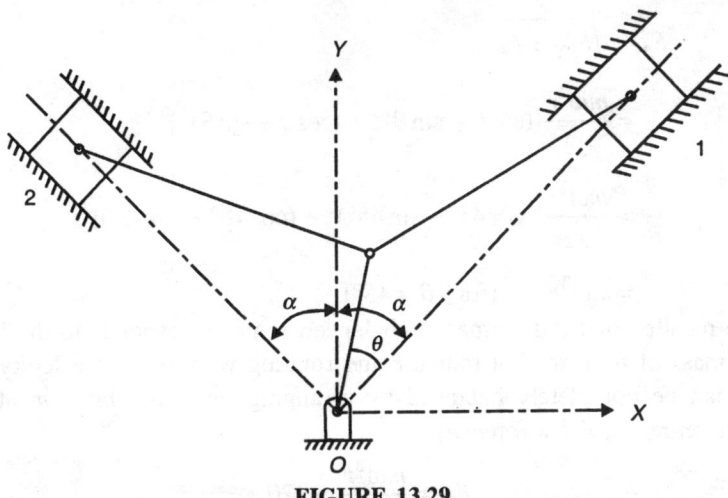

FIGURE 13.29

per cylinder is 10 kg, the crank is 70 mm and length of connecting rod is 350 mm. Show that the engine can be balanced for primary force by means of revolving balancing mass when crank angle for cylinder 1 is 45°. If the speed of engine is 500 rpm, what is maximum value of secondary unbalanced force?

Solution Primary forces for the two cylinders are:

$$F_{P1} = m\omega^2 r \cos\theta$$
$$F_{P2} = m\omega^2 r \cos(2\alpha - \theta)$$
$$\angle\theta = 45° \quad \text{and} \quad 2\alpha = 90° \text{ (given)}$$

Resolving forces along OY axis (refer Figure 1329),

$$F_{PY} = F_{P1} \cos\alpha + F_{P2} \cos\alpha$$
$$= m\omega^2 r \cos\alpha \, [\cos\theta + \cos(2\alpha - \theta)]$$
$$= m\omega^2 r \cos 45° \, [\cos\theta + \cos(90° - \theta)]$$
$$= \frac{m\omega^2 r}{\sqrt{2}} [\cos\theta + \sin\theta]$$

Resolving forces along OX axis,

$$F_{PX} = F_{P1} \sin\alpha - F_{P2} \sin\alpha$$
$$= m\omega^2 r \sin\alpha \, [\cos\theta - \cos(2\alpha - \theta)]$$
$$= m\omega^2 r \sin 45° \, [\cos\theta - \cos(90° - \theta)]$$
$$= \frac{m\omega^2 r}{\sqrt{2}} [\cos\theta - \sin\theta]$$

Resultant primary force:

$$F_P = \sqrt{F_{PX}^2 + F_{PY}^2}$$
$$= \frac{m\omega^2 r}{\sqrt{2}} [(\cos\theta - \sin\theta)^2 + (\cos\theta + \sin\theta)^2]^{0.5}$$
$$= \frac{m\omega^2 r}{\sqrt{2}} [(\cos 45° - \sin 45°)^2 + (\cos 45° + \sin 45°)]^{0.5}$$
$$= m\omega^2 r \quad \text{(for } \theta = 45°\text{)}$$

It may be recalled that this primary unbalanced force corresponds to the inertia force of an unbalanced mass of m, placed at radius r and rotating with angular velocity ω. Hence the primary force can be completely balanced by arranging revolving mass m at diametrically opposite to that represented by force F_P.

Secondary force

$$F_{S1} = \frac{m\omega^2 r}{n} \cos 2\theta$$

$$F_{S2} = \frac{m\omega^2 r}{n} \cos 2(2\alpha - \theta)$$

Resolving secondary forces along OY axis,

$$F_{SY} = F_{S1} \cos\alpha + F_{S2} \cos\alpha$$
$$= \frac{m\omega^2 r}{n} \cos\alpha \, [\cos 2\theta + \cos 2(2\alpha - \theta)]$$
$$= \frac{m\omega^2 r}{n} \cos\alpha \, [\cos 2\theta + \cos 2(90° - \theta)] \quad \text{for } 2\alpha = 90°$$
$$= \frac{m\omega^2 r}{n} \cos\alpha \, [\cos 2\theta + \cos(180° - 2\theta)] = 0$$

Similarly, resolving forces along OX axis,

$$F_{SX} = F_{S1}\sin\alpha - F_{S2}\sin\alpha$$

$$= \frac{m\omega^2 r}{n}\sin\alpha\,[\cos 2\theta - \cos 2(2\alpha - \theta)]$$

$$= \frac{m\omega^2 r}{n}\sin 45°\,[\cos 2\theta - \cos 2(90° - \theta)] \qquad \text{for } 2\alpha = 90°$$

$$= \frac{m\omega^2 r}{n} \times \frac{1}{\sqrt{2}}[\cos 2\theta - \cos(180° - 2\theta)]$$

$$= \frac{2m\omega^2 r}{n} \times \frac{1}{\sqrt{2}} \times \cos 2\theta$$

Since $F_{SY} = 0$, the maximum value of secondary force is possible when $\cos 2\theta = 1$, i.e. when angle $\theta = 0°$

or
$$F_{S\max} = \sqrt{2}\,\frac{m\omega^2 r}{n}$$

$$= \sqrt{2} \times 10 \times \left(\frac{2\pi \times 500}{60}\right)^2 \times \frac{70}{1000 \times 5}$$

$$= 542.8\ \text{N} \qquad\qquad\qquad\qquad \textbf{Ans.}$$

13.10 DIRECT AND REVERSE CRANK METHOD

The direct and reverse crank method is very useful to determine the primary and secondary unbalanced forces in radial or V-engines.

Let us consider a reciprocating engine in which crank OA rotates at constant angular velocity in clockwise direction (Figure 13.30). The crank OA makes an angle θ from the line of stroke at any instant. The primary force is:

$$F_p = m\omega^2 r\,\cos\theta$$

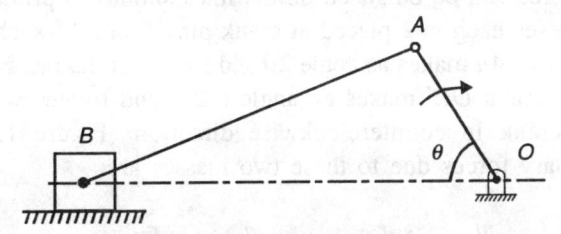

FIGURE 13.30 Reciprocating engine.

This force should be equal to horizontal component of force produced by mass m placed at the crank pin A. Let an imaginary crank OA' be drawn as shown in Figure 13.31 making an angle $(-\theta°)$. Then the crank OA is known as **direct crank** and imaginary crank OA' is

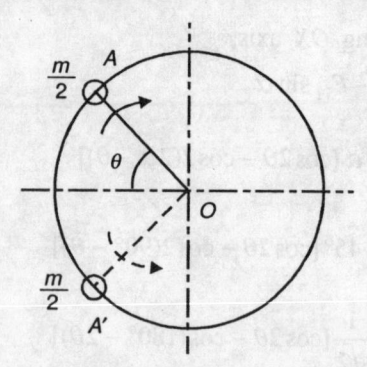

FIGURE 13.31 Direct and reverse crank (primary force).

known as **reverse crank.** This reverse crank rotates at same angular speed but in opposite direction, say counterclockwise direction.

Now the mass m, which otherwise is placed at crank pin A, is divided equally and placed at the crank pin of direct and reverse crank, i.e. at points A and A'. The horizontal components of force due to these two masses is:

$$\frac{m}{2}\omega^2 r \cos\theta + \frac{m}{2}\omega^2 r \cos\theta = m\omega^2 r \cos\theta \qquad (13.43)$$

which is equal to primary force. The vertical component of two masses will cancel each other. Thus by placing half of the reciprocating mass at the direct crank and another half at the reverse crank, the primary force can be completely balanced.

As regard to secondary force, we know that secondary force is expressed as

$$F_s = \frac{m\omega^2 r}{n}\cos 2\theta$$

or
$$F_s = m(2\omega)^2 \frac{r}{4n}\cos 2\theta \qquad (13.44)$$

The secondary force can be balanced on the lines similar to primary force. The mass m is replaced by two masses each $m/2$ placed at crank pins C and C' such that direct secondary crank OC of crank radius $r/4n$ makes as angle 2θ and rotates at 2ω rad/s in clockwise direction. The reverse secondary crank OC' makes an angle (-2θ) and rotates with same speed as that of direct secondary crank in counterclockwise direction (Figure 13.32). The horizontal components of secondary forces due to these two masses are:

$$\frac{m}{2}(2\omega)^2 \frac{r}{4n}\cos 2\theta + \frac{m}{2}(2\omega)^2 \frac{r}{4n}\cos 2\theta$$

$$= m(2\omega)^2 \frac{r}{4n}\cos 2\theta$$

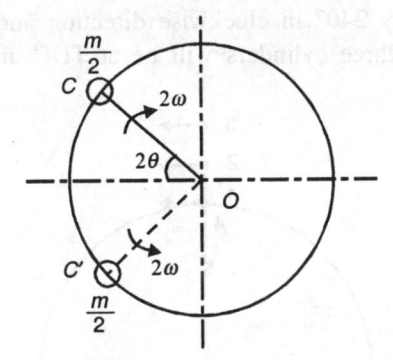

FIGURE 13.32 Direct and reverse crank (secondary force).

The application of above direct and reverse crank method is illustrated further through worked-out example.

EXAMPLE 13.12 A three cylinder radial engine has axes at 120° to one another and their connecting rods are coupled to a single common crank. The stroke length is 100 mm and length of each connecting rod is 150 mm. If the mass of reciprocating parts per cylinder is 1 kg, determine the primary and secondary force of the engine running at 2400 rpm. (See Figure 13.33.)

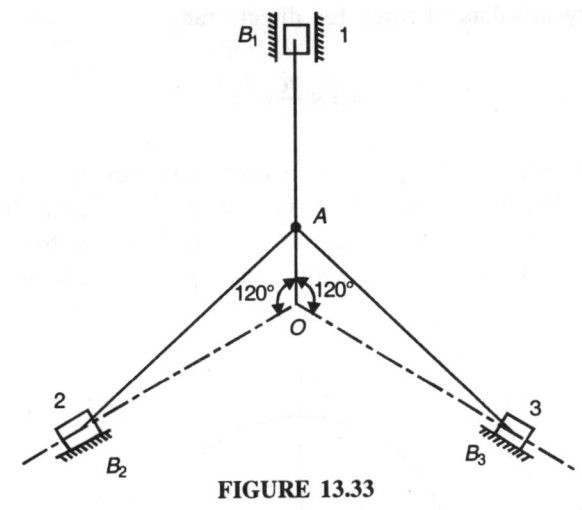

FIGURE 13.33

Solution

$$\text{Stroke length} = 2r = 100 \text{ mm}$$

Angular velocity:
$$\omega = \frac{2\pi N}{60} = \frac{2\pi \times 2400}{60} = 251.32 \text{ rad/s}$$

Primary direct crank: For cylinder 1, since crank angle θ is zero, direct crank will be at 0° from the top dead centre (TDC) rotating in clockwise direction. For cylinder 2, direct crank will rotate 120° clockwise and will coincide with direct crank of cylinder 1. For cylinder 3, the direct crank will rotate by 240° in clockwise direction and will also coincide at TDC. Thus the direct crank of all three cylinders will be at TDC and all shall rotate at ω rad/s [Figure 13.34(a)].

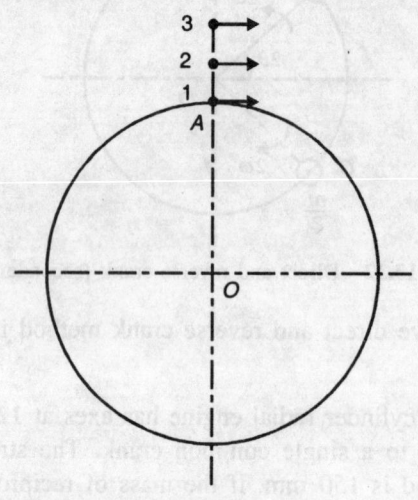

FIGURE 13.34(a) Primary direct crank.

Thus the primary unbalanced force for direct crank

$$= 3 \times \frac{m}{2} \omega^2 r$$

Primary reverse crank: The primary reverse crank for cylinder 1 will rotate 0° in counterclockwise direction. For cylinder 2, it will rotate 120° counterclockwise from axis of cylinder 2 and for cylinder 3, it will rotate 240 counterclockwise from the axis of cylinder 3 as shown in Figure 13.34(b), which shows that primary reverse crank is completely balanced.

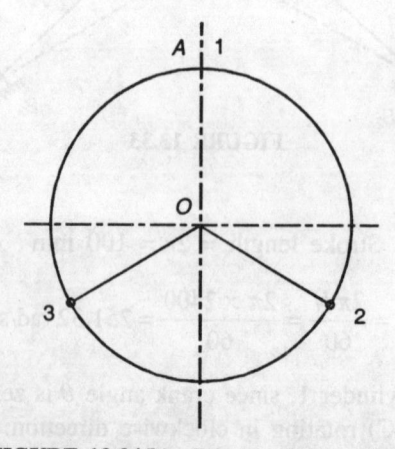

FIGURE 13.34(b) Primary reverse crank.

Therefore, total primary unbalanced force is equal to unbalanced force observed during direct crank:

$$= \frac{3}{2} m\omega^2 r$$

$$= \frac{3}{2} \times 1 \times \left(\frac{2\pi \times 2400}{60}\right)^2 \times \frac{50}{1000} = 4737.1 \text{ N} \qquad \text{Ans.}$$

Secondary direct crank: The secondary crank rotates at 2θ from initial position. Thus direct secondary crank for cylinder 1 rotates at 2θ in clockwise direction but $\theta = 0°$ So its position does not change. Direct crank for cylinder 2 is rotated by 240° (2 × 120°) in clockwise direction from the axis of cylinder 2 and for cylinder 3, it is rotated by 480° (2 × 240°) in clockwise direction from the axis of cylinder 3 as shown in Figure 13.34(c), which shows that secondary direct crank is completely balanced.

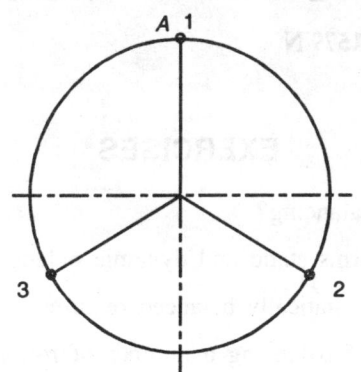

FIGURE 13.34(c) Secondary direct crank.

Secondary reverse crank: The secondary reverse crank for cylinder 1 is rotated by 0° in counterclockwise direction from the axis of cylinder 1. For cylinders 2 and 3, secondary reverse crank is rotated by 240° and 480° respectively from their axes. Thus all the cranks coincide at point A rotating at 2ω rad/s [Figure 13.34(d)].

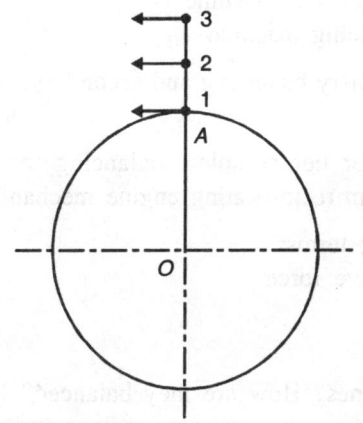

FIGURE 13.34(d) Secondary reverse crank.

Therefore, secondary force due to reverse crank is:

$$3 \times \frac{m}{2} \times (2\omega)^2 \times \frac{r}{4n}$$

where
$$r = \frac{l}{r} = \frac{150}{50} = 3$$

Total unbalanced secondary force is equal to secondary force due to reverse crank.

$$\therefore \quad F_s = 3 \times \frac{m}{2}(2\omega)^2 \times \frac{r}{4n}$$

$$= 3 \times \frac{1}{2} \times (2 \times 251.32)^2 \times \frac{50}{4 \times 3 \times 1000}$$

$$= 1579 \text{ N} \hspace{3cm} \text{Ans.}$$

EXERCISES

1. What is purpose of balancing?
2. Explain clearly the terms static and dynamic balancing.
3. Does a rotor which is statically balanced requires dynamic balancing?
4. Explain the method of balancing a number of rotating masses in one plane by a single mass in the same plane.
5. Describe the method and find out the masses for correction of unbalance for a rotor which has to be balanced at two planes.
6. What do you mean by balancing machine? Explain constructional features and working of the following types of balancing machine:
   (a) Pivoted-cradle balancing machine
   (b) Nodal point balancing machine
7. Explain the terms primary balancing and secondary balancing used for balancing of reciprocating masses.
8. Explain the reason for not complete balancing the disturbing effect due to the reciprocating masses in reciprocating engine mechanism.
9. Explain the following terms:
   (a) Variation in tractive force
   (b) Swaying couple
   (c) Hammer blow
10. What are in-line engines? How are they balanced? Is it possible to balance them completely?

11. Show that the primary unbalanced forces of a 90° V-twin engine can be balanced by a counter mass attached to the crank.

12. Four masses m_1, m_2, m_3 and m_4 having their radii of rotations as 20 mm, 15 mm, 25 mm and 30 mm are 20 kg, 30 kg, 24 kg and 26 kg in magnitudes respectively. The angles between the successive masses are 45°, 75° and 135° respectively. Find the position and magnitude of the balancing mass required to be placed at a radius of 20 mm.

[Ans: 11 kg at 201° from mass m_1]

13. Four masses A, B, C and D are required to be completely balanced. The planes containing masses B and C are 300 mm apart. The angle between planes containing masses B and C is 90°. The masses D and B make angles of 120° and 210° respectively with C in the same sense. Find: (i) the mass and angular position of mass A and (ii) the position of planes A and D.
The details of rotating masses are:

	Mass (kg)	Radius (mm)
A	–	180
B	30	240
C	50	120
D	40	150

[Ans: 20.83 kg, – 811.6 mm, 307 mm from mass B]

14. A shaft carries four masses in parallel planes A, B, C and D. The masses at B and C are 18 kg and 12.5 kg respectively and each has an eccentricity of 60 mm. The masses at A and D have an eccentricity of 80 mm. The angle between B and C is 100° and that between the masses B and A is 90° measured in the same direction. The axial distance between the planes A and B is 100 mm and that between B and C is 200 mm. If the shaft is in complete dynamic balance, determine:
 (i) masses at planes A and D
 (ii) distance between the planes A and D
 (iii) angular position of the mass at D

[Ans: 9.375 kg, 8.125 kg, 354 mm, 254°]

15. Three masses m_1, m_2 and m_3 located in the planes 1.2 and 3 are to be balanced by correcting masses in the plane 3 and are shown in Figure E13.1. Determine the magnitude of balancing masses. While balancing this rotor on a machine, if the maximum error in determining the magnitude of correcting masses is + 5 per cent, determine the bearing reactions left on the support as a percentage of the original reaction.

16. A rough casting of a rotor with a mass of 200 kg is mounted on two bearings 1100 mm apart. The rotor is statically balanced by attaching two masses 9 kg and 12 kg one in each plane A and B respectively. These are on opposite sides of the

FIGURE E13.1

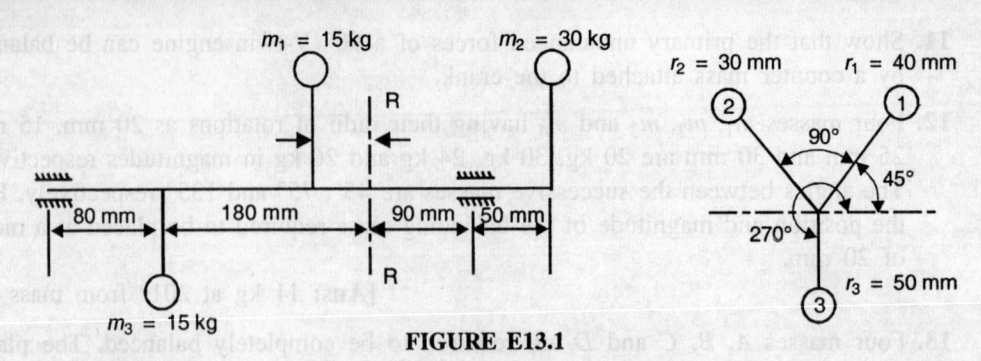

12 kg one in each plane A and B respectively. These are on opposite sides of the plane containing the mass centre of the rotor. The axial distances of the planes A and B from the mass centre are 500 mm and 300 mm respectively. The angle between the masses at A and B is 90°. The radial distances of the masses A and B from the axis of rotation are 370 mm and 440 mm respectively.
Determine the distance of the centre of gravity of the rotor from the axis of rotation and its angular location from the mass in the plane A. Also find the bearing forces if the rotor rotates at 100 rpm.

[Ans: 31.2 mm, θ = 122.2° counterclockwise from A, 277.2 N]

17. A horizontal shaft AB of length 1200 mm is rotating at a uniform speed of 250 rpm and carries two discs at A and B of masses 5 kg and 4 kg respectively. The centre of gravity of each disc is 2.5 mm away from the axis of rotation and the angle between the masses at A and B is 90°. The shaft is supported in the bearings at C and D such that AC = 300 mm and AD = 900 mm. It is desired to make the dynamic forces on the bearings equal and opposite and to have a minimum value for a given speed by means of a mass in the plane E, between C and D at a radius of 25 mm. Determine:
 (i) mass to be attached at plane E and angular position relative to the plane A
 (ii) the location of the plane E
 (iii) dynamic force on the bearings with the attached mass at E

[Ans: m_e = 0.64 kg 218° clockwise from A, 168.75 mm, F_e = 223 N]

18. In an experimental set-up to balance a thin disc, the following observations were obtained:
When the disc was run at 2000 rpm, vibration velocity level due to original imbalance was 2.6 mm/s. When a trial mass of log was fastened in position 1, the vibration level was V_1 = 6.5 mm/s. When the same mass was fastened at positions 2 and 3, the vibration level was V_2 = 1.9 mm/s and V_3 = 5.5 mm/s respectively. The angle between positions 1 and 2 is 180° and that between positions 1 and 3 is 90° measured in counterclockwise direction. Find the magnitude of balancing mass and its angular position.

19. A four crank engine has two outer cranks set at 120° to each other and their reciprocating masses are each 400 kg. The distances between the planes of rotation of adjacent crank are 450 mm, 750 mm and 600 mm. If the engine is to be in complete primary balance, find the reciprocating mass and the relative angular position for each of the inner crank.

If the length of each crank is 30 mm and length of each connecting rod is 120 mm, find the maximum secondary unbalanced force. The engine speed is 240 rpm.

[**Ans:** 892 kg at 159° in counterclockwise from A, 911 kg at 313° in counterclockwise from A, 987 N]

20. For an inside cylinder locomotive with two cranks at 90° angle, the mass of reciprocating part per cylinder is 300 kg. The distance between cylinder centre line is 600 mm and between the planes of rotation of wheel is 1.5 m. Each balanced mass is introduced in the planes of wheels to balance two-third of the reciprocating parts. The crank radius is 325 mm and diameter of driving wheels are 1.875 m.

Determine the maximum variation of (a) tractive force and (b) hammer blow when locomotive is running at a speed of 95.25 kmph.

[**Ans:** 55147 N, 39240 N]

21. The following data relate to a four cylinder coupled locomotive with two inside cylinders with crank at 60°:

Mass of revolving parts	= 260 kg/cylinder
Mass of reciprocating parts	= 300 kg/cylinder
Distance between wheels	= 1.5 m
Distance between cylinders	= 0.6 m
Distance between coupling rod cranks	= 1.9 m
Diameter of the wheels	= 1.9 m
Mass of revolving parts for coupling rod crank	= 125 kg
Length of coupling rod crank	= 220 mm
Angle between coupling rod crank and adjacent crank	= 180°
Radius of balanced masses placed in driving wheel	= 0.75 m

Determine:
(a) the position and magnitude of balanced masses in leading and trailing wheel to balance two-third of reciprocating and whole of revolving parts of masses. Assume that half of the reciprocating parts are to be balanced in each pair of coupled wheel.
(b) Maximum tractive force and hammer blow when locomotive is running at 100 kmph.

[**Ans:** (a) leading wheel 86.3 kg at 231.6°, 86.3 kg at 218.4° trailing wheel 21.25 kg at 145.4° and 21.25 kg at 304.6° (b) 3846.5 N, 8868 N]

22. Using the method of direct and reverse crank, prove that for a four stroke, five cylinder radial engine
 (i) primary force can be completely balanced by a rotating mass placed in the direction opposite to the crank.
 (ii) the secondary forces are completely balanced.

23. Find out the state of balance for 4th, 6th and 8th order force in a four stroke cycle five cylinders radial engine using the method of direct and reverse cranks.

24. A twin cylinder V-engine has the cylinder set at an angle of 40° with both pistons connected to a single crank. The crank radius is 60 mm and connecting rods are 300 mm long. The reciprocating mass is 1 kg per cylinder and total rotating mass is equivalent to 1.5 kg at the crank radius. A balance mass is fitted opposite to the crank equivalent to 1.8 kg at a radius of 80 mm.
    Determine, for an engine speed of 900 rpm, the maximum values of primary and secondary forces due to inertia of reciprocating and rotating masses.
    [**Ans: 3079 N, 116.41 N**]

25. The reciprocating mass per cylinder in 60° V-twin engine is 1.5 kg. The stroke and connecting rod lengths are 100 mm and 250 mm respectively. If the engine runs at 2500 rpm, determine the maximum and minimum values of the primary and secondary forces. Also find out the crank position corresponding to these values.
    [**Ans: 3427 N, at $\theta = 0°$ and 2570.2 N at $\theta = \pi/2$**]

26. An engine having five cylinders in-line has successive cranks 144° apart, the distance between cylinder centre lines being 450 mm. The reciprocating mass for each cylinder is 16 kg. The crank radius is 135 mm and length of the connecting rod is 540 mm. The engine runs at 600 rpm. Examine the engine for balance of primary and secondary forces and couples.
    [**Ans: 10150 Nm, 4100 Nm**]

MULTIPLE CHOICE QUESTIONS

1. If there are several unbalanced masses m in a rotor in different planes, the minimum number of balancing masses required is
    (a) 1 (b) 2
    (c) 3 (d) 4

2. In multi-cylinder in-line engine, the number of cylinders are chosen as even so that
    (a) primary forces are balanced (b) secondary forces are balanced
    (c) both (a) and (b) (d) none of the above

3. The entire reciprocating mass of an IC engine is never balanced because
    (a) it does not require balancing (b) it gives rise to unbalance in other direction
    (c) it give to unbalance couple (d) it is not possible to do so

4. Field balancing of a single plane rotor without measurement of phase angle can be accomplished by carrying out
    (a) one test run (b) two test runs
    (c) three test runs (d) four test runs

5. A change in the firing order of an in-line engine may reduce
    (a) primary force (b) secondary force
    (c) unbalanced couple (d) fluctuation in turning moment

6. Lanchester technique of balancing can be used for balancing of
   (a) primary forces
   (b) secondary forces
   (c) pitching moments
   (d) all of the above

7. The maximum total primary force in a three cylinder radial engine with axes 120° to one another, stroke length 100 mm, speed 3000 rpm, reciprocating mass/cylinder is
   (a) 100 kN
   (b) 12.05 kN
   (c) 11.1 kN
   (d) 15.2 kN

8. In dynamic balancing, the following conditions will hold
   (a) force polygon is closed
   (b) couple polygon is closed
   (c) force and couple polygon is closed
   (d) none of the above

9. In a symmetrical two cylinder V-engine, the resultant force is proportional to
   (a) $\cos \theta$
   (b) $\cos 2\theta$
   (c) $\sin \theta$
   (d) $\sin 2\theta$

10. When the primary direct crank of an IC engine makes an angle θ with the line of stroke, then what angle the secondary crank will make with the line of stroke?
    (a) $\theta/2$
    (b) θ
    (c) 2θ
    (d) 4θ

11. Mass m attached to a shaft rotating at ω rad/s at radius r from the axis of shaft is balanced by mass m_b at radius r_b from the axis. If the speed of the shaft is doubled for balance, the value of mass m_b is
    (a) doubled
    (b) quadruple
    (c) halved
    (d) unaffected

12. If the ratio of length of the connecting rod to the crank radius is increased
    (a) primary forces will be increased
    (b) primary forces will be decreased
    (c) secondary forces will be increased
    (d) secondary forces will be decreased

13. The frequency of secondary force as compared to primary force for ratio of connecting rod length to the crank radius of 4 is
    (a) half
    (b) twice
    (c) four times
    (d) sixteen times

CHAPTER 14

Mechanical Vibrations

14.1 INTRODUCTION

Machines are sometimes acted upon by oscillating forces which results into oscillating or to and fro type motion in a machine member about its mean position. Each type of motion which repeats itself after certain interval of time is called **vibration**. Mechanical vibration deals with the vibration of mechanical system. For a mechanical vibration to occur a minimum of two energy storage elements are required, namely a mass which stores kinetic energy and an elastic member which stores potential energy.

A pendulum swinging on either side of a mean position under the action of gravity is an example of mechanical vibration. When the pendulum swings through the mid-position, its mass is at the lowest position and it possesses only kinetic energy whereas when it is at extremity of its swing, it carries only potential energy. In the absence of damping, the motion of pendulum continues indefinitely.

In mechanical system, the main causes of vibration are the following:

(i) unbalanced forces produced from machine itself
(ii) external excitation which may be periodic, randam or of the nature of impact
(iii) self-excited such as machine tool chatter.

Vibration hampers the normal functioning of mechanical system by producing excessive stresses, undesirable noise, looseness of parts and fatigue failure of member due to its cyclic nature. Therefore, the study of vibration is aimed at either eliminating the sources of vibration or reducing its effect by proper design and utilizing improved material for vibration isolation.

14.2 ELEMENTS OF VIBRATORY SYSTEM

The main elements of a vibratory system are shown in Figure 14.1. These are:

(i) the mass
(ii) the spring
(iii) the damper
(iv) the excitation element.

The first three elements are called **passive parameters** which are used to define a system. Mass and spring are used to store energy whereas damper is for dissipation of energy. The excitation is called an **active element**.

Mechanical Vibrations 593

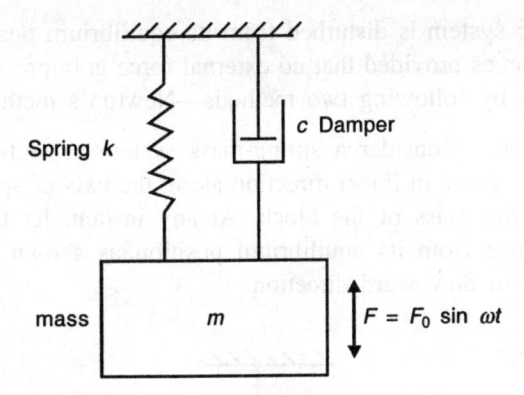

FIGURE 14.1 A typical vibratory system.

In a vibratory system, mass is a rigid body which executes vibration and can gain or loose kinetic energy which is proportional to the velocity change of the body. The spring has elasticity and is assumed to have negligible mass. The damping element neither has mass nor elasticity. The elasticity of the spring is measured in terms of stiffness k and damping capacity of the damper is measured by damping coefficient c, which is defined as force per unit velocity of the vibratory system. The following terms are used in vibratory system analysis:

(i) *Free vibration.* Vibration of a system in which periodic motion continues after the cause or the initial disturbance is removed is called **free vibration.**
(ii) *Damped vibration.* When the energy of a vibratory system is gradually dissipated due to resistance, such vibration is called **damped vibration.**
(iii) *Forced vibration.* If the vibratory motion persists because of repeated disturbing force acting on the system, then it is called **forced vibration.**
(iv) *Period of vibration.* It is time taken to complete one cycle of vibration.
(v) *Frequency.* The number of vibration cycles occuring in unit time is called **frequency.**
(vi) *Amplitude.* The maximum displacement of a vibratory body from the mean position is known as **amplitude of vibration.**
(vii) *Natural frequency.* It is a frequency of free vibrating system.
(viii) *Resonance.* It is a state of the system when forcing frequency is equal to the natural frequency of the system. The amplitude of vibration at resonance condition is maximum.
(ix) *Degree of freedom.* Number of independent motions which a vibratory system can perform during vibration is called **degree of freedom** (dof).

14.3 UNDAMPED FREE VIBRATION

In mechanical system, it is essential to estimate their natural frequencies of vibration to avoid resonance condition. In this section, a generalized model of undamped single dof system is discussed for determination of natural frequency.

When an elastic system is disturbed from its equilibrium position, it vibrates under the influence of inertia forces provided that no external force is impressed upon it. We shall solve this class of problem by following two methods—Newton's method and energy method.

Newton's method. Consider a spring-mass system as shown in Figure 14.2(a). The mass is constrained to move in linear direction along the axis of spring. Let k be the stiffness of spring and m be the mass of the block. At any instant, let the spring-mass system be displaced by x distance from its equilibrium position as shown in Figure 14.2(b). Let us assume x be positive in downward direction.

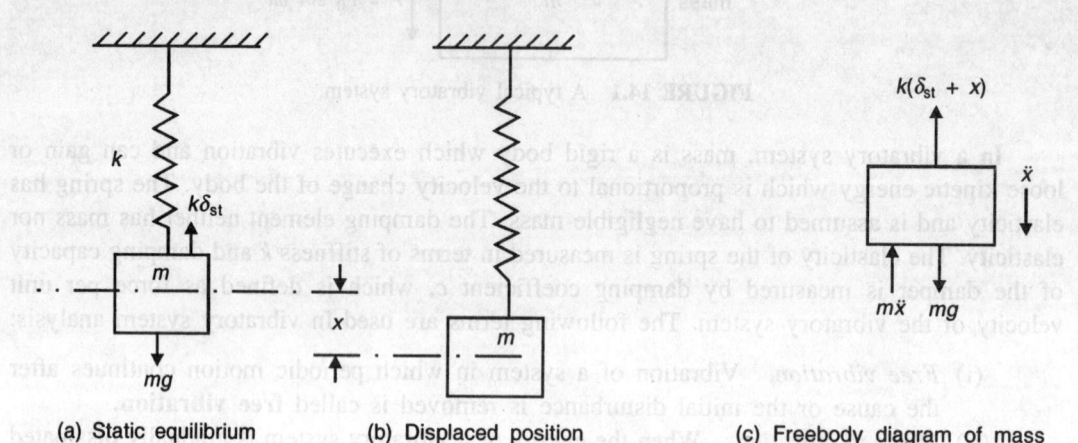

(a) Static equilibrium   (b) Displaced position   (c) Freebody diagram of mass

FIGURE 14.2 A Spring-mass vibratory system.

Considering the equilibrium position, the following forces are acting as shown in Figure 14.2(c):

(i) mg, acting vertical downward due to mass m
(ii) $m\ddot{x}$, an inertia force due to acceleration of mass, acting in vertical upward direction
(iii) kx, a spring force acting in vertical upward direction
(iv) $k\delta_{st}$, a vertical upward force due to deflection in spring due to weight mg, where δ_{st} is static deflection of the spring at static equilibrium position.

Applying Newton's law of motion for free undamped vibration, we get

$$m\ddot{x} = -k(\delta_{st} + x) + mg \qquad (14.1)$$

where $\ddot{x}$ is second derivative of displacement $x\left(=\dfrac{d^2x}{dt^2}\right)$

We know, for static equilibrium condition,

$$k\delta_{st} = mg \qquad (14.2)$$

Substituting Eq. (14.2) in Eq. (14.1), we get

$$m\ddot{x} + kx = 0 \qquad (14.3)$$

This is a differential equation of motion for a single dof spring-mass system having free vibration. Writing Eq. (14.3) as

$$\ddot{x} + \frac{k}{m}x = 0$$

or

$$\ddot{x} + \omega_n^2 x = 0 \qquad (14.4)$$

where ω_n is natural frequency of vibration ($= \sqrt{k/m}$)

Equation (14.3) is a standard second order differential equation for simple harmonic motion. The general solution of this equation is given as

$$x = A \sin \omega_n t + B \cos \omega_n t$$
$$= A_1 \sin(\omega_n t + \phi_1)$$

or

$$x = A_2 \cos(\omega_n t + \phi_2) \qquad (14.5)$$

where A, B, A_1, A_2 and ϕ_1, ϕ_2 are constants which can be determined from the initial conditions.

Energy method. When a conservative system is set in motion, the mechanical energy which is sum of kinetic energy and potential energy is constant and also the rate of change of total mechanical energy should be zero.

Mathematically,

$$T + U = 0$$

and

$$\frac{\partial}{\partial t}(T + U) = 0 \qquad (14.6)$$

where T = kinetic energy of the system
and U = potential energy of the system

Considering a spring-mass system, the kinetic energy is:

$$T = \frac{1}{2}m\dot{x}^2 \qquad (14.7)$$

The change in potential energy due to displacement x is equal to strain energy of spring minus potential energy due to change in the position of mass as shown in Figure 14.3.

Potential energy:
$$U = \int_0^x (\text{spring force} - mg)\, dx$$

$$= \int_0^x (mg + kx)\, dx - \int_0^x mg\, dx$$

or

$$U = \frac{1}{2}kx^2 \qquad (14.8)$$

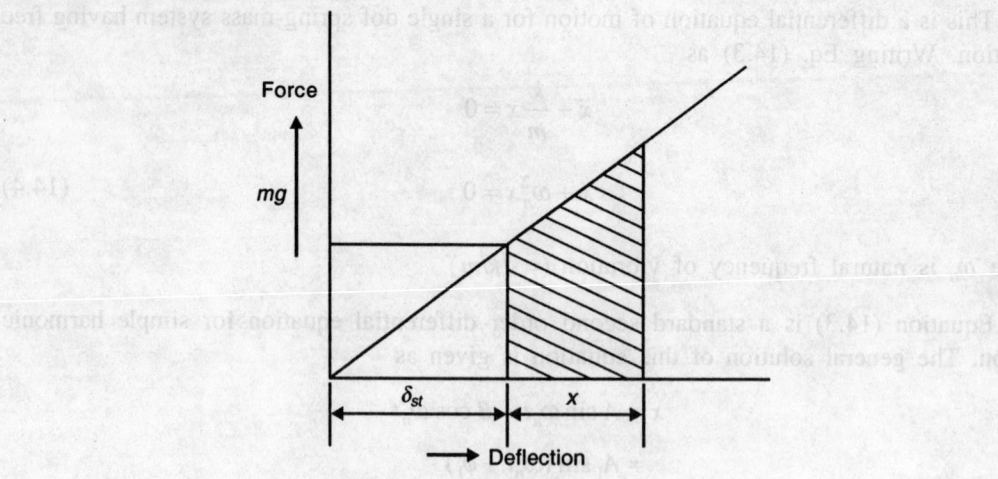

FIGURE 14.3 Force-deflection relation.

or rate of change of mechanical energy must be equal to zero, that is

$$\frac{\partial}{\partial t}(T+U) = 0$$

or

$$\frac{\partial}{\partial t}\left(\frac{1}{2}m\dot{x}^2 + \frac{1}{2}kx^2\right) = 0$$

or

$$\dot{x}(m\ddot{x} + kx) = 0$$

or

$$m\ddot{x} + kx = 0$$

which represents differential equation of motion for single dof undamped free vibrating system.

Some mechanical systems may have more than one springs. In such systems, springs are either connected in series or in parallel as shown in Figure 14.4. To determine the vibration characteristics of such a system, all springs are converted into a single spring having equivalent stiffness. The equivalent stiffness can be found by the following relations:

(a) Springs in series

$$\frac{1}{k_{eq}} = \frac{1}{k_1} + \frac{1}{k_2} + \cdots \qquad (14.9a)$$

(b) Springs in parallel

$$k_{eq} = k_1 + k_2 + \cdots \qquad (14.9b)$$

(a) Springs in series (b) Springs in parallel

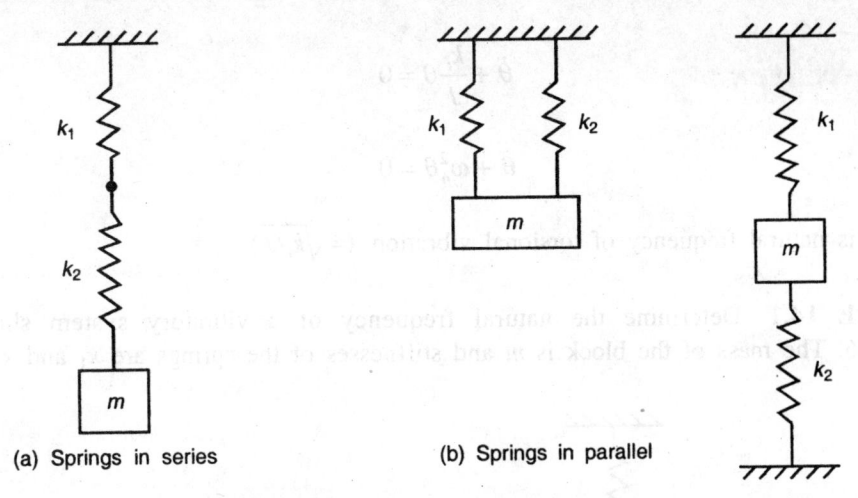

FIGURE 14.4 Springs in series and parallel.

14.4 TORSIONAL VIBRATION

Consider a circular disc of mass moment of inertia J mounted on a shaft of torsional stiffness k_t as shown in Figure 14.5. If the disc is twisted slightly, it executes torsional oscillation. The twisting torque acting on shaft when it is displaced through a small angle θ from its equilibrium position is $(-k_t\theta)$. The negative sign indicates that $k_t\theta$ acts in the direction opposite to rotation.

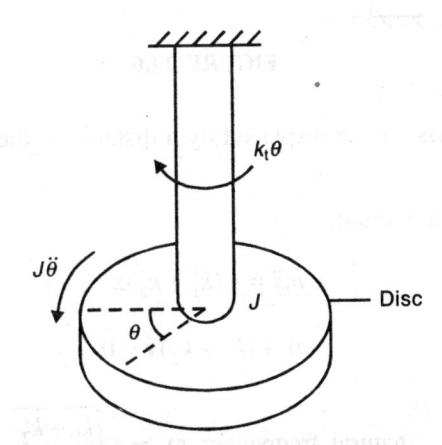

FIGURE 14.5 Torsional vibration.

Now applying Newton's law of motion,

$$J\ddot{\theta} = -k_t\theta$$

or

$$J\ddot{\theta} + k_t\theta = 0 \tag{14.10}$$

or
$$\ddot{\theta} + \frac{k_t}{J}\theta = 0$$

or
$$\ddot{\theta} + \omega_n^2 \theta = 0$$

where ω_n is natural frequency of torsional vibration $(= \sqrt{k_t/J})$

EXAMPLE 14.1 Determine the natural frequency of a vibratory system shown in Figure 14.6. The mass of the block is m and stiffnesses of the springs are k_1 and k_2.

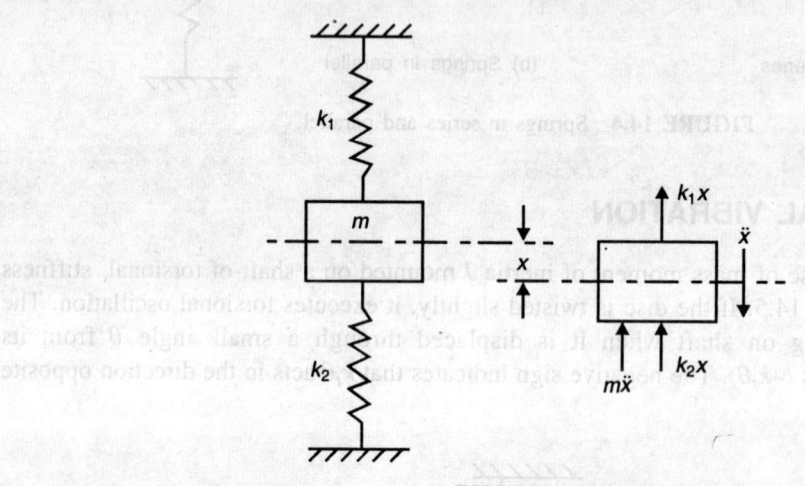

FIGURE 14.6

Solution Let the mass m be displaced by a distance x, the forces acting upon the mass are shown in the figure.

Applying Newton's law of motion,

$$m\ddot{x} = -(k_1 + k_2)x$$

or
$$m\ddot{x} + (k_1 + k_2)x = 0$$

or
$$\text{natural frequency: } \omega_n = \sqrt{\frac{k_1 + k_2}{m}}$$

or
$$f_n = \frac{\omega_n}{2\pi} = \frac{1}{2\pi}\sqrt{\frac{(k_1 + k_2)}{m}} \text{ Hz} \qquad \textbf{Ans.}$$

EXAMPLE 14.2 A cylinder partially immersed in riverwater is depressed slightly and released (Figure 14.7). Find its natural frequency assuming that it stays upright all the time.

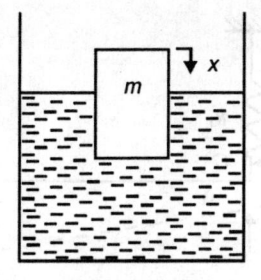

FIGURE 14.7

What is the change in its frequency if the riverwater is replaced by seawater of specific gravity equal to 1.1 times that of the riverwater?

Solution Let
x = displacement of the cylinder
A = area of cross section of the cylinder
m = mass of the cylinder
and ρ = specific gravity of riverwater

Applying Newton's law,

$$m\ddot{x} = -\rho A x$$

or
$$m\ddot{x} + \rho A x = 0$$

or
$$\ddot{x} + \frac{\rho A}{m} x = 0$$

Natural frequency:
$$f_n = \frac{1}{2\pi}\sqrt{\frac{\rho A}{m}} \text{ Hz}$$

If the riverwater is replaced by seawater, the natural frequency is:

$$f'_n = \frac{1}{2\pi}\sqrt{\frac{1.1\rho A}{m}}$$

Therefore, increase in frequency is:

$$\Delta f = f'_n - f_n = \frac{1}{2p}\sqrt{\frac{1.1\rho A}{m}} - \frac{1}{2p}\sqrt{\frac{\rho A}{m}} \text{ Hz} \qquad \text{Ans.}$$

EXAMPLE 14.3 Find the natural frequency of the system shown in Figure 14.8. The bar AB is assumed to be rigid and massless.

Solution To find the natural frequency of the system, Let us first find an equivalent spring at position B for replacement of spring at position C. Let a force F is applied at B, then the force at point C is:

600 Theory of Mechanisms and Machines

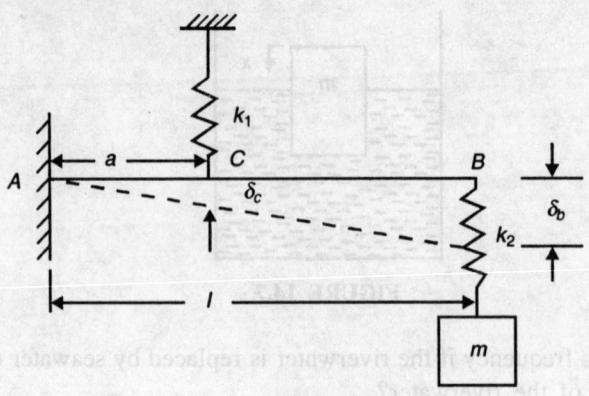

FIGURE 14.8

$$F_c = \frac{F \times l}{a}$$

Deflection at point C:
$$\delta_c = \frac{F_c}{k_1} = \frac{Fl}{ak_1}$$

Deflection at point B:
$$\delta_b = \delta_c \times \frac{l}{a}$$
$$= \frac{F}{k_1}\left(\frac{l}{a}\right)^2$$

Therefore, the stiffness of an equivalent spring to be placed at point B:
$$k_b = k_1 \left(\frac{a}{l}\right)^2$$

Thus the equivalent system is as shown in Figure 14.9.

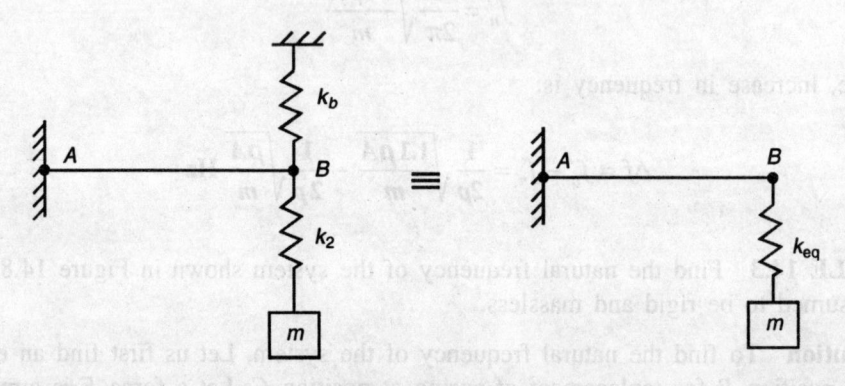

FIGURE 14.9

The overall equivalent stiffness of two springs in series is:

$$k_{eq} = \frac{k_b k_2}{k_b + k_2}$$

and natural frequency:
$$f_n = \frac{1}{2\pi}\sqrt{\frac{k_{eq}}{m}} \text{ Hz} \qquad \text{Ans.}$$

EXAMPLE 14.4 Determine the natural frequency of the mass-pulley-spring system as shown in Figure 14.10.

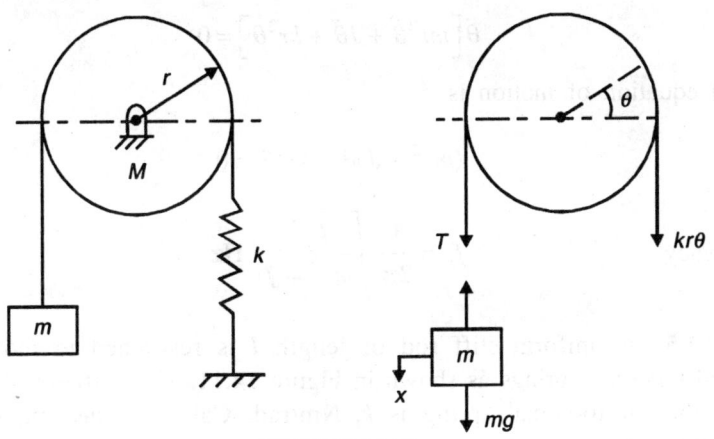

FIGURE 14.10

Solution
Let m = mass of the block
M = mass of the pulley
r = radius of the pulley
and x = displacement of the block

The total kinetic energy is:

T = kinetic energy of mass + kinetic energy of pulley

$$= \frac{1}{2}m\dot{x}^2 + \frac{1}{2}J\dot{\theta}^2$$

where J = polar moment of inertia of the pulley ($= \frac{1}{2}Mr^2$)
and θ = angular displacement of pulley

For small angular rotation of pulley, we can assume that linear displacement $x = r\theta$ and velocity $\dot{x} = r\dot{\theta}$. Therefore, the expression of kinetic energy is:

$$T = \frac{1}{2}mr^2\dot{\theta}^2 + \frac{1}{2}J\dot{\theta}^2$$

Potential energy of the spring is:

$$U = \frac{1}{2}kx^2 = \frac{1}{2}kr^2\theta^2$$

Therefore, according to energy method the rate of change of total energy must be equal to zero.

$$\frac{d}{dt}(T+U) = 0$$

or

$$\frac{d}{dt}\left[\frac{1}{2}mr^2\dot{\theta}^2 + \frac{1}{2}J\dot{\theta}^2 + \frac{1}{2}kr^2\theta^2\right] = 0$$

or

$$\dot{\theta}\left[mr^2\ddot{\theta} + J\ddot{\theta} + kr^2\theta\right] = 0$$

or differential equation of motion is:

$$(mr^2 + J)\ddot{\theta} + kr^2\theta = 0$$

Natural frequency:

$$f_n = \frac{1}{2\pi}\sqrt{\frac{kr^2}{mr^2 + J}} \text{ Hz}$$

Ans.

EXAMPLE 14.5 A uniform stiff rod of length l is restrained to move vertically by both linear and torsional springs as shown in Figure 14.11. The stiffness of linear springs is k N/mm and that of torsional spring is k_t Nm/rad. Calculate the frequency of vertical oscillation of the rod.

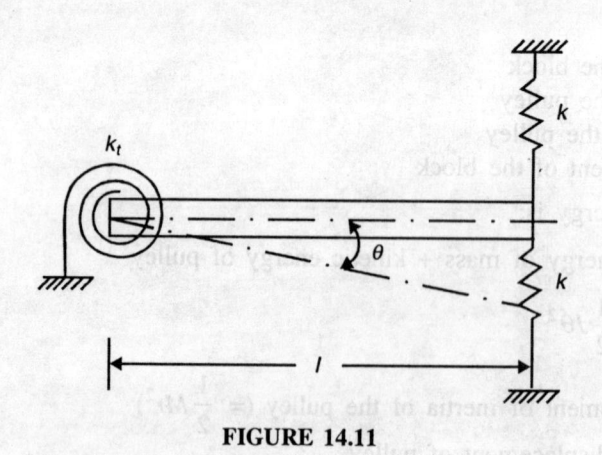

FIGURE 14.11

Solution Let the rod is displaced by angle θ. Then the resisting torque shall be

$$k_t\theta + 2kl^2\theta$$

Let J_0 = moment of inertia

Therefore, applying Newton's law,

$$\text{applied torque} = -\text{resisting torque}$$

or
$$J_0 \ddot{\theta} = -(k_t + 2kl^2)\theta$$

or
$$J_0 \ddot{\theta} + (k_t + 2kl^2)\theta = 0$$

Natural frequency:
$$f_n = \frac{1}{2\pi}\sqrt{\frac{k_t + 2kl^2}{J_0}} \text{ Hz} \qquad \text{Ans.}$$

EXAMPLE 14.6 A cylinder of mass m having radius r is connected to two springs of stiffness k as shown in Figure 14.12. Determine the natural frequency of vibration.

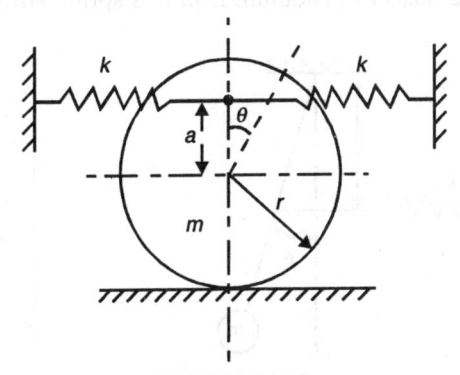

FIGURE 14.12

Solution Let x be the displacement of cylinder in translation plus rolling. Then $x = r\theta + a\theta$

where θ = angle turned by the cylinder

Kinetic energy:
$$T = \frac{1}{2}m(r\dot{\theta})^2 + \frac{1}{2}J_0\dot{\theta}^2$$

where J_0 is moment of inertia of the cylinder $\left(\frac{1}{2}mr^2\right)$

Potential energy:
$$U = 2\left(\frac{1}{2}kx^2\right)$$

or
$$U = k(r+a)^2\theta^2$$

Therefore,
$$\frac{\partial}{\partial t}(T+U) = 0$$

or
$$\frac{\partial}{\partial t}\left[\frac{1}{2}mr^2\dot{\theta}^2 + \frac{1}{4}mr^2\dot{\theta}^2 + k(r+a)^2\theta^2\right] = 0$$

or
$$\frac{3}{2}mr^2\ddot{\theta} + 2k(r+a)^2\theta = 0$$

or
$$\ddot{\theta} + \frac{4k(r+a)^2}{3mr^2}\theta = 0$$

Natural frequency: $\omega_n = \sqrt{\dfrac{4k(r+a)^2}{3mr^2}}$ rad/s   Ans.

EXAMPLE 14.7 Calculate the natural frequency of a spring connected pendulum system as shown in Figure 14.13. The mass of pendulum is m and spring stiffness is k. Neglect the mass of the rod.

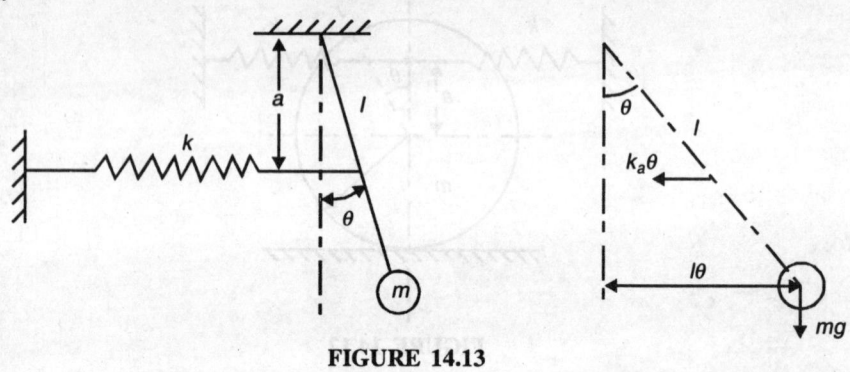

FIGURE 14.13

Solution At the equilibrium position of the pendulum at an angle θ, the various forces acting over it are shown in Figure 14.13.

Let J_0 = moment of inertia of the pendulum ($= ml^2$)
where l = length of the pendulum

The torque equation is:
$$J_0\ddot{\theta} = -mgl\theta - ka^2\theta$$

or
$$ml^2\ddot{\theta} + (mgl + ka^2)\theta = 0$$

Natural frequency: $f_n = \dfrac{1}{2\pi}\sqrt{\dfrac{mgl + ka^2}{ml^2}}$ Hz   Ans.

EXAMPLE 14.8 Write down the differential equation of motion and determine the expression for natural feqency of vibration for the system shown in Figure 14.14 given below.

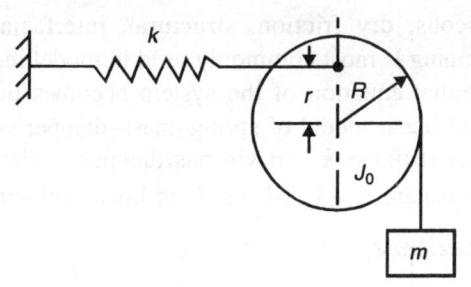

FIGURE 14.14

Solution:

Let m = mass of the block
 J_0 = moment of inertia of the pulley
 R = radius of the pulley
 k = stiffness of the spring
and x = displacement of the block ($= R\theta$)

Kinetic energy:
$$T = \frac{1}{2}m\dot{x}^2 + \frac{1}{2}J_0\dot{\theta}^2$$

and potential energy:
$$U = \frac{1}{2}k(r\theta)^2$$

Therefore,
$$\frac{\partial}{\partial t}(T + U) = 0$$

or
$$\frac{\partial}{\partial t}\left[\frac{1}{2}m\dot{x}^2 + \frac{1}{2}J_0\dot{\theta}^2 + \frac{1}{2}k(r\theta)^2\right] = 0$$

or
$$\frac{\partial}{\partial t}\left[\frac{1}{2}m(R\dot{\theta})^2 + \frac{1}{2}J_0\dot{\theta}^2 + \frac{1}{2}kr^2\theta^2\right] = 0$$

or differential equation of motion is:
$$(mR + J_0)\ddot{\theta} + kr^2\theta = 0$$

and natural frequency: $\omega_n = \sqrt{\dfrac{kr^2}{(mR + J_0)}}$ rad/s **Ans.**

14.5 FREE DAMPED VIBRATION

In general, all physical systems have one or other type of damping. In some cases, the amount of damping may be very small while in other cases, it may be large. The presence of damping decreases amplitude of free vibration with time and the rate of amplitude decay depends on the type and amount of damping. The damping in the physical system may be one of the several types namely viscous, dry friction, structural, interfacial damping and so forth. However, viscous type damping is most commonly used in modelling and analysis of vibratory system because the differential equation of the system becomes linear.

Consider a generalized linear model of spring–mass–damper system (Figure 14.15). This system contains a spring of stiffness k and viscous damper of damping coefficient 'c'. The damping resistance at any instant is $c\dot{x}$ where $\dot{x}$ is linear velocity at that time.

(a) A spring-mass-damper system (b) Free body diagram

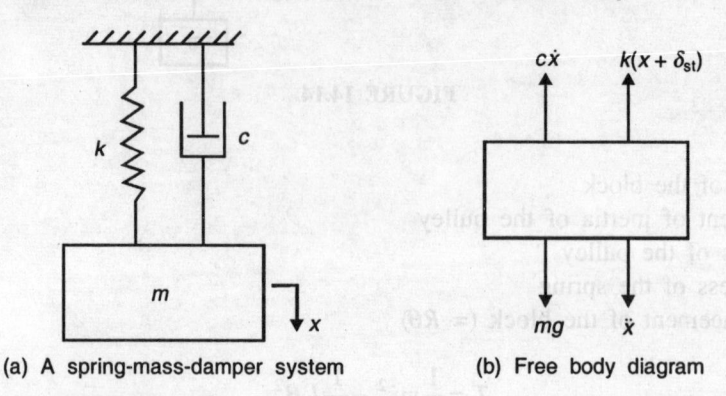

FIGURE 14.15 A linear model of damped free vibration system.

Suppose the system is displaced through a small displacement x from its static equilibrium position. The free body diagram showing various forces acting on the system is shown in Figure 14.15(b).

Using Newton's law of motion,

$$m\ddot{x} = -k(x + \delta_{st}) + mg - c\dot{x}$$

where for static equilibrium condition, $k\delta_{st} = mg$.

Therefore,
$$m\ddot{x} + c\dot{x} + kx = 0 \qquad (14.11)$$

Equation (14.11) is a fundamental linear second order differential equation of motion for a damped free vibration of single dof system. Let $x = e^{st}$ is solution of the Eq. (14.11). Substituting it in the same equation, we get the following characteristic equation of the system:

$$ms^2 + cs + k = 0 \qquad (14.12)$$

The roots of this quadratic equation are:

$$s_{1,2} = -\frac{c}{2m} \pm \sqrt{\left(\frac{c}{2m}\right)^2 - \frac{k}{m}} \qquad (14.13)$$

The most general solution of Eq. (14.11) can be written as

$$x = Ae^{s_1 t} + Be^{s_2 t} \qquad (14.14)$$

In vibration terminology that value of damping coefficient c which makes the expression within the radical sign of Eq. (14.13) vanish and thereby gives two equal roots of s is called **critical damping coefficient,** denoted by c_c.

Therefore,
$$\left(\frac{c_c}{2m}\right)^2 = \frac{k}{m} = \omega_n^2$$

or
$$c_c = 2m\omega_n \qquad (14.15)$$

The ratio of damping coefficient to critical damping coefficient c_c is called damping factor, denoted by ξ (Zeeta).

$$\xi = \frac{c}{c_c} = \frac{c}{2m\omega_n} \qquad (14.16)$$

Therefore, the roots of characteristics equation may be written as

$$s_{1,2} = \omega_n(-\xi \pm \sqrt{\xi^2 - 1}) \qquad (14.17)$$

Now we shall discuss the solution of differential equation for different conditions:

Overdamped system ($\xi > 1$). In this case, the damping factor is more than unity so both roots s_1 and s_2 are real but negative and hence the general solution for overdamped system is given as

$$x = Ae^{s_1 t} + Be^{s_2 t}$$

or
$$x = Ae^{[-\xi + \sqrt{\xi^2 - 1}]\omega_n t} + Be^{[-\xi - \sqrt{\xi^2 - 1}]\omega_n t} \qquad (14.18)$$

Figure 14.16(a) shows the general behaviour of the system. From Eq. (14.18), we find that power of both terms are negative and both decrease exponentially with time. In an overdamped system when the starting velocity is zero, the system comes to equilibrium position in exponential manner. Higher is the value of damping factor, slower is the response [See Figure 14.16(b)]. Theoretically such system will take infinite time to come back to the equilibrium position once it is disturbed from it.

Critically damped system ($\xi = 1$). In any system, if the value of damping factor (ξ) is unity, the system is said to be critically damped. The values of two roots s_1 and s_2 become equal to $-\omega_n$ and solution of the differential equation reduces to

$$x = (C + Dt)e^{-\omega_n t} \qquad (14.19)$$

The value of x decreases as time t increases and finally becomes zero as it approaches to infinity. This is shown in Figure 14.16(b) with dotted curve.

Underdamped system ($\xi < 1$). This situation generally occurs in those physical systems which have small amount of damping and is the most usual case that exists. Hence it is

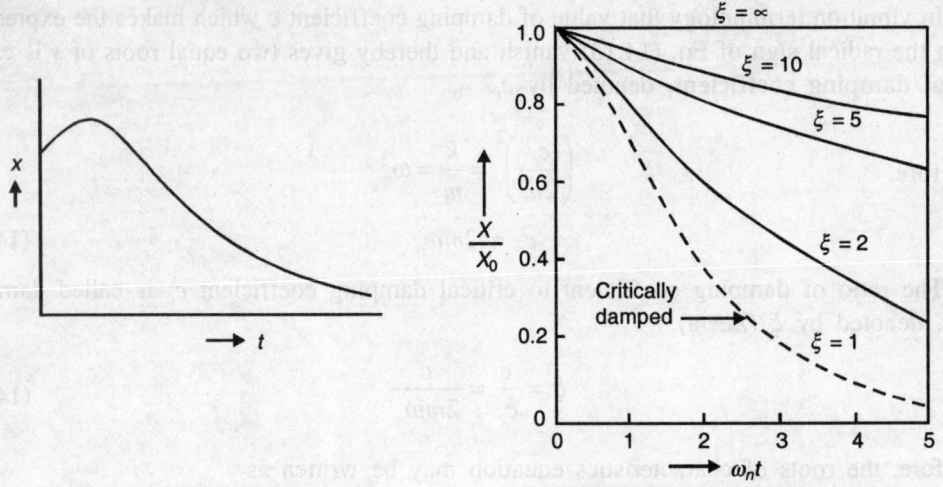

(a) Displacement-time curve for an over damped system

(b) Displacement-time curves of over damped and critically damped system

FIGURE 14.16 Displacement–time plot for a damped system.

important. In this case, values of two roots s_1 and s_2 are complex conjugates and written as

$$s_1 = \omega_n\left(-\xi + i\sqrt{1-\xi^2}\right)$$

$$s_2 = \omega_n\left(-\xi - i\sqrt{1-\xi^2}\right)$$

and damped natural frequency of a system is computed by the following expression:

$$\omega_d = \omega_n\sqrt{1-\xi^2} \qquad (14.20)$$

The solution of differential equation is given as

$$x = e^{-\xi\omega_n t}(A\cos\omega_d t + B\sin\omega_d t)$$

or
$$x = Ce^{-\xi\omega_n t}\sin(\omega_d t + \phi) \qquad (14.21)$$

where C = constant $(=\sqrt{A^2 + B^2})$

and ϕ = phase angle [= $\tan^{-1}(A/B)$]

The amplitude $Ce^{-\xi\omega_n t}$ decreases exponentially with time as shown in Figure 14.17.

14.5.1 Logarithmic Decrement

In order to measure the amount of damping, we must measure the rate of decay of oscillation in a free damped vibration. This rate of decay is expressed by logarithmic decrement. The

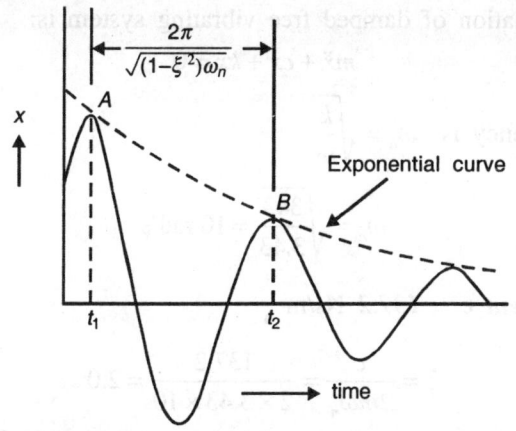

FIGURE 14.17 Displacement time curve for under-damped system.

solution of differential equation of underdamped free vibrating system is as given in Eq. (14.21) and its graphical representation is shown in the Figure 14.17.

Referring to Figure 14.17, Let t_1 and t_2 be the times at which two consecutive amplitudes occur, then the radio of amplitude to vibration is:

$$\frac{X_1}{X_2} = \frac{Ce^{-\xi\omega_n t_1}}{Ce^{-\xi\omega_n t_2}} = e^{\xi\omega_n(t_2 - t_1)}$$

where $(t_2 - t_1)$ is period of oscillation $(= 2\pi/\omega_d)$

$$\frac{X_1}{X_2} = e^{\xi\omega_n \times 2\pi/\omega_d}$$

or

$$\frac{X_1}{X_2} = e^{\frac{2\pi\xi}{\sqrt{1-\xi^2}}} \qquad (14.22)$$

The natural logarithm of this ratio is called **logarithmic decrement** and is represented by δ.

$$\delta = \ln\left(\frac{X_1}{X_2}\right) = \frac{2\pi\xi}{\sqrt{1-\xi^2}}$$

or

$$\delta = 2\pi\xi \quad \text{if} \quad \xi \ll 1 \qquad (14.23)$$

EXAMPLE 14.9 A spring–mass–damper system consists of a spring of stiffness 343 N/m. The mass is 3.43 kg. The mass is displaced by 20 mm beyond the equilibrium position and released. Find the equation of motion for the system if the damping coefficient of the damper is equal to (i) 137.2 Ns/m and (ii) 13.72 Ns/m.

Solution The equation of damped free vibrating system is:

$$m\ddot{x} + c\dot{x} + kx = 0$$

for which natural frequency is: $\omega_n = \sqrt{\dfrac{k}{m}}$

or
$$\omega_n = \sqrt{\dfrac{343}{3.43}} = 10 \text{ rad/s}$$

When damping coefficient c = 137.2 Ns/m

Damping factor: $\quad \xi = \dfrac{c}{2m\omega_n} = \dfrac{137.2}{2 \times 3.43 \times 10} = 2.0$

Therefore, the system is overdamped and its solution equation is:

$$x = Ae^{[-\xi + \sqrt{\xi^2 - 1}]\omega_n t} + Be^{[-\xi - \sqrt{\xi^2 - 1}]\omega_n t}$$

$$= Ae^{[-2 + \sqrt{2^2 - 1}]10t} + Be^{[-2 - \sqrt{2^2 - 1}]10t}$$

or
$$x = Ae^{-2.68t} + Be^{-37.32t} \qquad (i)$$

Differentiating Eq. (i) with respect to time, we get

$$\dot{x} = -2.68\, Ae^{-2.68t} - 37.32\, Be^{-37.32t} \qquad (ii)$$

For initial conditions: at $t = 0$, $x = 0.02$ m

and at $t = 0$, $\dot{x} = 0$

Substituting initial conditions in Eqs. (i) and (ii), we get

$$0.02 = A + B \qquad (iii)$$

and
$$0 = -2.68\, A - 37.32\, B \qquad (iv)$$

Solving the above eqs. for constants A and B, we get

$$A = 0.02155$$

and
$$B = -0.00155$$

The equation of motion is:

$$x = 0.02155 e^{-2.68t} - 0.00155 e^{-37.32t} \qquad \textbf{Ans.}$$

When damping coefficient c = 13.72 Ns/m

Natural frequency: $\quad \omega_n = 10$ rad/s

Damping factor: $\quad \xi = \dfrac{c}{2m\omega_n} = \dfrac{13.72}{2 \times 3.43 \times 10} = 0.2$

Mechanical Vibrations 611

Therefore, it is underdamped system.
The solution of equation of motion for underdamped system is:

$$x = Ce^{-\xi\omega_n t} \sin(\omega_d t + \phi)$$

where
$$\omega_d = \omega_n \times \sqrt{1-\xi^2} = 10 \times \sqrt{1-0.2^2} = 9.8$$

or
$$x = Ce^{-0.2 \times 10 \times t} \sin(9.8t + \phi)$$

or
$$x = Ce^{-2t} \sin(9.8t + \phi)$$

Differentiating the above equation, we get

$$\dot{x} = -2Ce^{-2t} \sin(9.8t + \phi) + 9.8 \times Ce^{-2t} \times \cos(9.8t + \phi)$$

Substituting initial conditions,

at $t = 0$, $x = 0.02$ m

and at $t = 0$, $\dot{x} = 0$,

$$0.02 = C \times \sin\phi \quad \text{(v)}$$
$$0 = -2C \times \sin\phi + 9.8\, C \times \cos\phi \quad \text{(vi)}$$

Solving Eqs. (v) and (vi), we get

$$\phi = 78.46° \quad \text{and} \quad C = 0.0204$$

Therefore, the equation for underdamped system is:

$$x = 0.0204\, e^{-2t} \sin(9.8t + 78.46°) \quad \text{Ans.}$$

EXAMPLE 14.10 A mass of 1 kg is supported on a spring damper having stiffness and damping coefficient 9800 N/m and 4.9 Ns/m. Determine the natural frequency of the system. Also determine the logarithmic decrement and amplitude after three cycles if the initial dispacement is 3 mm.

Solution Natural frequency:

$$\omega_n = \sqrt{\frac{k}{m}} = \sqrt{\frac{9800}{1}} = 99 \text{ rad/s}$$

Damping factor:
$$\xi = \frac{c}{2m\omega_n} = \frac{4.9}{2 \times 1 \times 99} = 0.0247$$

Therefore, the system is underdamped.

Logarithmic decrement:
$$\delta = \ln\left(\frac{X_1}{X_2}\right) = \frac{2\pi\xi}{\sqrt{1-\xi^2}}$$

$$= \frac{2\pi \times 0.0247}{\sqrt{1-0.0247^2}}$$

$$= \mathbf{0.155} \qquad \text{Ans.}$$

Let X_0 = initial amplitude of vibration
and X_n = amplitude after n cycles

$$\delta = \frac{1}{n}\ln\left(\frac{X_0}{X_n}\right)$$

or
$$0.155 = \frac{1}{3}\ln\left(\frac{3.0}{x_3}\right)$$

or
$$x_3 = \mathbf{1.884} \text{ mm} \qquad \text{Ans.}$$

EXAMPLE 14.11 The damped vibration records of a spring–mass–damper system show the following data:

  (i) amplitude on the second cycle = 12 mm
  (ii) amplitude on the third cycle = 10.5 mm
  (iii) stiffness of spring = 7840 N/m
  (iv) mass of the block = 2 kg

Determine the damping coefficient of the vibratory system.

Solution Natural frequency:

$$\omega_n = \frac{k}{m} = \sqrt{\frac{7840}{2}} = 62.6 \text{ rad/s}$$

Logarithmic decrement:
$$\delta = \ln\left(\frac{X_2}{X_3}\right) = \ln\left(\frac{12}{10.5}\right) = 0.1335$$

Also
$$\delta = \frac{2\pi\xi}{\sqrt{1-\xi^2}} = 2\pi\xi \quad \text{for} \quad \xi \ll 1$$

Therefore, Damping factor: $\xi = \dfrac{\delta}{2\pi} = \dfrac{0.1335}{2\pi} = 0.0212$

Also
$$\xi = \frac{c}{2m\omega_n}$$

or damping coefficient:
$$c = 2m\omega_n\xi$$
$$= 2 \times 2 \times 62.6 \times 0.0212$$
$$= 5.31 \text{ Ns/m} \qquad \textbf{Ans.}$$

EXAMPLE 14.12 A single pendulum is pivoted at a point O as shown in Figure 14.18. If the mass of the rod is negligible for small oscillations, find the damped natural frequency of the pendulum.

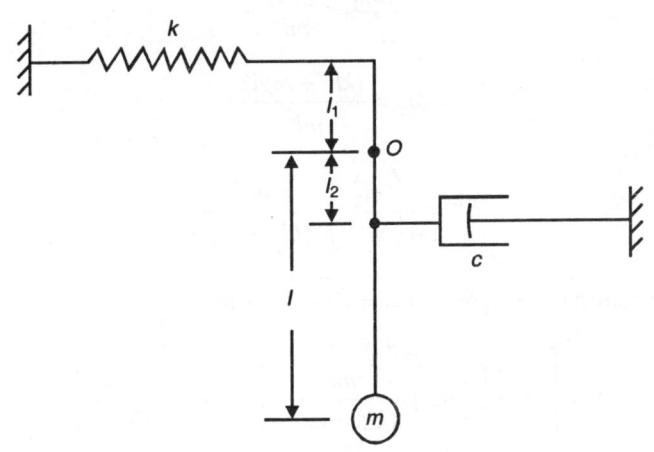

FIGURE 14.18

Solution Consider the position of the pendulum when it rotates by small angle θ. The various forces acting on the pendulum rod are shown in the Figure 14.19. Taking moment about fulcrum point O, we get

$$ml^2\ddot{\theta} + cl_2^2\dot{\theta} + kl_1^2\theta + mgl\sin\theta = 0.$$

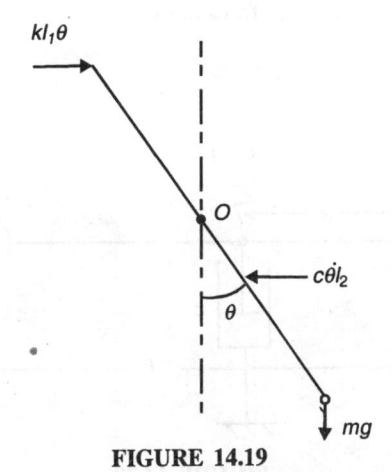

FIGURE 14.19

For small angle, $\sin\theta = \theta$.

$\therefore$
$$ml^2\ddot{\theta} + cl_2^2\dot{\theta} + (kl_1^2 + mgl)\theta = 0$$

or
$$\ddot{\theta} + \frac{cl_2^2}{ml^2}\dot{\theta} + \frac{(kl_1^2 + mgl)}{ml^2}\theta = 0 \qquad (i)$$

Comparing Eq. (i) with general equation of motion as given below:

$$\ddot{\theta} + 2\xi\omega_n\dot{\theta} + \omega_n^2\theta = 0$$

we find that
$$2\xi\omega_n = \frac{cl_2^2}{ml^2}$$

and
$$\omega_n^2 = \frac{(kl_1^2 + mgl)}{ml^2}$$

or
$$\xi^2 = \frac{1}{4}\left(\frac{cl_2^2}{ml^2}\right)^2 \frac{ml^2}{(kl_1^2 + mgl)}$$

Therefore, damped natural frequency: $\omega_d = \sqrt{1-\xi^2} \times \omega_n$

or
$$\omega_d = \left[1 - \frac{1}{4}\left(\frac{cl_2^2}{ml^2}\right)^2 \frac{ml^2}{(kl_1^2 + mgl)}\right]^{0.5} \times \left(\frac{kl_1^2 + mgl}{ml^2}\right)^{0.5}$$

or
$$\omega_d = \left[\frac{kl_1^2 + mgl}{ml^2} - \frac{cl_2^2}{2ml^2}\right]^{0.5} \qquad \textbf{Ans.}$$

EXAMPLE 14.13 Obtain the differential equation of motion for the system shown in Figure 14.20 given below and also find out the expression for (i) the critical damping coefficient and (ii) the damped natural frequency.

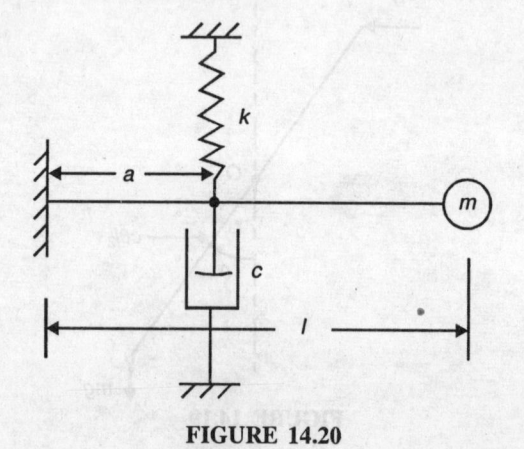

FIGURE 14.20

Solution For small angular displacement θ of the vibratory mass m, the vertical displacement at the spring is $a\theta$.

Therefore, spring force $= ka\theta$
and damping force $= ca\dot{\theta}$.
The equation of motion is:

$$J_0 \ddot{\theta} = -a(ka\theta + ca\dot{\theta})$$

where J_0 is moment of inertia of the mass $(= ml^2)$

or
$$ml^2 \ddot{\theta} + ca^2 \dot{\theta} + ka^2 \theta = 0$$

The solution of the equation is $\theta = e^{st}$ where s is the root of the following equation:

$$ml^2 s^2 + ca^2 s + ka^2 = 0$$

$$s_{1,2} = -\frac{ca^2}{2ml^2} \pm \sqrt{\left(\frac{ca^2}{2ml^2}\right)^2 - \frac{ka^2}{ml^2}}$$

For critical damping, the values inside the radical sign must be zero. Therefore,

$$\left(\frac{ca^2}{2ml^2}\right)^2 = \frac{ka^2}{ml^2}$$

or
$$c_c = \frac{2l}{a}\sqrt{km} \qquad \text{Ans.}$$

Damped natural frequency is radical value with negative sign.

$$\omega_d = \sqrt{\frac{ka^2}{ml^2} - \left(\frac{ca^2}{2ml^2}\right)^2} \qquad \text{Ans.}$$

EXAMPLE 14.14 A barrel of large gun on firing recoils against a spring. At the end of recoil, the damper is engaged to bring the barrel to its original position ready for the next firing within the minimum possible time. The barrel weighs 800 N and the recoil distance is 1.5 m. The gunfire shots weighing 15 N at 1.6 km/s. Find the required stiffness and damping coefficient.

Solution
Let m_1 = mass of the barrel
m_2 = mass of gunshell
v_1 = initial velocity with which the barrel recoils
and v_2 = velocity of the gunshell

From the principle of conservation of momentum, we know that

$$m_1 v_1 = m_2 v_2$$

or
$$\frac{800}{g} \times v_1 = \frac{15}{g} \times 1.6 \times 1000$$

or
$$v_1 = 30 \text{ m/s}$$

During the recoil the damper is inactive and therefore, the initial kinetic energy of the barrel must get absorbed as elastic energy in the spring.

Thus
$$\frac{1}{2} m_1 v_1^2 = \frac{1}{2} k d^2$$

where d = recoil distance
and k = required stiffness of the spring

or
$$\frac{1}{2} \times \frac{800}{9.81} \times 30^2 = \frac{1}{2} k \times 1.5^2$$

or
$$k = \mathbf{32.62 \text{ kN/m}} \qquad \textbf{Ans.}$$

In order to return the barrel to its original position ready for the next firing is shortest possible time, the damper should have critical damping. Therefore,

$$c_c = 2 m \omega_n = 2\sqrt{km}$$

$$= 2\sqrt{32.62 \times 10^3 \times 800/9.81}$$

$$= \mathbf{3.262 \text{ kNs/m}} \qquad \textbf{Ans.}$$

14.6 FORCED DAMPED VIBRATION

In free vibration, when a vibratory system is disturbed from its equilibrium position, it executes vibration due to elastic properties. The system comes to the rest position in a certain period of time depending upon the damping capacity. However, there is another category of vibration in which the system keeps it vibrating due to some externally impressed force such as vibration of IC engine, compressor, turbine and so forth. The impressed force may be periodic, impulsive or random in nature. In this section, we shall discuss about the classical harmonic forced vibration.

Consider a classical spring-mass-damper system which is excited by sinusoidal forcing function $F_0 \sin \omega t$ [see Figure 14.21(a)]. At any instant, if the mass is displaced through distance x in downward direction, the various forces acting on the mass are shown in the free body diagram [Figure 14.21(b)].

Mechanical Vibrations | **617**

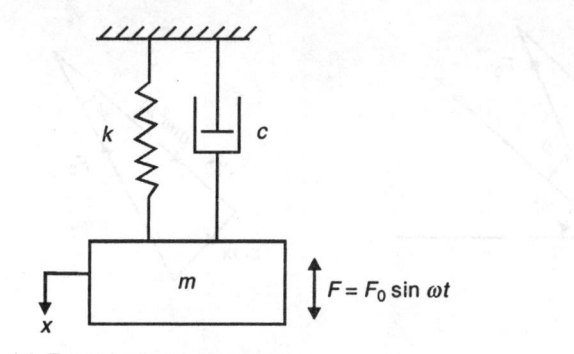

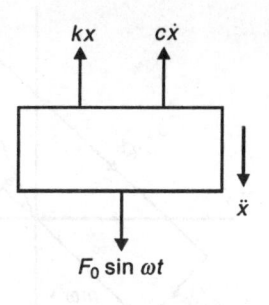

(a) Forced vibrating system. **(b)** Free body diagram

FIGURE 14.21 A forced damped vibrating system.

The differential equation of motion for the above system is:

$$m\ddot{x} + c\dot{x} + kx = F_0 \sin \omega t \tag{14.24}$$

The above equation is a linear, non-homogeneous second order differential equation which has solution in two parts—complimentary function and particular integral. The complimentary function is same as the solution of damped free vibration given by Eq. (14.21) for underdamped vibratory system.

The particular integral is a steady state harmonic vibration having a frequency equal to that of the excitation and displacement vector lags force vector by some angle. Let us assume that the particular integral is:

$$x_p = X \sin(\omega t - \phi) \tag{14.25}$$

where X = amplitude of vibration
and ϕ = angle by which displacement vector lags the force vector

Differentiating Eq. (14.25), we get

$$\dot{x}_p = \omega X \sin(\omega t - \phi + \pi/2)$$

and

$$\ddot{x}_p = -\omega^2 X \sin(\omega t - \phi + \pi)$$

Substituting the value of x_p, $\dot{x}_p$ and $\ddot{x}_p$ in Eq. (14.24), we get

$$F_0 \sin \omega t - kX \sin(\omega t - \phi) - c\omega X \sin(\omega t - \phi + \pi/2)$$

and

$$-m\omega^2 X \sin(\omega t - \phi + \pi) = 0 \tag{14.26}$$

The vector representation of Eq. (14.26) is shown in Figure 14.22, from which the following observations are derived:

(i) The displacement lags the impressed force by an angle ϕ.
(ii) Spring force is always opposite in direction to the displacement.
(iii) Damping force lags the displacement by 90° and is always opposite in direction to the velocity.
(iv) Inertia force is in phase with displacement but acts opposite to acceleration.

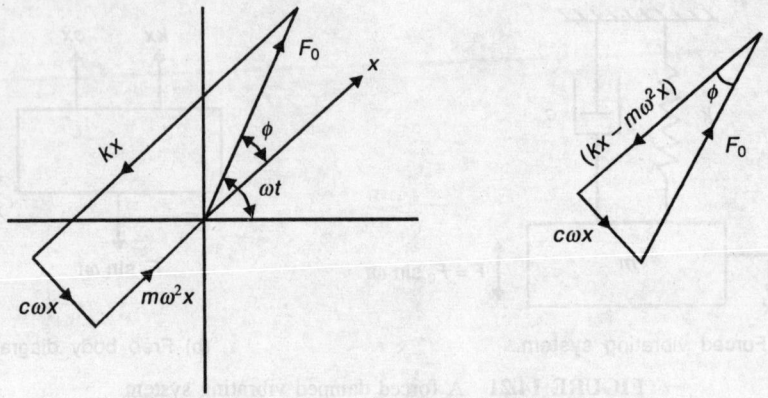

FIGURE 14.22 Vector representation of Eq. (14.26).

From Figure 14.22, the expressions for steady state amplitude X and phase angle ϕ are given as

$$X = \frac{F_0}{\sqrt{(k - m\omega^2)^2 + (c\omega)^2}} \tag{14.27}$$

and

$$\phi = \tan^{-1}\left(\frac{c\omega}{k - m\omega^2}\right) \tag{14.28}$$

The steady state amplitude X can be written as

$$X = \frac{F_0/k}{\sqrt{\left(1 - \frac{m\omega^2}{k}\right)^2 + \left(\frac{c\omega}{k}\right)^2}}$$

or

$$X = \frac{X_{st}}{\sqrt{(1 - r^2)^2 + (2\xi r)^2}} \tag{14.29}$$

and

phase angle: $\phi = \tan^{-1}\left(\frac{2\xi r}{1 - r^2}\right) \tag{14.30}$

where X_{st} = static deflection (= F_0/k)
and r = ratio of forcing frequency to natural frequency (= ω/ω_n)

Hence the particular integral is:

$$x_p = \frac{X_{st} \sin(\omega t - \phi)}{\sqrt{(1 - r^2)^2 + (2\xi r)^2}} \tag{14.31}$$

and the complete solution of the differential equation of motion is sum of Equations (14.21) and Eq. (14.31).

In practical system, the complimentary function, which is also called transient vibration, dies out after lapse of certain period of time leaving only steady state vibration. Therefore, it is important to know the behaviour of the system under steady state condition.

In vibration terminology, the ratio of steady state amplitude to the zero frequency deflection (static deflection), i.e. X/X_{st} is called **magnification factor**.

$$\frac{X}{X_{st}} = \frac{1}{\sqrt{(1-r^2)^2 + (2\xi r)^2}} \tag{14.32}$$

The non-dimensional plot of magnification factor versus frequency ratio and phase lag versus frequency ratio are shown in Figure 14.23 and Figure 14.24 respectively. From these plots, the following useful information is derived:

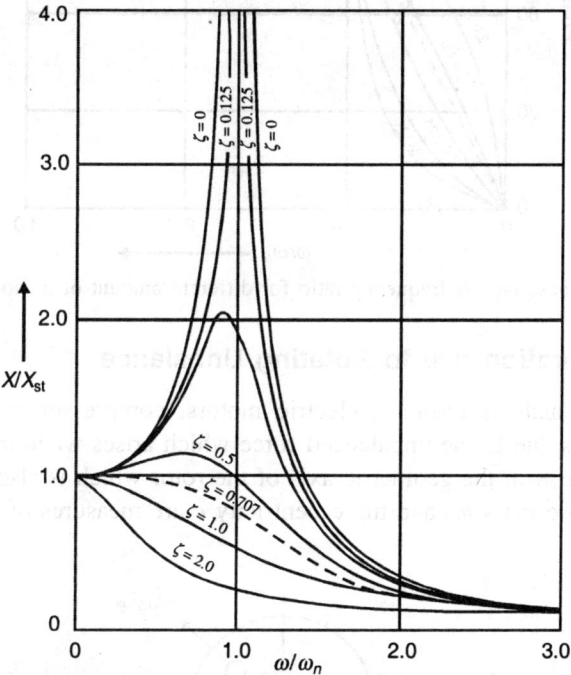

FIGURE 14.23 Amplitude ratio v/s frequency ratio for different amount of damping.

1. The steady state solution is independent of initial conditions.
2. At zero forcing frequency, the magnification factor is unity and is independent of damping factor.
3. At resonance condition, the amplitude of vibration becomes excessive for small amount of damping. It is maximum (theoretically infinite) when damping is zero.
4. The amplitude of vibration becomes small at very high forcing frequency.
5. For a given value of damping factor (ξ) and the ratio of frequency (r), the phase difference is constant. At resonance, the phase angle is 90°. However, for zero damping factor, it is zero degree upto resonance and 180° beyond that (Figure 14.24).

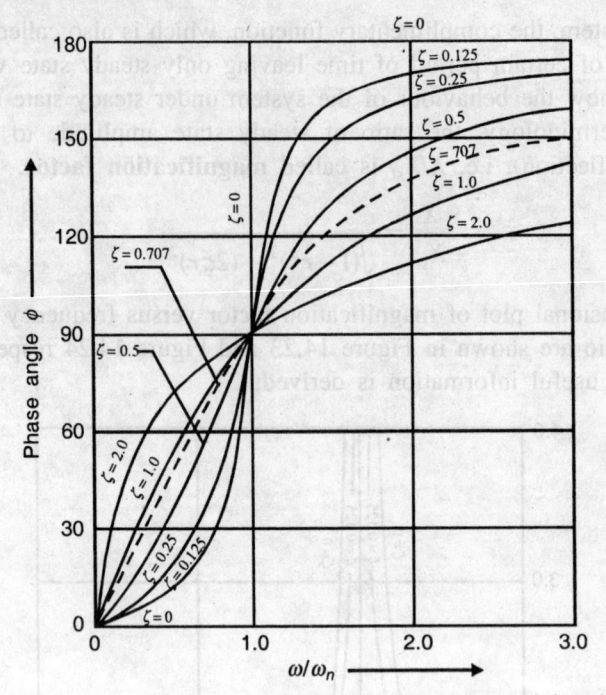

FIGURE 14.24 Phase lag v/s frequency ratio for different amount of damping.

14.6.1 Forced Vibration due to Rotating Unbalance

In rotating machines such as turbines, electric motors, compressor fans and so forth, the problem of vibration is due to the unbalanced force which arises when the mass centre of the rotor does not coincide with the geometric axis of the rotor which is also the axis of rotation. The equivalent eccentric mass m_0 and the eccentricity e are measures of the unbalanced force (Figure 14.25).

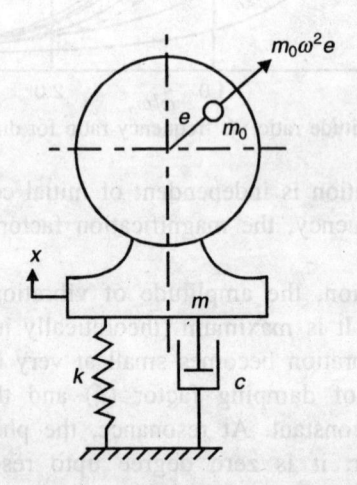

FIGURE 14.25 Vibration due to rotating unbalanced mass.

Refer to Figure 14.25.

Let m_0 = eccentric mass
m = mass of the machine including eccentric mass
e = eccentricity of mass m_0
ω = angular velocity of rotation
and x = vibratory displacement of the machine

For a single dof system, if the vertical displacement of the rotating machine is $(x + e \sin \omega t)$, the equation of motion of the vibratory rotating machine is given as

$$(m - m_0)\frac{d^2x}{dt^2} + m_0 \frac{d^2}{dt}(x + e\sin\omega t) = -kx - c\frac{dx}{dt}$$

or
$$m\ddot{x} + c\dot{x} + kx = m_0 \omega^2 e \sin \omega t \tag{14.33}$$

where $m_0 \omega^2 e \sin \omega t$ is the centrifugal force which excites the system.

The steady state amplitude of the system corresponding to Eq. (14.33) is

$$X = \frac{m_0 \omega^2 e / k}{\sqrt{(k - m\omega^2)^2 + (c\omega)^2}} \tag{14.34}$$

The non-dimensional amplitude ratio and phase angle are calculated by the following expressions:

$$\frac{X}{\frac{m_0 e}{m}} = \frac{r^2}{\sqrt{(1 - r^2)^2 + (2\xi r)^2}} \tag{14.35}$$

and
$$\text{phase angle: } \phi = \tan^{-1}\left(\frac{2\xi r}{1 - r^2}\right) \tag{14.36}$$

The frequency response curves for a rotating unbalanced system for different amounts of damping are shown in Figure 14.26.

14.6.2 Excitation of the Support

In some applications, the excitation of the system is through support instead of mass like suspension system of an automobile and seismic instruments. In these cases, the support is excited by sinusoidal motion as shown in Figure 14.27.

Let the support of vibratory system is displaced by

$$y = Y \sin \omega t \tag{14.37}$$

Suppose if x is absolute motion of the mass, then the differential equation of motion is:

$$m\ddot{x} = -c(\dot{x} - \dot{y}) - k(x - y) \tag{14.38}$$

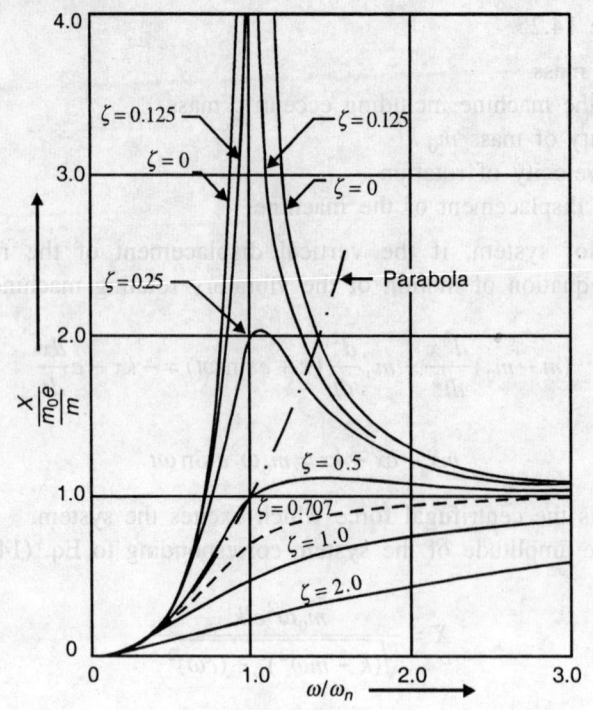

FIGURE 14.26 Amplitude ratio v/s frequency ratio for different amount of damping.

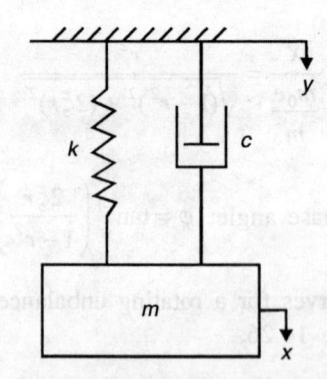

FIGURE 14.27 Forced vibration due to excitation of the support.

Substituting the value of y from Eq. (14.37), we have

$$m\ddot{x} + c\dot{x} + kx = Y[k \sin \omega t + c\omega \cos \omega t]$$

or
$$m\ddot{x} + c\dot{x} + kx = Y\sqrt{k^2 + (c\omega)^2} \sin(\omega t + \alpha) \qquad (14.39)$$

where
$$\alpha = \tan^{-1}\left(\frac{c\omega}{k}\right) = \tan^{-1}(2\xi r)$$

The steady state solution of Eq. (14.39) is given by

$$x = X \sin(\omega t + \alpha - \phi) \qquad (14.40)$$

where X = steady state amplitude.

$$X = \frac{Y\sqrt{k^2 + (c\omega)^2}}{\sqrt{(k^2 - m\omega^2)^2 + (c\omega)^2}}$$

or

$$X = \frac{Y\sqrt{1 + (2\xi r)^2}}{\sqrt{(1 - r^2)^2 + (2\xi r)^2}} \qquad (14.41)$$

and

$$\text{phase angle: } \phi = \tan^{-1}\left(\frac{2\xi r}{1 - r^2}\right) \qquad (14.42)$$

The ratio X/Y from Eq. (14.41) is called **motion transmissibility.**

14.6.3 Vibration Isolation and Transmissibility

Machines are often mounted on springs and dampers to reduce transmission of force to the foundation so that the adjoining structure or machine is not set into heavy vibration. In vibration, the ratio of force transmitted to the foundation to that impressed upon the machine is called **transmissibility,** where lesser the force is transmitted, greater is said to be isolation.

In classical single dof spring-mass-damper system as shown in Figure 14.28(a), the spring force kx and damper force $c\omega x$ are two forces acting on mass and foundation. Therefore, the amount of force transmitted to the foundation is vector sum of these two forces which are 90° out of phase with each other. The direction of transmitted force is opposite to that of the mass. Figure 14.28(b) shows the vector plot of these forces.

Let F_t = force transmitted to the foundation. Refer to Figure 14.28(c),

Force transmitted:
$$F_t = X\sqrt{k^2 + (c\omega)^2} \qquad (14.43)$$

Substituting the value of X from Eq. (14.27), we get:

$$F_t = \frac{F_0\sqrt{k^2 + (c\omega^2)}}{\sqrt{(k - m\omega^2)^2 + (c\omega)^2}}$$

Therefore,
$$\text{transmissibility: } T_R = \frac{F_t}{F_0}$$

or
$$T_R = \frac{\sqrt{1 + (2\xi r)^2}}{\sqrt{(1 - r^2)^2 + (2\xi r)^2}} \qquad (14.44)$$

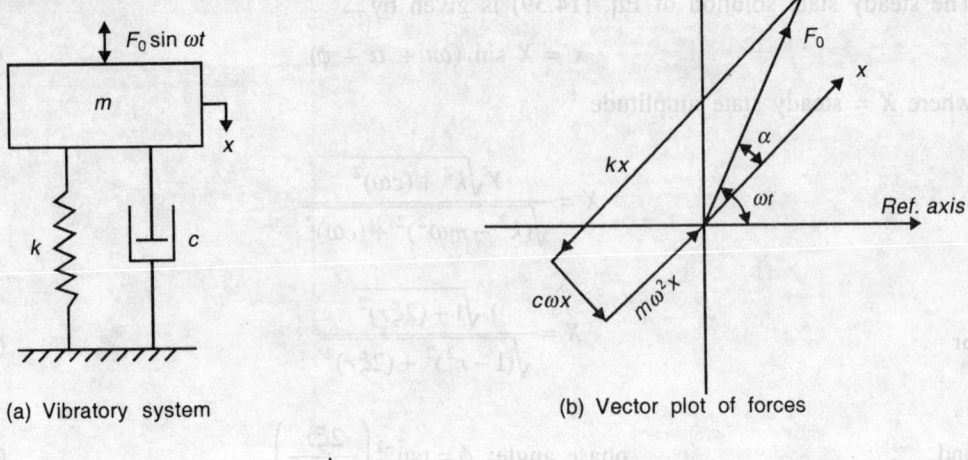

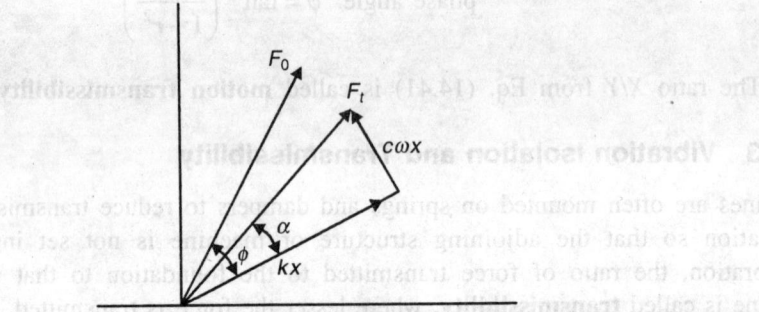

FIGURE 14.28 Determination of force transmissibility.

(a) Vibratory system

(b) Vector plot of forces

The transmitted force lags behind the impressed force by angle $(\phi - \alpha)$, where angle $\alpha = \tan^{-1}(2\xi r)$. [See Figure 14.28(c).]

The transmissibility and phase angle are plotted against frequency ratio as shown in Figure 14.29 and Figure 14.30 respectively.

The following useful information is derived from these figures:

1. The transmissibility curves start from unity value and pass through the unity transmissibility at frequency ratio $r = \sqrt{2}$ and after that they tend to zero as r approaches infinity.
2. In the first region of the curve that is upto r equal to $\sqrt{2}$, the greater amount of damping gives lower transmissibility but it is always greater than unity.
3. In second region that is when r is greater than $\sqrt{2}$, the lesser amount of damping gives lower transmissibility which is always less than unity. Therefore, in order to have low value of transmissibility, the operating frequency range is kept away in the second range when lower damping is suitable. However, vibrating system has to pass through resonance condition at which transmissibility will be maximum.

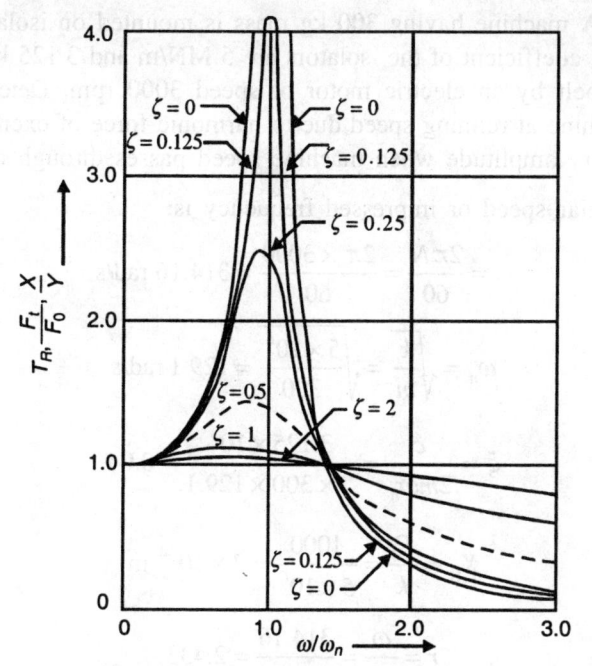

FIGURE 14.29 Transmissibility v/s frequency ratio curve for different amount of damping.

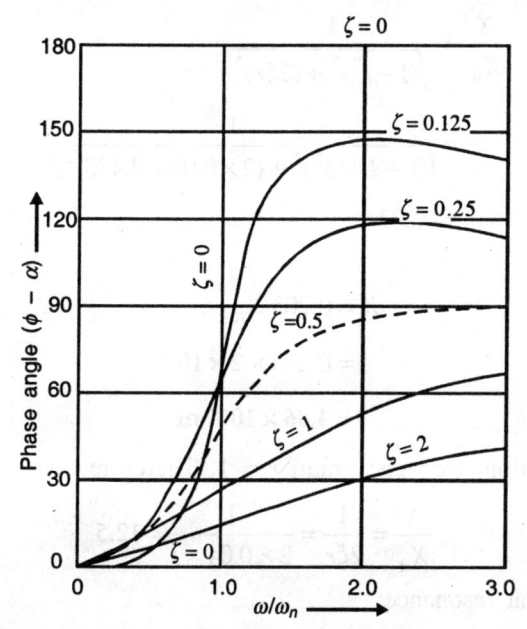

FIGURE 14.30 Phase angle v/s frequency ratio curve for different amount of damping.

4. The phase angle curves start from zero degree and approach to 90° when frequency ratio r tends to infinity. However, for damping factor less than 0.5, phase angle continues to increase to a maximum value and than gradually it becomes 90° when frequency ratio approaches infinity.

EXAMPLE 14.15 A machine having 300 kg mass is mounted on isolators. The combined stiffness and damping coefficient of the isolators are 5 MN/m and 3.125 kNs/m. The machine is driven through a belt by an electric motor of speed 3000 rpm. Determine the vibratory amplitude of the machine at running speed due to harmonic force of excitation of 1 kN. Also determine the vibratory amplitude when machine speed passes through resonance condition.

Solution Angular speed or impressed frequency is:

$$\omega = \frac{2\pi N}{60} = \frac{2\pi \times 3000}{60} = 314.16 \text{ rad/s}$$

Natural frequency:
$$\omega_n = \sqrt{\frac{k}{m}} = \sqrt{\frac{5 \times 10^6}{300}} = 129.1 \text{ rad/s}$$

Damping factor:
$$\xi = \frac{c}{2m\omega_n} = \frac{3.125 \times 10^3}{2 \times 300 \times 129.1} = 0.04$$

Static deflection:
$$X_{st} = \frac{F_0}{k} = \frac{1000}{5 \times 10^6} = 2 \times 10^{-4} \text{ m}$$

Frequency ratio:
$$r = \frac{\omega}{\omega_n} = \frac{314.16}{129.1} = 2.433$$

Magnification factor:
$$\frac{X}{X_{st}} = \frac{1}{\sqrt{(1-r^2)^2 + (2\xi r)^2}}$$

$$= \frac{1}{[(1-2.433^2)^2 + (2 \times 0.04 \times 2.433)^2]^{0.5}}$$

$$= 0.203$$

or

Steady state amplitude:
$$X = 0.203 \, X_{st}$$
$$= 0.203 \times 2 \times 10^{-4}$$
$$= \mathbf{4.06 \times 10^{-5} \text{ m}} \qquad \text{Ans.}$$

At the resonance condition, frequency ratio $r = 1$. Therefore,

$$\frac{X}{X_{st}} = \frac{1}{2\xi r} = \frac{1}{2 \times 0.04 \times 1} = 12.5$$

Steady state amplitude at resonance:

$$X = 12.5 \, X_{st}$$
$$= 12.5 \times 2 \times 10^{-4}$$
$$= \mathbf{2.5 \times 10^{-3} \text{ m}} \qquad \text{Ans.}$$

EXAMPLE 14.16 A machine part of 2kg mass vibrates in a viscous medium. Determine the damping coefficient when a harmonic excitation force of 25 N results in a resonant amplitude of 12.5 mm with a period of 0.1s, if the system is excited by a harmonic force of 4 Hz frequency, what will be the percentage increase in the amplitude of vibration when damper is removed as compared to that with damper?

Solution Natural frequency: $f_n = \dfrac{1}{t} = \dfrac{1}{0.1} = 10$ Hz **Ans.**

or $\omega_n = 2\pi f_n = 2\pi \times 10 = 62.83$ rad/s

Spring stiffness: $k = m\omega_n^2$

$$= 2 \times 62.83^2$$

$$= 7895.2 \text{ N/m}$$

Static deflection: $X_{st} = \dfrac{F_0}{k} = \dfrac{25}{7895.2} = 3.167 \times 10^{-3}$ m

We know that at resonance condition ($r = 1$), the magnification factor or ratio of amplitudes:

$$\dfrac{X}{X_{st}} = \dfrac{1}{2\xi}$$

or Damping factor: $\xi = \dfrac{X_{st}}{2X} = \dfrac{3.167 \times 10^{-3}}{2 \times 0.0125} = 0.126$

Damping coefficient: $c = 2m\omega_n\xi$

$$= 2 \times 2 \times 62.83 \times 0.126$$

$$= 31.6 \text{ Ns/m}$$ **Ans.**

If excitation frequency is 4Hz, i.e. $\omega = 2\pi \times 4 = 8\pi$ rad/s

Ratio of frequency: $r = \dfrac{\omega}{\omega_n} = \dfrac{8\pi}{20\pi} = 0.4$

The ratio of amplitudes: $\dfrac{X}{X_{st}} = \dfrac{1}{1-r^2} = \dfrac{1}{1-0.4^2} = 1.19$ (when $\xi = 0$)

or percentage increase in the amplitude is **19%**. **Ans.**

EXAMPLE 14.17 A mass of 100 kg is suspended on a spring having stiffness 19600 N/m and is acted upon by a harmonic force of 39.2 N at undamped natural frequency. If the damping coefficient is 98 Ns/m, determine the following:

(a) undamped natural frequency
(b) amplitude of vibration
(c) phase difference angle

Solution Natural frequency:

$$\omega_n = \sqrt{\frac{k}{m}} = \sqrt{\frac{19600}{100}} = 14 \text{ rad/s} \qquad \text{Ans.}$$

Static deflection:
$$X_{st} = \frac{F_0}{k} = \frac{39.2}{19600} = 2 \times 10^{-3} \text{ m}$$

Damping factor:
$$\xi = \frac{c}{2m\omega_n} = \frac{98}{2 \times 100 \times 14} = 0.035$$

At resonance condition ($r = 1$), the ratio of amplitude of vibration is:

$$\frac{X}{X_{st}} = \frac{1}{2\xi}$$

or Amplitude of vibration: $X = \dfrac{X_{st}}{2\xi} = \dfrac{2 \times 10^{-3}}{2 \times 0.035} = 0.0286 \text{ m}$ **Ans.**

Phase angle:
$$\phi = \tan^{-1}\left(\frac{2\xi r}{1 - r^2}\right) = \tan^{-1}\left(\frac{2 \times 0.035 \times 1}{1 - 1^2}\right) = 90° \qquad \text{Ans.}$$

EXAMPLE 14.18 A vertical single stage air compressor having a mass of 500 kg is mounted on springs having stiffness of 1.9×10^5 N/m and dashpot with a damping factor of 0.2. The rotating parts are completely balanced and equivalent reciprocating parts weigh 20 kg. The stroke is 0.2 m. Determine the dynamic amplitude of vibration and phase difference between motion and excited force if the compressor is operated at 200 rpm.

Solution Natural frequency:

$$\omega_n = \sqrt{\frac{k}{m}} = \sqrt{\frac{1.9 \times 10^5}{500}} = 19.5 \text{ rad/s}$$

Excitation frequency:
$$\omega = \frac{2\pi N}{60} = \frac{2\pi \times 200}{60} = 20.94 \text{ rad/s}$$

Frequency ratio:
$$r = \frac{\omega}{\omega_n} = \frac{20.94}{19.5} = 1.074$$

Eccentricity:
$$e = \frac{\text{stroke}}{2} = \frac{0.2}{2} = 0.1 \text{ m}$$

The ratio of amplitudes for unbalanced mass is:

$$\frac{X}{\frac{m_0 e}{m}} = \frac{r^2}{\sqrt{(1-r^2)^2 + (2\xi r)^2}}$$

$$\frac{X}{\frac{20 \times 0.1}{500}} = \frac{1.074^2}{\sqrt{(1-1.074^2)^2 + (2 \times 0.2 \times 1.074)^2}}$$

or Dynamic amplitude: $X = 0.01$ m   **Ans.**

Phase angle: $\phi = \tan^{-1}\left(\frac{2\xi r}{1-r^2}\right) = \tan^{-1}\left(\frac{2 \times 0.2 \times 1.074}{1-1.074^2}\right)$

or $\phi = 109.6°$ **Ans.**

EXAMPLE 14.19 A machine weighing 70 kg is mounted on springs of stiffness 1.2 MN/m with damping factor equal to 0.2. The piston within the machine weighing 2 kg has a reciprocating motion with a stroke of 80 mm and speed of 300 rpm. Determine:

 (i) the amplitude of motion
 (ii) its phase angle
 (iii) the transmissibility
 (iv) the phase angle of the transmissibility
 (v) the force transmitted to the foundation.

Solution Natural frequency:

$$\omega_n = \sqrt{\frac{k}{m}} = \sqrt{\frac{1.2 \times 10^6}{72}} = 129 \text{ rad/s}$$

Forcing frequency: $\omega = \frac{2\pi N}{60} = \frac{2\pi \times 3000}{60} = 314.16$ rad/s

Ratio of frequency: $r = \frac{\omega}{\omega_n} = \frac{314.16}{129} = 2.43$

Eccentricity: $e = \frac{\text{Stroke}}{2} = \frac{80}{2} = 40$ mm

The ratio of amplitude of vibration is:

$$\frac{X}{\frac{m_0 e}{m}} = \frac{r^2}{\sqrt{(1-r^2)^2 + (2\xi r)^2}}$$

$$= \frac{2.43^2}{\sqrt{(1-2.43^2)^2 + (2 \times 0.2 \times 2.43)^2}}$$

$$= 1.181$$

(i) Amplitude of vibration: $X = \dfrac{m_0 e}{m} \times 1.181 = \dfrac{2 \times 40}{70} \times 1.181$

$$= \mathbf{1.35 \text{ mm}} \qquad \textbf{Ans.}$$

(ii) Phase angle: $\phi = \tan^{-1}\left(\dfrac{2\xi r}{1-r^2}\right)$

$$= \tan^{-1}\left(\frac{2 \times 0.2 \times 2.43}{1 - 2.43^2}\right)$$

$$= \mathbf{168.8°} \qquad \textbf{Ans.}$$

(iii) Transmissibility: $T_R = \left[\dfrac{1 + (2\xi r)^2}{(1-r^2)^2 + (2\xi r)^2}\right]^{0.5}$

$$= \left[\frac{1 + (2 \times 0.2 \times 2.43)^2}{(1 - 2.43^2)^2 + (2 \times 0.2 \times 2.43)^2}\right]^{0.5}$$

$$= 0.2789$$

or Transmissibility = **27.89%**   **Ans.**

(iv) Force transmitted to the foundation:

$$F_t = T_R \times F_0$$

$$= T_R \times m_0 e \omega^2$$

$$= 0.2789 \times 2 \times 0.04 \times 314.16^2$$

$$= \mathbf{2202.1 \text{ N}} \qquad \textbf{Ans.}$$

(v) Phase angle:

$$\tan \alpha = 2\xi r$$

or $\alpha = \tan^{-1}(2\xi r)$

$\qquad = \tan^{-1}(2 \times 0.2 \times 2.43)$

$\qquad = 44.18°$

The angle by which transmitted force F_t leads by excitation force F_0

$\qquad = 360° - (180° - \phi) + \alpha$
$\qquad = 360° - (180° - 168.8°) + 44.18°$
$\qquad = 304.62°$ \hfill Ans.

EXAMPLE 14.20 A trailor moving over a road having an approximately sinusoidal profile with a wavelength of 5 m and an amplitude of 50 mm. The trailor weighs 500 kg and is pulled along the road with a velocity of 60 kmph. Determine the spring constant to give a vibration amplitude of 10 mm and the most unfavourable speed of the vehicle.

Solution The ratio of amplitude for support excitation problem is given as

$$\frac{X}{Y} = \frac{\sqrt{1+(2\xi r)^2}}{\sqrt{(1-r^2)^2 + (2\xi r)^2}}$$

when damping factor $\xi = 0$,

$$\frac{X}{Y} = \frac{1}{1-r^2}$$

The time period of forced vibration is equal to the time taken by trailor in moving a distance of 5 m, the wavelength of the road profile.

$$v = 60 \text{ kmph} = 16.67 \text{ m/s}$$

Time: $\qquad t = \dfrac{\text{Wave length}}{v} = \dfrac{5}{16.67} = 0.3 \text{ s}$

Frequency: $\qquad \omega = \dfrac{2\pi}{t} = \dfrac{2\pi}{0.3} = 20.94 \text{ rad/s}$

Therefore, for required amplitude of $X = 10$ mm and $Y = 50$ mm, the amplitude ratio

$$\frac{X}{Y} = \frac{1}{1-r^2}$$

$$\frac{10}{50} = \frac{1}{1-r^2}$$

or
or $\qquad r = 2.45$

Natural frequency: $\qquad \omega_n = \dfrac{\omega}{r} = \dfrac{20.94}{2.45} = 8.54 \text{ rad/s}$

Stiffness: $\quad k = m\omega_n^2 = 500 \times 8.54^2 = \mathbf{36466\ N/m}$ **Ans.**

Most unfavourable speed of the trailor is corresponding to natural frequency where resonance will occur.

$$\omega = \omega_n = 8.54 \text{ rad/s}$$

Speed: $\quad v = \dfrac{\text{distance} \times \omega_n}{2\pi} = \dfrac{5 \times 8.54}{2\pi} = 6.796 \text{ m/s}$

or $\quad v = \mathbf{24.46\ kmph}$ **Ans.**

EXAMPLE 14.21 Develop an equation of motion for the system as shown in Figure 14.31 and determine the steady state amplitude and phase angle.

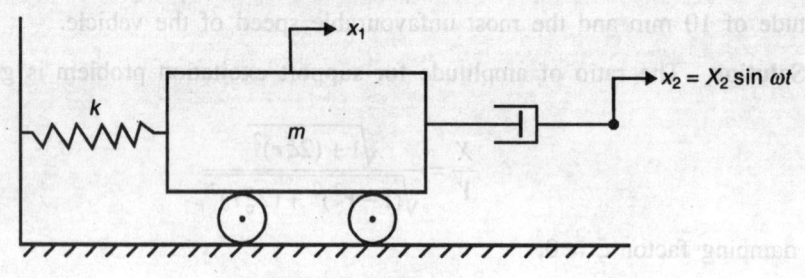

FIGURE 14.31

Solution: For a displacement x_1 of the mass m, the forces opposing the motion are:
(i) spring force = kx_1
(ii) damping force = $c(\dot{x}_1 - \dot{x}_2)$ assuming that $x_1 > x_2$

The equation of motion is:
$$m\ddot{x}_1 = -k_1 x_1 - c(\dot{x}_1 - \dot{x}_2)$$

or
$$m\ddot{x}_1 + c\dot{x}_1 + kx_1 = c\dot{x}_2 \qquad (i)$$

where $\quad x_2 = X_2 \sin \omega t \quad$ (given)

$$= \dfrac{X_2}{2i}(e^{i\omega t} - e^{-i\omega t})$$

and $\quad \dot{x}_2 = \dfrac{i\omega X_2}{2i}\left(e^{i\omega t} + e^{-i\omega t}\right)$

Let us assume that displacement x_1 lags behind x_2 by an angle ϕ. Therefore,

$$x_1 = \dfrac{X_1}{2i}\left[e^{i(\omega t - \phi)} - e^{-i(\omega t - \phi)}\right]$$

and $\quad \dot{x}_1 = \dfrac{i\omega X_1}{2i} e^{-\phi}\left(e^{i\omega t} + e^{-i\omega t}\right)$

or
$$\dot{x}_1 = \frac{\omega \overline{x}_1}{2}\left(e^{i\omega t} + e^{-i\omega t}\right)$$

and
$$\ddot{x}_1 = \frac{i\omega^2 \overline{x}_1}{2}\left(e^{i\omega t} - e^{-i\omega t}\right)$$

Putting these in Eq. (i), we get

$$im\omega^2 \overline{X}_1 + c\omega \overline{X}_1 + k\overline{X}_1 = c\omega X_2$$

where
$$\overline{X}_1 = \frac{i\omega c X_2}{(k - m\omega^2) - ic\omega}$$

or
$$\frac{X_1 e^{-i\omega}}{X_2} = \frac{c\omega}{\sqrt{(k - m\omega^2)^2 + (c\omega)^2}}$$

or
$$\frac{X_1}{X_2} = \frac{c\omega}{\sqrt{(k - m\omega^2)^2 + (c\omega)^2}}$$

$$= \frac{2\xi r}{\sqrt{(1 - r^2)^2 + (2\xi r)^2}} \qquad \text{Ans.}$$

and
$$\text{Phase angle: } \phi = \tan^{-1}\left(\frac{c\omega}{k - m\omega^2}\right)$$

$$= \tan^{-1}\left(\frac{2\xi r}{1 - r^2}\right) \qquad \text{Ans.}$$

14.7 TRANSVERSE VIBRATION

In many mechanical applications, member vibrates in the plane perpendicular to the longitudinal axis due to external excitation. Such vibrations are called **transverse vibration,** for example vibration in shaft and beams subjected to different types of load and end conditions. In this section, we shall discuss these types of vibrations.

14.7.1 Vibration due to Single Point Load

The shafts or beams of negligible mass carrying a point load is subjected to vibrating force proportional to the deflection of the load from the equilibrium position. The natural frequency f_n is computed by the following relation:

$$f_n = \frac{1}{2\pi}\sqrt{\frac{g}{\delta}} \qquad (14.45)$$

where δ = static deflection of the shaft (see Figure 14.32)

$\delta = \dfrac{Wl^3}{3EI}$ for cantilever beam subjected to a point load at free end [Figure 14.32(a)]

$= \dfrac{Wa^2b^2}{3EIl}$ for simply supported beam or a shaft supported in short bearings [Figure 14.32(b)]

$= \dfrac{Wa^3b^3}{3EIl^3}$ for fixed beam or a shaft supported in long bearings [Figure 14.32(c)].

(a) Cantilever beam   (b) Simply supported beam   (c) Fixed beam

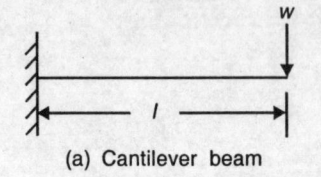

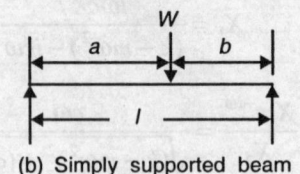

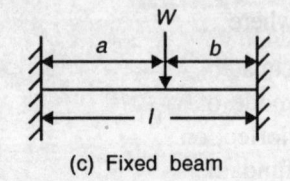

FIGURE 14.32 Transverse vibration of beam.

EXAMPLE 14.22 A hollow shaft 1.8 m long is supported in flexible bearings. It carries a wheel of 900 N weight at the centre. The external and internal diameters of the shaft are 60 mm and 40 mm respectively. Determine the transverse vibration if the elastic modulus of material is 210 GN/m^2.

Solution The moment of inertia of the hollow shaft:

$$I = \dfrac{\pi}{64}\left(d_0^4 - d_i^4\right) = \dfrac{\pi}{64}(0.06^4 - 0.04^4)$$

$$= 5.105 \times 10^{-7} \text{ m}^4$$

Static deflection of the shaft: $\delta = \dfrac{Wl^3}{48 EI}$

or $\delta = \dfrac{900 \times 1.8^3}{48 \times 210 \times 10^9 \times 5.105 \times 10^{-7}} = 1.02 \times 10^{-3}$ m

Natural frequency: $f_n = \dfrac{1}{2\pi}\sqrt{\dfrac{g}{\delta}} = \dfrac{1}{2\pi}\sqrt{\dfrac{9.81}{1.02 \times 10^{-3}}} = \mathbf{15.6\ Hz}$ **Ans.**

14.7.2 Several Point Loads

In practice, many mechanical members carry more than one discs or point loads, thus they form a case of several degree of freedom which can be solved for natural frequency by the following approximate methods:

Dunkerley's method. This method is based on algebraic property of the characteristic equation. Accordingly, if $\omega_{n1}, \omega_{n2}, \cdots \omega_{ni}$ are natural frequencies of a system subjected to one point load at a time, the frequency of multi-degree of freedom is computed by the following relation:

$$\frac{1}{\omega_n^2} = \frac{1}{\omega_{n1}^2} + \frac{1}{\omega_{n2}^2} + \cdots + \frac{1}{\omega_{ni}^2} \qquad (14.46)$$

where ω_n = fundamental natural frequency of multi-degree of freedom.

Rayleigh's method. This method is an extension of energy method used for single degree of freedom problem in which maximum kinetic energy and potential energy are equated. According to Rayleigh's principle, if any deflection curve is assumed for fundamental mode of vibration, the calculated frequency will approach natural frequency if the assumed deflection curve approaches the actual curve. Using this principle, a good estimate of fundamental natural frequency, generally within ±10 per cent error, can be obtained by assuming any reasonable deflection curve for the fundamental mode. Generally, static deflection curve is assumed to approximate the fundamental mode.

Consider a shaft with negligible mass carrying point loads w_1, w_2, w_3 as shown in Figure 14.33. Let y_1, y_2, y_3 be deflections under these loads.

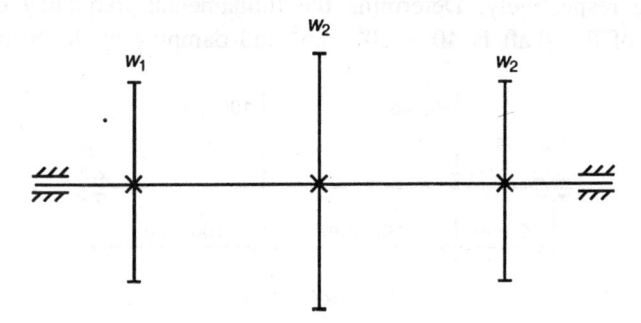

FIGURE 14.33 Transverse vibration of a three disc rotor.

The maximum potential energy is:

$$PE = \frac{1}{2}w_1 y_1 + \frac{1}{2}w_2 y_2 + \cdots$$

or

$$PE = \frac{g}{2}\sum_{i=1}^{n} m_i y_i \qquad (14.47)$$

and kinetic energy:

$$KE = \frac{1}{2}m_1 v_1^2 + \frac{1}{2}m_2 v_2^2 + \cdots$$

$$= \frac{1}{2}m_1(\omega_n y_1)^2 + \frac{1}{2}m_2(\omega_n y_2)^2 + \cdots$$

or
$$KE = \frac{\omega_n^2}{2}\sum_{i=1}^{n} m_i y_i^2 \qquad (14.48)$$

where ω_n is natural frequency of vibration.

According to Rayleigh's principle, equating potential energy and kinetic energy,

$$\frac{g}{2}\sum_{i=1}^{n} m_i y_i = \frac{\omega_n^2}{2}\sum_{i=1}^{n} m_i y_i^2$$

or
$$\text{Natural frequency: } \omega_n = \left[\frac{g\sum_{i=1}^{n} m_i y_i}{\sum_{i=1}^{n} m_i y_i^2}\right]^{0.5} \qquad (14.49)$$

EXAMPLE 14.23 A solid shaft of uniform diameter is supported by two short bearings as shown in Figure 14.34. The above shaft carries two discs having concentrated masses of 60 kg and 100 kg respectively. Determine the fundamental frequency of vibration if the moment of inertia of the shaft is 40×10^4 mm^4 and damping at the bearing is neglected.

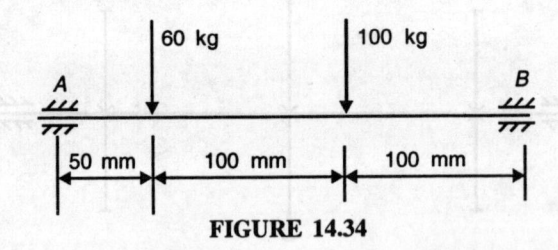

FIGURE 14.34

Solution The assumed deflection curve of the above shaft is shown in Figure 14.35.

FIGURE 14.35

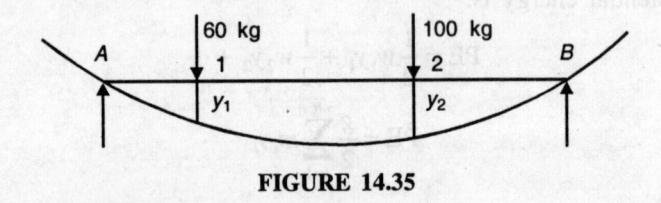

The deflection at point 1 due to unit load at point 1 is:

$$\alpha_{11} = \frac{wa^2b^2}{3EIl} = \frac{1 \times 50^2 \times 200^2}{3EI \times 250} = \frac{133.34 \times 10^3}{EI}$$

Deflection at point 1 due to unit load at point 2 is:

$$\alpha_{21} = \alpha_{12} = \frac{100 \times 50\,(250^2 - 50^2 - 100^2)}{6EI \times 250}$$

$$= \frac{166.67 \times 10^3}{EI}$$

Similarly, deflection at point 2 due to unit load at point 2 is:

$$\alpha_{22} = \frac{150^2 \times 100^2}{3EI \times 250} = \frac{300 \times 10^3}{EI}$$

Therefore, deflection at point 1:

$$y_1 = m_1 \alpha_{11} + m_2 \alpha_{12}$$

$$= \frac{60 \times 133.34 \times 10^3}{0.2 \times 10^6 \times 40 \times 10^4} + \frac{100 \times 166.67 \times 10^3}{0.2 \times 10^6 \times 40 \times 10^4}$$

$$= 3.1 \times 10^{-4} \text{ mm}$$

Similarly, deflection at point 2:

$$y_2 = m_1 \alpha_{21} + m_2 \alpha_{22}$$

$$= \frac{60 \times 166.67 \times 10^3}{0.2 \times 10^6 \times 40 \times 10^4} + \frac{100 \times 300 \times 10^3}{0.2 \times 10^6 \times 40 \times 10^4}$$

$$= 5.0 \times 10^{-4} \text{ mm}$$

$$\omega_n = \sqrt{\frac{g \sum_{i=1}^{n} m_i y_i}{\sum_{i=1}^{n} m_i y_i^2}}$$

$$= \left[\frac{9800[60 \times 3.1 \times 10^{-4} + 100 \times 5.0 \times 10^{-4}]}{60 \times (3.1 \times 10^{-4})^2 + 100 \times (5.0 \times 10^{-4})^2} \right]^{0.5}$$

$$= 4674.5 \text{ rad/s} \qquad \textbf{Ans.}$$

14.8 WHIRLING OF SHAFT

When speed of a rotating shaft is increased to such a value that it starts vibrating violently in the transverse direction, that speed of the shaft is called **whirling speed** or **critical speed**. In other words, critical speed of the shaft is that speed which is equal to its natural frequency of lateral vibration of the shaft. This phenomenon occurs due to many reasons namely unbalanced rotating mass, gyroscopic couple effect and so forth. The operation of shaft at its critical speed is very dangerous as the amplitude of vibration is maximum which may cause high fatigue stresses and may lead to failure.

Consider a shaft supporting a disc at its mid-span as shown in Figure 14.36. When the geometric centre of the disc C does not coincides with the mass centre G due to manufacturing

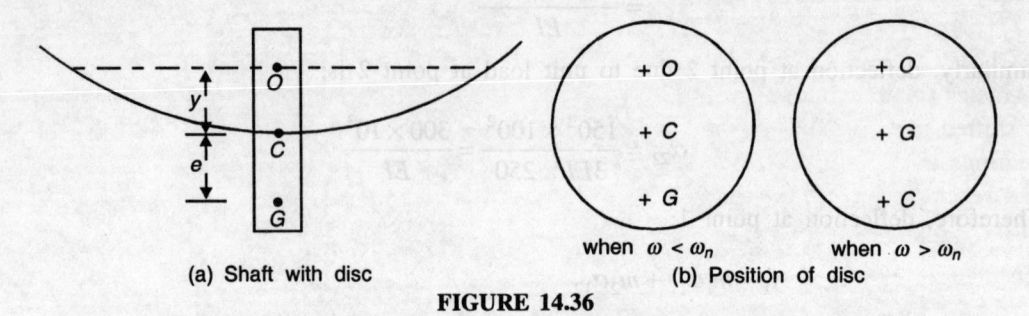

(a) Shaft with disc

when $\omega < \omega_n$ when $\omega > \omega_n$
(b) Position of disc

FIGURE 14.36

or other inaccuracies, it is subjected to centrifugal force which may act radially outward. For the equilibrium of the disc, a restoring force at point C acts in opposite direction.

Let m = mass of the disc
        k = transverse stiffness of the shaft
        y = transverse deflection of the shaft
        ω = angular velocity of the shaft
and e = eccentricity between geometric centre and the mass centre

Referring to Figure 14.36, the centrifugal force is:

$$F_c = m\omega^2(y + e)$$

and restoring force: $F_r = ky$
For the equilibrium of rotating disc,

$$F_c = F_r$$

or

$$m\omega^2(y + e) = ky$$

or

$$y = \frac{m\omega^2 e}{k - m\omega^2}$$

or

$$y = \frac{\left(\dfrac{\omega}{\omega_n}\right)^2 e}{\left[1 - \left(\dfrac{\omega}{\omega_n}\right)^2\right]} \qquad (14.50)$$

where ω_n is natural frequency (= $\sqrt{k/m}$).

At the condition of resonance ($\omega = \omega_n$), the deflection of the shaft is infinitely large. If the speed of shaft is increased beyond resonance condition ($\omega > \omega_n$), the light side (point C) will act outwardly [Figure 14.36(b)]. The deflection of the shaft at this condition will be negative which shows that shaft tries to deflect in opposite direction. As the deflection y approaches the value $(-e)$, the amplitude of vibration subsides and shaft starts operating at steady condition.

EXAMPLE 14.24 A circular disc having 10 kg mass is mounted at mid-span of 25 mm diameter shaft supported by two bearings. Due to manufacturing inaccuracy, the mass centre is shifted by 0.025 mm away from the geometric centre of the disc. If the span between two bearings is 500 mm and the shaft rotates at 4000 rpm, determine the amplitude of vibration and reactions on the bearings. Take modulus of elasticity $E = 2 \times 10^{11}$ N/m².

Solution Transverse stiffness:

$$k = \frac{48EI}{l^3}$$

or

$$k = \frac{48 \times 2 \times 10^{11} \times \frac{\pi}{64} \times 0.025^4}{0.5^3}$$

$$= 1.473 \times 10^6 \text{ N/m}$$

Natural frequency:

$$\omega_n = \sqrt{\frac{k}{m}} = \sqrt{\frac{1.473 \times 10^6}{10}}$$

$$= 383.8 \text{ rad/s}$$

The angular velocity or forcing frequency:

$$\omega = \frac{2\pi N}{60} = \frac{2\pi \times 4000}{60} = 418.88 \text{ rad/s}$$

and

Frequency ratio: $\dfrac{\omega}{\omega_n} = \dfrac{418.88}{383.8} = 1.091$

Amplitude of vibration: $y = \dfrac{\left(\dfrac{\omega}{\omega_n}\right)^2 e}{\left[1 - \left(\dfrac{\omega}{\omega_n}\right)^2\right]} = \dfrac{1.091 \times 0.025}{[1 - 1.091^2]} = -0.1433$ mm   **Ans.**

The negative sign indicates that amplitude of vibration is out of phase with centrifugal force.

Force on the bearing: $F = \dfrac{mg}{2} + \dfrac{ky}{2}$

$= \dfrac{10 \times 9.81}{2} + \dfrac{1.473 \times 10^6 \times 0.1433 \times 10^{-3}}{2}$

$= 155.6$ N **Ans.**

14.9 TORSIONAL VIBRATION OF ROTOR SYSTEM

The torsional vibration of a shaft carrying single rotor is discussed in Section 14.4 where the shaft is assumed to be massless and if the disc is given a twist about its vertical axis and then released, it will start vibrating (oscillating) about the axis. However, in practice, a shaft may have more than one rotor; the vibration analysis of such system is quite different than single rotor system. In this section, we shall discuss torsional vibration of shaft for the following cases:

(i) two rotor system
(ii) three rotor system
(iii) geared system

A detailed discussion about these three systems is given further.

14.9.1 Two Rotor System

A shaft held in bearings carrying one rotor at each end, as shown in Figure 14.37, is called two rotor system. This shaft vibrates in such a way that two rotors oscillate in opposite direction. Therefore, a portion of the shaft length near to the rotor is twisted in opposite direction. A point on the shaft which does not undergo any twist is called **nodal point.** The vibrating shaft in such a situation behaves as if it is clamped at the nodal point.

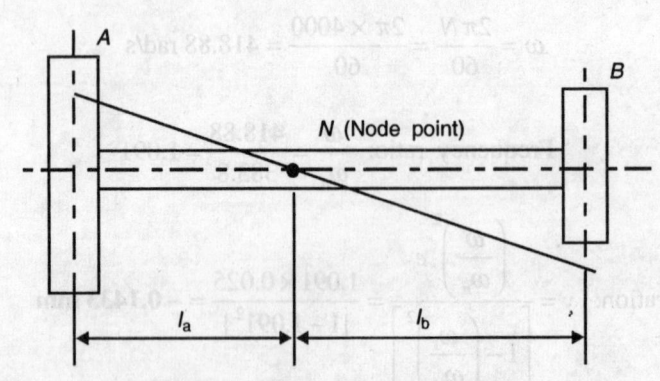

FIGURE 14.37 A two rotor system.

Let I_a and I_b = moments of inertia of rotors A and B respectively
l_a and l_b = lengths of two portions of the shaft

The above two rotor system can be composed of two separate shafts which are clamped at nodal point and vibrate at same frequency.

Therefore,
$$\omega_{na} = \omega_{nb}$$

$$\sqrt{\frac{k_{ta}}{I_a}} = \sqrt{\frac{k_{tb}}{I_b}}$$

or
$$\frac{k_{ta}}{I_a} = \frac{k_{tb}}{I_b} \tag{14.51}$$

where ω_{na} and ω_{nb} are natural frequencies of torsional vibration of rotors A and B respectively.

and k_{ta} and k_{tb} are torsional stiffnesses of the shaft corresponding to lengths l_a and l_b (= GJ/l)
with G = modulus of rigidity
J = polar moment of inertia of the shaft.

Equation (14.51) can be rewritten as
$$\frac{GJ}{I_a l_a} = \frac{GJ}{I_b l_b}$$

or
$$I_a l_a = I_b l_b \tag{14.52}$$

Therefore, the nodal point divides the length of shaft in the inverse of moment of inertia of two rotors.

$$\frac{l_a}{l_b} = \frac{I_b}{I_a} \tag{14.53}$$

The ratio of amplitude of vibration of rotor A and B is equal to the ratio corresponding length of shafts.

$$\frac{\text{Amplitude of vibration of rotor } A}{\text{Amplitude of vibration of rotor } B} = \frac{l_a}{l_b} \tag{14.54}$$

In certain applications, the rotors are mounted on the stepped shaft having different diameters at different sections. In such a case, it is easier to replace the step shaft with torsionally equivalent shaft having uniform diameter throughout its length. A torsionally equivalent shaft is one in which total angle of twist is equal to sum of angle of twist of each section.

Let θ be the total angle of twist and θ_1, θ_2, θ_3 be the angles of twist of individual steps having lengths l_1, l_2, l_3 respectively.

Therefore,
$$\theta = \theta_1 + \theta_2 + \theta_3$$

or
$$\frac{Tl}{GJ} = \frac{Tl_1}{GJ_1} + \frac{Tl_2}{GJ_2} + \frac{Tl_3}{GJ_3} \tag{14.55}$$

The length of equivalent shaft is:

$$l = l_1\left(\frac{d}{d_1}\right)^4 + l_2\left(\frac{d}{d_2}\right)^4 + l_3\left(\frac{d}{d_3}\right)^4 \tag{14.56}$$

where d is the diameter of equivalent shaft.

14.9.2 Three Rotor System

Consider a three rotor system as shown in Figure 14.38(a), in which two rotors A and B are placed at the ends of the shaft and third rotor C is placed in between. The possible mode shapes of torsional vibration of the three rotor system for the following conditions are shown in Figure 14.38.

(i) If the direction of rotation of rotors A and B are same and the rotor C rotates in opposite direction, the mode shape of the rotor has two nodal points and these are located at points D and E as shown in Figure 14.38(b).

(ii) The rotors A and C rotate in the same direction and rotor B rotates in the opposite direction. The mode shape of such system has a single node between rotor B and C as shown in Figure 14.38(c).

(iii) Similarly, when rotors B and C rotate in the same direction and rotor A rotates in the opposite direction, the mode shape of such system has a single node between rotor A and C as shown in Figure 14.38(d).

Let

I_a, I_b and I_c be mass moments of inertia of rotors A, B and C respectively

l_1, l_2, l_a, l_b, l_{c1}, and l_{c2} be distances on the shaft as shown in Figure 14.38

and k_{ta}, k_{tb} and k_{tc} be torsional stiffnesses of the shaft at different sections.

For analysis of a three rotor problem, let us consider the first case in which the rotors A and B rotate in the same direction and the rotor C rotates in the opposite direction. The mode shape of vibration have two nodal points, hence divide the shaft into three portions. Thus the above three rotor system is assumed to be composed of three separate shafts whose natural frequencies are same. Therefore,

$$\omega_{na} = \omega_{nb} = \omega_{nc}$$

or

$$\sqrt{\frac{k_{ta}}{I_a}} = \sqrt{\frac{k_{tb}}{I_b}} = \sqrt{\frac{k_{tc}}{I_c}}$$

or

$$\frac{k_{ta}}{I_a} = \frac{k_{tb}}{I_b} = \frac{k_{tc}}{I_c} \tag{14.57}$$

Since the length of the shaft or the portion of rotor C is divided into two parts, the torque required to produce unit twist of rotor C is the sum of the torques required to produce a unit twist in each of the lengths l_{c1} and l_{c2} [Figure 14.38(b)]. Therefore, Eq. (14.57) can be rewritten as

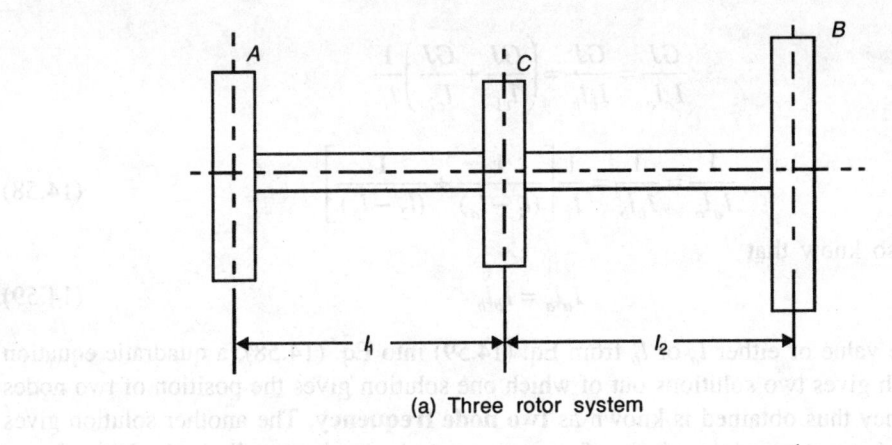

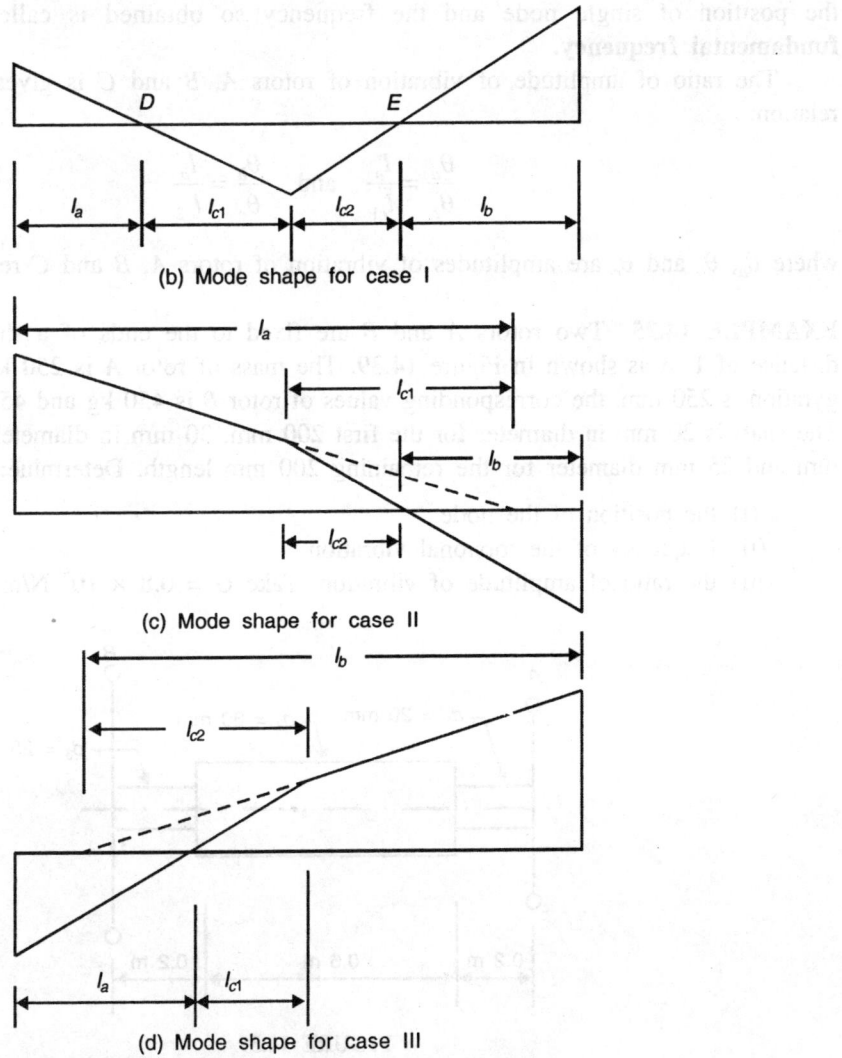

FIGURE 14.38 Vibration of a three rotor system.

644 Theory of Mechanisms and Machines

$$\frac{GJ}{I_a l_a} = \frac{GJ}{I_b l_b} = \left(\frac{GJ}{l_{c1}} + \frac{GJ}{l_{c2}}\right)\frac{1}{I_c}$$

or
$$\frac{1}{I_a l_a} = \frac{1}{I_b l_b} = \frac{1}{I_c}\left[\frac{1}{(l_1 - l_a)} + \frac{1}{(l_2 - l_b)}\right] \qquad (14.58)$$

Further, we also know that
$$I_a l_a = I_b l_b \qquad (14.59)$$

substituting the value of either l_a or l_b from Eq. (14.59) into Eq. (14.58), a quadratic equation is formed which gives two solutions out of which one solution gives the position of two nodes and the frequency thus obtained is known as **two node frequency**. The another solution gives the position of single node and the frequency so obtained is called **single node** or **fundamental frequency**.

The ratio of amplitude of vibration of rotors A, B and C is given by the following relation:

$$\frac{\theta_a}{\theta_c} = \frac{l_a}{l_{c1}} \quad \text{and} \quad \frac{\theta_b}{\theta_c} = \frac{l_b}{l_{c2}} \qquad (14.60)$$

where θ_a, θ_b and θ_c are amplitudes of vibration of rotors A, B and C respectively.

EXAMPLE 14.25 Two rotors A and B are fixed to the ends of a shaft which are at a distance of 1 m as shown in Figure 14.39. The mass of rotor A is 250 kg and its radius of gyration is 250 mm, the corresponding values of rotor B is 450 kg and 450 mm respectively. The shaft is 20 mm in diameter for the first 200 mm, 30 mm in diameter for the next 600 mm and 25 mm diameter for the remaining 200 mm length. Determine:

   (i) the position of the node
   (ii) frequency of the torsional vibration
   (iii) the ratio of amplitude of vibration. Take $G = 0.8 \times 10^5$ N/mm².

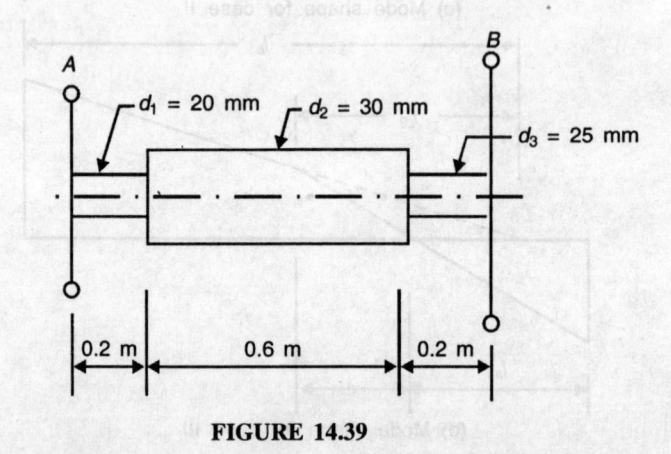

FIGURE 14.39

Mechanical Vibrations   645

Solution Moment of inertia of rotor A, $I_a = mk^2$

$$I_a = 250 \times 0.25^2 = 15.625 \text{ kgm}^2$$

Moment of Inertia rotor B: $I_b = 450 \times 0.45^2 = 91.125 \text{ kgm}^2$

Transforming the given two rotor step shaft system into an equivalent system having uniform diameter of 20 mm,

$$l_e = l_1\left(\frac{d}{d_1}\right)^4 + l_2\left(\frac{d}{d_2}\right)^4 + l_3\left(\frac{d}{d_3}\right)^4$$

where d is the uniform diameter of an equivalent system.

$$l_e = 0.2\left(\frac{20}{20}\right)^4 + 0.6\left(\frac{20}{30}\right)^4 + 0.2\left(\frac{20}{25}\right)^4$$

$$= 0.4 \text{ m}$$

To locate the nodal point, the condition of similar frequency is applied.

$$I_a l_a = I_b l_b$$

or $$\frac{l_a}{l_b} = \frac{I_b}{I_a} = \frac{91.125}{15.625} = 5.832 \qquad (i)$$

and Equivalent length: $l_e = l_a + l_b = 0.4$ (ii)

Solving Eqs. (i) and (ii), we get

$$l_a = \textbf{0.3415 m} \quad \text{and} \quad l_b = \textbf{0.0585 m} \qquad \textbf{Ans.}$$

Frequency of vibration: $$\omega_n = \sqrt{\frac{k_t}{I_a}} = \sqrt{\frac{GJ}{I_a l_a}}$$

where J is polar moment of inertia and is given by

$$\frac{\pi}{32}d^4 = \frac{\pi}{32} \times 0.02^4 = 1.57 \times 10^{-8} \text{ m}^4$$

$$\therefore \qquad \omega_n = \sqrt{\frac{0.8 \times 10^{11} \times 1.57 \times 10^{-8}}{15.625 \times 0.3415}}$$

$$= 15.342 \text{ rad/s}$$

or   Frequency: $f_n = \dfrac{\omega_n}{2\pi} = \dfrac{15.342}{2\pi} = \textbf{2.44 Hz}$ **Ans.**

Ratio of amplitude: $\dfrac{\theta_a}{\theta_b} = \dfrac{l_a}{l_b} = \textbf{5.832}$ **Ans.**

EXAMPLE 14.26 A 1.5 m long steel shaft AB has flywheels at its ends A and B. The mass of the flywheel at the end A is 600 kg and its radius of gyration is 400 mm; the corresponding values for the flywheel at the end B are 300 kg and 300 mm respectively. The diameter of shaft for the first 400 mm starting from the end A is 50 mm, the 60 mm diameter for the next portion of 500 mm long and the remaining portion of 600 mm long of unknown diameter. Determine:

    (i) diameter of shaft for the portion near to the end B so that the node of torsional vibration of the system will be at the centre of 500 mm long segment

    (ii) natural frequency of vibration

Solution The layout of the given shaft is as shown in Figure 14.40. Moments of inertia of the flywheels at both the ends are:

$$I_a = mk^2 = 600 \times 0.4^2 = 96 \text{ kg m}^2$$

$$I_b = 300 \times 0.3^2 = 27 \text{ kg m}^2$$

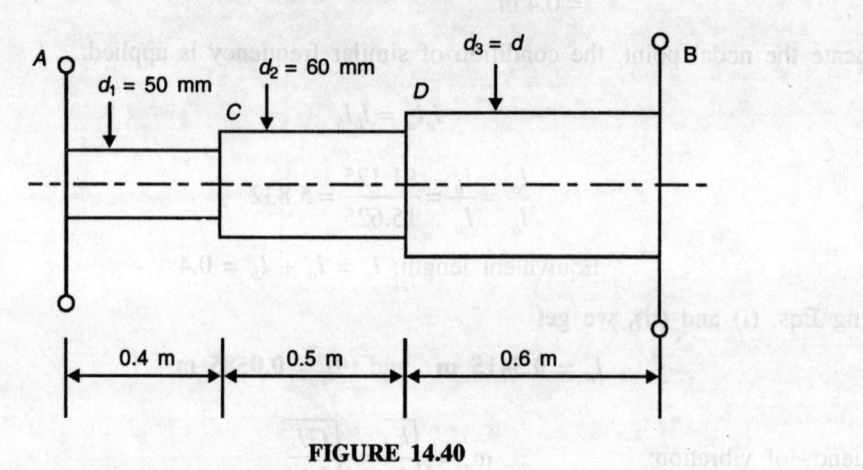

FIGURE 14.40

Assuming that the diameter of the equivalent shaft be equal to 50 mm, the length of torsionally equivalent shaft is:

$$l_e = l_1 \left(\frac{d}{d_1}\right)^4 + l_2 \left(\frac{d}{d_2}\right)^4 + l_3 \left(\frac{d}{d_3}\right)^4$$

$$= 0.4 + 0.5 \left(\frac{50}{60}\right)^4 + 0.6 \left(\frac{50}{d_3}\right)^4$$

$$= 0.64 + \frac{375 \times 10^4}{d_3^4} \text{ m}$$

(i) **Diameter of the shaft in BD portion.** Let nodal point lies at a distance of l_a from the flywheel A and at a distance of l_b from the flywheel B such that

$$I_a l_a = I_b l_b$$

or

$$l_a = \frac{I_b l_b}{I_a} = \frac{27}{96} l_b$$

Since the nodal point lies at the centre of the portion CD in the original system, its equivalent length from the rotor A is given by

$$l_a = l_1 + \frac{l_2}{2} \left(\frac{d_1}{d_2}\right)^4$$

$$= 0.4 + \frac{0.5}{2}\left(\frac{50}{60}\right)^4$$

$$= 0.52 \text{ m}$$

Therefore $\quad l_b = \dfrac{96}{27} l_a = \dfrac{96}{27} \times 0.52 = 1.85 \text{ m}$

or Equivalent length: $l_e = l_a + l_b = 1.85 + 0.52 = 2.37$ m

or

$$l_e = 2.37 = 0.64 + \frac{375 \times 10^4}{d_3^4}$$

or $\quad d_3 = \mathbf{38.37 \text{ mm}}$ **Ans.**

(ii) Natural frequency: $\quad f_n = \dfrac{1}{2\pi}\sqrt{\dfrac{GJ}{I_a l_a}}$

where $\quad J = \dfrac{\pi}{32} d^4 = \dfrac{\pi}{32} \times 0.05^4 = 6.1359 \times 10^{-7} \text{ mm}^4$

$\therefore \quad f_n = \dfrac{1}{2\pi}\sqrt{\dfrac{0.8 \times 10^{11} \times 6.1359 \times 10^{-7}}{96 \times 0.52}}$

$= \mathbf{5.0 \text{ Hz}}$ **Ans.**

EXAMPLE 14.27 A four cylinder engine with flywheel is coupled to a propellor. This assembly is approximated as a three rotor torsional vibration system in which the moment of inertia of the engine is 80 kg m² and that of flywheel and propellor are 35 kg m² and 40 kg m² respectively. The engine and flywheel rotors are connected by 40 mm diameter and 2m long shaft and the flywheel and propellor rotors are connected by 30 mm diameter and 2m long shaft. Neglecting the inertia of the shaft and taking modulus of rigidity as 0.8×10^5 N/mm², determine:

(i) Natural frequencies of the torsional vibration
(ii) the position of nodes

Solution The original three rotor system is as shown in Figure 14.41 given below:

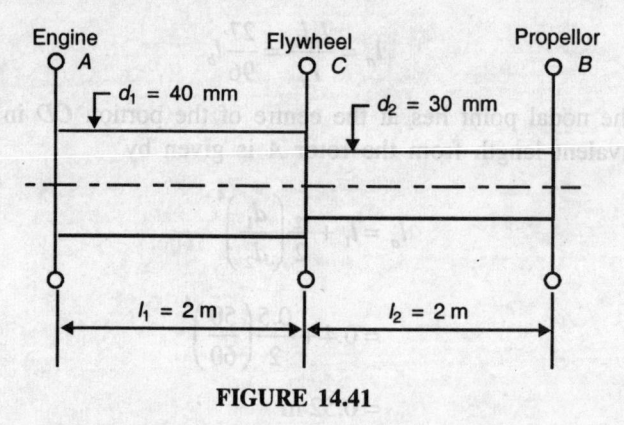

FIGURE 14.41

The length of equivalent system when the portion CB is transformed into a shaft of uniform diameter 40 mm:

$$l = l_1 + l_2 \left(\frac{d_1}{d_2}\right)^4$$

$$= 2 + 2\left(\frac{40}{30}\right)^4$$

$$= 8.32 \text{ m}$$

Therefore, the equivalent system is as shown in Figure 14.42 given below:

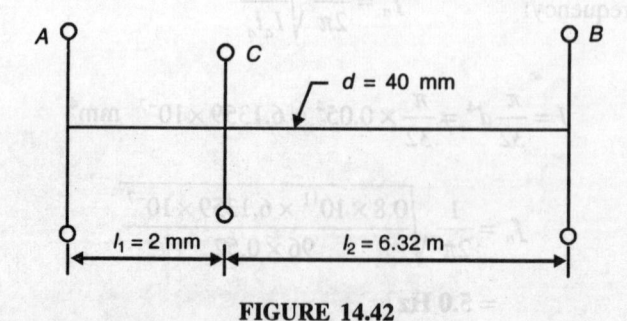

FIGURE 14.42

We know that for equal frequencies of two portions of the shaft:

$$I_a l_a = I_b l_b$$

or

$$\frac{l_a}{l_b} = \frac{I_b}{I_a} = \frac{40}{80} = 0.5 \qquad \text{(i)}$$

Also
$$\frac{1}{I_a l_a} = \frac{1}{I_b l_b} = \frac{1}{I_c}\left[\frac{1}{l_1 - l_a} + \frac{1}{l_2 - l_b}\right]$$

or
$$\frac{1}{I_a l_a} = \frac{1}{I_c}\left[\frac{(l_2 - l_b) + (l_1 - l_a)}{(l_1 - l_a)(l_2 - l_b)}\right]$$

or
$$(l_1 - l_a)(l_2 - l_b) = \frac{I_a l_a}{I_c}[(l_2 + l_1) - (l_a + l_b)]$$

For equivalent torsional system length: $l_2 = 6.32$ m. Therefore,

$$(2 - l_a)(6.32 - l_b) = \frac{80}{35} \times l_a[8.32 - (l_a + l_b)] \qquad (ii)$$

From Eq. (i), substituting the value of $l_a = 0.5\, l_b$ we get

$$(2 - 0.5 l_b)(6.32 - l_b) = \frac{80}{35} \times 0.5 l_b\,(8.32 - 1.5 l_b)$$

or
$$1.2145 l_b^2 - 4.35 l_b - 12.64 = 0$$

Solving above quadratic equation, we get

$$l_b = 5.48 \text{ m} \quad \text{and} \quad -1.9 \text{ m}$$

Case I When $l_b = 5.48$ m:
$$l_a = \frac{I_b}{I_a} \times l_b = 0.5 l_b = 2.74 \text{ m}$$

Torsional stiffness:
$$k_{ta} = \frac{GJ}{l_a}$$

or
$$k_{ta} = \frac{0.8 \times 10^{11} \times \frac{\pi}{32} \times 0.04^4}{2.74} = 7338 \text{ N/m}$$

Natural frequency: $f_n = \frac{1}{2\pi}\sqrt{\frac{k_{ta}}{I_a}} = \frac{1}{2\pi}\sqrt{\frac{7338}{80}} = 1.52$ Hz **Ans.**

Amplitude ratio:
$$\frac{\theta_a}{\theta_c} = \frac{l_a}{l_{c1}} = \frac{l_a}{l_1 - l_a}$$

or
$$\theta_c = \theta_a \times \frac{l_1 - l_a}{l_a} = 1 \times \frac{2 - 2.74}{2.74} = -0.27 \qquad \textbf{Ans.}$$

Similarly, $\theta_b = \theta_c \times \dfrac{l_b}{l_b - l_2} = -0.27 \times \dfrac{5.48}{5.48 - 6.32} = \mathbf{1.76}$ **Ans.**

The mode shape is as shown in Figure 14.43.

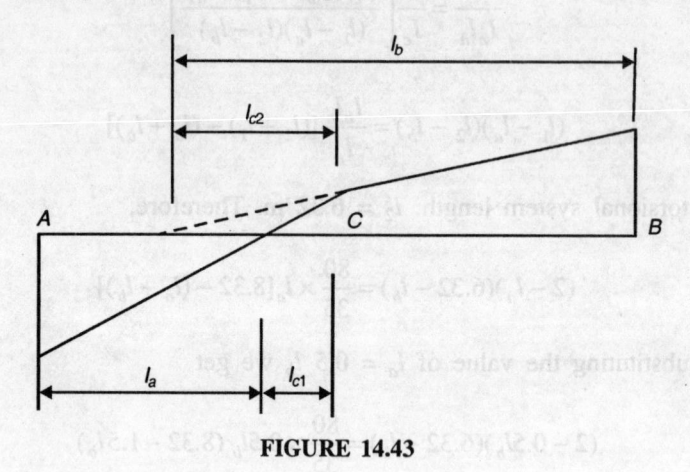

FIGURE 14.43

Case II When $l_b = |-1.9 \text{ m}| = 1.9 \text{ m}$:

$$l_a = 0.5 l_b = 0.5 \times 1.9 = 0.95 \text{ m}$$

Torsional stiffness: $k_t = \dfrac{0.8 \times 10^{11} \times \dfrac{\pi}{32} \times 0.04^4}{0.95}$

$$= 21164.4 \text{ N/m}$$

Natural frequency: $f_n = \dfrac{1}{2\pi}\sqrt{\dfrac{21164.4}{80}} = \mathbf{2.59 \text{ Hz}}$ **Ans.**

Amplitude ratio: $\dfrac{\theta_c}{\theta_a} = \dfrac{l_1 - l_a}{l_a}$

or $\theta_c = 1 \times \dfrac{(2 - 0.95)}{0.95} = \mathbf{1.1}$ **Ans.**

$$\theta_b = \theta_c \times \dfrac{l_b}{l_b - l_2} = 1.10 \times \dfrac{1.9}{1.9 - 6.32}$$

$$= \mathbf{-0.473}$$ **Ans.**

The mode shape of second case is shown in Figure 14.44.

Mechanical Vibrations 651

FIGURE 14.44

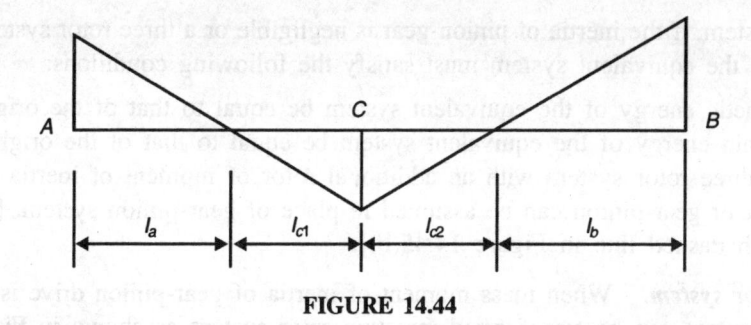

14.9.3 Geared System

In mechanical system, power is generally transmitted through geared system as shown in Figure 14.45(a). The shaft 1 carries a rotor A on one end and a pinion on the other end, while shaft 2 carries a gear meshing with pinion at one end and a rotor B on the other. For finding torsional vibration characteristics of such a system, it may be equivalently replaced by either

(a) Original geared system

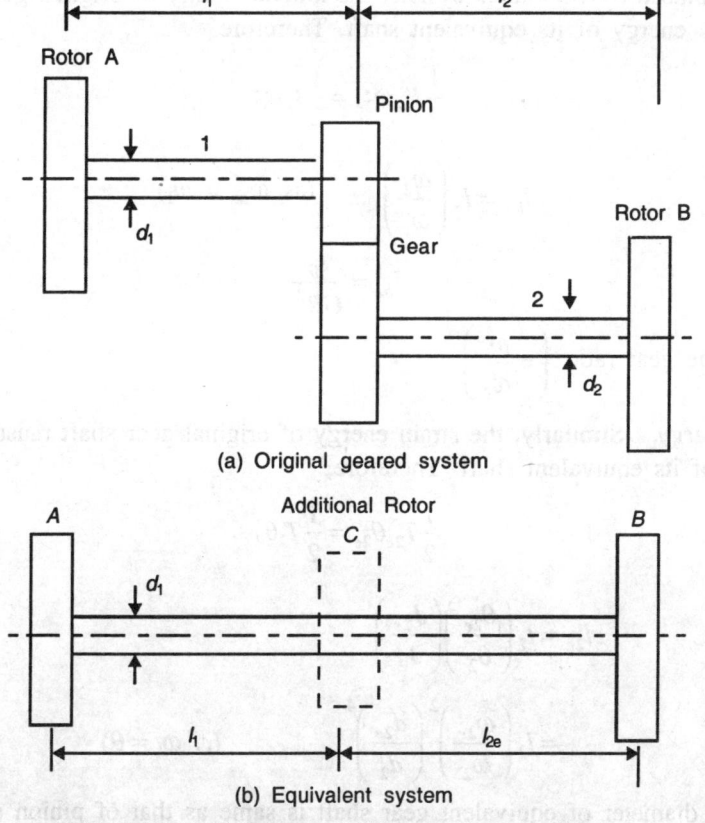

(b) Equivalent system

FIGURE 14.45 Torsional vibration of geared system.

a two rotor system, if the inertia of pinion-gear is negligible or a three rotor system. However, in either case, the equivalent system must satisfy the following conditions:

(i) Kinetic energy of the equivalent system be equal to that of the original system.
(ii) Strain energy of the equivalent system be equal to that of the original system
(iii) A three rotor system with an additional rotor of moment of inertia equivalent to that of gear-pinion can be assumed in place of gear-pinion system. [Rotor shown with dashed line in Figure 14.45(b)].

Two rotor system. When mass moment of inertia of gear-pinion drive is negligible, a geared rotor system can be transformed into two rotor system as shown in Figure 14.45(b) (without additional rotor C)

Let I_a, I_b = mass moment of inertia of rotors A and B respectively
l_1, l_2 = lengths of pinion and gear shaft as shown in Figure 14.45(a)
d_1, d_2 = diameters of the respective shafts
I_{be} = equivalent mass-moment of inertia of rotor B.
ω_1, ω_2 = angular velocities of pinion and gear shaft
and ω_{2e} = angular velocity of equivalent gear shaft

Kinetic energy. Assuming that the dimensions of the pinion shaft is maintained and gear shaft is replaced by equivalent system, the kinetic energy of original gear shaft must be equal to kinetic energy of its equivalent shaft. Therefore,

$$\frac{1}{2}I_{be}\omega_{2e}^2 = \frac{1}{2}I_b\omega_2^2$$

or $$I_{be} = I_b\left(\frac{\omega_2}{\omega_1}\right)^2 \quad (\text{as } \omega_{2e} = \omega_1)$$

or $$I_{be} = \frac{I_b}{GR^2} \qquad (14.61)$$

where GR is the gear ratio $\left(=\dfrac{\omega_1}{\omega_2}\right)$

Strain energy. Similarly, the strain energy of original gear shaft must be equal to the strain energy of its equivalent shaft. Therefore,

$$\frac{1}{2}T_{2e}\theta_{2e} = \frac{1}{2}T_2\theta_2$$

or $$l_{2e} = l_2\left(\frac{\theta_{2e}}{\theta_2}\right)\left(\frac{J_{2e}}{J_2}\right)$$

$$= l_2\left(\frac{\omega_{2e}}{\omega_2}\right)^2\left(\frac{d_{2e}}{d_2}\right)^4 \qquad (\because \omega t = \theta)$$

Assuming that diameter of equivalent gear shaft is same as that of pinion shaft ($d_{2e} = d_1$),

Mechanical Vibrations 653

$$l_{2e} = l_2 GR^2 \left(\frac{d_1}{d_2}\right)^4 \qquad (14.62)$$

In case of the inertia of the gear pinion is not negligible, an additional rotor C is assumed to be installed at a distance l_1 from the rotor A. The mass moment of inertia of this rotor is given by

$$I_c = I_{cg} + \frac{I_{cp}}{GR^2} \qquad (14.63)$$

where I_{cg} = mass moment of inertia of gear
and I_{cp} = mass moment of inertia of pinion

EXAMPLE 14.28 A motor drives a centrifugal pump through gearing. The speed of the pump is one-third of the speed of the motor. The diameter of pinion shaft is 40 mm and it is 300 mm long. The mass moment of inertia of motor is 400 kgm^2. The gear shaft, on which impeller of centrifugal pump is mounted, has 60 mm diameter and is 600 mm long. The mass moment of inertia of the impeller is 1500 m^2. Neglecting the inertia of gear and pinion, determine the natural frequency of torsional vibration.

Solution The equivalent length of impeller shaft keeping the diameter of the shaft as 40 mm:

$$l_{2e} = GR^2 l_2 \left(\frac{d_1}{d_2}\right)^4$$

$$= 3^2 \times 0.6 \times \left(\frac{40}{60}\right)^4 = 1.067 \text{ m}$$

The original geared rotor system can be transformed into two rotor systems as shown in Figure 14.46.

The equivalent moment of inertia of rotor B when gear drive is replaced:

$$I_{be} = \frac{I_b}{GR^2} = \frac{1500}{3^2} = 166.67 \text{ kg m}^2$$

Let the node of the equivalent system lies at point N, such that

$$I_a l_a = I_{be} l_b$$

or $\qquad I_a l_a = I_{be}(l_1 + l_{2e} - l_a)$

or $\qquad 400 \times l_a = 166.67 \times (0.3 + 1.067 - l_a)$

or $\qquad l_a = 0.4$ m   and $l_b = 0.967$ m

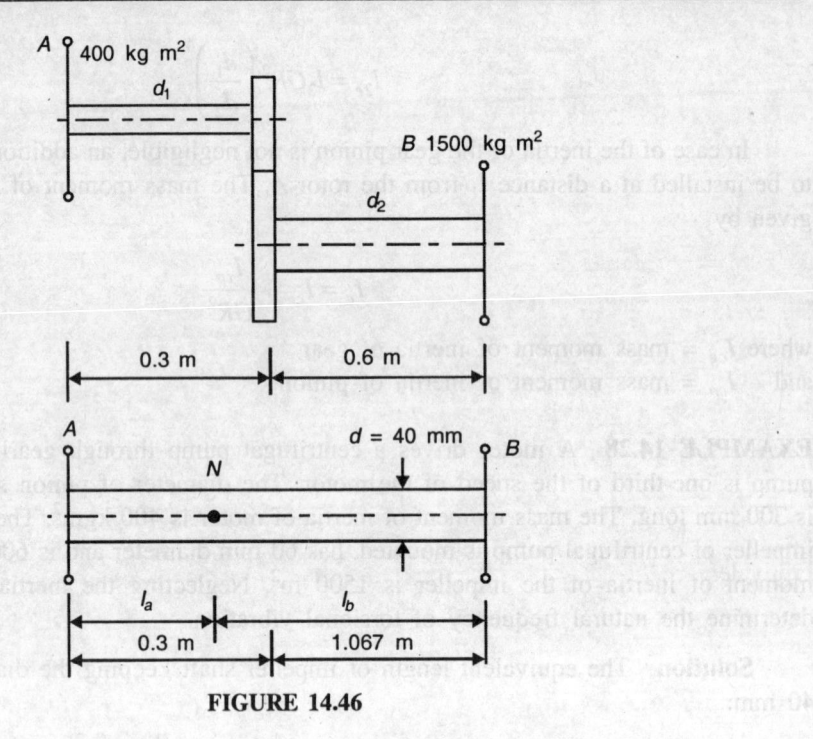

FIGURE 14.46

Frequency of torsional vibration:

$$f_n = \frac{1}{2\pi}\sqrt{\frac{k_t}{I_a}} = \frac{1}{2\pi}\sqrt{\frac{GJ}{I_a l_a}}$$

$$= \frac{1}{2\pi}\sqrt{\frac{0.8 \times 10^{11} \times \frac{\pi}{32} \times 0.04^4}{400 \times 0.4}}$$

$$= 1.784 \text{ Hz} \qquad \text{Ans.}$$

EXAMPLE 14.29 An IC engine drives a centrifugal compressor through a pair of pinion and gear. The compressor speed is one-fourth of the engine. The steel shaft from the flywheel of engine to the pinion is 50 mm in diameter and 0.4 m in length. The shaft from the gear to compressor is 60 mm in diameter and 0.5 m in length. If the mass moment of inertia of flywheel, pinion, gear and impeller are 500 kg m², 5 kg m², 15 kg m² and 500 kg m² respectively, determine the natural frequency of torsional vibration. Take $G = 0.8 \times 10^{11}$ N/m².

Solution The original geared system is as shown in Figure 14.47.
The equivalent moment of inertia of impeller is:

$$I_{be} = \frac{I_b}{GR^2} = \frac{500}{4^2} = 31.25 \text{ kg m}^2$$

Mechanical Vibrations 655

500 kg m² at A, $d_1 = 50$ mm, 5 kg m², $d_2 = 60$ mm, 15 kg m², 500 kg m² at B, 0.4 m, 0.5 m

FIGURE 14.47

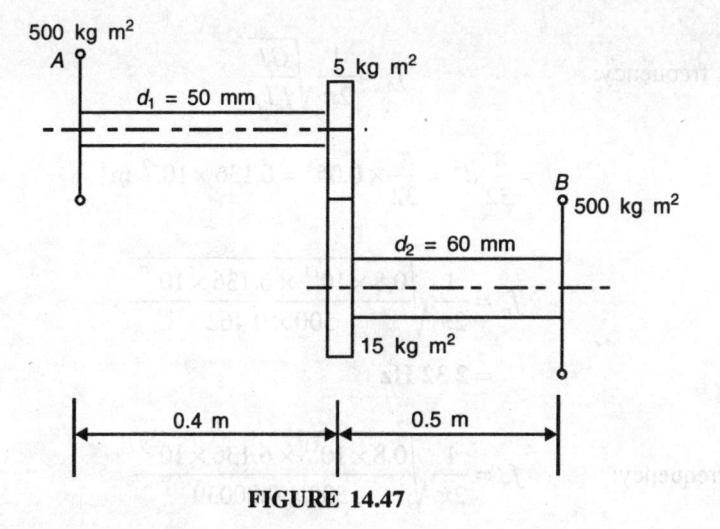

Equivalent moment of inertia of gear-pinion:

$$I_c = I_{cg} + \frac{I_{cp}}{GR^2}$$

$$= 15 + \frac{5}{4^2} = 15.3125 \text{ kg m}^2$$

Equivalent length of impeller shaft:

$$l_{2e} = l_2 \times GR^2 \left(\frac{d_1}{d_2}\right)^4$$

$$= 0.5 \times 4^2 \times \left(\frac{50}{60}\right)^2$$

$$= 3.858 \text{ m}$$

We know that $\qquad I_a l_a = I_{be} l_{be}$

or $\qquad 500\, l_a = 31.25\, l_{be}$

or $\qquad l_{be} = 16\, l_a$

Also

$$\frac{1}{I_a l_a} = \frac{1}{I_{be}} \left[\frac{1}{l_1 - l_a} + \frac{1}{l_{2e} - l_{be}} \right]$$

or $\qquad \dfrac{1}{500 \times l_a} = \dfrac{1}{15.3125} \left[\dfrac{1}{0.4 - l_a} + \dfrac{1}{3.858 - 16 l_a} \right]$

or $\qquad l_a = 0.462$ m and 0.00039 m

Single node frequency:
$$f_n = \frac{1}{2\pi}\sqrt{\frac{GJ}{I_a l_a}}$$

where
$$J = \frac{\pi}{32}d^4 = \frac{\pi}{32} \times 0.05^4 = 6.136 \times 10^{-7} \text{ m}^4$$

$\therefore$
$$f_n = \frac{1}{2\pi}\sqrt{\frac{0.8 \times 10^{11} \times 6.136 \times 10^{-7}}{500 \times 0.462}}$$
$$= 2.32 \text{ Hz} \qquad \text{Ans.}$$

Two node frequency:
$$f_n = \frac{1}{2\pi}\sqrt{\frac{0.8 \times 10^{11} \times 6.136 \times 10^{-7}}{500 \times 0.00039}}$$
$$= 79.85 \text{ Hz} \qquad \text{Ans.}$$

14.10 ELECTRICAL ANALOGY

The characteristics of mechanical vibrating system are defined by a set of differential equations. However, in many complex cases, the solution of these differential equations is cumbersome if not impossible. In such cases, these mechanical systems are transformed into an equivalent electrical circuits to study the characteristics of the system. This process is called **electrical analogy.** The basis of electrical analogy is that the differential equation of motion for two analogous systems should be the same. Under this condition, the terms in differential equation of mechanical system corresponds to the terms of differential equation of an analogous electrical circuit.

The problem of mechanical vibration can be solved by two types of electrical analogy—force-voltage analogy and force-current analogy. In this section, a brief discussion about force-voltage analogy, which is most widely used, is presented to limitation of scope.

In electrical circuits, two types of elements are used—active and passive. The voltage and current sources are known as **active elements,** whereas the resistor, capacitor and inductors are known as **passive elements.** These elements do not generate energy; on the contrary, they consume energy.

The equivalent electrical circuit of any mechanical system is constructed on the basis of Kirchhoff's law. Accordingly, the algebraic sum of all the voltage around any closed circuit is equal to zero in any network. While transforming mechanical system into an equivalent electrical circuit, the mass, springs and damper are replaced by inductor, capacitor, and resistor respectively. The equivalent mechanical-electrical components and voltage drop across them are given in Table 14.1. Further, one should note that if the mechanical system is such that the forces on the mass act parallel, then the equivalent electrical circuit will have the respective components in series. Similarly, if the mechanical system has its component in series then the analogous electrical circuit will have its components in parallel.

Mechanical Vibrations 657

Table 14.1 Electro-mechanical equivalence

Mechanical system	Electrical system	Voltage drop
F Force (N)	V Voltage	—
m Mass (kg)	L Inductance	$V = \dfrac{Ldi}{dt} = \dfrac{Ld^2q}{dt^2}$
c Damping coefficient	R Resistance	$V = iR = R\dfrac{dq}{dt}$
k Stiffness	$\dfrac{1}{c}$ Capacitance	$V = \dfrac{1}{c}\int i\,dt = \dfrac{q}{c}$
x Displacement	q Charge	—
$\dfrac{dx}{dt}$ Velocity	i Current	—

EXAMPLE 14.30 Convert the mechanical vibratory system as shown in Figure 14.48 into an analogous electrical circuit.

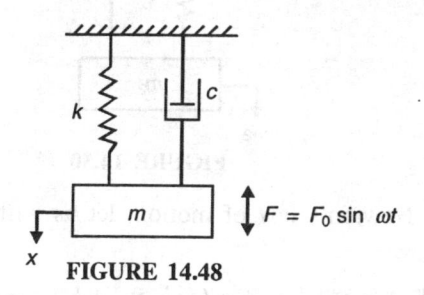

FIGURE 14.48

Solution The differential equation of motion is:

$$m\ddot{x} + c\dot{x} + kx = F_0 \sin \omega t \qquad (i)$$

The equivalent differential equation for electrical circuit is given as

$$L\frac{di}{dt} + Ri + \frac{1}{c}\int i\,dt = V_0 \sin \omega t \qquad (ii)$$

From Eq. (i), we see that the mass m, damper c, and spring stiffness k are associated with x and forces act on the mass parallel to each other. Therefore, equivalent electrical circuit will have all the components in series as shown in Figure 14.49.

FIGURE 14.49

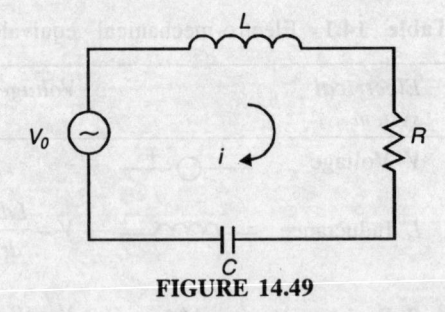

EXAMPLE 14.31 Convert the mechanical system shown in Figure 14.50 into an equivalent electrical circuit.

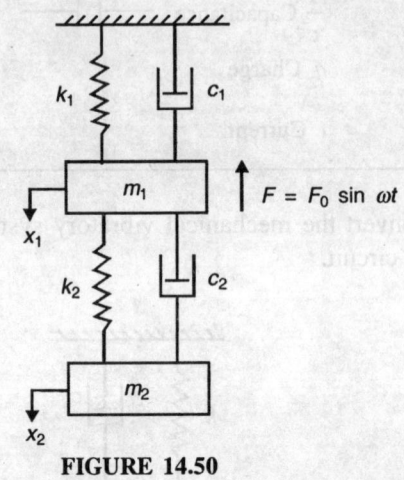

FIGURE 14.50

Solution Using Newton's law of motion, let us write down the following differential equations of motion:

$$m_1\ddot{x}_1 + c_1\dot{x}_1 + k_1 x_1 + c_2(\dot{x}_1 - \dot{x}_2) + k_2(x_1 - x_2) = F_0 \sin \omega t$$

$$m_2\ddot{x}_2 + c_2(\dot{x}_2 - \dot{x}_1) + k_2(x_2 - x_1) = 0$$

A close look to the above equation shows that m_1, c_1, and k_1 are associated with displacement x_1. Therefore, it will form one loop having i_1 current whereas mass m_2 is associated with displacement x_2 so it should be part of the second loop. The stiffness k_2 and damping coefficient c_2 are associated with x_1 and x_2 both, so they should be common element between these two loops. Thus the equivalent electrical circuit is as shown in Figure 14.51.

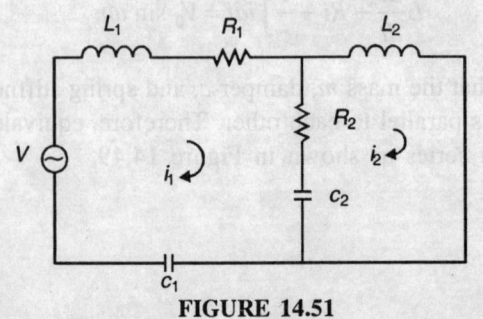

FIGURE 14.51

EXAMPLE 14.32 Draw the analogous electrical circuit for a mechanical vibratory system as shown in Figure 14.52 given below. Also write down the equation of motion.

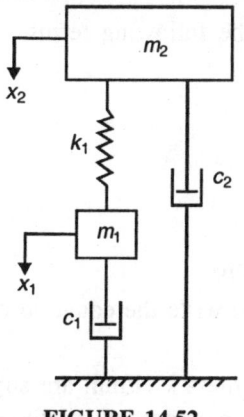

FIGURE 14.52

Solution The differential equation of motion for the above system is as given

$$m_1 \ddot{x}_1 + c_1 \dot{x}_1 + k_1 (x_1 - x_2) = F_1$$

$$m_2 \ddot{x}_2 + c_2 \dot{x}_2 + k_1 (x_1 - x_2) = 0$$

The corresponding differential equation for equivalent electrical circuit is:

$$L_1 \frac{di_1}{dt} + R_1 i_1 + \frac{1}{c_1} \int (i_1 - i_2) \, dt = V_1$$

and

$$L_2 \frac{di_2}{dt} + R_2 i_2 + \frac{1}{c_1} \int (i_1 - i_2) \, dt = 0$$

Comparing the differential equations of mechanical system and electrical circuit, the analogous electrical circuit can be drawn as shown in Figure 14.53.

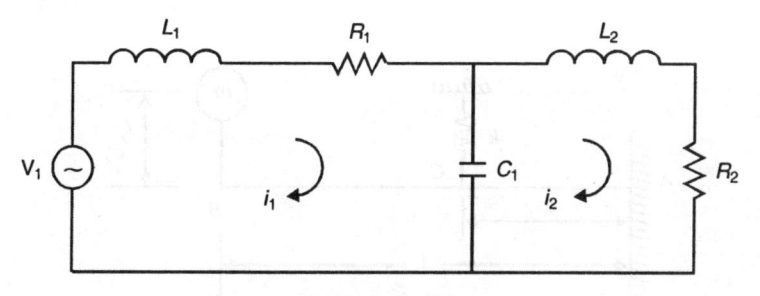

FIGURE 14.53

EXERCISES

1. What are sources of vibration in a mechanical system? Explain.

2. What do you mean by the following terms:
   (a) free vibration
   (b) forced vibration
   (c) damped vibration
   (d) period of vibration
   (e) resonance
   (f) natural frequency
   (g) amplitude of vibration

3. Explain energy method to write the equation of motion of a vibratory system with suitable example.

4. Four springs each of stiffness k N/mm are supporting an electric motor as shown in Figure E14.1. If the moment of inertia of the motor about central axis of vibration is J_0, determine the natural frequency.

$$\left[\text{Ans:} \ \frac{1}{2\pi} \sqrt{\frac{4ka^2}{J_0}} \ \text{Hz} \right]$$

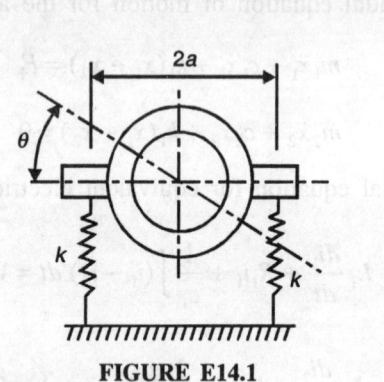

FIGURE E14.1

5. Determine the natural frequency of the system as shown in Figure E14.2

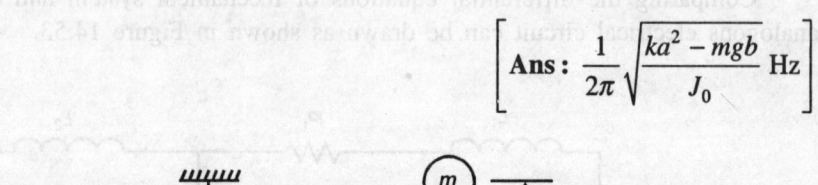

$$\left[\text{Ans:} \ \frac{1}{2\pi} \sqrt{\frac{ka^2 - mgb}{J_0}} \ \text{Hz} \right]$$

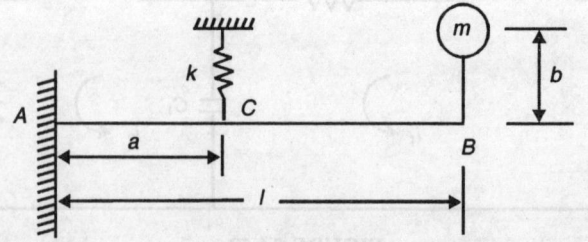

FIGURE E14.2

Mechanical Vibrations — 661

6. A circular cylinder of mass m and radius r is connected by a spring of stiffness k as shown in Figure E14.3. The cylinder is free to roll on the surface without slip. Find its natural frequency.

$$\left[\text{Ans:}\ \omega_n = \sqrt{\frac{2k}{3m}}\ \text{rad/s}\right]$$

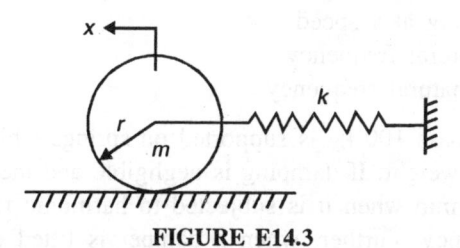

FIGURE E14.3

7. A pulley of mass m_0 is suspended through a spring stiffness k. A block of mass m kg passes over the pulley as shown in Figure E14.4. Determine the natural frequency of vibration.

$$\left[\text{Ans:}\ \omega_n = \sqrt{\frac{kr^2}{(J_0 + m_0 r^2 + 4mr^2)}}\right]$$

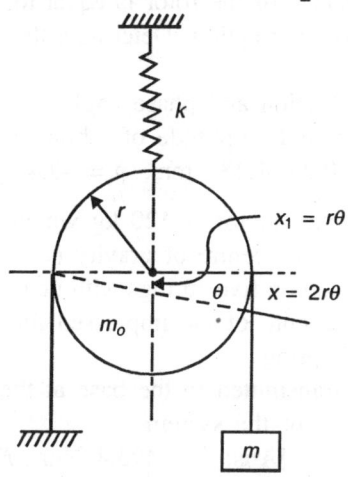

FIGURE E14.4

8. Explain the concept of overdamped, critically damped and underdamped system. Also show that in a free underdamped vibrating system, the amplitude of vibration decays exponentially.

9. Derive the expression for ratio of amplitude of steady state vibration to static deflection for a spring-mass-damper system subjected to forced vibration.

10. What do you mean by vibration transmissibility? Derive an expression of force transmissibility for forced damped vibrating system.

11. Draw and discuss the transmissibility versus ratio of frequency curve.

12. Explain the Rayleigh method for computing natural frequency of multi-degree of freedom vibratory system.

13. What do you mean by whirling of shaft? Derive an expression for amplitude of vibration of shaft supporting a disc at the mid-span. Discuss the behaviours of shaft when it is rotating at a speed
    (a) less than natural frequency
    (b) more than natural frequency

14. A machine of mass 100 kg is supported on springs which get deflected by 8 mm under the dead weight. If damping is negligible and the machine vibrates with an amplitude of 5 mm when it is subjected to harmonic force at 80 per cent of the resonant frequency. Further, when a damper is fitted on it, it is found that the resonant amplitude is 2 mm. Find:
    (i) the magnitude of damping force
    (ii) the damping coefficient
    [Ans: 2207.52 N, 31518 N s/m]

15. An electric motor of mass 30 kg is running at 500 rpm. The motor is supported by a spring of stiffness 7 kN/m and damper which offers a resistance of 60 N at 0.25 m/s. The unbalance of the rotor is equal to the mass of 0.8 kg m located at 50 mm from the axis of rotation. Determine the following for the above system:
    (i) damping factor
    (ii) amplitude of vibration and phase angle
    (iii) resonance speed and amplitude of vibration
    [Ans: 0.26, 1.435 mm, ϕ = –9.4°, f_n = 145.6 rpm, x = 2.56 mm]

16. The mass of an electric motor is 120 kg which runs at 1500 rpm. The mass of armature is 35 kg and its centre of gravity is 0.5 mm above the axis of rotation. If the motor is mounted on five springs with negligible damping such that the force transmitted is one-eleventh of the impressed force, determine:
    (i) the stiffness of spring
    (ii) dynamic force transmitted to the base at the operating speed
    (iii) natural frequency of the system
    [Ans: k = 493.4 N/m, F_t = 39.24 N, ω_n = 45.34 rad/s]

17. A gun barrel having 560 kg mass is designed to have initial recoil velocity 36 m/s and recoil distance on firing is 1.5 m. Calculate the following:
    (i) spring stiffness
    (ii) damping coefficient
    (iii) time required for the barrel to return to a position of 0.12 m from its initial position.
    [Ans: 322.5 kN/m, 26880 N s/m, 0.2354 s]

18. A pendulum with a disc of moment of inertia 0.05 kg m² is immersed in a viscous fluid. During the torsional vibration, the amplitude of vibration measured on same side of the neutral axis for successive cycles is decayed by 50 per cent. Determine:
    (i) damping factor
    (ii) damping torque
    (iii) periodic time of vibration
    (iv) natural frequency of vibration
    [Ans: 0.11, 45.8 Nm/rad, 1.5 × 10⁻³ s, 669.2 Hz]

19. A shock absorber is designed in such a way that its over shoot is 10 per cent of the initial displacement when released. Determine the damping factor. If the damping factor is reduced to one-half, what will be the overshoot?
    [Ans: 0.344, 33.45 per cent]

20. A single cylinder engine has an out of balance force of 500 N at an engine speed of 300 rpm. The complete mass of the engine is 150 kg and it is carried on a set of springs having total stiffness 30 N/mm. Find the amplitude of vibration and force transmitted to the foundation.
    If a viscous damper is interposed between the mass and the foundation with damping coefficient 1000 Ns/m, find the amplitude of vibration and its angle of lag with disturbing force.
    [Ans: 4.235 mm, 127 N, 4.093 mm, 165.1°]

21. A trailer has a mass of 1000 kg when fulley loaded and 250 kg when it is empty. The spring of suspension has stiffness of the order of 350 kN/m and damping factor 0.5. The speed of the trailer is 100 kmph. The road surface is sinusoidal with a wavelength of 5 m. Determine the ratio of amplitude of the trailer when it is fully loaded and empty.
    [Ans: 1.452, 0.619]

22. Figure E14.5 shows a three degree of freedom undamped system. Determine the first natural frequency of vibration using Dunkerley's principle.
    Stiffness: $k_1 = k_2 = k_3 = 100$ N/m and mass: $m_1 = m_2 = m_3 = 10$ kg
    [Ans: 4.082 rad/s]

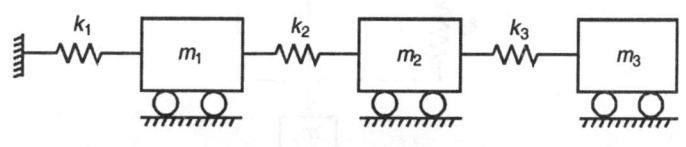

FIGURE E14.5

23. Two rotors A and B are attached to the end of a 500 mm long shaft as shown in Figure E14.6. The mass of the rotor A is 300 kg and its radius of gyration is 300 mm and the corresponding values of rotor B 500 kg and 450 mm respectively. Determine the position of node and frequency of torsional vibration. $G = 0.8 \times 10^5$ N/mm².
    [Ans: 274 Hz]

A ⌀ 70 mm   ⌀ 120 mm   ⌀ 100 mm   B

| 250 mm | 100 mm | 150 mm |

FIGURE E14.6

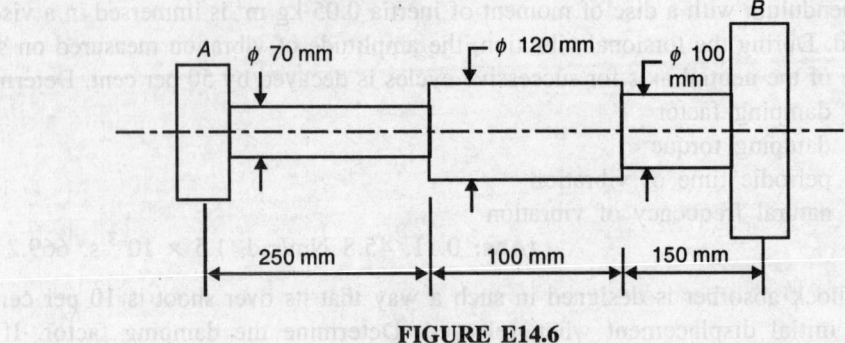

24. Draw the electrical analogous diagram of the following torsional vibratory system shown in Figure E14.7.

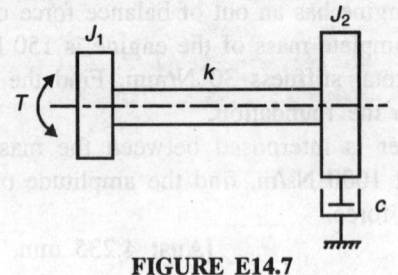

FIGURE E14.7

25. Write down the equation of motion of the mechanical system shown in Figure E14.8 and convert it into equivalent electrical circuit.

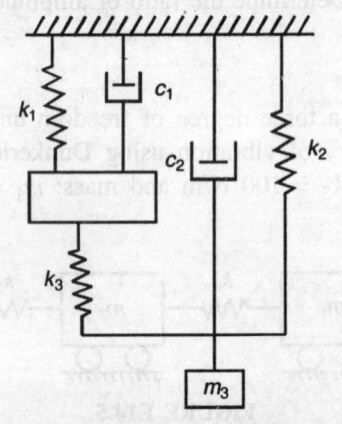

FIGURE E14.8

26. The springs of an automotive vehicle are compressed by 100 mm under its own weight. Find the critical speed when the vehicle is travelling over a road of profile approximated by a sine wave of amplitude 0.8 m and wavelength of 14 m. What will be the amplitude of vibration at 60 kmph?

[Ans: 79.5 kmph, 0.186 m]

MULTIPLE CHOICE QUESTIONS

1. Resonance is a phenomenon in which the frequency of the external exciting force
   (a) is half of the natural frequency of the system
   (b) is double of the natural frequency of the system
   (c) coincides with natural frequency of the system
   (d) does not coincide with the natural frequency of the system

2. The effective stiffness of two springs connected in series
   (a) is increased by 50 per cent   (b) is decreased by 50 per cent
   (c) remains same   (d) none of the above

3. A vibrating system comes back to rest as quickly as possible if the system is
   (a) overdamped   (b) underdamped
   (c) critically damped   (d) no damping is applied

4. In a spring-mass system if the mass of spring is taken into account while computing natural frequency, its natural frequency would be
   (a) increased   (b) decreased
   (c) remain same   (d) none of the above

5. The natural frequency of a system is a function of
   (a) the stiffness and mass of the system
   (b) the stiffness and damping coefficient
   (c) the mass and damping coefficient
   (d) the stiffness, mass and damping coefficient

6. In the spring-mass system, if the mass of the system is doubled and stiffness of spring is halved, the natural frequency of vibration
   (a) remains unchanged   (b) is doubled
   (c) is halved   (d) is four times

7. In damped free vibrating system.
   (a) the spring force vector acts in the direction opposite to the displacement
   (b) the damping force vector acts in the direction opposite to the velocity
   (c) the inertia force vector acts in the direction opposite to the acceleration
   (d) all of the above statements are true.

8. In damped force vibrating system
   (a) the spring force lags behind the displacement by 180°
   (b) the damping force leads the displacement by 90°
   (c) the inertia force leads the displacement by 180°
   (d) all of the above statements are true.

9. Three rotors connected by shafts when subjected to torsional vibration will have
   (a) one node   (b) two nodes
   (c) three nodes   (d) any number of nodes

10. The transmissibility is same for all values of damping factors at frequency ratio of
    (a) 1
    (b) 2
    (c) $\sqrt{2}$
    (d) $1/\sqrt{2}$

11. In the force transmissibility, the vibration isolation is possible only when frequency ratio is
    (a) less than unity
    (b) equal to unity
    (c) less than $\sqrt{2}$
    (d) greater than $\sqrt{2}$

12. In a forced vibration, the amplitude of vibration at very high forcing frequency becomes
    (a) large
    (b) small
    (c) no change
    (d) none of the above

13. In a force vibration at the resonance, the phase angle is
    (a) 0°
    (b) 45°
    (c) 90°
    (d) 180°

14. In a forced vibration due to rotating unbalance, at very high forcing frequency, the magnification factor approaches
    (a) less than unity
    (b) unity
    (c) greater than unity
    (d) zero

15. The transmissibility curves pass through unit at frequency ratio equal to
    (a) unity
    (b) $\sqrt{2}$
    (c) both (a) and (b)
    (d) greater than $\sqrt{2}$

CHAPTER 15

Gyroscope

15.1 INTRODUCTION

A gyroscope is a spatial mechanism which is generally employed for directional control namely gyrocompass used on aircraft, naval ship, control system of missiles and space shuttle. The gyroscopic effect is also felt in the bearings of an automotive vehicle while making a turn. A gyroscope consists of a rotor mounted in the inner gimbal. The inner gimbal is mounted in the outer gimbal which itself is mounted on a fixed frame as shown in Figure 15.1. When the rotor spins about X-axis with angular velocity ω rad/s and the inner gimbal precesses (rotates) about Y-axis, the spatial mechanism is forced to turn about Z-axis other than its own axis of rotation, and the gyroscopic effect is thus setup. The resistance to this motion is called **gyroscopic effect**.

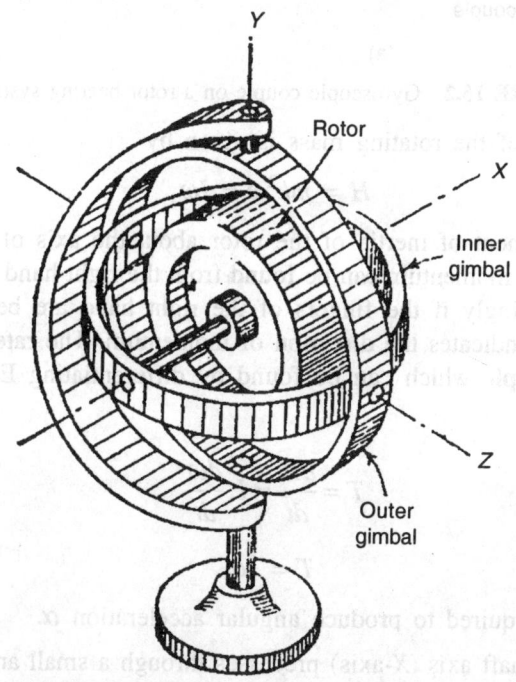

FIGURE 15.1 Gyroscope mechanism.

15.2 GYROSCOPIC COUPLE

Consider a rotor of mass m having radius of gyration k mounted on the shaft supported at two bearings. Let the rotor spins (rotates) about X-axis with constant angular velocity ω rad/s. The X-axis is, therefore, called axis of spin, Y-axis, the axis of precess and Z-axis, the axis of couple (Figure 15.2).

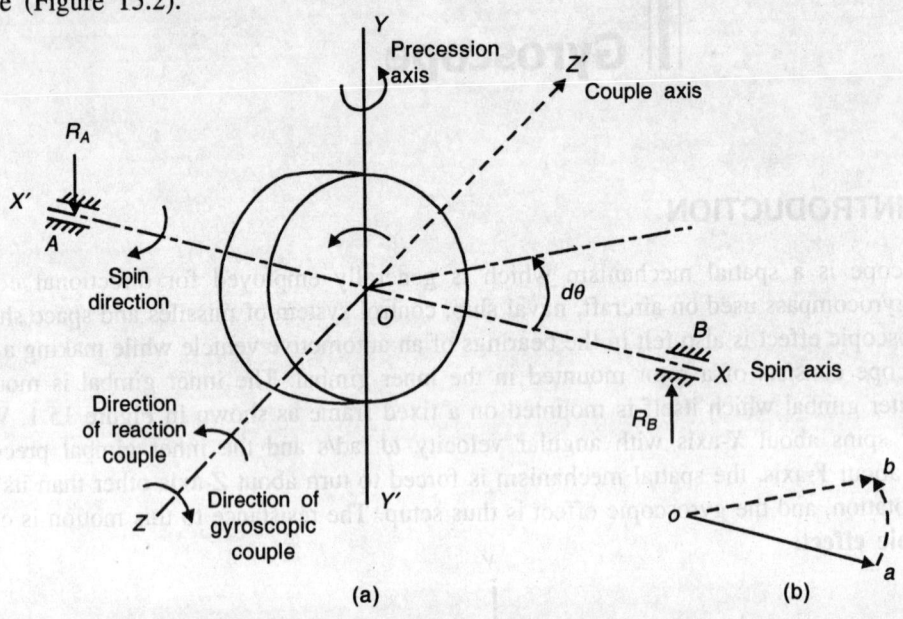

FIGURE 15.2 Gyroscopic couple on a rotor bearing system.

The angular momentum of the rotating mass is given by

$$H = mk^2\omega = I\omega \qquad (15.1)$$

where I is the mass moment of inertia of the rotor about the axis of spin.

The direction of the momentum can be found from the right hand screw rule or the right hand thumb rule, accordingly if the fingers of the right hand are bent in the direction of rotation then the thumb indicates the direction of momentum. The rate of change of angular momentum is called **couple** which can be found by differentiating Eq. (15.1).

Therefore

Couple: $$T = \frac{dH}{dt} = I \cdot \frac{d\omega}{dt}$$

or $$T = I\alpha \qquad (15.2)$$

where T is the couple required to produce angular acceleration α.

Now suppose the shaft axis (X-axis) precesses through a small angle $\delta\theta$ about Y-axis in the plane XOZ, then the angular momentum varies from H to $H + \delta H$, where δH is the change in the angular momentum, represented by vector $\vec{ab}$ [Figure 15.2(b)]. For the small value of angle of rotation $\delta\theta$, we can write

$$ab = oa \times \delta\theta$$

or
$$\delta H = H \times \delta\theta$$
$$= I\omega\delta\theta$$

Thus the rate of change of angular momentum is:

$$C = \frac{dH}{dt} = \lim_{\delta t \to 0}\left(\frac{I\omega\,\delta\theta}{\delta t}\right)$$

$$= I\omega\frac{d\theta}{dt}$$

or
$$C = I\omega\omega_p \qquad (15.3)$$

where C = gyroscopic couple
ω = angular velocity of spinning
and ω_p = angular velocity of precession

The direction of gyroscopic couple can be found from the right hand rule. Accordingly, if we put the thumb of our right hand towards the direction vector $\vec{ab}$ along the axis of couple (Z-axis), the direction of bent fingers indicate the direction of gyroscopic couple, which is clockwise for the present case (Figure 15.2). The resisting couple will act in the direction opposite to that of the gyroscopic couple. This means that whenever the axis of spin changes its direction, a gyroscopic couple is applied to it through the bearings which supports the spinning axis. In the case shown in the Figure 15.2, the reaction gyroscopic couple tends to raise the bearing B and lower the bearing A. Thus the reaction of the shaft on each bearing is to oppose the action of the bearing on the shaft.
Forces on the bearings A and B are:

$$R_A = \frac{C}{l}(\downarrow)$$

$$R_B = \frac{C}{l}(\uparrow) \qquad (15.4)$$

Figure 15.3 shows the direction of gyroscopic couple for the following two cases:

(a) When the disc spins in clockwise direction when viewed from the left end and precesses in clockwise direction [Figure 15.3(a)]
(b) When the disc spins in anticlockwise direction when viewed from the left end and precesses in clockwise direction [Figure 15.3(b)]

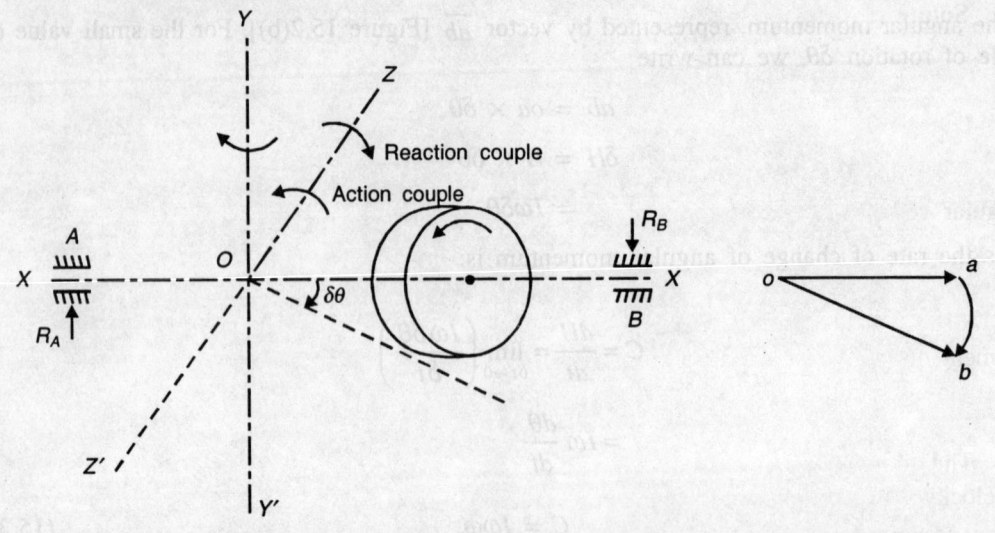

FIGURE 15.3(a) Disc spins in clockwise direction when viewed from left end and precesses in clockwise direction.

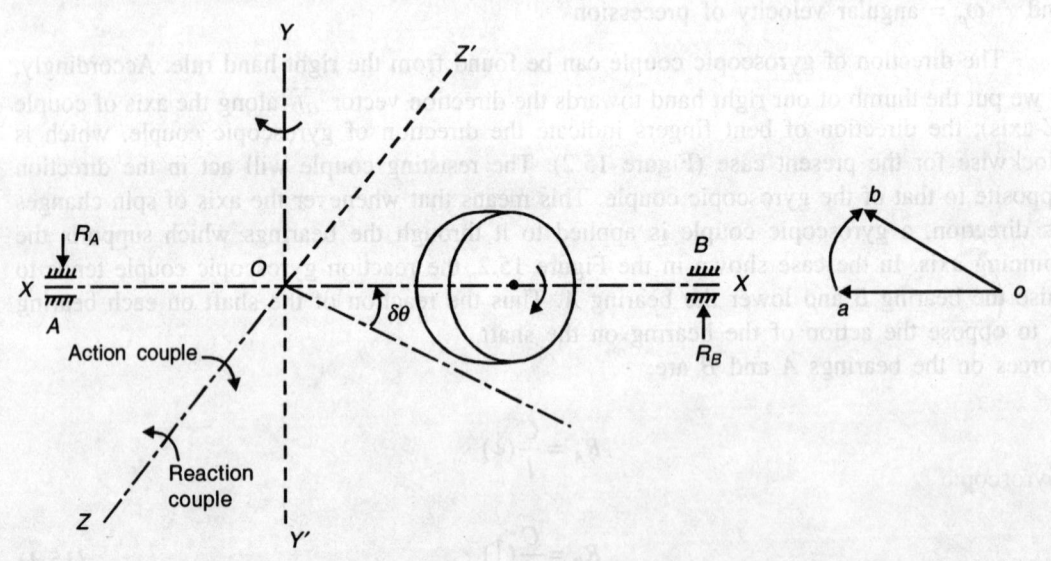

FIGURE 15.3(b) Disc spin in clockwise direction when viewed from right end and precesses in clockwise direction.

EXAMPLE 15.1 A disc of 5 kg mass with radius of gyration 70 mm is mounted at mid-span on a horizontal shaft of 120 mm length between the two bearings. The shaft spins at 720 rpm in clockwise direction when viewed from the right hand bearing. If the shaft precesses about the vertical axis at 30 rpm in clockwise direction when viewed from the above, determine the reactions at each bearing due to mass of the disc and gyroscopic effect.

Solution Angular velocity:
$$\omega = \frac{2\pi N}{60} = \frac{2\pi \times 720}{60}$$
$$= 75.4 \text{ rad/s}$$

Angular velocity of precession: $\omega_p = \frac{2\pi N_p}{60}$
$$= \frac{2\pi \times 30}{60} = 3.14 \text{ rad/s}$$

Moment of inertia: $I = mk^2$
$$= 5 \times 0.07^2 = 0.0245 \text{ kg m}^2$$

The angular momentum vector and induced reaction gyroscopic couple acting in anticlockwise direction is shown in Figure 15.4.

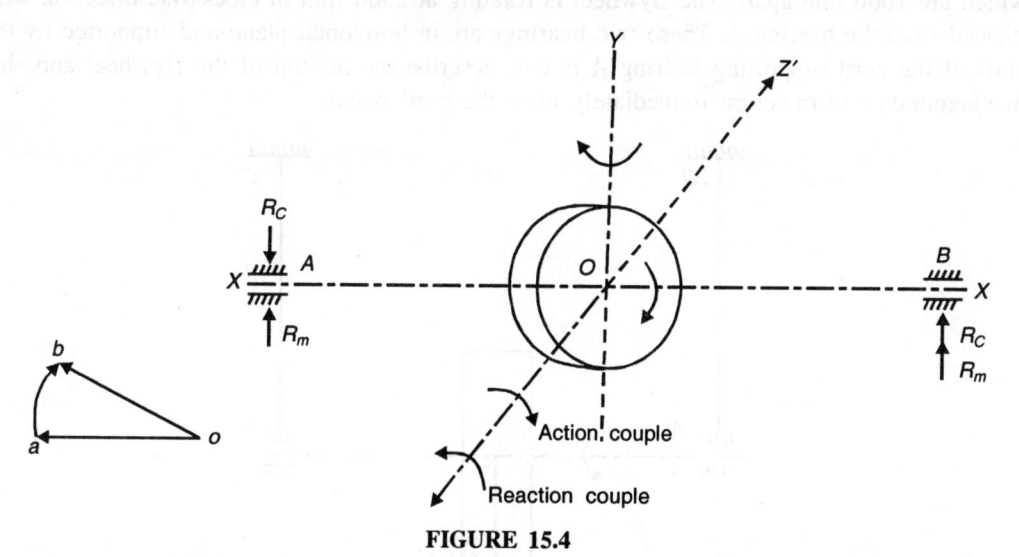

FIGURE 15.4

Gyroscopic couple:
$$C = I\omega\omega_p$$
$$= 0.0245 \times 75.4 \times 3.14$$
$$= 5.8 \text{ Nm}$$

This couple induces reaction R_c at the bearing support.

$$R_c \times \frac{120}{1000} = 5.8$$

or
$$R_c = 48.3 \text{ N}$$

The reaction R_c acts in upward direction at right hand bearing and in downward direction at left hand bearing.

The reaction due to weight of the disc acts in upward direction. Therefore,

Reaction at bearing A: $R_A = R_c - R_m$

$= 48.43 - 24.53$

$= 23.9 \text{ N}(\downarrow)$ **Ans.**

Reaction at bearing B: $R_B = R_c + R_m$

$= 48.43 + 24.53$

$= \mathbf{72.96 \text{ N}}(\uparrow)$ **Ans.**

EXAMPLE 15.2 A flywheel of 500 kg mass and 350 mm radius of gyration is mounted on the shaft at 400 mm away from bearing A. The shaft is supported on two bearings A and B which are 1000 mm apart. The flywheel is rotating at 2500 rpm in clockwise direction when viewed from the bearing A. These two bearings are in horizontal plane and supported by thin cord. If the cord supporting bearing A is cut, describe the motion of the flywheel and shaft in magnitude and direction immediately after the cord is cut.

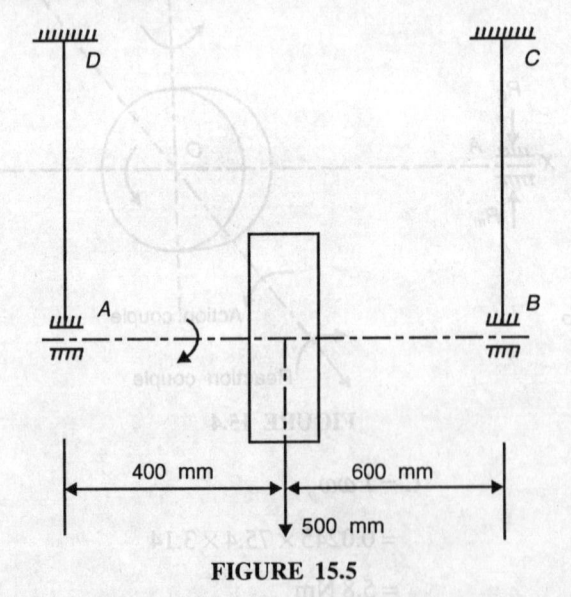

FIGURE 15.5

Solution Angular velocity:

$$\omega = \frac{2\pi N}{60} = \frac{2\pi \times 2500}{60}$$

$= 261.8 \text{ rad/s}$

Moment of inertia: $I = mk^2 = 500 \times 0.35^2 = 61.25 \text{ kgm}^2$

When cord A is cut, the flywheel rotates around the vertical cord BC in clockwise direction due to the couple formed.

Couple:
$$C = mg \times \text{distance}$$
$$= 500 \times 9.81 \times 0.6 = 2943 \text{ Nm}$$

This couple is resisted by reaction gyroscopic couple. Thus the angular momentum vector of spinning axis and precession axis are as shown in Figure 15.6.

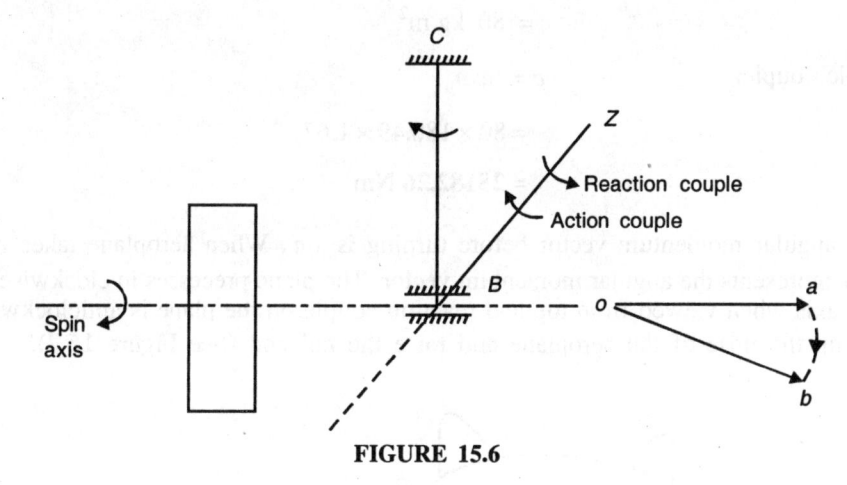

FIGURE 15.6

Thus reaction couple acting in anticlockwise direction will be balanced by gyroscopic couple.

or
$$I\omega\omega_p = C = 2943$$

or

Angular velocity of precession:
$$\omega_p = \frac{2943}{I \times \omega}$$

or
$$\omega_p = \frac{2943}{61.25 \times 261.8}$$
$$= 0.183 \text{ rad/s} \qquad \text{Ans.}$$

EXAMPLE 15.3 An aeroplane flying at a speed of 300 kmph takes right turn with a radius of 50 m. The mass of engine and propeller is 500 kg and radius of gyration is 400 mm. If the engine runs at 1800 rpm in clockwise direction when viewed from tail end, determine the gyroscople couple and state its effect on the aeroplane. What will be the effect if the aeroplane turns to its left instead of right?

Solution Angular velocity of aeroplane engine:
$$\omega = \frac{2\pi N}{60} = \frac{2\pi \times 1800}{60} = 188.49 \text{ rad/s}$$

Angular velocity of precession: $\omega_p = \dfrac{V}{R}$

or
$$\omega_p = \dfrac{300 \times 1000}{3600} \times \dfrac{1}{50}$$
$$= 1.67 \text{ rad/s}$$

Moment of inertia: $I = mk^2 = 500 \times 0.4^2$
$$= 80 \text{ kg m}^2$$

Gyroscopic couple: $c = I\omega\omega_p$
$$= 80 \times 188.49 \times 1.67$$
$$= 25182.26 \text{ Nm} \qquad \text{Ans.}$$

The angular momentum vector before turning is $\overrightarrow{oa}$. When aeroplane takes right turn, vector $\overrightarrow{ab}$ represents the angular momentum vector. The plane precesses in clockwise direction about OY axis when viewed from top and reaction couple on the plane is anticlockwise which tends to dip the nose of the aeroplane and raise the tail end (See Figure 15.7).

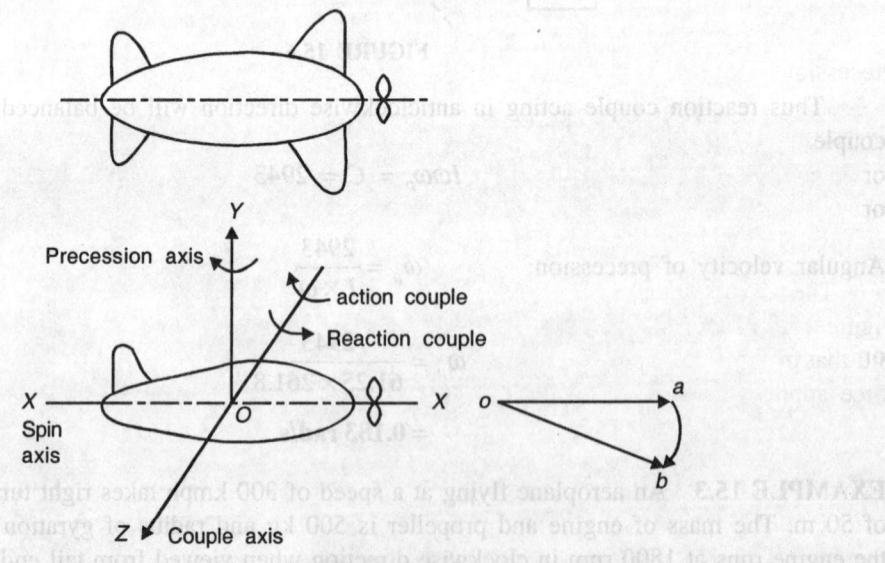

FIGURE 15.7

When aeroplane turns to its left, the magnitude of gyrocouple remains the same. However, the direction of reaction couple is reversed and it will raise the nose and dip the tail of the aeroplane.

EXAMPLE 15.4 For a single cylinder engine as shown in Figure 15.8, determine the bearing forces caused by the gyroscopic action of the flywheel as the engine vehicle traverses

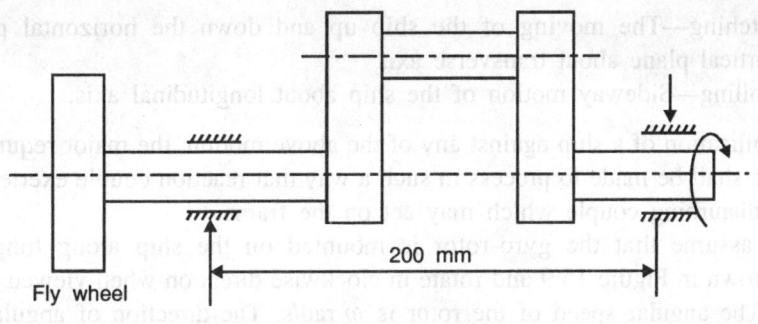

FIGURE 15.8

at 250 m radius at 85 kmph in a turn to its right side. The engine speed is 2500 rpm and is turning clockwise when viewed from the front of the engine. The moment of inertia of the flywheel is 0.35 kg m².

Solution Angular velocity:

$$\omega = \frac{2\pi N}{60} = \frac{2\pi \times 2500}{60} = 261.8 \text{ rad/s}$$

Linear velocity:

$$V = \frac{80 \times 1000}{3600} = 22.2 \text{ m/s}$$

Precession velocity:

$$\omega_p = \frac{V}{R} = \frac{22.5}{250} = 0.088 \text{ rad/s}$$

Gyroscopic couple:

$$C = I\omega\omega_p$$
$$= 0.35 \times 261.8 \times 0.088$$
$$= 8.063 \text{ Nm}$$

The angular momentum vector of the engine is shown in Figure 15.9. The direction of the gyroscopic couple is clockwise and that of the reaction couple is anticlockwise. Thus the bearing force applied through the bearings to the shaft is:

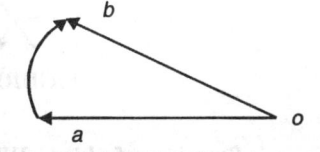

$$F = \frac{8.063}{0.2} = 40.3 \text{ N}$$ **Ans.**

15.3 GYROSCOPIC STABILIZATION OF SHIP

Gyroscope is used for stabilization and directional control of a ship sailing in the rough sea. A ship, while nevigating in the rough sea, may experience the following three different types of motion:

(i) Steering—The turning of ship in a curve while it moves forward

(ii) **Pitching**—The moving of the ship up and down the horizontal positions in a vertical plane about transverse axis
(iii) **Rolling**—Sideway motion of the ship about longitudinal axis.

For stabilization of a ship against any of the above motion, the major requirement is that the gyroscope shall be made to precess in such a way that reaction couple exerted by the rotor opposes the disturbing couple which may act on the frame.

Let us assume that the gyro-rotor is mounted on the ship along longitudinal axis (X-axis) as shown in Figure 15.9 and rotate in clockwise direction when viewed from rear end of the ship. The angular speed of the rotor is ω rad/s. The direction of angular momentum vector $\overline{oa}$, based on direction of rotation of rotor, is decided using right hand thumb rule (Figure 15.9). The gyroscopic effect during the above three types of motion is discussed further.

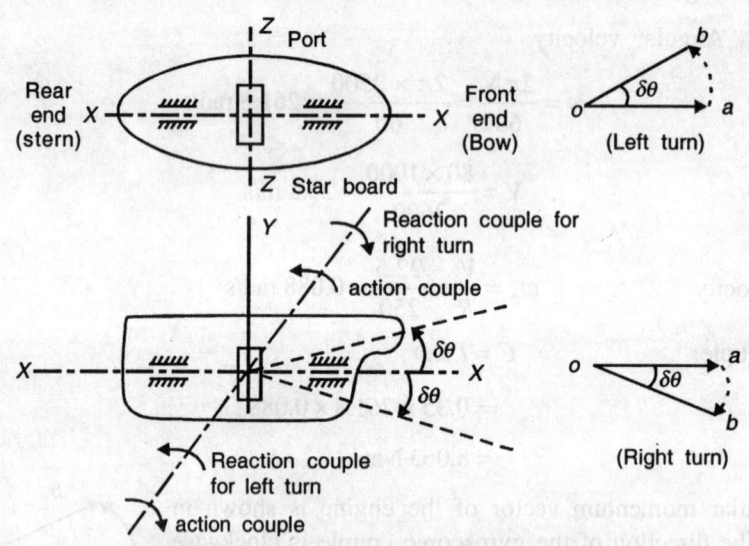

FIGURE 15.9 Gyroscopic stabilization of a ship (steering).

Steering of ship. When a ship takes a left turn, the gyro-rotor precesses about Y-axis in XZ-plane as shown in Figure 15.9 and the gyroscopic couple which is calculated by Eq. (15.3), acts in clockwise direction on the Z-axis. The reaction couple, which is equal in magnitude acts in opposite direction (anticlockwise). This reaction couple tends to raise the front end and lower the rear end of the ship. However, when the ship takes right turn, the direction of reaction couple is reversed so that the front end of the ship is lowered and rear end is raised.

Pitching of ship. The pitching motion of a ship generally occurs due to waves which can be approximated as sine wave. During pitching, the ship moves up and down the horizontal position in vertical plane by an angle θ from the axis of spin due to precessing of rotor about transverse axis ZZ (Figure 15.10).

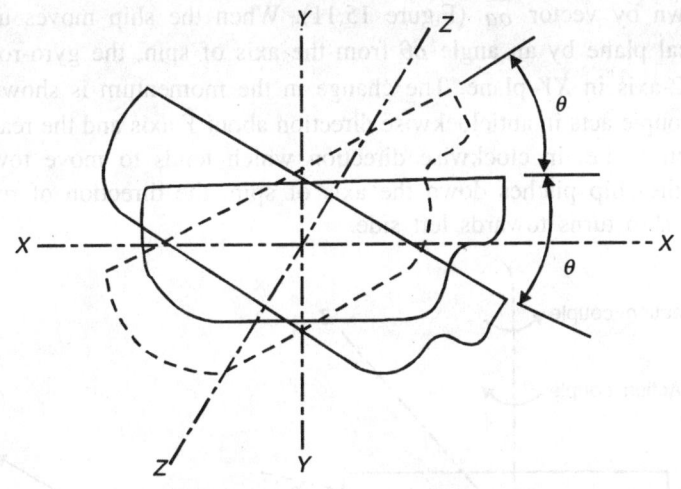

FIGURE 15.10 Pitching action of ship.

Let θ = angular displacement of spin axis from its mean equilibrium position
A = amplitude of swing
$$\left(= \text{angle in degree} \times \frac{2\pi}{360°}\right)$$
and ω_0 = angular velocity of simple hormonic motion $\left(= \frac{2\pi}{\text{time period}}\right)$

The angular motion of the rotor is given as
$$\theta = A \sin \omega_0 t$$

Angular velocity of precess:
$$\omega_p = \frac{d\theta}{dt}$$
$$= \frac{d}{dt}(A \sin \omega_0 t)$$

or
$$\omega_p = A\omega_0 \cos \omega_0 t \qquad (15.5)$$

The angular velocity of precess will be maximum when $\cos \omega_0 t = 1$

or
$$\omega_{p\max} = A\omega_0$$
$$= A \times \frac{2\pi}{t}$$

Thus the gyroscopic couple:
$$C = I \omega \omega_p \qquad (15.6)$$

Let the gyro-rotor is mounted along the longitudinal axis and rotates in clockwise direction when viewed from the rear end of the ship. The direction of momentum for this condition is shown by vector $\overrightarrow{oa}$ (Figure 15.11). When the ship moves up the horizontal position in vertical plane by an angle $\delta\theta$ from the axis of spin, the gyro-rotor axis (X-axis) precesses about Z-axis in XY-plane. The change in the momentum is shown by vector $\overrightarrow{ob}$. The gyroscopic couple acts in anticlockwise direction about Y-axis and the reaction couple acts in opposite direction, i.e. in clockwise direction which tends to move towards right side. However, when the ship pitches down the axis of spin, the direction of reaction couple is reversed and the ship turns towards left side.

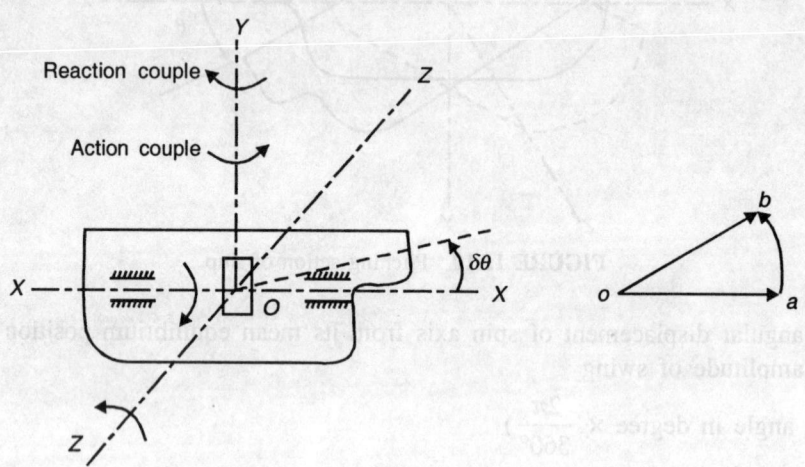

FIGURE 15.11 Ship stabilization during pitching.

Rolling of ship. In a ship when axis of spin is along the logitudinal axis, there is no precession of this axis, thus there is no effect of gyroscopic couple on the ship frame when the ship rolls.

EXAMPLE 15.5 A ship is propelled by a turbine rotor which has a mass of 3500 kg and rotates at a speed of 2000 rpm. The rotor has a radius of gyration of 0.5 m and rotates in clockwise direction when viewed from the stern (rear) end. Find the magnitude of gyroscopic couple and its direction for the following:

(i) When the ship runs at a speed of 12 knots and steers to the left in a curve of 70 m radius

(ii) When the ship pitches 6° above and 6° below the horizontal position and the bow (Front) end is lowered. The pitching motion is simple harmonic with pereodic time 30s

(iii) When the ship rolls and at a certain instant, it has an angular velocity of 0.05 rad/s clockwise when viewed from the stern

Also find the maximum angular acceleration during pitching.

Solution Since 1 knot = 1.86 kmph, the linear velocity of the ship:

$$V = 1.86 \times 12 = 22.32 \text{ kmph}$$

$$= \frac{22.32 \times 1000}{3600} = 6.2 \text{ m/s}$$

Angular velocity of the rotor: $\omega = \dfrac{2\pi N}{60} = \dfrac{2\pi \times 2000}{60}$

$= 209.44$ rad/s

(i) When ship steers to the left, the reaction gyroscopic couple action is in anticlockwise direction and the bow of the ship is raised and stern is lowered, as shown in Figure 15.12.

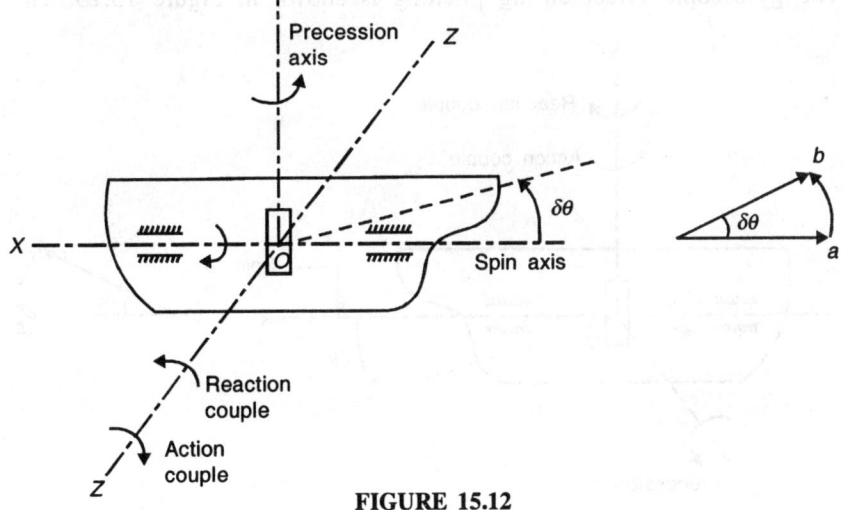

FIGURE 15.12

Precession velocity: $\omega_p = \dfrac{V}{R} = \dfrac{6.2}{70} = 0.08857$ rad/s

Moment of inertia: $I = mk^2 = 3500 \times 0.5^2 = 875$ kg m^2

Gyroscopic couple: $C = I\omega\omega_p$

$= 875 \times 209.44 \times 0.08857$

$= \mathbf{16231.34}$ **Nm**   Ans.

(ii) Amplitude of swing: $A = \dfrac{6° \times 2\pi}{360°} = 0.1047$ rad

Angular displacement: $\theta = A \sin \omega_0 t$

Angular velocity of precession: $\omega_p = \dfrac{d\theta}{dt} = A\omega_0 \cos \omega_0 t$

Maximum angular velocity of precession:

$$\omega_{pmax} = \omega_0 A$$

where $\omega_0 = \dfrac{2\pi}{\text{time period of oscillation}} = \dfrac{2\pi}{30}$

$= 0.2094$ rad/s

$$\omega_{pmax} = 0.2094 \times 0.1047 = 0.022 \text{ rad/s}$$

Maximum couple for pitching:

$$C_{max} = I\omega\omega_{pmax}$$
$$= 875 \times 209.44 \times 0.022$$
$$= 4031.72 \text{ Nm} \qquad \text{Ans.}$$

The gyroscopic effect during pitching is shown in Figure 15.13. The reaction

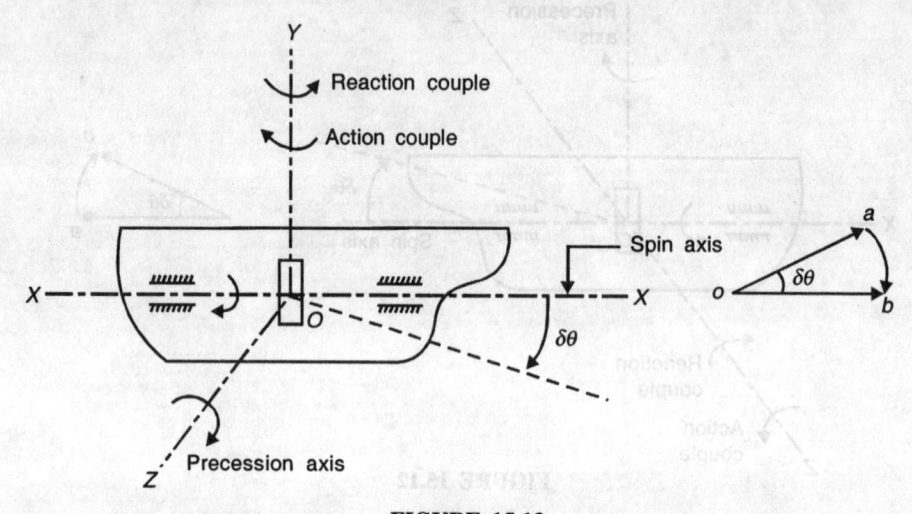

FIGURE 15.13

gyroscopic couple will act in anticlockwise direction during the bow descending and will turn ship towards the left side.

(iii) Angular velocity of precession while the ship rolls is:

$$\omega_p = 0.05 \text{ rad/s}$$

and gyroscopic couple: $C = I\omega\omega_p$

$$= 875 \times 209.44 \times 0.05$$
$$= 9163 \text{ Nm} \qquad \text{Ans.}$$

Since the ship rolls in the same plane as the plane of spin, there is no gyroscopic effect. Angular velocity of precess during pitching is:

$$\omega_p = \frac{d\theta}{dt} = A\omega_0 \cos \omega_0 t$$

Therefore, angular acceleration:

$$\alpha = \frac{d^2\theta}{dt^2} = -A\omega_0^2 \sin \omega_0 t$$

Maximum angular acceleration:

$$\alpha_{max} = -A\omega_0^2$$
$$= 0.1047 \times 0.2094^2$$
$$= \mathbf{0.00459 \text{ rad/s}^2} \qquad \textbf{Ans.}$$

EXAMPLE 15.6 The rotor of a ship has a mass of 2000 kg and its radius of gyration is 0.4 m. The rotor rotates at 2400 rpm in clockwise direction when viewed from front end. Determine the gyroscopic couple and its effect when:

(i) the ship takes left turn at a radius of 350 m with a speed of 35 kmph
(ii) the ship pitches with the bow rising at an angular velocity of 1 rad/s
(iii) the ship rolls at an angular velocity of 0.15 rad/s

Solution Angular velocity:

$$\omega = \frac{2\pi N}{60} = \frac{2\pi \times 2400}{60} = 251.33 \text{ rad/s}$$

Linear velocity: $V = 35 \text{ kmph} = \dfrac{35 \times 1000}{3600} = 9.72 \text{ m/s}$

Moment of inertia: $I = mk^2 = 2000 \times 0.4^2 = 320 \text{ kg m}^2$

Steering towards left

Angular velocity of precession: $\omega_p = \dfrac{V}{R} = \dfrac{9.72}{350} = 0.0278 \text{ rad/s}$

Gyroscopic couple:
$$C = I\omega\omega_p$$
$$= 320 \times 251.33 \times 0.0278$$
$$= 2235.8 \text{ Nm}$$

The reaction gyroscopic couple will act in anticlockwise and will tend to lower the bow as shown in Figure 15.14.

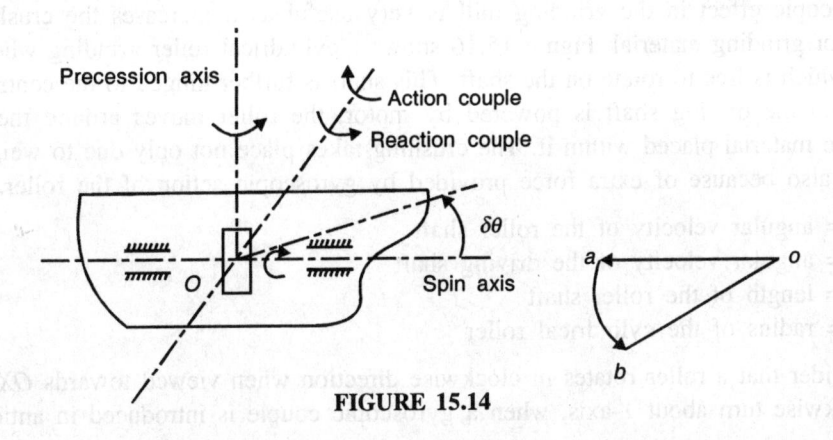

FIGURE 15.14

Pitching. Angular velocity of precession during pitching $\omega_p = 1.0$ rad/s

Gyroscopic couple: $C = 320 \times 251.33 \times 1.0$

$$= 80425.6 \text{ Nm} \qquad \textbf{Ans.}$$

The reaction gyroscopic couple acting in anticlockwise direction will tend to turn the bow towards the left side as shown in Figure 15.15.

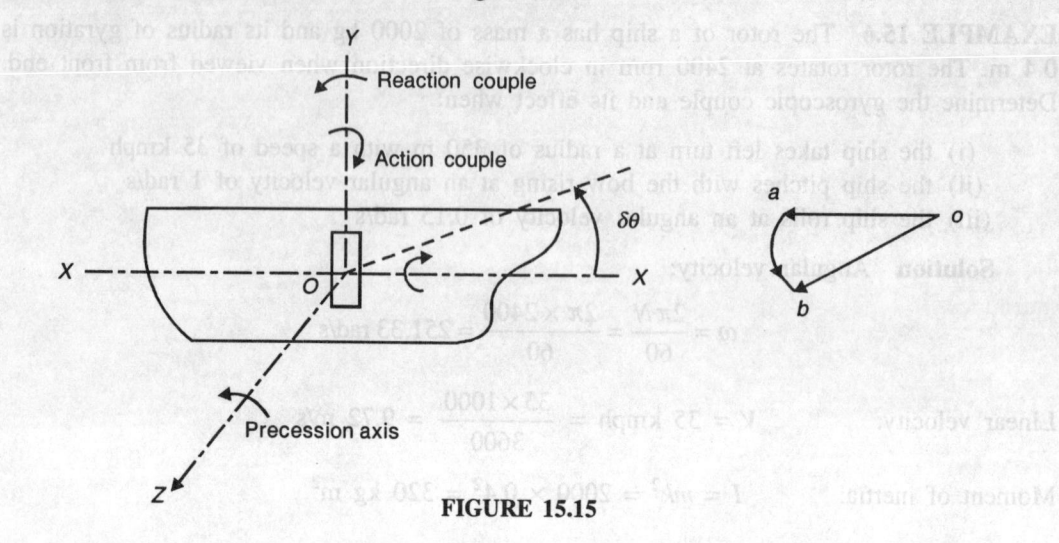

FIGURE 15.15

Rolling. Gyroscopic couple: $C = I\omega\omega_p$

$$= 320 \times 251.33 \times 0.15$$

$$= 12063.84 \text{ Nm}$$

During rolling, the ship rolls in the same plane as the plane of spin and there will be no gyroscopic effect.

15.4 GYROSCOPIC EFFECT IN GRINDING MILL

The gyroscopic effect in the grinding mill is very useful as it increases the crushing force required for grinding material. Figure 15.16 shows a cylindrical roller grinding wheel placed in a pan which is free to rotate on the shaft. This shaft is further hinged to the central driving shaft. When the driving shaft is powered by motor, the roller moves around the pan and crushes the material placed within it. The crushing takes place not only due to weight of the roller but also because of extra force provided by gyroscopic action of the roller.

Let ω = angular velocity of the roller shaft
ω_1 = angular velocity of the driving shaft
l = length of the roller shaft
and r = radius of the cylindrical roller

Consider that a roller rotates in clockwise direction when viewed towards OX-axis and takes clockwise turn about Y-axis, when a gyroscopic couple is introduced in anticlockwise

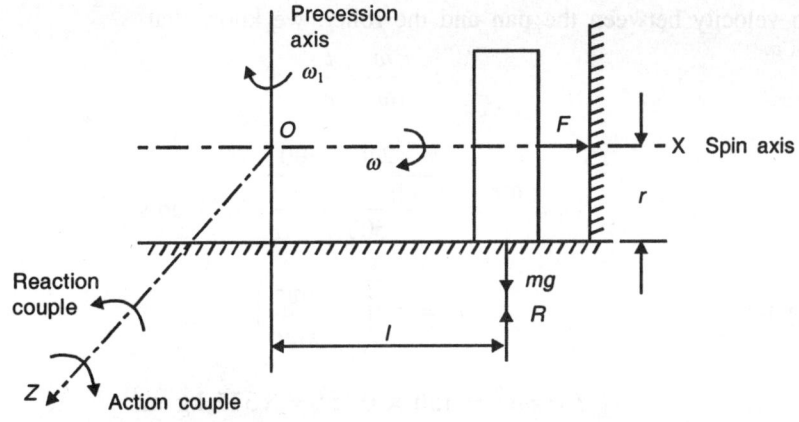

FIGURE 15.16 Gyroscopic effect in grinding mill.

direction. The reaction couple acts in the direction opposite to the gyroscopic couple which introduces a resultant crushing force R. The magnitude of gyroscopic couple is:

$$C = I\omega\omega_1 \qquad (15.7)$$

For the condition that relative velocity between the roller and the pan floor at a point it is zero, we get

$$\frac{\omega}{\omega_1} = \frac{l}{r} \qquad (15.8)$$

Therefore, gyroscopic couple: $C = I\omega_1^2 \dfrac{l}{r}$ \qquad (15.9)

This gyroscopic couple is balanced by the moment of forces acting on the roller. Taking moment about O,

$$M = R \times l - mgl \qquad (15.10)$$

Equating Eqs. (15.9) and (15.10), we get

$$R \times l - mgl = I\omega_1^2 \times \frac{l}{r}$$

or

$$\frac{R}{mg} = 1 + \frac{I\omega_1^2}{mgr} \qquad (15.11)$$

Equation (15.11) gives the ratio of resultant crushing force to the weight of roller which is always greater than unity. Hence the gyroscopic effect increases crushing force in the grinding mill.

EXAMPLE 15.7 In a roller grinding wheel, a roller having 300 mm diameter is of mass 150 kg and rotates at 240 rpm by the driving shaft. If the effect of length of roller arm revolving about central driving shaft is 500 mm, determine the total crushing force.

Solution Refer to Figure 15.16. The ratio of resultant crushing force R to weight of the roller is given by

$$\frac{R}{mg} = 1 + \frac{I\omega_1^2}{mgr}$$

For zero velocity between the pan and the roller, we know that

$$\frac{\omega}{\omega_1} = \frac{l}{r}$$

or

$$\omega_1 = \frac{\omega r}{l} = \frac{\frac{2\pi \times 240}{60} \times \frac{300}{2}}{500} = 7.54 \text{ rad/s}$$

Crushing force:

$$R = mg\left(1 + \frac{I\omega_1^2}{mgr}\right)$$

where

$$I = mk^2 = 150 \times 0.15^2 = 3.375 \text{ kg m}^2$$

or

$$R = 150 \times 9.81\left[1 + \frac{3.375 \times 7.54^2}{150 \times 9.81 \times 0.15}\right]$$

$$= 2750.6 \text{ N} \qquad \text{Ans.}$$

So it is seen that crushing force is 86.93 per cent greater than weight of the roller.

15.5 STABILITY OF AUTOMOTIVE VEHICLE

An automotive vehicle running on the road is said to be stable when no wheel is supposed to leave the road surface. In other words, the resultant reactions by the road surface on wheels should act in upward direction. For a moving vehicle, one of the reaction is due to gyroscopic couple produced by the rotating wheels and rotating parts of the engine. In the following subsection, we shall discuss stability of four wheels and two wheels vehicle.

15.5.1 Stability of Four Wheels Vehicle

Consider a four wheels automotive vehicle as shown in Figure 15.17. The engine is mounted at the rear with its crank shaft parallel to the rear axle. The centre of gravity of the vehicle lies vertically above the ground where total weight of the vehicle is assumed to be acted upon.

Let m = mass of the vehicle
b = distance between two wheels
h = distance of centre of gravity above the road surface
r = radius of wheel
R = radius of curvature while taking turn
ω = angular velocity of wheels
ω_p = angular velocity of precession
v = linear velocity of the vehicle
ω_e = angular velocity of the rotating parts of the engine
I_w = moment of inertia of the wheel
and I_e = moment of inertia of the rotating parts of the engine.

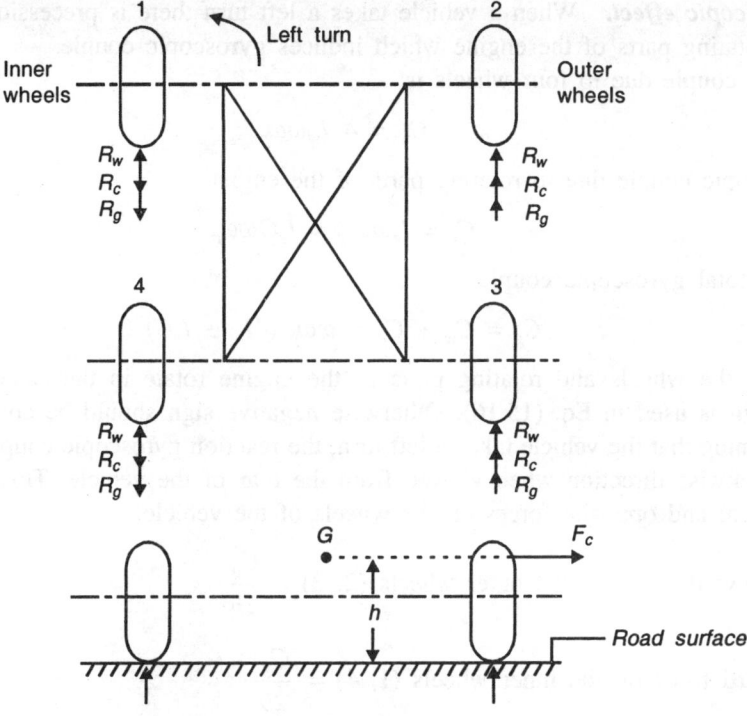

FIGURE 15.17 Stability of four wheels vehicle.

The reaction by the road surface on the wheels acts due to the following forces:

Weight of the vehicle. Assuming that weight of the vehicle (mg) is equally distributed over four wheels. Therefore, the force on each wheel acting downward is $mg/4$ and the reaction by the road surface on the wheel acts in upward direction.

$$R_w = \frac{mg}{4} \tag{15.12}$$

Centrifugal force. When a vehicle moves on a curved path, a centrifugal force acts on the vehicle in outward direction at the centre of gravity of the vehicle (Figure 15.17).

Centrifugal force: $\quad F_c = m\omega_p^2 R = \dfrac{mv^2}{R}$

This force forms a overturning couple.

$$C_c = \frac{mv^2 h}{R}$$

The overturning couple is balanced by vertical reaction which is upward on the outer wheels (2, 3) and downward on the inner wheels (1, 4).
Let R_c = reaction on each wheel

$$\therefore \qquad R_c = \frac{C_c}{2b} \quad \text{or} \quad R_c = \frac{mv^2 h}{2Rb} \tag{15.13}$$

Gyroscopic effect. When a vehicle takes a left turn there is precession of wheel spin axis and rotating parts of the engine which induces gyroscopic couple.
Gyroscopic couple due to four wheels is:

$$C_w = 4\, I_w \omega \omega_p \tag{15.14}$$

and gyroscopic couple due to rotating parts of the engine

$$C_e = I_e \omega_e \omega_p = I_e G \omega \omega_p \tag{15.15}$$

Therefore, total gyroscopic couple:

$$C_g = C_w + C_e = \omega \omega_p\, (4I_w \pm I_e G) \tag{15.16}$$

When the wheels and rotating parts of the engine rotate in the same direction, then positive sign is used in Eq. (15.16). Otherwise negative sign should be considered.

Assuming that the vehicle takes a left turn, the reaction gyroscopic couple on the vehicle acts in clockwise direction when viewed from the rear of the vehicle. This reaction couple induces equal and opposite forces on the wheels of the vehicle.

Downward force on the outer wheels (2, 3) = $\dfrac{C_g}{2b}$

Upward force on the inner wheels (1, 4) = $\dfrac{C_g}{2b}$

To balance this reaction couple, road surface introduces reaction forces which act vertically upward on the outerwheels and vertically downward on the inner wheels.

Upward reaction on each outer wheel: $R_g = \dfrac{C_g}{2b}(\uparrow)$ \hfill (15.17)

Downward reaction on each inner wheel: $R_g = \dfrac{C_g}{2b}(\downarrow)$ \hfill (15.18)

Hence the total vertical reaction on the outer wheel is:

$$R_O = R_w + R_c + R_g \tag{15.19}$$

and the total vertical reaction on the inner wheel is:

$$R_i = R_w - R_c - R_g \tag{15.20}$$

When the vehicle runs at high speed, the total vertical reaction on the inner wheel may become zero or negative. This may cause the inner wheel to leave the road surface and overturn the vehicle. Thus to avoid this situation,

$$R_i = R_w - R_c - R_g > 0$$

or
$$R_w > R_c + R_g$$

or
$$\frac{mg}{4} > \frac{mv^2 h}{2Rb} + \frac{\omega \omega_p (4I_w \pm I_e G)}{2b} \tag{15.21}$$

EXAMPLE 15.8 An automobile car is travelling along a track of 100 m mean radius. The moment of inertia of 500 mm diameter wheel is 1.8 kg m². The engine axis is parallel to the rear axle and crank shaft rotates in the same sense as the wheel. The moment of inertia of rotating parts of the engine is 1 kg m². The gear ratio is 4 and the mass of the vehicle is 1500 kg. If the centre of gravity of the vehicle is 450 mm above the road level and width of the track of the vehicle is 1.4 m, determine the limiting speed of the vehicle for condition that all four wheels maintain contact with the road surface.

Solution Let v = limiting velocity of the vehicle.

Angular velocity: $\omega = \dfrac{v}{r} = \dfrac{v}{0.25}$ rad/s

Precession velocity: $\omega_p = \dfrac{v}{R} = \dfrac{v}{100}$ rad/s

(i) Reaction due to gyroscopic couple:
   (a) Gyroscopic couple due to four wheels:

$$C_w = 4I_w \omega \omega_p$$

$$= 4 \times 2 \times \dfrac{v}{0.25} \times \dfrac{v}{100} = 0.32 v^2 \text{ Nm}$$

   (b) Gyroscopic couple due to engine parts:

$$C_e = I_e G \omega \omega_p$$

$$= 1 \times 4 \times \dfrac{v}{0.25} \times \dfrac{v}{100} = 0.16 v^2 \text{ Nm}$$

Total gyroscopic couple:

$$C_g = C_w + C_e$$

$$= 0.32v^2 + 0.16v^2 = 0.48v^2 \text{ Nm}$$

Reaction due to total gyroscopic couple on each outer wheel:

$$R_g = \dfrac{C_g}{2b} = \dfrac{0.48v^2}{2 \times 1.5} = 0.16 v^2 \text{N} (\uparrow)$$

Reaction due to total gyroscopic couple on each inner wheel:

$$C_g = 0.16 \ v^2 \text{N} \ (\downarrow)$$

(ii) Reaction due to centrifugal couple:

Centrifugal force: $F_c = \dfrac{mv^2}{R} = \dfrac{1500 \times v^2}{100} = 15v^2$ N

Overturning couple due to centrifugal force:

$$C_c = F_c \times h$$

$$= 15 v^2 \times 0.45 = 6.75 v^2 \text{ Nm}$$

Vertical downward reaction on each inner wheel is:

$$R_c = \frac{C_c}{2b} = \frac{6.75 v^2}{2 \times 1.5} = 2.25 v^2 \text{ N} (\downarrow)$$

(iii) Reaction due to weight of the vehicle:

$$R_w = \frac{mg}{4} = \frac{1500 \times 9.81}{4} = 3678.75 \text{ N} (\uparrow)$$

The limiting condition to avoid lifting of inner wheels from the road surface is:

$$R_i = R_w - R_c - R_g > 0$$

or

$$R_w > R_c + R_g$$

$$3678.75 \geq 2.25 v^2 + 0.16 v^2$$

or

$$v = 39.07 \text{ m/s} \quad \text{or} \quad \mathbf{140.65 \text{ kmph}} \qquad \text{Ans.}$$

EXAMPLE 15.9 A section of an electric rail track of gauge 1.5 m has a left hand curve of radius 300 m, the superelevation of the outer rail being 260 mm. The approach to the curve is along a straight length of track, over the last 50 m there is a uniform increase in elevation of the outer rail from level track to the super elevation of 260 mm. Each motor used for traction has a rotor of mass 550 kg and radius of gyration 300 mm. The motor shaft is parallel to the axes of the running wheels. It is supported in bearings 780 mm apart and runs at four times the wheel speed but in opposite direction. The diameter of running wheel is 1.2 m. Determine the forces on the bearings due to gyroscopic action when the train is travelling at 90 kmph (a) on the last 50 m of approach track (b) on the curve track.

Solution Angular velocity:

$$\omega = \frac{\text{Gear ratio} \times v}{r}$$

$$= \frac{4 \times 90 \times 1000}{3600 \times 0.6} = 166.67 \text{ rad/s}$$

Let ω_p = angular velocity of precession.

Moment of inertia: $I = mk^2 = 550 \times 0.3^2 = 49.5 \text{ kg m}^2$

Gyroscopic couple: $C = I\omega\omega_p$
$= 49.5 \times 166.67 \times \omega_p$
$= 8250.16\, \omega_p$ Nm

Forces on bearings: $P = \dfrac{8250.16\, \omega_p}{0.78}$
$= 10577.1\, \omega_p$ N

(a) Angle turned by engine shaft in the last 50 m track
$= \dfrac{0.26}{1.5} = 0.1734$ rad

Time taken to cover this distance $= \dfrac{50}{90/3.6} = 2$ sec

Velocity of precession: $\omega_p = \dfrac{0.1734}{2} = 0.0867$ rad/s

Forces on bearings: $P = 10577.1 \times 0.0867 = 917.03$ N

The change in momentum is represented by vector $\overrightarrow{oa}$ and $\overrightarrow{ob}$ as shown in Figure 15.18.

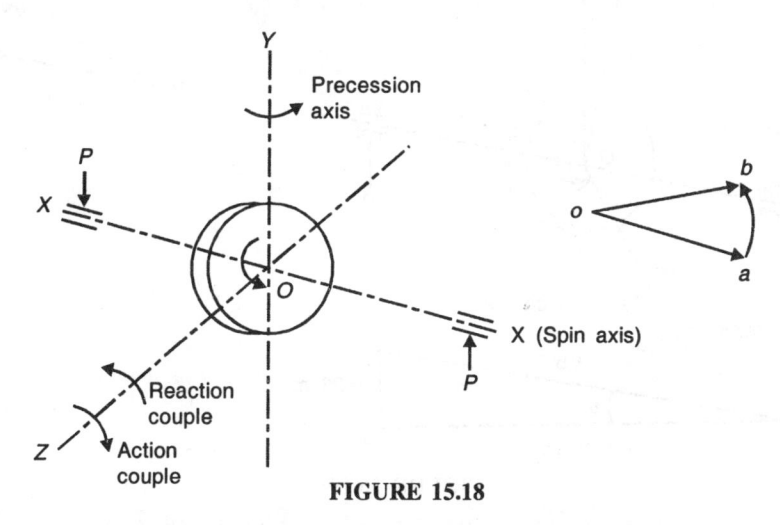

FIGURE 15.18

The couple required for precession is, therefore, acting in clockwise looking upward direction. The reaction couple acts in anticlockwise direction looking downward as the forces on the bearings are in the directions shown in Figure 15.18.

(b) When electric rail moves on curved path, the effective angular velocity of precession about the axis perpendicular to the axis of rotation is:

$$\omega_p = \dfrac{v}{R}\cos\theta$$

where θ is angle due to superelevation of outer rail. Referring to Figure 15.19.

$$\cos\theta = \frac{AB}{AC} = \frac{1.4773}{1.5} = 0.9848$$

or

$$\omega_p = \frac{90 \times 1000}{3600 \times 300} \times 0.9848 = 0.08206 \text{ rad/s}$$

Effective angular velocity of spin = $\omega - \omega_p \sin\theta \approx \omega$
Therefore,
Forces on bearings: $\quad P = 10577.1\ \omega_p$

$$= 10577.1 \times 0.08206$$

$$= 867.95 \text{ N} \qquad\qquad \textbf{Ans.}$$

The change in angular momentum vector and reaction couple shown in Figure 15.19 shows direction of forces on the bearings.

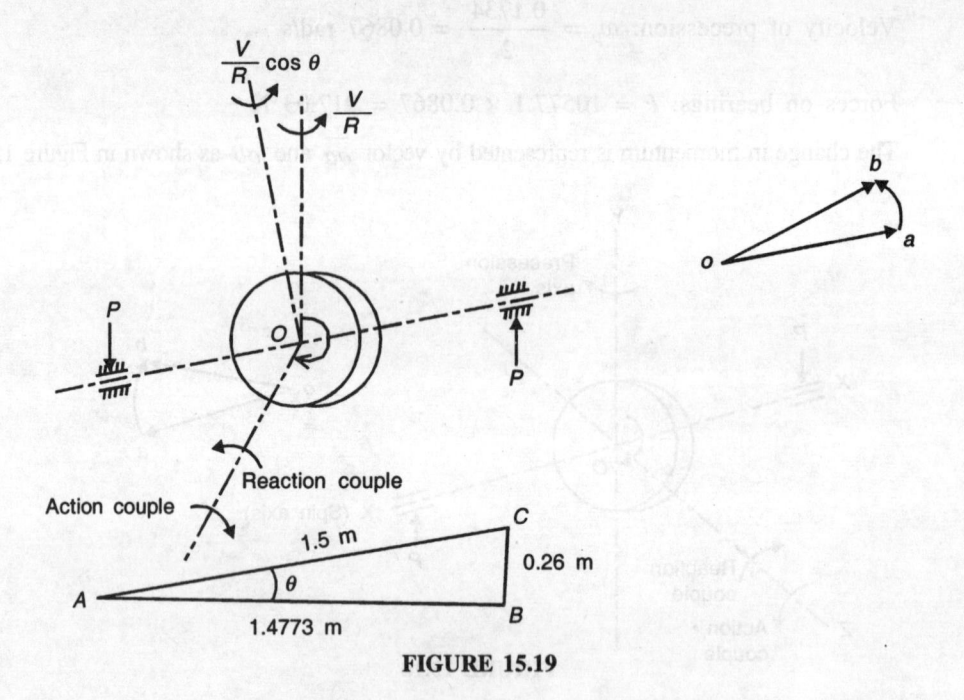

FIGURE 15.19

15.5.2 Stability of Two Wheeler Vehicle

A two wheeler vehicle is inherently unstable compared to a four wheel vehicle. However, a two wheeler vehicle can be easily lifted inward to negate the effect of overturning couple and can be brought to equilibrium while taking a turn.

Figure 15.20(a) shows a two wheeler vehicle taking left turn over a curved path. The vehicle is inclined to the vertical for equilibrium by an angle θ known as **angle of heel**. The vehicle is subjected to overturning couple due to the following:

(i) *Centrifugal force*—When vehicle moves a curved path centrifugal force acts horizontally in outward direction at the centre of gravity of the vehicle wheel [Figure 15.20(b)].

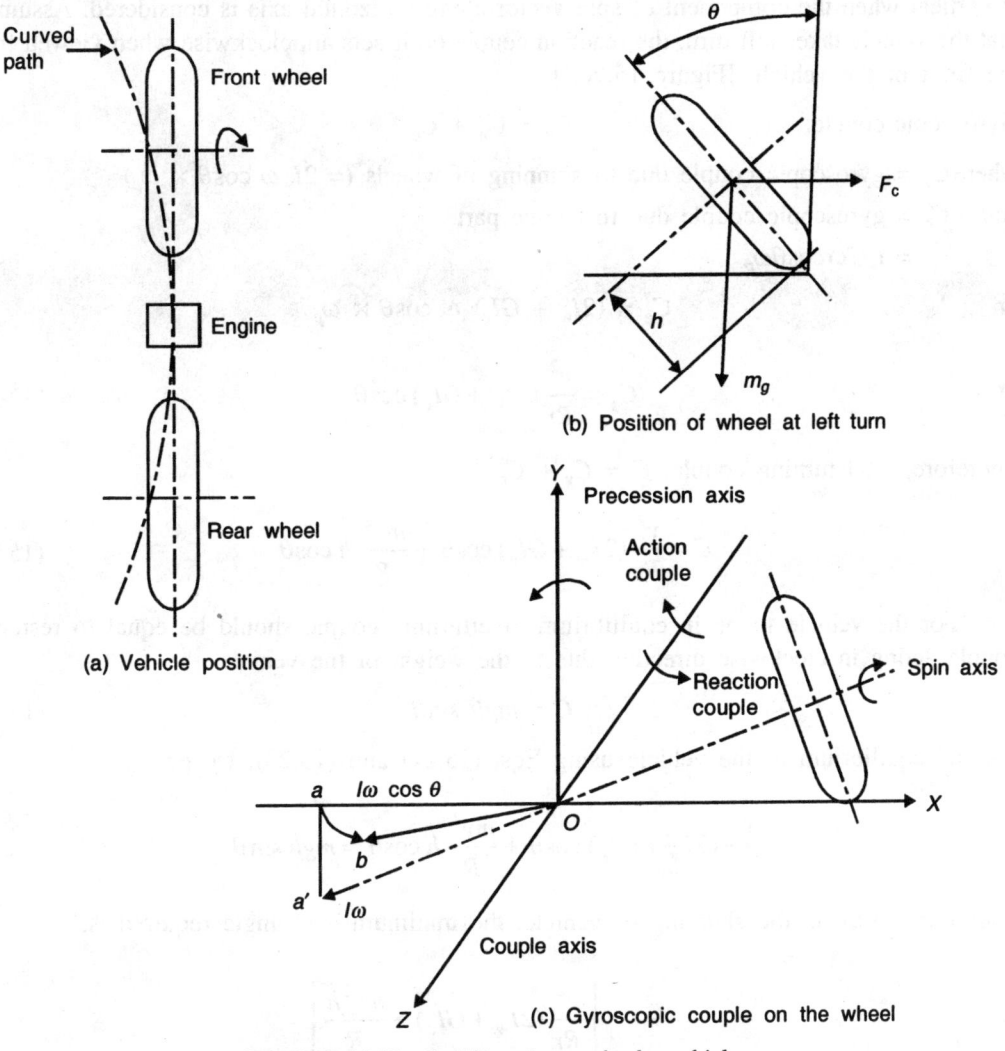

FIGURE 15.20 Stability of two wheeler vehicle.

(a) Vehicle position
(b) Position of wheel at left turn
(c) Gyroscopic couple on the wheel

Centrifugal force:
$$F_c = \frac{mv^2}{R}$$

or

Couple:
$$C_c = F_c \times h \cos\theta$$
$$= \frac{mv^2}{R} h \cos\theta \qquad (15.22)$$

(ii) *Gyroscopic effect*—When a vehicle moves over the curved path, the axis of spin is inclined to horizontal at the angle of heel. Thus the angular momentum of combined wheel and engine part due to spinning is represented by vector $\overrightarrow{oa'}$. However the axis of precession is vertical when the component of spin vector along horizontal axis is considered. Assuming that the vehicle takes left turn, the reaction couple on it acts anticlockwise when viewed from the front of the vehicle [Figure 15.9(c)].

Gyroscopic couple: $\qquad C_g = C_w + C_e$

where C_w = gyroscopic couple due to spinning of wheels ($= 2I_w\omega \cos\theta \times \omega_p$)

and $\quad C_e$ = gyroscopic couple due to engine part

$\qquad = I_e G\omega \cos\theta \omega_p$

or $\qquad C_g = (2I_w + GI_e)\, \omega \cos\theta \times \omega_p$

or $\qquad C_g = \dfrac{v^2}{Rr}(2I_w + GI_e)\cos\theta \qquad (15.23)$

Therefore, total turning couple: $C = C_g + C_c$

$$C = \dfrac{v^2}{Rr}(2I_w + GI_e)\cos\theta + \dfrac{mv^2}{R} h\cos\theta \qquad (15.24)$$

For the vehicle to be in equilibrium, overturning couple should be equal to restoring couple acting in clockwise direction due to the weight of the vehicle.

$\therefore \qquad C = mgh\,\sin\theta \qquad (15.25)$

For the equilibrium of the vehicle using Eqs. (15.24) and (15.25), we get

$$\dfrac{v^2}{Rr}(2I_w + GI_e)\cos\theta + \dfrac{mv^2}{R} h\cos\theta = mgh\sin\theta$$

Therefore, to avoid the skidding of vehicle, the minimum heel angle required is:

$$\tan\theta = \dfrac{\left[\dfrac{v^2}{Rr}(2I_w + GI_e) + \dfrac{mv^2 h}{R}\right]}{mgh} \qquad (15.26)$$

EXAMPLE 15.10 A motorcycle and its rider together weighs 2000 N and their combined centre of gravity is 550 mm above the road when motorcycle is upright. Each wheel is of 580 mm diameter and has a moment of inertia of 1.0 kgm². The moment of inertia of rotating parts of engine is 0.15 kg m². The engine rotates at 5 times the speed of the vehicle and in the same sense. Determine the angle of heel necessary when motorcycle is taking a turn over a track of 35 m radius at a speed of 60 kmph.

Solution Velocity of vehicle:

$$v = \frac{60 \times 1000}{3600} = 16.67 \text{ m/s}$$

Angular velocity of wheel: $\omega = \dfrac{2v}{d} = \dfrac{2 \times 16.67}{0.58} = 57.48$ rad/s

Angular velocity of precession: $\omega_p = \dfrac{v}{R} = \dfrac{16.67}{35} = 0.476$ rad/s

(i) Gyroscopic couple due to rotating parts of two wheels:

$$C_w = 2I_w \omega \omega_p \cos\theta$$
$$= 2 \times 1.0 \times 57.48 \times 0.476 \times \cos\theta$$
$$= 54.72 \cos\theta \text{ Nm}$$

(ii) Gyroscopic couple due to rotating parts of engine:

$$C_e = I_e G \omega \omega_p \cos\theta$$
$$= 0.15 \times 5 \times 57.48 \times 0.476 \times \cos\theta$$
$$= 20.52 \cos\theta \text{ Nm}$$

(iii) Centrifugal force due to angular velocity of the wheel:

$$F_c = \frac{mv^2}{R} = \frac{2000 \times 16.67^2}{9.81 \times 35} = 1618.7 \text{ N}$$

Centrifugal couple: $C_c = 1618.7 \times 0.55 \cos\theta$
$$= 890.28 \cos\theta \text{ Nm}$$

Total overturning couple: $C = C_w + C_e + C_c$
$$= (54.72 + 20.52 + 890.28) \cos\theta$$
$$= 965.52 \cos\theta \text{ Nm}$$

Resisting couple due to weight of the motorcycle
$$= mgh \sin\theta$$
$$= \frac{2000}{9.81} \times 9.81 \times 0.55 \sin\theta$$
$$= 1100 \sin\theta \text{ Nm}$$

For the equilibrium of the motorcycle, overturning couple should be equal to resisting couple.

∴ $$1100 \sin\theta = 965.52 \cos\theta$$

or $$\tan\theta = \frac{965.52}{1100} = 0.877$$

or heel angle: $\theta = \mathbf{41.27°}$ **Ans.**

EXERCISES

1. Explain the following terms:
   (a) Axis of spin
   (b) Axis of precession
   (c) Precession
   (d) Gyroscope
   (e) Gyroscopic effect

2. What is meant by gyroscopic couple? Prove that the value of this couple is $I\omega\omega_p$, where I, ω and ω_p are moments of inertia, angular velocity of spin and angular velocity of precession respectively.

3. Explain the terms: steering, pitching and rolling in a ship. About which axis do these take place?

4. What do you mean by gyroscopic stabilization of ship? Explain with suitable sketches how it is carried out for steering, pitching and rolling motions.

5. How is the stability of a moving vehicle, which is taking a turn, ascertained? Explain, what factors affect the stability of vehicle?

6. A rotor of the marine ship having a mass of 750 kg and radius of gyration 300 mm at 1500 rpm clockwise when viewed from stern (rear end). Determine the gyroscopic couple and its effect on the ship in the following conditions:

   (i) When the ship pitches with angular velocity 1 rad/s and the bow (front end) is (a) rising (b) falling
   (ii) When the ship is moving at 30 kmph and takes left turn of 200 m radius.
   (iii) When the ship rolls and at a certain instant, it has an angular velocity of 0.04 rad/s.

   [Ans: 10602.8 Nm, 441.78 Nm, 530.14 Nm]

7. A propeller of a steamer has mass 1500 kg and radius of gyration of 1.2 m. The steamer turns left in a circle of 150 m radius at 25 kmph. The speed of propeller being 100 rpm in clockwise direction when viewed from the rear. Determine the magnitude and effect of the gyroscopic couple on the steamer.

   [Ans: 1047.2 Nm, bow raised]

8. A turbine rotor of a ship rotates at 1000 rpm clockwise when viewed from the rear. Its mass is 600 kg and radius of gyration is 350 mm. If the ship pitches with an angular velocity of 0.8 rad/s, determine the gyroscopic couple transmitted to the hull when the bow is rising. Also show in what direction the couple acts on hull.
[Ans: 6157.5 Nm, CW]

9. A propeller of an aircraft engine has a mass of 450 kg with radius of gyration 1m. The propeller shaft rotates at 300 rpm clockwise when viewed from the tail end. If the plane takes a left turn at a radius of 150 m with speed of 400 kmph, determine the gyroscopic couple and its effect on the air craft. Also determine the forces on the bearings if the distance between two bearings is 0.8 m.
[Ans: 10472 Nm, 13090 N]

10. A disc with radius of gyration 60 mm and a mass of 4 kg is mounted centrally on a horizontal axle of 80 mm length between the bearings. It spins about the axle at 800 rpm anticlockwise when viewed from the right hand side. The axle precesses about a vertical axis at 50 rpm in the clockwise direction when viewed from above. Determine the resultant reaction at each bearing due to the mass and gyroscopic effect.
[Ans: 25.94 N, 13.3 N]

11. A two wheeler vehicle of 350 mm wheel radius is negotiating a turn of radius 70 m at a speed of 100 kmph. The combined mass of the vehicle with rider is 250 kg, the centre of gravity of the combined system is 0.6 m above the road and moments of inertia of flywheel and road wheels are 0.3 kg m^2 and 1.0 kg m^2 respectively. If the speed of the engine flywheel is five times that of the wheel and in the same direction, find the angle of heel of the vehicle.
[Ans: 50.16°]

12. The driving axle of a locomotive with two wheels has a mass moment of inertia 350 kg m^2. The wheels are 1.8 m in diameter. The distance between the planes of the wheels is 1.5 m. While travelling at 100 kmph, the locomotive passes over a defective rail which causes the right hand wheel to fall 10 mm and raise again, in a total time of 0.1s, the vertical movement of wheel in simple harmonic motion. Determine the maximum gyroscopic couple caused and direction in which it acts when wheel is falling. Also determine the reaction between the wheel and rail.
[Ans: 2260.3 Nm, CCW, 1506.9 N]

13. A four wheeled car of total mass 2000 kg running on the rail of 1.6 m gauge, rounds a curve of 30 m radius at 54 kmph. The track is super elevated by 8°. The diameter of wheel is 0.7 m with mass 200 kg per wheel. The radius of gyration being 1 m, determine the pressure on rails due to centrifugal force and gyroscopic couple.
[Ans: 18574.3 N, 2942.4 N]

14. A racing car weighs 20 kN. It has a track width 1 m and height of 300 mm above the ground level and lies midway between front and rear axles. The flywheel rotates at 3000 rpm clockwise when viewed from the front. Moment of inertia of the flywheel is 4 kg m^2. The moment of inertia of each wheel is 3 kg m^2. Find the reactions between the wheels and ground when the car takes a curve of 15 m radius

at 30 kmph, taking into account the effect of gyroscopic effect and centrifugal force. Wheel radius is 400 mm.

[Ans: 6836.9 N, 3163.1 N]

15. A rear engine automobile is travelling along a track of 100 mm mean radius. The moment of inertia of wheels is 2 kg m^2 and that of engine parts is 1.2 kg m^2. The engine axis is parallel to the rear axle and crank shaft rotates in the same direction as the wheel. The diameter of the wheel is 600 mm. The gear ratio is 3. The mass of automobile is 1600 kg with centre of gravity lying 500 mm above the road level. The width of the vehicle is 1.5 m. Determine the limiting speed of the vehicle when road surface is not cambered.

[Ans: 96.48 kmph]

16. A horizontal axle AB 1 m long is pivoted at its centre. It carries a weight of 20 N at B and a rotor weighing 50 N at A. The rotor rotates at 600 rpm in clockwise direction looking from its front. Calculate the angular velocity of precession taking the radius of gyration of the rotor to be 300 mm.

[Ans: 0.32 rad/s]

MULTIPLE CHOICE QUESTIONS

1. The axis of spin, the axis of precession and axis of applied gyroscopic torque are contained in
   (a) one plane
   (b) two planes perpendicular to each other
   (c) three planes perpendicular to one another
   (d) none of the above

2. If the rotor of aeroplane is rotating clockwise when looking from front and makes a right turn, the gyroscopic effect will
   (a) tend to depress the nose and raise the tail
   (b) tend to raise the nose and depress the tail
   (c) tend to rotate aeroplane about spin axis
   (d) none of the above

3. Net reaction of ground on wheel due to gyroscopic couple due to wheels and the dead weight and centrifugal force of a vehicle negotiating a curve is
   (a) increased on inner wheels and decreased on outer wheels
   (b) decreased on inner wheels and increased on outer wheels
   (c) decreased on all the wheels
   (d) increased on all the wheels

4. Gyroscopic effect on a grinding mill
   (a) cause no change in crushing force
   (b) decrease crushing force
   (c) increase crushing force
   (d) none of the above

5. A spinning top does not immediately fall on the ground because of
   (a) gyroscopic couple
   (b) centrifugal force
   (c) self weight
   (d) external moments

6. The rotor of a ship rotates in clockwise direction when viewed from the stern and the ship takes a left turn. The effect of gyroscopic couple is
   (a) to raise the bow and stern
   (b) to move the ship towards star board
   (c) to raise the bow and lower the stern
   (d) to lower the bow and raise the stern

7. When a naval ship is pitching with the bow rising, rotor rotating clockwise when seen from stern the gyroscopic effect acting on it will be
   (a) to move it towards the port
   (b) to raise the bow and lower the stern
   (c) to move it towards the start board
   (d) to raise the stern and lower the bow.

8. The gyroscopic effects due to rotating parts of a turbojet engine of an aircraft on a curved course depend on
   (a) flight velocity
   (b) flight altitude
   (c) radius of the curve
   (d) flight velocity and radius of the curve

Bibliography

Barnacle, H.E. and Walker, G.E., *Mechanics of Machines*, Pergamon Press, London, 1965.

Black, P., *Mechanics of Machines*, Pergamon Press, London, 1967.

Ballaney, P.L., *Theory of Machines*, Khanna Publishers, Delhi, 1994.

Bansal, R.K., *Theory of Machines*, Laxmi Publications, New Delhi, 2004.

Bevan, Thomas, *The Theory of Machines*, CBS Publishers and Distributers, New Delhi, 1984.

Church, A.H., *Mechanical Vibrations*, John Wiley & Sons, New York, 1983.

Doughtie, V.L. and James, W.L., *Elements of Mechanism*, John Wiley & Sons, New York, 1964.

Doughtie, V.L. and Vallance, A., *Design of Machine Elements*, John Wiley & Sons, New York, 1964.

Dukkipati, Rao V. and Srinivas, J., *Textbook of Mechanical Vibrations*, Prentice-Hall of India, New Delhi, 2004.

Dyson, F., *Principles of Mechanisms*, Oxford University Press, Oxford, 1964.

Fairs, V.M. and Keown, *Mechanism*, McGraw-Hill, New York, 1960.

Ghosh, A. and Mallik, A.K., *Theory of Mechanisms and Machines*, Affiliated East-West Press, New Delhi, 1990.

Green, W.G., *Theory of Machines*, Blackie & Sons, 1963.

Grover, G.K., *Mechanical Vibrations*, Nemichand & Brothers, Roorkee, 1983.

Hannah, J. and Stephens, R.C., *Mechanics of Machines—Elementary Theory and Examples*, Viva Books, New Delhi, 1999.

Hannah, J. and Stephens, R.C., *Mechanics of Machines: Advanced Theory and Examples*, Viva Books, New Delhi, 1999.

Harland, Billings J., *Applied Kinematics*, Affiliated East-West Press, New Delhi, 1969.

Hartenberg, R.S. and Denavit, J., *Kinematic Synthesis of Linkages*, McGraw-Hill, New York, 1964.

Hinkle, R.T., *Kinematics of Machines*, Prentice-Hall, Inc., Englewood Cliffs, New Jersey, 1960.

Hirschhorn, J., *Kinematics and Dynamics of Plane Mechanism*, McGraw-Hill, New York, 1962.

Howard, P.J., *Theory of Machines*, MacDonalds, London, 1966.

Khurmi, R.S. and Gupta, J.K., *Theory of Machines*, S. Chand & Co., New Delhi, 2004.

Lal, Jadish, *Theory of Machines*, Metropolitan Book Co., Delhi, 1978.

Lent, D., *Analysis and Design of Mechanism*, Prentice-Hall Inc., Englewood Cliffs, New Jersey, 1961.

Levitskii, N.I., *Analysis and Synthesis of Mechanism*, Amerind Publishing Co., New Delhi, 1975.

Marbie, H.H. and Ocvirk, F.W., *Dynamics of Machinery*, John Wiley & Sons, New York, 1963.

Martin, G.H., *Kinematics and Dynamics of Machines*, McGraw-Hill, New York, 1982.

Maxwell, R.L., *Kinematics and Dynamics of Machinery*, Prentice-Hall Inc., Englewood Cliffs, New Jersay, 1960.

Paul, B., *Kinematics and Dynamics of Planar Machinery*, Prentice-Hall Inc., Englewood Cliffs, New Jersey, 1979.

Ramamurti, V., *Mechanics of Machines*, Narosa Publishing House, New Delhi, 2002.

Rao, J.S. and Dukkipati, Rao V., *Mechanism and Machine Theory*, New Age International, New Delhi, 1995.

Rao, J.S. and Gupta, K., *Theory and Practice of Mechanical Vibrations*, Wiley Eastern, New Delhi, 1984.

Rattan, S.S., *Theory of Machines*, Tata McGraw-Hill, New Delhi, 2004.

Sha, J.M. and Jadvani, H.M., *Theory of Machines*, Dhanpat Rai & Sons, Delhi, 1990.

Shariff, Abdulla, *Theory of Machines*, Dhanpat Rai & Sons, Delhi, 1984.

Sharma, C.S., *Mechanical Vibrations Analysis*, Khanna Publishers, Delhi, 1983.

Sharma, C.S. and Purohit, Kamlesh, *Design of Machine Elements*, Prentice-Hall of India, New Delhi, 2005.

Shigley, J.E. and Uicker, J.J., *Theory of Machines and Mechanisms*, McGraw-Hill International Book Co., 1981.

Singh, Sadhu, *Theory of Machines*, Pearson Education (Singapore), New Delhi, 2002.

Singh, V.P., *Theory of Machines*, Dhanpat Rai & Sons, Delhi, 2001.

Soni, A.H., *Mechanism Synthesis and Analysis*, McGraw-Hill, New York, 1974.

Vierck, R.K., *Vibration Analysis*, Harper & Row Publisher, New York, 1979.

Zimmerman, J.R., *Elementary Kinematics of Machines*, John Wiley & Sons, New York, 1962.

Answers to Multiple Choice Questions

Chapter 2
1. (c) 2. (b) 3. (a) 4. (c) 5. (b) 6. (d) 7. (c) 8. (a) 9. (d) 10. (a) 11. (a)
12. (c) 13. (c) 14. (b)

Chapter 3
1. (d) 2. (b) 3. (b) 4. (c) 5. (b) 6. (c) 7. (d) 8. (a) 9. (c) 10. (c) 11. (b)
12. (b) 13. (a) 14. (c) 15. (c)

Chapter 5
1. (b) 2. (c) 3. (a) 4. (b) 5. (c) 6. (a) 7. (d) 8. (c) 9. (d) 10. (c)
11. (a) 12. (b)

Chapter 6
1. (b) 2. (a) 3. (b) 4. (a) 5. (b) 6. (d) 7. (c) 8. (a) 9. (d) 10. (a) 11. (c)

Chapter 7
1. (a) 2. (c) 3. (a) 4. (c) 5. (d) 6. (b) 7. (c)

Chapter 8
1. (b) 2. (a) 3. (b) 4. (d) 5. (c) 6. (b) 7. (d) 8. (a) 9. (d) 10. (c)

Chapter 9
1. (a) 2. (c) 3. (a) 4. (b) 5. (c) 6. (b) 7. (b) 8. (a) 9. (d) 10. (a) 11. (a)

Chapter 10
1. (d) 2. (c) 3. (b) 4. (c) 5. (c) 6. (b) 7. (a) 8. (d) 9. (c) 10. (c) 11. (b)
12. (c) 13. (a) 14. (b)

Chapter 11
1. (b) 2. (a) 3. (c) 4. (d) 5. (d) 6. (b) 7. (a)

Chapter 12

1. (d) 2. (d) 3. (b) 4. (a) 5. (c) 6. (a)

Chapter 13

1. (b) 2. (c) 3. (b) 4. (c) 5. (a) 6. (d) 7. (c) 8. (c) 9. (d) 10. (c) 11. (d)
12. (d) 13. (b)

Chapter 14

1. (c) 2. (b) 3. (c) 4. (b) 5. (a) 6. (c) 7. (d) 8. (a) 9. (b) 10. (c) 11. (d)
12. (b) 13. (c) 14. (b) 15. (c)

Chapter 15

1. (b) 2. (b) 3. (b) 4. (c) 5. (a) 6. (d) 7. (c) 8. (d)

Index

Absorption dynamometer, 274
Acceleration analysis, 40
  analytical method, 38
  centripetal acceleration, 40
  Coriolis acceleration, 51
  Klein's construction, 49
  tangential acceleration, 41
  polygon, 42
Angle of friction, 137
Arnold Kennedy theorem, 34

Balancing, 542
  direct and reverse crank, 581
  dynamic, 547
  in-line engine, 571
  locomotive, 567
  reciprocating mass, 564
  static, 543
  two planes, 548
  V engine, 577
Balancing machine, 558
  field balancing, 561
  nodal point, 560
  pivot cradle, 559
Band brake, 250
  simple, 251
  differential, 252
Band and block brake, 262
Belt drive, 184
  centrifugal force, 199
  length, 192
  materials, 204
  ratio of tension, 195
  slip, 190
  stresses, 204
Bevis–Gibson dynamometer, 285
Bevel gear, 432
Bloch's method, 94

Block brakes, 235
  small angle, 236
  large angle, 239
Brakes, 235

Cams, 331
  followers, 333
  motion, 335
  pressure angle, 359
  synthesis, 346
Cams with specified profile, 362
  circular cam, 372
  tangent cam, 362
Chain drive, 224
  types, 224
  length, 228
  chordal action, 229
Chebysheve spacing, 91
Collar bearing, 160
Compound gear train, 443
Cone clutch, 168
Constrained mechanism, 73
  Coriolis acceleration, 51
  crank effort, 503
  crank and slotted lever, 12
  critically damped system, 605

Damping coefficient, 605
Damped free vibration, 605
Dead centre, 81
Dimensional synthesis, 82
Dynamics, 2
Direct and reverse crank, 581
Dynamic equivalent link, 500
Dynamic force analysis, 492
Dynamic of slider crank, 495
  connecting rod, 497
  piston acceleration, 495
  piston velocity, 496

Index

Dynamometer, 274
  absorption, 274
  epicyclic train, 278
  transmission, 278
  torsion, 284

Effect of braking, 270
Effect of governor, 320
Electrical analogy, 656
Epicyclic gear train, 445
  tabulation method, 445
  relative velocity method, 446
  torque transmitted, 447
Energy method, 595
Engine indicator mechanism, 114
  crossby, 115
  Dobbie McInne, 117
  simplex, 114
  Thomson, 116

Field balancing, 561
Flywheel, 517
Followers, 333
  motion, 335
  constant velocity, 336
  simple harmonic motion, 338
  cycloidal, 343
  constant acceleration/deceleration, 339
  polynomial, 344
Force analysis, 471
  crank effort, 498
  dynamic, 486
  free body diagram, 474
  static, 471
  turning moment, 515
Forced damped vibration, 616
Free damped vibration, 605
Friction, 136
  guided friction, 145
  inclined plane, 138
  pivot and collars, 158
  screw, 153
  wedge, 147
Friction axis, 176
Friction circle, 175
Friction clutch, 165
  cone clutch, 168
  plate clutch, 166

Gears 388
  terminology, 389
  law of gearing, 391
  form of teeth, 394
  arc of contact, 397
  Bevel, 432
  force analysis, 415
  helical, 417
  interference, 405
  spiral, 420
  worm gear, 427
Gear strain, 441
  compound, 443
  epicyclic, 445
  simple, 441
Governor, 292
  controlling force, 322
  effort, 320
  hunting, 320
  gravity, 316
  Hartnell, 305
  isochronism, 319
  Porter, 295
  Proell, 300
  sensitiveness, 318
  watt, 292
Grashof's criterion, 77
Grasshopper mechanism, 112
Grubler's equation, 5, 73
Gyroscope, 667
  couple, 668
  automotive vehicle, 684
  grinding mill, 682
  ship stabilization, 675

Hart mechanism, 106
Helical gear, 427
  force analysis, 419
  formative teeth, 419
Higher pair, 2
Hooke's joint, 126
Hydraulic dynamometer, 276

Inclined plane friction, 138
Inertia force analysis, 488
Internal expanding brake, 265
Interference, 405
  minimum teeth, 406

Index

Inversion, 9
   double slider crank, 13
   four bar, 9
   slider crank, 10
Instantaneous centre, 32
   Arnold Kennedy theorem, 34
Involute profile, 394

Kinematics, 2
Kinematic chain, 5
   inversion of, 9
Kinematic link, 2
Kinematic pairs, 2
Klien's construction, 49

Law of belting, 189
Law of dry friction, 137
Law of gearing, 391
Least square technique, 93
Length of belt, 192
   cross belt, 194
   open belt, 192
Limit position, 80
Logarithmic decrement, 608
Lower pairs, 2
Lower pair mechanism, 104

Magnification factor, 619
Mechanical vibration, 592
   undamped, 593
   torsional, 597
Mechanism, 6
   constrained, 73

Natural frequency, 593
Newtorn's method, 594
Nodal point balancing, 560

Overdamped system, 607

Path of approach, 398
Path of recess, 398
Peaucellier mechanism, 105
Pentograph, 112
Pickering governor, 317
Pivot bearing, 163
Pivot cradle balancing machine, 558
Plate clutch, 166
Porter clutch, 295
Power of governor, 321

Proel governor, 300
Prony brake dynamometer, 275

Quick return ratio, 90

Ratio of tension, 195
Rayleigh principle, 635
Relative pole method, 82
Relative velocity method, 17
   crank and slotted lever, 20
   four bar mechanism, 19
   slider crank mechanism, 19
Reverted gear train, 444
Robert's mechanism, 112
Rolling pair, 2
Rope brake dynamometer, 275
Rope drive, 218
   rope construction, 218

Scott Russel mechanism, 108
Screw friction, 153
Simple gear train, 441
Sliding pair, 2
Slider crank mechanism, 7
Spherical pair, 2
Spiral gear, 420
   efficiency, 422
Spur gear, 415
Stability of vehicle, 684
   four wheels, 684
   two wheeler, 690
Static balancing, 542
Static force analysis, 471
Stepped pulley drive, 214
Straight line mechanism, 104
Support excitation, 621
Synthesis of cam, 346
   analytical method, 355
   graphical method, 346
Synthesis of mechanism, 72
   Grashof's criterion, 77
   Grubler's equation, 73
   inversion method, 87
   number synthesis, 72
   relative pole method, 82
   three position method, 85, 87
   two position method, 84, 86

Tatham dynamometer, 279
Techebicheff mechanism, 110

Index

Torsion dynamometer, 284
Torsional vibration, 597
Train value, 441
Transmission angle, 81
Transmission dynamometer, 278
Transverse vibration, 633
Turning moment diagram, 515
Turning pair, 2
Two plane balancing, 548

Undamped free vibration, 593
Underdamped system, 607

Velocity analysis, 17
    analytical method, 58
    relative velocity method, 17
    instantaneous centre method, 32
Velocity polygon, 18
Vibration, 599
    damped free, 605
    electrical analogy, 656
    forced damped, 616

isolation, 623
logarithmic decrement, 608
rotating unbalance, 620
support excitation, 621
transmissibility, 623
transverse, 633
undamped, 593
whirling of shaft, 637
Vibration of rotor system, 640
    geared system, 651
    three rotor, 642
    two rotor, 640
Virtual work, 476
Von-Hefner dynamometer, 280

Watt governor, 292
Watt mechanism, 109
Wedge friction, 147
Willson–Hartnell governor, 312
Worm gear, 427
    efficiency, 430
Whirling of Shaft, 637